Studies in Systems, Decision and Control

Volume 301

The series "Studies in Systems, Decision and Control" (SSDC) covers both new developments and advances, as well as the state of the art, in the various areas of broadly perceived systems, decision making and control–quickly, up to date and with a high quality. The intent is to cover the theory, applications, and perspectives on the state of the art and future developments relevant to systems, decision making, control, complex processes and related areas, as embedded in the fields of engineering, computer science, physics, economics, social and life sciences, as well as the paradigms and methodologies behind them. The series contains monographs, textbooks, lecture notes and edited volumes in systems, decision making and control spanning the areas of Cyber-Physical Systems, Autonomous Systems, Sensor Networks, Control Systems, Energy Systems, Automotive Systems, Biological Systems, Vehicular Networking and Connected Vehicles, Aerospace Systems, Automation, Manufacturing, Smart Grids, Nonlinear Systems, Power Systems, Robotics, Social Systems, Economic Systems and other. Of particular value to both the contributors and the readership are the short publication timeframe and the world-wide distribution and exposure which enable both a wide and rapid dissemination of research output.

** Indexing: The books of this series are submitted to ISI, SCOPUS, DBLP, Ulrichs, MathSciNet, Current Mathematical Publications, Mathematical Reviews, Zentralblatt Math: MetaPress and Springerlink.

More information about this series at http://www.springer.com/series/13304

Ju H. Park
Editor

Recent Advances in Control Problems of Dynamical Systems and Networks

Springer

Editor
Ju H. Park
Department of Electrical Engineering
Yeungnam University
Kyongsan, Korea (Republic of)

ISSN 2198-4182 ISSN 2198-4190 (electronic)
Studies in Systems, Decision and Control
ISBN 978-3-030-49122-2 ISBN 978-3-030-49123-9 (eBook)
https://doi.org/10.1007/978-3-030-49123-9

This Springer imprint is published by the registered company Springer Nature Switzerland AG
The registered company address is: Gewerbestrasse 11, 6330 Cham, Switzerland

To my parents

Preface

In the last decade, there have been rapid developments and changes in the field of control engineering. Departing from the study of classical linear/nonlinear control systems, tremendous efforts have been made to tackle more complex systems or networks and many real-world application problems. In particular, with the more powerful IT technology, the components of the control systems are connected to a wireless network, and the concept of a cyber-physical system in which physical layers and cyberspaces coexist has also emerged. Besides, under the influence of the 4th Industrial Revolution, advanced technologies such as artificial intelligence, machine learning, Internet of Things, virtual reality, wearable devices, and big data are grafting with control systems.

In addition to the progress of this innovative technology, the field of control theory has been creating a stepping stone to solve the limitations of reality one by one that has not been addressed in the past. From this point of view, even in the same research subject in control engineering, the intention of planning this book is to cover with issues such as (i) control theories that weren't developed, (ii) constraints of systems that weren't addressed, or (iii) control systems of new structures that could not be dealt with when readers look at books or papers published 10 years ago.

Aim of the Book

The aim of this book is (1) to provide an introduction to recent research trend on various topics on control problems of dynamical systems and networks and (2) to present novel analysis and synthesis frameworks for investigating the stability and designing controllers of the systems and networks.

To show the progress of research on these topics from various perspectives, in this book, several dynamic system models in the continuous or discrete-time domain such as traditional linear and nonlinear systems, networked control systems, fractional-order systems, switched systems, large-scale interconnected systems, neural networks, and complex networks have been considered for various control problems.

In the system analysis, certain system phenomena including uncertainties, time delays, actuator faults, switching, asynchronous sampling, event-triggered scheme, impulse, and cyber-attacks are considered to deal with the stability and stabilization problems in the systems and networks. In order to show applicability to real problems, it is noted that some examples are chosen from practical industrial models.

From a detailed view of the contents of the book, it collects works carried out recently by promising experts in this field. By their contribution, this book covers the problems on observer/estimator design, observer-based control, finite-time stability and stabilization, eigenvector-centrality-based consensus protocol, sampled-data control, fuzzy-based control, sliding mode control, fault alarm, fault detection/estimation/reconstruction, fault-tolerant control, event-triggered control, asynchronous quantized control, synchronization, positivity, stochastic stability, and so on.

Some preliminary results of the material presented in the book may have been published in eminent international journals in recent years by contributors. Therefore, if readers want to know additional basic knowledge and advanced one related to the contents of each chapter, please refer to the previously published results of each contributor, which can be easily found through academic DB search.

It is noted that this book attempts to consolidate the recent new results or extended results of previous studies from contributors on the topics mentioned above. In this respect, the book as a tutorial reference as well as an advanced monograph is likely to be of use for the wide and heterogeneous group of graduate students and doctoral researchers in science and engineering who are focusing on control theory and corresponding real applications for various dynamical systems and networks.

Main Features of This Book

This book presents both theoretical development and applications in the real world in the important research areas in control engineering including linear systems, nonlinear systems, time-delay systems, positive systems, multi-agent systems, fractional-order systems, networked control systems, hybrid and switching systems, stochastic systems, fuzzy systems, descriptor systems, parabolic systems, impulsive systems, complex networks, and neural networks.

As can be seen from the system classification mentioned above, this book covers the latest control problems for almost all systems studied in control engineering.

For dealing with these systems and networks, some new control strategies are applied based on the following control schemes including sampled-data control, fault-tolerant control, event-triggered control, sliding mode control, observer-based control, and $\mathcal{H}_\infty$ control.

Appropriate features suggested by this book include the following:

1. Providing various mathematical and engineering basic and advanced knowledge.
2. Introduction of various control systems and networks.
3. Presenting recent research trends on control theory.
4. State-of-the-art controller design techniques.
5. Constraints in the control system to be considered in the real environment.
6. Examples to help readers understand how the theories and methods presented apply.
7. Accompanying detailed simulation results.

Prerequisites and Intended Audience

This book should be of interest to both graduate students and researchers in the area of control theory and applications, as well as those new to the area. The reporting on the latest developments from various points of view in this area is certainly to attract intense interest to relevant researchers who are concerned with the newest trends and topics.

To understand the research flow and contents on the various topics covered in this book, the background required of the readers is basic and advanced knowledge of linear algebra, calculus, linear and nonlinear control system theory, Lyapunov stability theory, probability theory, and simulation techniques.

In particular, for the fractional-order systems discussed in Part, additional knowledge of the Lyapunov function theory associated with fractional derivatives is required.

Outline of the Book

The book is composed of four parts, with the system classification ranging from Linear and Nonlinear Dynamic Systems, Switched Systems and Hybrid Control, Fractional-Order Systems, and Dynamical Networks.

In soliciting the chapter contributions from experts on related fields, maximum effort has been attempted to cover the hottest topics in four above research categories.

Part, "**Linear and Nonlinear Dynamic Systems**", starts in chapter "Partial State Observers for Linear Time-Delay Systems" and ends with chapter "Fault Estimator for Parabolic Systems with Distributed Inputs and Outputs", which introduces the recent control problems on a variety of topics on linear and nonlinear dynamic systems.

As we know well, a dynamic system is a system in which the states (or variables) are expressed as a function of time in physical space. Many people believe that the concept of such dynamic systems was invented by French mathematician Henri Poincare in the late 1800s. The most basic classification of dynamic systems is to distinguish between linear and nonlinear systems. A linear dynamic system is a mathematical model of a system based on the use of a linear operator. Nonlinear systems refer to the type of dynamic systems where the output from the system does not vary directly with respect to input to the system.

Starting from the nineteenth century, control problems of linear or nonlinear dynamic systems in various areas we encounter result from system modeling to stability analysis and controller design. First, there has been a tremendous amount of research into the stability of dynamic systems over two centuries. Various stability concepts including asymptotic stability, exponential stability, bounded-input bounded-output stability, Nyquist stability, uniform stability, stochastic stability, and absolute stability have been introduced through stability theories. In addition, control theories for stabilizing such systems have evolved. For example, PID control, optimal control, adaptive control, robust control, fuzzy control, neural control, stochastic control, model-predictive control, and so on. In general, systems are analyzed in the time domain or in the frequency domain by considering the characteristics of the system to deal with control problems. As is well known, in the frequency domain analysis of classical control theory, the Fourier transform, Laplace transform, or Z-transform is used to deal with the various analyses of linear systems in depth. On the other hand, in the time domain for modern control theory, the analysis of real-world nonlinear systems including linear systems is carried out using state equations, another form of differential equations.

For stability and stabilization of dynamic systems, in this Part, for various linear systems and nonlinear systems having impulse, fault, delay, or uncertain factors in continuous-time or discrete-time space, the controller design techniques of a number of ways for the stabilization of closed-loop systems are introduced.

Part, "**Switched Systems and Hybrid Control**", starts in chapter "Event-Triggered Control of Discrete-Time Switched Linear Systems Via Communication Channels with Limited Capacity" and ends with chapter "Asynchronous Quantized Control for Markov Switching Systems with Channel Fading", which introduces hottest topics on switched systems and hybrid control.

A hybrid system is capable of exhibiting simultaneously several kinds of dynamic behavior in different parts of a system like continuous-time dynamics, discrete-time dynamics, jump phenomena, operating mode changes, and logic-based switching. From the mathematical standpoint, these hybrid systems have mathematical models of different forms, such as algebraic equations, difference equations, ordinary differential equations, and/or partial differential equations.

Switched systems are an important class of hybrid dynamical systems, which are widely used to model many real physical systems with multi-mode switching, such as, in electronic circuits field, power electronics, chemical processes, mechanical systems, automotive industry, aircraft and air traffic control, and so on. In general, a switched system consists of a finite number of continuous (or discrete)-time subsystems and a logical rule that orchestrates switching between them. The switching rule, determined by time or system state, both time and state, other supervisory logic decision, or even stochastic processes, yields different switching signals. The switching signals, which have a great influence on the system performances, can be classified into two categories: arbitrary and constrained. It is known that the constrained switching compared with arbitrary switching arises quite naturally from the physical constraints of a real switched system.

Naturally, stability and stabilization are very important issues in the study of switched systems. A great number of efficient methods have been proposed to solve these problems for switched systems in the past few decades. It is known that the stability of switched systems depends not only on the dynamics of each subsystem but also on the properties of switching signals. When all the subsystems are stable, it usually cannot ensure the stability of the switched system under arbitrary switching. There is always the possibility, however, that the stability of the whole system is maintained if the switching is sufficiently slow, which means that the interval between consecutive switchings is large enough. In fact, it really does not matter if the system occasionally has fast switchings, provided this does not occur too frequently. The idea is proved to be reasonable and captured by the concept of average dwell-time switching proposed by J. P. Hespanha and A. S. Morse from the paper "Stability of switched systems with average dwell-time, in: Proceedings of the 38th IEEE Conference on Decision and Control, Vol. 3, pp. 2655–2660, 1999" which has showed that switching among stable linear systems results in a stable system provided that the average dwell time is sufficiently large.

Until now, several approaches have been proposed for the control problems for switched systems, such as arbitrary switching, restricted switching (like dwell time and average dwell time), multiple Lyapunov functions method, and piecewise quadratic Lyapunov functions. Especially, the average dwell-time approach has been acknowledged to be more flexible and efficient in the analysis of switched systems. While there are many books and papers that can be consulted on these switched systems, the following two references (R1. D. Liberzon, Switching in Systems and Control, Birkhauser, Boston, 2003; R2. H. Lin, P. J. Antsaklis, Stability and stabilizability of switched linear systems: A survey of recent results, IEEE Trans. Automatic Control, Vol. 54, pp. 308–322, 2009) are recommended for readers to get the basic knowledge of the system.

Markov jump systems are a type of stochastic hybrid systems with finite operation modes, where the jumps among these modes are governed by a Markov process taking values in a finite set. Markov jump systems have the advantage of modeling dynamical systems subject to random abrupt changes in their parameters or structures, and have been applied to many practical plants, such as powers systems, manufactory processes, fault-tolerant systems, telecommunications systems, immunology, etc.

When dealing with this system, transition probabilities determining the transition between different modes have a crucial influence on system stability and performance. In some cases, the transition probability obeys the exponential distribution, making the sojourn time memoryless and resulting in a constant index distribution for Markov jump systems. However, in the actual process, the sojourn time does not follow the exponential distribution, which describes semi-Markov jump systems of more general cases.

Meanwhile, in recent years, a remarkable development of networked control systems in which components are networked by wire or wireless has been progressed. In addition to this, a change in the control structure is taking place for efficient control of the system, away from the classical control method. For instance, a hybrid-triggered scheme consisting of time-triggered scheme and event-triggered scheme is developed recently to alleviate the load of network bandwidth. The representative new type of control method is the event-triggered control method. This method can save resources and cost by changing the control input only when a certain trigger condition is satisfied.

As another special control method, the hybrid-triggered scheme contains a random switch between the time-triggered scheme and the event-triggered scheme. For a good application of this hybrid control scheme, refer to a recent paper by J. Liu, Y. Gu, X. Xie, D. Yue, J. H. Park, Hybrid-driven-based $\mathcal{H}_\infty$ control for networked cascade control systems with actuator saturations and stochastic cyber attacks, IEEE Transactions on Systems, Man, and Cybernetics: Systems, Vol. 49, No. 12, pp. 2452–2463, 2019.

In any case, the purpose of Part, "**Switched Systems and Hybrid Control**", is to explore the latest trends in research on hybrid characteristics of systems or controllers.

Part, "**Fractional-Order Systems**", starts in chapter "Observer-Based Controller Design for Fractional-Order Neutral-Type Systems" and ends with chapter "New Results on Stability of Coupled Impulsive Fractional-Order Systems on Networks", which presents some recent hot topics on fractional-order dynamic systems or networks.

As an extension and generalization of the traditional integer-order differential and integral calculus, it is known that the fractional calculus is a mathematical branch of the analysis that studies the several different possibilities of defining real number powers or complex number powers of the differentiation operator and of the integration operator.

Since its first appearance in 1695, the fractional calculus was introduced in one of Niels Henrik Abel's early papers in 1823. Especially, from the twenty-first century, the fractional calculus becomes a dynamic research topic in applied science and engineering and has appealed more consideration because the fractional-order characterization is established to be more enough to explain some real-time problems than the integer-order representation. Some notable contributions on fractional calculus have been generated to the implementation of fractional differential equations. Compared to the integer-order models, the main desirability of fractional models is that it provides a valuable implementation for the account of remembrance and hereditary effects of many resources and procedures. Therefore, fractional differential equations occur in many notable applications such as viscoelasticity, electrochemistry, control, electromagnetic theory, aerodynamics, polymer rheology, etc. Notably, due to the memory effect, the fractional calculus has been intensively used in neural networks and complex dynamic systems in the last decade because the feature of the fractional calculus plays a crucial role in the description of neural networks having memory. For instance, it was reported that a generalized capacitor can be more accurately modeled by fractional operators and the circuit systems were developed with the use of fractional calculus.

As an application of fractional-order systems, let us look at it in the field of neural networks. A model of fractional Hopfield neural network was proposed by E. Kaslik and S. Sivasundaram in the paper titled "Nonlinear dynamics and chaos in fractional-order neural networks," Neural Networks, Vol. 32, pp. 245–256, 2012, which has the following description:

$$ {}_{t_0}^{C}D_t^{\alpha}x(t) = -Ax(t) + Bf(x(t)) + d $$

where $x = (x_1, x_2, \ldots, x_n)^T \in \mathbb{R}^n$ is the state vector, the diagonal matrix

$A = \operatorname{diag}(a_1, a_2, \ldots, a_n)$ with $0 < a_i$ is the state feedback coefficients, $B = (b_{ij})_{n\times n}$ is the network's interconnection matrix, $f(x)$ is the activation function, and d denotes the constant external input. This neural network has the memory and storage information from initial time t_0 to t.

Also, very recently, a fractional discrete-time neural network is proposed by L. Huang, J. H. Park, G.-C. Wu, and Z.-W. Mo from the paper, "Variable-order fractional discrete-time recurrent neural networks," *Journal of Computational and Applied Mathematics*, Vol. 370, Article ID 112622, 2020, which are described by

$$ {}^{C}\Delta_a^{v}x(t) = -Ax(t+v-1) + Bf(x(t+v-1)), $$

$$ 0 < v \leq 1,\ x(a) = x_a, t \in \mathbb{N}_{a+1-v} $$

where f is a continuous function and $f : \ell_a^{\infty} \to \mathbb{R}^n$.

It is clear that the above equation has a zero solution if $f(0) = 0$. For $v = 1$, it can be reduced to a standard recurrent neural network by the use of Euler method $\dot{x}(t) \approx \Delta x(t)$:

$$ x(k+1) = (I - A)x(k) + Bf(x(k)) + d $$

if $d = 0$.

Further, in control engineering, the stability and stabilization problems on linear or nonlinear fractional-order systems in continuous- or discrete-time domain become one of the hottest topics in recent 5 years. Some real applications such fractional dynamic systems include chaotic systems with applications to image encryption, color image denoising, enhancing and decomposition, secure communications, pattern recognition applications, battery external short circuit fault diagnosis, tumor-immune system interaction, financial systems, population dynamic models, human respiratory syncytial virus infection, and so on.

Part, "**Dynamical Networks**", starts with the positivity and stability problems in neural networks in chapter "Positivity and Stability of Nonlinear Time-Delay Systems in Neural Networks" and ends with a synchronization problem in chapter "Finite-Time Synchronization Control for Markovian Jump Memristive Neural Networks with Reaction-Diffusion Terms".

This part introduces recent progress of some spotlighting topics of dynamic networks including neural networks and complex networks which are the areas of research that are of most interest in the era of the 4th industrial revolution along with cyber-physical systems, artificial intelligence, and big data.

Artificial neural networks are computing systems vaguely inspired by the biological neural networks that constitute animal brains. Usually, a practical neural network that can perform some complex tasks has a very large number of neurons and layers, since more neurons and layers bring higher accuracy output results. There are some popular models of neural networks including Hopfield neural networks, cellular neural networks, convolutional neural networks, Cohen-Grossberg neural networks, and bidirectional associative memory neural network, which have found many successful applications in various fields, such as signal processing, pattern recognition, image processing, parallel computation, and combinatorial optimization. Especially, convolutional neural networks have particular strengths in these areas such as image and video recognition, image classification, medical image analysis, and natural language processing. Thus, in the recent study of intelligent autonomous vehicles, convolutional neural networks are widely used to find patterns to recognize objects, faces, and scenes in images because the networks learn features directly, extract features automatically, and then provide the highest level of recognition results.

For the research on neural networks from the control engineering perspective, the problems on mathematical modeling and their stability of developed neural networks have been active in the last two decades since the developed neural network is regarded as a dynamic system in the model-based control problems.

Since the past two decades, complex networks have drawn increasing attention from various scientific fields, such as mathematics, engineering, physics, economics, ecology, and even sociology. A complex network is a large set of interconnected nodes, in which every node represents an individual agent of the system, while edges represent relations between nodes. In this regard, complex dynamical networks have garnered wide-scale attention from many fields due to their broad applications in various fields, such as electrical power grids, sensor networks, communication networks, Internet, coupled oscillators, World Wide Web, epidemic spreading networks, a platoon of unmanned aerial vehicles or mobile robots, and so on. In particular, in recent years, several types of flu or epidemics have occurred in various regions and spread to the whole world, which has a great negative effect on the health of many people. The spread of these viruses and infectious diseases is also a problem of a complex network.

The nature of complex networks, such as topological structures, dynamical evolution, and node diversities, has been fully investigated in the existing literature. Also, the study on geometric features, collective behaviors, and control and synchronization of complex networks has received another significant concern. Among these topics, a lot of interest and related research has been recently focused on synchronization by control engineering researchers. Some types of synchronization have been discovered, such as complete synchronization, cluster synchronization, impulsive synchronization, phase synchronization, lag synchronization, quasi-synchronization, and cluster synchronization. It is noted that the cluster synchronization needs the coupled complex networks split into several subgroups, called cluster, and the systems in the same clusters should be synchronized. That is, each cluster has its own synchronization goal which could be distinct for different clusters.

In this Part, we will look at stability, fault detection, control, and synchronization issues in various types of dynamic networks.

Linear and Nonlinear Dynamic Systems

Chapter "Partial State Observers for Linear Time-Delay Systems".

This chapter considers the design problem of state observers for linear systems with time delay. It is well known that state estimation is an essential part of any model-based control systems. In particular, partial state estimation is very useful when the full set of the states of a system is not observable or detectable, or whenever the system is large scale and only a few states of the system need to be monitored. Recent advancements in functional observer design algorithms for linear time-invariant systems will be reviewed first in this chapter. Then, delay-free and delay-dependent functional observer design schemes for linear time-delay systems with state and input delays are illustrated.

Chapter "Finite-Time Stability and Control of Impulsive Positive Systems with Interval Uncertainty".

This chapter addresses the problem of finite-time stability and stabilization of positive dynamic systems including impulses and interval uncertainties. By considering a few types of impulsive effects, the finite-time stability conditions are proposed for interval impulsive positive systems on the basis of a time-varying copositive Lyapunov function and an average impulsive interval method. Also, the finite-time stability guaranteeing $\mathcal{L}_1$-gain performance is further studied as an application work.

Chapter "An Eigenvector-Centrality Based Consensus Protocol Design for Discrete-Time Multi-agent Systems with Communication Delays".

This chapter investigates a leader-follower consensus protocol design problem for a class of multi-agent systems in the discrete-time domain with time-varying communication delay. A consensus protocol is proposed by applying the concept of centrality measure for the agent which is determined by the structure of information flow among multiple agents. In order to find the most central node within the connected graph, the various centrality measures have been attempted in the field of social network analysis. To improve the consensus performance in two aspects of $\mathcal{H}_\infty$ performance and the stability in consideration of the time-delay, a novel weighted consensus protocol is proposed by calculating the eigenvector centrality for agents in a network by building simple Lyapunov-Krasovskii functional combined with a linear matrix inequality framework.

Chapter "Fault-Tolerant Sampled-Data Synchronization of Chaotic Systems with Random Occurring Uncertainties: A Semi-Markov Jump Model Approach".

This chapter focuses on addressing the issue of sampled-data synchronization for nonlinear chaotic systems with uncertainties by fault-tolerant control techniques, where a semi-Markov jump chain is utilized to describe the switching characteristics of the system. In this approach, it is assumed that the randomly occurring uncertainties in the chaotic systems are subject to the Bernoulli distributed white noise sequences. A mean-square synchronization condition of the system by means of employing appropriate Lyapunov-Krasovskii functional and reasonable integral inequality methods is presented.

Chapter "Fuzzy-Based Sliding Mode Control Design for Stability Analysis of Nonlinear Interconnected Systems".

This chapter investigates the problem of the stability and stabilization of interconnected large-scale systems. The nonlinear large-scale interconnected system is equivalently expressed into a T-S fuzzy-model-based system with time delays. Via a sliding mode control scheme having a suitable sliding surface combined with the techniques such as Wirtinger inequality approach, Lyapunov stability theory, and linear matrix inequality framework, delay-dependent stability and stabilization conditions for guaranteeing asymptotic stability of the system are derived.

Chapter "Observer-Based Fault-Tolerant Control for Non-Infinitely Observable Descriptor Systems".

This chapter is concerned with a fault-tolerant control for a class of non-infinitely observable descriptor systems with disturbances. The proposed control scheme consists of a sliding mode observer to estimate the states and unknown inputs and a controller to perform fault-tolerant stabilization. Then, the closed-loop system can be expressed as an infinitely observable reduced-order system by treating some states as unknown inputs. A linear matrix inequality optimization technique is used to design the sliding mode observer and stabilizing controller such that the estimates from the observer and the output of the closed-loop system converge asymptotically. A practical system is considered so that the proposed method can demonstrate its effectiveness.

Chapter "Fault Estimator for Parabolic Systems with Distributed Inputs and Outputs".

This chapter investigates the problem of the $\mathcal{H}_\infty$ fault estimation of a class of parabolic systems with distributed inputs and outputs. The fault in the considered parabolic system is characterized by a minimal state-space description of a low-pass filter, which incorporates fault's prior knowledge. First, the fault estimator is designed to estimate both the fault and the states of the considered system. Then, exponential stability is considered for the system without external disturbance.

Next, the results are extended to guarantee the improved robustness and ensure that the system is the $\mathcal{H}_\infty$ performance or mixed $\mathcal{H}_\infty$ and passivity performance in terms of an optimization problem based on the Lyapunov theory with the framework of the linear matrix inequality approach.

Switched Systems and Hybrid Control

Chapter "Event-Triggered Control of Discrete-Time Switched Linear Systems via Communication Channels with Limited Capacity".

This chapter considers a state feedback control problem for discrete-time switched linear systems. The switched system with a control input is assumed to be connected via a limited capacity communication channel and the controller can only access the successfully event-triggered transmitted information of system state and mode. Considering the signal transmission delays and packet losses, respectively, mode-dependent event-triggered transmission schemes are proposed to mitigate unnecessary data transmission. Based on the augmented switching signal, which is generated by merging the switching signal with its transmitted switching signal, the closed-loop systems are modeled as switched systems with the delayed state. Then, by using the multiple Lyapunov functional and average dwell-time methods, sufficient exponential stability conditions for the closed-loop systems are obtained.

Chapter "Fault Alarm-Based Hybrid Control Design for Periodic Piecewise Time-Delay Systems".

This chapter presents a novel fault alarm-based hybrid control scheme which is about to stabilize the periodic piecewise systems with time delay and disturbances. The hybrid controller has devised to handle both the normal and faulty systems. By an alarm signal in the system, the controller automatically switches to the reliable control mode from conventional control mode. The asymptotic stability of the concerned system is ensured with a minimum disturbance attenuation level by the controller.

Chapter "Event-Triggered Sliding Mode Control for Stochastic Markov Jump Systems".

This chapter examines the design problem of sliding mode controller for the stabilization of stochastic Markov jump systems via a novel event-triggered communication scheme as a tool of saving computational resources in the system. Through stochastic Lyapunov functional and observer design theories, a stabilizing sliding mode control law is constructed to guarantee that the closed-loop system's trajectories can arrive at the specified desired sliding surface. To demonstrate the effectiveness and applicability of the proposed control method, an example of a practical model is utilized.

Chapter "Asynchronous Quantized Control for Markov Switching Systems with Channel Fading".

This chapter deals with an asynchronous control problem for Markov switching systems with channel fading, where a more general class of discrete-time Markov switching system is developed by exploring the channel fading. To analyze the system as a hidden Markov model, the original systems and designed controller are characterized by stochastic variables. As an effective method to reduce energy consumption in the controller, a mode-dependent event-triggered communication mechanism for guaranteeing stochastic stability of a closed-loop system is proposed.

Fractional-Order Systems

Chapter "Observer-Based Controller Design for Fractional-Order Neutral-Type Systems".

This chapter studies the design problem of the observer-based controller for a class of fractional-order neutral-type systems. Then, based on the Lyapunov stability theory, an observer-based guaranteed cost control problem of the fractional-order systems is formulated. A stability criterion for guaranteeing the stability of the closed-loop system is given, and the design of the observer and the feedback controller is attributed to the feasible solution problem of a class of linear matrix inequalities.

Chapter "Adaptive Synchronization of Fractional-Order Delayed Memristive Neural Networks".

This chapter deals with an adaptive synchronization problem for fractional-order memristor-based neural networks with a constant time delay. It is noted that this work is the first attempt to study the synchronization of fractional-order delayed memristive neural networks using a novel adaptive delay feedback controller.

Chapter "New Results on Stability of Coupled Impulsive Fractional-Order Systems on Networks".

This chapter is devoted to researching the stability issue of fractional-order impulsive coupled systems over networks. Five classes of criteria on stability, uniform stability, uniform asymptotic stability, finite-time stability, and Mittag-Leffler stability, respectively, of the fractional-order impulsive coupled systems are derived by the graph theory and Lyapunov second technique.

Part Dynamical Networks

Chapter "Positivity and Stability of Nonlinear Time-Delay Systems in Neural Networks".

This chapter deals with the problems of positivity and stability of nonlinear time-delay systems in the area of neural network models including Hopfield neural networks, bidirectional associative memory neural networks, and inertial neural networks. A systematic approach involving comparison techniques via differential and integral inequalities in combination with nonlinear analysis has been presented.

Then, linear-programming-based conditions for the existence and exponential stability of positive equilibriums have been derived.

Chapter "Fault Detection and Isolation Methodology in Finite Frequency Domain for Constrained Networked systems".

This chapter investigates the transmission-dependent fault detection and isolation strategy in a finite frequency domain for networked systems with constrained communication capacities. A switched system with multiple stochastic parameters is firstly derived to model the whole system. Then, drawing support from the geometric method and the finite frequency stochastic performance indices, sufficient conditions are established to guarantee the desired performances and characterize the filter gains based on the proposed transmission-dependent fault detection and isolation scheme.

Chapter "Mean-Square Stochastic Stability of Delayed Hybrid Stochastic Inertial Neural Networks".

This chapter considers the problem of the mean-square stochastic stability analysis for hybrid stochastic inertial neural networks with time delays. The hybrid stochastic inertial neural networks are represented as the combination of a two-level system in which the first level is directed by the system of second-order differential equations and the second level is directed by the discrete set representing switching jump nodes. By considering a few types of time delays, sufficient delay-dependent mean-square stability conditions for the networks are established using variable transformation techniques and some integral inequalities based on the Lyapunov theory and linear matrix inequality framework.

Chapter "Exponential Stability and Stabilization of Stochastic Neural Network Systems via Switching and Impulsive Control".

This chapter deals with the exponential stability and stabilization problem of stochastic neural network systems via a new switching and impulsive control scheme. The main characteristics of this proposed controller lie in the asynchronous behavior between switchings and impulses. Also, the concepts of dwell time and impulsive interval are employed to analyze the system. Then, based on multiple Lyapunov functions approach, sufficient conditions for mean-square exponential stability are established for stability and controller design.

Chapter "Hybrid-Triggered Synchronization of Delayed Complex Dynamical Networks Subject to Stochastic Cyber-Attacks".

This chapter deals with a synchronization problem for a class of delayed complex dynamical networks with randomly occurring cyber-attacks. A hybrid event-triggered communication scheme, which includes a new event-triggered scheme and the conventional time-triggered scheme in one framework, is proposed to reduce the transmission load of networks and control resources. Then, a stability criterion of an error dynamic system for the synchronization problem is obtained such that the synchronization error is asymptotically stable in the mean-square sense.

Chapter "Cluster Synchronization on Derivative Coupled Lur'e Networks: Impulsive Pinning Strategy".

This chapter is concerned with the global and exponential synchronization issue for a class of nonidentically delay coupled Lur'e networks with stochastic disturbance and derivative couplings. By thinking the cluster topological structure of the Lur'e networks, a novel impulsive pinning controller is proposed. Then, the stability criteria and estimate of the exponential convergence velocity for the cluster synchronization are derived.

Chapter "Finite-Time Synchronization Control for Markovian Jump Memristive Neural Networks with Reaction-Diffusion Terms".

This chapter investigates the issue of finite-time synchronization for a class of memristive neural networks, where the reaction-diffusion items and Markovian jump parameters are both considered. In the analysis, it is assumed that the drive and response systems are set to obey inconsistent Markovian jumping laws, which is more practical in real systems. A discontinuous controller is designed so that the influence of the inconsistent parameters can be avoided by combining the novel Lyapunov-Krasovskii functional with a canonical Bessel-Legendre inequality and free weighted matrix method.

Kyongsan, Korea (Republic of) Ju H. Park
April 2020 Chumna Chair Professor

Acknowledgements

The editor of this edited book, I, Ju H. (Jessie) Park, has been studying various topics on nonlinear dynamical systems and networks for about 25 years. In particular, in recent years, my research team named "Laboratory for Advanced Nonlinear Dynamics and Control" in Yeungnam University, Republic of Korea has made remarkable achievements in the areas of control theory, signal processing, and artificial intelligence. So, in connection with these recent achievements, I have found consensus with my members of the team that the latest work on various hot topics on dynamic systems and networks needs to be introduced to as many readers as possible. In addition, I invited some of the experts in the related research field to increase the completeness of this project. This led to the high quality and diversity of this book, which consisted of 21 chapters. The material presented in this book has been the recent outcome of research efforts by contributors. Hence, I would like to thank all contributors for their dedicated participation to complete each chapter.

I as the editor of this edited volume would like to express my gratitude to the Editor of the book series *Studies in Systems, Decision and Control* of Springer-Nature, Prof. Dr. Janusz Kacprzyk (Polish Academy of Sciences); Executive Editor of the series, Dr. Thomas Ditzinger; and Project Coordinator, Ms. Saranya Kalidoss, Project Manager, Viju Falgon at the Springer-Nature for their professional and powerful handling of this book. Without their editorial comments and detailed examination, the publication of the book would not have gone so smoothly.

Personally, there are no words that suffice to thank my acquaintances, Eunkyung, Jaeho, Hoyoul, Michelle, Claude, Melvin, Jeonghee, Hyunjik, William, Jinguk, Jiyeon, Maurice, Woocheol, George, Brian, Johanne, Achim, Kevin, James, Johnny, and Helena for their constant encouragement, support, and sacrifice.

Finally, this work of writing the book was supported by the National Research Foundation of Korea (NRF) grant funded by the Korea government (MSIT) (No. 2020R1A2B5B02002002).

Kyongsan, Korea (Republic of) Ju H. Park

Contents

Switched Systems and Hybrid Control

Fractional-Order Systems

Dynamical Networks

Editor and Contributors

About the Editor

Ju H. Park (Jessie Park) received the Ph.D. degree in Electronics and Electrical Engineering from Pohang University of Science and Technology (POSTECH), Pohang, Republic of Korea, in 1997. From May 1997 to February 2000, he was a Research Associate Fellow in Engineering Research Center—Automation Research Center, POSTECH. He joined Yeungnam University, Kyongsan, Republic of Korea, in March 2000, where he is currently the Chuma Chair Professor.

From 2006 to 2007, he was a Visiting Professor in the Department of Mechanical Engineering, Georgia Institute of Technology. His research interests include robust control and filtering, neural networks, complex networks, fuzzy systems, multi-agent systems, chaotic systems, embedded systems, signal processing, intelligent vehicles, and applied mathematics. He has published a number of papers more than 600 in top leading journals and two books titled Recent Advances in Control and Filtering of Dynamic Systems with Constrained Signals and Dynamic Systems with Time Delays: Stability and Control in these areas.

Professor Park serves as an Editor of the International Journal of Control, Automation and Systems, Springer-Nature. He is also a Subject Editor/Advisory Editor/Associate Editor/Editorial Board member for several international journals, including IET Control Theory and Applications, Applied Mathematics and Computation, Journal of The Franklin Institute, Nonlinear Dynamics, Cogent Engineering, Engineering Reports, IEEE Transactions on Fuzzy Systems, IEEE Transactions on Neural Networks and Learning Systems, IEEE Transactions on Cybernetics, and so on.

Since 2015, he is a recipient of Highly Cited Researcher Award by Clarivate Analytics (formerly, Thomson Reuters) and listed in three fields, engineering, computer sciences, and mathematics in 2019. He is a Fellow of the Korean Academy of Science and Technology (KAST).

Further information about Prof. Park and his laboratory is available from the web page: http://yu.ac.kr/~rcl/.

Contributors

Haibo Bao School of Mathematics and Statistics, Southwest University, Chongqing, People's Republic of China

G. Bhuvaneshwari Division of Mathematics, School of Advanced Sciences, Vellore Institute of Technology-Chennai Campus, Chennai, India

Jinde Cao School of Mathematics, Southeast University, Nanjing, People's Republic of China

Joseph Chang Lun Chan Division of Electronic Engineering, Jeonbuk National University, Jeonju, Republic of Korea

Jun Cheng College of Mathematics and Statistics, Guangxi Normal University, Guilin, People's Republic of China

S. Dhanasekar Division of Mathematics, School of Advanced Sciences, Vellore Institute of Technology-Chennai Campus, Chennai, India

Dong Ding Engineering Research Center of Internet of Things Technology Applications (Ministry of Education), Jiangnan University, Wuxi, People's Republic of China

Cunchen Gao School of Mathematical Science, Ocean University of China, Qingdao, People's Republic of China

Raju K. George Department of Mathematics, Indian Institute of Space Science and Technology, Thiruvananthapuram, Kerala, India

S. Harshavarthini Department of Applied Mathematics, Bharathiar University, Coimbatore, India

Le Van Hien Faculty of Mathematics and Informatics, Hanoi National University of Education, Hanoi, Vietnam

Mengjie Hu Department of Electrical Engineering, Yeungnam University, Gyeongsan, Republic of Korea

Ticao Jiao School of Electrical and Electronic Engineering, Shandong University of Technology, Shandong, People's Republic of China

Youggui Kao Department of Mathematics, Harbin Institute of Technology (Weihai), Weihai, People's Republic of China

R. Krishnasamy Department of Applied Mathematics and Computational Sciences, PSG College of Technology, Coimbatore, Tamil Nadu, India

O. M. Kwon School of Electrical Engineering, Chungbuk National University, Seowon-gu, Cheongju, Republic of Korea

S. H. Lee School of Electrical Engineering, Chungbuk National University, Seowon-gu, Cheongju, Republic of Korea

Tae H. Lee Division of Electronic Engineering, Jeonbuk National University, Jeonju, Republic of Korea

Yajuan Liu School of Control and Computer Engineering, North China Electric Power University, Beijing, People's Republic of China

Yue Long School of Automation Engineering, University of Electronic Science and Technology of China, Chengdu, People's Republic of China

Jingtao Man School of Information Engineering, Henan University of Science and Technology, Luoyang, People's Republic of China

A. Manivannan School of Advanced Sciences - Division of Mathematics, Vellore Institute of Technology-Chennai Campus, Chennai, India

K. Mathiyalagan Department of Applied Mathematics, Bharathiar University, Coimbatore, India

Reza Mohajerpoor Data 61, CSIRO, Sydney, Australia

Ju H. Park Department of Electrical Engineering, Yeungnam University, Gyeongsan, Republic of Korea

M. J. Park Center for Global Converging Humanities, Kyung Hee University, Giheung-gu, Yongin, Republic of Korea

V. Parthiban Division of Mathematics, School of Advanced Sciences, Vellore Institute of Technology-Chennai Campus, Chennai, India

Wenhai Qi School of Engineering, Qufu Normal University, Rizhao, People's Republic of China

Yongxia Qu School of Mathematical Science, Ocean University of China, Qingdao, People's Republic of China

R. Sakthivel Department of Applied Mathematics, Bharathiar University, Coimbatore, India

Yaonan Shan University of Electronic Science and Technology of China, Chengdu, Sichuan, People's Republic of China

Hao Shen School of Electrical and Information Engineering, Anhui University of Technology, Ma'anshan, People's Republic of China

Shuai Song School of Automation, Nanjing University of Science and Technology, Nanjing, People's Republic of China

Xiaona Song School of Information Engineering, Henan University of Science and Technology, Luoyang, People's Republic of China

Chee Pin Tan School of Engineering and Advanced Engineering Platform, Monash University Malaysia, Selangor, Malaysia

Ze Tang Engineering Research Center of Internet of Things Technology Applications (Ministry of Education), Jiangnan University, Wuxi, People's Republic of China

Jing Wang School of Electrical and Information Engineering, Anhui University of Technology, Ma'anshan, People's Republic of China

Yudong Wang School of Metallurgy Engineering, Anhui University of Technology, Ma'anshan, People's Republic of China

Yude Xia School of Electrical and Information Engineering, Anhui University of Technology, Ma'anshan, People's Republic of China

Xiaoqing Xiao School of Information Science and Technology, Nantong University, Jiangsu, People's Republic of China

Xiaojian Yi School of Mechanical Engineering, Beijing Institute of Technology, Beijing, People's Republic of China

Li Zhang School of Mathematical Science, Ocean University of China, Qingdao, People's Republic of China

Lei Zhou School of Information Science and Technology, Nantong University, Jiangsu, People's Republic of China

List of Figures

Partial State Observers for LinearTime-Delay Systems

Finite-Time Stability and Control of Impulsive Positive Systems with Interval Uncertainty

An Eigenvector-Centrality Based Consensus Protocol Design for Discrete-Time Multi-agent Systems with Communication Delays

Fault-Tolerant Sampled-Data Synchronization of Chaotic Systems with Random Occurring Uncertainties: A Semi-Markov Jump Model Approach

Fuzzy-Based Sliding Mode Control Design for Stability Analysis of Nonlinear Interconnected Systems

Adaptive Synchronization of Fractional-Order Delayed Memristive Neural Networks

New Results on Stability of Coupled Impulsive Fractional-Order Systems on Networks

Positivity and Stability of Nonlinear Time-Delay Systems in Neural Networks

Fault Detection and Isolation Methodology in Finite Frequency Domain for Constrained Networked Systems

Mean-Square Stochastic Stability of Delayed Hybrid Stochastic Inertial Neural Networks

Exponential Stability and Stabilization of Stochastic Neural Network Systems via Switching and Impulsive Control

Hybrid-Triggered Synchronization of Delayed Complex Dynamical Networks Subject to Stochastic Cyber-Attacks

Cluster Synchronization on Derivative Coupled Lur'e Networks: Impulsive Pinning Strategy

Finite-Time Synchronization Control for Markovian Jump Memristive Neural Networks with Reaction-Diffusion Terms

List of Tables

Linear and Nonlinear Dynamic Systems

Partial State Observers for Linear Time-Delay Systems

Reza Mohajerpoor

Abstract State estimation is an essential component of any model-based control system. In particular, partial state estimation is favorable when the full set of the states of a system is not observable or detectable, or for large-scale systems that need only a few functions of the states to be monitored. This chapter overviews recent advancements in functional observer design algorithms for linear time-invariant systems. The concept of functional observability is explained for systems with and without perturbations, as a generalization of observability or detectability perceptions. Moreover, delay-free and delay-dependent functional observer design schemes for linear time-delay systems with state and input delays are illustrated. Since the exact delay values are often not measurable in practice, a methodology is demonstrated to deal with uncertain state delays in designing delay-dependent functional observers. In light of that, an auxiliary time-varying delay with arbitrary bounds is exploited in the observer, which effectively relaxes the conservative nature of delay-free observers.

Keywords State estimation · Time-delay · Functional observability · Observer

1 Introduction

A broad range of nonlinear dynamic systems can be linearized around desired set points that maximize the performance of the system. To maintain the operating condition of the system around a desired set point, linear control laws can be constructed by real-time feedback from the states of the system. However, monitoring/measuring all the states of a system is often not possible or extremely costly. To lift this burden the un-measured states of the system can be estimated using model-based state observers such as Luenberger observers [1]. If the linearized dynamics of the system

R. Mohajerpoor (✉)
Data 61, CSIRO, Sydney, Australia
e-mail: reza.mohajerpoor@data61.csiro.au

J. H. Park (ed.), *Recent Advances in Control Problems of Dynamical Systems and Networks*, Studies in Systems, Decision and Control 301,
https://doi.org/10.1007/978-3-030-49123-9_1

is accurate, then the error due to estimation is zero amid the steady-state operations. Otherwise, bounded errors can be expected depending on the accuracy of the modeled dynamics.

Another problem facing with model-based state estimators is that there might be a number of states that do not have any connection with the measured states, making the system *un-observable*. If the un-observable states decay to zero asymptotically over time, then the system is *detectable*. Traditional Luenberger-type observers can only estimate the states of the system if the system is observable or detectable. Otherwise, asymptotic state estimation cannot be achieved for aims such as system monitoring, fault detection and isolation, and control of the observable states of the system. Moreover, in large-scale systems, such as interconnected power systems or chemical reactors [2–4], it is often required to estimate a small set of linear functions of the states to achieve system monitoring, while estimating the whole states due to the implied numerical costs and complexities.

To resolve the issues related to conventional observers, functional observers (FOs) were introduced [5, 6] that aim to estimate a single or multiple linear functions of the states of the system, instead of the full set of the states. FO design for time-delay systems has been studied since 1999. Trinh [7] treated state delay terms as *unknown inputs* and designed a delay-free observer for the system. Later, Darouach [8] designed delay-dependent FOs for linear time-invariant (LTI) systems with *constant and known* state delays. The state delay is employed in the observer structure proposed in the paper. Several contributions were made thereafter to address delay-dependent FO design for state delay systems, though, the majority made the same limiting assumption of accurately measuring/estimating the state delays in real-time [9–12]. Seuret et al. [13] addressed the unknown state delay issue for full-order observers by proposing a sliding mode observer. Mohajerpoor et al. [14], later on, proposed a new robust functional observer structure and design algorithm to relax the problem of uncertain state delays.

This chapter provides an overview of the structure and design procedures of functional observers for LTI systems. We first introduce functional observers for LTI systems, and layout the existence conditions for FOs in Sect. 2. Thereafter, recent advancements on the design of functional observer for systems with input-delay and state-delay given that the state-delay is accurately measurable in real-time are comprehended in Sects. 3 and 4, respectively. Then, the restrictive assumption on the delay measurement for delay-dependent observer structure is relaxed in Sect. 5. Finally, the chapter is summarized and a few foreseeable directions are sketched in Sect. 6.

Notations: In this chapter, $\mathbb{R}$ is the space of real numbers; $\mathbb{C}^+$ is the space of complex numbers with positive real values; $\mathbb{R}^{n\times m}$ is the space of $n \times m$ matrices; $\mathbb{S}^{n\times n}$ is the space of $n \times n$ symmetric matrices; $\mathbb{R}^n$ is the n-dimensional Euclidean space; $[X_1; X_2]$ is equivalent to $\left[X_1^T, X_2^T\right]^T$; rows$(X)$ shows the number of rows of the matrix X; I_k the $k \times k$ identity matrices, and $\mathbf{0}$ is the zero matrix with appropriate dimension. Furthermore, $\otimes$ is the Kronecker product; $*$ in a symmetric matrix stands for the symmetric element; $\mathrm{sym}\,(X) = X + X^T$; $X^\dagger$ indicates the pseudo-inverse or the generalized inverse of matrix X; $X^{\perp}$ represents the right orthogonal matrix of X, wherein $XX^{\perp} = \mathbf{0}$; $X \prec 0\,(X \preceq 0)$ means that matrix X is negative definite (negative semi-definite), and $X \succ 0\,(X \succeq 0)$ indicates that X is positive definite (positive semi-definite). In addition, $\mathcal{R}(X)$ and $\mathcal{N}(X)$ are the row space and the null space of matrix X, respectively. $[[\mathcal{S}]]$ demonstrates a matrix constructed from the row basis vectors for subspace $\mathcal{S}$. Finally, $\mathcal{C}_n(\Omega)$ pinpoints the space of continuous functions mapping from Ω to $\mathbb{R}^n$ with the topology of uniform convergence.

2 Functional Observer Design for LTI Systems

Consider the following LTI dynamic system:

$$\begin{aligned} \dot{x}(t) &= Ax(t) + Bu(t), \\ y(t) &= Cx(t), \\ z(t) &= Lx(t), \end{aligned} \tag{1}$$

where $x(\cdot) \in \mathbb{R}^n$ is the vector of the states of the system, $u(\cdot) \in \mathbb{R}^m$ is the control input, and $y(\cdot) \in \mathbb{R}^p$ and $z(\cdot) \in \mathbb{R}^l$ are the vector of measurement outputs and the desired functions to be constructed. To add, $A \in \mathbb{R}^{n\times n}$, $B \in \mathbb{R}^{n\times m}$, $C \in \mathbb{R}^{p\times n}$, and $L \in \mathbb{R}^{l\times n}$. Moreover, we assume $\operatorname{rank}([C; L]) = p + l \leq n$, i.e. matrices C and L are full row rank and span independent spaces, and the number of functions to be estimated is not greater than the detectable number of states $n - p$.

The aim of the functional observer is to construct $\hat{z}(t) \in \mathbb{R}^l$ in a way that the error $e(t) = \hat{z}(t) - z(t)$ approaches to zero asymptotically. To this aim, the below lth order dynamic structure can be considered for the observer:

$$\begin{aligned} \dot{\omega}(t) &= F\omega(t) + Gu(t) + Hy(t), \\ \hat{z}(t) &= \omega(t) + Vy(t), \end{aligned} \tag{2}$$

where $\omega(\cdot) \in \mathbb{R}^l$ is the internal states of the observer, and $F \in \mathbb{R}^{l\times l}$, $G \in \mathbb{R}^{l\times m}$, $H \in \mathbb{R}^{l\times p}$, and $V \in \mathbb{R}^{l\times p}$ are observer design parameters. The following theorem summarizes the conditions required for the FO error dynamics to be (globally) asymptotically stable.

Theorem 1 ([15]) *The observer error $e(t)$ converges to zero asymptotically if and only if the following conditions are satisfied:*

I. *there exists a matrix $T \in \mathbb{R}^{l\times n}$ that complies with the following matrix equations:*

$$FT - TA + HC = \mathbf{0}, \tag{3a}$$

$$G = TB, \tag{3b}$$

$$L - T - VC = \mathbf{0}. \tag{3c}$$

II. *Matrix F is stable, i.e. $F \prec 0$.*

Nonlinear Eq. (3a) is called Sylvester equation and appears in numerous practical problems. Designing matrices F, T, H, V, and G is not straightforward due to the nonlinear term FT in the Sylvester equation. There are a few algorithms that solve (3) in a way that F is Hurwitz. It has been shown that each method implies numerical conservatism and carries limitations in solving the equation effectively and efficiently [6]. *Transformation based* frameworks employ matrix transformations to linearize the nonlinear equation, and they are found to be more effective than the other solution methods. There are multiple transformation based solution algorithms in the literature. A recent less conservative solution method in this category proposed in [6] is presented in the coming section.

2.1 A FO Design Algorithm

It is obtained from (3c) that

$$\begin{bmatrix} T & V \end{bmatrix} = LM^{\dagger} + Z(I_{n+p} - MM^{\dagger}) \tag{4}$$

where $M \triangleq [I_n; C] \in \mathbb{R}^{(n+p)\times n}$, and $Z \in \mathbb{R}^{l\times(n+p)}$ is an arbitrary matrix that works as a design parameter. By partitioning $M^{\dagger}$ and $I_{n+p} - MM^{\dagger}$ as $[\, M_1 \; M_2 \,] \triangleq M^{\dagger}$ and $[\, \bar{M}_1 \; \bar{M}_2 \,] := I_{n+p} - MM^{\dagger}$, where $M_1 \in \mathbb{R}^{n\times n}$, $M_2 \in \mathbb{R}^{n\times p}$, $\bar{M}_1 \in \mathbb{R}^{(n+p)\times n}$, and $\bar{M}_2 \in \mathbb{R}^{(n+p)\times p}$. Therefore, (4) can be written as

$$T = LM_1 + Z\bar{M}_1, \tag{5a}$$

$$V = LM_2 + Z\bar{M}_2. \tag{5b}$$

Moreover, post-multiplying (3a) by $\bar{C} \triangleq C(I_n - L^{\dagger}L)$, one gets

$$H = -FTC^{\dagger} + TAC^{\dagger}. \tag{6}$$

$$FTC^{\perp} = TAC^{\perp}. \tag{7}$$

Now, from (5a) and (7) we have

$$FLM_1C^{\perp} + FZ\bar{M}_1C^{\perp} = LM_1AC^{\perp} + Z\bar{M}_1AC^{\perp}. \tag{8}$$

Since it can be shown that $\bar{M}_1$ and $C^{\perp}$ are orthogonal, i.e. $\bar{M}_1C^{\perp} = \mathbf{0}$ [6], (8) can be written as

$$\begin{bmatrix} F, & -Z \end{bmatrix} \Omega = \Phi, \tag{9}$$

where $\Omega \triangleq [LM_1C^\perp; \bar{M}_1AC^\perp] \in \mathbb{R}^{(n+p+l)\times(n-p)}$ and $\Phi \triangleq LM_1AC^\perp \in \mathbb{R}^{l\times(n-p)}$. If there is a solution for (9), one obtains

$$F = N_{11} + \tilde{Z}N_{21}, \tag{10}$$

$$Z = -N_{12} - \tilde{Z}N_{22}, \tag{11}$$

where $\begin{bmatrix} N_{11} & N_{12} \end{bmatrix} = N_1 \triangleq \Phi\Omega^\dagger \in \mathbb{R}^{l\times(n+p+l)}$, $\begin{bmatrix} N_{21} & N_{22} \end{bmatrix} = N_2 \triangleq (I_{n+p+l} - \Omega\Omega^\dagger) \in \mathbb{R}^{(n+p+l)\times(n+p+l)}$, $N_{11} \in \mathbb{R}^{l\times l}$, $N_{12} \in \mathbb{R}^{l\times(n+p)}$, $N_{21} \in \mathbb{R}^{(n+p+l)\times l}$, $N_{22} \in \mathbb{R}^{(n+p+l)\times(n+p)}$, and $\tilde{Z} \in \mathbb{R}^{l\times(n+p+l)}$ is a free matrix that is designed in a way that matrix F is Hurwitz.

Note that $\tilde{Z}$ can be designed using multiple numerical methods such as pole-placement [16], provided that the pair (N_{11}, N_{21}) is observable or detectable.

2.2 Existence of a Solution

Solving the set of matrix equations (3) in a way that conditions of Theorem 1 are fulfilled implies that (9) can be solved and the pair (N_{11}, N_{21}) is at least detectable. These two conditions are necessary and sufficient for the presence of an stable functional observer (2) for System (1), and it has been proven that they can be translated into the following conditions [17]:

$$\text{rank}\left(\begin{bmatrix} LA \\ CA \\ C \\ L \end{bmatrix}\right) = \text{rank}\left(\begin{bmatrix} CA \\ C \\ L \end{bmatrix}\right), \tag{12}$$

$$\text{rank}\left(\begin{bmatrix} sL - LA \\ CA \\ C \end{bmatrix}\right) = \text{rank}\left(\begin{bmatrix} CA \\ C \\ L \end{bmatrix}\right) \quad \forall s \in \mathbb{C}^+. \tag{13}$$

It can be shown that Conditions (12) and (13) are less conservative than conventional observability/detectability conditions.

A beautiful aspect of functional observers is their potential flexibility, such that even if Conditions (12) and (13) are not realized for a system, there is still an opportunity to design a functional observer to estimate $z(t) = Lx(t)$. This can be achieved via augmenting extra functions to $z(t)$ to generate a new functional $z_a(t) = L_a x(t)$, where $L_a = [L; R_1; R_2]$. Functional vector $R_1x(t)$ is added to fulfill Condition (12) and $R_2x(t)$ is added to satisfy (13) in a way that (12) is not violated.

Prior constructing additional row vectors R_1 and R_2, we should ensure that there exist an stable functional observer for System (1). The answer lies in the concept of *functional observability/detectability* defined in theorem below.

Theorem 2 ([18]) *The triple* (A, C, L) *is functional detectable if and only if*

$$\text{rank}\left(\begin{bmatrix} sI_n - A \\ C \\ L \end{bmatrix}\right) = \text{rank}\left(\begin{bmatrix} sI_n - A \\ C \end{bmatrix}\right), \quad \forall s \in \mathbb{C}^+ \tag{14}$$

To add, System (1) is functional observable if and only if (14) holds for all $s \in \mathbb{C}$. *Equivalently, the triple* (A, C, L) *is functional observable if and only if*

$$\text{rank}\left(\begin{bmatrix} C \\ CA \\ \vdots \\ CA^{n-1} \\ L \\ LA \\ \vdots \\ LA^{n-1} \end{bmatrix}\right) = \text{rank}\left(\begin{bmatrix} C \\ CA \\ \vdots \\ CA^{n-1} \end{bmatrix}\right) \tag{15}$$

Remark 1 Condition (14) implies that the complex *Rosenbrock* matrices on both sides of the equation have the same *invariant zeros*. This condition can be readily testified using numerical matrix libraries such as Python libraries or MATLAB.

2.3 Constructing Minimal Order Functional Observers

Establishing the row vectors R_1 and R_2 such that minimum possible order FO (*minimal* FO) can be designed is not straightforward. A powerful recursive algorithm called *MOI-FO* (minimum order increase for the functional observer) algorithm is presented below to achieve this aim (refer to [6] for further insights).

MOI-FO Algorithm:

1. If $L_0 = L$ fulfills (12), define $L_\beta = L_0$, and proceed to Step 3. Otherwise, set $i = 0$ proceed to the next step.
2. Define $\Pi_i \triangleq [[\mathcal{R}(\Theta_i)^\perp]]\Phi_i H_2^i$, where $H_1^i \triangleq [[\mathcal{R}([C; CA; L_i])]]$, $H_2^i \triangleq [[\mathcal{R}([H_1^i; L_i A])]]$, $\begin{bmatrix} \Phi_i & \Psi_i \end{bmatrix} \triangleq \left[\left[\mathcal{N}\left(\begin{bmatrix} H_2^j A; & H_2^j \end{bmatrix}\right)\right]\right]$, and $\begin{bmatrix} \Theta_i & \Gamma_i \end{bmatrix} \triangleq \left[\left[\mathcal{N}\left(\begin{bmatrix} \Phi_i H_2^i; & H_1^i \end{bmatrix}\right)\right]\right]$. For $j = \{1, \ldots, \text{rows}(\Pi_i)\}$ define $L_i^j = [L_i^{j-1}; q_j]$ with $L_i^1 = [L_i; q_1]$, where q_j is the jth row of Π_i. If (12) is complied by L_i^j, then $L_\beta = L_i^j$ and proceed to Step 4. Otherwise, apply $j \leftarrow j + 1$ and repeat this step. If augmenting the whole Π_i does not satisfy condition (12), then move to the next step.

3. Establish $\Lambda_i \triangleq \left[\left[\mathcal{R}(H_2^i) \cap \mathcal{R}\left(\left[\, H_1^i;\ \Pi_i \,\right]\right)^{\perp}\right]\right]$, define $L_{i+1} = [L_i;\ \Pi_i;\ \Lambda_i]$, set $i \leftarrow i+1$, and return to Step 2.
4. If (13) is satisfied by L_β, the algorithm is terminated successfully by $L_\mathrm{a} = L_\beta$. Otherwise, define $\left[\,\Xi\ \Delta\,\right] \triangleq [[\mathcal{N}\left(\left[\, CA^2;\ H_1^\beta \,\right]\right)]]$, where Ξ has p columns. Furthermore, define $\Pi \triangleq \mathcal{R}(\Xi CA)$, and $L_\beta^j = [L_\beta^{j-1};\ q_j]$ for $j = \{1, \ldots,$ rows$(\Pi)\}$, where $L_\beta^1 = [L_\beta;\ q_1]$ and q_j is the jth row of Π. If L_β^j satisfies (13), then $L_\mathrm{a} \leftarrow L_\beta^j$, and the algorithm is terminated. Otherwise, the step is repeated by $j \leftarrow j+1$.

Example 1 Consider the following system

$$A = \begin{bmatrix} -5 & -2 & 5 & 0 & 1 \\ 2 & -6 & 1 & 0 & -3 \\ 0 & -2 & -8 & 0 & 0 \\ -6 & 5 & 7 & -5 & -5 \\ -2 & 0 & -4 & 0 & 0 \end{bmatrix}, \quad B = \begin{bmatrix} 1 \\ -3 \\ -5 \\ 5 \\ -2 \end{bmatrix},$$

$$C = \left[\begin{bmatrix} 0 & 1 \\ 1 & 1 \end{bmatrix}, \mathbf{0}_{2\times 3}\right], \quad L = \begin{bmatrix} 0 & 0 & 1 & 0 & 0 \\ 0 & 0 & 0 & 1 & 0 \end{bmatrix}.$$

The pair (A, C) is *not* observable, nor detectable. Moreover, the triple (A, C, L) is *not* functional observable, however, it is functional detectable according to Condition (14) in Theorem 2.

It can be realized that L meets Condition (12), whereas Condition (13) is not satisfied indicating that the system is *not* functional detectable under this setting. Therefore, following the MOI-FO algorithm a row vector $q = \left[\, -3\ -8\ 6\ 0\ -2 \,\right]$ is obtained to be augmented to L. Hence, $L_\mathrm{a} = [L;\ q]$ meets both Conditions (12) and (13). The invariant zeros of the matrix pencil $S = \left[\, sI_n - A;\ C;\ L \,\right]$ is $\lambda = -5$, which is the undetectable eigenvalue of the observer and should be incorporated as desired FO's eigenvalue.

Assigning the FO's poles at $\{-10.2648, -5, -9\}$ ($\{-10.2648, -9\}$ are arbitrary and adjust the performance of the observer) the observer parameters were calculated based on the presented method in Sect. 2.1 and another transformation-based framework proposed in [19].

Thereafter, the performance of the designed FOs were examined through numerical simulation of the system excited with an arbitrary input $u(t) = 2 + 10e^{-0.4t}\cos(2t)$, $(t \geq 0)$, as shown in Fig. 1. The results indicate that the presented observer design algorithm is performing strong in estimating functions $[z_1;\ z_2;\ z_3] = L_\mathrm{a} x(t)$, while the solution algorithm proposed in [19] fails to estimate the functions with a desirable accuracy. The reason for such failure lies in the numerical issues confronted with the majority of solution algorithms for the nonlinear Sylvester equation.

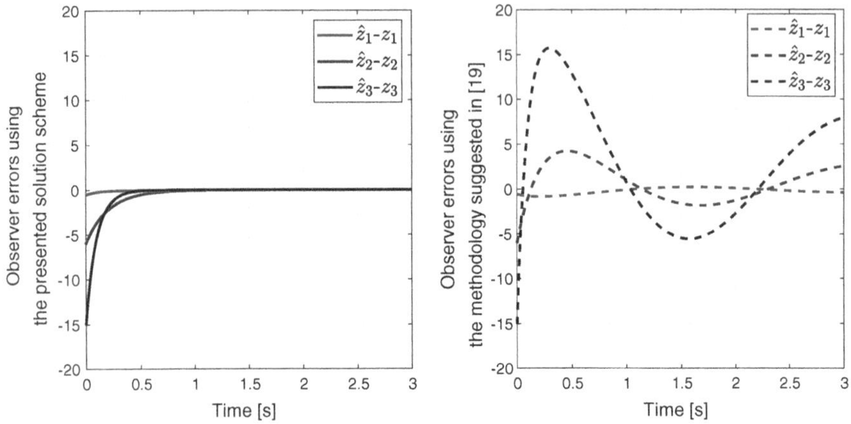

Fig. 1 The errors due to estimating functions $z(t) = L_a x(t)$ using: (i) presented solution in Sect. 2.1 (left figure) and (ii) the solution proposed in [19] (right figure)

3 Delay-Free FO Design for Systems with Unknown Input Delay

Time delay can appear in the states (internal delay), the input (actuator lag), or even the output (measurement lag) of a system [20].

This section focuses on LTI systems with input delay, and the delay terms are treated as unknown-inputs and a delay-free FO structure with minimal order is designed for the system.

Consider an LTI system with time-varying input delay described below:

$$\begin{aligned} \dot{x}(t) &= Ax(t) + Bu(t) + B_d u_d(t - d_u(t)), \\ y(t) &= Cx(t), \\ z(t) &= Lx(t), \end{aligned} \tag{16}$$

where A, B, C, and L are as defined in (1), $B_d \in \mathbb{R}^{n \times m_d}$, and $d_u(\cdot)$ is the time-varying input delay. We design an observer with structure (2) for the system such that the error $e(t) = \hat{z}(t) - z(t)$ asymptotically converges to zero, irrespective of the delay term $B_d u_d(t - d_u(t))$.

Indeed, the observer parameters are designed to decouple the delay term from the error dynamics. This aim can be achieved if conditions given in Proposition 1 below are assured.

Proposition 1 *The lth order FO (2) is an asymptotically stable FO for system (16) if and only if conditions I and II of Theorem 1 are satisfied, and in addition parameter T satisfies the following equality constraint:*

$$TB_d = \mathbf{0}. \tag{17}$$

Additional condition (17) is added to the conditions of Theorem 2 to decouple the observer dynamics from the input delay terms. Similar to the FO design procedure, solving (3) and (17) such that matrix F is stable is obscured and faces numerical difficulties. Therefore, a solution algorithm adopted from [21] is suggested in the next section that is less conservative than a number of other algorithms. This class of observers are categorized as *unknown-input functional observers* (UIFOs) in the literature [22–24].

3.1 A Solution Algorithm for UIFOs

Let us first define a non-singular transformation matrix $U \triangleq \left[C^{\dagger}\ C^{\perp}\right]$, and $\tilde{x}(t) \triangleq U^{-1}x(t)$. Thus, system (16) is transformed into

$$\begin{aligned} \dot{\tilde{x}}(t) &= \tilde{A}\tilde{x}(t) + \tilde{B}u(t) + \tilde{B}_d u_d(t - d_u(t)), \\ y(t) &= \tilde{C}x(t), \\ z(t) &= \tilde{L}x(t), \end{aligned} \tag{18}$$

where $\tilde{A} \triangleq U^{-1}AU$, $\tilde{B} \triangleq U^{-1}B$, $\tilde{B}_d \triangleq U^{-1}B_d$, $\tilde{C} \triangleq CU$, and $\tilde{L} \triangleq LU$.

As a result of the system transformation, conditions (3) and (17) also undergo the transformation, which results in the following set of equations (here $\tilde{T} \triangleq TU$):

$$F\tilde{T} + H\tilde{C} - \tilde{T}\tilde{A} = \mathbf{0}, \tag{19a}$$

$$\tilde{T} + V\tilde{C} - \tilde{L} = \mathbf{0}, \tag{19b}$$

$$\tilde{T}\tilde{B}_d = \mathbf{0}, \tag{19c}$$

$$G = \tilde{T}\tilde{B}. \tag{19d}$$

Combining (19b) and (19b), one obtains:

$$\tilde{T} = \tilde{L}_m\tilde{M}_1 + Z_u\bar{\tilde{M}}_1, \tag{20}$$

$$V = \tilde{L}_m\tilde{M}_2 + Z_u\bar{\tilde{M}}_2, \tag{21}$$

where $\tilde{M} \triangleq \left[I_n, \tilde{B}_d; \tilde{C}, \mathbf{0}_{p\times m_d}\right]$, $\tilde{L}_m \triangleq \left[\tilde{L}, \mathbf{0}_{l\times m_d}\right]$, $\left[\tilde{M}_1\ \tilde{M}_2\right] \triangleq \tilde{M}^{\dagger}$, and $\left[\bar{\tilde{M}}_1\ \bar{\tilde{M}}_2\right] \triangleq I_{n+p} - \tilde{M}\tilde{M}^{\dagger}$, with $\tilde{M}_1 \in \mathbb{R}^{(n+m_d)\times n}$, $\tilde{M}_2 \in \mathbb{R}^{(n+m_d)\times p}$, $\bar{\tilde{M}}_1 \in \mathbb{R}^{(n+p)\times n}$, $\bar{\tilde{M}}_2 \in \mathbb{R}^{(n+p)\times p}$, and $Z_u \in \mathbb{R}^{l\times(n+p)}$ is an arbitrary matrix to be designed.

Now, post-multiplying Sylvester equation (19a) by $\left[\, \tilde{C}^{\dagger} \; \tilde{C}^{\perp} \,\right]$, we get

$$H = -F\tilde{T}\tilde{C}^{\dagger} + \tilde{T}\tilde{A}\tilde{C}^{\dagger}. \tag{22}$$

$$F\tilde{T}\tilde{C}^{\perp} = \tilde{T}\tilde{A}\tilde{C}^{\perp}. \tag{23}$$

Putting $\tilde{T}$ from (20) into (23) yields

$$\left[\, F, \, -Z_u \,\right] \Omega_u = \Phi_u, \tag{24}$$

where $\Omega_u \triangleq \left[\, \tilde{L}_m \tilde{M}_1 \tilde{C}^{\perp}; \; \bar{\tilde{M}}_1 \tilde{A} \tilde{C}^{\perp} \,\right]$, and $\Phi_u \triangleq \tilde{L}_m \tilde{M}_1 \tilde{A} \tilde{C}^{\perp}$. Equation (24) is similar to (9) when designing ordinary FOs.

Assuming that (24) can be solved, the following can be obtained

$$F = \tilde{N}_{11} + \tilde{Z}_u \tilde{N}_{21}, \tag{25}$$

$$Z_u = -\tilde{N}_{12} - \tilde{Z}_u \tilde{N}_{22}, \tag{26}$$

where $\tilde{N}_1 \triangleq \Phi_u \Omega_u^{\dagger}$, $\tilde{N}_2 \triangleq (I_{n+p+l} - \Omega_u \Omega_u^{\dagger})$, $\left[\, \tilde{N}_{11} \; \tilde{N}_{12} \,\right] \triangleq \tilde{N}_1$, and $\left[\, \tilde{N}_{21} \; \tilde{N}_{22} \,\right] \triangleq \tilde{N}_2$, with $\tilde{N}_{11} \in \mathbb{R}^{l \times l}$, $\tilde{N}_{12} \in \mathbb{R}^{l \times (n+p)}$, $\tilde{N}_{21} \in \mathbb{R}^{(n+p+l) \times l}$, and $\tilde{N}_{22} \in \mathbb{R}^{(n+p) \times (n+p+l)}$. Moreover, $\tilde{Z}_u \in \mathbb{R}^{l \times (n+p+l)}$ is a free matrix that should be designed in a way that matrix F is Hurwitz. This is doable if and only if the pair $(\tilde{N}_{11}, \tilde{N}_{21})$ is observable or detectable.

3.2 Existence of a Solution for UIFOs

The following theorem explains easy to check conditions for the existence of an asymptotically stable and delay-free FO (2) for System (16).

Theorem 3 ([22]) *There exists an asymptotically stable lth order and proper FO with structure (2) for System (16) if and only if the below conditions are satisfied:*

$$\operatorname{rank}\left(\begin{bmatrix} LA & LB_d \\ CA & CB_d \\ C & \mathbf{0} \\ L & \mathbf{0} \end{bmatrix}\right) = \operatorname{rank}\left(\begin{bmatrix} CA & CB_d \\ C & \mathbf{0} \\ L & \mathbf{0} \end{bmatrix}\right). \tag{27}$$

$$\operatorname{rank}\left(\begin{bmatrix} sL - LA & LB_d \\ CA & CB_d \\ C & \mathbf{0} \end{bmatrix}\right) = \operatorname{rank}\left(\begin{bmatrix} CA & CB_d \\ C & \mathbf{0} \\ L & \mathbf{0} \end{bmatrix}\right) \quad \forall s \in \mathbb{C}^{+}. \tag{28}$$

It can be shown that condition (27) ensures that there is a solution for (24), and condition (28) guarantees the detectability of the pair $(\tilde{N}_{11}, \tilde{N}_{21})$.

It can be shown that conditions of Theorem 3 can be relaxed by adding extra auxiliary rows to matrix L, if one or both of conditions (27) and (28) are not satisfied, provided that the quadruple (A, C, L, B_d) is *unknown-input functional observable* or *unknown-input functional detectable*.

Theorem 4 ([25]) *The quadruple (A, C, L, B_d) is unknown-input functional detectable if and only if the following conditions are attained,*

$$\text{rank}\left(\begin{bmatrix} sI_n - A & B_d \\ C & \mathbf{0} \\ L & \mathbf{0} \end{bmatrix}\right) = \text{rank}\left(\begin{bmatrix} sI_n - A & B_d \\ C & \mathbf{0} \end{bmatrix}\right), \quad \forall s \in \mathbb{C}^+ \tag{29}$$

$$\text{rank}(CB_d) = \text{rank}\left(\begin{bmatrix} CB_d \\ LB_d \end{bmatrix}\right) \tag{30}$$

The quadruple is unknown-input functional observable if and only if condition (30) is realized, and condition (29) holds for all $s \in \mathbb{C}$.

Remark 2 Condition (30) is called the matching condition for a proper UIFO. It is dictated by this condition that $\mathcal{R}(LB_d)$ must lie in the row space spanned by CB_d. However, this condition is still less conservative than the matching condition for designing unknown-input observers, where it is essential for the number of perturbations to be less than or equal to the number of outputs implied by $\text{rank}(CB_d) = \text{rank}(B_d)$ [21]. By the way, the matching condition can be further relaxed, if the derivatives of the outputs are employed in the observer structure, resulting in *improper* observers [26]. In addition, condition (29) insinuates that L must not generate additional invariant zeros in the right half complex plane than what C imposes on the system's matrix.

A systematic and recursive algorithm to find minimum number of additional row vectors to be appended to L to satisfy conditions (27) and (28) is presented below [21]. The algorithm is called *MOI-UIFO* (minimum order increase for unknown-input functional observer), and it essentially builds minimum number of auxiliary functions $R_1x(t)$ and $R_2x(t)$, such that $L_a = [L; R_1; R_2]$ fulfills conditions (27) and (28). This enables the design of a minimal order UIFO for the delayed system (16). R_1 literally is added to assure (27), and R_2 is added to satisfy (28) without breaching (27). Unknown-input functional detectability guarantees the existence or nonexistence of the row vectors R_1 and R_2. Derivation of the algorithm may look intricate, though it follows the same philosophy of the MOI-FO algorithm.

MOI-UIFO Algorithm:

1. If L satisfies (27), define $L_\beta = L$, $\beta = 0$, and proceed to Step 3. Otherwise, set $i = 0$, $L_0 = L$, and proceed to the next step.

2. Define $\Pi_i^d \triangleq \mathcal{W}_i\,[[\mathcal{R}(\Theta_i^d)^\perp]]\,\bar{\Phi}_i^d \Phi_i^d \bar{H}_2^{d,i}$, where
$H_1^{d,i} \triangleq [[\mathcal{R}([\,CA,\ CB_d;\ C,\ \mathbf{0};\ L_i,\ \mathbf{0}\,])]]$; $H_2^{d,i} \triangleq \left[\left[\mathcal{R}\left(\left[\,H_1^{d,i};\ L_iA,\ L_iB_d\,\right]\right)\right]\right]$;
$[\,\Phi_i^d,\ \Psi_i^d\,] \triangleq \left[\left[\mathcal{N}\left(\left[\,\bar{H}_2^{d,i}[\,A,\ B_d\,];\ H_2^{d,i}\,\right]\right)\right]\right]$;
$[\,\bar{\Phi}_i^d,\ \bar{\Psi}_i^d\,] \triangleq [[\mathcal{N}([\,\Phi_i^d\bar{H}_2^{d,i}[\,I_n\ \mathbf{0}_{n\times m_d}\,];\ H_2^{d,i}\,])]]$;
$[\,\Theta_i^d\ \Gamma_i^d\,] \triangleq \left[\left[\mathcal{N}\left(\left[\,[\,\bar{\Phi}_i^d\Phi_i\bar{H}_2^{d,i},\ \mathbf{0}_{\kappa\times m_d}\,];\ H_1^{d,i}\,\right]\right)\right]\right]$; $\kappa \triangleq \text{rows}(\Phi_i^d)$; and
$[\,\mathcal{W}_i\ \mathcal{G}_i\,] \triangleq [[\mathcal{N}([\,[[\mathcal{R}(\Theta_i^d)^\perp]]\,\bar{\Phi}_i^d\Phi_i^d\bar{H}_2^{d,i}B_d;\ CB_d\,])]]$. Moreover, $\bar{H}_1^{d,i}$ and $\bar{H}_2^{d,i}$ are any n independent columns of $H_1^{d,i}$ and $H_2^{d,i}$, respectively. Define $L_i^j = [L_i^{j-1};\ q_j]$ for $j = \{1, \ldots, \text{rows}(\Pi_i^d)\}$, with $L_i^1 = [L_i;\ q_1]$, where q_j is the jth row of Π_i^d. If (27) is met by L_i^j, then $L_\beta = L_i^j$, $\beta = i$, and proceed to Step 4. Contrarily, if (27) is not realized even by augmenting the whole Π_i^d to L_i, then the next step should be carried forward.
3. Define $\Lambda_i^d \triangleq \left[\left[\mathcal{R}(\bar{H}_2^{d,i}) \cap \mathcal{R}\left([\,\bar{H}_1^i;\ \Pi_i^d\,]\right)^\perp\right]\right]$ and
$[\,\bar{\mathcal{W}}_i\ \bar{\mathcal{G}}_i\,] \triangleq [[\mathcal{N}([\,\Lambda_i^d B_d;\ CB_d\,])]]$. Then, set $L_{i+1} \leftarrow [L_i;\ \Pi_i^d;\ \bar{\mathcal{W}}_i\Lambda_i^d]$, $i \leftarrow i+1$, and return to Step 2.
4. If the detectability condition (28) is fulfilled by $L = L_\beta$, then $L_\text{a} = L_\beta$ and the algorithm stops. Otherwise, define $\Pi^d \triangleq \mathcal{R}(\tilde{\mathcal{W}}\Xi CA)$, where $\left[\,\Xi^d\ \Delta^d\,\right] \triangleq \left[\left[\mathcal{N}\left(\left[\,CA[\,A\ B_d\,];\ H_1^{d,\beta}\,\right]\right)\right]\right]$, and $[\,\tilde{\mathcal{W}}\ \tilde{\mathcal{G}}\,] \triangleq [[\mathcal{N}([\,\Xi^d CAB_d;\ CB_d\,])]]$. Now, for $j = \{1, \ldots, \text{rows}(\Pi^d)\}$ define $L_\beta^j = [L_\beta^{j-1};\ q_j]$, with $L_\beta^1 = [L_\beta;\ q_1]$. If (28) is satisfied by $L = L_\beta^j$, then we have $L_\text{a} = L_\beta^j$. Otherwise, this step is repeated after an increment in $j \leftarrow j+1$.

Example 2 Consider a system with the following parameters:

$$A = \begin{bmatrix} -2.51 & 0.33 & 0.68 & 1.12 & -0.25 \\ 0.14 & -0.23 & -0.31 & 0.91 & 0.36 \\ 0.51 & -1.18 & 0.41 & 0.63 & -0.77 \\ 0.22 & 0.33 & 0.46 & 0.65 & -0.77 \\ 0.23 & 0.33 & 3.97 & 0.06 & 0.69 \end{bmatrix},$$

$$B = \begin{bmatrix} 0.43 \\ 0 \\ 0.92 \\ 1.2 \\ -1.27 \end{bmatrix}, \quad B_d = \begin{bmatrix} 1 & 0 \\ -3 & -1 \\ 0 & 0.5 \\ 0.45 & 0 \\ 0 & 0 \end{bmatrix},$$

$$C = \begin{bmatrix} 1 & 0 & 0 & 0 & 0.6 \\ 0 & 0 & 0 & 1 & 0 \end{bmatrix}, \quad L = \begin{bmatrix} 2 & 0 & 0 & 9 & 0.3 \end{bmatrix}.$$

After checking conditions (29) and (30), it is acknowledged that the quadruple (A, C, L, B_d) is unknown-input *functional detectable*.

Nevertheless, the system is not unknown-input functional observable, as the system's matrix $\left[\,sI_n - A,\ B_d;\ C,\ \mathbf{0}\,\right]$, has two invariant zeros at $-2.1622 \pm 2.0253i$, while the augmented system's matrix $\left[\,sI_n - A,\ B_d;\ \left[\,C;\ L\,\right],\ \mathbf{0}\,\right]$ has no invariant zeros. The two invariant zeros correspond to the stable un-observable modes of the system that should be included in the set of desired observer's eigenvalues.

In addition, it can be simply found that $\text{rank}(CB_d) \neq \text{rank}(B_d)$. Conclusively, there is *no* proper unknown-input observer for the system.

Condition (27) is not satisfied by L, thus, the order of observer is increased by applying the MOI-UIFO algorithm that results in an additional row vector $q_1 = \begin{bmatrix} 0.0603 \; 0.1620 \; 0.3239 \; -0.0439 \; -0.01 \end{bmatrix}$ to be appended to L. In addition, Condition (28) is not met by $L_\beta = [L; q_1]$.

Accordingly, $q_2 = \begin{bmatrix} -0.0045 \; 0.1041 \; 0.2082 \; 0.2082 \; -0.2050 \end{bmatrix}$ is obtained from Step 4 of the MOI-UIFO algorithm, such that $L_a = [L; q_1; q_2]$ fulfills both conditions of Theorem 4.

Adhering to the framework outlined in Sect. 3.1 appropriate observer parameters are south to assign the poles of F at $(-2.1622 \pm 2.0253i, -5)$. However, due to a near-singular numerical composition involved in this particular example and numerous matrix operations such as inverse and multiplications that propagate numerical errors, the algorithm proposed in [22] fails to address appropriate parameters. This highlights the numerical superiority of the proposed algorithm in this chapter over comparable frameworks.

4 Delay-Dependent FO Design for Systems with Measurable State Delay

When a number of the states of system (1) are lagged, instead of the actuator inputs, the dynamics of the system is written as

$$\begin{aligned} \dot{x}(t) &= Ax(t) + A_d x(t - d(t)) + Bu(t), \\ y(t) &= Cx(t), \\ z(t) &= Lx(t), \\ x(\theta) &= \phi(\theta) \qquad \forall \theta \in [-d_2, 0], \end{aligned} \tag{31}$$

where $A_d \in \mathbb{R}^{n \times n}$, $\phi(\cdot) \in \mathcal{C}_n([-d_2, 0])$ is the initial function, and A, B, C, and L are as defined in (1).

The delay is assumed to be time-varying, bounded and with arbitrary upper-bound on its derivative [27]:

$$0 < d_1 \leq d(t) \leq d_2, \qquad \dot{d}(t) \leq \mu, \tag{32}$$

where $d_2 > d_1 \geq 0$, and $\mu \geq 0$ is the upper-bound of the rate of change of the delay.

Let us denote $d_{12} \triangleq d_2 - d_1$. The aim of the FO design algorithm is to achieve the asymptotic stability in the sense formulated below:

$$\lim_{t \to \infty} \sup_{-d_2 \leq \theta \leq 0} |e(t)| = 0 \qquad \forall \theta \in [-d_2, 0]. \tag{33}$$

The following lth order delay-dependent proper functional observer structure is appraised that incorporates the real-time delay value:

$$\begin{aligned} \dot{\omega}(t) &= F\omega(t) + F_d\omega(t-d(t)) + Gu(t) + Hy(t) + H_d y(t-d(t)) \\ \hat{z}(t) &= \omega(t) + Vy(t) \\ \omega(\theta) &= 0 \qquad \forall\theta \in [-d_2, 0], \end{aligned} \tag{34}$$

where $F_d \in \mathbb{R}^{l\times l}$, $G_d \in \mathbb{R}^{l\times m}$, and $H_d \in \mathbb{R}^{l\times p}$.

A number of lemmas that are employed in deriving the stability criterion for the observer are given below.

Lemma 1 (Wirtinger-based single integral inequality [28]) *Given $Y \succ 0 \in \mathbb{S}^{n\times n}$ and any function $f : [a, b] \to \mathbb{R}^n$, if $f(\cdot)$ can be integrated on $[a, b]$, the following inequality holds:*

$$\int_a^b f^T(s)Yf(s)\mathrm{d}s \geq \frac{2}{b-a}\chi_1^T \begin{bmatrix} 2Y & \frac{-3}{b-a}Y \\ \frac{-3}{b-a}Y & \frac{6}{(b-a)^2}Y \end{bmatrix} \chi_1, \tag{35}$$

where $\chi_1^T \triangleq \left[\int_a^b f^T(s)\mathrm{d}s,\ \int_a^b \int_u^b f^T(s)\mathrm{d}s\mathrm{d}u\right]$. Jensen's inequality that is a more conservative estimation of the upper-limit of the integral can be written as

$$\int_a^b f^T(s)Yf(s)\mathrm{d}s \geq \frac{1}{b-a}\left(\int_a^b f^T(s)\mathrm{d}s\right)Y\left(\int_a^b f(s)\mathrm{d}s\right).$$

Lemma 2 ([29]) *The following inequalities hold for a differentiable function $f : [a, b] \to \mathbb{R}^n$, given real scalars a, b, and c, such that $a < c$ and $b - a > b - c$, and $Y \succ 0 \in \mathbb{S}^{n\times n}$,*

$$\int_a^c \int_u^b f^T(s)Yf(s)\mathrm{d}s\mathrm{d}u \geq \frac{6\zeta_4(a,b)}{\tilde{\zeta}(a,b)}\chi_2^T \begin{bmatrix} Y & -\frac{2\zeta_3(a,b)}{\zeta_4(a,b)}Y \\ -\frac{2\zeta_3(a,b)}{\zeta_4(a,b)}Y & \frac{6\zeta_2(a,b)}{\zeta_4(a,b)}Y \end{bmatrix} \chi_2, \tag{36}$$

$$\int_a^c \int_u^b \dot{f}^T(s)Y\dot{f}(s)\mathrm{d}s\mathrm{d}u \geq \frac{6\zeta_4(a,b)}{\tilde{\zeta}(a,b)}\chi_3^T \begin{bmatrix} Y & -\frac{2\zeta_3(a,b)}{\zeta_4(a,b)}Y \\ -\frac{2\zeta_3(a,b)}{\zeta_4(a,b)}Y & \frac{6\zeta_2(a,b)}{\zeta_4(a,b)}Y \end{bmatrix} \chi_3, \tag{37}$$

where

$$\begin{aligned} \chi_2^T &\triangleq \left[\int_a^c \int_u^b f^T(s)\mathrm{d}s\mathrm{d}u,\ \int_a^c \int_u^b \int_{u_1}^b f^T(s)\mathrm{d}s\mathrm{d}u_1\mathrm{d}u\right], \\ \chi_3^T &\triangleq \left[(c-a)f^T(b) - \int_a^c f^T(s)\mathrm{d}s,\ \frac{\zeta_2(a,b)}{2}f^T(b) - \int_a^c \int_u^b f^T(s)\mathrm{d}s\mathrm{d}u\right], \end{aligned}$$

$\zeta_i(a,b) \triangleq (b-a)^i - (b-c)^i$, and $\tilde{\zeta}(a,b) \triangleq 3\zeta_2(a,b)\zeta_4(a,b) - 2\zeta_3^2(a,b)$.

Lemma 3 (Reciprocally convex combination lemma [30]) *Given $f_1, f_2, \ldots, f_N$ are functions mapping from $\mathbb{R}^n$ to $\mathbb{R}$, and positive in an open subset $\mathcal{D}$ of $\mathbb{R}^n$, then* *the reciprocally convex combination of the functions over $\mathcal{D}$ fulfills the following equation*

$$\min_{\{\alpha_i|\alpha_i>0,\sum_i \alpha_i=1\}} \sum_i \frac{1}{\alpha_i} f_i(t) = \sum_i f_i(t) + \max_{g_{i,j}(t)} \sum_{i \neq j} g_{i,j}(t),$$

subject to:

$$\left\{ g_{i,j}(t) : \mathbb{R}^n \to \mathbb{R}, \begin{bmatrix} f_i(t) & g_{i,j}(t) \\ g_{i,j}(t) & f_j(t) \end{bmatrix} \succeq 0 \right\}.$$

Lemma 4 ([29]) *Suppose $G : [a, b] \to \mathbb{R}^{n\times n}$ is a convex function. Then, $G(t)$ is negative for all $t \in [a, b]$, if and only if $G(a) \prec 0$ and $G(b) \prec 0$.*

Let us define a supplementary error vector $\epsilon(t) \triangleq \omega(t) - Tx(t)$.

The following theorem discusses the stability conditions of the observer error dynamics.

Theorem 5 ([31]) *Functional observer (34) for System (31) is globally asymptotically stable, if and only if*

i. the auxiliary error dynamics comprehended below is asymptotically stable

$$\begin{aligned} \dot{\epsilon}(t) &= F\epsilon(t) + F_d\epsilon(t - d(t)) \\ \epsilon(\theta) &= -T\phi(\theta)\ \forall\theta \in [-d_2, 0] \end{aligned} \tag{38}$$

ii. there exist a matrix T that satisfies the following constrained Sylvester equations

$$FT - TA + HC = \mathbf{0}, \tag{39a}$$

$$F_dT - TA_d + H_dC = \mathbf{0}, \tag{39b}$$

$$T + VC - L = \mathbf{0}, \tag{39c}$$

$$G = TB. \tag{39d}$$

The delay-dependent observer design algorithm thus confronts with two steps:
(i) finding a sufficient condition on the asymptotic stability of the delayed (and therefore infinite-dimensional) dynamics (38) (studied in Sect. 4.1), and
(ii) solving the set of interconnected equations (39) in a way to comply with the stability condition established in the former step (addressed in Sect. 4.2).

Remark 3 It is possible to design a delay-free observer for system (31) in a similar way that we designed the observer for the system with input delays, (16), as demonstrated in Sect. 3, by treating the delay term as an unknown-input. This study has been covered in the literature to some extent (see e.g. [7]). Moreover, LTI systems with multiple state and input delays can be dealt with in a similar way as comprehended in this section (see [10, 11]).

4.1 Developing a Stability Criterion

Establishing delay-dependent sufficient conditions for stability of delay differential equation (38) has been broadly studied in the literature.

The following theorem summarizes one of the latest stability criterion for the system. A sketch of the proof is also given, though a more comprehensive proof can be found in [31].

Theorem 6 *Observer error dynamics (38) is asymptotically stable for any delay function that complies with (32), if there exist* $l \times l$ *symmetric matrices* $P_i \succ 0$, $i = \{1, 4, 6\}$, $R_j \succ 0$, $j = \{1, \ldots, 4\}$, $Q_m \succ 0$, $m = \{1, 2, 3\}$, *and* $S_k \succ 0$, $k = \{1, 2\}$, *and any* $l \times l$ *matrices* $P_i, i = \{2, 3, 5\}$ *and* U_j, $j = \{1, \ldots, 4\}$ *such that the following matrix inequalities are satisfied*

$$\Psi_i \prec 0 \quad i = \{1, 2\}, \tag{40}$$

where $\Psi_1 \triangleq \Psi(d(t))|_{d(t)=d_1}$, $\Psi_2 \triangleq \Psi(d(t))|_{d(t)=d_2}$, *and*

$$\Psi(d(t)) \triangleq \begin{bmatrix} \Gamma(t) & \Gamma_9 & \Gamma_{10}(t) & \Gamma_{11}(t) & \Gamma_{12} \\ * & -72\frac{\zeta_2^2(d_1,d_2)}{\tilde{\zeta}(d_1,d_2)}S_2 & \mathbf{0} & \mathbf{0} & \mathbf{0} \\ * & * & -12R_2 & \mathbf{0} & \mathbf{0} \\ * & * & * & -12R_2 & \mathbf{0} \\ * & * & * & * & \tilde{\Gamma}_{12} \end{bmatrix}.$$

In addition, $\Gamma(t) = [\Gamma_{i,j}]_{8\times 8}$ *consisting non-zero elements as below*

$\Gamma_{1,1} \triangleq \mathrm{sym}(P_3) + Q_1 + d_1^2 R_1 + d_{12}^4 R_2 - 6d_1^2 S_1 - 12\frac{\zeta_2(d_1,d_2)}{\tilde{\zeta}(d_1,d_2)}\eta_{11} S_2 - \frac{4}{d_1} R_3$
$+ \mathrm{sym}(U_1^T F_1)$,
$\Gamma_{2,1} \triangleq F_d^T U_1$, $\quad \Gamma_{2,2} \triangleq -(1-\mu)Q_3 - 4(d_2 - d(t))^2 R_4 - 4(d(t) - d_1)^2 R_4$,
$\Gamma_{3,1} \triangleq P_2^T - P_3^T - \frac{2}{d_1} R_3$, $\quad \Gamma_{3,2} \triangleq -2(d(t) - d_1)^2 R_4$,
$\Gamma_{3,3} \triangleq Q_2 - Q_1 + Q_3 - \frac{4}{d_1} R_3 - 4(d(t) - d_1)^2 R_4$, $\quad \Gamma_{4,1} \triangleq -P_2^T$,
$\Gamma_{4,2} \triangleq -2(d_2 - d(t))^2 R_4$, $\quad \Gamma_{4,4} \triangleq -Q_2 - 4(d(t) - d_1)^2 R_4$,
$\Gamma_{5,1} \triangleq P_6 + \frac{6}{d_1^2} R_3 + U_3^T F_1$, $\quad \Gamma_{5,2} \triangleq U_3^T F_d$, $\quad \Gamma_{5,3} \triangleq -P_6 + \frac{6}{d_1^2} R_3 + P_5^T$,
$\Gamma_{5,4} \triangleq -P_5^T$, $\quad \Gamma_{5,5} \triangleq -4R_1 - \frac{12}{d_1^3} R_3 - 12S_1$,
$\Gamma_{6,1} \triangleq 12S_1$, $\quad \Gamma_{6,5} \triangleq \frac{6}{d_1} R_1 + \frac{24}{d_1} S_1$,
$\Gamma_{6,6} \triangleq -\frac{12}{d_1^2} R_1 - \frac{72}{d_1^2} S_1$, $\quad \Gamma_{7,1} \triangleq -\frac{12\zeta_2(d_1,d_2)}{\tilde{\zeta}(d_1,d_2)}\eta_{17} S_2 + P_5 + U_2^T F_1$,
$\Gamma_{7,2} \triangleq 6(d_2 - d(t))R_4 + U_2^T F_d$, $\quad \Gamma_{7,3} \triangleq P_4 - P_5$,
$\Gamma_{7,4} \triangleq -P_4 + 6(d_2 - d(t))R_4$,
$\Gamma_{7,7} \triangleq -12R_4 - 4(d_2 - d(t))^2 R_2 - 12\frac{\zeta_2(d_1,d_2)\zeta_4(d_1,d_2)}{\tilde{\zeta}(d_1,d_2)} S_2$,
$\Pi_{8,1} \triangleq -\frac{12\zeta_2(d_1,d_2)}{\tilde{\zeta}(d_1,d_2)}\eta_{17} S_2 + P_5 + U_2^T F_1$, $\quad \Gamma_{8,2} \triangleq 6(d(t) - d_1)R_4 + U_2^T F_d$,
$\Gamma_{8,3} \triangleq 6(d(t) - d_1)R_4 + P_4 - P_5$, $\quad \Gamma_{8,4} \triangleq -P_4$,
$\Gamma_{8,7} \triangleq -12\frac{\zeta_2(d_1,d_2)}{\tilde{\zeta}(d_1,d_2)}\zeta_4(d_1,d_2) S_2$,

$\Gamma_{8,8} \triangleq -12\frac{\zeta_2(d_1,d_2)\zeta_4(d_1,d_2)}{\tilde{\zeta}(d_1,d_2)}S_2 - 12R_4 - 4(d(t)-d_1)^2R_2,$
$\Gamma_9 \triangleq \left[12\frac{\zeta_2(d_1,d_2)}{\tilde{\zeta}(d_1,d_2)}\eta_{19}S_2, \mathbf{0}, \mathbf{0}, \mathbf{0}, \mathbf{0}, \mathbf{0}, 24\frac{\zeta_2(d_1,d_2)\zeta_3(d_1,d_2)}{\tilde{\zeta}(d_1,d_2)}S_2, 24\frac{\zeta_2(d_1,d_2)\zeta_3(d_1,d_2)}{\tilde{\zeta}(d_1,d_2)}S_2\right]^T,$
$\Gamma_{10}(t) \triangleq [\mathbf{0}, \mathbf{0}, \mathbf{0}, \mathbf{0}, \mathbf{0}, \mathbf{0}, 6(d_2-d(t))R_2, \mathbf{0}]^T,$
$\Gamma_{11}(t) \triangleq [\mathbf{0}, \mathbf{0}, \mathbf{0}, \mathbf{0}, \mathbf{0}, \mathbf{0}, \mathbf{0}, 6(d_2-d(t))R_2]^T,$
$\Gamma_{12} \triangleq \left[P_1 - U_1 + U_4^TF_1, U_4^TF_d, \mathbf{0}, \mathbf{0}, P_3 - U_3, \mathbf{0}, P_2 - U_2, P_2 - U_2\right]^T$, *and*
$\tilde{\Gamma}_{12} \triangleq d_1R_3 + d_{12}^4R_4 + d_1^4S_1 + \zeta_2^2(d_1,d_2)S_2 - \mathrm{sym}(U_4).$
Furthermore,
$\eta_{11} \triangleq d_{12}^2\zeta_4(d_1,d_2) - 2d_{12}\zeta_2(d_1,d_2)\zeta_3(d_1,d_2) + 1.5\zeta_2^3(d_1,d_2),$
$\eta_{17} \triangleq \zeta_2(d_1,d_2)\zeta_3(d_1,d_2) - d_{12}\zeta_4(d_1,d_2),$
and $\eta_{19} \triangleq 3\zeta_2^2(d_1,d_2) - 2d_{12}\zeta_3(d_1,d_2).$

Proof Consider the following Lyapunov–Krasovskii functional candidate

$$V(t,d(t)) = \sum_{i=1}^{4} V_i(t), \tag{41}$$

where

$$\begin{aligned}
V_1(t,d(t)) &= \tilde{\epsilon}^T(t)EP\tilde{\epsilon}(t),\\
V_2(t,d(t)) &= \int_{t-d_1}^{t} \epsilon^T(s)Q_1\epsilon(s)ds + \int_{t-d_2}^{t-d_1} \epsilon^T(s)Q_2\epsilon(s)\mathrm{d}s\\
&\quad + \int_{t-d(t)}^{t-d_1} \epsilon^T(s)Q_3\epsilon(s)\mathrm{d}s,\\
V_3(t,d(t)) &= h_1\int_{-d_1}^{0}\int_{t+s_1}^{t} \epsilon^T(s)R_1\epsilon(s)\mathrm{d}s\mathrm{d}s_1 + \int_{-d_1}^{0}\int_{t+s_1}^{t} \dot{\epsilon}^T(s)R_3\dot{\epsilon}(s)\mathrm{d}s\mathrm{d}s_1\\
&\quad + d_{12}^3\int_{-d_2}^{-d_1}\int_{t+s_1}^{t} \epsilon^T(s)R_2\epsilon(s)\mathrm{d}s\mathrm{d}s_1\\
&\quad + d_{12}^3\int_{-d_2}^{-d_1}\int_{t+s_1}^{t} \dot{\epsilon}^T(s)R_4\dot{\epsilon}(s)\mathrm{d}s\mathrm{d}s_1,\\
V_4(t,d(t)) &= 2d_1^2\int_{-d_1}^{0}\int_{s_1}^{0}\int_{t+s_2}^{t} \dot{\epsilon}^T(s)S_1\dot{\epsilon}(s)\mathrm{d}s\mathrm{d}s_2\mathrm{d}s_1\\
&\quad + 2(d_2^2 - d_1^2)\int_{-d_2}^{-d_1}\int_{s_1}^{0}\int_{t+s_2}^{t} \dot{\epsilon}^T(s)S_2\dot{\epsilon}(s)\mathrm{d}s\mathrm{d}s_2\mathrm{d}s_1,
\end{aligned}$$

with

$$\tilde{\epsilon}(t) = \left[\epsilon^T(t), \int_{t-d_2}^{t-d_1} \epsilon^T(s)\mathrm{d}s, \int_{t-d_1}^{t} \epsilon^T(s)\mathrm{d}s, \dot{\epsilon}^T(t)\right]^T,$$
$$P \triangleq \begin{bmatrix} P_1 & P_2 & P_3 & \mathbf{0} \\ * & P_4 & P_5 & \mathbf{0} \\ * & * & P_6 & \mathbf{0} \\ U_1 & U_2 & U_3 & U_4 \end{bmatrix},$$
$$E \triangleq \left[I_{3l}, \mathbf{0}; \mathbf{0}, \mathbf{0} \right].$$

Moreover, let us define the following variables with omitted arguments $(t, d(t))$: $z_1 \triangleq x(t)$, $z_2 \triangleq x(t-d(t))$, $z_3 \triangleq x(t-d_1)$, $z_4 \triangleq x(t-d_2)$, $z_5 \triangleq \int_{t-d_1}^{t} x(s)\mathrm{d}s$, $z_6 \triangleq \int_{-d_1}^{0}\int_{s}^{0} x(t+u)\mathrm{d}u\mathrm{d}s$, $z_7 \triangleq \int_{-d_2}^{-d(t)} x(t+u)\mathrm{d}u$, $z_8 \triangleq \int_{-d(t)}^{-d_1} x(t+u)\mathrm{d}u$, $z_9 \triangleq \int_{-d_2}^{-d_1}\int_{s}^{0} x(t+u)\mathrm{d}u\mathrm{d}s$, $z_{10} \triangleq \int_{-d_2}^{-d(t)}\int_{s}^{-d(t)} x(t+u)\mathrm{d}u\mathrm{d}s$, $z_{11} \triangleq \int_{-d(t)}^{-d_1}\int_{s}^{-d_1} x(t+u)\mathrm{d}u\mathrm{d}s$, $z_{12} \triangleq \dot{x}(t)$, and $\delta(t) \triangleq (d_2 - d(t))^3 + (d(t) - d_1)^3$.

Differentiating (41) along the solution of (38), noting that $\int_{t-d_2}^{t-d_1} f(s)\mathrm{d}s = \int_{t-d_2}^{t-d(t)} f(s)\mathrm{d}s + \int_{t-d(t)}^{t-d_1} f(s)\mathrm{d}s$ holds for any integrable function $f : [t-d_2, t-d_1] \to \mathbb{R}$, using Lemmas 1, 2 and 3, and the fact that $\delta(t) \le d_{12}^3$, results in the sequel expressions (arguments of $\zeta_i(\cdot,\cdot)$, $i = \{2, 3, 4\}$, and $\tilde{\zeta}(\cdot,\cdot)$ are removed for simplicity):

$$\dot{V}_1(t, d(t)) = \mathrm{sym}\left(\begin{bmatrix} z_1 \\ z_7 + z_8 \\ z_5 \\ z_{12} \end{bmatrix}^T \begin{bmatrix} P_1 & P_2 & P_3 & \mathbf{0} \\ * & P_4 & P_5 & \mathbf{0} \\ * & * & P_6 & \mathbf{0} \\ U_1 & U_2 & U_3 & U_4 \end{bmatrix}^T \begin{bmatrix} z_{12} \\ z_3 - z_4 \\ z_1 - z_3 \\ F z_1 + F_d z_2 - z_{12} \end{bmatrix}\right), \tag{42}$$

$$\begin{aligned}
\dot{V}_2(t, d(t)) \le\, & z_1^T Q_1 z_1 - z_3^T Q_1 z_3 + z_3^T Q_2 z_3 - z_4^T Q_2 z_4 \\
& + z_3^T Q_3 z_3 - (1-\mu) z_2^T Q_3 z_2, \qquad (43)\\
\dot{V}_3(t, d(t)) \le\, & d_1^2 z_1^T R_1 z_1 + d_1 z_{12}^T R_3 z_{12} + d_{12}^4 z_1^T R_2 z_1 + d_{12}^4 z_{12}^T R_4 z_{12} \\
& - 2\begin{bmatrix} z_5 \\ z_6 \end{bmatrix}^T \begin{bmatrix} 2 & \frac{-3}{d_1} \\ \frac{-3}{d_1} & \frac{6}{d_1^2} \end{bmatrix} \otimes R_1 \begin{bmatrix} z_5 \\ z_6 \end{bmatrix} - 2\tilde{z}_1^T \Upsilon(t) \otimes R_2 \tilde{z}_1 \qquad (44)\\
& - \frac{2}{d_1}\begin{bmatrix} z_1 - z_3 \\ d_1 z_1 - z_5 \end{bmatrix}^T \begin{bmatrix} 2 & \frac{-3}{d_1} \\ \frac{-3}{d_1} & \frac{6}{d_1^2} \end{bmatrix} \otimes R_3 \begin{bmatrix} z_1 - z_3 \\ d_1 z_1 - z_5 \end{bmatrix} \\
& - 2\tilde{z}_2^T \Upsilon(t) \otimes R_4 \tilde{z}_2, \\
\dot{V}_4(t, d(t)) \le\, & d_1^4 z_{12}^T S_1 z_{12} + \zeta_2^2 z_{12}^T S_2 z_{12} \qquad (45)\\
& - 12\begin{bmatrix} d_1 z_1 - z_5 \\ \frac{d_1^2}{2} z_1 - z_6 \end{bmatrix}^T \begin{bmatrix} 1 & \frac{-2}{d_1} \\ \frac{-2}{d_1} & \frac{6}{d_1^2} \end{bmatrix} \otimes S_1 \begin{bmatrix} d_1 z_1 - z_5 \\ \frac{d_1^2}{2} z_1 - z_6 \end{bmatrix}
\end{aligned}$$

$$- 12\frac{\zeta_2}{\bar{\zeta}} \begin{bmatrix} d_{12}z_1 - z_7 - z_8 \\ \frac{\zeta_2}{2}z_1 - z_9 \end{bmatrix}^T \begin{bmatrix} \zeta_4 & -2\zeta_3 \\ -2\zeta_3 & 6\zeta_2 \end{bmatrix} \otimes S_2$$
$$\times \begin{bmatrix} d_{12}z_1 - z_7 - z_8 \\ \frac{\zeta_2}{2}z_1 - z_9 \end{bmatrix}, \tag{46}$$

where $\tilde{z}_1 \triangleq [z_7; z_{10}; z_8; z_{11}]$, $\tilde{z}_2 \triangleq [(z_2 - z_4); ((d_2 - d(t))z_2 - z_7); (z_3 - z_2); ((d(t) - d_1)z_3 - z_8)]$, $\Theta(a, b) \triangleq \left[2(a-b)^2, -3(a-b); -3(a-b), 6\right]$, and $\Upsilon(t) \triangleq \left[\Theta(d_2, d(t)), \mathbf{0}; \mathbf{0}, \Theta(d(t), d_1)\right]$.

Accordingly, it is observed from (46)–(42) that

$$\dot{V}(t, d(t)) \leq z^T \Psi(d(t))z, \tag{47}$$

where $z^T \triangleq \left[z_1^T, z_2^T, \ldots, z_{12}^T\right]$.

Since $\Psi(\cdot)$ is constructed by weighted sum and vector composition of convex functions $d(t) - d_1$, $d_2 - d(t)$, and $(d(t) - d_i)^2$, $i = \{1, 2\}$, it is a convex function of $d(t)$ [32]. Therefore, according to the convex combination technique (Lemma 4), criterion (47) is satisfied if and only if matrix inequalities (40) are achievable. The proof thus follows from applying the Lyapunov–Krasovskii stability theorem [33]. □

4.2 Observer Design Scheme

The observer design procedure can be called an extension of the framework outlined in Sect. 2.1.

The aim is to solve equations (39), while adhering to the stability criterion developed in Theorem 6.

Retaining (4) to (8), it is followed from the same process that

$$F_d T C^{\perp} = T A_d C^{\perp}. \tag{48}$$

$$F_d L M_1 C^{\perp} + F_d Z \bar{M}_1 C^{\perp} = L M_1 A_d C^{\perp} + Z \bar{M}_1 A_d C^{\perp}. \tag{49}$$

Conclusively, we have

$$\left[F, F_d, -Z\right] \Omega_d = \Phi_d, \tag{50}$$

where $\Omega_d \triangleq \left[LM_1C^{\perp}, \mathbf{0}; \mathbf{0}, LM_1C^{\perp}; \bar{M}_1AC^{\perp}, \bar{M}_1A_dC^{\perp}\right] \in \mathbb{R}^{(n+p+2l)\times 2(n-p)}$ and $\Phi_d \triangleq \left[LM_1AC^{\perp}, LM_1A_dC^{\perp}\right] \in \mathbb{R}^{l\times 2(n-p)}$.

If there is a solution for (50), one obtains

$$F = N_{d,11} + \tilde{Z} N_{d,21}, \tag{51}$$

$$F_d = N_{d,12} + \tilde{Z} N_{d,22}, \tag{52}$$

$$Z = -N_{d,13} - \tilde{Z} N_{d,23}, \tag{53}$$

where $\left[N_{d,11} \; N_{d,12} \; N_{d,13} \right] \triangleq \Phi_d \Omega_d^\dagger$, $\left[N_{d,21} \; N_{d,22} \; N_{d,23} \right] \triangleq (I_{n+p+2l} - \Omega_d \Omega_d^\dagger) \in \mathbb{R}^{(n+p+2l)\times(n+p+2l)}$, $N_{d,11} \in \mathbb{R}^{l\times l}$, $N_{d,12} \in \mathbb{R}^{l\times l}$, $N_{d,13} \in \mathbb{R}^{l\times(n+p)}$, $N_{d,21} \in \mathbb{R}^{(n+p+2l)\times l}$, $N_{d,22} \in \mathbb{R}^{(n+p+2l)\times l}$, $N_{d,23} \in \mathbb{R}^{(n+p+2l)\times(n+p)}$, and $\tilde{Z} \in \mathbb{R}^{l\times(n+p+2l)}$ is a free matrix.

The necessary and sufficient condition for the existence of a solution for (50) can be expressed as [31]

$$\operatorname{rank}\left(\begin{bmatrix} LM_1 A & LM_1 A_d \\ LM_1 & \mathbf{0} \\ \mathbf{0} & LM_1 \\ \bar{M}_1 A & \bar{M}_1 A_d \\ C & \mathbf{0} \\ \mathbf{0} & C \end{bmatrix}\right) = \operatorname{rank}\left(\begin{bmatrix} LM_1 & \mathbf{0} \\ \mathbf{0} & LM_1 \\ \bar{M}_1 A & \bar{M}_1 A_d \\ C & \mathbf{0} \\ \mathbf{0} & C \end{bmatrix}\right) \tag{54}$$

Given that condition (54) is met, the free matrix $\tilde{Z}$ should be found in a way that matrix inequalities (40) are satisfied. The inequalities are nonlinear due to the multiplicative terms $U_i^T \tilde{Z}$, $i = \{1, \ldots, 4\}$, which make it hard to solve the matrix inequalities as the problem is non-convex and thus $\mathcal{NP}$-hard.

Various algorithms are proposed in the literature to resolve this prevalent obstacle. A resolution to this dilemma is simplifying the problem by defining $U_i = \alpha_i U_4$, $i = \{1, 2, 3\}$, where α_i are free scalar parameters. Next, the following corollary provides a guideline to design $\tilde{Z}$ assuming that α_i are given.

Corollary 1 *Given scalars α_i, $i = \{1, 2, 3\}$, functional observer (34) is a globally asymptotically stable functional state estimator for System (31) with delay function complying with (32), if condition (54) is realized, and there exist symmetric positive definite $l \times l$ matrices P_i, $i = \{1, 4, 6\}$, Q_j, $j = \{1, 2, 3\}$, R_k, $k = \{1, \ldots, 4\}$, and S_m, $m = \{1, 2\}$, and any matrices $P_i \in \mathbb{R}^{l\times l}$, $i = \{2, 3, 5\}$, $U_4 \in \mathbb{R}^{l\times l}$, and $\Lambda \in \mathbb{R}^{l\times(2l+n+p)}$, such that the following linear matrix inequalities (LMIs) are feasible*

$$\tilde{\Psi}_i \prec 0 \quad i = \{1, 2\}, \tag{55}$$

where $\bar{\Psi}_1 \triangleq \bar{\Psi}(t)|_{d(t)=d_1}$, $\bar{\Psi}_2 \triangleq \tilde{\Psi}(t)|_{d(t)=d_2}$, and

$$\tilde{\Psi}(t) = \begin{bmatrix} \bar{\Gamma}(t) & \Gamma_9 & \Gamma_{10}(t) & \Gamma_{11}(t) & \bar{\Gamma}_{12} \\ * & -72\frac{\zeta_2^2}{\zeta} S_2 & \mathbf{0} & \mathbf{0} & \mathbf{0} \\ * & * & -12R_2 & \mathbf{0} & \mathbf{0} \\ * & * & * & -12R_2 & \mathbf{0} \\ * & * & * & * & \tilde{\Gamma}_{12} \end{bmatrix},$$

wherein Γ_9, Γ_{10}, Γ_{11}, *and* $\tilde{\Gamma}_{12}$ *are as defined in Theorem 6,*

$$\bar{\Gamma}_{12} = \big[P_1 - \alpha_1 U_4 + U_4^T N_{d,11} + \Lambda N_{d,21}, U_4^T N_{d,12} + \Lambda N_{d,22}, \\ \mathbf{0}, \mathbf{0}, P_3 - \alpha_3 U_4, \mathbf{0}, P_2 - \alpha_2 U_4, P_2 - \alpha_2 U_4]^T ,$$

and all the components of $\bar{\Gamma}(t) = [\bar{\Gamma}_{i,j}]_{8\times 8}$ *are identical to* $\Gamma(t)$ *defined in Theorem 6 other than the following elements:*
$\bar{\Gamma}_{1,1} = \mathrm{sym}(P_3) + Q_1 + d_1^2 R_1 + d_{12}^4 R_2 - 6d_1^2 S_1 - 12\frac{\zeta_2}{\zeta}\eta_{11} S_2 - \frac{4}{d_1} R_3 + \alpha_1 \mathrm{sym}$ $(U_4 N_{d,11} + \Lambda N_{d,21})$, $\bar{\Gamma}_{2,1} = \alpha_1 N_{d,22}^T \Lambda^T + \alpha_1 N_{d,12}^T U_4$, $\bar{\Gamma}_{5,1} = P_6 + \frac{6}{d_1^2} R_3 + \alpha_3$ $U_4^T N_{d,11} + \alpha_3 \Lambda N_{d,21}$, $\bar{\Gamma}_{5,2} = \alpha_3 U_4^T N_{d,12} + \alpha_3 \Lambda N_{d,22}$, $\bar{\Gamma}_{7,1} = -\frac{12\zeta_2}{\zeta}\eta_{17} S_2 + P_5 +$ $\alpha_2 U_4^T N_{d,11} + \alpha_2 \Lambda N_{d,21}$, $\bar{\Gamma}_{7,2} = 6(d_2 - d(t))R_4 + \alpha_2 U_4^T N_{d,12} + \alpha_2 \Lambda N_{d,22}$, $\bar{\Gamma}_{8,1}$ $= -\frac{12\zeta_2}{\zeta}\eta_{17} S_2 + P_5 + \alpha_2 U_4^T N_{d,11} + \alpha_2 \Lambda N_{d,21}$, *and* $\bar{\Gamma}_{8,2} = 6(d(t) - d_1)R_4 + \alpha_2$ $U_4^T N_{d,12} + \alpha_2 \Lambda N_{d,22}$.

In addition,

$$\tilde{Z} = U_4^{-T} \Lambda. \tag{56}$$

Proof The proof follows from Theorems 5 and 6. Firstly, Condition (ii) of Theorem 5 is met, only upon the fulfillment of rank condition (54). Secondly, by substituting for F and F_d from (51) and (52), setting $U_i = \alpha_i U_4$, $i = \{1, 2, 3\}$, and defining $\Lambda \triangleq U_4^{-1} \tilde{Z}$, it can be shown that inequalities (55) and (40) are equivalent. Therefore, from Theorem 6 Condition (i) of Theorem 5 is also satisfied. □

The final dilemma is how to find the scalar weightings α_i, $i = \{1, 2, 3\}$, that appear in nonlinear forms $\alpha_i U_4$. To resolve this issue an artificial intelligence (AI) based approach introduced in [31] is summarized below that is named *GABT* (genetic algorithm based tuning) scheme.

GABT Algorithm:

1. Consider $f = t_{\min}$ as the fitness function constrained by

$$\bar{\Psi}_i \prec t_{\min} \quad i = \{1, 2\}, \tag{57}$$

   where $t_{\min} < 0$ is the value to be minimized, and α_i are the variables to be tuned. Moreover, select arbitrary and feasible bounds for the variables as $\alpha_i \in [\alpha_i^{\min}, \alpha_i^{\max}]$.
2. Generate N_p arbitrary sets of α_is to be the initial population's chromosomes, let $\bar{\alpha}^j \triangleq [\alpha_1^j, \alpha_2^j, \alpha_3^j]$, $j = \{1, \ldots, N_p\}$, and define an arbitrarily large integer number $N_g > 0$.
3. Employ the genetic algorithm to efficiently reach a chromosome that complies with LMIs (57). If even after N_g generations the GA algorithm was unable to find an appropriate set of values $\bar{\alpha}^j$, then the algorithm terminates without any solution.

5 Addressing Uncertainties in State Delays

The main shortcoming of the delay-dependent FO structure and design framework comprehended in Sect. 4 is using the actual real-time values of the state delay, which generally can be erroneously estimated in the majority of practical problems.

To resolve the sustainability problem of FO (34), an alternative observer structure that employs an auxiliary delay function instead of the actual delay values is demonstrated in this section.

A sliding mode robust scheme to design the observer parameters is also illustrated.

Let us augment the output vector and the functional $z(t)$ as $z_a(t) = L_a x(t)$, where $L_a \triangleq [L; C]$. Consider the following FO structure

$$\begin{aligned}\dot{\omega}(t) &= F\omega(t) + F_d\omega(t-\hat{d}(t)) + Gu(t) + Hy(t) + H_d y(t-\hat{d}(t))\\ &\quad - \begin{bmatrix} X_2 \\ F_{d,22}\end{bmatrix}\epsilon_2(t-\hat{d}(t)) - \begin{bmatrix} X_1 \\ F_{22}-G_\epsilon\end{bmatrix}\epsilon_2(t) + \begin{bmatrix} R \\ -I_p\end{bmatrix}\nu(t) \qquad (58)\\ \hat{z}_a(t) &= \omega(t) + Vy(t)\\ \omega(\theta) &= \mathbf{0} \qquad \forall\theta\in[-\hat{d}_2, 0],\end{aligned}$$

where $\omega(\cdot) \in \mathbb{R}^{l+p}$ is the states of the observer dynamics; $\hat{z}_a(\cdot)$ is the estimation of $z_a(\cdot)$; $X_1 \in \mathbb{R}^{l\times p}$, $X_2 \in \mathbb{R}^{l\times p}$, $R \in \mathbb{R}^{l\times p}$, and $G_\epsilon \in \mathbb{R}^{p\times p}$ are additional observer parameters; and F and F_d are partitioned as $[F_{11}, F_{12}; F_{21}, F_{22}] \triangleq F$ and $\left[F_{d,11}, F_{d,12}; F_{d,21}, F_{d,22}\right] \triangleq F_d$, respectively.

Moreover, for a free parameter $T \in \mathbb{R}^{(l+p)\times n}$, $\epsilon(\cdot) \triangleq \omega(\cdot) - Tx(\cdot) \in \mathbb{R}^{l+p}$, $[\epsilon_1(\cdot); \epsilon_2(\cdot)] \triangleq \epsilon(\cdot)$, $\epsilon_2(\cdot) \in \mathbb{R}^p$, and $\nu(t)$ is the norm-bounded uncertainty vector function defined as

$$\nu(t) = \begin{cases} \rho(t)\frac{\tilde{P}\epsilon_2(t)}{|\tilde{P}\epsilon_2(t)|} & \epsilon_2(t) \neq \mathbf{0} \\ \mathbf{0} & \epsilon_2(t) = \mathbf{0} \end{cases} \qquad (59)$$

where $\tilde{P} \in \mathbb{S}^{p\times p}$ is a positive-definite matrix and $\rho(t)$ is a time-dependent upper-bound of the perturbation term. To add, $\hat{d}(t)$ is the ancillary delay term that complies with the below conditions

$$0 \le \hat{d}_1 \le \hat{d}(t) \le \hat{d}_2, \ \dot{\hat{d}}(t) \le \hat{\mu} < 1, \qquad (60)$$

where $\hat{d}_1$, $\hat{d}_2$, and $\hat{\mu}$ are constant scalars.

Note that the delay bounds of the auxiliary delay function can be set at the actual delay bounds d_1, d_2, and μ. However, being capable of adjusting these parameters empowers the FO design algorithm to come up with less conservative stability criterion for the observer error dynamics, and at the same time it mitigates the structural restrictions that face delay-free observers.

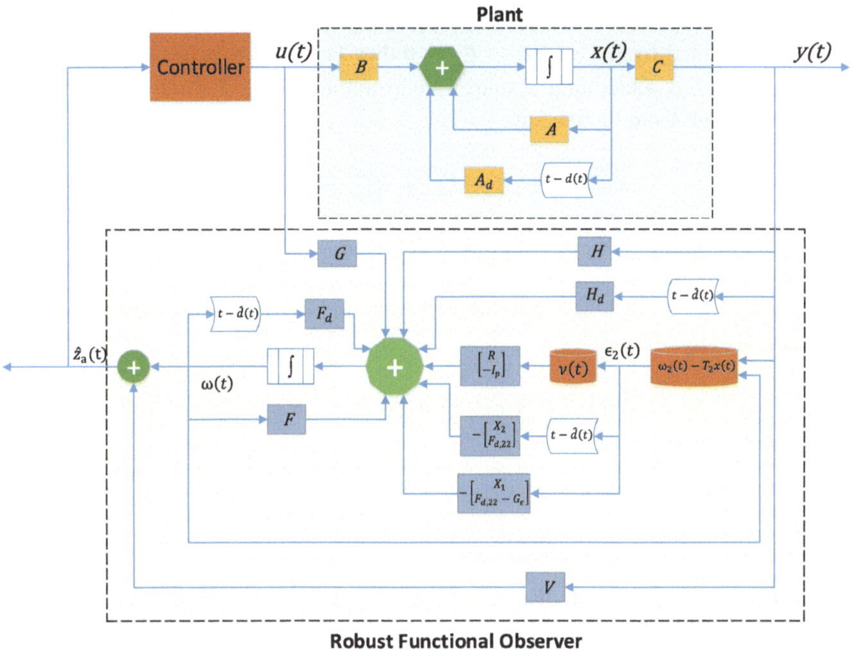

Fig. 2 The block diagram of the robust closed-loop functional observer-based controlled system. Observer parameters, system parameters, and functional processes are highlighted in gray, yellow, and orange, respectively

Figure 2 depicts the block diagram of a closed-loop control system, wherein the system behaves according to (31) and the observer dynamics complies with (58).

The observed functions here could be for instance $z(t) = u(t) = K_p x(t)$, where K_p is the proportional control gain.

The following theorem outlines the necessary and sufficient conditions for the stability of FO (58) for System (31) in the sense of (33), where $e(\cdot) = \hat{z}_{\mathrm{a}}(\cdot) - z_{\mathrm{a}}(\cdot)$.

Theorem 7 ([14]) *Consider System (31). FO (58) is a globally asymptotically stable functional observer for the system in the sense of (33), if and only if*

i. *for any initial function $\phi(\cdot) \in \mathcal{C}_n([-d_2, 0])$, the following error dynamics is globally asymptotically stable*

$$\begin{aligned}\dot{\epsilon}(t) = {} & F\epsilon(t) + F_d\epsilon(t - d(t)) - H_d C \int_{t-\hat{d}(t)}^{t-d(t)} \dot{x}(s)\mathrm{d}s - F_d \int_{t-\hat{d}(t)}^{t-d(t)} \dot{\omega}(s)\mathrm{d}s \\ & - \begin{bmatrix} X_2 \\ F_{d,22} \end{bmatrix} \epsilon_2(t - \hat{d}(t)) - \begin{bmatrix} X_1 \\ F_{22} - G_\epsilon \end{bmatrix} \epsilon_2(t) + \begin{bmatrix} R \\ -I_p \end{bmatrix} v(t), \\ \epsilon(\theta) = {} & -T\phi(\theta) \quad \forall\theta \in [-d_2, 0], \end{aligned} \tag{61}$$

ii. *there exists a matrix T that satisfies constrained equations (39) with L in (39c) replaced by L_{a}.*

Let us partition T, V, and $\omega(\cdot)$ as $[T_1; T_2] \triangleq T$, $[V_1; V_2] \triangleq V$, and $[\omega_1(\cdot); \omega_2(\cdot)] \triangleq \omega(\cdot)$. It is obtained from $T + VC - L_a = \mathbf{0}$ that $T_2 = (I_p - V_2)C$. Replacing $\omega(\cdot)$ by $\epsilon(\cdot) + Tx(\cdot)$, conducting a system transformation using $\mathcal{T} \triangleq \left[I_l, R; \mathbf{0}, I_p\right]$ and dictating X_1 and X_2 to be fixed at

$$X_1 = -F_{11}R + F_{12} + RG_\epsilon - RF_{21}R, \tag{62}$$

$$X_2 = -F_{d,11}R - RF_{d,21}R + F_{d,12}, \tag{63}$$

in an attempt to decouple the dynamics of $\epsilon_1(\cdot)$ and $\epsilon_2(\cdot)$, error dynamics (61) can be represented as follows:

$$\begin{aligned}\dot{\bar{\epsilon}}_1(t) &= \bar{F}_{11}\bar{\epsilon}_1(t) + \bar{F}_{d,11}\bar{\epsilon}_1(t - \hat{d}(t)) \\ &\quad - (H_{d,1}C + F_{d,1}T + RH_{d,2}C + RF_{d,2}T)\int_{t-\hat{d}(t)}^{t-d(t)} \dot{x}(s)\mathrm{d}s,\end{aligned} \tag{64}$$

$$\dot{\bar{\epsilon}}_2(t) = (G_\epsilon - F_{21}R)\bar{\epsilon}_2(t) + \varepsilon(t, \bar{\epsilon}(t), x(t)) - \nu(t), \tag{65}$$

where $F_{d,1} \triangleq [F_{d,11}, F_{d,12}]$, $F_{d,2} \triangleq [F_{d,21}, F_{d,22}]$, $\bar{\epsilon}(t) = \mathcal{T}\epsilon(t)$, $[\bar{\epsilon}_1(\cdot); \bar{\epsilon}_2(\cdot)] \triangleq \bar{\epsilon}(\cdot)$, $\bar{F}_{11} \triangleq F_{11} + RF_{21}$, $\bar{F}_{d,11} \triangleq F_{d,11} + RF_{d,21}$, and $\varepsilon(\cdot, \cdot, \cdot)$ is the perturbation term defined as follows

$$\begin{aligned}\varepsilon(t, \bar{\epsilon}(t), x(t)) \triangleq\, & F_{21}\bar{\epsilon}_1(t) + F_{d,21}\bar{\epsilon}_1(t - \hat{d}(t)) - F_{d,21}R\bar{\epsilon}_2(t - \hat{d}(t)) \\ & - (H_{d,2}C + F_{d,2}T)\int_{t-\hat{d}(t)}^{t-d(t)} \dot{x}(s)\mathrm{d}s.\end{aligned} \tag{66}$$

Conclusively, after appropriately partitioning $[H_1, H_2] \triangleq H$ and $\left[H_{d,1}, H_{d,2}\right] \triangleq H_d$, Theorem 7 can be re-written as below.

Corollary 2 *FO (58) is a globally asymptotically stable functional observer for system (31) if and only if Condition (i) of Theorem 7 is appeased, and*

a. for any initial function $\phi(\cdot) \in \mathcal{C}_n([-d_2, 0])$, the following error dynamics is globally asymptotically stable

$$\dot{\bar{\epsilon}}_1(t) = \bar{F}_{11}\bar{\epsilon}_1(t) + \bar{F}_{d,11}\bar{\epsilon}_1(t - \hat{d}(t)), \tag{67a}$$

$$\dot{\bar{\epsilon}}_2(t) = (G_\epsilon - F_{21}R)\bar{\epsilon}_2(t) + \varepsilon(t, \bar{\epsilon}(t), x(t)) - \nu(t), \tag{67b}$$

$$\bar{\epsilon}(\theta) = -\mathcal{T}T\phi(\theta).$$

b. the following constraint equation holds

$$H_{d,1}C + F_{d,11}T_1 + F_{d,12}T_2 + RH_{d,2}C + RF_{d,21}T_1 + RF_{d,22}T_2 = \mathbf{0}. \tag{68}$$

5.1 Establishing the Stability Criteria

It is deduced from (66) that the perturbation term $\varepsilon(\cdot,\cdot,\cdot)$ is essentially bounded if the error terms $\epsilon_1(\cdot)$ and $\epsilon_2(\cdot)$, as well as $x(t-d(t)) - x(t-\hat{d}(t))$ are bounded. The dynamic upper-bound of $\varepsilon(\cdot,\cdot,\cdot)$ is expressed as

$$\varepsilon(t, \bar{\epsilon}(t), x(t)) \leq \rho(t). \tag{69}$$

We first prove the finite time stability of the error dynamics $\bar{\epsilon}_2(t)$ (67b) in Theorem 8, and then establish sufficient conditions for the asymptotic stability of $\bar{\epsilon}_1(t)$ in Theorem 9.

Theorem 8 *The ancillary error function $\bar{\epsilon}_2(\cdot)$ comprehended in (67b) is finite time stable subject to applying the sliding control (59).*

Proof Consider the following Lyapunov candidate function:

$$V_s(t) = \bar{\epsilon}_2^T(t)\tilde{P}\bar{\epsilon}_2(t), \tag{70}$$

where $\tilde{P}$ is any symmetric positive definite matrix.

Moreover, let us define $\tilde{G}_\epsilon \prec 0$ as an arbitrary matrix with negative eigenvalues, and determine $G_\epsilon = \tilde{G}_\epsilon + F_{21}R$. Now, differentiating (70) along (67b), taking the upper-bound of the perturbation term into account according to (69), and substituting from (59) we have

$$\begin{aligned} \dot{V}_s(t) &= 2\bar{\epsilon}_2^T(t)\tilde{P}\left(\tilde{G}_\epsilon\bar{\epsilon}_2(t) + \varepsilon(t,\bar{\epsilon}(t),x(t)) - \rho(t)\text{sign}\left(\tilde{P}\bar{\epsilon}_2(t)\right)\right) \\ &\leq 2 \parallel \tilde{P} \parallel_2 \bar{\epsilon}_2^T(t)\tilde{G}_\epsilon\bar{\epsilon}_2(t) \\ &\leq -2\bar{\rho}V_s(t), \end{aligned} \tag{71}$$

where sign$(\cdot)$ is the signum function, and $\bar{\rho} > 0$ is a scalar that fits the following criterion

$$\parallel \tilde{P} \parallel_2 \tilde{G}_\epsilon \leq -\bar{\rho}\tilde{P}.$$

Accordingly, $V_s(t)$ is exponentially stable, and this results in the ideal sliding of $\bar{\epsilon}_2(t)$ to the origin [34]. This concludes the finite time convergence of the error vector $\bar{\epsilon}_2(t)$. □

Theorem 9 *The error dynamics (67) with auxiliary delay function complying with (60) is globally asymptotically stable, if there exist $l \times l$ symmetric positive definite matrices P_{ii}, Q_i, $i = \{1, 2, 3\}$, R_j, $j = \{1, \ldots, 4\}$, and arbitrary $l \times l$ real matrices P_{12}, P_{13}, P_{23}, and D_i, $i = \{1, \ldots, 4\}$ that satisfy the following matrix inequality:*

$$\begin{bmatrix} \tilde{\Omega}_{1,1} & * & * & * & * & * & * \\ P_{12}^T - P_{13}^T + R_3 & \tilde{\Omega}_{2,2} & * & * & * & * & * \\ -P_{12}^T & R_4 & -Q_2 - R_4 & * & * & * & * \\ P_{33} + D_3^T \bar{F}_{11} & P_{23}^T - P_{33} & -P_{23}^T & -R_1 & * & * & * \\ P_{23} + D_2^T \bar{F}_{11} & P_{22} - P_{23} & -P_{22} & \mathbf{0} & -R_2 & * & * \\ \bar{F}_{d,11}^T D_1 & \mathbf{0} & \mathbf{0} & \bar{F}_{d,11}^T D_3 & \bar{F}_{d,11}^T D_2 & -(1-\hat{\mu})Q_3 & * \\ \tilde{\Omega}_{7,1} & \mathbf{0} & \mathbf{0} & P_{13} - D_3 & P_{12} - D_2 & D_4^T \bar{F}_{d,11} & \tilde{\Omega}_{7,7} \end{bmatrix} \prec 0 \quad (72)$$

where $\tilde{\Omega}_{1,1} \triangleq \mathrm{sym}(P_{13}) + Q_1 + \hat{d}_1^2 R_1 + \hat{d}_{12}^2 R_2 + \mathrm{sym}(D_1^T \bar{F}_{11})$, $\tilde{\Omega}_{2,2} \triangleq Q_2 + Q_3 - Q_1 - R_3 - R_4$, $\tilde{\Omega}_{7,1} \triangleq D_4^T \bar{F}_{11} + P_{11} - D_1$, *and* $\tilde{\Omega}_{7,7} \triangleq -\mathrm{sym}(D_4) + \hat{d}_1^2 R_3 + \hat{d}_{12}^2 R_4$.

Proof The proof follows a similar procedure undertaken for Theorem 6, considering the Lyapunov–Krasovskii functional described below and employing Jensen's inequality instructed in Lemma 1.

For further details the reader can consult with [14].

$$V(t) = \sum_{i=1}^{3} V_i(t), \quad (73)$$

$$V_1(t) = \tilde{\epsilon}_1^T(t) E P \tilde{\epsilon}_1(t), \quad (74)$$

$$V_2(t) = \int_{t-\hat{d}_1}^{t} \bar{\epsilon}_1^T(s) Q_1 \bar{\epsilon}_1(s)\mathrm{d}s + \int_{t-\hat{d}_2}^{t-\hat{d}_1} \bar{\epsilon}_1^T(s) Q_2 \bar{\epsilon}_1(s)\mathrm{d}s + \int_{t-\hat{d}(t)}^{t-\hat{d}_1} \bar{\epsilon}_1^T(s) Q_3 \bar{\epsilon}_1(s)\mathrm{d}s, \quad (75)$$

$$V_3(t) = \hat{d}_1 \int_{-\hat{d}_1}^{0} \int_{t+s_1}^{t} \bar{\epsilon}_1^T(s) R_1 \bar{\epsilon}_1(s)\mathrm{d}s\mathrm{d}s_1 + \hat{d}_{12} \int_{-\hat{d}_2}^{-\hat{d}_1} \int_{t+s_1}^{t} \bar{\epsilon}_1^T(s) R_2 \bar{\epsilon}_1(s)\mathrm{d}s\mathrm{d}s_1 + \hat{d}_1 \int_{-\hat{d}_1}^{0} \int_{t+s_1}^{t} \dot{\bar{\epsilon}}_1^T(s) R_3 \dot{\bar{\epsilon}}_1(s)\mathrm{d}s\mathrm{d}s_1 + \hat{d}_{12} \int_{-\hat{d}_2}^{-\hat{d}_1} \int_{t+s_1}^{t} \dot{\bar{\epsilon}}_1^T(s) R_4 \dot{\bar{\epsilon}}_1(s)\mathrm{d}s\mathrm{d}s_1, \quad (76)$$

where $\tilde{\epsilon}_1(t) = \left[\bar{\epsilon}_1^T(t), \int_{t-\hat{d}_2}^{t-\hat{d}_1} \bar{\epsilon}_1^T(s)\mathrm{d}s, \int_{t-\hat{d}_1}^{t} \bar{\epsilon}_1^T(s)\mathrm{d}s, \dot{\bar{\epsilon}}_1^T(t)\right]^T$, $E \triangleq [I_{3l}, \mathbf{0}; \mathbf{0}, \mathbf{0}]$, and

$$P = \begin{bmatrix} P_{11} & P_{12} & P_{13} & \mathbf{0} \\ * & P_{22} & P_{23} & \mathbf{0} \\ * & * & P_{33} & \mathbf{0} \\ D_1 & D_2 & D_3 & D_4 \end{bmatrix}.$$

□

5.2 Observer Design Algorithm

The observer design policy is similar to the strategy explained in Sect. 2.1. Considering (4)–(5b) with L_a instead of L, $\bar{M}_1 C^{\perp} = \mathbf{0}$, and a system transformation using $[C^{\dagger}, C^{\perp}]$, Sylvester equations (39a) and (39b) can be broken into the following equations:

$$H_1 = T_1 A C^{\dagger} - F_{12} T_2 C^{\dagger} - F_{11} T_1 C^{\dagger}. \tag{77a}$$

$$H_2 = T_2 A C^{\dagger} - F_{22} T_2 C^{\dagger} - F_{21} T_1 C^{\dagger}. \tag{77b}$$

$$H_{d,1} = T_1 A_d C^{\dagger} - F_{d,12} T_2 C^{\dagger} - F_{d,11} T_1 C^{\dagger}. \tag{78a}$$

$$H_{d,2} = T_2 A_d C^{\dagger} - F_{d,22} T_2 C^{\dagger} - F_{d,21} T_1 C^{\dagger}. \tag{78b}$$

$$\begin{bmatrix} F_{11} & F_{12} & F_{d,11} & F_{d,12} & Z_1 \end{bmatrix} \tilde{\Omega}_d = \tilde{\Phi}_{1,d}, \tag{79}$$

$$\begin{bmatrix} F_{21} & F_{22} & F_{d,21} & F_{d,22} & Z_2 \end{bmatrix} \tilde{\Omega}_d = \tilde{\Phi}_{2,d}, \tag{80}$$

where $[Z_1, Z_2] \triangleq Z \in \mathbb{R}^{(l+p)\times(n+p)}$ is a free parameter,

$$\tilde{\Omega}_d \triangleq \begin{bmatrix} L M_1 C^{\perp} & \mathbf{0} \\ C M_1 C^{\perp} & \mathbf{0} \\ \mathbf{0} & L M_1 C^{\perp} \\ \mathbf{0} & C M_1 C^{\perp} \\ -\bar{M}_1 A C^{\perp} & -\bar{M}_1 A_d C^{\perp} \end{bmatrix},$$

$\tilde{\Phi}_{1,d} \triangleq \begin{bmatrix} L M_1 A C^{\perp} & L M_1 A_d C^{\perp} \end{bmatrix}$, and $\tilde{\Phi}_{2,d} \triangleq \begin{bmatrix} C M_1 A C^{\perp} & C M_1 A_d C^{\perp} \end{bmatrix}$. In addition, it can be shown that the constraint equation (68) is equivalent to

$$\begin{bmatrix} Z_1 & R Z_2 \end{bmatrix} \begin{bmatrix} \bar{M}_1 A_d \\ \bar{M}_1 A_d \end{bmatrix} = -(L + RC) M_1 A_d. \tag{81}$$

Furthermore, it can be verified that (80) has at least one nontrivial solution [14], which can be expressed as

$$\begin{bmatrix} F_{21} & F_{22} & F_{d,21} & F_{d,22} & Z_2 \end{bmatrix} = \tilde{N}_{d,1,2} + \tilde{Z}_2 \tilde{N}_{d,2}, \tag{82}$$

where $\tilde{N}_{d,1,2} \triangleq \tilde{\Phi}_{2,d} \tilde{\Omega}_d^{\dagger}$, $\tilde{N}_{d,2} \triangleq (I_{\varrho} - \tilde{\Omega}_d \tilde{\Omega}_d^{\dagger})$, $\tilde{Z}_2 \in \mathbb{R}^{p\times\varrho}$ is an arbitrary matrix, and $\varrho \triangleq 2(l+p) + n + p$.

Unlike (80) that always has a solution, Eq. (79) has a solution if and only if the following rank condition is satisfied [14]:

$$\mathrm{rank}\left(\begin{bmatrix} LM_1A & LM_1A_d \\ LM_1 & \mathbf{0} \\ CM_1 & \mathbf{0} \\ \mathbf{0} & LM_1 \\ \mathbf{0} & CM_1 \\ -\bar{M}_1A & -\bar{M}_1A_d \\ C & \mathbf{0} \\ \mathbf{0} & C \end{bmatrix}\right) = \mathrm{rank}\left(\begin{bmatrix} LM_1 & \mathbf{0} \\ CM_1 & \mathbf{0} \\ \mathbf{0} & LM_1 \\ \mathbf{0} & CM_1 \\ -\bar{M}_1A & -\bar{M}_1A_d \\ C & \mathbf{0} \\ \mathbf{0} & C \end{bmatrix}\right). \tag{83}$$

The solution of (79) is expressed as

$$\begin{bmatrix} F_{11} & F_{12} & F_{d,11} & F_{d,12} & Z_1 \end{bmatrix} = \tilde{N}_{d,1,1} + \tilde{Z}_1\tilde{N}_{d,2}, \tag{84}$$

where $\tilde{N}_{d,1,1} \triangleq \tilde{\Phi}_{1,d}\tilde{\Omega}_d^{\dagger}$, and $\tilde{Z}_1 \in \mathbb{R}^{l\times\varrho}$ is an arbitrary matrix.

Furthermore, given that $\mathrm{rank}([L;\, C]) = l + p$, it can be shown that (81) has a solution if and only if the following matching condition is realized [14]:

$$\mathrm{rank}\left(\begin{bmatrix} LA_d \\ CA_d \end{bmatrix}\right) = \mathrm{rank}(CA_d). \tag{85}$$

However, if L is not in the orthogonal space spanned by C, matrix R can be designed to reduce the rank of $(L + RC)M_1A_d$, and in such a circumstance a less conservative condition than matching condition (85) can be achieved for the existence of a solution for (81):

$$\mathrm{rank}\left(\begin{bmatrix} \bar{M}_1A_d \\ (L + RC)M_1A_d \end{bmatrix}\right) = \mathrm{rank}(\bar{M}_1A_d). \tag{86}$$

It highlights that the design parameter R can contribute as a leverage to relax the restrictive matching condition (85) to a less conservative condition. Next, via partitioning $\begin{bmatrix} \tilde{N}_{d,1,i,1}, \ldots, \tilde{N}_{d,1,i,5} \end{bmatrix} \triangleq \tilde{N}_{d,1,i}, i = \{1, 2\}$, and $\begin{bmatrix} \tilde{N}_{d,2,1}, \ldots, \tilde{N}_{d,2,5} \end{bmatrix} \triangleq \tilde{N}_{d,2}$, (82) and (84) read as $(i, j = \{1, 2\})$

$$F_{ij} = \tilde{N}_{d,1,i,j} + \tilde{Z}_i\tilde{N}_{d,2,j}, \tag{87}$$

$$F_{d,ij} = \tilde{N}_{d,1,i,j+2} + \tilde{Z}_i\tilde{N}_{d,2,j+2}, \tag{88}$$

$$Z_i = \tilde{N}_{d,1,i,5} + \tilde{Z}_i\tilde{N}_{d,2,5}. \tag{89}$$

Accordingly, from (89) and assuming that (86) holds, (81) reads

$$\tilde{Z}_1 + R\tilde{Z}_2 = W + J\bar{W}, \tag{90}$$

where $W \triangleq \left(-(L+RC)M_1 A_d - (\tilde{N}_{d,1,1,5} + R\tilde{N}_{d,1,2,5})\bar{M}_1 A_d\right)\left(\tilde{N}_{d,2,5}\bar{M}_1 A_d\right)^{\dagger}$, $\bar{W} \triangleq \left(I_{\varrho} - \left(\tilde{N}_{d,2,5}\bar{M}_1 A_d\right)\left(\tilde{N}_{d,2,5}\bar{M}_1 A_d\right)^{\dagger}\right)$, and $J \in \mathbb{R}^{l\times\varrho}$ is an arbitrary parameter.

Hence, one can find from (87), (88), and (90) that

$$\bar{F}_{11} = W_f + J\bar{W}\tilde{N}_{d,2,1}, \tag{91}$$

$$\bar{F}_{d,11} = W_n + J\bar{W}\tilde{N}_{d,2,3}, \tag{92}$$

where $W_f \triangleq \tilde{N}_{d,1,1,1} + R\tilde{N}_{d,1,2,1} + W\tilde{N}_{d,2,1}$ and $W_n \triangleq \tilde{N}_{d,1,1,3} + R\tilde{N}_{d,1,2,3} + W\tilde{N}_{d,2,3}$.

Now, the whole procedure of the FO design scheme has ended in finding an appropriate matrix J that fulfills the matrix inequality in Theorem 9. However, apparently the inequality suffers from nonlinear terms $D_i^T J$, similar to what occurred in Theorem 6. Therefore, an analogous trick to linearize matrix inequality (72) is to first assume that $D_i = \alpha_i D_4$, $i = \{1, 2, 3\}$, where α_i are scalar parameters, secondly fix the scalar parameters and derive an LMI equivalent to (72), and finally adjusting those fixed parameters using an AI empowered framework like the GABT algorithm.

The following corollary that follows from Corollary 2 and Theorem 9 comprehends on finding the key parameter J.

Corollary 3 *Given α_i, $i = \{1, 2, 3\}$, functional observer (58) for System (31) is globally asymptotically stable with any ancillary delay function complying with (60), if rank Conditions (83) and (86) are satisfied, and there exist $l \times l$ symmetric matrices $P_{ii} \succ 0$, $Q_i \succ 0$, $i = \{1, 2, 3\}$, and R_j, $j = \{1, \ldots, 4\}$, and arbitrary $l \times l$ real matrices P_{12}, P_{13}, P_{23}, D_4, and $\Lambda \in \mathbb{R}^{l\times\varrho}$, such that the following LMI is feasible:*

$$\begin{bmatrix} \bar{\tilde{\Omega}}_{1,1} & * & * & * & * & * & * \\ \bar{\tilde{\Omega}}_{2,1} & \bar{\tilde{\Omega}}_{2,2} & * & * & * & * & * \\ -P_{12}^T & R_4 & -Q_2 - R_4 & * & * & * & * \\ \bar{\tilde{\Omega}}_{4,1} & P_{23}^T - P_{33} & -P_{23}^T & -R_1 & * & * & * \\ \bar{\tilde{\Omega}}_{5,1} & P_{22} - P_{23} & -P_{22} & \mathbf{0} & -R_2 & * & * \\ \bar{\tilde{\Omega}}_{6,1} & \mathbf{0} & \mathbf{0} & \bar{\tilde{\Omega}}_{6,4} & \bar{\tilde{\Omega}}_{6,5} & -(1-\hat{\mu})Q_3 & * \\ \bar{\tilde{\Omega}}_{7,1} & \mathbf{0} & \mathbf{0} & P_{13} - \alpha_3 D_4 & P_{12} - \alpha_2 D_4 & \bar{\tilde{\Omega}}_{7,6} & \bar{\tilde{\Omega}}_{7,7} \end{bmatrix} \prec 0, \tag{93}$$

where $\bar{\tilde{\Omega}}_{1,1} \triangleq \mathrm{sym}(P_{13}) + Q_1 + \hat{d}_1^2 R_1 + \hat{d}_{12}^2 R_1 + \alpha_1 \mathrm{sym}(D_4^T W_f) + \alpha_1 \mathrm{sym}(\Lambda\bar{W}\tilde{N}_{d,2,1})$, $\bar{\tilde{\Omega}}_{2,1} \triangleq P_{12}^T - P_{13}^T + R_3$, $\bar{\tilde{\Omega}}_{2,2} \triangleq Q_2 + Q_3 - Q_1 - R_3 - R_4$, $\bar{\tilde{\Omega}}_{4,1} \triangleq P_{33} + \alpha_3 D_4^T W_f + \alpha_3\Lambda\bar{W}\tilde{N}_{d,2,1}$, $\bar{\tilde{\Omega}}_{5,1} \triangleq P_{23} + \alpha_2 D_4^T W_f + \alpha_2\Lambda\bar{W}\tilde{N}_{d,2,1}$, $\bar{\tilde{\Omega}}_{6,1} \triangleq \alpha_1 W_n^T E_4 + \alpha_1 \tilde{N}_{d,2,3}^T \bar{W}^T\Lambda^T$, $\bar{\tilde{\Omega}}_{6,4} \triangleq \alpha_3 W_n^T D_4 + \alpha_3\tilde{N}_{d,2,3}^T\bar{W}^T\Lambda^T$, $\bar{\tilde{\Omega}}_{6,5} \triangleq \alpha_2 W_n^T D_4 + \alpha_2\tilde{N}_{d,2,3}^T\bar{W}^T\Lambda^T$, $\bar{\tilde{\Omega}}_{7,1} \triangleq D_4^T W_f + \Lambda\bar{W}\tilde{N}_{d,2,1} + P_{11} - \alpha_1 D_4$, $\bar{\tilde{\Omega}}_{7,6} \triangleq D_4^T W_n + \Lambda\bar{W}\tilde{N}_{d,2,3}$, *and* $\bar{\tilde{\Omega}}_{7,7} \triangleq -\mathrm{sym}(D_4) + \hat{d}_1^2 R_3 + \hat{d}_{12}^2 R_4$.

In addition, we have

$$J = \Lambda D_4^{-T}. \tag{94}$$

Example 3 Consider a delayed LTI system (31) with parameters described below:

$$A = \begin{bmatrix} -10 & 1 & 0 & 0 \\ -41 & -2 & 4 & 0 \\ 0 & 0 & -2 & 1 \\ 2 & 0 & -20 & -6 \end{bmatrix}, \; A_d = \begin{bmatrix} 2 & 0 & 0 & -5 \\ 0 & 0 & 0 & 9 \\ 0 & 0 & 0 & 0 \\ 0 & 0 & 0 & 0 \end{bmatrix}, \; C = \begin{bmatrix} 1.5 & 0 & 0 & 0 \\ 0 & -2 & 0 & 0 \end{bmatrix},$$

$B = \begin{bmatrix} 1 & 2 & -1 & 0.2 \end{bmatrix}^T$, and $L = \begin{bmatrix} \mathbf{0}_2 & I_2 \end{bmatrix}$.

The actual time-varying unknown delay function is assumed to behave as $d(t) = 0.5(d_2 + d_1) + 0.5(d_2 - d_1)\sin\left(\frac{2\mu}{d_2 - d_1}t\right)$ with delay bounds arbitrarily fixed at $d_1 = 0.1$[s], $d_2 = 3$[s], and $\mu = 1.2$.

It can be examined that Conditions (83) and (86) are satisfied. Interestingly, it can be shown that by setting $\hat{d}(\cdot) \equiv 0$, i.e. a delay-free FO, a necessary condition for the stability of the observer dynamics outlined below is not met [14]:

$$\text{rank}\left(\begin{bmatrix} sI_l - W_f - W_n \\ \bar{W}(\tilde{N}_{d,2,1} + \tilde{N}_{d,2,3}) \end{bmatrix}\right) = l \quad \forall s \in \mathbb{C}^+. \tag{95}$$

This phenomenon emphasizes that adding auxiliary delay terms into the observer dynamics can effectively reduce from conservatism structurally implied by unknown-input functional observers (see Sect. 3).

We further assume that only the upper-bound of delay d_2 is known. Hence, $\hat{d}_2 = d_2 = 3$[s], $\hat{d}_1 = 0$ and $\hat{\mu} = 0.4$ are arbitrarily assigned, and the ancillary delay function is chosen as $\hat{d}(t) = 0.5(\hat{d}_2 + \hat{d}_1) + 0.5(\hat{d}_2 - \hat{d}_1)\cos\left(\frac{2\hat{\mu}}{\hat{d}_2 - \hat{d}_1}t\right)$. Fixing $\alpha_i = 1$, $i = \{1, 2, 3\}$, returns infeasible LMI (93). Therefore, the GABT algorithm is used to obtain $[\alpha_1, \alpha_2, \alpha_3] = \begin{bmatrix} 25 & 4 & 5 \end{bmatrix}$ as a feasible set that results in appropriate observer parameters.

Simulations were carried out in the MATLAB/Simulink environment using an arbitrary input signal that reads as $u(t) = 5 + 10e^{-0.4t}\cos(2t)$, and assuming $\phi(\theta) = \begin{bmatrix} -2 & 0.01 & 0 & 10 \end{bmatrix}^T$, $\theta \in [-3, 0]$, and $\tilde{P} = I_2$.

Two alternative values were assigned for the sliding gain to emphasize the crucial role of the estimation of the upper-bound of disturbance terms: (i) $\rho(t) = 0.1$ and (ii) $\rho(t) = 5$. The results are depicted in Figs. 3 and 4. It is clear from Fig. 3 that $\rho(t) = 0.1$ is not an upper-bound of the disturbance term $\varepsilon(t, \bar{\epsilon}(t), x(t))$, thus finite-time convergence of the outputs estimation errors does not occur, which causes deviations in the estimation of the desired functional $z(t)$.

On the other hand, Fig. 4 demonstrates that sufficiently enlarging the sliding gain to $\bar{\rho} = 5$ the outputs estimations converge in a finite time, and the estimated functionals converge asymptotically.

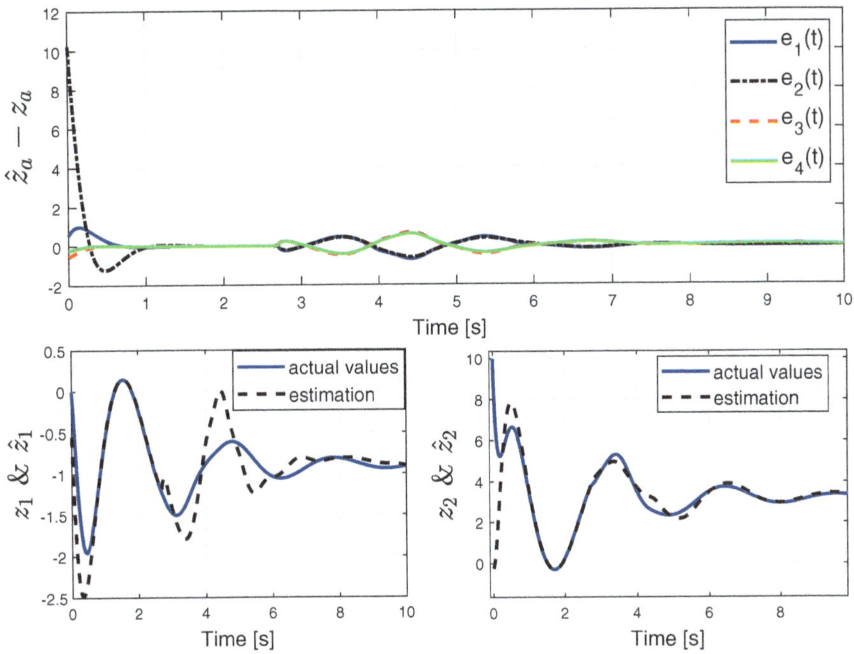

Fig. 3 Estimation of the functions with insufficient estimated upper-bound on the disturbance term $\varepsilon(t, \bar{\epsilon}(t), x(t))$ ($\rho(t) = 0.1$). z_i and $\hat{z}_i$ indicate the ith element of z_a and $\hat{z}_a$, respectively

To emphasize the effectiveness of the delayed terms and the robustifying sliding term, the estimation were carried out by dropping parameters F_d, H_d, X_2, and function $\nu(\cdot)$ from the observer structure, thus replacing observer (58) with a conventional FO similar to (2). The results shown in Fig. 5 clarifies that the simplified FO structure fails to satisfy the stability requirements.

6 Summary and Future Directions

Fundamental concepts and recent advancements in partial state estimation of LTI systems with time-varying delays have been reviewed in this chapter. The chapter sheds light on the import aspects of studying (unknown-input) functional observers and the main challenges confronting the current state of the practice. It has been shown this class of generic Luenberger observers can relax the observability/detectability requirement of the system, as well as the matching condition that is invoked for designing unknown-input observers.

Sophisticated order-increase algorithms illustrated in this chapter can be readily practiced using matrix programming languages available as Phyton/R libraries or MATLAB language. It has been shown that delay-free and delay-dependent FOs can

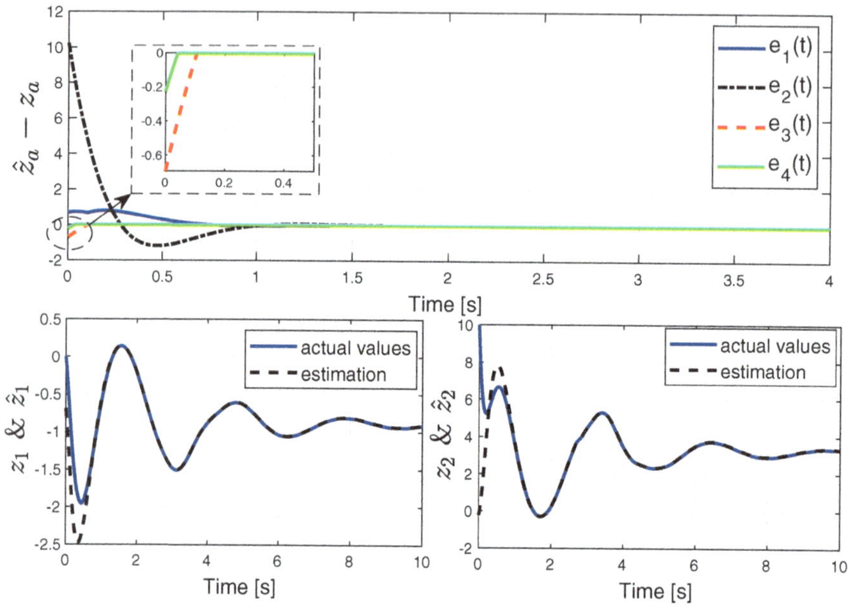

Fig. 4 Estimation of the functions with sufficiently large estimated upper-bound on the disturbance term $\varepsilon(t, \bar{\epsilon}(t), x(t))$ ($\rho(t) = 5$). z_i and $\hat{z}_i$ indicate the ith element of z_a and $\hat{z}_a$, respectively

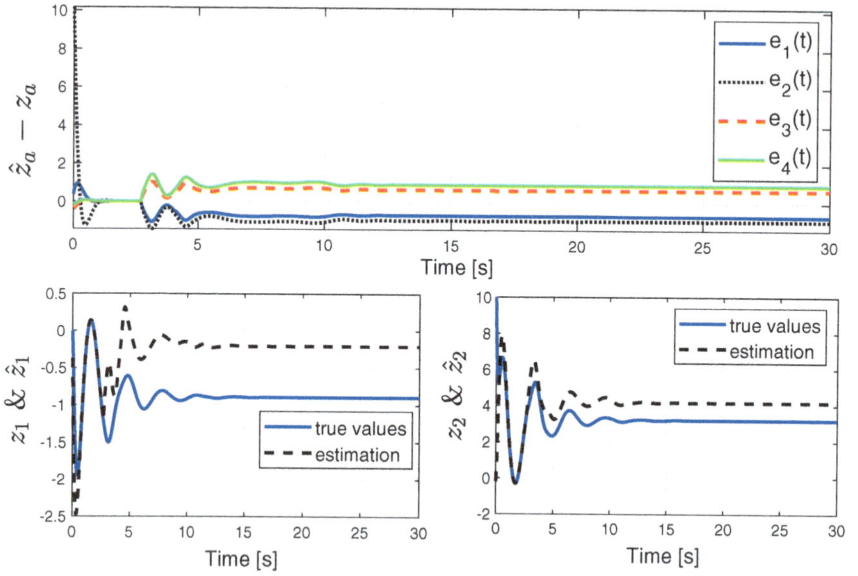

Fig. 5 Estimation results of the delay-free observer obtained from omitting the delayed terms and removing the sliding term $\nu(t)$. z_i and $\hat{z}_i$ indicate the ith element of z_a and $\hat{z}_a$, respectively

be designed for time-delay systems with measurable or uncertain input and state time-varying delays. Lyapunov–Krasovskii stability analysis techniques have empowered the delay-dependent observer design schemes.

There are numerous fields that the developed theories for functional observers can be employed to cope with restrictions implied by conventional full-order observers. This includes system monitoring, fault detection and isolation, and feedback control of large-scale dynamic systems, such as power systems [35], chemical processes [2], and traffic networks [36]. On the theoretical side of this field, the most interesting challenge to the author's mind is designing optimum FO-based control algorithms for (delayed) dynamic systems that are not observable nor detectable.

References

1. Luenberger, D.: Observers for multivariable systems. IEEE Trans. Autom. Control **11**(2), 190–197 (1966)
2. Nazmi, S., Mohajerpoor, R., Abdi, H.: Functional observer design with application to pre-compensated multi-variable systems. In: IEEE Conference on Control Applications (CCA2015), Sydney, Australia (2015)
3. Pham, T.N., Trinh, H., Hien, L.V.: Load frequency control of power systems with electric vehicles and diverse transmission links using distributed functional observers. IEEE Trans. Smart Grid **7**(1), 238–252 (2016)
4. Ashraf-Modarres, A., Momeni, H., Yazdizadeh, A.: A new decentralized sliding mode observer design for interconnected power systems with unknown and time-varying delay. In: 4th Conference on Thermal Power Plants (CTPP2012), Tehran, Iran, pp. 1–6 (2012)
5. Sirisena, H.R.: Minimal-order observers for linear functions of a state vector. Int. J. Control **29**(2), 235–254 (1979)
6. Mohajerpoor, R., Abdi, H., Nahavandi, S.: A new algorithm to design minimal multi-functional observers for linear systems. Asian J. Control **18**(3), 842–857 (2016)
7. Trinh, H.: Linear functional state observer for time-delay systems. Int. J. Control **72**(18), 1642–1658 (1999)
8. Darouach, M.: Linear functional observers for systems with delays in state variables. IEEE Trans. Autom. Control **46**(3), 491–496 (2001)
9. Mohajerpoor, R., Leong, W.Y., Abdi, H., Nahavandi, S.: A delay-dependent functional observer for linear time-invariant systems with input delay. In: Proceedings of the 13th International Conference on Control, Automation, Robotics and Vision (ICARCV2014). IEEE, Singapore, pp. 746–751 (2014)
10. Mohajerpoor, R., Abdi, H., Nahavandi, S.: Reduced-order functional observers with application to partial state estimation of linear systems with input-delays. J. Control Decis. **2**(4), 233–256 (2015)
11. Mohajerpoor, R., Abdi, H., Shanmugam, L., Nahavandi, S.: Minimum-order filter design for partial state estimation of linear systems with multiple time-varying state delays and unknown input delays. Int. J. Robust Nonlinear Control **27**(3), 393–409 (2017)
12. Mohajerpoor, R., Abdi, H., SHanmugam, L., Nahavandi, S.: Partial state estimation of lti systems with multiple constant time-delays. J. Franklin Inst. **353**(2), 541–560 (2016)
13. Seuret, A., Floquet, T., Richard, J.P., Spurgeon, S.K.: A sliding mode observer for linear systems with unknown time varying delay. In: American Control Conference (ACC2007), New York, USA, pp. 4558–4563 (2007)
14. Mohajerpoor, R., Shanmugam, L., Abdi, H., Nahavandi, S., Park, J.H.: Delay-dependent functional observer design for linear systems with unknown time-varying state delays. IEEE Trans. Cybern. **48**(7), 2036–2048 (2018)

15. Mohajerpoor, R., Abdi, H., Nahavandi, S.: Partial state estimation: a new design approach. In: 2014 13th International Conference on Control Automation Robotics Vision (ICARCV), Singapore, pp. 752–757 (2014)
16. Brogan, W.L.: Modern Control Theory. Pearson Education India, Noida (1982)
17. Darouach, M.: Existence and design of functional observers for linear systems. IEEE Trans. Autom. Control **45**(5), 940–943 (2000)
18. Jennings, L.S., Fernando, T.L., Trinh, H.M.: Existence conditions for functional observability from an eigenspace perspective. IEEE Trans. Autom. Control **56**(12), 2957–2961 (2011)
19. Trinh, H., Nahavandi, S., Tran, T.D.: Algorithms for designing reduced-order functional observers of linear systems. Int. J. Innov. Comput. Inf. Control **4**(2), 321–334 (2008)
20. Park, J.H., Lee, T.H., Liu, Y., Chen, J.: Dynamic Systems with Time Delays: Stability and Control. Springer-Nature, Berlin (2019)
21. Mohajerpoor, R., Abdi, H., Nahavandi, S.: Minimal unknown-input functional observers for multi-input multi-output lti systems. J. Process Control **35**, 143–153 (2015)
22. Darouach, M.: Functional observers for systems with unknown inputs. In: 16th International Symposium on Mathematical Theory of Networks and Systems (MTNS2004), Leuven, Belgium (2004)
23. Hou, M., Pugh, A.C., Muller, P.C.: Disturbance decoupled functional observers. IEEE Trans. Autom. Control **44**(2), 382–386 (1999)
24. Bejarano, F.J., Fridman, L., Poznyak, A.: Unknown input and state estimation for unobservable systems. SIAM J. Control Optim. **48**(2), 1155–1178 (2009)
25. Mohajerpoor, R., Abdi, H., Nahavandi, S.: On unknown-input functional observability of linear systems. In: American Control Conference (ACC2015), Chicago, USA
26. Angulo, M.T., Moreno, J.A., Fridman, L.: On functional observers for linear systems with unknown inputs and hosm differentiators. J. Franklin Inst. **351**(4), 1982–1994 (2014)
27. Chen, J., Park, J.H., Xu, S.: Stability analysis of continuous-time systems with time-varying delay using new lyapunov-krasovskii functionals. J. Franklin Inst. **355**(13), 5957–5967 (2018)
28. Seuret, A., Gouaisbaut, F.: Wirtinger-based integral inequality: application to time-delay systems. Automatica **49**(9), 2860–2866 (2013)
29. Mohajerpoor, R., Shanmugam, L., Abdi, H., Rakkiyappan, R., Nahavandi, S., Shi, P.: New delay range-dependent stability criteria for interval time-varying delay systems via wirtinger-based inequalities. Int. J. Robust Nonlinear Control **28**(2), 661–677 (2018)
30. Park, P., Ko, J.W., Jeong, C.: Reciprocally convex approach to stability of systems with time-varying delays. Automatica **47**(1), 235–238 (2011)
31. Mohajerpoor, R., Shanmugam, L., Abdi, H., Nahavandi, S., Saif, M.: Functional observer design for retarded system with interval time-varying delays. Int. J. Syst. Sci. **48**(5), 1060–1070 (2017)
32. Boyd, S., Vandenberghe, L.: Convex Optimization. Cambridge University Press, Cambridge (2004)
33. Hale, J.K.: Introduction to Functional Differential Equations, vol. 99. Springer, Berlin (1993)
34. Edwards, C., Spurgeon, S.: Sliding Mode Control, Theory and Applications. CRC Press, Boca Raton (1998)
35. Emami, K., Ariakia, H., Fernando, T.: A functional observer based dynamic state estimation technique for grid connected solid oxide fuel cells. IEEE Trans. Energy Convers. **33**(1), 96–105 (2018)
36. Mohajerpoor, R., Saberi, M., Vu, H.L., Garoni, T.M., Ramezani. M.: $\mathcal{H}_\infty$ robust perimeter flow control in urban networks with partial information feedback. Transportation Research Part B: Methodological (2019)

Finite-Time Stability and Control of Impulsive Positive Systems with Interval Uncertainty

Mengjie Hu and Ju H. Park

Abstract This chapter addresses finite-time stability and control issues of impulsive positive systems (IPSs) with interval uncertainty. In terms of different types of impulsive effects, a time-varying copositive Lyapunov function is constructed and the average impulsive interval method is applied for studying the finite-time stability of interval IPSs. The relevance between the impulsive effect and the finite time period is disclosed by theoretical results and simulations. Sufficient conditions for the existence of the finite-time controller are established to ensure the finite-time stability of the closed-loop system while maintaining the positivity. The conditions in terms of vector inequalities are easily solvable. Numerical examples are presented to show the usefulness of the obtained theorems.

Keywords Positive systems · Finite-time stability · Interval uncertainty · Robustness

1 Problem Formulation and Preliminaries

In nature and engineering applications, there exists a class of dynamic systems which can keep the system state and output variables non-negative under a non-negative initial state [1]. It is widely used in many fields, such as the metabolic kinetics of drugs in the biological field [2], the industrial process of distillation columns in the chemical field [3], and the probability problem of stochastic models in economics and

---

M. Hu (✉) · J. H. Park
Department of Electrical Engineering, Yeungnam University, Gyeongsan 38541, Republic of Korea
e-mail: humjynu@gmail.com

J. H. Park
e-mail: jessie@ynu.ac.kr

J. H. Park (ed.), *Recent Advances in Control Problems of Dynamical Systems and Networks*, Studies in Systems, Decision and Control 301,
https://doi.org/10.1007/978-3-030-49123-9_2

sociology [4]. It is worth noting that the state of positive systems is constrained within the convex polyhedron in the first quadrant instead of the entire state space, which makes the method for studying general dynamic systems often not applicable to the analysis of positive systems. In 1912, Frobenius proposed the Perron–Frobenius theorem and a series of results on non-negative matrices [5], which provides a powerful mathematical tool for later research works. Since then, positive systems have attracted attentions of many scholars and achieved rapid development [6–10].

In practice, the system dynamic evolution often exhibits the transient state, that is, impulsive phenomenon. Dynamic systems with positive constraints and impulsive effects are called impulsive positive systems (IPSs). IPSs include positive continuous-time dynamics and positive discrete-time dynamics, and possess typical hybrid characteristics and rich dynamic behaviors. In recent years, some researches have been conducted on IPSs, please see [11–13] as references.

As is well known, Lyapunov stability is defined in limitless time intervals and utilized to indicate the steady-state characteristics. From the perspective of practical applications, some systems may be defined only within a finite time interval, or their states need to be restricted to a specific range. If the system state does not overstep a preset threshold within a specified time interval under a prescribed initial condition, then the system is finite-time stable (FTS). [14] summarized the development of finite-time stability research, and pointed out that the difference between the concepts of finite-time stability and Lyapunov stability is mainly in two aspects:

(1) The finite-time stability means the system will be analysed within a limited time interval.

(2) A threshold for the system state is needed to be set in advance. Results on the finite-time stability of general impulsive systems can refer to [15, 16], and this problem has also made much progress in positive systems [17–20]. Among them, [19] established sufficient and necessary conditions for the finite-time stability of linear positive systems. By the average dwell time method, [20] discussed the finite-time $\mathcal{L}_1$ control issue of positive switched systems with delay.

Based on the above discussion, although there are some interesting results on the finite-time stability for positive systems and impulsive systems, there is a lack of research on the finite-time stability of interval IPSs. This chapter extends the definition of finite-time stability to IPSs. The finite-time stability and control issues of interval IPSs are studied. With the help of the time-varying co-positive Lyapunov function and the average impulsive interval method, under different impulsive effects (disturbance impulses, stabilizing impulses and "neutral" impulses), a finite-time stability criterion for interval IPSs is established. The relationship between the impulses and the finite time interval is disclosed. The finite-time controller is designed to achieve the positive stabilization of interval IPSs. Numerical examples are presented to show the usefulness of the obtained theorems.

Notations: $\mathcal{R}^n$ is the n-dimensional Euclidean space. N is the natural number set, $\mathcal{R}^n_+$ denotes the positive quadrant of $\mathcal{R}^n$. We denote $\mathbf{1} = [1, 1, \ldots, 1]^T$. A matrix

$A \succ 0$ ($\prec 0$, $\succeq 0$, $\preceq 0$) denotes all entries of A are positive (negative, nonnegative, non-positive).

Consider the impulsive system with interval uncertainty:

$$\begin{cases} \dot{x}(t) = \mathcal{A}x(t), t \neq t_k, \\ x(t_k) = (I + \mathcal{H})x(t^-), k \in \mathbb{N}, \\ x(t_0) = x_0, t_0 = 0, \end{cases} \tag{1}$$

where $x(t) \in \mathcal{R}^n$ is state vector. $\mathcal{A}$ and $\mathcal{H}$ are constant matrices with appropriate dimensions.

Define the set of interval uncertainty matrices as follows:

$$\mathcal{A} \in [\mathcal{A}_m, \mathcal{A}_M], \mathcal{H} \in [\mathcal{H}_m, \mathcal{H}_M].$$

$x(t_k^-) = \lim\limits_{h \to 0^-} x(t_k + h)$. x_0 is the initial state.

Assume $x(t_k) = x(t_k^+) = \lim\limits_{h \to 0^+} x(t_k + h), k \in \mathbb{N}$. The impulse time sequence $\sigma = \{t_1, t_2, t_3 ...\}$ satisfies $0 < t_1 < t_2 < t_3 ... < t_{k-1} < t_k < ...$ with $\lim_{k \to \infty} t_k \to \infty$.

Remark 1 Note that the uncertainty refers to changes within a possible range, which is unavoidable in practice. There are usually two types of system uncertainty. One is the integral quadratic constraint description. The other is a structured description, which usually involves interval uncertainty.

Before presenting the main results, the following definitions and lemmas are provided.

Definition 1 ([1]) If for $x_0 \succeq 0$, the system state $x(t) \succeq 0$ holds $\forall t \geq t_0$, then system (1) is positive.

Definition 2 ([11]) If there exist $\tau_a > 0$ and integer $N_0 \geq 0$, such that

$$\frac{T_2 - T_1}{\tau_a} - N_0 \leq N_\sigma(T_2, T_1) \leq \frac{T_2 - T_1}{\tau_a} + N_0, \forall T_2 \geq T_1 \geq 0 \tag{2}$$

holds, where $N_\sigma(T_2, T_1)$ refers to the number of impulsive instants over (T_1, T_2), then the average impulsive interval of impulse time sequence σ is τ_a.

Lemma 1 ([12]) *If $\mathcal{A}_m$ is Metzler matrix and $(I + \mathcal{H}_m) \succeq 0$, then system (1) is positive.*

2 Finite-Time Stability Analysis

In this section, we will establish sufficient conditions to make interval impulsive positive systems (IPSs) finite-time stable (FTS), specially, different impulsive effects are considered.

In order to ensure the positivity of system (1), suppose the system matrices satisfy $\mathcal{A}_m$ is Metzler and $(I + \mathcal{H}_m) \succeq 0$.

First, the definition of finite-time stability is presented below.

Definition 3 ([21]) For a vector $v \succ 0$, and a vector-valued function $r(\bullet) : t \in [t_0, t_f] \mapsto \mathcal{R}_+^n$ with $r(t_0) \prec v$, IPS (1) is FTS w.r.t. $(t_0, t_f, v, r(\cdot))$ if

$$x_0^T v \leq 1 \Rightarrow x^T(t)r(t) < 1, \forall t \in [t_0, t_f]. \tag{3}$$

Now, we have the first main result.

Theorem 1 *For prescribed positive parameters τ_a, t_f, λ_1 and λ_2, an integer $N_0 \geq 0$, a vector $v \in \mathcal{R}_+^n$, and a vector function $r(t) \succ 0$ with $r(t_0) \prec v$, set $\phi = \frac{\lambda_1}{\lambda_2}\beta^{-N_0}$ and $\varphi = \frac{\lambda_1}{\lambda_2}\beta^{N_0}$, if there exist parameters $\beta > 0$ and ε, and a continuously differentiable vector function $\delta(t) \succ 0$ such that the following inequalities hold,*

$$\dot{\delta}(t) + \mathcal{A}_M^T \delta(t) - \varepsilon\delta(t) \preceq 0, \tag{4}$$

$$((1 - \beta)I + \mathcal{H}_M^T)\delta(t_k) \preceq 0, k \in \mathbb{N}, \tag{5}$$

$$\lambda_1 r(t) \prec \delta(t), \tag{6}$$

$$\delta(t_0) \preceq \lambda_2 v, \tag{7}$$

(a) When $\beta \geq 1$,

(i) if τ_a and t_f satisfy

$$0 < t_f \leq \frac{\ln \phi}{\ln \beta + \varepsilon\tau_a}\tau_a, \tag{8}$$

$$ln\beta + \varepsilon\tau_a > 0, \tag{9}$$

then system (1) is FTS w.r.t. $(t_0, t_f, v, r(\cdot))$.

(ii) if τ_a satisfies

$$ln\beta + \varepsilon\tau_a \leq 0, \tag{10}$$

and $\phi \geq 1$, then system (1) is FTS w.r.t. $(t_0, t_f, v, r(\cdot))$ $\forall t_f > t_0$.

(b) When $0 < \beta < 1$,

(iii) if τ_a and t_f satisfy

$$0 < t_f \leq \frac{\ln \varphi}{\ln \beta + \varepsilon\tau_a}\tau_a, \tag{11}$$

and τ_a satisfies (9), then system (1) is FTS w.r.t. $(t_0, t_f, v, r(\cdot))$.

(iv) if τ_a satisfies (10) and $\varphi \geq 1$, then system (1) is FTS w.r.t. $(t_0, t_f, v, r(\cdot))$ $\forall t_f > t_0$.

Remark 2 Note that in the two special cases of Theorem 1, namely, Case (ii) and Case (iv), under certain conditions, IPS (1) is FTS w.r.t. $(t_0, t_f, v, r(\cdot))$ regardless of the value of t_f.

Proof Define a time-varying copositive Lyapunov function:

$$V(t) = x^T(t)\delta(t), \tag{12}$$

where $\delta(t) \succ 0$ is to be determined.

When $t \in [t_{k-1}, t_k)$, $k \in \mathbb{N}$, we have

$$\begin{aligned} D^+V(t) &= \dot{x}^T(t)\delta(t) + x^T(t)\dot{\delta}(t) \\ &= x^T(t)(\mathcal{A}^T\delta(t) + \dot{\delta}(t)). \end{aligned} \tag{13}$$

By (4), we can derive

$$\mathcal{A}^T\delta(t) + \dot{\delta}(t) \preceq \dot{\delta}(t) + \mathcal{A}_M^T\delta(t) \preceq \varepsilon\delta(t), \tag{14}$$

then,

$$D^+V(t) - \varepsilon V(t) = x^T(t)(\dot{\delta}(t) + \mathcal{A}^T\delta(t) - \varepsilon\delta(t)) \leq 0. \tag{15}$$

Integrating (15) from t_{k-1} to t, $t \in [t_{k-1}, t_k)$, $k \in \mathbb{N}$, we can obtain

$$V(t) \leq e^{\varepsilon(t-t_{k-1})}V(t_{k-1}, x(t_{k-1})). \tag{16}$$

For $t = t_k$,$k \in \mathbb{N}$,

$$\begin{aligned} V(t_k, x(t_k)) &= x^T(t_k)\delta(t_k) \\ &= x^T(t_k^-)(I + \mathcal{H})^T\delta(t_k). \end{aligned} \tag{17}$$

From (5), we have

$$(I + \mathcal{H})^T\delta(t_k) \preceq (I + \mathcal{H}_M)^T\delta(t_k) \preceq \beta\delta(t_k), \tag{18}$$

then,

$$V(t_k) \leq \beta V(t_k^-). \tag{19}$$

Then, we can obtain

$$V(t) \leq \beta^{N_\sigma(t,t_0)}e^{\varepsilon(t-t_0)}V(t_0, x(t_0)), t \geq t_0, \tag{20}$$

where $N_\sigma(t, t_0)$ is the number of impulsive instants in $[t_0, t)$.

In what follows, two cases of $\beta \geq 1$ and $0 < \beta < 1$ are considered respectively.

(A) For $\beta \geq 1$:
From (2), we derive

$$V(t) \leq \beta^{N_0} e^{(\frac{ln\beta}{\tau_a}+\varepsilon)(t-t_0)} V(t_0, x(t_0)). \tag{21}$$

By (6), (7) and (21), we have

$$\begin{aligned}\lambda_1 x^T(t) r(t) = V(t) &\leq \beta^{N_0} e^{(\frac{\ln\beta}{\tau_a}+\varepsilon)(t-t_0)} x^T(t_0)\delta(t_0) \\ &\leq \lambda_2 \beta^{N_0} e^{(\frac{\ln\beta}{\tau_a}+\varepsilon)(t-t_0)} x^T(t_0) v. \end{aligned} \tag{22}$$

(i) When $ln\beta + \varepsilon\tau_a > 0$, by (8), we have the following inequality

$$\frac{\lambda_2}{\lambda_1}\beta^{N_0} e^{(\frac{\ln\beta}{\tau_a}+\varepsilon)(t-t_0)} \leq \frac{\lambda_2}{\lambda_1}\beta^{N_0} e^{(\frac{\ln\beta}{\tau_a}+\varepsilon)(t_f-t_0)} \leq 1, t \in [t_0, t_f], \tag{23}$$

holds, then from (22), we can derive

$$x^T(t) r(t) < \frac{\lambda_2}{\lambda_1}\beta^{N_0} e^{(\frac{\ln\beta}{\tau_a}+\varepsilon)(t_f-t_0)} x^T(t_0) v \leq x^T(t_0) v. \tag{24}$$

(ii) When $ln\beta + \varepsilon\tau_a \leq 0$, by $\phi \geq 1$, we can obtain

$$\frac{\lambda_2}{\lambda_1}\beta^{N_0} e^{(\frac{\ln\beta}{\tau_a}+\varepsilon)(t-t_0)} \leq \frac{\lambda_2}{\lambda_1}\beta^{N_0} e^{(\frac{\ln\beta}{\tau_a}+\varepsilon)(t_0-t_0)} \leq 1, t \in [t_0, t_f]. \tag{25}$$

From (22), we have

$$x^T(t) r(t) < \frac{\lambda_2}{\lambda_1}\beta^{N_0} e^{(\frac{\ln\beta}{\tau_a}+\varepsilon)(t_0-t_0)} x^T(t_0) v \leq x^T(t_0) v. \tag{26}$$

By (24) and (26), we can conclude that when $\beta \geq 1$,

$$x^T(t) r(t) < x^T(t_0) v, t \in [t_0, t_f], \tag{27}$$

then,

$$x^T(t_0) v \leq 1 \Rightarrow x^T(t) r(t) < 1 \tag{28}$$

holds.

(B) For $0 < \beta < 1$:
(iii) When $ln\beta + \varepsilon\tau_a > 0$, by (2), (65) and (20), we have

$$x^T(t) r(t) < \frac{\lambda_2}{\lambda_1}\beta^{-N_0} e^{(\frac{\ln\beta}{\tau_a}+\varepsilon)(t_f-t_0)} x^T(t_0) v \leq x^T(t_0) v, t \in [t_0, t_f]. \tag{29}$$

(iv) When $ln\beta + \varepsilon\tau_a \leq 0$, by $\varphi \geq 1$ and (20), we have

$$x^T(t)r(t) < \frac{\lambda_2}{\lambda_1}\beta^{-N_0}e^{(\frac{\ln\beta}{\tau_a}+\varepsilon)(t_0-t_0)}x^T(t_0)v \leq x^T(t_0)v, t \in [t_0, t_f]. \tag{30}$$

Then, from (29) and (30), we can obtain that when $0 < \beta < 1$,

$$x^T(t_0)v \leq 1 \Rightarrow x^T(t)r(t) < 1, \forall t \in [t_0, t_f]. \tag{31}$$

Thus, combining with Case (A) and Case (B), system (1) is FTS w.r.t. $(t_0, t_f, v, r(\cdot))$. This completes the proof. □

According to the above proof, we can observe from (16) and (19) that the parameters ε and β denote the continuous dynamics and impulsive effects respectively. Specifically,

(a) The parameter ε refers to the evolution of Lyapunov function in $t \in [t_{k-1}, t_k)$, $k \in \mathbb{N}$ between the adjacent impulsive instants. The Lyapunov function $V(t)$ may increase over $t \in [t_{k-1}, t_k)$, $k \in \mathbb{N}$ if $\varepsilon \geq 0$. The considered $V(t)$ will decrease over $t \in [t_{k-1}, t_k)$, $k \in \mathbb{N}$ if $\varepsilon < 0$.

(b) The parameter β characterizes different impulsive effects. The Lyapunov function $V(t)$ may jump up at t_k if $\beta > 1$, for this case, the impulses are disturbance impulses. The impulses are "neutral" if $\beta = 1$. The considered $V(t)$ will jump down at t_k if $0 < \beta < 1$, for this case, the impulses are stabilizing impulses.

In the following, for different ε and β, according to the results in Theorem 1, we can obtain the bounds of the average impulsive interval τ_a and the final time t_f, which are arranged in Table 1. Table 2 shows the correlation among key parameters.

Remark 3 For different types of impulsive effects, from Tables 1 and 2, we can obtain:

(a) When $\beta \geq 1$, i.e., disturbance impulses and "neutral" impulses. From Table 1, there always exists the lower bound of τ_a, which suggests that in order to keep the finite-time stability, the frequency of disturbance impulses should not be too high. There always exists the upper bound of t_f if $\tau_a < \kappa_2$.

In addition, by Table 2, we can see the positive correlation between $\underline{\tau_a}$&β, $\underline{\tau_a}$&t_f and $\overline{t_f}$&τ_a when disturbance impulses are taken into account.

(b) When $0 < \beta < 1$, i.e., stabilizing impulses. From Table 1, we can observe if $\tau_a \leq \kappa_2$ holds, i.e., if the stabilizing impulses occur frequently enough, the IPS is FTS for any t_f. If $t_f \leq \upsilon_2$ holds, the IPS is FTS for any τ_a, i.e., if the wanted finite time interval is short sufficiently, the IPS is FTS regardless of the frequency of stabilizing impulses.

Besides, by Table 2, we can see the negative correlation between $\overline{\tau_a}$&β, $\overline{\tau_a}$&t_f and $\overline{t_f}$&τ_a when stabilizing impulses are taken into account.

Next, we will study the finite-time stability of IPSs under $\mathcal{L}_1$ performance. Consider the IPS with exogenous disturbance,

$$\begin{cases} \dot{x}(t) = \mathcal{A}x(t) + \mathcal{F}\omega(t), t \neq t_k, \\ \Delta x(t) = \mathcal{H}x(t^-), t = t_k, k \in \mathbb{N}, \\ y(t) = \mathcal{C}x(t) + \mathcal{E}\omega(t), \\ x(t_0) = x_0, t_0 = 0, \end{cases} \tag{32}$$

where system matrices are belong to the set of interval uncertainty matrices as follows:

$$\begin{aligned} &\mathcal{A} \in [\mathcal{A}_m, \mathcal{A}_M], \mathcal{H} \in [\mathcal{H}_m, \mathcal{H}_M], \mathcal{F} \in [\mathcal{F}_m, \mathcal{F}_M], \\ &\mathcal{C} \in [\mathcal{C}_m, \mathcal{C}_M], \mathcal{E} \in [\mathcal{E}_m, \mathcal{E}_M], \end{aligned}$$

in which $\mathcal{A}_m \in \mathcal{R}^{n\times n}$ is Metzler, $I + \mathcal{H}_m \succeq 0$, $\mathcal{F}_m \succeq 0$, $\mathcal{C}_m \succeq 0$ and $\mathcal{E}_m \succeq 0$. $y(t) \in \mathcal{R}^q$ is the system output.
Here, it is assumed that the disturbance input $\omega(t) \in \mathcal{R}^p_+$ satisfies

$$\int_{t_0}^{t_f} \|\omega(t)\|\, dt \leq h,\ h \geq 0. \tag{33}$$

Definition 4 ([11]) For a vector $\upsilon \succ 0$, and a vector function $r(\bullet) : t \in [t_0, t_f] \mapsto \mathcal{R}^n_+$ with $r(t_0) \prec \upsilon$, system (32) is finite-time bounded w.r.t. $(t_0, t_f, \upsilon, r(\cdot), h)$, if

$$x_0^T \upsilon \leq 1 \Rightarrow x^T(t) r(t) < 1, \forall t \in [t_0, t_f]. \tag{34}$$

Table 1 The bounds of τ_a and t_f

ε	β		
	$\beta \geq 1$		$0 < \beta < 1 (\varepsilon \neq 0)$
$\varepsilon \geq 0$	τ_a	$\tau_a > \kappa_1 0$	• $\tau_a \leq \kappa_3$ if $\upsilon_2 < t_f \leq \upsilon_3$ • τ_a has no bounds if $t_f \leq \upsilon_2$ • $\tau_a \leq \kappa_2$ if $t_f \geq \upsilon_3$
	t_f	$t_f < \upsilon_1$	• $t_f \leq \upsilon_3$ if $\tau_a > \kappa_2$ • t_f has no bounds if $\tau_a \leq \kappa_2$
$\varepsilon < 0$	τ_a	• $\tau_a > \kappa_1$ if $t_f < \upsilon_1$ • $\tau_a \geq \kappa_2$ if $t_f \geq \upsilon_1$	τ_a has no bounds
	t_f	• $t_f < \upsilon_1$ if $\tau_a \in (\kappa_1, \kappa_2)$ • t_f has no bounds if $\tau_a \geq \kappa_2$	t_f is has no bounds

[1] $\kappa_1 = \frac{\ln\beta}{\ln\phi - \varepsilon t_f} t_f$, $\kappa_2 = -\frac{\ln\beta}{\varepsilon}$, $\kappa_3 = \frac{\ln\beta}{\ln\varphi - \varepsilon t_f} t_f$, $\upsilon_1 = \frac{\ln\phi}{\ln\beta + \varepsilon\tau_a}\tau_a$, $\upsilon_2 = \frac{\ln\varphi}{\varepsilon}$ and $\upsilon_3 = \frac{\ln\varphi}{\ln\beta + \varepsilon\tau_a}\tau_a$

Table 2 The correlations among key parameters

ε	β	
	$\beta \geq 1$	$0 < \beta < 1$ ($\varepsilon \neq 0$)
$\varepsilon \geq 0$	$\underline{\tau_a} \uparrow$ when $\beta \uparrow$.	$\overline{\tau_a} \downarrow$ when $\beta \uparrow$.
	$\underline{\tau_a} \uparrow$ when $t_f \uparrow$.	$\overline{\tau_a} \downarrow$ when $t_f \uparrow$ if $\tau_a > \kappa_2$ and $t_f > \upsilon_2$.
	$\overline{t_f} \uparrow$ when $\tau_a \uparrow$.	$\overline{t_f} \downarrow$ when $\tau_a \uparrow$ if $\tau_a > \kappa_2$.
$\varepsilon < 0$	$\underline{\tau_a} \uparrow$ when $\beta \uparrow$.	\
	$\underline{\tau_a} \uparrow$ when $t_f \uparrow$ if $\tau_a < \kappa_2$.	
	$\overline{t_f} \uparrow$ when $\tau_a \uparrow$ if $\tau_a < \kappa_2$.	

[1] $\uparrow$ denotes the parameter increases, $\downarrow$ denotes the parameter decreases
[2] $\overline{\tau_a}$ means the upper bound of τ_a, $\underline{\tau_a}$ means the lower bound of τ_a
[3] $\kappa_2 = -\frac{\ln \beta}{\varepsilon}$ and $\upsilon_2 = \frac{\ln \varphi}{\varepsilon}$

Definition 5 ([11]) For a given constant t_f, the IPS (32) is finite-time bounded w.r.t. $(t_0, t_f, v, r(\cdot), h)$ under $\mathcal{L}_1$-gain performance if

(1) System (32) is finite-time bounded w.r.t. $(t_0, t_f, v, r(\cdot), h)$;

(2) Inequality

$$\theta \int_{t_0}^{t_f} e^{-\eta t} \|y(t)\| dt \leq \gamma \int_{t_0}^{t_f} \|\omega(t)\| dt, \tag{35}$$

holds under zero initial conditions, where $\theta > 0$, $\gamma > 0$, $\eta > 0$ and $\omega(t)$ satisfies (33).

Without loss of generality, with the help of Theorem 1, we present the theoretical criterion for the case of disturbance impulses.

Theorem 2 *For prescribed positive constants τ_a, t_f, λ_1, λ_2, λ and h, an integer $N_0 \in N$, a vector $v \in \mathcal{R}_+^n$, and a vector function $r(t) \succ 0$ defined over $[t_0, t_f]$ with $r(t_0) \prec v$, set $\psi = \frac{\lambda_1}{\lambda_2 + \lambda h} \beta^{-N_0}$, if there exist parameters $\varepsilon > 0$, $\beta > 1$ and $0 < \gamma \leq \lambda$, and a continuously differentiable vector function $\delta(t) \succ 0$ such that*

$$\dot{\delta}(t) + \mathcal{A}_M^T \delta(t) - \varepsilon \delta(t) + \mathcal{C}^T \mathbf{1}_q \preceq 0, \tag{36}$$
$$\mathcal{F}_M^T \delta(t) + \mathcal{E}_M^T \mathbf{1}_q - \gamma \mathbf{1}_p \preceq 0, \tag{37}$$
$$((1-\beta)I + \mathcal{H}_M^T)\delta(t_k) \preceq 0, \ k \in \mathbb{N}, \tag{38}$$
$$\lambda_1 r(t) \prec \delta(t), \tag{39}$$
$$\delta(t_0) \preceq \lambda_2 v, \tag{40}$$

hold and τ_a satisfies

$$\tau_a \geq \max\left\{\frac{\ln\beta}{\varepsilon}, \frac{\ln\beta t_f}{\ln\psi - \varepsilon t_f}\right\}, \tag{41}$$

then system (32) is finite-time bounded under $\mathcal{L}_1$-gain performance w.r.t. $(t_0, t_f, v, r(\cdot), h)$.

Proof The proof is conducted by two steps:

(1) Finite-time boundedness.

Consider the Lyapunov function in (12).

When $t \in [t_{k-1}, t_k), k \in \mathbb{N}$, we derive

$$\begin{aligned} D^+V(t) &= \dot{x}^T(t)\delta(t) + x^T(t)\dot{\delta}(t) \\ &= x^T(t)(\mathcal{A}^T\delta(t) + \dot{\delta}(t)) + \omega^T(t)\mathcal{F}^T\delta(t). \end{aligned} \tag{42}$$

By (36) and (37), we obtain

$$D^+V(t) - \varepsilon V(t) \leq \lambda\omega^T(t)\mathbf{1}_p. \tag{43}$$

Integrating (43) from t_{k-1} to t, $t \in [t_{k-1}, t_k), k \in \mathbb{N}$, we can obtain

$$V(t) \leq e^{\varepsilon(t-t_{k-1})}V(t_{k-1}, x(t_{k-1})) + \lambda\int_{t_{k-1}}^{t} e^{\varepsilon(t-s)}\,\|\omega(t)\|\,ds. \tag{44}$$

For $t = t_k, k \in \mathbb{N}$, by (38), we obtain

$$V(t_k, x(t_k)) \leq \beta V(t_k^-, x(t_k^-)). \tag{45}$$

Combining (44) and (45), we derive

$$\begin{aligned} V(t) \leq\ & \beta^{N_\sigma(t,t_0)}e^{\varepsilon(t-t_0)}V(t_0, x(t_0)) \\ & + \lambda\int_{t_0}^{t} e^{\varepsilon(t-s)}\beta^{N_\sigma(t,s)}\,\|\omega(t)\|\,ds, t \geq t_0. \end{aligned} \tag{46}$$

Note that $t \leq t_f$ and $\int_{t_0}^{t_f}\|\omega(t)\|\,ds \leq h$, by (2), we have

$$\begin{aligned} V(t) &\leq \beta^{N_\sigma(t,t_0)}e^{\varepsilon(t-t_0)}(V(t_0, x(t_0)) + \lambda h) \\ &\leq \beta^{N_0}e^{(\frac{\ln\beta}{\tau_a}+\varepsilon)(t-t_0)}(V(t_0, x(t_0)) + \lambda h). \end{aligned} \tag{47}$$

By (39)–(41), we have $x^T(t_0)v \leq 1 \Rightarrow x^T(t)r(t) < 1$. Then, system (32) is finite-time bounded w.r.t. $(t_0, t_f, v, r(\cdot), h)$.

(2) $\mathcal{L}_1$-gain performance of system (32).

Substituting (36) and (37) to (42), we have

$$D^{+}V(t) \le \varepsilon V(t) + \gamma \|\omega(t)\| - \|y(t)\| . \tag{48}$$

Let $\Gamma(t) = \gamma \|\omega(t)\| - \|y(t)\|$. Integrating (48) from t_{k-1} to t, $t \in [t_{k-1}, t_k)$, $k \in \mathbb{N}$, we can derive

$$V(t) \le e^{\varepsilon(t-t_{k-1})} V(t_{k-1}, x(t_{k-1})) + \int_{t_{k-1}}^{t} e^{\varepsilon(t-s)} \Gamma(s) ds. \tag{49}$$

For $t = t_k, k \in \mathbb{N}$, by (38), we have (45) holds, then

$$V(t) \le \beta^{N_\sigma(t,t_0)} e^{\varepsilon(t-t_0)} V(t_0, x(t_0)) + \int_{t_0}^{t} e^{\varepsilon(t-s)} \beta^{N_\sigma(t,s)} \Gamma(s) ds, t \ge t_0. \tag{50}$$

Since the zero initial condition,

$$\int_{t_0}^{t} e^{\varepsilon(t-s)} \beta^{N_\sigma(t,s)} \|y(s)\| ds \le \gamma \int_{t_0}^{t} e^{\varepsilon(t-s)} \beta^{N_\sigma(t,s)} \|\omega(t)\| ds, \tag{51}$$

holds, then,

$$\int_{t_0}^{t} e^{\varepsilon(t-s)} \beta^{-N_\sigma(s,0)} \|y(s)\| ds \le \gamma \int_{t_0}^{t} e^{\varepsilon(t-s)} \beta^{-N_\sigma(s,0)} \|\omega(t)\| ds. \tag{52}$$

Substituting (2) and (41) to (52), we obtain

$$\int_{t_0}^{t} e^{\varepsilon(t-s)-\varepsilon s} \beta^{-N_0} \|y(s)\| ds \le \gamma \int_{t_0}^{t} e^{\varepsilon(t-s)} \|\omega(t)\| ds. \tag{53}$$

Thus,

$$\begin{aligned} \beta^{-N_0} \int_{t_0}^{t_f} e^{-2\varepsilon s} \|y(s)\| ds &\le \gamma \int_{t_0}^{t_f} e^{-\varepsilon s} \|\omega(t)\| ds \\ &\le \gamma \int_{t_0}^{t_f} \|\omega(t)\| ds. \end{aligned} \tag{54}$$

From Definition 5, we can conclude that system (32) is finite-time bounded under $\mathcal{L}_1$-gain performance w.r.t. $(t_0, t_f, v, r(\cdot), h)$. This completes the proof. □

3 Finite-Time Stabilization of Interval IPSs

In this section, the finite-time control problem of interval IPSs is addressed.

Based on the finite-time stability result of Theorem 1, the state feedback controller for system (1) is designed.

Consider the following system:

$$\begin{cases} \dot{x}(t) = \mathcal{A}x(t) + \mathcal{D}u(t), t \neq t_k, \\ \Delta x(t) = \mathcal{H}x(t^-), t = t_k, k \in \mathbb{N}, \\ x(t_0) = x_0, t_0 = 0, \end{cases} \tag{55}$$

where

$$u(t) = Kx(t), \tag{56}$$

matrix $K \preceq 0$ is to be determined.

System matrices are belong to the set of interval uncertainty matrices as follows:

$$\mathcal{A} \in [\mathcal{A}_m, \mathcal{A}_M], \quad \mathcal{H} \in [\mathcal{H}_m, \mathcal{H}_M], \quad \mathcal{D} \in [\mathcal{D}_m, \mathcal{D}_M],$$

in which $\mathcal{A}_m$ is Metzler, $I + \mathcal{H}_m \succeq 0$ and $\mathcal{D}_m \succeq 0$.

Lemma 2 ([7]) *For matrix $\mathcal{A} \in \mathcal{R}^{n\times n}$, $\mathcal{A}$ is Metzler iff there exists a constant ς such that $\mathcal{A} + \varsigma I \succeq 0$ holds.*

Theorem 3 *For preset positive scalars τ_a, t_f, λ_1 and λ_2, an integer $N_0 \in N$, vectors $v \in \mathcal{R}^n_+$, $\tilde{\delta} \in \mathcal{R}^r_+$, and a vector function $r(t) \succ 0$ with $r(t_0) \prec v$, set $\phi = \frac{\lambda_1}{\lambda_2}\beta^{-N_0}$, if there exist scalars $\beta > 0$, ε and ς, and vectors $\delta \in \mathcal{R}^n_+$ and $w \in \mathcal{R}^n$ such that the following inequalities hold,*

$$\tilde{\delta}^T \mathcal{D}_m^T \delta \mathcal{A}_m + \mathcal{D}_m \tilde{\delta} w^T + \varsigma I \succeq 0, \tag{57}$$

$$\mathcal{A}_M^T \delta - \varepsilon\delta + w \preceq 0, \tag{58}$$

$$((1-\beta)I + \mathcal{H}_M^T)\delta \preceq 0, \tag{59}$$

$$\lambda_1 r(t) \prec \delta \preceq \lambda_2 v, \tag{60}$$

(A) When $\beta \geq 1$,
(i) if τ_a and t_f satisfy

$$0 < t_f \leq \frac{\ln\phi}{\ln\beta + \varepsilon\tau_a}\tau_a, \tag{61}$$

$$\varepsilon\tau_a + \ln\beta > 0, \tag{62}$$

then with the controller

$$u(t) = Kx(t) = \frac{1}{\tilde{\delta}^T \mathcal{D}_m^T \delta}\tilde{\delta} w^T x(t). \tag{63}$$

Then, system (55) is positive and FTS w.r.t. $(t_0, t_f, v, r(\cdot))$.
(ii) if $\phi \geq 1$ and τ_a satisfies

$$\varepsilon\tau_a + \ln\beta \leq 0, \tag{64}$$

then with the controller (63), system (55) is positive and FTS w.r.t. $(t_0, t_f, v, r(\cdot))$ $\forall t_f > t_0$.

(B) When $0 < \beta < 1$,
(iii) if τ_a *satisfies (9) and* t_f *satisfies*

$$0 < t_f \leq \frac{\ln\varphi}{\ln\beta + \varepsilon\tau_a}\tau_a, \tag{65}$$

then with the controller (63), system (55) is positive and FTS w.r.t. $(t_0, t_f, v, r(\cdot))$.
(iv) if $\varphi \geq 1$ *and* τ_a *satisfies (10), then with the controller (63), system (55) is FTS w.r.t.* $(t_0, t_f, v, r(\cdot))$ $\forall t_f > t_0$.

Remark 4 In Theorem 3, inequality (57) makes the closed-loop system positive. Inequalities (58)–(61) and (64) ensure the closed-loop system be FTS.

Remark 5 The relationships between τ_a/t_f and K are discussed here.
(1) From (61) or (64), when τ_a decreases, the parameter ε may be smaller, then w will decrease by (58), which will cause matrix K to decrease from (63), i.e., in order to ensure the finite-time stability of the closed-loop system, the intensity of the controller should increase as the disturbance impulses occur more frequently.
(2) From (61), we have ε may be smaller when t_f increases, then w will decrease by (58), which will cause matrix K to decrease from (63), i.e., when the expected finite time interval is expanded, the intensity of the controller need to increase.

Proof Since $\tilde{\delta} \succ 0$, $\mathcal{D}_m \succeq 0$, $\tilde{\delta} \in \mathcal{R}^r$, $\mathcal{D}_m \in \mathcal{R}^{n\times r}$ and $\delta \in \mathcal{R}^n$, then we have $\tilde{\delta}^T\mathcal{D}_m^T\delta$ is a positive constant.
Divide (57) by $\tilde{\delta}^T\mathcal{D}_m^T\delta$, we can derive

$$\mathcal{A}_m + \frac{1}{\tilde{\delta}^T\mathcal{D}_m^T\delta}\mathcal{D}_m\tilde{\delta}w^T + \frac{\varsigma}{\tilde{\delta}^T\mathcal{D}_m^T\delta}I \succeq 0. \tag{66}$$

From Lemma 2, we have that $\mathcal{A}_m + \frac{1}{\tilde{\delta}^T\mathcal{D}_m^T\delta}\mathcal{D}_m\tilde{\delta}w^T$ is Metzler.
From (63), we also have the matrix $\mathcal{A}_m + \mathcal{D}_m K$ is Metzler.
Then, we conclude that system (55) is positive.
By Theorem 1, let the vector function $\delta(t)$ be a constant vector δ. Then, the finite-time stability conditions are presented as follows:

$$\mathcal{A}_M^T\delta - \varepsilon\delta + K^T\mathcal{D}_m^T\delta \preceq 0, \tag{67}$$

$$((1-\beta)I + \mathcal{H}_M^T)\delta \preceq 0, \tag{68}$$

$$\lambda_1 r(t) \prec \delta \preceq \lambda_2 v. \tag{69}$$

It is easy to see that inequalities (58)–(60) are equivalent to inequalities (67)–(69). In addition, Cases (i)–(iv) of Theorem 3 are the same as Cases (i)–(iv) of Theorem 1. Then, system (55) is FTS. This completes the proof. □

4 Illustrative Example

In this section, three examples are given to exhibit the usefulness of obtained results. The first one based on the drug metabolism is for the finite-time stability of IPSs, and the other two are for the positive finite-time stabilization.

Example 1 Consider the drug metabolism model in Fig. 1 [22], which is widely applied to analyze the drug distribution in the body after injections.

This example is to show that the concentrations of drugs in plasma and extravascular space begin from a specified range, then they are kept within a permissible range over a finite time period after impulse injection.

The system model is

$$\begin{cases} \dot{x}(t) = \begin{pmatrix} -(a_{11}+a_{21}) & a_{12} \\ a_{21} & -(a_{22}+a_{12}) \end{pmatrix} x(t), t \neq t_k, \\ \Delta x(t) = \begin{pmatrix} b_{11} & 0 \\ 0 & b_{22} \end{pmatrix} x(t^-), t = t_k, k \in \mathbb{N}, \\ \mathrm{x}(\mathrm{t}_0) = \mathrm{x}_0, t_0 = 0, \end{cases}$$

where system state x_1 and x_2 denote the concentration of drugs in blood plasma and extravascular space respectively. $a_{11} > 0$ and $a_{22} > 0$ are the ratios of drug excretion in the two spaces respectively. a_{12} and a_{21} are the ratios of drug transmission between the two compartments. b_{11} and b_{22} are the amounts of drug injection in the two compartments respectively.

We consider the system parameter uncertainty, the following system matrices are taken into consideration:

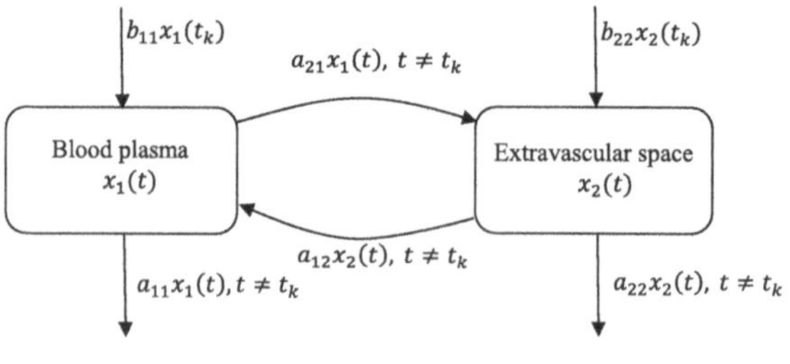

Fig. 1 Drug metabolism model

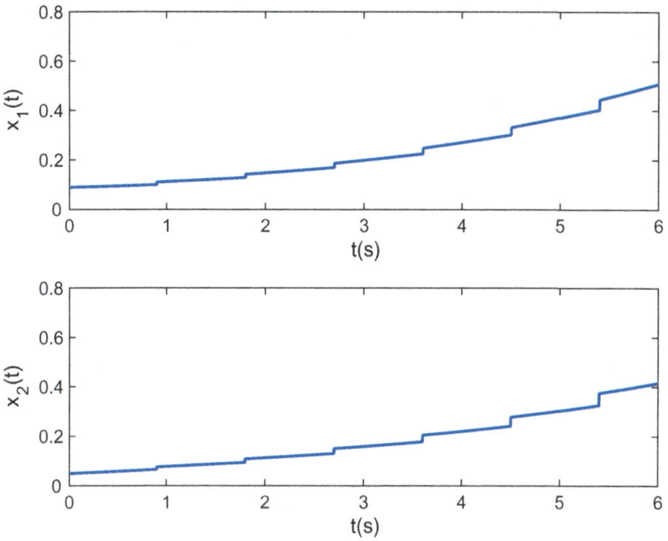

Fig. 2 State response in Example 1

Fig. 3 Evolution of $x^T(t)r(t)$ in Example 1

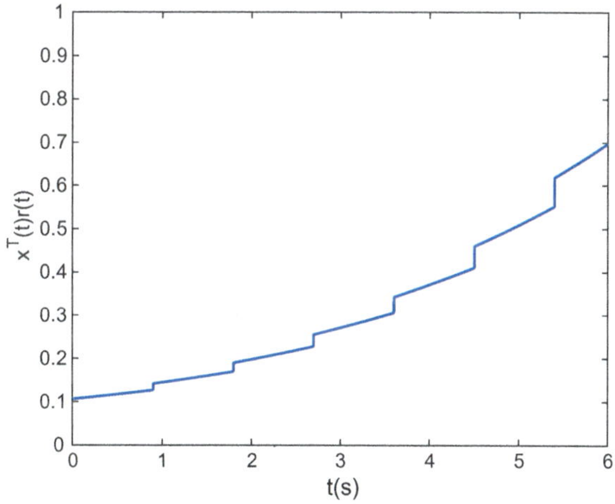

$$\mathcal{A}_M = \begin{bmatrix} -0.15 & 0.45 \\ 0.35 & -0.25 \end{bmatrix}, \quad \mathcal{H}_M = \begin{bmatrix} 0.1 & 0 \\ 0 & 0.15 \end{bmatrix},$$

$$\mathcal{A}_m = \begin{bmatrix} -0.2 & 0.4 \\ 0.3 & -0.3 \end{bmatrix}, \quad \mathcal{H}_m = \begin{bmatrix} 0.05 & 0 \\ 0 & 0.15 \end{bmatrix}.$$

Suppose that the concentration of drugs starts from $6x_1 + 6x_2 \leq 1$. After impulse injection, the permissible range is $0.8x_1 + 0.7x_2 < 1$ over a finite time period $[0, 4s]$. Then, we have vectors $r(t) = [0.8, 0.7]^T$ and $\upsilon = [6, 6]^T$.

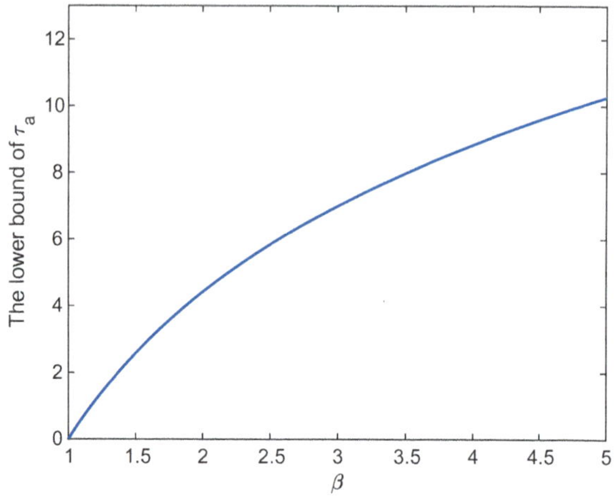

Fig. 4 The bound of τ_a with varying β of Example 1

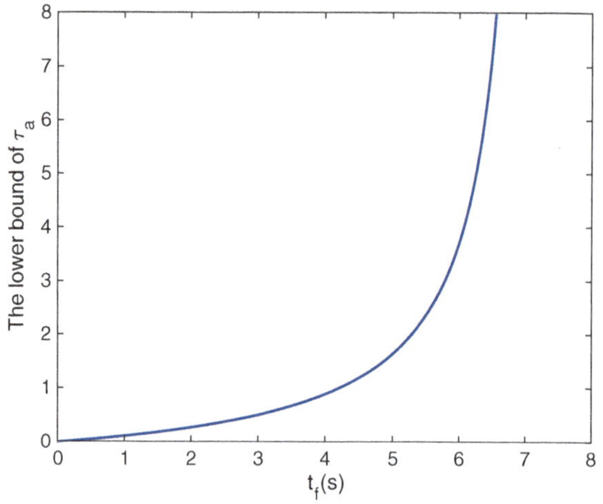

Fig. 5 The bound of τ_a with varying t_f of Example 1

The initial state is given as $[0.09, 0.05]^T$. The parameters are chosen as $\tau_a = 0.9s$, $\varepsilon = 0.2$, $\beta = 1.15$, $\lambda_1 = 1$ and $\lambda_2 = 0.24$.

By applying the results in Theorem 1, the feasible solution of inequalities (4)–(7) is

$$\delta(t) = \begin{bmatrix} 1.3412 - 0.0817 * t \\ 1.3689 - 0.0748 * t \end{bmatrix}.$$

As simulation results, Fig. 2 presents state response of system (1). Figure 3 depicts the curve of $x^T(t)r(t)$ of system (1). It is easy to see that the IPS (1) with disturbance impulses is FTS w.r.t. $(t_0, t_f, v, r(\cdot))$.

When other parameters being settled, the relationships between β and $\underline{\tau_a}$, and between $\underline{\tau_a}$ and t_f are discussed here.

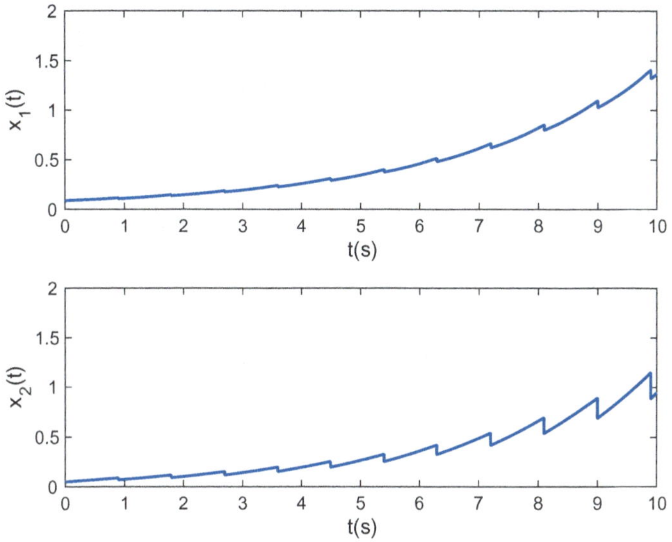

Fig. 6 State evolution of the open-loop system of Example 2

From Case (i) in Theorem 1, the lower bound of τ_a with varying β is depicted in Fig. 4. We can observe that the lower bound of τ_a will be larger if β increases. By (8), the relationship between $\underline{\tau_a}$ and t_f is shown in Fig. 5. We can see the lower bound of τ_a increases as t_f increases. The simulations verify the analysis in Table 2.

Example 2 This example is to show the case of finite-time control with stabilizing disturbance.

Consider system (55) with matrices

$$\mathcal{A}_M = \begin{bmatrix} -0.1 & 0.6 \\ 0.5 & -0.1 \end{bmatrix}, \quad \mathcal{H}_M = \begin{bmatrix} -0.17 & 0.13 \\ 0.06 & -0.3 \end{bmatrix}, \quad \mathcal{D}_M = \begin{bmatrix} 1.1 \\ 1.1 \end{bmatrix},$$
$$\mathcal{A}_m = \begin{bmatrix} -0.12 & 0.6 \\ 0.45 & -0.1 \end{bmatrix}, \quad \mathcal{H}_m = \begin{bmatrix} -0.2 & 0.1 \\ 0.05 & -0.3 \end{bmatrix}, \quad \mathcal{D}_M = \begin{bmatrix} 1 \\ 1 \end{bmatrix}.$$

The initial state is $x(t_0) = [0.09 \;\; 0.05]^T$. Parameters are $t_f = 9$, $v = [6.5 \;\; 6.5]^T$, $r(t) = [0.7 \;\; 0.7]^T$, and $\tau_a = 0.9$.
The state response of the open-loop system is in Fig. 6. We can observe from Fig. 7 that $x^T(t)r(t)$ is larger than 1 when $t \geq 7.917$, which suggests that the system is not FTS over $[t_0, t_f]$.

Choose scalars $\lambda_1 = 1$, $\lambda_2 = 0.23$, $\beta = 0.89$, $\varepsilon = 0.28$, $\varsigma = 0.4$ and $N_0 = 0$, and vector $\tilde{\delta} = 1$.
By Theorem 3, one feasible solution of inequalities (57)–(60) is

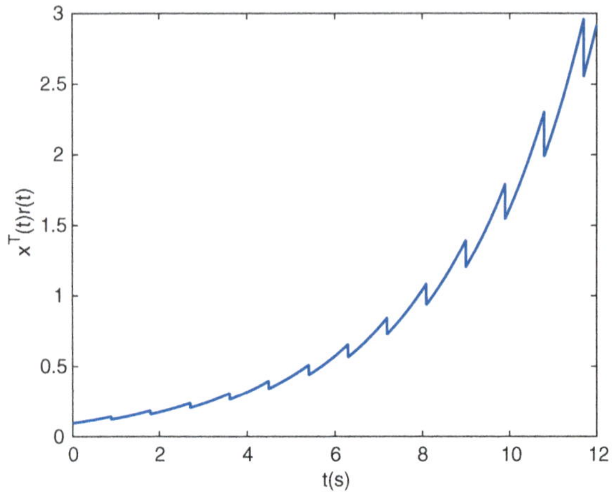

Fig. 7 Evolution of $x^T(t)r(t)$ of the open-loop system of Example 2

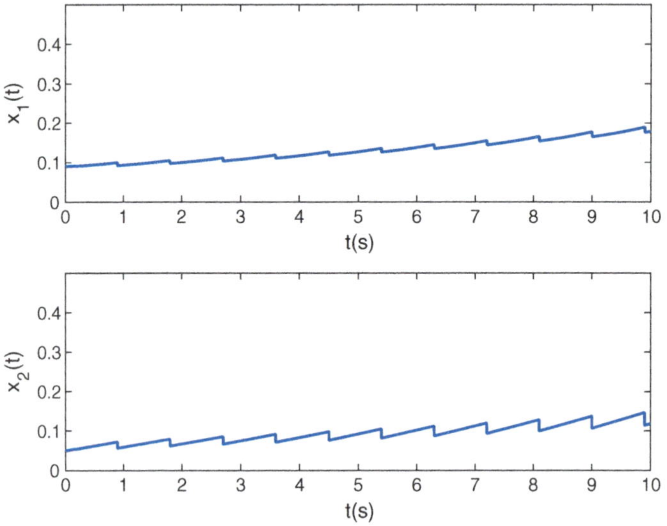

Fig. 8 State evolution of the closed-loop system of Example 2

$$\delta = \begin{bmatrix} 0.8072 \\ 0.8033 \end{bmatrix}, \quad w = \begin{bmatrix} -0.1344 \\ -0.2082 \end{bmatrix}.$$

Further, from (63), we have the matrix

$$K = \begin{bmatrix} -0.0834 & -0.1293 \end{bmatrix}.$$

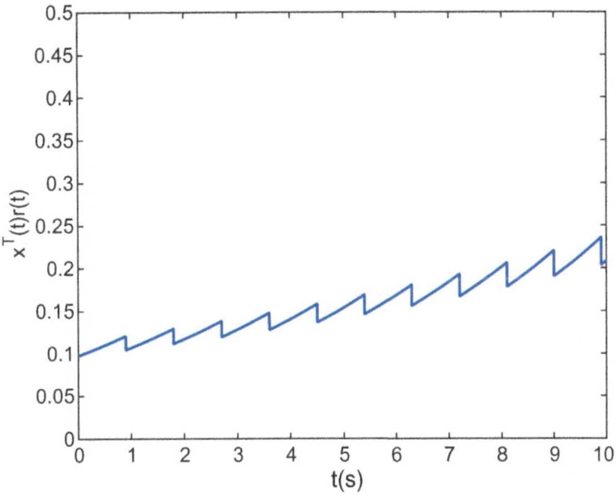

Fig. 9 Evolution of $x^T(t)r(t)$ of the closed-loop system of Example 2

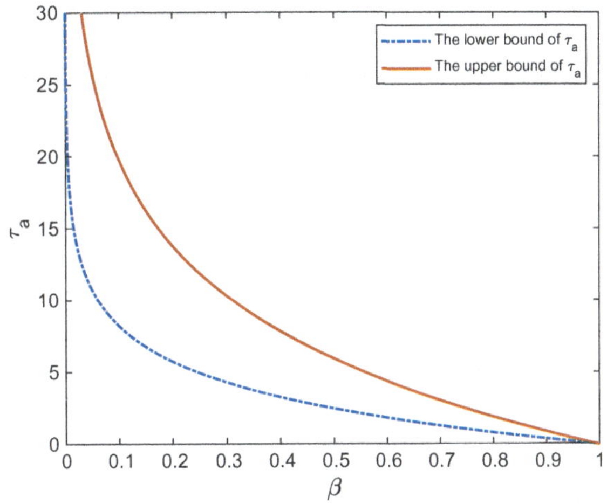

Fig. 10 The bound of τ_a with varying β of Example 2

Figure 8 presents the state trajectory of the closed-loop system. The curve of $x^T(t)r(t)$ of the closed-loop system is in Fig. 9. We can observe that the closed-loop system (55) is positive and FTS w.r.t. $(t_0, t_f, v, r(\cdot))$ under the proposed controller.

When other parameters being selected, the relationships between β and τ_a, and between $\overline{\tau_a}$ and t_f are discussed. From Case (iii) in Theorem 3, the bounds of τ_a with varying β is in Fig. 10. We can see that the upper bound of τ_a will decrease if β increases. By (65), the relationship between $\overline{\tau_a}$ and t_f is shown in Fig. 11. We can see the upper bound of τ_a will decrease if t_f increases. The simulation results can confirm the analysis of Table 2.

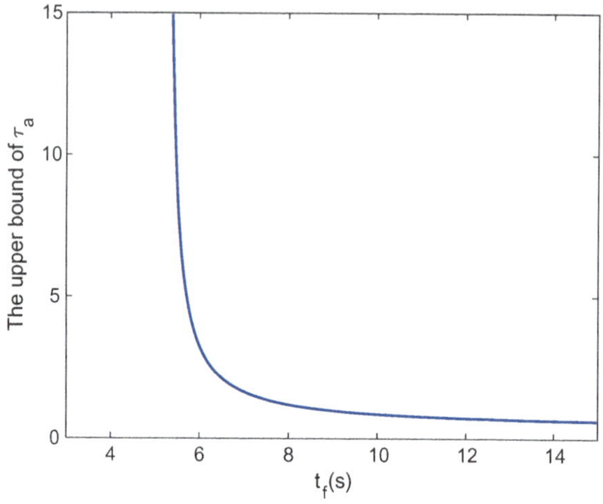

Fig. 11 The bound of τ_a with varying t_f of Example 2

Example 3 This example is to present the case of finite-time control with disturbance impulses.

Consider system (55) with

$$\mathcal{A}_M = \begin{bmatrix} -0.3 & 0.15 \\ 0.1 & -0.3 \end{bmatrix}, \quad \mathcal{H}_M = \begin{bmatrix} 0.05 & 0.25 \\ 0.2 & -0.1 \end{bmatrix}, \quad \mathcal{D}_M = \begin{bmatrix} 0.5 \\ 0.5 \end{bmatrix}.$$
$$\mathcal{A}_m = \begin{bmatrix} -0.3 & 0.15 \\ 0.08 & -0.32 \end{bmatrix}, \quad \mathcal{H}_m = \begin{bmatrix} 0.04 & 0.25 \\ 0.2 & -0.1 \end{bmatrix}, \quad \mathcal{D}_m = \begin{bmatrix} 0.4 \\ 0.4 \end{bmatrix}.$$

The initial state is set as $x(t_0) = [0.2 \;\; 0.3]^T$. Parameters are chosen as $t_f = 10$, $v = [2 \;\; 1.9]^T$, $r(t) = [0.4 \;\; 0.4]^T$, and $\tau_a = 0.5$.

The state response of the open-loop system is in Fig. 12. We can observe from Fig. 13 that $x^T(t)r(t)$ is larger than 1 when $t \geq 1.834$, so the open-loop system is not FTS over $[t_0, t_f]$.

Let $\lambda_1 = 2.3$, $\lambda_2 = 0.6$, $\beta = 1.23$, $\varepsilon = -0.28$, $\varsigma = 10$, $N_0 = 0$, and $\tilde{\delta} = 1$.

By Theorem 3, the inequalities (57)–(60) are feasible with the solution:

$$\delta = \begin{bmatrix} 1.1334 \\ 1.0096 \end{bmatrix}, w = \begin{bmatrix} -0.1209 \\ -0.2381 \end{bmatrix}.$$

Further, by simple calculation, we have

$$K = \begin{bmatrix} -0.1410 & -0.2777 \end{bmatrix}.$$

Figure 14 presents the state trajectory of the closed-loop system. The curve of $x^T(t)r(t)$ of the closed-loop system (55) is shown in Fig. 15. We can observe that the

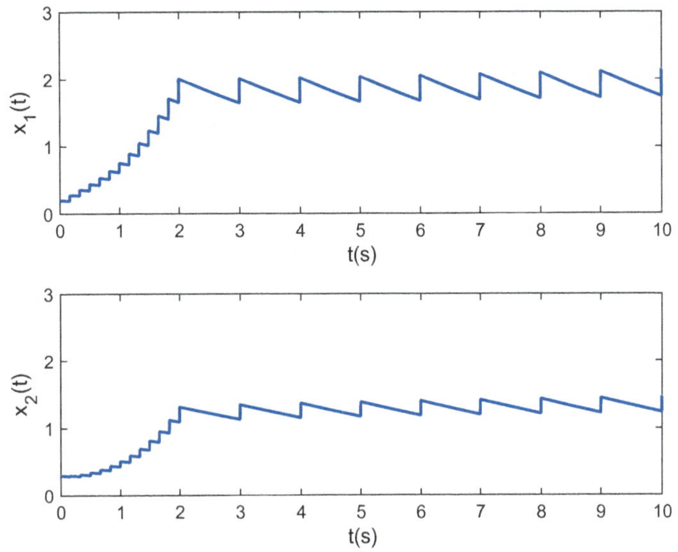

Fig. 12 State evolution of the open-loop system of Example 3

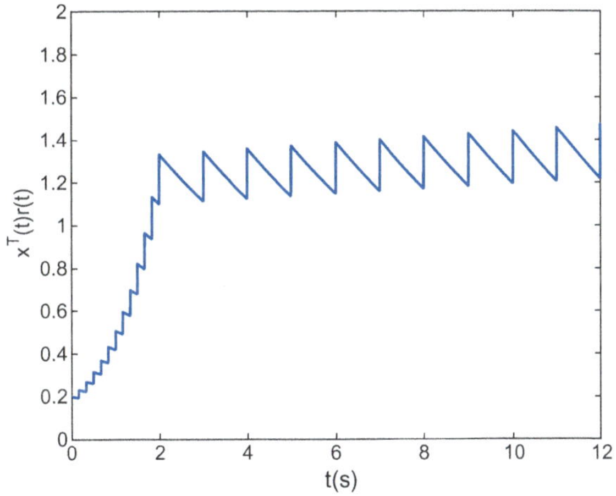

Fig. 13 Evolution of $x^T(t)r(t)$ of the open-loop system of Example 3

closed-loop system (55) is positive and FTS w.r.t. $(t_0, t_f, \upsilon, r(\cdot))$ under the proposed controller.

When other parameters being settled, the relationships between β and τ_a, and between $\underline{\tau_a}$ and t_f are discussed. From Case i) in Theorem 3, the bounds of τ_a with varying β is shown in Fig. 16. We can see that the lower bound of τ_a will increase if β increases. The relationship between $\underline{\tau_a}$ and t_f is shown in Fig. 17. We can observe that the lower bound of τ_a will increase if t_f increases. The simulations can support the analysis in Table 2.

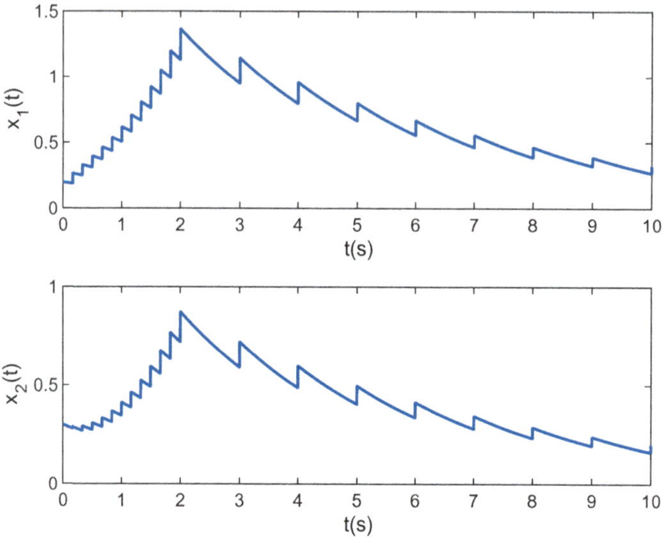

Fig. 14 State evolution of the closed-loop system of Example 3

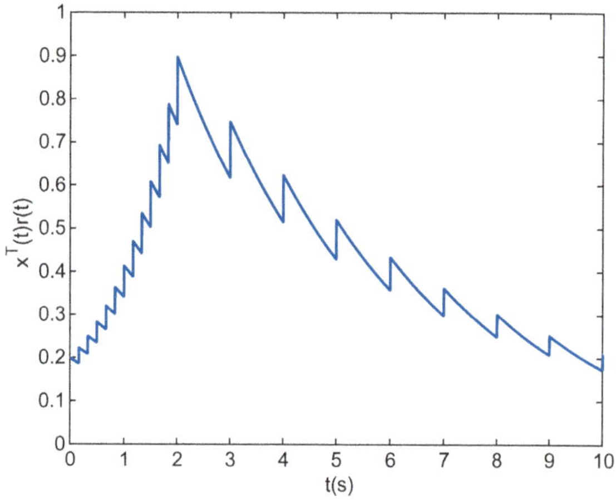

Fig. 15 Evolution of $x^T(t)r(t)$ of the closed-loop system of Example 3

5 Conclusion

In this chapter, the finite-time stability and control issues of interval IPSs have been addressed. For different impulsive effects, the finite-time stability criterion has been proposed for interval IPSs by applying a time-varying copositive Lyapunov function. The association between the frequency of impulses and the finite time interval has been disclosed thoroughly by theoretical and simulation analysis. The finite-time controller of interval IPSs has been designed with the help of the finite-time stability

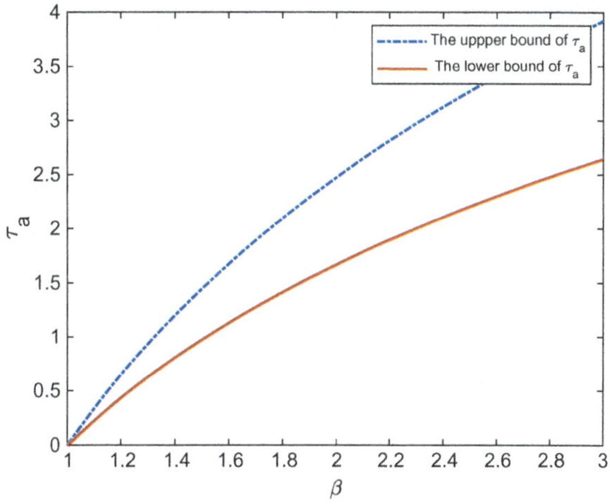

Fig. 16 The bound of τ_a with varying β of Example 3

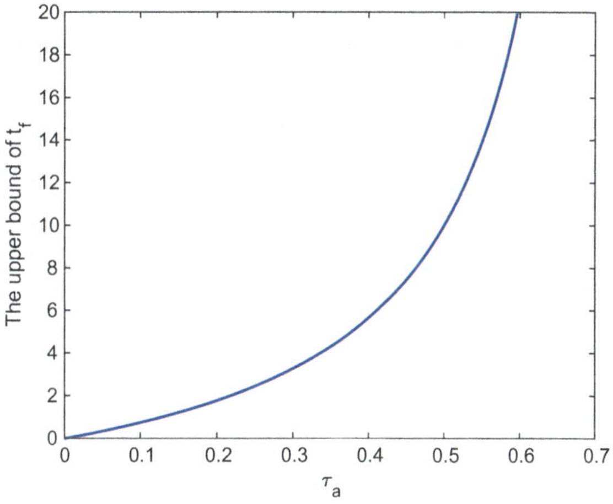

Fig. 17 The bound of τ_a with varying t_f of Example 3

results. Finally, numerical examples have been demonstrated to show the usefulness of the obtained results.

Acknowledgements This work was supported by the National Research Foundation of Korea (NRF) grant funded by the Korea government (Ministry of Science and ICT) (Grant No. 2019R1A5A 8080290 11).

References

1. Farina, L., Rinaldi, S.: Positive Linear Systems: Theory and Applications. Wiley, Hoboken (2011)
2. Hernandez-Vargas, E., Colaneri, P., Middleton, R., et al.: Discrete-time control for switched positive systems with application to mitigating viral escape. Int. J. Robust Nonlinear Control **21**(10), 1093–1111 (2011)
3. Silva-Navarro, G., Alvarez-Gallegos, J.: On the property sign-stability of equilibria in quasi-monotone positive nonlinear systems. In: Proceedings of IEEE 33rd Conference on Decision and Control, Lake Buena Vista, USA, 14–16 Dec 1994, pp. 4043–4048
4. Johnson, C.R.: Sufficient conditions for D-stability. J. Econ. Theory **9**(1), 53–62 (2006)
5. Berman, A., Neumann, M., Stern, R.J.: Nonnegative Matrices in Dynamic Systems. Wiley, New York (1989)
6. Rami, M.A., Tadeo, F.: Controller synthesis for positive linear systems with bounded controls. IEEE Trans. Circuits Syst. II Express Briefs **54**(2), 151–155 (2007)
7. Liu, X., Yu, W., Wang, L.: Stability analysis for continuous-time positive systems with time-varying delays. IEEE Trans. Autom. Control **55**(4), 1024–1028 (2010)
8. Zhao, X., Zhang, L., Shi, P., et al.: Stability of switched positive linear systems with average dwell time switching. Automatica **48**(6), 1132–1137 (2012)
9. Qi, W., Park, J.H., Cheng, J., Chen, X.: Stochastic stability and $\mathcal{L}_1$-gain analysis for positive nonlinear semi-Markov jump systems with time-varying delay via TS fuzzy model approach. Fuzzy Sets Syst. **371**, 110–122 (2019)
10. Qi, W., Park, J.H., Cheng, J., Kao, Y., Gao, X.: Exponential stability and $\mathcal{L}_1$-gain analysis for positive time-delay Markovian jump systems with switching transition rates subject to average dwell time. Inf. Sci. **424**, 224–234 (2018)
11. Hu, M., Wang, Y., Xiao, J.: On finite-time stability and stabilization of positive systems with impulses. Nonlinear Anal.: Hybrid Syst. **31**, 275–291 (2019)
12. Hu, M., Xiao, J., Xiao, R., Chen, W.: Impulsive effects on the stability and stabilization of positive systems with delays. J. Franklin Inst. **354**(10), 4034–4054 (2017)
13. Yang, H., Zhang, Y.: Impulsive control of continuous-time homogeneous positive delay systems of degree one. Int. J. Robust Nonlinear Control **29**(11), 3341–3362 (2019)
14. Dorato P (2006) An overview of finite-time stability. Current Trends in Nonlinear Systems and Control, pp. 185–194
15. Amato, F., Ambrosino, R., Ariola, M., et al.: Robust finite-time stability of impulsive dynamical linear systems subject to norm-bounded uncertainties. Int. J. Robust Nonlinear Control **21**(10), 1080–1092 (2011)
16. Amato, F., De Tommasi, G., Pironti, A.: Necessary and sufficient conditions for finite-time stability of impulsive dynamical linear systems. Automatica **49**(8), 2546–2550 (2013)
17. Qi, W., Zong, G., Cheng, J., et al.: Robust finite-time stabilization for positive delayed semi-Markovian switching systems. Appl. Math. Comput. **351**, 139–152 (2019)
18. Han, Z., Wu, H., Zhang, J.: Robust finite-time stability and stabilisation of switched positive systems. IET Control. Theory Appl. **8**(1), 67–75 (2013)
19. Chen, G., Yang, Y.: Finite-time stability of switched positive linear systems. Int. J. Robust Nonlinear Control **24**(1), 179–190 (2014)
20. Xiang, M., Xiang, Z.: Finite-time $\mathcal{L}_1$ control for positive switched linear systems with time-varying delay. Commun. Nonlinear Sci. Numer. Simul. **18**(11), 3158–3166 (2013)
21. Amato, F., Ambrosino, R., Cosentino, C., De Tommasi, G.: Finite-time stabilization of impulsive dynamical linear systems. Nonlinear Anal.: Hybrid Syst. **5**(1), 89–101 (2011)
22. Haddad, W.M., Chellaboina, V.S., Hui, Q.: Nonnegative and Compartmental Dynamical Systems. Princeton University Press, Princeton (2010)

An Eigenvector-Centrality Based Consensus Protocol Design for Discrete-Time Multi-agent Systems with Communication Delays

M. J. Park, S. H. Lee, and O. M. Kwon

Abstract The leader-follower consensus protocol design problem for a class of discrete-time multi-agent systems (DT-MASs) with the communication delay is addressed in this work, wherein it is assumed that the communication delay is time-varying function. The proposed consensus protocol is substituted by applying the concept of network-centrality for agent, from now on, the protocol will be called to as the substitute consensus protocol. Here, the network-centrality is determined by the flow structure of the information among multiple agents. So, in analyzing the flow structure of the information between each agent, the graph representation is used. In order to find the most influential node, which means agent, within the graph, the various network-centralities have been attempted among the methods of analyzing the social networks. Through this work, we try to take the eigenvector-centrality which is one of network-centralities. Here, its advantage is that the node's influence is determined by the network-centrality of the node as well as the number of other nodes to which the node is connected. The substitute consensus protocol based on the eigenvector-centrality will be applied to the leader-follower consensus problem for second-order DT-MASs with time-varying communication delay. To do this, by utilizing the Lyapunov method and some mathematical techniques, sufficient conditions for the concerned problem will be proposed in terms of linear matrix inequality. Finally, a point-mass kinematics for aircraft modeled with the DT-MASs

M. J. Park
Center for Global Converging Humanities, Kyung Hee University, 1732 Deogyeong-daero, Giheung-gu, Yongin 17104, Republic of Korea
e-mail: netgauss@khu.ac.kr

S. H. Lee · O. M. Kwon (✉)
School of Electrical Engineering, Chungbuk National University, 1 Chungdae-ro, Seowon-gu, Cheongju 28644, Republic of Korea
e-mail: madwind@chungbuk.ac.kr

S. H. Lee
e-mail: acafela@chungbuk.ac.kr

J. H. Park (ed.), *Recent Advances in Control Problems of Dynamical Systems and Networks*, Studies in Systems, Decision and Control 301,
https://doi.org/10.1007/978-3-030-49123-9_3

and its simulation results are given to verify the advantages of the proposed consensus protocol from a variety of perspective; as an example, the stability in consideration of time-delay.

Keywords Eigenvector centrality · Discrete-time multi-agent system · Lyapunov method

1 Introduction

Multi-agent systems have received considerable attention for their extensive application in many areas, including biology, physics, robotics, control engineering, and so on [1–3]. The major concern in the multi-agent systems is the issue of the consensus problem between the leader and the followers. In recent years, this issue has been used in many research areas: for examples, vehicle systems [4], intelligent decision support system for power grid dispatching [5] and networked control systems [6]. Because of the finite speed of signal processing in the flow of information between the two agents in the multi-agent systems, the multi-agent systems are used in the issue of the leader-follower consensus problem on communication time-delay.

It is well known that time-delay often causes unwelcome dynamic behaviors such as poor performance and instability of the system. It should be pointed out that the issue of consensus between the two agents in the multi-agent systems with time-delay can be addressed by investigating the stability of the multi-agent systems.

In addition, the issue of consensus between the two agents is one of prime research topics on the applications of the multi-agent systems. In this regard, various approaches to the issue of consensus problem on the multi-agent systems with time-delay are in the literature [7–14]. However, the above-mentioned literature mainly addressed the terms of consensus in the continuous-time multi-agent systems. In other words, the issue of consensus on the DT-MASs with time-delay has yet to be fully investigated.

In this work, motivated by the above mentioned, a new delay-dependent consensus protocol design method for the DT-MASs with time-delay is studied. The main contribution of this work lies in:

- The main question in this work is *How can we weight the node to improve the stability criterion?*. We pay attention to the *eigenvector-centrality* for each node as one answer to the question.

By building a suitable Lyapunov–Krasovskii functional and using the matrix theory, the consensus criteria are derived from the aspect of of linear matrix inequality (LMI). There are some numerical simulations that show the effectiveness of the proposed method.

1.1 Graph Theory and Notations

In the multi-agent systems , the flow of information between each agent is portrayed by employing the directed graph $\mathcal{G} = (\mathcal{V}, \mathcal{E})$, where it represents the sets of edge $\mathcal{E} = \{(i, j) : i, j \in \mathcal{V}\} \subset \mathcal{V} \times \mathcal{V}$ and node $\mathcal{V} = \{1, 2, \ldots, N\}$. From the graph $\mathcal{G}$, the matrix containing elements that satisfy $a_{ii} = 0$ and $a_{ij} \geq 0$ is called as the adjacency matrix $\mathcal{A} = [a_{ij}]_{N \times N}$.

If there is the flow of information between node i and node j, the elements of matrix $\mathcal{A}$ are described by $a_{ij} > 0 \Leftrightarrow (i, j) \in \mathcal{E}$. $\mathcal{N}_i = \{j \in \mathcal{V} : (i, j) \in \mathcal{E}\}$ denotes the set of neighbors of agent i.

The diagonal matrix defined as $\mathcal{D} = \begin{bmatrix} d_1 & & \\ & \ddots & \\ & & d_N \end{bmatrix}$, where $d_i = \sum_{k \in \mathcal{N}_i} a_{ik}$ is the degree of node i, is the degree matrix of the graph $\mathcal{G}$. The Laplacian matrix $\mathcal{L}$ of the graph $\mathcal{G}$ is obtained by $\mathcal{L} = \mathcal{D} - \mathcal{A}$. More details can be seen in [15].

Throughout the work, the used notations are standard. $\mathbb{R}^n$ and $\mathbb{R}^{m \times n}$ are, respectively, the sets of n-vectors and $m \times n$ matrices. I_n, and 0 are $n \times n$ identity matrix, and zero scalar or zero matrix of appropriate dimension. $X_\perp$ denotes a basis for the null space of X.

1.2 Advantage Claimed for the Use of Network-Centrality

In the well-used simple consensus protocol:

$$u_i(t) = \sum_{j=1, j \neq i}^{N} a_{ij}[p_j(t) - p_i(t)], \tag{1}$$

nodes directly connected by the largest number of nodes are considered to be more important. The consensus protocol based on (1) applies to almost all existing work on the multi-agent systems; hereinafter, the protocol (1) should be referred to as the existing protocol.

However, having more nodes in real networks in itself does not guarantee that someone is important. For example, from Fig. 1, the betweenness-centrality of edge between the two nodes is proportional to the thickness of the edges, while node colors indicate the degrees (white is the minimum degree). There, node 6 has the largest center-to-edge value from betweenness-centrality of edge between nodes 5 and 6, and nodes 6 and 7, while node 5 has the largest value of degree. According to the work [16], by reflecting the betweenness-centrality of the two nodes, the substitute protocol:

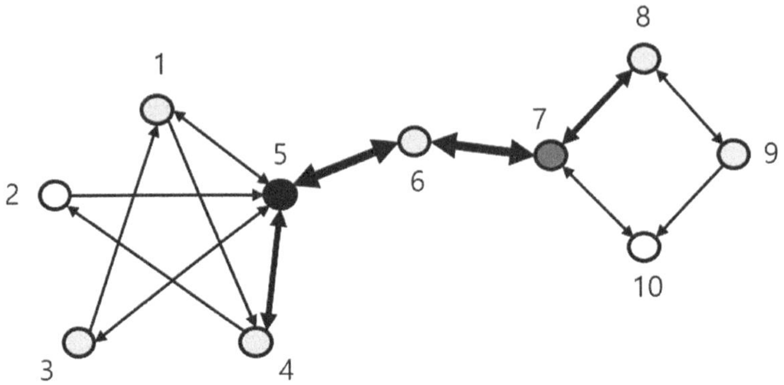

Fig. 1 Structure example for information flow

Table 1 Local information about each node of unknown 10-node graph

Node 1	(1, 4) and (1, 5)	Node 6	(6, 5) and (6, 7)
Node 2	(2, 5)	Node 7	(7, 6), (7, 8) and (7, 10)
Node 3	(3, 1) and (3, 5)	Node 8	(8, 7) and (8, 9)
Node 4	(4, 2) and (4, 5)	Node 9	(9, 8) and (9, 10)
Node 5	(5, 1), (5, 3) and (5, 4)	Node 10	(10, 7)

(i, j) corresponds to the information flow: $i \rightarrow j$

$$u_{\epsilon,i}(t) = \sum_{j=1, j\neq i}^{N} \epsilon_j a_{ij}[p_j(t) - p_i(t)] \tag{2}$$

can be used not only by the local information for the interconnection between each node, but also by the effect, ϵ_j, as the intermediary between each agent of edges.

When viewing network-centrality, the protocol (1) considers only the flow structure of the local information between each node because the protocol (1) is obtained by the degree-centrality, that is, the number of nodes adjacent to the node. In degree-centrality, we consider that direct interconnection places more importance on information from more nodes. However, considering more important nodes in the actual networks provides stronger information for nodes.

To obtain the value of the network-centrality of all the nodes, the overall structure of the flow of information between each node in the network should be identified beforehand. The answer is very simple: if we have the information on the value of a_{ij} used in the existing protocol (1), we can easily draw the flow structure of the information between each node in the network that is needed for the substitute protocol (2). For example, suppose we only know the flow of information between each node listed in Table 1. By considering the flow structure of the local information for the interconnection between each node listed in Table 1, the overall structure of the flow of information between each node in the network can be easily constructed as shown in Fig. 1. Therefore, knowing the flow structure of the local information for the interconnection between each node means that the value of the network-centrality can be computed from the overall structure of the flow of information between each node in the network. In this sense, the next step is the procedure of forming the substitute protocol. From the above-mentioned point of view, the steps are as follows:

- **Step 1**: The flow structure of the local information between each node created by the use of the local information about the direct neighbors of each node can be drawn to the overall flow structure of the information between each node $\mathcal{G}$.
- **Step 2**: From the graph $\mathcal{G}$ representing the local information, we can obtain the value of the network-centrality of each node. This value is used for the flow structure of the global information between each node because the information from both direct and indirect neighbors of each node is used to obtain the node and is defined as ϵ_j.
- **Step 3**: By multiplying a_{ij} and ϵ_j, the graph $\mathcal{G}$ is recreated as the weighted graph $\mathcal{G}_\epsilon$ reflecting the global flow of information between each node. Based on the weighted graph $\mathcal{G}_\epsilon$, the existing protocol (1) can be converted to the substitute protocol (2).

At that time, the weight ϵ_j corresponds to the value of the network-centrality of each node. Thus, because the design concept presented in this work does not alter the flow structure of the information between each node, the substitute protocol does not harm the distributed design. In addition, the value of a_{ij} as well as the network-centrality ϵ_j is easily obtained using NodeXL, which is one of the social network analysis tools. Therefore, not only the flow structure of the local information for the interconnection between each node but also its global flow of information reflected in a_{ij} and ϵ_j, respectively, can be considered.

In this respect, by adding the main role of each node based on the eigenvector-centrality, a new consensus protocol will be proposed to improve the consensus performance in two aspects of the stability in consideration of time-delay and $\mathcal{H}_\infty$ performance. The advantage of using the eigenvector-centrality centered is that the centrality is proportional to the sum of the centralities of the adjacent nodes, in other words, the eigenvector-centrality incorporates the importance of adjacent nodes to generalize the degree. This is the major contribution of this work. Noted that the concept of the network-centrality can be imposed in [17].

2 Problem Formation

The DT-MASs to be focused in this work is chosen in the form given below:

$$\begin{aligned} p_i(t_{k+1}) &= p_i(t_k) + \tau v_i(t_k), \\ v_i(t_{k+1}) &= v_i(t_k) + \tau u_i(t_k) + w_i(t_k), \quad i \in \mathcal{I} = \{1, 2, \ldots, N\}. \end{aligned} \tag{3}$$

In the above equations, N stands for the number of agents, $p_i(t_k) \in \mathbb{R}^n$ and $v_i(t_k) \in \mathbb{R}^n$ are the position and the velocity vectors, respectively. $u_i(t_k) \in \mathbb{R}^n$ is the consensus control input vector (the consensus protocol). $w_i(t_k) \in \mathbb{R}^n$ is the external disturbance which belongs to $l_2[0, \infty)$. τ is sampling period with assumption that $0 < \tau < 1$ between instant t_k and t_{k+1}.

The leader is considered as follows:

$$\begin{aligned} p_0(t_{k+1}) &= p_0(t_k) + \tau v_0(t_k), \\ v_0(t_{k+1}) &= v_0(t_k) + \tau u_0(t_k), \end{aligned} \tag{4}$$

where $p_0(t_k) \in \mathbb{R}^n$ and $v_0(t_k) \in \mathbb{R}^n$ represent the position and velocity vectors, and $u_0(t_k) \in \mathbb{R}^n$ is the given control input vector.

The output vector is defined as follows:

$$z_i(t_k) = \begin{bmatrix} p_i(t_k) - \frac{1}{N}\sum_{j=1}^{N} p_j(t_k) \\ v_i(t_k) - \frac{1}{N}\sum_{j=1}^{N} v_j(t_k) \end{bmatrix}, \quad i \in \mathcal{I}. \tag{5}$$

This means the relative displacement mean for all agents.

Before giving the substitute consensus protocol based the eigenvector-centrality, let us introduce the consensus protocol of the existing form:

$$\begin{aligned} u_i(t_k) = {} & u_0(t_k) + u_i^p(t_k) + u_i^v(t_k) \\ & - b_i[p_i(t_k) - p_0(t_k) + v_i(t_k) - v_0(t_k)] \end{aligned} \tag{6}$$

for $i \in \mathcal{I}$, where

$$\begin{aligned} u_i^p(t_k) &= \sum_{j=1, j\neq i}^{N} a_{ij}[p_j(t_k - \mu(t_k)) - p_i(t_k - \mu(t_k))], \\ u_i^v(t_k) &= \sum_{j=1, j\neq i}^{N} a_{ij}[v_j(t_k - \mu(t_k)) - v_i(t_k - \mu(t_k))]. \end{aligned}$$

Here, the interconnection weights a_{ij} and b_i are defined as:

$$\begin{cases} a_{ij} = 1 \text{ if agent } i \text{ is connected to agent} j, & a_{ij} = 0 \text{ otherwise}, \\ b_i = 1 \text{ if leader is connected to agent} i, & b_i = 0 \text{ otherwise}. \end{cases}$$

And, the flow of the information is considered to be delayed by $\mu(t_k)$ with

$$1 \leq \mu(t_k) \leq \mu_M, \tag{7}$$

where μ_M is the provided scalar value.

Then, the considered system (3) can be transformed as

$$
\begin{aligned}
\rho_i(t_{k+1}) &= \rho_i(t_k) + \tau \nu_i(t_k), \\
\nu_i(t_{k+1}) &= \nu_i(t_k) - \tau \sum_{j=1}^{N} l_{ij}[\rho_j(t_k - \mu(t_k)) + \nu_j(t_k - \mu(t_k))] \\
&\quad -\tau b_i[\rho_i(t_k) + \nu_i(t_k)] + w_i(t_k),
\end{aligned} \tag{8}
$$

for $i \in \mathcal{I}$, where $\rho_i(t_k) = p_i(t_k) - p_0(t_k)$, $\nu_i(t_k) = v_i(t_k) - v_0(t_k)$ and l_{ij} defined as $l_{ij} = -a_{ij}$ $(i \neq j)$ and $l_{ii} = -\sum_{k=1, j\neq i}^{N} l_{ik}$ is the elements of the Laplacian matrix $\mathcal{L} = [l_{ij}]_{N\times N}$.

Then, the matrix forms of system (8) and output (5) can be written by

$$
\begin{aligned}
x(t_{k+1}) &= \mathbf{A}x(t_k) + \mathbf{B}|_{\mathcal{L}}\, x(t_k - \mu(t_k)) + \mathbf{W}w(t_k), \\
z(t_k) &= \mathbf{D}x(t_k)
\end{aligned} \tag{9}
$$

with

$$
x(t_k) = \begin{bmatrix} \rho(t_k) \\ \nu(t_k) \end{bmatrix}, \ \rho(t_k) = \begin{bmatrix} \rho_1(t_k) \\ \vdots \\ \rho_N(t_k) \end{bmatrix},
$$

$$
\nu(t_k) = \begin{bmatrix} \nu_1(t_k) \\ \vdots \\ \nu_N(t_k) \end{bmatrix}, \ w(t_k) = \begin{bmatrix} w_1(t_k) \\ \vdots \\ w_N(t_k) \end{bmatrix},
$$

where

$$
\mathbf{A} = \begin{bmatrix} I_N & \tau I_N \\ -\tau\mathcal{B} & I_N - \tau\mathcal{B} \end{bmatrix} \otimes I_n, \ \mathcal{B} = \begin{bmatrix} b_1 & & \\ & \ddots & \\ & & b_N \end{bmatrix},
$$

$$
\mathbf{B}|_{\mathcal{L}} = \begin{bmatrix} 0 & 0 \\ -\tau\mathcal{L} & -\tau\mathcal{L} \end{bmatrix} \otimes I_n, \ \mathbf{W} = \begin{bmatrix} 0 \\ I_N \end{bmatrix} \otimes I_n,
$$

$$
\mathbf{D} = \begin{bmatrix} I_N - \frac{1}{N}\mathbf{1}_N\mathbf{1}_N^T & 0 \\ 0 & I_N - \frac{1}{N}\mathbf{1}_N\mathbf{1}_N^T \end{bmatrix} \otimes I_n.
$$

Here, $\mathbf{B}|_{\mathcal{L}}$ means to include the matrix $\mathcal{L}$.

The objective of this work is to configure the weighted consensus protocol for the DT-MASs (9). That is, by the property of the eigenvector-centrality, the substitute protocol will be weighted to investigate the leader-follower consensus problem for any initial condition if and only if

$$
\begin{aligned}
&\lim_{t_k\to\infty} (p_i(t_k) - p_0(t_k)) = 0, \\
&\lim_{t_k\to\infty} (v_i(t_k) - v_0(t_k)) = 0, \quad i \in \mathcal{I}.
\end{aligned} \tag{10}
$$

Note that for any initial condition, the consensus protocol is known to resolve the leader-follower consensus problem asymptotically if and only if the positions and velocities of agents satisfy (10).

3 Main Results

In this section, the substitute protocol is proposed by weighting the main role of each agent based on eigenvector-centrality.

3.1 Existing Leader-Follower Consensus Protocol

In this subsection, the leader-follower consensus condition for the multi-agent systems (9) with the external disturbance is proposed with the existing protocol (6) first.

Now, the following theorem is given as the first result.

Proposition 1 *Through the protocol (6), for the provided integer* μ_M *equal to or greater than 1, the multi-agent systems (9) is asymptotically stable with* $\mathcal{H}_\infty$ *attenuation* γ *if positive definite matrices* $P,\ Q,\ R \in \mathbb{R}^{2N\times 2N}$ *fulfilling the following LMI:*

$$(\Upsilon|_{\mathcal{L}} + \mathbf{W}e_5^T)_\perp^T(\Xi + \Sigma)(\Upsilon|_{\mathcal{L}} + \mathbf{W}e_5^T)_\perp < 0, \tag{11}$$

where

$$\begin{aligned}
e_i &= [0_{2\nu\times(i-1)2\nu}, I_{2\nu}, 0_{2\nu\times(4-i)2\nu}, 0_\nu]^T,\ i = 1, 2, \ldots, 4,\\
e_5 &= [0_{2\nu\times 8\nu}, I_\nu]^T,\ \nu = Nn,\\
\Xi &= e_1(P\otimes I_n)e_1^T + e_4(P\otimes I_n)e_1^T + [e_4(P\otimes I_n)e_1^T]^T\\
&\quad + e_1(Q\otimes I_n)e_1^T - e_3(Q\otimes I_n)e_3^T\\
&\quad + \mu_M^2 e_4(R\otimes I_n)e_4^T - (e_1 - e_2)(R\otimes I_n)(e_1 - e_2)^T\\
&\quad - (e_2 - e_3)(R\otimes I_n)(e_2 - e_3)^T,\\
\Sigma &= e_1\mathbf{D}^T\mathbf{D}e_1^T - \gamma^2 e_5 e_5^T,\\
\Upsilon|_{\mathcal{L}} &= (\mathbf{A} - I_{2N}\otimes I_n)e_1^T + \mathbf{B}|_{\mathcal{L}}\, e_2^T + \mathbf{W}e_5^T - I_{2\nu}\, e_4^T,
\end{aligned}$$

are present. Then, consensus of the multi-agent systems (9) can be made with $\mathcal{H}_\infty$ *disturbance attenuation* γ *under the existing protocol (6).*

Proof The Lyapunov–Krasovskii functional candidate is chosen as follows:

$$V(t_k) = x^T(t_k)(P \otimes I_n)x(t_k) + \sum_{s=t_k-\mu_M}^{t_k-1} x^T(s)(Q \otimes I_n)x(s) + \mu_M \sum_{s=-\mu_M}^{-1} \sum_{u=t_k+s}^{t_k-1} \Delta x^T(u)(R \otimes I_n)\Delta x(u). \tag{12}$$

The $\Delta V(t_k)$ defined as $V(t_{k+1}) - V(t_k)$, which is called as the forward difference of $V(t_k)$, is calculated as

$$\begin{aligned}\Delta V(t_k) &= x^T(t_{k+1})(P \otimes I_n)x(t_{k+1}) - x^T(t_k)(P \otimes I_n)x(t_k) \\ &+x^T(t_k)(Q \otimes I_n)x(t_k) - x^T(t_k - \mu_M)(Q \otimes I_n)x(t_k - \mu_M) \\ &+\mu_M^2 \Delta x^T(t_k)(R \otimes I_n)\Delta x(t_k) \\ &-\mu_M \sum_{s=t_k-\mu_M}^{t_k-1} \Delta x^T(s)(R \otimes I_n)\Delta x(s). \end{aligned} \tag{13}$$

Its upper bound is obtained as by Jensen's inequality

$$\begin{aligned}\Delta V(t_k) &\le \Delta x^T(t_k)(P \otimes I_n)\Delta x(t_k) + 2\Delta x^T(t_k)(P \otimes I_n)x(t_k) \\ &+x^T(t_k)(Q \otimes I_n)x(t_k) - x^T(t_k - \mu_M)(Q \otimes I_n)x(t_k - \mu_M) \\ &+\mu_M^2 \Delta x^T(t_k)(R \otimes I_n)\Delta x(t_k) \\ &-\left(\sum_{s=t_k-\mu(t_k)}^{t_k-1} \Delta x(s)\right)^T (R \otimes I_n) \left(\sum_{s=t_k-\mu(t_k)}^{t_k-1} \Delta x(s)\right) \\ &-\left(\sum_{s=t_k-\mu_M}^{t_k-\mu(t_k)-1} \Delta x(s)\right)^T (R \otimes I_n) \left(\sum_{s=t_k-\mu_M}^{t_k-\mu(t_k)-1} \Delta x(s)\right). \end{aligned} \tag{14}$$

As a result, a new upper bound of $\Delta V(t_k)$ is

$$\Delta V(t_k) \le \zeta^T(t_k)\Xi\zeta(t_k), \tag{15}$$

where

$$\zeta(t_k) = [x^T(t_k), x^T(t_k - \mu(t_k)), x^T(t_k - \mu_M), \Delta x^T(t_k), w^T(t_k)]^T.$$

It should be noted that if (15)< 0 is maintained, then there is a scalar $\epsilon > 0$ such that

$$\Delta V(t_k) + z^T(t_k)z(t_k) - \gamma^2 w^T(t_k)w(t_k) < -\epsilon x^T(t_k)x(t_k).$$

From the above inequality, the following relationship is obtained:

$$\Delta V(t_k) + z^T(t_k)z(t_k) - \gamma^2 w^T(t_k)w(t_k) < 0. \tag{16}$$

According to [18], the multi-agent systems (9) with an initial condition of zero and a given constant $\gamma > 0$ should be asymptotically stable in view of $\mathcal{H}_\infty$ attenuation γ if the following inequality is met:

$$\Psi = \sum_{t_k=0}^{\infty}(z^T(t_k)z(t_k) - \gamma^2 w^T(t_k)w(t_k)) < 0. \tag{17}$$

In this regard, taking summation from 0 to ∞ on both sides of (16) yields

$$V(\infty) - V(0) + \sum_{t_k=0}^{\infty}(z^T(t_k)z(t_k) - \gamma^2 w^T(t_k)w(t_k)) < 0.$$

Because of $V(0) = 0$ with the initial condition of zero and $V(\infty) \to 0$, we obtain the inequality (17). Hence, it can be ensured by the Lyapunov stability theory that the system (3) is stable. The Ψ can be represented as follows:

$$\Psi = \sum_{t_k=0}^{\infty}\big(z^T(t_k)z(t_k) - \gamma^2 w^T(t_k)w(t_k) + \Delta V(t_k)\big) - V(t_{k+1}). \tag{18}$$

Considering $V(t_{k+1}) \geq 0$, the bound of Ψ is

$$\begin{aligned}\Psi &\leq \sum_{t_k=0}^{\infty}\big(x^T(t_k)\mathbf{D}^T\mathbf{D}x(t_k) - \gamma^2 w^T(t_k)w(t_k) + \Delta V(t_k)\big)\\ &\leq \sum_{t_k=0}^{\infty}\zeta^T(t_k)(\Sigma + \Xi)\zeta(t_k).\end{aligned} \tag{19}$$

Here, the $\Psi < 0$ is equivalent to

$$\Xi + \Sigma < 0. \tag{20}$$

Therefore, if (20) satisfies, then system (9) is asymptotically stable in view of $\mathcal{H}_\infty$ attenuation γ.

Moreover, the following zero equality can be obtained from the system equation (9)

$$(\Upsilon|_{\mathcal{L}} + \mathbf{W}e_5^T)\zeta(t_k) = 0. \tag{21}$$

From Finsler's lemma [19], if (20) subject to (21) holds, then the LMI (11) satisfies. It can be concluded that if (11) holds, then system (9) is asymptotically stable. This completes our proof. □

From Proposition 1, we make the stability condition for the system (9). The system without the external disturbance is described here as

$$x(t_{k+1}) = \mathbf{A}x(t_k) + \mathbf{B}|_{\mathcal{L}}\ x(t_k - \mu(t_k)). \tag{22}$$

Now, the second criterion is given as follows:

Proposition 2 *Through the protocol (6), for the provided integer μ_M equal to or greater than 1, the multi-agent systems (22) is asymptotically stable with $\mathcal{H}_\infty$ disturbance attenuation γ if positive definite matrices P, Q, $R \in \mathbb{R}^{2N\times 2N}$ fulfilling the following LMI:*

$$(\Upsilon|_{\mathcal{L}})_\perp^T \Xi (\Upsilon|_{\mathcal{L}})_\perp < 0, \tag{23}$$

are present. Then, consensus of the multi-agent systems (22) can be made under the existing protocol (6).

Proof By taking the place of e_i $(i = 1, 2, \dots, 4)$ in Proposition 1 with

$$[0_{\nu\times(i-1)\nu}, I_\nu, 0_{\nu\times(4-i)\nu}]^T,$$

the LMI (23) can be easily induced. □

3.2 Good Substitute Leader-Follower Consensus Protocol Based on the Eigenvector-Centrality

By weighting the values of the eigenvector-centrality of each agent, the substitute protocol is introduced in this subsection.

The substitute protocol based on the eigenvector-centrality to be proposed in this work is constructed as the form given below:

$$\begin{aligned} u_{\text{eigen},i}(t_k) = u_0(t_k) &+ u^p_{\text{eigen},i}(t_k) + u^v_{\text{eigen},i}(t_k) \\ &- b_i[p_i(t_k) - p_0(t_k) + v_i(t_k) - v_0(t_k)], \end{aligned} \tag{24}$$

where

$$u^p_{\text{eigen},i}(t_k) = \sum_{j=1, j\neq i}^{N} c_{\text{eigen},j}\ a_{ij}[p_j(t_k - \mu(t_k)) - p_i(t_k - \mu(t_k))],$$

$$u^v_{\text{eigen},i}(t_k) = \sum_{j=1, j\neq i}^{N} c_{\text{eigen},j}\ a_{ij}[v_j(t_k - \mu(t_k)) - v_i(t_k - \mu(t_k))].$$

Here, the eigenvector of the adjacency matrix $\mathcal{A}$ of the structure of information flow of network $\mathcal{G}$ obtained as $\mathbf{C}_{\text{eigen}} = \begin{bmatrix} c_{\text{eigen},1} \\ \vdots \\ c_{\text{eigen},N} \end{bmatrix}$, where $c_{\text{eigen},j}$ is the value of the eigenvector-centrality of agent j.

Remark 1 The graph $\mathcal{G}$ should be strongly connected to give weight to the eigenvector-centrality. Perron-Frobenius theorem [20] allows us to calculate the eigenvalues of $\mathcal{A}$ and then select the largest eigenvalues. Then, by Perron-Frobenius theorem, all elements of the eigenvector $\mathbf{C}_{\text{eigen}}$ are positive, and the element of this vector corresponds to the value of eigenvector-centrality for the graph $\mathcal{G}$. As a result, the weight of each edge in the substitute protocol (24) will be the tail agent's eigenvector-centrality value.

For the substitute protocol (24), the following results are obtained from Propositions 1 and 2.

Theorem 1 *Through the protocol (24), for the provided integer μ_M equal to or greater than* 1, *the multi-agent systems (9) is asymptotically stable with $\mathcal{H}_\infty$ attenuation γ if positive definite matrices P, Q, $R \in \mathbb{R}^{2N \times 2N}$ fulfilling the following LMI:*

$$(\Upsilon|_{\mathcal{L}_{\text{eigen}}} + \mathbf{W}e_5^T)_\perp^T(\Xi + \Sigma)(\Upsilon|_{\mathcal{L}_{\text{eigen}}} + \mathbf{W}e_5^T)_\perp < 0. \tag{25}$$

are present. Then, consensus of the multi-agent systems (9) can be made with $\mathcal{H}_\infty$ disturbance attenuation γ under the substitute protocol (24).

Proof By taking the place of $\mathcal{L}$ in Proposition 1 with $\mathcal{L}_{\text{eigen}} = [l_{\text{eigen},ij}]_{N \times N}$, where $l_{\text{eigen},ij} = -c_{\text{eigen},j}a_{ij}$ for $i \neq j$ and $l_{\text{eigen},ii} = -\sum_{j=1,j\neq i}^{N} l_{\text{eigen},ij}$, the condition (25) can be easily derived. □

Theorem 2 *Through the protocol (24), for the provided integer μ_M equal to or greater than* 1, *the multi-agent systems (22) is asymptotically stable if positive definite matrices P, Q, $R \in \mathbb{R}^{2N \times 2N}$ fulfilling the following LMI:*

$$(\Upsilon|_{\mathcal{L}_{\text{eigen}}})_\perp^T \Xi (\Upsilon|_{\mathcal{L}_{\text{eigen}}})_\perp < 0. \tag{26}$$

are present. Then, consensus of the multi-agent systems (22) can be made under the substitute protocol (24).

Proof By taking the place of $\mathcal{L}$ in Proposition 2 with $\mathcal{L}_{\text{eigen}}$, the condition (26) can be easily derived. □

4 Numerical Simulations

In this section, the advantages of the eigenvector-centrality based consensus protocol are shown by comparing Propositions (the existing case) and Theorems (the eigenvector-centrality-based case).

Consider the network consisting of 3-aircrafts with the flow of the information shown in Fig. 2a. For each aircraft, its point-mass kinematics can be formulated from the two frames [21]: the inertial frame fixed to inertial space (subscript I) and the body frame fixed to body (subscript b), shown in Fig. 2b. By choosing the former, the point-mass kinematics of each aircraft can be expressed by

$$\begin{aligned}\dot{\mathbf{p}}_I &= \mathbf{v}_I,\\ \dot{\mathbf{v}}_I &= \frac{1}{m}\mathbf{F}_I,\end{aligned} \tag{27}$$

where $\mathbf{p}_I = \mathtt{col}\big[p_{x,I},\ p_{y,I},\ p_{z,I}\big]$ and $\mathbf{v}_I = \mathtt{col}\big[v_{x,I},\ v_{y,I},\ v_{z,I}\big]$ are, respectively, the translational position and the translational velocity, $\mathbf{F}_I$ and m, respectively, the force and the mass of each aircraft.

With the multi-agent systems (3) of the discrete-time form, the above kinematics of each aircraft can be reformed as follows:

$$\begin{aligned}\mathbf{p}_{I,i}(t_{k+1}) &= \mathbf{p}_{I,i}(t_k) + 0.1\mathbf{v}_{I,i}(t_k),\\ \mathbf{v}_{I,i}(t_{k+1}) &= \mathbf{v}_{I,i}(t_k) + 0.1\mathbf{F}_{I,i}(t_k) + w_i(t_k),\quad i \in \mathcal{I},\end{aligned} \tag{28}$$

where the values of m and τ are assumed as $m = 1$ and $\tau = 0.1$.

In order to apply the consensus protocol, $\mathbf{F}_{I,i}(t_k)$ is replaced with $u_i(t_k)$ in (6) or $u_{\text{eigen},i}(t_k)$ in (24). From Fig. 2a, the adjacency matrix can be obtained as

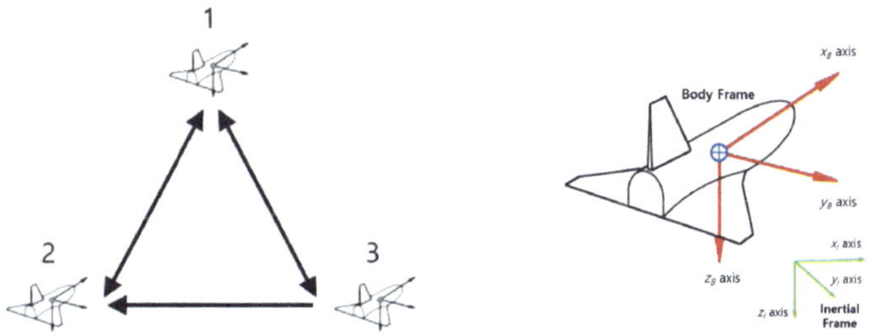

(a) Information flow. (b) Two frames.

Fig. 2 3-Aircrafts

Table 2 Laplacian matrices

$\mathcal{L}$	$\mathcal{L}_{\text{eigen}}$
$\begin{bmatrix} 2 & -1 & -1 \\ -1 & 2 & -1 \\ -1 & 0 & 1 \end{bmatrix}$	$\begin{bmatrix} 1.2959 & -0.6479 & -0.6479 \\ -0.4004 & 0.8009 & -0.4004 \\ -0.6479 & 0 & 0.6479 \end{bmatrix}$

Table 3 Minimum values of γ with $\mu_M = 3$

Methods	Proposition 1	Theorem 1
$\gamma_{\min}$	36.4153	19.7765

$$\mathcal{A} = \begin{bmatrix} 0 & 1 & 1 \\ 1 & 0 & 1 \\ 1 & 0 & 0 \end{bmatrix}.$$

Then, the eigenvector to obtain the values of the eigenvector-centrality of all aircrafts is computed as

$$\mathbf{C}_{\text{eigen}} = \begin{bmatrix} 0.6479 \\ 0.4004 \\ 0.6479 \end{bmatrix}.$$

Thus, the adjacency matrix with the eigenvector-centrality can be obtained as

$$\mathcal{A}_{\text{eigen}} = \begin{bmatrix} 0.6479 & 0 & 0 \\ 0 & 0.4004 & 0 \\ 0 & 0 & 0.6479 \end{bmatrix} \underbrace{\begin{bmatrix} 0 & 1 & 1 \\ 1 & 0 & 1 \\ 1 & 0 & 0 \end{bmatrix}}_{\mathcal{A}} = \begin{bmatrix} 0 & 0.6479 & 0.6479 \\ 0.4004 & 0 & 0.4004 \\ 0.6479 & 0 & 0 \end{bmatrix}.$$

In addition, the corresponding Laplacian matrices are listed in Table 2. To rule out the impact of choosing the connection between the leader and any agent, the leader assumes that it is connected to all agents, that is, $\mathcal{B} = \begin{bmatrix} 1 & 0 & 0 \\ 0 & 1 & 0 \\ 0 & 0 & 1 \end{bmatrix}$.

The followings are assumed as conditions for simulation:

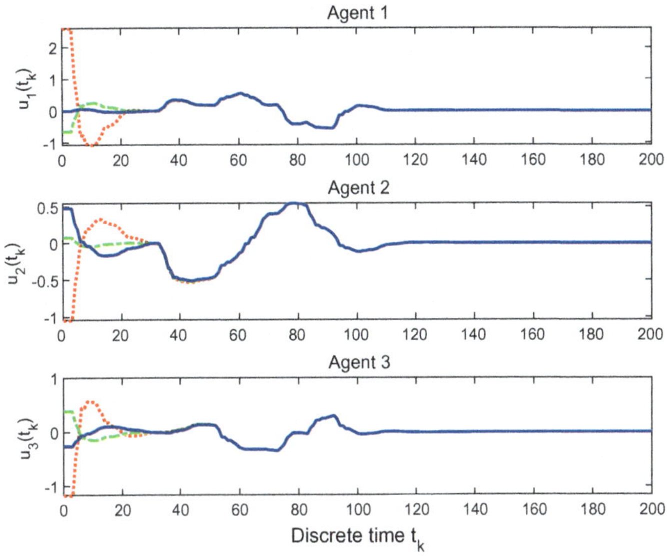

Fig. 3 Trajectories of the consensus protocol based on the eigenvector-centrality

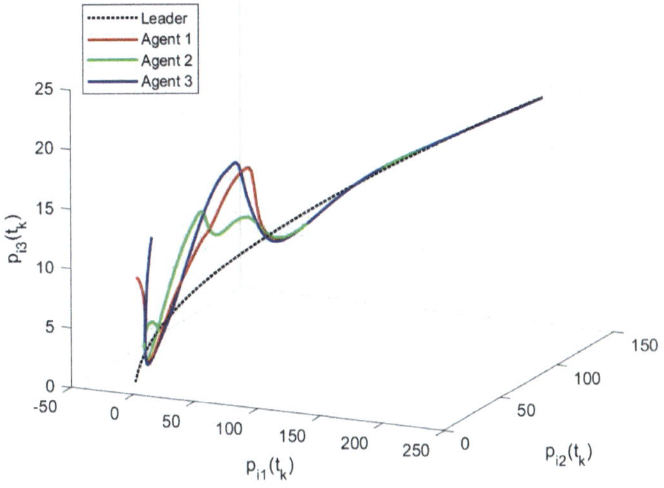

Fig. 4 The position trajectories of the three aircrafts

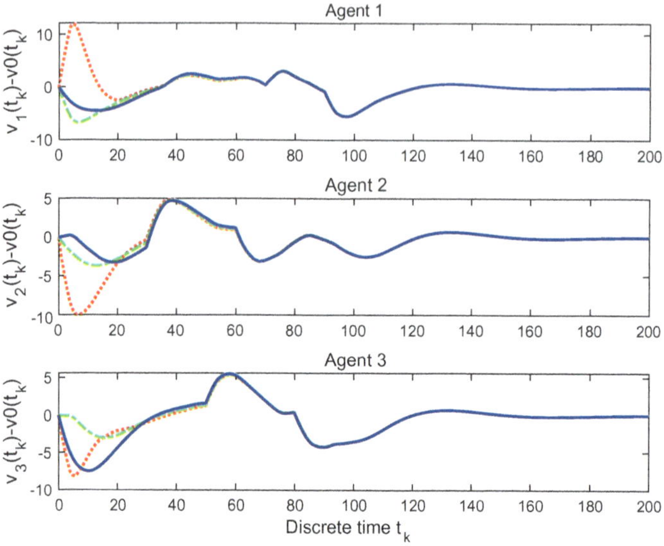

Fig. 5 The velocity error trajectories of the three aircrafts

$$\text{Agents}: \begin{cases} \mathbf{p}_{I,1}(0) = \begin{bmatrix} -8 \\ 10 \\ 8 \end{bmatrix}, \ \mathbf{v}_{I,1}(0) = \begin{bmatrix} 0 \\ 0 \\ 0 \end{bmatrix}, \ w_1(t_k) = \begin{bmatrix} 1 \\ 1 \\ 1 \end{bmatrix} \\ \text{when } t_k \in [70, 90], \\ \mathbf{p}_{I,2}(0) = \begin{bmatrix} 14 \\ 6 \\ 4 \end{bmatrix}, \ \mathbf{v}_{I,2}(0) = \begin{bmatrix} 0 \\ 0 \\ 0 \end{bmatrix}, \ w_2(t_k) = \begin{bmatrix} 1 \\ 1 \\ 1 \end{bmatrix} \\ \text{when } t_k \in [30, 60], \\ \mathbf{p}_{I,3}(0) = \begin{bmatrix} 10 \\ 4 \\ 12 \end{bmatrix}, \ \mathbf{v}_{I,3}(0) = \begin{bmatrix} 0 \\ 0 \\ 0 \end{bmatrix}, \ w_3(t_k) = \begin{bmatrix} 1 \\ 1 \\ 1 \end{bmatrix} \\ \text{when } t_k \in [50, 80], \end{cases}$$

$$\text{Leader}: \ \mathbf{p}_{I,0}(0) = \begin{bmatrix} 1 \\ 1 \\ 1 \end{bmatrix}, \ \mathbf{v}_{I,0}(0) = \begin{bmatrix} 1 \\ 1 \\ 1 \end{bmatrix}, \ u_0(0) = \begin{bmatrix} 1 \\ 0.5 \\ 0 \end{bmatrix}.$$

The advantages of the substitute protocol (24) based on the eigenvector-centrality are illustrated from the following two perspectives:

Table 4 Maximum bounds

Methods	Proposition 2	Theorem 2
μ_M	3	5

Table 5 Maximum bounds on μ_M

Methods	Proposition 2	Theorem 2
μ_M	2	4

- *$\mathcal{H}_\infty$ performance*
The minimum values of γ for $\mathcal{H}_\infty$ performance are given in Table 3 and computed by programming the following optimization problem: for given $\mu_M = 3$, if

$$\min_{P,Q,R} \quad \gamma \quad \text{s.t. LMI (3.11)}$$

has solutions with P, Q, R and γ, then, the protocol (6) is the leader-follower $\mathcal{H}_\infty$ consensus protocol for the multi-agent systems (9) and $\mathcal{H}_\infty$ disturbance attenuation is minimized by γ. Thus, from a view of performing $\mathcal{H}_\infty$ in Table 3, the substitute protocol plotted by Fig. 3 can be more effective than the existing protocol. Moreover, Figs. 4 and 5 show the positions and velocities of three aircraft follow the leader's trajectory, even if through having the external disturbance.

- *Stability in consideration of time-delay*
In Table 4, the value of maximum bound of time-delay obtained using the eigenvector-centrality based consensus protocol (24) is greater than the one with the existing protocol (6). This means that the value of maximum bound of time-delay under the substitute protocol based on the eigenvector-centrality for each agent that guarantees the stability of multi-agent systems can be further increased. To verify the results listed in Table 4, the root mean square of position error between the leader and each follower defined as in an equation:

$$\mathsf{rmse}(t_k) = \sqrt{\frac{\sum_{i=1}^{3}(p_i(t_k) - p_0(t_k))^2}{3}},$$

according to each protocol are drawn in Fig. 6. The detail is that when there is a time-varying delay that satisfies $1 \leq \mu(t_k) \leq 5$ in this figure, the substitute protocol based on the eigenvector-centrality ensures the stability of the multi-agent system, but the existing protocol does not. Fig. 7 shows the position trajectories of the three aircrafts system (28) under the substitute protocol. We can see three aircraft are following the leader's trajectory.

In addition, consider the network consisting of 5-nodes with the flow of the information shown in Fig. 8. From this graph, the adjacency matrix can be known as

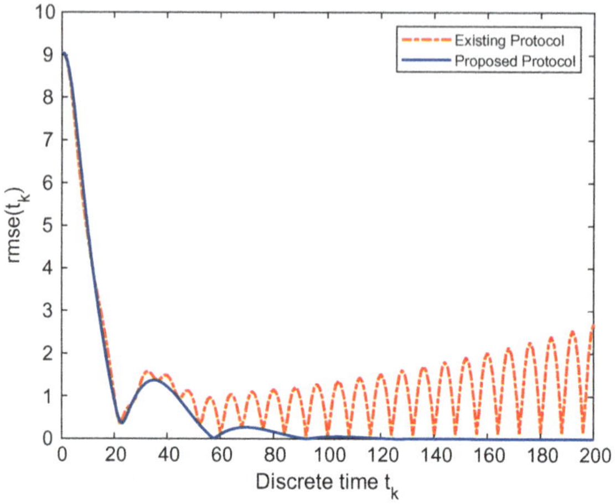

Fig. 6 The trajectories of $\mathsf{rmse}(t_k)$ with each protocol

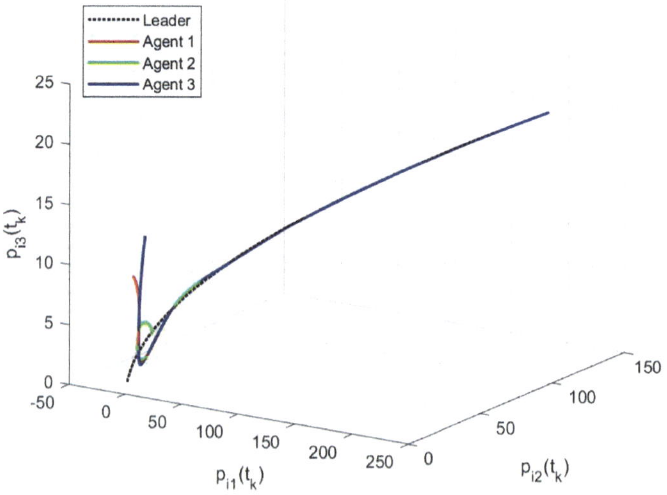

Fig. 7 The position trajectories of the three aircrafts

$$\mathcal{A} = \begin{bmatrix} 0\ 1\ 1\ 0\ 1 \\ 1\ 0\ 1\ 0\ 1 \\ 1\ 0\ 0\ 1\ 1 \\ 0\ 0\ 1\ 0\ 0 \\ 1\ 0\ 1\ 0\ 0 \end{bmatrix}.$$

Then, the eigenvector to calculate the values of the eigenvector-centrality of all nodes is obtained as

Fig. 8 5-nodes network

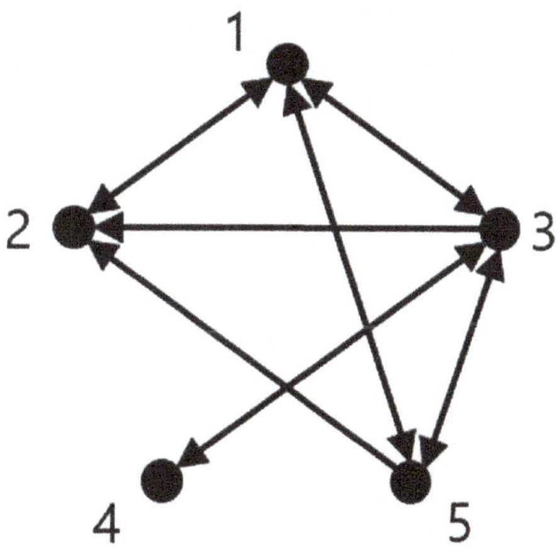

Fig. 9 The trajectories of $\mathrm{rmse}(t_k) = \sqrt{\frac{\sum_{i=1}^{5}(p_i(t_k)-p_0(t_k))^2}{5}}$ with each protocol

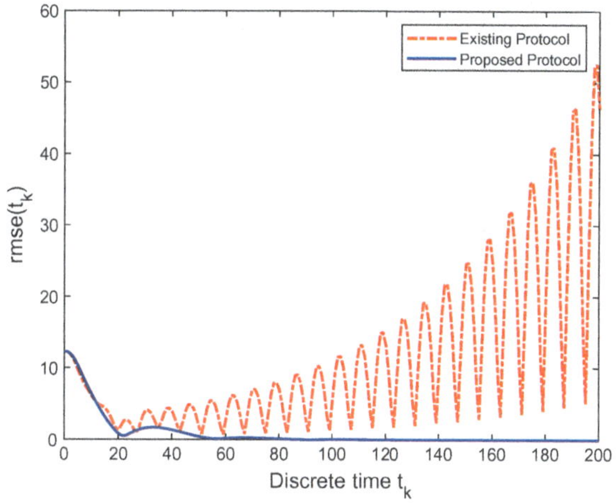

$$\mathbf{C}_{\mathrm{eigen}} = \begin{bmatrix} 0.5251 \\ 0.2081 \\ 0.5917 \\ 0.2345 \\ 0.5251 \end{bmatrix}.$$

From this, the corresponding Laplacian matrix can be computed as

$$\mathcal{L}_{\mathrm{eigen}} = \begin{bmatrix} 1.5754 & -0.5251 & -0.5251 & 0 & -0.5251 \\ -0.2081 & 0.6244 & -0.2081 & 0 & -0.2081 \\ -0.5917 & 0 & 1.7751 & -0.5917 & -0.5917 \\ 0 & 0 & -0.2345 & 0.2345 & 0 \\ -0.5251 & 0 & -0.5251 & 0 & 1.0503 \end{bmatrix}.$$

From the perspective of the stability in consideration of time-delay, the maximum bounds of time-delay under two protocols are listed in Table 5. From Fig. 9, the results listed in Table 5 can be confirmed when $1 \leq \mu(t_k) \leq 4$. Also, the results of Table 5 and Fig. 9 show that the substitute protocol has better performance. Note that the simulation conditions, such as the initial value of the leader, were randomly set.

5 Conclusion

The issue of the leader-follower consensus problem for the DT-MASs was been investigated in this work. To improve the consensus performance in two aspects of $\mathcal{H}_\infty$ performance and the stability in consideration of time-delay, the weighted consensus protocols were proposed by calculating the eigenvector-centrality for agents in a network. To achieve this, by building simple Lyapunov–Krasovskii functional, sufficient conditions for such problems for the proposed protocols were derived from the LMI perspective. Some numerical simulations were provided to demonstrate the effectiveness of the proposed protocols. Moreover, for the secure communication between the leader and any the selected agent, our future works will focus on grafting a new pinning control concept onto this work.

Acknowledgements This work was supported by the National Research Foundation of Korea(NRF) grant funded by the Korea government(MSIT; Ministry of Science and ICT) (NRF-2017R1C1B5076 878).

References

1. Saber, R.O., Fax, J.A., Murray, R.M.: Consensus and cooperation in networked multi-agent systems. Proc. IEEE **95**, 215–233 (2007)
2. Ren, W., Beard, R.W.: Consensus seeking in multi-agent systems under dynamically changing interaction topologies. IEEE Trans. Autom. Control **50**, 655–661 (2005)
3. Wang, J., Cheng, D., Hu, X.: Consensus of multi-agent linear dynamic systems. Asian J. Control **10**, 144–155 (2008)
4. Ren, W.: Consensus strategies for cooperative control of vehicle formations. IET Control Theory Appl. **1**, 505–512 (2007)
5. Qiong, W., Wenyin, L., Yihan, Y., Chuan, Z., Yong, L.: Intelligent decision support system for power grid dispatching base on multi-agent system. In: International Conference on Power System Technology (PowerCon 2006), pp. 1–5 (2006)
6. DeLellis, P., di Bernardo, M., Garofalo, F., Liuzza, D.: Analysis and stability of consensus in networked control systems. Appl. Math. Comput. **217**, 988–1000 (2010)
7. Yu, W., Ghen, G., Gao, M.: Some necessary and sufficient conditions for second-order consensus in multi-agent dynamical systems. Automatica **46**, 1089–1095 (2010)
8. Song, Q., Cao, J., Yu, W.: Second-order leader-following consensus of nonlinear multi-agent systems via pinning control. Syst. Control Lett. **59**, 553–562 (2010)
9. Xue, D., Yao, J., Wang, J., Guo, Y., Han, X.: Formation control of multi-agent systems with stochastic switching topology and time-varying communication delays. IET Control Theroy Appl. **7**, 1689–1698 (2013)

10. Lu, X., Austin, F., Chen, S.: Formation control for second-order multi-agent systems with time-varying delays under directed topology. Commun. Nonlinear Sci. Numer. Simulat. **17**, 1382–1391
11. He, P., Li, Y., Park, J.H.: Noise tolerance leader-following of high-order nonlinear dynamical multi-agent systems with switching topology and communication delay. J. Frankl. Inst. **353**, 108–143
12. Liu, X., Dou, L., Sun, J.: Consensus for networked multi-agent systems with unknown communication delays. J. Franklin Inst. **353**, 4176–4190 (2016)
13. Ding, L., Guo, G.: Sampled-data leader-following consensus for nonlinear multi-agent systems with Markovian switching topologies and communication delay. J. Franklin Inst. **352**, 369–383 (2015)
14. Park, M.J., Kwon, O.M., Park, J.H., Lee, S.M., Cha, E.J.: Randomly changing leader-following consensus control for Markovian switching multi-agent systems with interval time-varying delays. Nonlinear Anal.-Hybrid Syst. **12**, 117–131 (2014)
15. Godsil, C., Royle, G.: Algebraic Graph Theory. Springer, New York (2001)
16. Park, M.J., Lee, S.H., Kwon, O.M., Park, J.H., Choi, S.G.: Betweenness centrality-based consensus protocol for second-order multiagent systems with sampled-data. IEEE Trans. Cybern. **47**(8), 2067–2078 (2017)
17. Newman, M.E.J.: Networks: An Introduction. Oxford University Press, Oxford (2010)
18. Anton, S.: The $\mathcal{H}_\infty$ Control Problem. Prentice Hall, New York (1992)
19. de Oliveira, M.C., Selton, R.E.: Stability Tests for Constrained Linear Systems. Springer, Berlin (2001)
20. Horn, R.A., Johnson, C.R.: Matrix Analysis. Cambride University Press, New York (1990)
21. Stengel, R.F.: Flight Dynamics. Princeton University Press, New Jersey (2004)

Fault-Tolerant Sampled-Data Synchronization of Chaotic Systems with Random Occurring Uncertainties: A Semi-Markov Jump Model Approach

Yude Xia, Yudong Wang, Jing Wang, and Hao Shen

Abstract This work focuses on addressing the issue of sampled-data synchronization for nonlinear chaotic systems with uncertainties by fault-tolerant control techniques, where a semi-Markov jump chain is utilized to describe the switching characteristic. The random occurring uncertainties in chaotic systems are subject to the Bernoulli distributed white noise sequences therein. For such a synchronization control problem, the sampled-data control strategy, capable of alleviating the bandwidth pressure is applied, in which the random time-varying delays are adequately cogitated. Then, by establishing a proper Lyapunov–Krasovskii functional, some constrained criteria are obtained via the rational use of integral inequality technique. The validity and effectiveness of the developed methods are, at last, substantiated by a numerical example.

Keywords Chaotic systems · Sampled-data fault-tolerant synchronization · Semi-Markov jump model · Randomly occurring uncertainties

1 Introduction

Stochastic jumping systems (SJSs), revealing the hybrid dynamic properties of systems, have gotten extensive applications in many practical situations for its great capability in modeling [1–5]. The SJSs consist of some subsystems and the switching among them are governed by some stochastic processes, such as Markov process (MP) and semi-Markov process (SMP) [6, 7]. What is important in the long and arduous process of the analysis and synthesis on Markov jump systems (MJS) is an ideal

Y. Xia · J. Wang · H. Shen (✉)
School of Electrical and Information Engineering, Anhui University of Technology, Ma'anshan 243002, People's Republic of China
e-mail: haoshen10@gmail.com

Y. Wang
School of Metallurgy Engineering, Anhui University of Technology, Ma'anshan 243002, People's Republic of China

J. H. Park (ed.), *Recent Advances in Control Problems of Dynamical Systems and Networks*, Studies in Systems, Decision and Control 301,
https://doi.org/10.1007/978-3-030-49123-9_4

hypothesis that the sojourn-time subject to exponential distribution in MP. More precisely, in this hypothesis, the transition rate is represented by a time-invariant matrix and is not only known but also memoryless. Subsequently, two questions arise, one is that the transition rate is too hard to acquire in advance, and the other one is that memoryless transition rate is untoward to accommodate the practical situations. The features mentioned above bring many restrictions to the applications of MJSs. Compared with the MP, the SMP is with the property of a transition rate matrix and a probability density function (PDF), which are changeable. By right of relaxing the restrictions on the condition that the sojourn-time obeys to an exponential distribution, the SMP has been widely used in numerous situations. Consequently, the researches on semi-Markov jump systems have enchanted a crowd of scholars close attention [8, 9].

In parallel, the discovery of chaotic phenomena is one of the most important achievements in nonlinear disciplines. Chaos is considered to be the complex form of motion in the dynamic systems, which contains various states of dynamic systems [10, 11]. As an important research direction of chaos dynamics, chaos synchronization theories have become the key to open the gate of theoretical researches in chaotic systems, and its research progress has become a hot spot in many fields [12–14]. The work in [15] studied the synchronization condition of chaotic systems. As stated in [16], the pertinence of chaos synchronization in physiology, nonlinear optics and dynamics has been concerned. However, on account of the interference of external factors including imprecise modeling and changeable operating point, which result in the performance of the system may become deteriorating degenerate and even the instability of the system may occur. Hence, a good many researchers focus on the study of robustness analysis [17, 18]. Recently, the authors in [19] investigated the random occurring uncertainties (ROUs) on account of a fact that during the system operation may be suffered from probabilistic uncertainties. In this regard, the issue for synchronization of semi-Markov jump chaotic systems (SMJCSs) with ROUs is worth exploring.

When designing controllers to realize synchronization of forced-unforced systems, the conventional control scheme commonly considered that system information at any time will be transferred to the controller, and a large number of network resources are occupied in this case. This is, whereas, unreasonable due to limited network resources. Facing this problem, the controller in the sampled-data techniques was designed, in which the control signals can only be changeable at the sampling instant, while remaining constant otherwise. Therefore, on this occasion, the limited network resources could effectively economize.
Relatively speaking, the sampled-data controller has the advantages of high accuracy, effective suppression of interference and favorable commonality. By the aid of these merits, some accomplishments have been acquired [20, 21]. As concerned in [22] , the neural networks reached the synchronous state driven by a fault-tolerant controller using sampled-data.

Furthermore, influenced by some practical conditions, time delays phenomena [29], which often makes the system property decline or even causes instability, constantly appear in industrial processes and engineering systems [23–25]. Meanwhile,

the delay parameters in the systems are usually unknown or time-varying. And the existence of the time delays verge to be random, which may make the comprehension and analysis of the system more difficult. Therefore, the researches of random time-varying delays (TVDs) systems belonging to the main domain of control engineering, have been conducted for various systems in recent years.
To name just a few, networked control systems [26], neural networks [27, 28], etc. It should be pointed out that the works aforesaid do not with a view to the condition that the actuator may fail during the execution. Howbeit, in the actual control process, these failures, which may result in the instability of the systems, are sometimes difficult to avoid. Based on this reason, the strategy of fault-tolerance control is introduced in this chapter, which allows or tolerates the occurrence of errors within certain limits in the systems, and making the systems more stable and reliable. In the regard, how to handle the fundamental but essential problem for time-varying delayed systems becomes the motivation of this work.

Inspired by the above discussions, the attention is focused on disposing of the fault-tolerant synchronization issue of SMJCSs with ROUs, where the sampled-data control scheme is concerned to make the proposed method more pragmatic. To extract the core works from three aspects:

(1) As the first attempt, the fault-tolerant sampled-data synchronization of switched chaotic systems with ROUs is investigated, and the switching among these subsystems is governed by the SMP.
(2) By utilizing applicable integral inequalities to acquire the less conservative results, some constrained criteria are proposed to force the error system to be synchronously stable.
(3) The cost-efficient sampled-data controller, which involves the randomness of TVDs, is designed to reduce a mass of load pressure on transferring data in the SMJCSs.

Notations: The notations employed can refer to [30], hence it is omitted here.

2 Preliminaries

In this section, by modifying the model in [30], a kind of unforced and forced SMJCSs with ROUs are concerned as follow

$$\dot{x}(t) = (\hat{A}(\varsigma(t)) + \delta(t)\Delta\hat{A}(t))\hat{x}(t) + (\hat{B}(\varsigma(t)) + \delta(t)\Delta\hat{B}(t))\hat{f}\left(\hat{x}(t)\right), \quad (1)$$

$$\begin{aligned}\dot{y}(t) = {} & (\hat{A}(\varsigma(t)) + \delta(t)\Delta\hat{A}(t))\hat{y}(t) \\ & + (\hat{B}(\varsigma(t)) + \delta(t)\Delta\hat{B}(t))\hat{f}\left(\hat{y}(t)\right) + u(t),\end{aligned} \quad (2)$$

in which $\hat{x}(t) \in \mathbb{R}^n$ and $\hat{y}(t) \in \mathbb{R}^n$ are state vectors of unforced and forced systems, separately; $u(t) \in \mathbb{R}^n$ expresses the control input. $\hat{A}(\varsigma(t))$ and $\hat{B}(\varsigma(t))$ are system matrices. $\varsigma(t)$ stands for the SMP which take values from a denumerable aggregate $\check{S} \triangleq \{1, 2, \ldots, r\}$.

Denote $\hat{A}_v \triangleq \hat{A}(\varsigma(t))$ for each $\varsigma(t) \triangleq v \in \check{S}$ and the others are similarly denoted. Furthermore, its transition rate matrix $\Pi(\epsilon) \triangleq \{\pi_{vn}(\epsilon)\}$ is formulated as

$$\Pr\{\varsigma(t+\epsilon) = n | \varsigma(t) = v\} = \begin{cases} \pi_{vn}(\epsilon)\epsilon + o(\epsilon), v \neq n, \\ 1 + \pi_{vv}(\epsilon)\epsilon + o(\epsilon), v = n, \end{cases} \tag{3}$$

where $\epsilon > 0$ is the sojourn time, $\lim_{\epsilon \to 0}(o(\epsilon)/\epsilon) = 0$ and $\pi_{vn}(\epsilon) \geq 0$, (for $v \neq n$), is the transition rate from mode $v \xrightarrow[t]{t+\epsilon} n$ and $\pi_{vv}(\epsilon) = -\sum_{n \in \check{S}, n \neq v} \pi_{vn}(\epsilon)$.

And, $\hat{f} : \mathbb{R}^n \to \mathbb{R}^n$ is assumed that as a nonlinear vector which satisfies the global Lipschitz condition, that is

$$\left\| \hat{f}(\grave{a}) - \hat{f}\left(\grave{b}\right) \right\| \leq \grave{l} \left\| \grave{a} - \grave{b} \right\|, \tag{4}$$

with respect to a positive scalar $\grave{l}$.

Meanwhile, $\Delta\hat{A}(t)$ and $\Delta\hat{B}(t)$ are the uncertainties satisfying

$$\left[\Delta\hat{A}(t)\ \Delta\hat{B}(t)\right] = \hat{D}\hat{\mathcal{F}}(t)\left[E_a\ E_b\right], \tag{5}$$

where $\hat{D}$, E_a and E_b are given real matrices and $\hat{\mathcal{F}}(t)$ is restrained by

$$\hat{\mathcal{F}}^T(t)\hat{\mathcal{F}}(t) \leq I,\ \forall t \geq 0. \tag{6}$$

For the synchronization purpose, the synchronization error can be marked as $\mathrm{e}(t) \triangleq \hat{y}(t) - \hat{x}(t)$ and the following error system is embodied as follow naturally

$$\dot{\mathrm{e}}(t) = (\hat{A}_v + \hat{\delta}(t)\Delta\hat{A}(t))\mathrm{e}(t) + (\hat{B}_v + \hat{\delta}(t)\Delta\hat{B}(t))\tilde{f}(\mathrm{e}(t)) + u(t), \tag{7}$$

in which $\tilde{f}(\mathrm{e}(t)) \triangleq \hat{f}(y(t)) - \hat{f}(x(t))$.

To illustrate the ROUs, ones adopt a stochastic variable $\hat{\delta}(t)$ obeying Bernoulli distribution and complying with the following stochastic distribution rules

$$\Pr\left\{\hat{\delta}(t) = 1\right\} = \hat{\delta},\ \Pr\left\{\hat{\delta}(t) = 0\right\} = 1 - \hat{\delta}, \tag{8}$$

where $0 \leq \hat{\delta} \leq 1$.

Then, the error system (7) can be expressed as

$$\begin{cases} \dot{\mathrm{e}}(t) = \hat{A}_v \mathrm{e}(t) + \hat{B}_v \tilde{f}(\mathrm{e}(t)) + u(t) + \hat{\delta}(t) D p(t), \\ p(t) = F(t) q(t), \\ q(t) = E_a e(t) + E_b \tilde{f}(e(t)). \end{cases} \tag{9}$$

It is worthy noting that the sampled-data controller plays a key role in modern networked systems. In this regard, it is of great importance to show solicitude for sample-data control method. Hence, in this work, the sampled signal of feedback controller is generated by making use of a zero-order hold function with a set of hold-times $0 = t_0 < t_1 < \cdots < t_k < \cdots$, and its concrete expression is shown as follow

$$u(t) = K\mathrm{e}(t_k),\ t_k \leq t < t_{k+1},\ k = 0, 1, 2, \ldots, N, \tag{10}$$

here, K is the gain parameter to build a sampled-data controller and t_k is the instantaneous time which updates by the zero order holder.

Meanwhile, $\mathrm{e}(t_k)$ is the measured value of $\mathrm{e}(t)$ at sampling instantaneous time t_k, $\lim_{k\to\infty} t_k = +\infty$.
Defining $t_{k+1} - t_k \triangleq d_k$, and $d_k > 0$ means the sampling interval.
Assume that there exist m sampling intervals $d_0, d_1, \ldots, d_m$ with $0 < d_0 < d_1 < \cdots < d_m$.
Then, setting $\tau(t) \triangleq t - t_k$, $t_k \leq t < t_{k+1}$, expression (10) is rewritten as follow

$$u(t) = K\mathrm{e}(t_k) = K\mathrm{e}(t - \tau(t)),\ t_k \leq t < t_{k+1}, \tag{11}$$

and the time-varying delays $\tau(t)$ contents $\dot{\tau}(t) = 1$.
For the purpose of designing a controller owns the property of random TVDs , the stochastic variable $\alpha_i(t)$ can be defined as follow

$$\alpha_i(t) = \begin{cases} 1\ d_{i-1} \leq \tau_i(t) \leq d_i, \\ 0 \text{ otherwise.} \end{cases}\ i = 1, 2, \ldots, m.$$

Here, one supposes the stochastic variable $\alpha_i(t)$ subjects to the Bernoulli distribution as well. Thus, we have

$$\Pr\{\alpha_i(t) = 1\} = \alpha_i,\ \Pr\{\alpha_i(t) = 0\} = 1 - \alpha_i,\ i = 1, 2, \ldots, m.$$

Remark 1 TVDs are ubiquitous in the practical systems, which interferes with the system performance. Therefore, in this chapter, the Bernoulli distribution is employed to govern the occurrence of the TVDs, which reflects the randomness of the TVDs.

Remark 2 As far as our information goes, when studying the controller design of the SMJCSs, the majority of results are based on an assumption that the complete mode information of the systems can be obtained by the mode-dependent controller on time. However, this assumption may sometimes be unsatisfied with actual applications. So as to surmount the defect of the mode-dependent controllers, we introduce the mode-independent one, which is more powerful and has a wide range of applications when the jump mode is not applied to the controller.

Under some practical conditions, it is frequent that the actuator fails on account of the chaotic systems. Taking this feature into account, we derive the actuator fault model from [31].

For the control input, more concretely, let $u_f(t)$ be the signal sent from actuators, obeying

$$u_f(t) = F_\upsilon u(t), \tag{12}$$

in which

$$F_\upsilon = diag\left\{f_\upsilon^1, f_\upsilon^2, \ldots, f_\upsilon^n\right\},$$

is actuator faults matrix, with the known constants $f_{\upsilon-}^l$ and $f_{\upsilon+}^l$ and satisfying $0 \leq f_{\upsilon-}^l \leq f_\upsilon^l \leq f_{\upsilon+}^l \leq 1, l = 1, 2, \ldots, n$.

By setting

$$f_{0\upsilon}^l = \frac{f_{\upsilon-}^l + f_{\upsilon+}^l}{2}, \; m_\upsilon^l = \frac{f_\upsilon^l - f_{0\upsilon}^l}{2}, \; j^l = \frac{f_{\upsilon+}^l - f_{\upsilon-}^l}{f_{\upsilon+}^l + f_{\upsilon-}^l}, \tag{13}$$

then, it is clear that

$$F_\upsilon = F_{0\upsilon}\left(I + M_\upsilon\right), \; \left|M_\upsilon\right| \leq J_\upsilon \leq I, \tag{14}$$

where

$$F_{0\upsilon} \triangleq diag\left\{f_{0\upsilon}^1, \ldots, f_{0\upsilon}^n\right\}, \left|M_\upsilon\right| \triangleq diag\left\{\left|m_\upsilon^1\right|, \ldots, \left|m_\upsilon^n\right|\right\}, J_\upsilon \triangleq diag\left\{j_\upsilon^0, \ldots, j_\upsilon^n\right\}.$$

Thus, the system (9) with m sampling intervals can be expressed as

$$\begin{cases} \dot{e}(t) = \hat{A}_\upsilon e(t) + \hat{B}_\upsilon \tilde{f}\left(e(t)\right) + \sum\limits_{i=1}^{m} \alpha_i\left(t\right) F_{0\upsilon}\left(I + M_\upsilon\right) K e(t - \tau_i(t)) \\ \qquad + \hat{\delta}(t) D p\left(t\right), \\ p\left(t\right) = F\left(t\right) q\left(t\right), \\ q\left(t\right) = E_a e(t) + E_b \tilde{f}\left(e(t)\right), \end{cases} \tag{15}$$

in which $d_{i-1} \leq \tau_i(t) \leq d_i$.

Remark 3 In theoretical research, it is generally considered that the system is operating in an ideal environment. On the contrary, the operation of the system will be interfered by external factors such as parameter errors and component aging. Hence, it is spontaneous to hypothesize system uncertainties for the sake of overcoming this situation. Moreover, the uncertainties sometimes exist in a random way during the system's operation, such as, random failures caused by the network and sudden environmental disturbances. Therefore, it is worthy to concern about ROUs for reality.

Remark 4 The random variable $\hat{\delta}(t)$ is applied to modeling the probability distribution of ROUs, which meets conditions that $\mathcal{E}\left\{\hat{\delta}(t)\right\} = \hat{\delta}$ and $\mathcal{E}\left\{\left(\hat{\delta}(t) - \hat{\delta}\right)^2\right\} = \hat{\delta}\left(1 - \hat{\delta}\right)$. The description like this can trace to [19] and has never been concerned in the control problem of SMJCSs.

Lemma 1 ([32]) *For any given differentiable function σ in $\left[\acute{\alpha}, \acute{\beta}\right] \to \mathbb{R}^n$, and a symmetric matrix $\varpi > 0$, it is confirmed that:*

$$\int_{\acute{\alpha}}^{\acute{\beta}} \dot{\sigma}^T(s)\,\varpi\,\dot{\sigma}(s)\,ds \geq \frac{W}{\acute{\beta} - \acute{\alpha}} + \frac{3\tilde{W}}{\acute{\beta} - \acute{\alpha}}, \tag{16}$$

where

$$W \triangleq \left(\sigma\left(\acute{\beta}\right) - \sigma\left(\acute{\alpha}\right)\right)^T \varpi \left(\sigma\left(\acute{\beta}\right) - \sigma\left(\acute{\alpha}\right)\right), \quad \tilde{W} \triangleq \mathring{\mathcal{W}}^T \varpi \mathring{\mathcal{W}},$$

$$\mathring{\mathcal{W}} \triangleq \sigma\left(\acute{\beta}\right) + \sigma\left(\acute{\alpha}\right) - \frac{2}{\acute{\beta} - \acute{\alpha}} \int_{\acute{\alpha}}^{\acute{\beta}} \sigma(s)\,ds^T.$$

3 Stability Analysis

This part is aimed at designing a sampled-data feedback controller to realize the synchronization of the unforced and forced systems. Thereinto, the switching among aforesaid two systems are described by the SMP.

Now, we have the following main result.

Theorem 1 *For given positive constants γ, $\hat{\delta} \in [0, 1]$, ε λ and known matrices $\tilde{F}$, $\tilde{S}$, E_a, E_b, $\hat{D}$, $\hat{A}$, $\hat{B}$ with appropriate dimensions, if there exists matrices $P_\upsilon > 0$, $Q_i > 0$, $R_i > 0$, $Z_{i\upsilon} > 0$, $T_i > 0$, any matrices S_i, $\tilde{S}_i$, and a symmetric matrices G satisfying the inequalities for $i = 1, 2, \ldots, m$, $\upsilon \in \check{S}$*

$$\Xi_{1\upsilon} \triangleq \begin{bmatrix} \tilde{\Xi}_{1\upsilon} & \Phi_1 & \Phi_2 \\ \star & \Phi_3 & \Phi_4 \\ \star & \star & \Phi_5 \end{bmatrix} < 0, \tag{17}$$

$$\Xi_2 \triangleq -\begin{bmatrix} \mathcal{R}_i & S_i \\ \star & \mathcal{R}_i \end{bmatrix} < 0, \tag{18}$$

$$\Xi_{3\upsilon} \triangleq -\begin{bmatrix} \mathcal{Z}_{i\upsilon} & \tilde{S}_i \\ \star & \mathcal{Z}_{i\upsilon} \end{bmatrix} < 0, \tag{19}$$

$$\Xi_{4\upsilon} \triangleq h_i \sum_{n \in \check{S}} \mathring{\pi}_{\upsilon n} Z_{in} - T_i < 0, \tag{20}$$

where

$$\tilde{\Xi}_{1v} \triangleq \sum_{n\in\check{S}} \pi_{vn}(\epsilon) P_n + Q_1 - 4R_1 - 4Z_{1v} + \lambda l^2 + G\hat{A}_v + \varepsilon E_a^T E_a + \hat{A}_v^T G^T,$$

$$\Phi_1 \triangleq \begin{cases} (1,1) \triangleq -2R_1 - S_{11} - S_{13} - S_{12} - S_{14} - 2Z_{1v} - \tilde{S}_{11} - \tilde{S}_{13} \\ \quad - \tilde{S}_{12} - \tilde{S}_{14} + GF_{0v}(I+m_v)K, \\ (1,2) \triangleq S_{11} + S_{13} - S_{12} - S_{14} + \tilde{S}_{11} + \tilde{S}_{13} - \tilde{S}_{12} - \tilde{S}_{14}, \\ (1,3) \triangleq 6R_1 + 6Z_{1v}, \\ (1,4) \triangleq 2S_{12} + 2S_{14} + 2\tilde{S}_{12} + 2\tilde{S}_{14}, \\ (1,i) \triangleq \alpha_i GF_{0v}(I+m_v)K,\ i = 5, 9, 13, \ldots, 4m+1, \\ otherwise \triangleq 0, \end{cases}$$

$$\mathcal{R}_i \triangleq \begin{bmatrix} R_i & 0 \\ 0 & 3R_i \end{bmatrix},\ \mathcal{Z}_{iv} \triangleq \begin{bmatrix} Z_{iv} & 0 \\ 0 & 3Z_{iv} \end{bmatrix},\ \Phi_2 \triangleq \left[\Phi_{21}\ \Phi_{31}\ \delta G\hat{D}\right],$$

$$\Phi_{21} \triangleq GB + \varepsilon E_a^T E_b,\ \Phi_{31} \triangleq -G + \gamma\hat{A}_v^T G^T + P_v,\ h_i \triangleq d_i - d_{i-1},$$

$$\Phi_4 \triangleq [0\ \Phi_{42}\ 0],\ \Phi_{42}(i,1) \triangleq \begin{cases} \alpha_i \gamma K^T (I+m_v)^T F_{0v}^T G^T,\ i = 1, 5, \ldots, 4m-3, \\ otherwise \triangleq 0, \end{cases}$$

$$\Phi_3 \triangleq \begin{cases} (i,i) \triangleq sym\{-4\left(R_{\frac{i+3}{4}} + Z_{\frac{i+3}{4}v}\right) + S_{\frac{i+3}{4}1} + S_{\frac{i+3}{4}2} - S_{\frac{i+3}{4}3} - S_{\frac{i+3}{4}4} \\ \quad + \tilde{S}_{\frac{i+3}{4}1} + \tilde{S}_{\frac{i+3}{4}2} - \tilde{S}_{\frac{i+3}{4}3} - \tilde{S}_{\frac{i+3}{4}4},\ i = 1, 5, \ldots, 4m-3, \\ (i,i+1) \triangleq -S_{\frac{i+3}{4}1} + S_{\frac{i+3}{4}3} - 2\left(R_{\frac{i+3}{4}} + Z_{\frac{i+3}{4}v}\right) + S_{\frac{i+3}{4}2} - S_{\frac{i+3}{4}4} \\ \quad - \tilde{S}_{\frac{i+3}{4}1} + \tilde{S}_{\frac{i+3}{4}3} + \tilde{S}_{\frac{i+3}{4}2} - \tilde{S}_{\frac{i+3}{4}4},\ i = 1, 5, \ldots, 4m-3, \\ (i,i+2) \triangleq 6R_{\frac{i+3}{4}} + 2S^T_{\frac{i+3}{4}3} + 2S^T_{\frac{i+3}{4}4} + 6Z_{\frac{i+3}{4}v} + 2\tilde{S}^T_{\frac{i+3}{4}3} + 2\tilde{S}^T_{\frac{i+3}{4}4}, \\ \quad i = 1, 5, \ldots, 4m-3, \\ (i,i+3) \triangleq -2S_{\frac{i+3}{4}2} + 2S_{\frac{i+3}{4}4} + 6R_1 - 2\tilde{S}_{\frac{i+3}{4}2} + 2\tilde{S}_{\frac{i+3}{4}4} + 6Z_{\frac{i+3}{4}v}, \\ \quad i = 1, 5, \ldots, 4m-3, \\ (i,i) \triangleq -4R_{\frac{i+2}{4}} - 4R_{\frac{i+2}{4}+1} + Q_{\frac{i+2}{4}+1} - Q_{\frac{i+2}{4}} - 4Z_{\frac{i+2}{4}v} - 4Z_{(\frac{i+2}{4}+1)v}, \\ \quad i = 2, 6, \ldots, 4m-2, \\ (i,i+1) \triangleq -2S^T_{\frac{i+2}{4}3} + 2S^T_{\frac{i+2}{4}4} - 2\tilde{S}^T_{\frac{i+2}{4}3} + 2\tilde{S}^T_{\frac{i+2}{4}4}, \\ \quad i = 2, 6, \ldots, 4m-2, \\ (i,i+2) \triangleq 6R_{\frac{i+2}{4}} + 6Z_{\frac{i+2}{4}},\ i = 2, 6, \ldots, 4m-2, \\ (i-4,i-1) \triangleq -2\left(R_{\frac{i+2}{4}} + Z_{\frac{i+2}{4}v}\right) - S_{\frac{i+2}{4}1} - S_{\frac{i+2}{4}3} - S_{\frac{i+2}{4}2} \\ \quad - S_{\frac{i+2}{4}4} - \tilde{S}_{\frac{i+2}{4}1} - \tilde{S}_{\frac{i+2}{4}3} - \tilde{S}_{\frac{i+2}{4}2} - \tilde{S}_{\frac{i+2}{4}4}, \\ \quad i = 6, 10, \ldots, 4m-2, \\ (i-4,i) \triangleq S_{\frac{i+2}{4}1} + S_{\frac{i+2}{4}3} - S_{\frac{i+2}{4}2} - S_{\frac{i+2}{4}4} + \tilde{S}_{\frac{i+2}{4}1} + \tilde{S}_{\frac{i+2}{4}3} \\ \quad - \tilde{S}_{\frac{i+2}{4}2} - \tilde{S}_{\frac{i+2}{4}4},\ i = 6, 10, \ldots, 4m-2, \\ (i-4,i+1) \triangleq 6R_{\frac{i+2}{4}} + 6Z_{\frac{i+2}{4}v},\ i = 6, 10, \ldots, 4m+2, \\ (i-4,i+2) \triangleq 2S_{\frac{i+2}{4}2} + 2S_{\frac{i+2}{4}4} + 2\tilde{S}_{\frac{i+2}{4}2} + 2\tilde{S}_{\frac{i+2}{4}4}, \\ \quad i = 6, 10, \ldots, 4m+2, \\ (i,i) \triangleq -12R_{\frac{i+1}{4}} - 12Z_{\frac{i+1}{4}v},\ i = 3, 7, \ldots, 4m-1, \\ (i,i+1) \triangleq -4S_{\frac{i+1}{4}4} - 4\tilde{S}_{\frac{i+1}{4}4},\ i = 3, 7, \ldots, 4m-1, \\ (i,i) \triangleq -12R_{\frac{i}{4}} - 12Z_{\frac{i}{4}v},\ i = 4, 8, \ldots, 4m, \\ otherwise \triangleq 0, \end{cases}$$

$$\Phi_5 \triangleq \begin{bmatrix} \Phi_{51} & \Phi_{52} & 0 \\ \star & \Phi_{53} & \Phi_{54} \\ \star & \star & -\varepsilon I \end{bmatrix}, \ \Phi_{51} \triangleq \varepsilon E_b^T E_b - \lambda I, \ \Phi_{52} \triangleq \gamma \hat{B}_v^T G^T, \ \Phi_{54} \triangleq \gamma \hat{\delta} G \hat{D},$$

$$\Phi_{53} \triangleq \sum_{i=1}^{m} \left[\left(h_i^2 Z_{iv} + \frac{d_i^2 - d_{i-1}^2}{2} T_i \right) + h_i^2 R_i \right] - \gamma G - \gamma G^T,$$

$$\mathring{\pi}_{vn} \triangleq \mathcal{E}\{\pi_{vn}(\epsilon)\} \triangleq \int_0^{\infty} \pi_{vn}(\epsilon) f_v(\epsilon) d\epsilon.$$

Then, the robust synchronization between the system (1) and system (2) is guaranteed by the sampled-data controller (12).

Proof For the error system (15), the following Lyapunov–Krasovskii functional can be considered

$$V(\mathrm{e}_t) = V_1(\mathrm{e}_t) + V_2(\mathrm{e}_t) + V_3(\mathrm{e}_t), \tag{21}$$

where

$$\begin{aligned} V_1(\mathrm{e}_t) &= \mathrm{e}^T(t) P_v \mathrm{e}(t), \\ V_2(\mathrm{e}_t) &= \sum_{i=1}^{m} \left(\int_{t-d_i}^{t-d_{i-1}} \mathrm{e}^T(s) Q_i \mathrm{e}(s) ds \right. \\ &\quad \left. + h_i \int_{-d_i}^{-d_{i-1}} \int_{t+\theta}^{t} \dot{\mathrm{e}}^T(s) R_i \dot{\mathrm{e}}(s) ds d\theta \right), \\ V_3(\mathrm{e}_t) &= \sum_{i=1}^{m} \left(h_i \int_{-d_i}^{-d_{i-1}} \int_{t+\beta}^{t} \dot{\mathrm{e}}^T(s) Z_{iv} \dot{\mathrm{e}}(s) ds d\beta \right. \\ &\quad \left. + \int_{-d_i}^{-d_{i-1}} \int_{r}^{0} \int_{t+\beta}^{t} \dot{\mathrm{e}}^T(s) Z_{iv} \dot{\mathrm{e}}(s) ds d\beta dr \right). \end{aligned}$$

Regard $\mathcal{L}$ as the weak infinitesimal operator and $f_i(\epsilon)$ is the PDF of sojourn time ϵ staying at mode i [33], then we defined $\mathcal{L}$ of $V(\mathrm{e}_t)$ as follow:

$$\mathcal{L}\{V(\mathrm{e}_t)\} = \lim_{h \to 0^+} \frac{1}{h} \{\mathcal{E}\{V(\mathrm{e}_{t+h}) / \mathrm{e}_t\} - V(\mathrm{e}_t)\}. \tag{22}$$

Then, taking (21) and (22) into account, ones can be achieved that

$$\mathcal{L}\{V_1(\mathrm{e}_t)\} = sym\left\{\mathrm{e}^T(t) P_v \dot{e}(t)\right\} + \mathrm{e}^T(t) \sum_{n \in \check{S}} \pi_{vn}(\epsilon) P_n \mathrm{e}(t),$$

$$\begin{aligned} \mathcal{L}\{V_2(\mathrm{e}_t)\} &= \sum_{i=1}^{m} (\mathrm{e}^T(t - d_{i-1}) Q_i \mathrm{e}(t - d_{i-1}) - \mathrm{e}^T(t - d_i) Q_i \mathrm{e}(t - d_i) \\ &\quad - h_i \int_{t-\tau_i(t)}^{t-d_{i-1}} \dot{\mathrm{e}}^T(s) R_i \dot{\mathrm{e}}(s) ds - h_i \int_{t-d_i}^{t-\tau_i(t)} \dot{\mathrm{e}}^T(s) R_i \dot{\mathrm{e}}(s) ds) \end{aligned}$$

$$+h_i^2\dot{e}^T(t)R_i\dot{e}(t),$$

$$\mathcal{L}\{V_3(\mathrm{e}_t)\} = \sum_{i=1}^{m}(h_i^2\dot{\mathrm{e}}^T(t)Z_{iv}\dot{\mathrm{e}}(t) + \dot{\mathrm{e}}^T(t)\left(\frac{d_i^2 - d_{i-1}^2}{2}\right)\dot{\mathrm{e}}(t)T_i$$
$$-h_i\int_{t-d_i}^{t-\tau_i(t)}\dot{\mathrm{e}}^T(s)Z_{iv}\dot{\mathrm{e}}(s)ds - h_i\int_{t-\tau_i(t)}^{t-d_{i-1}}\dot{\mathrm{e}}^T(s)Z_{iv}\dot{\mathrm{e}}(s)ds)$$
$$+h_i\int_{-d_i}^{-d_{i-1}}\int_{t+\beta}^{t}\dot{\mathrm{e}}^T(s)\Xi_{4v}\dot{\mathrm{e}}(s)dsd\beta.$$

In the light of Lemma 1 and the results in [34], One may get that

$$-h_i\int_{t-\tau_i(t)}^{t-d_{i-1}}\dot{\mathrm{e}}^T(s)R_i\dot{\mathrm{e}}(s)ds - h_i\int_{t-d_i}^{t-\tau_i(t)}\dot{\mathrm{e}}^T(s)R_i\dot{\mathrm{e}}(s)ds)$$
$$\leq -\frac{h_i}{\tau_i(t) - d_{i-1}}\Theta_i^T(t)\mathcal{R}_i\Theta_i(t) - \frac{h_i}{d_i - \tau_i(t)}\Psi_i^T(t)\mathcal{R}_i\Psi_i(t)$$
$$\leq -\Lambda_i(t)^T\Xi_2\Lambda_i(t),$$
$$-h_i\int_{t-\tau_i(t)}^{t-d_{i-1}}\dot{\mathrm{e}}^T(s)Z_{iv}\dot{\mathrm{e}}(s)ds - h_i\int_{t-d_i}^{t-\tau_i(t)}\dot{\mathrm{e}}^T(s)Z_{iv}\dot{\mathrm{e}}(s)ds)$$
$$\leq -\frac{h_i}{\tau_i(t) - d_{i-1}}\Theta_i^T(t)\mathcal{Z}_{iv}\Theta_i(t) - \frac{h_i}{d_i - \tau_i(t)}\Psi_i^T(t)\mathcal{Z}_{iv}\Psi_i(t)$$
$$\leq -\Lambda_i(t)^T\Xi_2\Lambda_i(t),$$

where

$$\Theta_i(t) = \begin{bmatrix} \mathrm{e}(t - d_{i-1}) - \mathrm{e}(t - \tau_i(t)) \\ \mathrm{e}(t - d_{i-1}) + \mathrm{e}(t - \tau_i(t)) - \frac{2}{\tau_i(t) - d_{i-1}}\int_{t-\tau_i(t)}^{t-d_{i-1}}\mathrm{e}(s)ds \end{bmatrix},$$
$$\Psi_i(t) = \begin{bmatrix} \mathrm{e}(t - \tau_i(t)) - \mathrm{e}(t - d_i) \\ \mathrm{e}(t - \tau_i(t)) + \mathrm{e}(t - d_i) - \frac{2}{d_i - \tau_i(t)}\int_{t-d_i}^{t-\tau_i(t)}\mathrm{e}(s)ds \end{bmatrix},$$
$$\Lambda_i(t) = \left[\Theta_i^T(t)\ \Psi_i^T(t)\right]^T.$$

According to the property of nonlinear function $\hat{f}$, one naturally find that

$$\lambda\tilde{f}^T(\mathrm{e}(t))\tilde{f}(\mathrm{e}(t)) \leq \lambda\mathrm{e}^T(t)l^2\mathrm{e}(t). \tag{23}$$

From the error system (15), for G, which is an appropriately dimensioned symmetrical matrix, satisfies the following equation

$$sym\{\left[\mathrm{e}^T(t)G + \gamma\dot{\mathrm{e}}^T(t)G\right] \times [-\dot{\mathrm{e}}(t) + \hat{A}_v\mathrm{e}(t) + \hat{B}_v\tilde{f}(\mathrm{e}(t))$$
$$+F_{0v}(I + M_v)K\mathrm{e}(t - \tau_i(t)) + \hat{\delta}\hat{D}p(t)]\} = 0. \tag{24}$$

Also, from (5) and (6), one can get

$$p^T(t)\,p(t) \le q^T(t)\,q(t)\,.$$

Then, ε, as a positive constant, satisfies the following equation

$$\varepsilon\left[\Gamma - p^T(t)\,p(t)\right] \ge 0, \tag{25}$$

where

$$\begin{aligned}
\Gamma &\triangleq \zeta^T(t)\,(\mathrm{e}_1^T E_a^T E_a \mathrm{e}_1 + \mathrm{e}_1^T E_a^T E_b \mathrm{e}_{4m+2} + \mathrm{e}_{4m+2}^T E_b^T E_a \mathrm{e}_1 \\
&\quad + \mathrm{e}_{4m+2}^T E_b^T E_b \mathrm{e}_{4m+2})\zeta(t), \\
\zeta(t) &\triangleq \left[\mathrm{e}^T(t)\ \mathrm{e}_m^T(t)\ \tilde{f}^T(\mathrm{e}(t))\ \dot{\mathrm{e}}^T(t)\ p^T(t)\right]^T,
\end{aligned}$$

with

$$\begin{aligned}
\mathrm{e}_m(t) \triangleq \Bigg[& \mathrm{e}^T(t-\tau_1(t))\ \mathrm{e}^T(t-d_1)\ \frac{\int_{t-\tau_1(t)}^{t} \mathrm{e}^T(s)ds}{\tau_1(t)}\ \frac{\int_{t-d_1}^{t-\tau_1(t)} \mathrm{e}^T(s)ds}{d_1-\tau_1(t)} \\
& \mathrm{e}^T(t-\tau_2(t))\ \mathrm{e}^T(t-d_2)\ \frac{\int_{t-\tau_2(t)}^{t-d_1} \mathrm{e}^T(s)ds}{\tau_2(t)}\ \frac{\int_{t-d_2}^{t-\tau_2(t)} \mathrm{e}^T(s)ds}{d_2-\tau_2(t)}\ \cdots \\
& \mathrm{e}^T(t-\tau_m(t))\ \mathrm{e}^T(t-d_m)\ \frac{\int_{t-\tau_m(t)}^{t-d_{m-1}} \mathrm{e}^T(s)ds}{\tau_m(t)}\ \frac{\int_{t-d_m}^{t-\tau_m(t)} \mathrm{e}^T(s)ds}{d_m-\tau_m(t)} \Bigg]^T,
\end{aligned}$$

and e_i $(i = 1, 2, \ldots, 4m+3) \in \mathbb{R}^{(4mn+4n)*n}$, for example, $\mathrm{e}_3 = \left[0\ 0\ I\ 0_{4mn*n}\right]^T$.

Based on the above derivations, adding left sides of (23), (24) and (25) to $\mathcal{L}\{V(\mathrm{e}_t)\}$, the following new upper bound of $\mathcal{L}\{V(\mathrm{e}_t)\}$ can be acquired

$$\mathcal{L}\{V(\mathrm{e}_t)\} \le \zeta^T(t)\,\Xi_{1\upsilon}\zeta(t)\,.$$

Therefore, in view of the explanation for the globally mean square stability in [30], the error system (15) is mean square stable if conditions in Theorem 1 hold. This illustrates that the synchronization between the drive system (1) and response system (2) is implemented by the designed sampled-data controller (12). □

It is noted that some unknown matrix coupling terms exist in the conditions of Theorem 1, which lead to the difficulty of getting the controller gains.
To solve such a problem, we present the theorem below.

Theorem 2 *For given positive constants γ, $\hat{\delta} \in [0, 1]$, ε, λ and known matrices $\tilde{F}$, $\tilde{S}$, E_a, E_b, $\hat{D}$, $\hat{A}_\upsilon$, $\hat{B}_\upsilon$, J_υ with appropriate dimensions, if there exists matrices $P_\upsilon > 0$, $Q_i > 0$, $R_i > 0$, $Z_{i\upsilon} > 0$, $T_i > 0$, $\eta_\upsilon > 0$, any matrices S_i, $\tilde{S}_i$, H,and a symmetric matrices G satisfying the conditions (18)–(20) and the following inequalities for $i = 1, \ldots, m$, $\upsilon \in \check{S}$*

$$\check{\Xi}_{1v} \triangleq \begin{bmatrix} \hat{\Xi}_{1v} & \hat{\Phi}_1 & \hat{\Phi}_2 \\ \star & \Phi_3 & \hat{\Phi}_4 \\ \star & \star & \hat{\Phi}_5 \end{bmatrix} < 0, \tag{26}$$

where

$$\hat{\Xi}_{1v} \triangleq \sum_{n\in\check{S}} \pi_{vn} p_n + Q_1 - 4(R_1 + Z_{1v}) + \lambda l^2 + sym\{GA\}$$
$$+\varepsilon E_a^T E_a + \eta_v F_{0v} J_v J_v^T F_{0v}^T,$$

$$\hat{\Phi}_1 \triangleq \begin{cases} (1,1) \triangleq -2R_1 - S_{11} - S_{13} - S_{12} - S_{14} - 2Z_{1v} - \tilde{S}_{11} - \tilde{S}_{13} \\ \qquad - \tilde{S}_{12} - \tilde{S}_{14} + F_{0v} H, \\ (1,2) \triangleq S_{11} + S_{13} - S_{12} - S_{14} + \tilde{S}_{11} + \tilde{S}_{13} - \tilde{S}_{12} - \tilde{S}_{14}, \\ (1,3) \triangleq 6R_1 + 6Z_{1v}, \\ (1,4) \triangleq 2S_{12} + 2S_{14} + 2\tilde{S}_{12} + 2\tilde{S}_{14}, \\ (1,i) \triangleq \alpha_i F_{0v} H,\ i = 5, 9, 13, \ldots, 4m+1, \\ otherwise \triangleq 0, \end{cases}$$

$$\hat{\Phi}_2 \triangleq \left[\Phi_{21}\ \Phi_{31}\ \hat{\delta} GD + \eta_v \gamma F_{0v} J_v J_v^T F_{0v}^T\ 0 \right],$$

$$\hat{\Phi}_4 \triangleq \left[0_{4mn\times n}\ \hat{\Phi}_{42}\ 0_{4mn\times n}\ \Phi_{44} \right],$$

$$\hat{\Phi}_{42}(i,1) \triangleq \begin{cases} \alpha_i \gamma H^T F_{0v}^T,\ i = 1, 5, \ldots, 4m-3, \\ otherwise = 0, \end{cases}$$

$$\Phi_{44}(i,1) \triangleq \begin{cases} \alpha_i H^T,\ i = 1, 5, \ldots, 4m-3, \\ otherwise = 0, \end{cases}$$

$$\hat{\Phi}_5 \triangleq \begin{bmatrix} \Phi_{51} & \Phi_{52} & 0 & 0 \\ \star & \Phi_{53} & \Phi_{54} & 0 \\ \star & \star & \Phi_{55} & 0 \\ \star & \star & \star & -\eta_v I \end{bmatrix},$$

$$\Phi_{55} \triangleq -\varepsilon I + \eta_v \gamma^2 F_{0v} J_v J_v^T F_{0v}^T.$$

Then, the robust synchronization between the system (1) and system (2) is guaranteed by the sampled-data controller (12). In this regard, the expression of the controller gain (12) is presented as $K = G^{-1} H$.

Proof Letting $H \triangleq GK$, it is clear that $\Xi_{1v} < 0$ can be rewritten as the following form

$$\begin{bmatrix} \tilde{\Xi}_{1v} & \tilde{\Phi}_1 & \Phi_2 \\ \star & \Phi_3 & 0 \\ \star & \star & \Phi_5 \end{bmatrix} + \mathcal{R} M_v \mathcal{N}^T + \mathcal{N} M_v^T \mathcal{R}^T < 0, \tag{27}$$

where

$$\tilde{\Phi}_1 \triangleq \begin{cases} (1,1) \triangleq -2R_1 - S_{11} - S_{13} - S_{12} - S_{14} - 2Z_{1\upsilon} \\ \quad -\tilde{S}_{11} - \tilde{S}_{13} - \tilde{S}_{12} - \tilde{S}_{14}, \\ (1,2) \triangleq S_{11} + S_{13} - S_{12} - S_{14} + \tilde{S}_{11} + \tilde{S}_{13} - \tilde{S}_{12} - \tilde{S}_{14}, \\ (1,3) \triangleq 6R_1 + 6Z_{1\upsilon}, \\ (1,4) \triangleq 2S_{12} + 2S_{14} + 2\tilde{S}_{12} + 2\tilde{S}_{14}, \\ otherwise \triangleq 0, \end{cases}$$

$$\mathcal{R} \triangleq \begin{cases} (1,1) \triangleq F_{0\upsilon}, \\ (4m+3,1) \triangleq \gamma F_{0\upsilon}, \\ otherwise \triangleq 0, \end{cases}$$

$$\mathcal{N} \triangleq \begin{cases} (i,1) \triangleq \alpha_i H^T, \ i = 2, 6, \ldots, 4m-2, \\ otherwise \triangleq 0. \end{cases}$$

In addition, one readily has

$$\mathcal{R}M_\upsilon \mathcal{N}^T + \mathcal{N}M_\upsilon^T \mathcal{R}^T \leq \eta_\upsilon \mathcal{R}J_\upsilon J_\upsilon^T \mathcal{R}^T + \eta_\upsilon^{-1} \mathcal{N}\mathcal{N}^T. \tag{28}$$

After that, combining (27) and (28), and using Schur complement for (27), the conditions (17)–(20) can be obtained easily. This completes the proof.

Remark 5 In practical situations, it is necessary to economize the limited network bandwidth resources due to the limitation of the communication load. Based on this, this chapter designs a sampling controller, which is used for SMJCSs. In addition, the randomly occurring time delays are considered in the controller which is tally with reality. Furthermore, control signals with the sampled-data process can be transmitted at the sampling instant, which decreases the amount of data transmission such that, the bandwidth resource can be economized efficiently.

4 Numerical Example

This section explores a numerical example, which for the sake of demonstrating the feasibility of the main results expressed above.

It is about the synchronization of hyperchaotic [35] which is described by SMP with the following parameters with two modes ($\upsilon = 1, 2$)

$$\hat{A}_1 = \begin{bmatrix} 0 & -2 & -2 & 0 \\ 2 & 0.5 & 0 & 2 \\ 0 & 0 & 0 & 0 \\ 0 & 0 & -1 & 0.1 \end{bmatrix}, \ \hat{A}_2 = \begin{bmatrix} 0 & -1 & -1 & 0 \\ 1 & 0.25 & 0 & 1 \\ 0 & 0 & 0 & 0 \\ 0 & 0 & -0.5 & 0.05 \end{bmatrix},$$

$$\hat{B}_1 = \text{diag}\{1, 1, 1, 1\},$$
$$\hat{B}_2 = \text{diag}\{2, 2, 2, 2\}, \ F_{01} = \text{diag}\{0.5, 0.5, 0.5, 0.5\},$$
$$F_{02} = \text{diag}\{0.7, 0.7, 0.7, 0.7\}, \ H_1 = \text{diag}\{0, 0, 0, 0\},$$

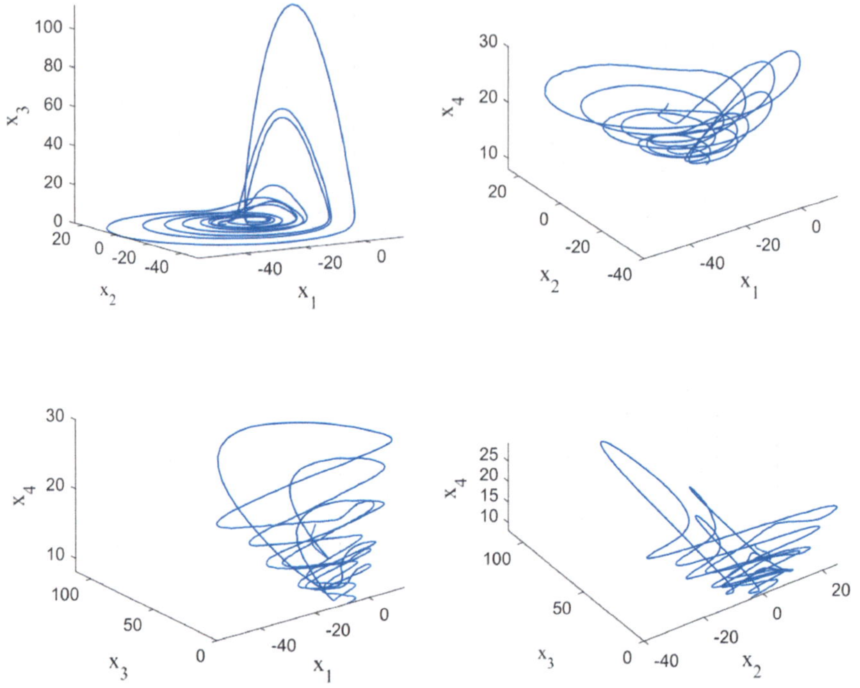

Fig. 1 The chaotic behavior of the hyperchaotic Rossler system

$$H_2 = \text{diag}\{0.5, 0.5, 0.5, 0.5\}, \ \hat{f}\left(\hat{x}(t)\right) = \left[0 \ 0 \ 3 + \hat{x}_1(t)\,\hat{x}_3(t) \ 0\right]^T.$$

Let the Lipschitz constant $l=1$ in the nonlinear function $\hat{f}$ and choose initial values of each system as $x(0) = \left[-7.2 \ -3.6 \ 1.35 \ 8.1\right]^T$, $y(0) = \left[-5.5 \ -6.5 \ 5 \ -10\right]^T$.

Then, the sampling intervals are considering as $d_1 = 0.04$, $d_2 = 0.08$, $d_3 = 0.15$, respectively.

In view of the conditions in Theorem 2, the mode-independent controller gain matrix may be calculated according to the procedure given below.

In addition, the switching between the two modes is described by a semi-Markov process, which obeys the transfer rate matrix shown below:

$$\Pi(\epsilon) = \begin{bmatrix} -\frac{1}{2}\epsilon & \frac{1}{2}\epsilon \\ 3(\epsilon)^2 & -3(\epsilon)^2 \end{bmatrix}.$$

On the basis of Weibull distribution [33], the probability density functions of sojourn time are concerned as $f_1(\epsilon) = \frac{1}{2}(\epsilon)\,e^{-0.25(\epsilon)^2}$ and $f_2(\epsilon) = 3(\epsilon)^2 e^{-(\epsilon)^3}$. Therefore, the mathematical expectation of $\Pi(\epsilon)$ can be computed easily:

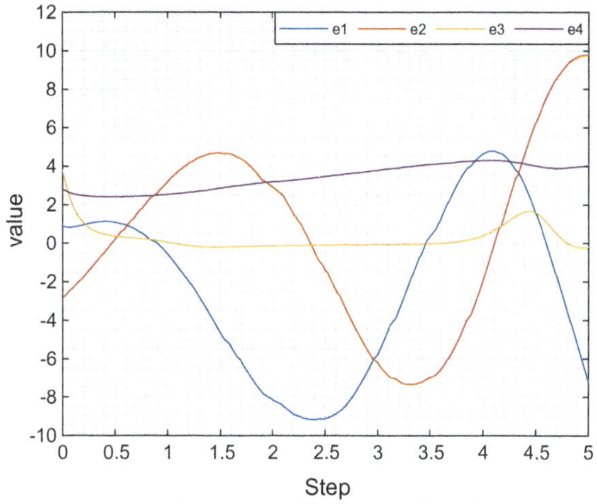

(a) uncontrolled error signals

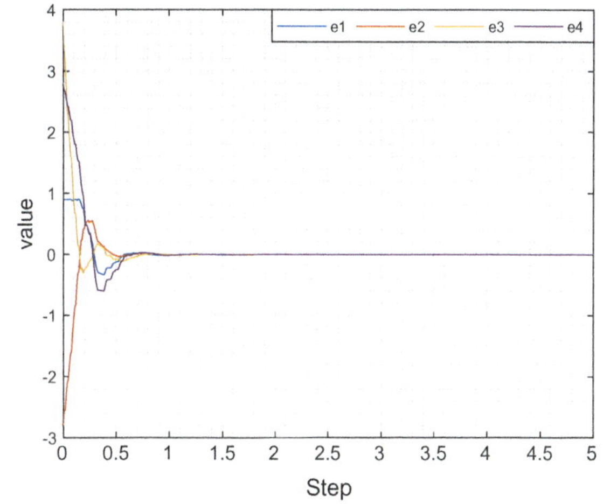

(b) controlled error signals

Fig. 2 Uncontrolled and controlled error signals

Table 1 The minimum value of $\breve{f}_{-}$ for different stochastic variable $\hat{\delta}$

$\hat{\delta}$	0.1	0.3	0.5	0.7	0.9
$\breve{f}_{-}$	0.1328	0.1354	0.1381	0.1408	0.1436

Fig. 3 Uncontrolled and controlled state response

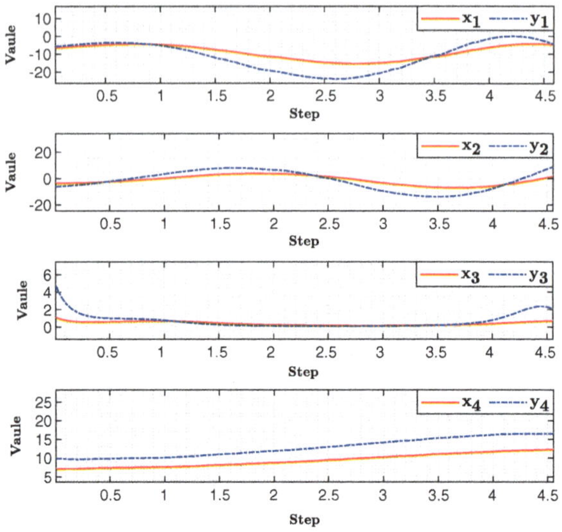

(a) Uncontrolled state responses

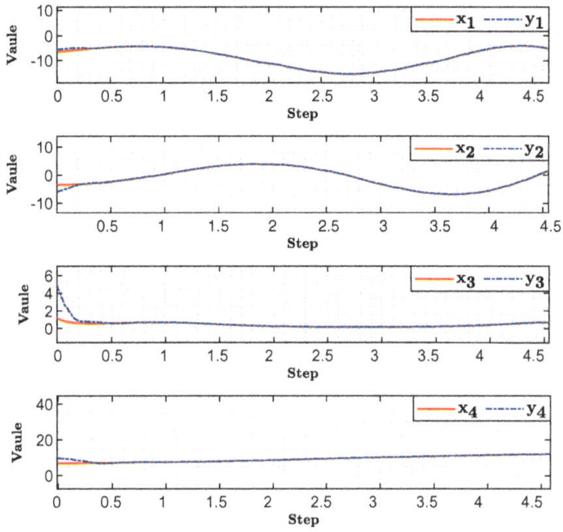

(b) Controlled state responses

$$\varepsilon\left\{\Pi\left(\epsilon\right)\right\} = \begin{bmatrix} -0.8862 & 0.8862 \\ 2.7082 & -2.7082 \end{bmatrix}.$$

The other parameters are listed as: $E_a = 0.5I_n$, $\hat{D} = I_n$, $\hat{\mathcal{F}}(t) = 0.4 + 0.2\sin t$, $E_b = 0.2I_n$ and $\hat{\delta} = 0.3$. Three sampling intervals are taken into account as d_1, d_2, d_3 in this example.

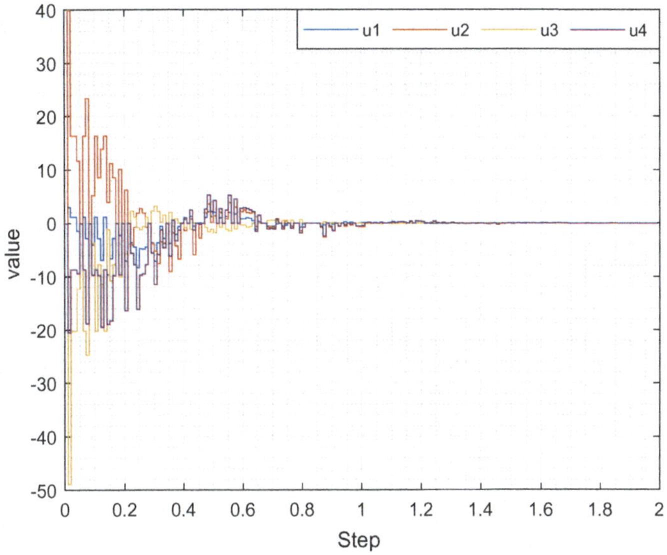

Fig. 4 The sampled-data control input with three sampling intervals

Figure 1 manifests the chaotic behavior. According to the above simulation settings, we can assume that $f_{1-}^{l} = f_{2-}^{l} = \check{f}_{-}$. Then, from the information obtained in Table 1 which is calculated by the Theorem 2. We can know that as the $\hat{\delta}$ decrease, the $\check{f}_{-}$ also decrease. After that, Remark 6 can be verified.

$$K = \begin{bmatrix} -8.7681 & 0.0874 & 2.9331 & -0.4818 \\ 0.0874 & -10.4061 & -0.1717 & -2.8332 \\ 2.9331 & 0.1717 & -8.5967 & 1.3147 \\ -0.4818 & -2.8332 & 1.3147 & -7.8816 \end{bmatrix}.$$

Remark 6 For this example, the system becomes unstable as the uncertainty magnitude increases. And it is not difficult to gain the information from the table that when the uncertainty increases, the fault tolerance of the system decreases $\left(\check{f}_{-} \text{ increases}\right)$. Then, the minimum value of $\check{f}_{-}$ which implies the maximum fault tolerance of the system is obtained.

The uncontrolled and controlled error signals are revealed in Fig. 2a,b, respectively. Comparing these two figures, one can find that the error signals controlled by the sampled-data controller tend to zero when time goes on.

Meanwhile, as desired, the error signals do not converge to zero for the uncontrolled system, which means that the controller we proposed realizes the synchronization of SMJCSs.

Fig. 3a shows the open-loop system, which expresses that the error system (15) is not synchronization. And, by observing the curve trace of Fig. 3b, we can comprehend the error system (15) becomes synchronization, which means the controller we designed is valid. Finally, Fig. 4 shows the sampled-data control input with three sampling intervals.

5 Conclusion

The design of a fault-tolerant sampled-data controller, in this work, has been put forward to accomplish the robust synchronization of SMJCSs. On the basis of a semi-Markov jump model, some available conditions have been established, which guarantee the mean-square synchronization of chaotic systems by means of employing appropriate Lyapunov–Krasovskii functional and reasonable integral inequality methods. Furthermore, ROUs have been taken into account in normal situations. In addition, by solving a group of LMIs, the anticipated controller gain can be acquired. An explained example has been employed to elucidate the preponderance of the method proposed. Note that, the aforementioned results can be generalized to the fuzzy model with extended dissipative performance, which is worthy of exploring.

Acknowledgements This work was supported by the National Natural Science Foundation of China under Grants 61703004, 61873002, the Natural Science Foundation of Anhui Province under grant 1808085QA18.

References

1. Qi, W., Zong, G., Karimi, H.R.: Observer-based adaptive SMC for nonlinear uncertain singular semi-Markov jump systems with applications to DC motor. IEEE Trans. Circuits Syst. I: Regul. Pap. **65**(9), 2951–2960 (2018)
2. Wang, J., Chen, M., Shen, H.: Event-triggered dissipative filtering for networked semi-Markov jump systems and its applications in a mass-spring system model. Nonlinear Dyn. **87**(4), 2741–2753 (2017)
3. Shen, H., Li, F., Xu, S., et al.: Slow state variables feedback stabilization for semi-Markov jump systems with singular perturbations. IEEE Trans. Autom. Control **63**(8), 2709–2714 (2017)
4. Shen, H., Li, F., Yan, H., et al.: Finite-time event-triggered $\mathcal{H}_\infty$ control for T-S fuzzy Markov jump systems. IEEE Trans. Fuzzy Syst. **26**(5), 3122–3135 (2018)
5. Park, J.H., Shen, H., Chang, X.H., Lee, T.H.: Recent Advances in Control and Filtering of Dynamic Systems with Constrained Signals. Springer, Cham, Switzerland (2018). https://doi.org/10.1007/978-3-319-96202-3
6. Shen, H., Men, Y., Wu, Z.Q., et al.: Network-based quantized control for fuzzy singularly perturbed semi-Markov jump systems and its application. IEEE Trans. Circuits Syst. I: Regul. Pap. **66**(3), 1130–1140 (2019)
7. Zhang, L., Leng, Y., Colaneri, P.: Stability and stabilization of discrete time semi-Markov jump linear systems via semi-Markov kernel approach. IEEE Trans. Autom. Control **61**(2), 503–508 (2015)

8. Liang, K., Dai, M., Shen, H., et al.: $\mathcal{L}_2 - \mathcal{L}_\infty$ synchronization for singularly perturbed complex networks with semi-Markov jump topology. Appl. Math. Comput. **321**, 450–462 (2018)
9. Wang, Z., Wu, H.: Global synchronization in fixed time for semi-Markovian switching complex dynamical networks with hybrid couplings and time-varying delays. Nonlinear Dyn. **95**(3), 2031–2062 (2019)
10. Joo, Y.H., Shieh, L.S., Chen, G.: Hybrid state-space fuzzy model-based controller with dual-rate sampling for digital control of chaotic systems. IEEE Trans. Fuzzy Syst. **7**(4), 394–408 (1999)
11. Aihara, K., Takabe, T., Toyoda, M.: Chaotic neural networks. Phys. Lett. A **144**(6–7), 333–340 (1990)
12. Voss, H.U.: Anticipating chaotic synchronization. Phys. Rev. E **61**(5), 5115 (2000)
13. Boutayeb, M., Darouach, M., Rafaralahy, H.: Generalized state-space observers for chaotic synchronization and secure communication. IEEE Trans. Circuits Syst. I: Fundam. Theory Appl. **49**(3), 345–349 (2002)
14. Park, J.H.: Synchronization of Genesio chaotic system via backstepping approach. Chaos, Solitons & Fractals **27**(5), 1369–1375 (2006)
15. Pecora, L.M., Carroll, T.L.: Synchronization in chaotic systems. Phys. Rev. Lett. **64**(8), 821 (1990)
16. Boccaletti, S., Kurths, J., Osipov, G., et al.: The synchronization of chaotic systems. Phys. Rep. **366**(1–2), 1–101 (2002)
17. Mohammad-Hoseini, S., Farrokhi, M., Koshkouei, A.: Robust adaptive control of uncertain non-linear systems using neural networks. Int. J. control **81**(8), 1319–1330 (2008)
18. Balasubramaniam, P., Vembarasan, V., Rakkiyappan, R.: Delay-dependent robust exponential state estimation of Markovian jumping fuzzy Hopfield neural networks with mixed random time-varying delays. Commun. Nonlinear Sci. Numer. Simul. **16**(4), 2109–2129 (2011)
19. Hu, J., Wang, Z., Gao, H., et al.: Robust sliding mode control for discrete stochastic systems with mixed time delays, randomly occurring uncertainties, and randomly occurring nonlinearities. IEEE Trans. Ind. Electron. **59**(7), 3008–3015 (2011)
20. Fridman, E., Seuret, A., Richard, J.P.: Robust sampled-data stabilization of linear systems: an input delay approach. Automatica **40**(8), 1441–1446 (2004)
21. Yamamoto, Y.: A function space approach to sampled data control systems and tracking problems. IEEE Trans. Autom. Control **39**(4), 703–713 (1994)
22. Su, L., Shen, H.: Fault-tolerant mixed $\mathcal{H}_\infty$/passive synchronization for delayed chaotic neural networks with sampled-data control. Complexity **21**(6), 246–259 (2016)
23. Lee, T.H., Park, J.H.: Stability analysis of sampled-data systems via freematrix- based time-dependent discontinuous Lyapunov approach. IEEE Trans. Autom. Control **62**(7), 3653–3657 (2017)
24. Lee, T.H., Park, J.H.: A novel Lyapunov functional for stability of timevarying delay systems via matrix-refined-function. Automatica **80**(80), 239–242 (2017)
25. Wu, Z.G., Park, J.H., Su, H., et al.: Discontinuous Lyapunov functional approach to synchronization of time-delay neural networks using sampled-data. Nonlinear Dyn. **69**(4), 2021–2030 (2012)
26. Sun, Y., Qin, S.: Stability analysis and controller design for networked control systems with random time delay. Int. J. Syst. Sci. **42**(3), 359–367 (2011)
27. Hu, X., Xia, J., Wei, Y., et al.: Passivity-based state synchronization for semi-Markov jump coupled chaotic neural networks with randomly occurring time delays. Appl. Math. Comput. **361**, 32–41 (2019)
28. Xing, M., Shen, H., Wang, Z.: Synchronization of semi-Markovian jump neural networks with randomly occurring time-varying delays. Complexity **2018**(Article ID 8094292), 1–16 (2018)
29. Park, J.H., Lee, T.H., Liu, Y., Chen, J.: Dynamic Systems with Time Delays: Stability and Control. Springer-Nature, Singapore (2019). https://doi.org/10.1007/978-981-13-9254-2
30. Lee, T.H., Park, J.H., Lee, S.M., et al.: Robust synchronisation of chaotic systems with randomly occurring uncertainties via stochastic sampled-data control. Int. J. Control **86**(1), 107–119 (2013)

31. Yang, G.H., Zhang, S.Y., Lam, J., et al.: Reliable control using redundant controllers. IEEE Trans. Autom. Control **43**(11), 1588–1593 (1998)
32. Seuret, A., Gouaisbaut, F.: Wirtinger-based integral inequality: application to time-delay systems. Automatica **49**(9), 2860–2866 (2013)
33. Wei, Y., Park, J.H., Karimi, H.R., et al.: Improved stability and stabilization results for stochastic synchronization of continuous-time semi-Markovian jump neural networks with time-varying delay. IEEE Trans. Neural Netw. Learn. Syst. **29**(6), 2488–2501 (2017)
34. Park, P., Lee, W.I., Lee, S.Y.: Auxiliary function-based integral inequalities for quadratic functions and their applications to time-delay systems. J. Frankl. Inst. **352**(4), 1378–1396 (2015)
35. Rossler, O.: An equation for hyperchaos. Phys. Lett. A **71**(2–3), 155–157 (1979)

Fuzzy-Based Sliding Mode Control Design for Stability Analysis of Nonlinear Interconnected Systems

A. Manivannan, G. Bhuvaneshwari, R. Krishnasamy, V. Parthiban, and S. Dhanasekar

Abstract In general, the nonlinear interconnected systems are highly interconnected with N number of sub-systems, therefore, it is difficult to a design decentralized control scheme for analyzing the stability of the system. This study is mainly concerned with fuzzy based sliding mode control design to analyze the stability and stabilization of nonlinear interconnected systems. In order to derive the stabilization condition, we proposed a fuzzy-based- sliding model controller and the sufficient condition is derived based on suitable Lyapunov function and LMI approach. The sufficient condition contains the information on the sliding motions, some tuning parameters and slack variables. The condition ensures that the asymptotic stability of the states of nonlinear interconnected systems under the fuzzy-based sliding mode controller. Finally, the numerical example and their simulation results are given to demonstrate the effectiveness of the proposed conditions.

Keywords Interconnected systems · Lyapunov stability · T-S fuzzy · Sliding mode control · Linear matrix inequality

---

A. Manivannan (✉) · G. Bhuvaneshwari · V. Parthiban · S. Dhanasekar
Division of Mathematics, School of Advanced Sciences,
Vellore Institute of Technology-Chennai Campus, Chennai 600127, India
e-mail: manivannan.a@vit.ac.in

G. Bhuvaneshwari
e-mail: bhuvaneshwari129@gmail.com

V. Parthiban
e-mail: parthiban.v@vit.ac.in

S. Dhanasekar
e-mail: dhanasekar.sundaram@vit.ac.in

R. Krishnasamy
Department of Applied Mathematics and Computational Sciences,
PSG College of Technology, Coimbatore, Tamil Nadu, India
e-mail: rksamy123@gmail.com

J. H. Park (ed.), *Recent Advances in Control Problems of Dynamical Systems and Networks*, Studies in Systems, Decision and Control 301,
https://doi.org/10.1007/978-3-030-49123-9_5

1 Introduction

The large-scale interconnected system plays a significant role in the various industrial applications such as power systems, transportation systems, socio-economic systems, haptic-enabled systems, and wind turbine systems. The main advantage of large-scale systems is that it communicates the viable information between the subsystems. For an instance, the wind turbines are connected over a tie-lines, in which through interconnections the data available in each wind turbine system can be shared over a network that results in better optimization of large-scale wind farms. Therefore, many researchers have paid great attention on the stability and stabilization issues of large-scale interconnected systems in the recent years [1–11].

In general, the control setup for each sub-system, say, centralized control does not reflect the overall stable performance of the complete large-scale system. To an alternative, a decentralized controller [12], a shared controller among the subsystems will provide stable performance as well as user reliability. Notice that, different types of decentralized control strategies are frequently encountered in the literature to guarantee the stabilization of large-scale systems.

To list a few, adaptive neural network control [13, 14], fault-tolerant control [15], sampled-data control [16] and integral sliding mode control (ISMC) [17, 18]. Among these controllers, ISMC has the ability to cope with the system having incomplete information and uncertainties. To an evident, authors in [17] have proposed a linear sliding variable-based decentralized sliding mode control strategy for interconnected systems. In [18], a reduced-order observer-based integral sliding mode control scheme has been proposed by the authors to stabilize a class of interconnected systems. Hence, this study focuses on designing the ISMC for nonlinear large-scale interconnected systems.

Technically, the industrial problems that can be framed into nonlinear differential systems cannot be solved directly. To solve those nonlinear systems, different types of theoretical approaches are utilized such as the Jacobian method, pseudo-linearization method and Takagi-Sugeno (T-S) fuzzy approach. Among these techniques, T-S fuzzy is considered to be a convenient method to express the nonlinear system into T-S fuzzy-based model [19–22]. In the view of literature, significant works have published regarding the T-S fuzzy-model based large scale interconnected system. To list a few, authors in [23] proposed the decentralized fault-tolerant control for a class of large-scale fuzzy systems with actuator faults containing stuck, outage and loss of effectiveness. Also, the authors in [24] investigated the large-scale system involving the certainties. Based on fuzzy SMC, the global stability of the closed-loop system was ensured.

Besides, as each subsystem is connected over the network, the communication between the subsystems will realize the time-delay over transmission. The ignorance of these types of communication time-delays may degrade the overall efficacy of the system. Therefore, the consideration of time-delays while formulating the large-scale system becomes necessary. For instance, authors in [25] proposed a new method to stabilize the system with control design. In [26], authors have been discussed the decentralized control of nonlinear interconnected time-delayed systems using T-S fuzzy mixed $\mathcal{H}_2/\mathcal{H}_\infty$ optimization with smith predictor is established. While investigating of stability of the system, the exogenous disturbances become a part of the analysis. In order to diminish the effect of the disturbance, $\mathcal{H}_2/\mathcal{H}_\infty$ performance came into the picture.

Theoretically, based on Lyapunov stability theory, remarkable stability and stabilization results have proposed on nonlinear dynamical systems [26–30].

In this chapter, we utilized the Lyapunov stability theory to derive the sufficient stability conditions in terms of LMIs, which can be solved with the help of MATLAB LMI control toolbox.

Inspired by the above works, the overall contributions of the present study can be listed in the following.

(i) The nonlinear large-scale interconnected is equivalently expressed into a T-S fuzzy-model-based system. We have utilized the double and triple integral terms with information of time-delays.
(ii) Wirtinger inequality approach is used to estimate the double integral terms and utilizing a slack variable with a tuning parameter, then delay-depended stability and stabilization conditions are derived.
(iii) The suitable controller is designed under the fuzzy integral sliding surface and their reachability condition is validated
(iv) A numerical example is provided to validate the effectiveness of the proposed theoretical conditions.

Notations. $\mathbb{R}^n$ and $\mathbb{R}^{n\times m}$ denote n-dimensional Euclidean space and the set of all $n \times m$ matrices with real entries respectively; $P_1 > 0$ ($P_1 \geq 0$) represent the positive definite (positive semi-definite) matrix. $P_2 < 0$ ($P_2 \leq 0$) represent the negative definite (negative semi-definite) matrix. P^T indicates the transpose and the symbol $*$ denotes the transposed elements in the symmetric matrix. I denotes an identity matrix; $\|\cdot\|$ is Euclidean norm.

2 Preliminaries

Now, let us consider the T-S fuzzy model of a nonlinear interconnected system with N-number of subsystems [10], as given below,

Plant rule l:
IF $\theta_{i1}^1(t)$ is M_{i1}^{1l} and $\cdots$ and $\theta_{i1}^q(t)$ is M_{i1}^{ql} and $\cdots$ and $\theta_{ij}^1(t)$ is M_{ij}^{1l} and $\cdots$ and $\theta_{ij}^q(t)$ is M_{ij}^{ql}

THEN

$$\dot{x}_i(t) = A_{il}x_i(t) + D_{il}x_i(t-\tau_i(t)) + B_{il}u_i(t) + \sum_{j=1,j\neq i}^{N} E_{ij,l}x_j(t) + F_{il}w_i(t),$$
$$y_i(t) = C_{il}x_i(t), \qquad i = 1,2,\dots,N. \quad l = 1,2,\dots,r. \tag{1}$$

Here, N is the total number of subsystems, $x_i(t),\ x_j(t) \in \mathbb{R}^n$ represent the i-th and jth subsystem respectively, $u_i(t) \in \mathbb{R}^m$ is control input, and $w_i(t) \in \mathbb{R}^m$ stands for external disturbance.

The coefficient matrices with appropriate dimensions are given as $A_{il}, B_{il}, D_{il}, F_{il}$, $E_{ij,l}$, and C_{il}; M_{ij}^{ql}, $(i, j, p, l) \in \{\mathcal{I}_N := \{1,2,\dots,N\}\} \times \{\mathcal{I}_J := \{1,2,\dots,J\}\} \times \{\mathcal{I}_q := \{1,2,\dots,q\}\} \times \{\mathcal{I}_r := \{1,2,\dots,r\}\}$ is the fuzzy set for $\theta_{ij}^p(t)$; $\mathcal{I}_N \times \mathcal{I}_J$ denotes all pairs $(i, j) \in \mathbb{Z}_{>0} \times \mathbb{Z}_{>0}$ with $1 \le i, j \le N$ and $i \neq j$; $\theta_{ij}^p(t) = \left[\theta_{i1}^1(t), \cdots \theta_{i1}^q(t)\ \cdots \theta_{ij}^1(t) \cdots \theta_{ij}^q(t)\right]^T$ are fuzzy premise variables.

Moreover, $\tau_i(t)$ denotes the time-varying delays for the ith subsystem, which needs to satisfy the following condition

$$0 \le \tau_i(t) \le \tau_i \quad \dot{\tau}_i(t) \le \mu_i < \infty,$$

for real constants $\tau_i > 0$, $\mu_i > 0$. $\varphi_i(t)$ denotes the initial value on $[-\tau_i, 0]$.

Finally, T-S fuzzy model is used to represent the each subsystem with r plant rules. Then, the overall fuzzy inferred system is derived as in (2)

$$\begin{aligned}\dot{x}_i(t) &= \sum_{l=1}^{r} h_i^l(\theta_i(t))\Big[A_{il}x_i(t) + D_{il}x_i(t-\tau_i(t)) + B_{il}u_i(t)\\ &\quad + \sum_{j=1,j\neq i}^{N} E_{ij,l}x_j(t) + F_{il}w_i(t)\Big],\\ y_i(t) &= \sum_{l=1}^{r} h_i^l(\theta_i(t))C_{il}x_i(t), \qquad i = 1,2,\dots,N,\end{aligned} \tag{2}$$

where

$$h_i^l(\theta_i(t)) = \frac{v_i^l(\theta_i(t))}{\sum_{l=1}^{r} v_i^l(\theta_i(t))}, \qquad v_i^l(\theta_i(t)) = \prod_{p=1}^{q} M_i^{pl}(\theta_i^p).$$

and $M_i^{pl} : U_{\theta_i^p} \subset \mathbb{R} \to \mathbb{R}_{[0,1]}$denotes the membership function of θ_{ij}^p on the compact set $U_{\theta_i^p}$.

Remark 1 The number of fuzzy plant rules depends on the number of fuzzy premise variables. For instance, for two fuzzy premise variables, it requires four fuzzy membership rules to express the nonlinear model into linear local submodels. In general, for $n-$ premise variables, 2^n fuzzy membership rules are required.

The sliding mode operation based on fuzzy sliding surface function and the reachability of the sliding motion is discussed in the following sections to ensure the stabilization of the considered system.

3 Sliding Surface Design

Sliding surface design is chosen as

$$\begin{aligned} s_i(t) = \mathcal{F}_i x_i(t) - \int_0^t \mathcal{F}_i \Big[\sum_{l=1}^{r} h_i^l(\theta_i(t)) \Big((A_{il} + B_{il} K_{il}) x_i(\sigma) \\ + D_{il} x_i(\sigma - \tau_i(\sigma)) \Big) + \sum_{j=1, j \neq i}^{N} E_{ij,l}\, x_j(\sigma) \Big] d\sigma. \end{aligned} \tag{3}$$

Here, if the state trajectories reach the sliding mode, then the sliding surface function must satisfy the conditions $s_i(t) = 0, \quad \dot{s}_i(t) = 0$.

In this aspect, the derivative of the Eq. (3) with respect to time yields

$$\dot{s}_i(t) = \mathcal{F}_i \sum_{l=1}^{r} h_i^l(\theta_i(t)) \Big(B_{il} u_i(t) + F_{il} w_i(t) - B_{il} K_{il} x_i(t) \Big).$$

Now, the equivalent controller is derived from the above equation as follows

$$\begin{aligned} u_{ie}(t) = (\mathcal{F}_i \sum_{l=1}^{r} h_i^l(\theta_i(t)) B_{il})^{-1} \mathcal{F}_i \sum_{l=1}^{r} h_i^l(\theta_i(t)) \\ \times \Big(- F_{il} w_i(t) + B_{il} K_{il} x_i(t) \Big). \end{aligned} \tag{4}$$

If the system state enters into sliding mode dynamics, then the state trajectories can be controlled through a designed equivalent controller (4).

In this regard, substituting the equation (4) into (2) can lead to the sliding mode dynamical state equations, which are given below

$$\dot{x}_i(t) = \sum_{l=1}^{r} h_i^l(\theta_i(t))\Big[(A_{il} + B_{il}K_{il})x_i(t) + D_{il}x_i(t-\tau_i(t)) + \mathcal{B}_{il}\mathcal{F}_i w_i(t) + \sum_{j=1,j\neq i}^{N} E_{ij,l}\, x_j(t)\Big], \tag{5}$$

$$y_i(t) = \sum_{l=1}^{r} h_i^l(\theta_i(t))C_{il}x_i(t), \qquad i = 1, 2, \ldots, N,$$

where $\mathcal{B}_{il} = I_n - B_{il}(\mathcal{F}_i B_{il})^{-1}\mathcal{F}_i$ is the transition matrix and I_n denotes the identity matrix.

Also, the given condition must hold $det(\mathcal{F}_i\mathcal{B}_{il}) \neq 0$, and $\mathcal{F}_i \in \mathbb{R}^{m_i \times n_i}$ and K_{il} are control gain matrices $(i = 1, 2, \ldots, N,$ and $l = 1, 2, \ldots, r)$.

The useful lemmas required for the analysis are given in the following:

Lemma 1 ([31]) *For a given matrix $M_1 > 0$, the following inequality holds for all continuously given scalars a and b satisfying $a < b$,the following inequality holds for all continuously differentiable function $x(t) : [a, b] \to R^n$*

$$\frac{(b-a)^2}{2}\int_a^b\int_s^b x^T(u)M_1x(u)duds \geq \Big(\int_a^b\int_s^b x^T(u)duds\Big)M_1\Big(\int_a^b\int_s^b x(u)duds\Big) + 2\Theta^T M_1\Theta$$

where $\Theta = -\int_a^b\int_s^b x(u)duds + \frac{3}{b-a}\int_a^b\int_s^b\int_u^b x(v)dvduds.$

Lemma 2 ([32]) *For any constant matrix $M_2 > 0$, any scalars a and b with $a < b$ and a vector function $x(t) : [a, b] \longrightarrow \mathbb{R}^n$ such that the integrals concerned are well defined, the following holds*

$$\Big(\int_a^b x(s)ds\Big)^T M_2\Big(\int_a^b x(s)ds\Big) \leq (b-a)\Big(\int_a^b x^T(s)M_2x(s)ds\Big).$$

Lemma 3 (Schur Complement) [33] *Given constant matrices Ω_1, Ω_2 and Ω_3 with appropriate dimensions, where $\Omega_1^T = \Omega_1$ and $\Omega_2^T = \Omega_2 > 0$, the inequality*

$$\Omega_1 + \Omega_3^T\Omega_2^{-1}\Omega_3 < 0$$

holds, if and only if

$$\begin{bmatrix} \Omega_1 & \Omega_3^T \\ * & -\Omega_2 \end{bmatrix} < 0, \quad or \quad \begin{bmatrix} -\Omega_2 & \Omega_3 \\ * & \Omega_1 \end{bmatrix} < 0.$$

The following section discusses about the derivation of sufficient stability conditions with respect to the derived closed-loop system (5).

4 Main Results

This section comprises the stability and stabilization of derived closed-loop system (5) under ISMC and suitable Lyapunov functionals.

The following section discusses the stability conditions which contain the non-convex terms that cannot be solved directly.

In this regard, through simple calculations, the stabilization conditions are derived in terms of solvable LMIs.

4.1 Stability and Stabilization Results

The sufficient condition for stability of the system (5) is summarized in the following theorem.

Theorem 1 *For given gain matrices* K_{il}, *if there exist the symmetric positive definite matrices* P_i, R_{1i}, R_{2i}, Q_i, *and* T_i, *which satisfy the following LMI, then the system (5) is said to be asymptotically stable*

$$\Gamma_{ij}^{l} = \begin{bmatrix} (\phi_{ij}^{l})_{1,1} & \tau_i Q_i & G_i D_{il} & G_i E_{ij,l} & (\phi_{ij}^{l})_{1,5} & (\phi_{ij}^{l})_{1,6} & (\phi_{ij}^{l})_{1,7} & (\phi_{ij}^{l})_{1,8} \\ * & (\phi_{ij}^{l})_{2,2} & 0 & 0 & 0 & 0 & 0 & 0 \\ * & * & (\phi_{ij}^{l})_{3,3} & 0 & (\phi_{ij}^{l})_{3,5} & 0 & 0 & 0 \\ * & * & * & (\phi_{ij}^{l})_{4,4} & (\phi_{ij}^{l})_{4,5} & 0 & 0 & 0 \\ * & * & * & * & (\phi_{ij}^{l})_{5,5} & (\phi_{ij}^{l})_{5,6} & 0 & 0 \\ * & * & * & * & * & (\phi_{ij}^{l})_{6,6} & 0 & 0 \\ * & * & * & * & * & * & (\phi_{ij}^{l})_{7,7} & (\phi_{ij}^{l})_{7,8} \\ * & * & * & * & * & * & * & (\phi_{ij}^{l})_{8,8} \end{bmatrix} < 0, \tag{6}$$

where

$$(\phi_{ij}^{l})_{1,1} = R_{1i} + R_{2i} - \tau_i^2 T_i - \tau_i Q_i - 2\Big(\frac{\tau_i^2}{2}\Big) T_i + 2G_i A_{il} + 2G_i B_{il} K_{il} + M_i + C_{il}^T C_{il}$$

$$(\phi_{ij}^{l})_{1,5} = P_i - G_i + b_1 (G_i A_{il})^T + b_1 (G_i B_{il} K_{il})^T, \ (\phi_{ij}^{l})_{1,6} = G_i \mathcal{B}_{il} F_{il},$$

$$(\phi_{ij}^{l})_{1,7} = -\tau_i T_i, \ (\phi_{ij}^{l})_{1,8} = 3T_i, \ (\phi_{ij}^{l})_{2,2} = -R_{1i} - \tau_i Q_i,$$

$$(\phi_{ij}^{l})_{3,3} = -(1 - \mu_i) R_{2i}, \ (\phi_{ij}^{l})_{3,5} = b_1 (G_i D_{il})^T,$$

$$(\phi_{ij}^{l})_{4,4} = -M_j, \ (\phi_{ij}^{l})_{4,5} = b_1 \Big(G_i E_{ij,l}\Big)^T,$$

$$(\phi_{ij}^{l})_{5,5} = \tau_i^2 Q_i + \Big(\frac{\tau_i^2}{2}\Big)^2 T_i - 2b_1 G_i, \ (\phi_{ij}^{l})_{5,6} = b_1 G_i \mathcal{B}_{il} F_{il},$$

$$(\phi_{ij}^{l})_{6,6} = -\gamma^2 I, \ (\phi_{ij}^{l})_{7,7} = -3T_i,$$

$$(\phi_{ij}^{l})_{7,8} = \frac{6T_i}{\tau_i}, \ (\phi_{ij}^{l})_{8,8} = \frac{-18T_i}{\tau_i^2}.$$

Proof Consider the following LKF

$$V(x_i(t)) = \sum_{i=1}^{N} \Big(V_{1i}(x_i(t)) + V_{2i}(x_i(t)) + V_{3i}(x_i(t)) \tag{7}$$

$$+ V_{4i}(x_i(t)) + V_{5i}(x_i(t)) \Big) \tag{8}$$

where

$$V_{1i}(x_i(t)) = x_i^T(t) P_i x_i(t),$$
$$V_{2i}(x_i(t)) = \int_{t-\tau_i}^{t} x_i^T(s) R_{1i} x_i(s) ds,$$
$$V_{3i}(x_i(t)) = \int_{t-\tau_i(t)}^{t} x_i^T(s) R_{2i} x_i(s) ds,$$
$$V_{4i}(x_i(t)) = \tau_i \int_{-\tau_i}^{0} \int_{t+\theta}^{t} \dot{x}_i^T(s) Q_i \dot{x}_i(s) ds d\theta,$$
$$V_{5i}(x_i(t)) = \frac{\tau_i^2}{2} \int_{t-\tau_i}^{t} \int_{s}^{t} \int_{u}^{t} \dot{x}_i^T(v) T_i \dot{x}_i(v) dv du ds.$$

Evaluating the time derivative of $V(x_i(t)$ on the state trajectories of the Eq. (5), one can obtain

$$\dot{V}(x_i(t)) = \sum_{i=1}^{N} \Big(\dot{V}_{1i}(x_i(t)) + \dot{V}_{2i}(x_i(t)) + \dot{V}_{3i}(x_i(t)) + \dot{V}_{4i}(x_i(t)) + \dot{V}_{5i}(x_i(t)) \Big)$$

where

$$\dot{V}_{1i}(x_i(t)) = 2x_i^T(t) P_i \dot{x}_i^T(t),$$
$$\dot{V}_{2i}(x_i(t)) = \Big(x_i^T(t) R_{1i} x_i(t) - x_i^T(t-\tau_i) R_{1i} x_i(t-\tau_i) \Big),$$
$$\dot{V}_{3i}(x_i(t)) = \Big(x_i^T(t) R_{2i} x_i(t) - (1-\mu_i) x_i^T(t-\tau_i(t)) R_{2i} x_i(t-\tau_i(t)) \Big),$$
$$\dot{V}_{4i}(x_i(t)) = \tau_i^2 \dot{x}_i^T(t) Q_i \dot{x}_i(t) - \tau_i \int_{t-\tau_i}^{t} \dot{x}_i^T(s) Q_i \dot{x}_i(s) ds,$$
$$\dot{V}_{5i}(x_i(t)) = \Big(\frac{\tau_i^2}{2} \int_{t-\tau_i}^{t} \int_{s}^{t} \dot{x}_i^T(t) T_i \dot{x}_i(t) du ds - \frac{\tau_i^2}{2} \int_{t-\tau_i}^{t} \int_{s}^{t} \dot{x}_i^T(u) T_i \dot{x}_i(u) du ds \Big).$$

From $\dot{V}_{4i}(t)$, it can be seen that the single integral term cannot be solved.

In this regard, the most frequently used Jensen's inequality (Lemma 1) can be utilized and integral term $-\tau_i \int_{t-\tau_i}^{t} \dot{x}_i^T(s) Q_i \dot{x}_i(s) ds$ is approximated as

$$\dot{V}_{4i}(x_i(t)) \leq \tau_i^2 \dot{x}_i^T(t) Q_i \dot{x}_i(t) \tag{9}$$

$$- \tau_i \Big(\Big[x_i^T(t) - x_i^T(t - \tau_i) \Big] Q_i \Big[x_i^T(t) - x_i^T(t - \tau_i) \Big] \Big). \tag{10}$$

Similarly, the Wirtinger-based inequalities are employed to solve the complex double integral term. For more details, refer the Lemma 2. In detail, with lower bound of double integral, one can solve the integral $-\frac{\tau_i^2}{2} \int_{t-\tau_i}^{t} \int_{s}^{t} \dot{x}_i^T(s) T_i \dot{x}_i(s) duds$ into the following inequality $\dot{V}_{5i}(x_i(t))$:

$$\begin{aligned} \dot{V}_{5i}(x_i(t)) \leq \; & \frac{\tau_i^4}{4} \dot{x}_i^T(t) T_i \dot{x}_i(t) \\ & - \Big[\tau_i x_i(t) - \int_{t-\tau_i}^{t} x_i(s) ds \Big]^T T_i \Big[\tau_i x_i(t) - \int_{t-\tau_i}^{t} x_i(s) ds \Big] \\ & - 2 \Big[\frac{\tau_i}{2} x(t) - \int_{t-\tau_i}^{t} x_i(s) ds + \frac{3}{\tau_i} \int_{t-\tau_i}^{t} \int_{s}^{t} x(u) duds \Big]^T \\ & \times T_i \Big[\frac{\tau_i}{2} x(t) - \int_{t-\tau_i}^{t} x_i(s) ds + \frac{3}{\tau_i} \int_{t-\tau_i}^{t} \int_{s}^{t} x(u) duds \Big]. \end{aligned} \tag{11}$$

Instead of accompanying the closed-loop system (3), the most convenient way is to employ the closed-loop system through the zero-equation approach.

In this regard, for any matrix G_i and a scalar b_1, the following zero equation can be derived

$$\begin{aligned} 0 = 2 \Big[x_i^T(t) + b_1 \dot{x}_i^T] G_i \Big[- \dot{x}_i(t) + \sum_{l=1}^{r} h_i^l(\theta_i(t)) \Big(A_{il} + B_{il} K_{il} x_i(t) \\ + D_{il} x_i(t - \tau_i(t)) + \mathcal{B}_{il} \mathcal{F}_i w_i(t) \Big) + \sum_{j=1, j \neq i}^{N} E_{ij,l} x_j(t) \Big]. \end{aligned} \tag{12}$$

Obviously, we can get

$$\sum_{i=1}^{N} \sum_{i=1, j \neq i}^{N} x_i^T(t) M_i x_i(t) = \sum_{i=1}^{N} \sum_{j=1, j \neq i}^{N} x_i^T(t) M_j x_i(t). \tag{13}$$

Now, $\mathcal{H}_\infty$ attenuation performance for a given scalar γ under the zero initial condition can be considered as

$$\sum_{i=1}^{N}\int_{0}^{\infty} y_i^T(\tau)y_i(\tau)d\tau < \gamma^2\sum_{i=1}^{N}\int_{0}^{\infty} w_i^T(\tau)w_i(\tau)d\tau. \tag{14}$$

Summing up the derivatives $\dot{V}_{1i}(x_i(t))$, $\dot{V}_{2i}(x_i(t))$, $\dot{V}_{3i}(x_i(t))$ and the equations (9) – (14), we can obtain that

$$\dot{V}(x_i(t)) + \sum_{i=1}^{N}\Big(y_i^T(t)y_i(t) - \gamma^2 w_i^T(t)w_i(t)\Big) \leq 0. \tag{15}$$

And, with (15), one can have

$$\dot{V}(x_i(t)) + \sum_{i=1}^{N}\Big(y_i^T(t)y_i(t) - \gamma^2 w_i^T(t)w_i(t)\Big) \leq \sum_{i=1}^{N}\sum_{j=1,j\neq i}^{N}\sum_{l=1}^{r} h_i^l(\theta_i(t))\xi_{ij}^T(t)\Gamma_{ij}^l\xi_{ij}(t) \tag{16}$$

where

$$\xi_{ij}^T(t) = \Big[x_i^T(t) \quad x_i^T(t-\tau_i) \quad x_i^T(t-\tau_i(t)) \quad x_j^T(t) \quad \dot{x}_i^T(t) \quad w_i^T(t) \quad \int_{t-\tau_i}^{t} x_i(s)ds \quad \int_{t-\tau_i}^{t}\int_{s}^{t} x_i^T(u)duds\Big].$$

Based on Lyapunov stability condition and $\mathcal{H}_\infty$ performance (15) and LMI (6), one can have $\dot{V}(x_i(t)) < 0$ which implies that the proposed T-S fuzzy system is asymptotically stable with $\mathcal{H}_\infty$ performance. □

Remark 2 Here, it should be mentioned that the control gain matrices are assumed to be known matrices in the sufficient of Theorem 1. To find the control gain matrices, let us consider, $K_{il}\hat{G}_i = L_{il}$ where $\hat{G}_i = G_i^{-1}$, $\hat{R}_{1i} = \hat{G}_i R_{1i}\hat{G}_i$, $\hat{R}_{2i} = \hat{G}_i R_{2i}\hat{G}_i$, $\hat{Q}_i = \hat{G}_i Q_i \hat{G}_i$, $\hat{T}_i = \hat{G}_i T_i \hat{G}_i$, $\hat{M}_{ij} = \hat{G}_i M_j \hat{G}_i$ and pre-and-post multiplication of $diag\{\hat{G}_i, \hat{G}_i, \hat{G}_i, \hat{G}_i, \hat{G}_i, I, \hat{G}_i, \hat{G}_i\}^T$, and $diag\{\hat{G}_i, \hat{G}_i, \hat{G}_i, \hat{G}_i, \hat{G}_i, I, \hat{G}_i, \hat{G}_i\}$ in (6), one can get the following LMI

$$\hat{\Gamma}^l_{ij} = \begin{bmatrix} (\hat{\phi}^l_{ij})_{1,1} & \tau_i\hat{Q}_i & D_{il}\hat{G}_i & E_{ij,l}\hat{G}_i & (\hat{\phi}^l_{ij})_{1,5} & (\hat{\phi}^l_{ij})_{1,6} & (\hat{\phi}^l_{ij})_{1,7} & (\hat{\phi}^l_{ij})_{1,8} \\ * & (\hat{\phi}^l_{ij})_{2,2} & 0 & 0 & 0 & 0 & 0 & 0 \\ * & * & (\hat{\phi}^l_{ij})_{3,3} & 0 & (\hat{\phi}^l_{ij})_{3,5} & 0 & 0 & 0 \\ * & * & * & (\hat{\phi}^l_{ij})_{4,4} & (\hat{\phi}^l_{ij})_{4,5} & 0 & 0 & 0 \\ * & * & * & * & (\hat{\phi}^l_{ij})_{5,5} & (\hat{\phi}^l_{ij})_{5,6} & 0 & 0 \\ * & * & * & * & * & (\hat{\phi}^l_{ij})_{6,6} & 0 & 0 \\ * & * & * & * & * & * & (\hat{\phi}^l_{ij})_{7,7} & (\hat{\phi}^l_{ij})_{7,8} \\ * & * & * & * & * & * & * & (\hat{\phi}^l_{ij})_{8,8} \end{bmatrix} < 0, \tag{17}$$

where

$$(\hat{\phi}^l_{ij})_{1,1} = \hat{R}_{1i} + \hat{R}_{2i} - \tau_i^2 T_i - \tau_i\hat{Q}_i - 2\Big(\frac{\tau_i^2}{2}\Big)\hat{T}_i + 2A_{1il}\hat{G}_i$$
$$+ 2B_{il}L_{il} + \hat{M}_i + \hat{G}_i C_{il}^T C_{il}\hat{G}_i$$

$$(\hat{\phi}^l_{ij})_{1,5} = \hat{P}_i - \hat{G}_i + b_1(A_{il}\hat{G}_i)^T + b_1(B_{il}L_{il})^T, \quad (\hat{\phi}^l_{ij})_{1,6} = \mathcal{B}_{il}F_{il}\hat{G}_i,$$

$$(\hat{\phi}^l_{ij})_{1,7} = -\tau_i\hat{T}_i,\ (\hat{\phi}^l_{ij})_{1,8} = 3\hat{T}_i,\ (\hat{\phi}^l_{ij})_{2,2} = -\hat{R}_{1i} - \tau_i\hat{Q}_i,$$
$$(\hat{\phi}^l_{ij})_{3,3} = -(1-\mu_i)\hat{R}_{2i},\ (\hat{\phi}^l_{ij})_{3,5} = b_1(D_{il}\hat{G}_i)^T,$$

$$(\hat{\phi}^l_{ij})_{4,4} = -\hat{M}_{ij},\ (\hat{\phi}^l_{ij})_{4,5} = b_1\Big(E_{ij,l}\hat{G}_i\Big)^T,$$

$$(\hat{\phi}^l_{ij})_{5,5} = \tau_i^2\hat{Q}_i + \Big(\frac{\tau_i^2}{2}\Big)^2\hat{T}_i - 2b_1\hat{G}_i,\ (\hat{\phi}^l_{ij})_{5,6} = b_1\mathcal{B}_{il}F_{il}\hat{G}_i,$$

$$(\hat{\phi}^l_{ij})_{6,6} = -\gamma^2 I,\ (\hat{\phi}^l_{ij})_{7,7} = -3\hat{T}_i,$$
$$(\hat{\phi}^l_{ij})_{7,8} = \frac{6\hat{T}_i}{\tau_i},\ (\hat{\phi}^l_{ij})_{8,8} = -\frac{-18\hat{T}_i}{\tau_i^2}.$$

Applying Lemma 3 (Schur complement lemma) to solve the Eq. (17) and their corresponding stabilization result can be summarized in the following theorem.

Theorem 2 *If there exist the symmetric positive definite matrices $\hat{P}_i$, $\hat{R}_{1i}$, $\hat{R}_{2i}$, $\hat{Q}_i$, and $\hat{T}_i$, which satisfy the following LMI condition*

$$\hat{\Gamma}^l_{ij} = \begin{bmatrix} (\hat{\phi}^l_{ij})_{1,1} & \tau_i\hat{Q}_i & D_{il}\hat{G}_i & E_{ij,l}\hat{G}_i & (\hat{\phi}^l_{ij})_{1,5} & (\hat{\phi}^l_{ij})_{1,6} & (\hat{\phi}^l_{ij})_{1,7} & (\hat{\phi}^l_{ij})_{1,8} & C^T_{il}\hat{G}_i \\ * & (\hat{\phi}^l_{ij})_{2,2} & 0 & 0 & 0 & 0 & 0 & 0 & 0 \\ * & * & (\hat{\phi}^l_{ij})_{3,3} & 0 & (\hat{\phi}^l_{ij})_{3,5} & 0 & 0 & 0 & 0 \\ * & * & * & (\hat{\phi}^l_{ij})_{4,4} & (\hat{\phi}^l_{ij})_{4,5} & 0 & 0 & 0 & 0 \\ * & * & * & * & (\hat{\phi}^l_{ij})_{5,5} & (\hat{\phi}^l_{ij})_{5,6} & 0 & 0 & 0 \\ * & * & * & * & * & (\hat{\phi}^l_{ij})_{6,6} & 0 & 0 & 0 \\ * & * & * & * & * & * & (\hat{\phi}^l_{ij})_{7,7} & (\hat{\phi}^l_{ij})_{7,8} & 0 \\ * & * & * & * & * & * & * & (\hat{\phi}^l_{ij})_{8,8} & 0 \\ * & * & * & * & * & * & * & * & -I \end{bmatrix} < 0, \tag{18}$$

then the solutions of the considered system (5) are said to be asymptotically stable under the control gain matrices $K_{il} = L_{il}G_i^{-1}$.

4.2 Reachability Condition

This subsection discusses the synthesis of T-S fuzzy sliding mode control law, in which the state trajectories of the T-S fuzzy system (5) can be driven onto the proposed sliding surface.

In this regard, the appropriate T-S fuzzy sliding mode control law is designed as follows:

Control Rule: U_i^l:
IF $\theta^1_{i1}(t)$ is M^{1l}_{i1} and $\cdots$ and $\theta^q_{i1}(t)$ is M^{ql}_{i1} THEN

$$u_i(t) = \sum_{l=1}^{r} h^l_i(\theta_i(t))K_{il} - \delta_i(t)(\mathcal{F}_i B_{il})^{-1}sgn(s_i(t)) \tag{19}$$

where

$$\delta_i(t) = \lambda_i + \alpha_i\|x_i(t)\|\|\mathcal{F}_i F_{il}\|,$$

λ_i and α_i are positive scalars.

Remark 3 Notice that, in the selection of membership functions for plant and control rules, different types of strategies have followed. For example, matched and mismatched membership for plant and control rule. In this chapter, we have considered the same membership functions for both the control and plant rules.

Theorem 3 *For the given system* (5), *if sliding surface is designed as in* (3) *with the unknown control gains K_{il}, then the state trajectories of the system are driven onto the sliding mode dynamics under the designed control rule* (19).

Proof Consider the appropriate Lyapunov function as follows

$$U_i(t) = \frac{1}{2}s_i(t)s_i^T(t). \tag{20}$$

Then, taking the derivative of (20) with respect to time, one may get

$$\begin{aligned}
\dot{U}_i(t) &= s_i^T(t)\Big[\mathcal{F}_i B_{il} u_i(t) + \mathcal{F}_i F_i w_i - \sum_{l=1}^{r} \mathcal{F}_i B_{il} K_{il} x_i(t)\Big] \\
&\leq s_i^T(t)\Big[\mathcal{F}_i B_{il}\Big(\sum_{l=1}^{r} h_i^l(\theta_i(t)) K_{il} x_i(t) - \delta_i(t)\Big(\mathcal{F}_i B_{il}\Big)^{-1} sgn(s_i(t))\Big) \\
&\quad + \mathcal{F}_i F_i w_i(t) - \sum_{l=1}^{r} h_i^l(\theta_i(t)) \mathcal{F}_i B_{il} K_{il} x_i(t)\Big] \\
&\leq s_i^T(t)\Big[-\mathcal{F}_i B_{il}\delta_i(t)\Big(\mathcal{F}_i B_{il}\Big)^{-1} sgn(s_i(t)) + \mathcal{F}_i F_i w_i(t)\Big] \\
&\leq s_i^T(t)\Big[-\delta_i(t)\frac{s_i(t)}{||s_i(t)||} + \mathcal{F}_i F_i w_i(t)\Big] \\
&\leq \Big[-\delta_i(t)||s_i(t)|| + ||s_i(t)||||\mathcal{F}_i||||F_i||||w_i(t)||\Big] \\
&\leq \Big[-\delta_i(t)\|s_i(t)\| + \alpha_i||s_i(t)||||\mathcal{F}_i||||F_i||||x_i(t)||\Big] \\
&\leq -\lambda_i(t)||s_i(t)||.
\end{aligned}$$

Hence, it is clear that $\dot{U}_i(t) \leq 0$, which guarantees the global stability of the sliding surface and ensure the reachability of the sliding mode dynamics. □

Therefore, based on the theoretical sufficient condition, the following section contains a numerical evaluation of the closed-loop system that helps to validate the efficiency of the proposed sufficient conditions.

5 Numerical Simulations

In this section, the numerical simulations are performed under the theoretical stability conditions. In this regard, the closed-loop system (5) is simulated with the following constant matrices:

$$A_{11} = \begin{bmatrix} 3.0000 & -4.2667 \\ 1.0000 & 0 \end{bmatrix}, \quad A_{12} = \begin{bmatrix} 4.0000 & -2.2667 \\ 1.0000 & 0 \end{bmatrix},$$

$$A_{21} = \begin{bmatrix} -4.0000 & -0.3333 \\ 1.0000 & 0 \end{bmatrix}, \quad A_{22} = \begin{bmatrix} -4.0002 & -0.3333 \\ 1.0000 & 0 \end{bmatrix},$$

$$E_{12,1} = \begin{bmatrix} 0 & -1.5 \\ 0 & 0 \end{bmatrix}, \quad E_{21,1} = \begin{bmatrix} 0 & -0.5 \\ 0 & 0 \end{bmatrix}, \quad D_{11} = \begin{bmatrix} 0 & 5.05 \\ 0 & 0 \end{bmatrix},$$

$$D_{12} = \begin{bmatrix} 0 & 5.05 \\ 0 & 0 \end{bmatrix}, \quad D_{21} = \begin{bmatrix} 0 & 1.05 \\ 0 & 0 \end{bmatrix}, \quad D_{22} = \begin{bmatrix} 0 & 1.05 \\ 0 & 0 \end{bmatrix},$$

$$F_{11} = \begin{bmatrix} 0.003 \\ 0 \end{bmatrix}, \quad F_{21} = \begin{bmatrix} -0.01 \\ 0 \end{bmatrix}, \quad B_{11} = \begin{bmatrix} 0.001 \\ 0 \end{bmatrix},$$

$$B_{12} = \begin{bmatrix} 0.0005 \\ 0 \end{bmatrix}, \quad B_{21} = \begin{bmatrix} 0.2 \\ 0 \end{bmatrix}, \quad B_{22} = \begin{bmatrix} 0.12 \\ 0 \end{bmatrix},$$

$$C_{11} = C_{12} = \begin{bmatrix} 1 \\ 0 \end{bmatrix}, \quad C_{21} = C_{22} = \begin{bmatrix} 1 \\ 0 \end{bmatrix}.$$

The membership functions $h_i^1(\theta_i(t)),\ h_i^2(\theta_i(t))$ are chosen as

$$h_i^1(\theta_i(t)) = 0.5\left(\frac{1 + x_{i2}}{d_{i1}}\right)$$

and

$$h_i^2(\theta_i(t)) = 1 - h_i^1(\theta_i(t)).$$

And, assuming that $x_{i2} \in [-d_{i1}, d_{i1}]$, where $d_{i1} = -0.03$. The remaining parameters required for the derivation of control gain matrices are chosen as the time-varying delay is $\tau_i(t) = 0.3\sin(t)$, $b_1 = 0.2$, $\mu_i = 0.5$, and $\gamma = 0.2$.

With all the given constant coefficient matrices and parameters, the MATLAB LMI control toolbox is utilized to evaluate the feasibility of the solutions and the corresponding control gain matrices are derived as follows:

$$\begin{aligned} K_{11} &= 1.0e + 04 \times \begin{bmatrix} -8.8331 & -8.1036 \end{bmatrix}, \\ K_{12} &= 1.0e + 04 \times \begin{bmatrix} -8.8331 & -8.1036 \end{bmatrix}, \\ K_{21} &= \begin{bmatrix} -724.5071 & -849.2530 \end{bmatrix}, \\ K_{22} &= \begin{bmatrix} -724.5071 & -849.2530 \end{bmatrix}. \end{aligned}$$

Along with control gain matrices, the solution behavior of the closed-loop system is evaluated through Runge–Kutta fourth-order numerical approximation technique. And the corresponding results are demonstrated for more understanding of solution behavior. In this regard, Figs. 1 and 2 demonstrate the state responses of Subsystem-1. Figures 3 and 4 represent the solution behavior of Subsystem-2. The sliding surface with respect to state responses is depicted in Fig. 5. In addition, Figs. 6, 7 and 8 represent the state responses for various initial conditions and their corresponding sliding surface.

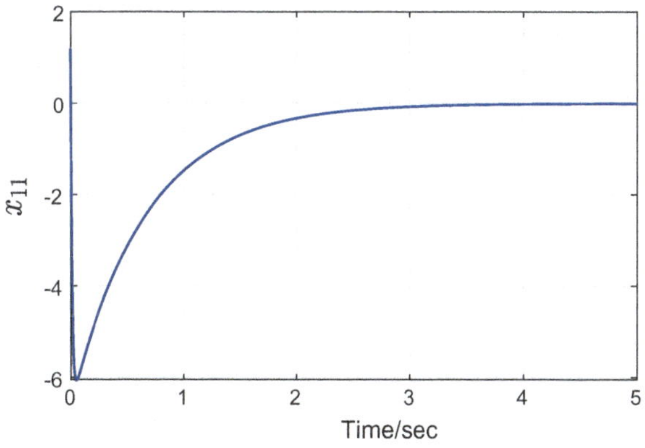

Fig. 1 State response x_1 of subsystem-1

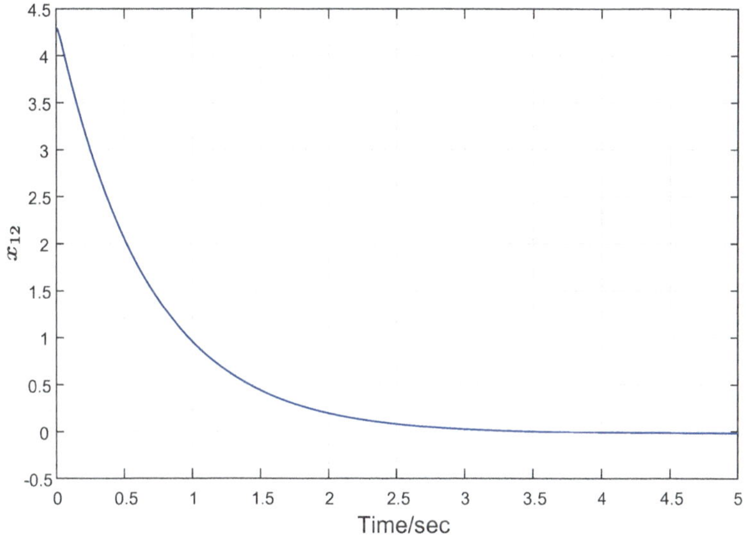

Fig. 2 State response x_2 of subsystem-1

6 Conclusion

In this chapter, the stability and stabilization of T-S fuzzy-based interconnected large-scale systems were investigated. Initially, the nonlinear system is equivalently expressed into a T-S fuzzy system via membership rules.

A suitable sliding surface is designed as a control law to ensure the global stability of the closed-loop system. Based on the Lyapunov stability theory, sufficient conditions are derived in terms of LMIs, which can be solved with the help of MATLAB LMI control toolbox. As a result, the appropriate control gain matrices are obtained

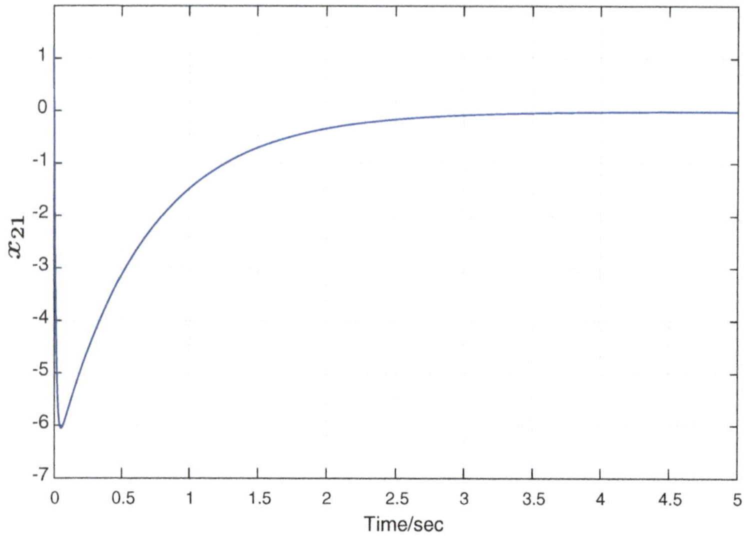

Fig. 3 State response x_1 of subsystem-2

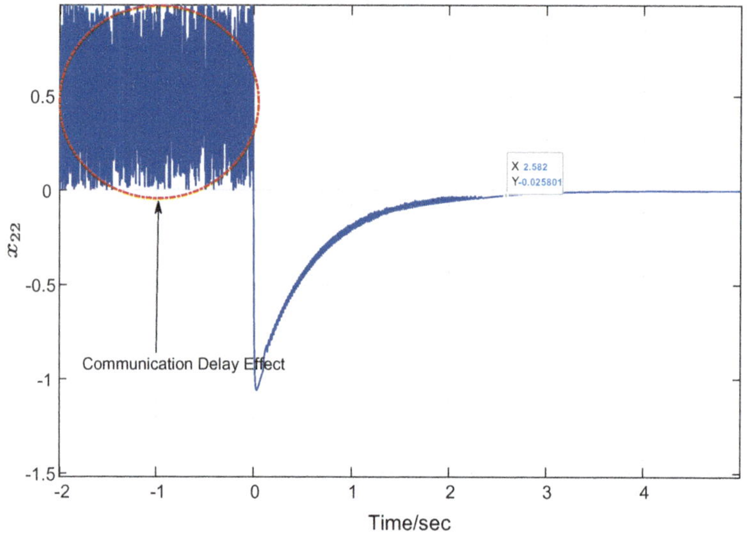

Fig. 4 State response x_2 of subsystem-2

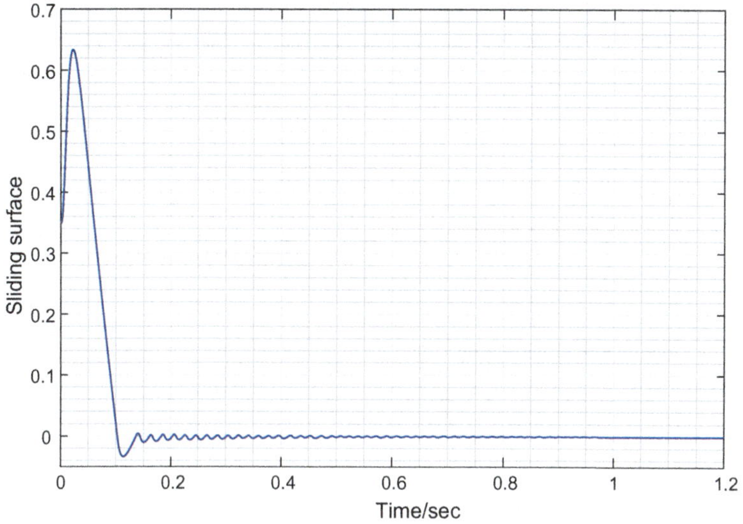

Fig. 5 Sliding surface

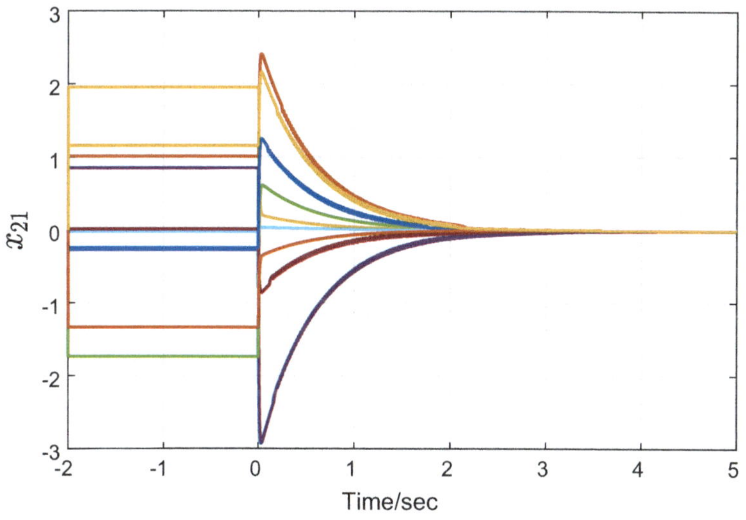

Fig. 6 State under different initial conditions

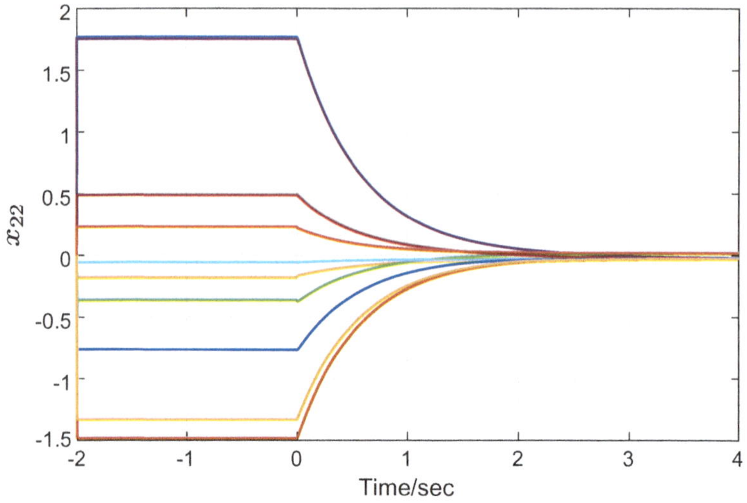

Fig. 7 State under different initial conditions

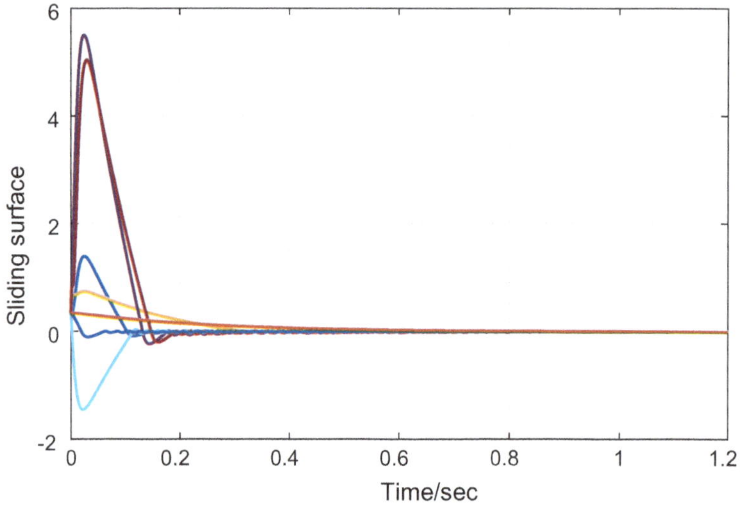

Fig. 8 Sliding motion under different initial conditions

for the constant coefficients and corresponding results are illustrated to demonstrate the efficiency of the proposed results.

References

1. Edwards, C., Spurgeon, S.: Sliding Mode Control: Theory and Applications. CRC Press, Boca Raton (1998)
2. Liu, J., Gao, Y., Yin, Y., Wang, J., Luo, W., Sun, G.: Basic Theory of Sliding Mode Control. Sliding Mode Control Methodology in the Applications of Industrial Power Systems, pp. 11–25. Springer, Cham, Switzerland (2020)
3. Tan, Y., Liu, Q., Du, D., Niu, B., Fei, S.: Observer-based Finite-time $\mathcal{H}_\infty$ control for interconnected fuzzy systems with quantization and random network attacks. IEEE Trans. Fuzzy Syst. (2019). https://doi.org/10.1109/TFUZZ.2019.2960719
4. Le, H.V., Chu, K.T.: Robust control of positive fractional-order interconnected systems with heterogeneous delays. Asian J. Control **21**(1), 596–608 (2019)
5. Xue, H., Zhang, J.: Robust exponential stability of interconnected switched systems with mixed delays and impulsive effect. Nonlinear Dyn. **97**(1), 679–696 (2019)
6. Gu, Z., Park, J.H., Yue, D., Wu, Z.G., Xie, X.: Event-triggered security output feedback control for networked interconnected systems subject to cyber-attacks. IEEE Trans. Syst. Man Cybern. Syst. (2019). https://doi.org/10.1109/TSMC.2019.2960115
7. Liu, Y., Fang, F., Park, J.H.: Decentralized dissipative filtering for delayed nonlinear interconnected systems based on T-S fuzzy model. IEEE Trans. Fuzzy Syst. **27**(4), 790–801 (2019)
8. Park, J.H.: Robust non-fragile decentralized controller design for uncertain large-scale interconnected systems with time-delays. ASME J. Dyn. Syst. Meas. Control **124**(2), 332–336 (2002)
9. Park, J.H.: Robust non-fragile guaranteed cost control of uncertain large-scale systems with time-delays in subsystem interconnections. Int. J. Syst. Sci. **35**(4), 233–241 (2004)
10. Kim, H.J., Koo, G.B., Park, J.B., Joo, Y.H.: Decentralized sampled-data $\mathcal{H}_\infty$ fuzzy filter for nonlinear large-scale systems. Fuzzy Sets Syst. **273**, 68–86 (2015)
11. Kim, H.J., Park, J.B., Joo, Y.H.: Decentralized $\mathcal{H}_\infty$ sampled-data fuzzy filter for nonlinear interconnected oscillating systems with uncertain interconnections. IEEE Trans. Fuzzy Syst. **28**(3), 487–498 (2020)
12. Zhao, B., Wang, D., Shi, G., Liu, D., Li, Y.: Decentralized control for large-scale nonlinear systems with unknown mismatched interconnections via policy iteration. IEEE Trans. Syst. Man Cybern. Syst. **48**(10), 1725–1735 (2017)
13. Wang, H., Liu, X., Liu, K.: Robust adaptive neural tracking control for a class of stochastic nonlinear interconnected systems. IEEE Trans. Neural Netw. Learn. Syst. **27**(3), 510–523 (2015)
14. Li, T., Wang, D., Li, J., Li, Y.: Adaptive decentralized NN control of nonlinear interconnected time-delay systems with input saturation. Asian J. Control **15**(2), 533–542 (2013)
15. Zhao, B., Li, Y., Liu, D.: Self-tuned local feedback gain based decentralized fault tolerant control for a class of large-scale nonlinear systems. Neurocomputing **235**, 147–156 (2017)
16. Li, S., Xiang, Z.: Sampled-data decentralized output feedback control for a class of switched large-scale stochastic nonlinear systems. IEEE Syst. J. (2019). https://doi.org/10.1109/JSYST.2019.2934512
17. Gao, R., Zhai, D., Cheng, J.: Decentralized static output feedback sliding mode control for interconnected descriptor systems via linear sliding variable. Appl. Math. Comput. **357**, 185–198 (2019)
18. Gao, R., Zhang, Q., Li, J., Zhai, D.: Integral sliding mode control for interconnected descriptor systems based on a reduced-order observer. Int. J. Syst. Sci. **50**(10), 1947–1960 (2019)

19. Takagi, T., Sugeno, M.: Fuzzy identification of systems and its applications to modeling and control. IEEE Trans. Syst. Man Cybern. **15**(1), 116–132 (1985)
20. Sugeno, M.: An introductory survey of fuzzy control. Inf. Sci. **36**(1–2), 59–83 (1985)
21. Yoneyama, J., Nishikawa, M., Katayama, H., Ichikawa, A.: Output stabilization of Takagi–Sugeno fuzzy systems. Fuzzy Sets Syst. **111**(2), 253–266 (2000)
22. Tseng, C.S.: A novel approach to $\mathcal{H}_\infty$ decentralized fuzzy-observer-based fuzzy control design for nonlinear interconnected systems. IEEE Trans. Fuzzy Syst. **16**(5), 1337–1350 (2008)
23. Deng, C., Yang, G.H.: Decentralized fault-tolerant control for a class of nonlinear large-scale systems with actuator faults. Inf. Sci. **382**, 334–349 (2017)
24. Chiang, C.C., Kuo, Z.H.: Decentralized adaptive fuzzy controller design of large-scale nonlinear systems with unmatched uncertainties. In 2002 IEEE World Congress on Computational Intelligence. 2002 IEEE International Conference on Fuzzy Systems. FUZZ-IEEE'02. Proceedings (Cat. No. 02CH37291), vol. 1, pp. 668–673. IEEE, Honolulu, HI (2002)
25. Hua, C.C., Wang, Q.G., Guan, X.P.: Exponential stabilization controller design for interconnected time delay systems. Automatica **44**(10), 2600–2606 (2008)
26. Hwang, C.L.: Decentralized fuzzy control of nonlinear interconnected dynamic delay systems via mixed $\mathcal{H}_2/\mathcal{H}_\infty$ optimization with Smith predictor. IEEE Trans. Fuzzy Syst. **19**(2), 276–290 (2010)
27. Zhang, K., Zhang, H., Mu, Y., Liu, C.: Decentralized tracking optimization control for partially unknown fuzzy interconnected systems via reinforcement learning method. IEEE Trans. Fuzzy Syst. (2020). https://doi.org/10.1109/TFUZZ.2020.2966418
28. Kharitonov, V.L., Zhabko, A.P.: Robust stability of time-delay systems. IEEE Trans. Autom. Control **39**(12), 2388–2397 (1994)
29. Zhang, H., Zhong, H., Dang, C.: Delay-dependent decentralized $\mathcal{H}_\infty$ filtering for discrete-time nonlinear interconnected systems with time-varying delay based on the T-S fuzzy model. IEEE Trans. Fuzzy Syst. **20**(3), 431–443 (2011)
30. Kim, H.J., Park, J.B., Joo, Y.H.: $\mathcal{H}_\infty$ fuzzy filter for non-linear sampled-data systems under imperfect premise matching. IET Control Theory Appl. **11**(5), 747–755 (2017)
31. Park, M., Kwon, O., Park, J.H., Lee, S., Cha, E.: Stability of time-delay systems via Wirtinger-based double integral inequality. Automatica **55**, 204–208 (2015)
32. Velmurugan, G., Rakkiyappan, R., Lakshmanan, S.: Passivity analysis of memristor-based complex-valued neural networks with time-varying delays. Neural Process. Lett. **42**(3), 517–540 (2015)
33. Gu, K., Chen, J., Kharitonov, V.L.: Stability of time-delay systems. Springer Science & Business Media, Berlin (2003)

Observer-Based Fault-Tolerant Control for Non-Infinitely Observable Descriptor Systems

Joseph Chang Lun Chan, Tae H. Lee, and Chee Pin Tan

Abstract This chapter considers a scheme to perform fault-tolerant control (FTC) in a class of non-infinitely observable descriptor systems (NIODS) that are affected by disturbances. The proposed scheme consists of a sliding mode observer (SMO) to estimate the states and unknown inputs, and a controller to perform FTC. The system is re-expressed as an infinitely observable reduced-order system by treating some states as unknown inputs. A SMO is implemented onto the reduced-order system to estimate the states and unknown inputs of the original system. These estimates are then passed into a controller that counteracts the effects of the unknown inputs. Linear matrix inequality techniques are used to design the SMO and controller such that the estimates from the SMO and the output of the system converge asymptotically. The necessary and sufficient conditions for the existence of the scheme are investigated and expressed in terms of the original system matrices. Finally, the scheme is implemented on a simulation of a practical system, and the results demonstrate its effectiveness.

Keywords Descriptor systems · Sliding mode observers · Fault reconstruction · Fault-tolerant control · Uncertain systems · Infinite observability

J. C. L. Chan · T. H. Lee (✉)
Division of Electronic Engineering, Jeonbuk National University, 567 Baekje-daero, Jeonju 54896, Republic of Korea
e-mail: thlee@jbnu.ac.kr

J. C. L. Chan
e-mail: pfjoseph@jbnu.ac.kr

C. P. Tan
School of Engineering and Advanced Engineering Platform, Monash University Malaysia, Selangor 47500, Malaysia
e-mail: tan.chee.pin@monash.edu

J. H. Park (ed.), *Recent Advances in Control Problems of Dynamical Systems and Networks*, Studies in Systems, Decision and Control 301,
https://doi.org/10.1007/978-3-030-49123-9_6

1 Introduction

Practical systems often experience faults, as well as unknown influences arising from modelling mismatches or parasitic dynamics (referred to hereafter as *disturbances*) [1, 2]. Faults and disturbances (both of which would be referred to collectively as *unknown inputs*) can potentially cause unacceptable deviations in performance, leading to costly downtime [3, 4]. Thus, it is important for corrective action to be taken to counter the effects of these unknown inputs. Observer-based controllers are a popular method for fault-tolerant control (FTC) [5], where the states and faults affecting the system are first estimated using an observer [4, 6–8]. These estimates are then passed into a controller which would counteract the faults' influence on the system.

Systems containing subsystems of differential equations linked together with algebraic equations can be represented as *descriptor systems* [9]. Descriptor systems (also referred to as *singular systems* [10, 11]) are found in many different domains; some examples include biological systems [12], chemical mixing tanks [13], RLC electronic circuits [14], and constrained mechanical systems [15]. A trait unique to descriptor systems is that they have multiple measures of observability - one of which is *infinite observability*, which implies all infinite modes of the system are measurable outputs [16]. Many observer schemes for FTC rely on the system being infinitely observable [7, 8, 10, 15, 17, 18], or make assumptions such that the performance of the observer can be expressed as an infinitely observable descriptor system [19–23].

Descriptor systems where some (or all) of the infinite modes are not measured are called *non-infinitely observable descriptor systems* (NIODS). There have been investigations on NIODS: Ooi et al. treated certain states as unknown inputs [24] and re-expressed some states as linear combinations of other signals [25], thereby reformulating the NIODS as an infinitely observable reduced-order system compatible with existing observer schemes. Chan et al. [3, 9] later extended these findings by considering the effects of disturbances on the fault, and designing the observer such that these effect are bounded, achieving robust fault reconstruction. Tan et al. [13, 14, 26] then considered the use of observers in cascade to relax the conditions required for the feasibility of the schemes. These works however only considered state and fault estimation in NIODS, but not FTC.

Motivated by these limitations, we propose an FTC scheme for NIODS, which consists of a sliding mode observer (SMO), and a state and unknown input feedback controller. Firstly, the system would be transformed into a form easing further analysis. Some states would be treated as unknown inputs to re-express the system as an infinitely observable reduced-order system compatible with existing SMOs for descriptor systems. The SMO is then designed to estimate the states and unknown inputs affecting the system. The estimates are then fed into the controller, which is designed such that the original descriptor system converges asymptotically, thereby achieving *fault-tolerant control*. Linear matrix inequalities (LMIs) would be used to design the SMO and controller such that they are stable.

The chapter is structured as follows: Sect. 2 presents the problem, and the preliminary transformations applied to the system. Section 3 then introduces the SMO to estimate the states and unknown inputs in the original system, and studies its performance. The controller for FTC is then investigated in Sect. 4. Necessary and sufficient conditions for the existence of the FTC scheme, as well as its design procedures are detailed in Sect. 5. A simulated example is then given in Sect. 6 to verify the performance of the proposed scheme, and lastly Sect. 7 draws some conclusions. The following notation is used in this chapter: the norm of a vector $x \in \mathbb{R}^a$ is defined as $\|x\| = \sqrt{\sum_{i=1}^{a} x_i^2}$, and the spectrum of a matrix A is denoted as $\lambda(A)$.

2 Preliminaries

Consider the following non-infinitely observable descriptor system (NIODS):

$$E\dot{x}_0 = Ax_0 + Bu + Mf + Q\xi, \tag{1}$$

$$y = Cx_0, \tag{2}$$

where $E, A \in \mathbb{R}^{n\times n}$, $B \in \mathbb{R}^{n\times m}$, $M \in \mathbb{R}^{n\times q}$, $Q \in \mathbb{R}^{n\times h}$, and $C \in \mathbb{R}^{p\times n}$ are known and constant, whilst $x_0 \in \mathbb{R}^n$, $u \in \mathbb{R}^m$, $y \in \mathbb{R}^p$, $f \in \mathbb{R}^q$, and $\xi \in \mathbb{R}^h$ represent the states, inputs, outputs, faults, and disturbances, respectively.

Without loss of generality, E and C are assumed to be rank-deficient and full row rank, respectively, i.e., $\text{rank}\,(E) = r < n$, $\text{rank}\,(C) = p$, and the system is *non-infinitely observable*, i.e., $\text{rank} \begin{bmatrix} E \\ C \end{bmatrix} = \bar{n} < n$ [16].

The objective is to design u such that the output y converges asymptotically, i.e., as $t \to \infty$, $y \to 0$.

The input u is designed based on the state and unknown input-feedback, which would require knowledge of x, f, and ξ.

To estimate these signals in real-time, a SMO scheme similar to that by Chan et al. [3] is utilized.

System (1)–(2) is first re-expressed into a form that facilitates further analysis using the following lemma and proposition.

Lemma 1 *Let* (1) *be pre-multiplied by* T_1 *and state* x_0 *be transformed such that* $x_0 \mapsto T_2x_0$, *where* T_1 *and* T_2 *are non-singular matrices (referred to hereafter as a* state equation transformation *and* state transformation, *respectively). These transformations preserve the dynamic properties of system* (1)–(2). ♯

Proof System (1)–(2) will be transformed into the form

$$\left(T_1ET_2^{-1}\right)(T_2\dot{x}_0) = \left(T_1AT_2^{-1}\right)(T_2x_0) + (T_1B)\,u + (T_1M)\,f + (T_1Q)\,\xi, \tag{3}$$

$$y = \left(CT_2^{-1}\right)(T_2x_0). \tag{4}$$

It can be seen that the matrices (E, A, B, M, Q, C) have been transformed. These transformations change these matrices into forms that are more convenient for later design and analysis, but preserve the dynamic properties of the original system (1)–(2).

The inputs and outputs to the system are unaffected by the transformations as well. The transfer function $G(s)$ of a descriptor system is given by $G(s) = C(sE - A)^{-1}B$. Therefore, if T_1 and T_2 are non-singular, the transfer functions of systems (1)–(2) and (3)–(4) are identical.

The dynamic properties of the systems, e.g., inputs, outputs, finite poles (the finite values of $s \in \mathbb{C}$ satisfying $|sE - A| = 0$ [16]), and dynamical order (the ranks of E and $T_1 E T_2^{-1}$ [15]) are thus said to be *equivalent*. □

Introduce a non-singular matrix $T_a = \begin{bmatrix} N_C^T \\ C \end{bmatrix}$, where $CN_C = 0$. Hence, we have

$$E \mapsto ET_a^{-1} = \begin{bmatrix} E_{a1} & E_{a2} \end{bmatrix}, \quad C \mapsto CT_a^{-1} = \begin{bmatrix} 0 & I_p \end{bmatrix}. \tag{5}$$

Since $\operatorname{rank} \begin{bmatrix} E \\ C \end{bmatrix} = \bar{n}$, (5) implies $\operatorname{rank}(E_{a1}) = \bar{n} - p$.

Apply singular-value decomposition (SVD) on E_{a1} to obtain $T_b E_{a1} T_{c1} = \begin{bmatrix} 0 & I_{\bar{n}-p} \\ 0 & 0 \end{bmatrix}$, where T_b and T_{c1} are non-singular.

Define another non-singular matrix $T_c = \begin{bmatrix} T_{c1}^{-1} & 0 \\ 0 & I_p \end{bmatrix}$, and apply the state transformation $T_c T_a$ such that

$$x_0 \mapsto (T_c T_a)\, x_0 = x = \begin{bmatrix} x_1 \\ x_2 \\ y \end{bmatrix} \begin{matrix} \updownarrow n - \bar{n} \\ \updownarrow \bar{n} - p \\ \updownarrow p \end{matrix}. \tag{6}$$

In the coordinates of (6), (E, A, M, Q) would be be transformed into

$$\begin{aligned} \begin{bmatrix} E & A & M & Q \end{bmatrix} &= \begin{bmatrix} T_b E\,(T_c T_a)^{-1} & T_b A\,(T_c T_a)^{-1} & T_b M & T_b Q \end{bmatrix} \\ &= \left[\begin{array}{ccc|ccc|c|c} 0 & I_{\bar{n}-p} & E_{c1} & A_{c1} & A_{c2} & A_{c3} & M_{c1} & Q_{c1} \\ 0 & 0 & E_{c2} & A_{c4} & A_{c5} & A_{c6} & M_{c2} & Q_{c2} \end{array}\right], \end{aligned} \tag{7}$$

where $A_{c2} \in \mathbb{R}^{(\bar{n}-p)\times(\bar{n}-p)}$, $A_{c4} \in \mathbb{R}^{(n-\bar{n}+p)\times(n-\bar{n})}$, and $M_2 \in \mathbb{R}^{(n-\bar{n}+p)\times q}$.

Proposition 1 *Assume there exists a non-singular matrix* $T_{d1} \in \mathbb{R}^{(n-\bar{n}+p)\times(n-\bar{n}+p)}$ *such that*

$$T_{d1} \begin{bmatrix} M_{c2} & A_{c4} & Q_{c2} \end{bmatrix} = \left[\begin{array}{c|c|c} I_q & 0 & 0 \\ 0 & I_{n-\bar{n}} & 0 \\ 0 & 0 & I_h \\ 0 & 0 & 0 \end{array}\right] \begin{matrix} \updownarrow q \\ \updownarrow n - \bar{n} \\ \updownarrow h \\ \updownarrow p - q - h \end{matrix}, \tag{8}$$

where the satisfaction of (8) *would be addressed later in Proposition 3.*

Then, there exists a set of transformations for system (1)–(2) *such that it would have the structure*

$$\begin{bmatrix} 0 & I_{\bar{n}-p} & E_1 \\ 0 & 0 & E_2 \end{bmatrix} \dot{x} = \begin{bmatrix} 0 & A_2 & A_3 \\ A_4 & A_5 & A_6 \end{bmatrix} x + \begin{bmatrix} B_1 \\ B_2 \end{bmatrix} u + \begin{bmatrix} 0 \\ M_2 \end{bmatrix} f + \begin{bmatrix} 0 \\ Q_2 \end{bmatrix}, \tag{9}$$

$$y = \begin{bmatrix} 0 & 0 & I_p \end{bmatrix} x, \tag{10}$$

$$\begin{bmatrix} M_2 & A_4 & A_5 & A_6 & Q_2 \end{bmatrix} = \left[\begin{array}{c|c|c|c|c} I_q & 0 & A_{51} & A_{61} & 0 \\ 0 & I_{n-\bar{n}} & A_{52} & A_{62} & 0 \\ 0 & 0 & A_{53} & A_{63} & I_h \\ 0 & 0 & A_{54} & A_{64} & 0 \end{array}\right], \tag{11}$$

where $A_{54} \in \mathbb{R}^{(p-q-h)\times(\bar{n}-p)}$. ♯

Proof Equation (8) implies that a matrix $T_{d2} \in \mathbb{R}^{(\bar{n}-p)\times(n-\bar{n}+p)}$ exists such that $\begin{bmatrix} A_{c1} & M_{c1} & Q_{c1} \end{bmatrix} = T_{d2} \begin{bmatrix} A_{c4} & M_{c2} & Q_{c2} \end{bmatrix}$. Hence, define the non-singular matrix $T_d = \begin{bmatrix} I_{\bar{n}-p} & -T_{d2} \\ 0 & T_{d1} \end{bmatrix}$.

Apply the state equation transformation $T_d T_b$ and state transformation $T_c T_a$ onto system (1)–(2) to obtain the structures in (9)–(11), thus completing the proof. □

Remark 1 Assumption (8) is a *matching condition*, and is used to decouple the dynamics of the unknown inputs from the states. It is not restrictive, and has been commonly assumed in various works for SMOs [3, 9, 13, 14, 17, 27–29]. ♯

Define a non-singular matrix T, where

$$T = \begin{bmatrix} I_{\bar{n}-p} & 0 \\ 0 & T_p \\ 0 & \bar{T}_p \end{bmatrix}, \quad T_p = \begin{bmatrix} I_{q+n-\bar{n}+h} & 0 \\ 0 & T_{44} \end{bmatrix}, \tag{12}$$

where $T_{44} \in \mathbb{R}^{(p-q-n+\bar{n}-h)\times(p-q-h)}$ is chosen to be full-row rank.

Apply a state equation transformation onto (9), and treat x_1 as an unknown input. Define $\bar{x} = \begin{bmatrix} x_2 \\ y \end{bmatrix}$ and $\bar{f} = \begin{bmatrix} f \\ x_1 \end{bmatrix}$. System (9)–(11) can then be reformulated as

$$\underbrace{\begin{bmatrix} I_{\bar{n}-p} & E_1 \\ 0 & T_p E_2 \end{bmatrix}}_{\bar{E}} \dot{\bar{x}} = \underbrace{\begin{bmatrix} A_2 & A_3 \\ T_p A_5 & T_p A_6 \end{bmatrix}}_{\bar{A}} \bar{x} + \underbrace{\begin{bmatrix} B_1 \\ T_p B_2 \end{bmatrix}}_{\bar{B}} u + \underbrace{\begin{bmatrix} 0 & 0 & 0 \\ T_p M_2 & T_p A_4 & T_p Q_2 \end{bmatrix}}_{\bar{M}} \bar{f}, \tag{13}$$

$$y = \underbrace{\begin{bmatrix} 0 & I_p \end{bmatrix}}_{\bar{C}} \bar{x}. \tag{14}$$

It can be seen that

$$\operatorname{rank} \begin{bmatrix} \bar{E} \\ \bar{C} \end{bmatrix} = \bar{n}, \tag{15}$$

i.e., $\begin{bmatrix} \bar{E} \\ \bar{C} \end{bmatrix}$ is full-column rank, which means that system (13)–(14) is *infinitely observable*. The SMO by Yeu et al. [17] can then be implemented on system (13)–(14) to estimate its states and the unknown inputs.

3 The SMO for State and Unknown Input Estimation

The SMO by Yeu et al. for system (13)–(14) has the following structure [17]:

$$\dot{z} = \left(R\bar{A} - G_l\bar{C}\right) z - R\bar{B}u - \left(G_l\left(I_p - \bar{C}V\right) + R\bar{A}V\right) y - G_n\nu, \tag{16}$$

$$\hat{\bar{x}} = Vy - z, \tag{17}$$

$$\nu = -\rho\frac{e_y}{\|e_y\|}, \; e_y = \bar{C}\hat{\bar{x}} - y, \tag{18}$$

where $\hat{x} = \begin{bmatrix} \hat{x}_2 \\ \hat{y} \end{bmatrix} \begin{matrix} \updownarrow \bar{n} - p \\ \updownarrow p \end{matrix}$.

The matrices $R \in \mathbb{R}^{\bar{n}\times\bar{n}}$ and $V, G_l, G_n \in \mathbb{R}^{\bar{n}\times p}$ are chosen to have the structures

$$R\bar{E} + V\bar{C} = I_{\bar{n}}, \; R = \begin{bmatrix} I_{\bar{n}-p} & R_2 \\ 0 & I_p \end{bmatrix}, \; \begin{bmatrix} G_l & G_n \end{bmatrix} = \left[\begin{array}{c|c} A_3 + R_4T_pA_6 & 0 \\ A_o + T_pA_6 & G_{n2} \end{array}\right], \tag{19}$$

where $R_2 = \begin{bmatrix} 0 & R_{24} \end{bmatrix}$, $R_{24} \in \mathbb{R}^{(\bar{n}-p)\times(p-q-n+\bar{n}-h)}$, $A_o \in \mathbb{R}^{p\times p} = A_o^T > 0$, and $G_{n2} \in \mathbb{R}^{p\times p} = G_{n2}^T > 0$.

3.1 Observer Error Analysis

Substituting (17) into (16) and rearranging gives

$$\dot{\hat{\bar{x}}} = \left(R\bar{A} - G_l\bar{C}\right)\hat{\bar{x}} + R\bar{B}u + G_l\bar{C}\bar{x} + G_n\nu + V\dot{y}. \tag{20}$$

Pre-multiply (13) with R and add $V\dot{y}$, and rearrange using (15) to get

$$R\bar{E}\dot{\bar{x}} + V\dot{y} = \left(R\bar{E} + V\bar{C}\right)\dot{\bar{x}} = R\bar{A}\bar{x} + R\bar{B}u + R\bar{M}\bar{f} + V\dot{y}. \tag{21}$$

Define $e = \hat{\bar{x}} - \bar{x} = \begin{bmatrix} e_2 \\ e_y \end{bmatrix} \begin{matrix} \updownarrow \bar{n} - p \\ \updownarrow p \end{matrix}$, and subtract (21) from (20) to obtain the error equation (which characterizes the performance of the observer):

$$\dot{e} = \left(R\bar{A} - G_l\bar{C}\right) e - R\bar{M}\bar{f} + G_n\nu. \tag{22}$$

Define $\bar{A}_2 = A_2 + R_{24}T_{44}A_{54}$. By substituting for $(\bar{A}, \bar{M}, \bar{C})$ from (13)–(14), R, G_l, and G_n from (19), and T_p from (12), Eq. (22) becomes

$$\begin{bmatrix} \dot{e}_2 \\ \dot{e}_y \end{bmatrix} = \begin{bmatrix} \bar{A}_2 & 0 \\ T_pA_5 & -A_o \end{bmatrix} \begin{bmatrix} e_2 \\ e_y \end{bmatrix} - \begin{bmatrix} 0 & 0 & 0 \\ T_pM_2 & T_pA_4 & T_pQ_2 \end{bmatrix} \bar{f} + \begin{bmatrix} 0 \\ G_{n2} \end{bmatrix} \nu. \tag{23}$$

Proposition 2 *Define a matrix $P_2 \in \mathbb{R}^{(\bar{n}-p)\times(\bar{n}-p)} = P_2^T > 0$, and a positive scalar γ. Let $\bar{P}_2 = P_2R_{24}$. Suppose for a given γ there exists a set of values for P_2 and $\bar{P}_2$ that satisfies the following LMI:*

$$P_2A_2 + \bar{P}_2T_{44}A_{54} + \left(P_2A_2 + \bar{P}_2T_{44}A_{54}\right)^T + \gamma P_2 < 0. \tag{24}$$

Also, let ρ from (18) *satisfy*

$$\rho \geq \|G_{n2}^{-1}T_pA_5\|\alpha + \|G_{n2}^{-1}T_p\left[M_2 \; A_4 \; Q_2\right]\|\beta + \eta, \tag{25}$$

where $\alpha \geq \|e_2(0)\|e^{-\frac{\gamma t}{2}}\sqrt{\frac{\lambda_{\max}(P_2)}{\lambda_{\min}(P_2)}}$, $\beta > \|f\|_{\max} + \|x_1\|_{\max} + \|\xi\|_{\max}$, and $\eta \in \mathbb{R}^+$.

Then, an ideal sliding motion would take place on surface $\mathcal{S} = \{e : \bar{C}e = 0\}$ in finite time. ♯

Proof The proof consists of two parts: the first portion aims to show how e_2 is bounded by α.

Define the Lyapunov candidate function $W_2 = e_2^TP_2e_2 > 0$, and differentiate it with respect to time to obtain

$$\dot{W}_2 = e_2^T\left(P_2A_2 + P_2R_{24}T_{44}A_{54} + (P_2A_2 + P_2R_{24}T_{44}A_{54})^T\right)e_2. \tag{26}$$

Substituting for $\bar{P}_2 = P_2R_{24}$ into (24) yields

$$P_2A_2 + P_2R_{24}T_{44}A_{54} + (P_2A_2 + P_2R_{24}T_{44}A_{54})^T + \gamma P_2 < 0. \tag{27}$$

It can be seen from comparing (26)–(27) that satisfying LMI (24) implies $\dot{W}_2 < -\gamma W_2$. This in turn implies $\|e_2(t)\| \leq \|e_2(0)\|e^{-\frac{\gamma t}{2}}\sqrt{\frac{\lambda_{\max}(P_2)}{\lambda_{\min}(P_2)}} \leq \alpha$ [29], completing the first part of the proof.

The next and final part of the proof shows how setting ρ to satisfy (25) would result in sliding motion on $\mathcal{S}$ in finite time.

Define another Lyapunov candidate function $W_y = e_y^TG_{n2}^{-1}e_y$, and differentiate it with respect to time to obtain

$$\begin{aligned} \dot{W}_y = & -2e_y\left(G_{n2}^{-1}A_o + A_o^TG_{n2}^{-1}\right)e_y \\ & +2e_y^T\left(G_{n2}^{-1}T_pA_5e_2 - G_{n2}^{-1}T_p\left[M_2 \; A_4 \; Q_2\right]\bar{f} + \nu\right). \end{aligned} \tag{28}$$

Since $G_{n2}^{-1}A_o > 0$ is symmetrical and $\|e_2\| \leq \alpha$, (28) can be re-expressed as

$$\dot{W}_y \leq -2\|e_y\| \left(\rho - \|G_{n2}^{-1} T_p A_5\|\alpha - \|G_{n2}^{-1} T_p \left[M_2 \; A_4 \; Q_2\right] \|\beta\right). \tag{29}$$

Also, notice that

$$\begin{aligned}\|e_y\|^2 &= \left(e_y\sqrt{G_{n2}^{-1}}\right)^T G_{n2} \left(e_y\sqrt{G_{n2}^{-1}}\right) \\ &\geq \lambda_{\min}\left(G_{n2}\right) \left\|e_y\sqrt{G_{n2}^{-1}}\right\|^2 = \lambda_{\min}\left(G_{n2}\right) W_y.\end{aligned}$$

Therefore, by setting ρ to satisfy (25), Eq. (29) becomes

$$\dot{W}_y \leq -2\eta\|e_y\| \leq -2\eta\sqrt{\lambda_{\min}\left(G_{n2}\right)}\sqrt{W}_y, \tag{30}$$

which is the *reachability condition* [28]. This results in $e_y = 0$ (and sliding motion taking place) in finite time, thus completing the proof. □

After sliding motion takes place, $e_y, \dot{e}_y = 0$, and error system (23) becomes

$$\dot{e}_2 = \bar{A}_2 e_2, \tag{31}$$

$$0 = T_p A_5 e_2 - T_p \left[M_2 \; A_4 \; Q_2\right] \bar{f} + G_{n2}\nu_{eq}, \tag{32}$$

where ν_{eq} is the *equivalent output error injection* which is needed to maintain the sliding motion on $\mathcal{S}$.

Then, define a measurable signal $\bar{\nu} = G_{n2}\nu_{eq}$, and substitute for A_5 and Q_2 from (11) and T_p from (12) to obtain

$$\bar{\nu} = \begin{bmatrix}\nu_1\\ \nu_2\\ \nu_3\\ \nu_4\end{bmatrix} = -\begin{bmatrix}A_{51}\\ A_{52}\\ A_{53}\\ A_{54}\end{bmatrix} e_2 + \begin{bmatrix}I_q & 0 & 0\\ 0 & I_{n-\bar{n}} & 0\\ 0 & 0 & I_h\\ 0 & 0 & 0\end{bmatrix}\begin{bmatrix}f\\ x_1\\ \xi\end{bmatrix}, \tag{33}$$

where $\nu_4 \in \mathbb{R}^{p-q-n+\bar{n}-h}$.

Define the estimates for f, x_1, and ξ as

$$\hat{f} = \nu_1 + L_f\nu_4, \; \hat{x}_1 = \nu_2 + L_1\nu_4, \; \hat{\xi} = \nu_3 + L_x\nu_4, \tag{34}$$

respectively, where $L_f \in \mathbb{R}^{q\times(p-q-n+\bar{n}-h)}$, $L_1 \in \mathbb{R}^{(n-\bar{n})\times(p-q-n+\bar{n}-h)}$, and $L_x \in \mathbb{R}^{h\times(p-q-n+\bar{n}-h)}$.

Next, define the estimation errors $e_f = \hat{f} - f$, $e_1 = \hat{x}_1 - x_1$, and $e_\xi = \hat{\xi}_1 - \xi$. Then, from (31) and (33) it follows that

$$\dot{e}_2 = \bar{A}_2 e_2, \tag{35}$$

$$\begin{bmatrix} e_1 \\ e_2 \\ e_f \\ e_\xi \end{bmatrix} = \begin{bmatrix} -A_{52} - L_1 T_{44} A_{54} \\ I_{\bar{n}-p} \\ -A_{51} - L_f T_{44} A_{54} \\ -A_{53} - L_x T_{44} A_{54} \end{bmatrix} e_2. \tag{36}$$

From Proposition 2, e_2 is asymptotically stable (i.e., as $t \to \infty,\ e_2 \to 0$ and $\hat{x}_2 \to x_2$), which implies $e_1,\ e_f$, and e_ξ are also asymptotically stable, and therefore $\hat{x}_1 \to x_1,\ \hat{f} \to f$, and $\hat{\xi} \to \xi$. Hence, one possible choice for $L_1,\ L_f$, and L_x is

$$A_{51} + L_f T_{44} A_{54} \approx 0,\ A_{52} + L_1 T_{44} A_{54} \approx 0,\ A_{53} + L_x T_{44} A_{54} \approx 0. \tag{37}$$

Thus, the design of the SMO to estimate states and unknown inputs in system (1)–(2) has been completed.

4 The Controller for Fault-Tolerant Control

In this section, the state and unknown input estimates obtained from the SMO are fed into a controller, which would generate the control input u.

The gains of the controller would be designed such that the states of the system converge asymptotically, rejecting the influence of the unknown inputs.

Let the input u be given by

$$u = -K_x \hat{x}_0 - K_f \hat{f} - K_\xi \hat{\xi}, \tag{38}$$

where $K_x \in \mathbb{R}^{m\times n},\ K_f \in \mathbb{R}^{m\times q}$, and $K_\xi \in \mathbb{R}^{m\times h}$ are the gains of the controller (which are to be designed), and $\hat{x}_0$ are state estimates for the system in the coordinates of (1)–(2).

4.1 Error Analysis and Controller Design

Define $e_x = \hat{x}_0 - x_0$, and suppose K_f and K_ξ designed to satisfy

$$K_f = B^\dagger M,\ K_\xi = B^\dagger Q, \tag{39}$$

where $BB^\dagger \begin{bmatrix} M & Q \end{bmatrix} = \begin{bmatrix} M & Q \end{bmatrix}$.

Applying u onto system (1)–(2) yields

$$E\dot{x}_0 = (A - BK_x)\,x_0 + \begin{bmatrix} -BK_x & -M & -Q \end{bmatrix} \begin{bmatrix} e_x \\ e_f \\ e_\xi \end{bmatrix}, \tag{40}$$

$$y = Cx_0. \tag{41}$$

Recall from (6) that $x = T_c T_a x_0 = \begin{bmatrix} x_1 \\ x_2 \\ y \end{bmatrix}$, where $T_c T_a$ is non-singular. Then, define $e_{xca} = \hat{x} - x = T_c T_a \left(\hat{x}_0 - x_0\right) = \begin{bmatrix} e_1 \\ e_2 \\ e_y \end{bmatrix}$.

After sliding motion takes place, $e_x = (T_c T_a)^{-1} \begin{bmatrix} e_1 \\ e_2 \\ 0 \end{bmatrix}$.

Then, using (36), e_x, e_f, and e_ξ can be expressed in terms of e_2 as follows:

$$\begin{bmatrix} e_x \\ e_f \\ e_\xi \end{bmatrix} = \begin{bmatrix} (T_c T_a)^{-1} & 0 \\ 0 & I_{q+h} \end{bmatrix} \begin{bmatrix} e_1 \\ e_2 \\ 0 \\ \hline e_f \\ e_\xi \end{bmatrix} = \underbrace{\begin{bmatrix} (T_c T_a)^{-1} & 0 \\ 0 & I_{q+h} \end{bmatrix} \begin{bmatrix} -A_{52} - L_1 T_{44} A_{54} \\ I_{\bar{n}-p} \\ 0 \\ \hline -A_{51} - L_f T_{44} A_{54} \\ -A_{53} - L_x T_{44} A_{54} \end{bmatrix}}_{B_{x2}} e_2. \tag{42}$$

Substituting e_x, e_f, and e_ξ from (42) into (40) yields

$$E\dot{x}_0 = (A - BK_x)\,x_0 + \underbrace{\begin{bmatrix} -BK_x & -M & -Q \end{bmatrix} B_{x2}}_{B_x}\, e_2, \tag{43}$$

$$y = Cx_0. \tag{44}$$

The aim is to now design the controller gain K_x such that system (43)–(44) is asymptotically stable.

The following lemma and theorem presents the design for the controller:

Lemma 2 (Lemma 2 [7]) *All non-singular $Z \in \mathbb{R}^{a \times a}$ satisfying $Z^T Y^T = YZ \geq 0$ can be parameterized as $Z = Z_1 Y^T + N_Y Z_2$, where $Z_1 \in \mathbb{R}^{a \times a} > 0$ and $Z_2 \in \mathbb{R}^{(a-b) \times a}$, and $N_Y \in \mathbb{R}^{a \times (a-b)}$ is a matrix satisfying $YN_Y = 0$ and $\mathrm{rank}\,(N_Y) = a - \mathrm{rank}(Y) = a - b$.* ♯

Theorem 1 *Define matrices $H \in \mathbb{R}^{n \times n} > 0$, $J \in \mathbb{R}^{(n-r) \times n}$, and $S \in \mathbb{R}^{m \times n}$. Suppose there exists a set of values for H, J, and S satisfying the following LMI:*

$$AHE^T + AN_E J - BS + \left(AHE^T + AN_E J - BS\right)^T < 0, \tag{45}$$

and K_f and K_ξ are designed to satisfy (39), *and*

$$K_x = S\left(HE^T + N_E J\right)^{-1}, \tag{46}$$

where $EN_E = 0$. *Then, system* (1)–(2) *is asymptotically stable.* ♯

Proof The proof is divided into two portions: the first part re-expresses LMI (45) into a form that facilitates analysis.

Using Lemma 2, there exists a non-singular matrix $U_0 \in \mathbb{R}^{n\times n}$ satisfying

$$U_0 = HE^T + N_E J,\ U_0^T E^T = EU_0 \geq 0. \tag{47}$$

Substitute for S from (46) and U_0 into LMI (45) to obtain

$$\begin{aligned}(A - BK_x)\,U_0 + U_0^T\,(A - BK_x)^T &< 0,\\ U_0^T E^T &= EU_0 \geq 0.\end{aligned} \tag{48}$$

Define $U = U_0^{-1}$, and then pre-multiply (48) with U^T and post-multiply with U to obtain

$$\Delta_0 = U^T\,(A - BK_x) + (A - BK_x)^T\,U < 0, \tag{49}$$

$$E^T U = U^T E \geq 0. \tag{50}$$

LMI (45) has been re-expressed in a form that is similar to that for Theorem 2 in [7], concluding the first part of the proof.

The next and final portion of the proof shows how asymptotic stability is achieved by satisfying LMI (45).

Define the Lyapunov candidate functions

$$W_0 = x_0^T U^T E x_0,\ W_2 = e_2^T N e_2,\ W_c = W_0 + \delta W_2, \tag{51}$$

where $N \in \mathbb{R}^{(\bar{n}-p)\times(\bar{n}-p)} = N^T > 0$ and $\delta \in \mathbb{R}^+$.

Using (51), W_c can be re-expressed as $W_c = \begin{bmatrix} x_0^T & e_2^T \end{bmatrix}\begin{bmatrix} U^T E & 0 \\ 0 & \delta N \end{bmatrix}\begin{bmatrix} x_0 \\ e_2 \end{bmatrix}$, which implies

$$W_c \leq \underbrace{\lambda_{\max}\left(\begin{bmatrix} U^T E & 0 \\ 0 & \delta N \end{bmatrix}\right)}_{\theta_1}\left\|\begin{bmatrix} x_0 \\ e_2 \end{bmatrix}\right\|^2. \tag{52}$$

where $\theta_1 \in \mathbb{R}^+$.

Differentiating W_0 and W_2 with respect to time, substituting for $E\dot{x}_0$ from (43), and using (35) and (49)–(50) yields

$$\dot{W}_0 = x_0^T \Delta_0 x_0 + 2x_0^T U^T B_x e_2,\ \dot{W}_2 = e_2^T \underbrace{\left(N\bar{A}_2 + \bar{A}_2^T N\right)}_{\Delta_2} e_2. \tag{53}$$

It can be seen that LMI (45) implies (through (49)–(50)) that $\Delta_0 < 0$.

Furthermore, since $\bar{A}_2$ has been designed to be stable in Proposition 6, $\Delta_2 < 0$ as well. Therefore, define $\mu_0 = \lambda_{\min}(-\Delta_0)$ and $\mu_2 = \lambda_{\min}(-\Delta_2)$, where $\mu_0, \mu_2 \in \mathbb{R}^+$.

Differentiating W_c with respect to time yields

$$\begin{aligned}\dot{W}_c &= x_0^T \Delta_0 x_0 + 2x_0^T U^T B_x e_2 + \delta e_2^T \Delta_2 e_2 \\ &\leq -\mu_0 \|x_0\|^2 + \underbrace{2\|U^T B_x\|}_{\mu_x} \|x_0\| \|e_2\| - \delta \mu_2 \|e_2\|^2.\end{aligned} \tag{54}$$

Then, set $\delta > \frac{\mu_x^2}{\mu_0 \mu_2}$, and inequality $-\frac{1}{2}\left(\mu_0 \|x_0\|^2 + \delta \mu_2 \|e_2\|^2\right) \leq -\sqrt{\mu_0 \mu_2 \delta}\|x_0\| \|e_2\|$ implies (54) can be re-expressed as

$$\dot{W}_c \leq -\frac{\mu_0}{2}\|x_0\|^2 - \frac{\mu_x^2}{2\mu_0}\|e_2\|^2. \tag{55}$$

Let $\theta_2 = \min\left\{\frac{\mu_0}{2}, \frac{\mu_x^2}{2\mu_0}\right\}$, which implies $\dot{W}_c \leq -\theta_2 \left\| \begin{bmatrix} x_0 \\ e_2 \end{bmatrix} \right\|^2$.
From (52), $\dot{W}_c$ can then be re-expressed as $\dot{W}_c \leq -\frac{\theta_2}{\theta_1} W_c$, which in turn implies that as $t \to \infty$, $x_0 \to 0$ [7]. Hence, the satisfaction of LMI (45) results in the asymptotic stability of x_0, completing the proof. □

The controller for state and unknown input feedback has also been designed, concluding the error analysis of the FTC scheme.

5 Design of the Fault-Tolerant Control Scheme

Several assumptions were made during the analysis of the FTC scheme in previous sections. These assumptions are made in terms of design matrices, and are difficult to verify without undertaking the design process. Hence, it is important to re-express them in terms of the original system matrices, so that the designer knows from the outset if the proposed scheme is applicable to their system.

5.1 Existence Conditions

Theorem 2 *The proposed FTC scheme can be applied onto system* (1)–(2) *such that it is asymptotically stable if and only if the following conditions hold:*

A1. $rank \begin{bmatrix} E & A & M & Q \\ C & 0 & 0 & 0 \\ 0 & E & 0 & 0 \\ 0 & C & 0 & 0 \end{bmatrix} = n + \bar{n} + q + h.$

A2. $p \geq q + n - \bar{n} + h.$

A3. $rank \begin{bmatrix} B & M & Q \end{bmatrix} = rank(B).$

The following condition is necessary, and also sufficient if $p > q + n - \bar{n} + h$*:*

A4. $rank \begin{bmatrix} sE - A & M \\ C & 0 \end{bmatrix} = n + q \; \forall \; s \in \mathbb{C}^+.$

If $p = q + n - \bar{n} + h$*, the sufficient condition is that*

A5. The eigenvalues of A_2 *as defined in* (9) *are stable.* ♯

Proof The constructive proof is presented by the remainder of this subsection. □

Several assumptions were made during the prior analysis of the FTC scheme:

B1. There exists a non-singular matrix T_{d2} that transforms $\begin{bmatrix} M_{c2} & A_{c4} & Q_{c2} \end{bmatrix}$ to have the structures in (8).

B2. The matrix T_{44} has the dimensions given in (12).

B3. A matrix $B^\dagger$ exists satisfying (39).

B4. Error system (31)–(32) is stable.

The following propositions show the equivalence of A1–A5 and B1–B4.

Proposition 3 *Condition A1 is necessary and sufficient to satisfy B1.* ♯

Proof Assumption B1 implies rank $\begin{bmatrix} M_{c1} & A_{c1} & Q_{c1} \\ M_{c2} & A_{c4} & Q_{c2} \end{bmatrix} = \text{rank} \begin{bmatrix} M_{c2} & A_{c4} & Q_{c2} \end{bmatrix}$. The left-hand side of A1 can be simplified using the structures in (5) and (7) into

$$\text{rank} \begin{bmatrix} E & A & M & Q \\ C & 0 & 0 & 0 \\ 0 & E & 0 & 0 \\ 0 & C & 0 & 0 \end{bmatrix} = \text{rank} \begin{bmatrix} M_{c2} & A_{c4} & Q_{c2} \end{bmatrix} + 2\bar{n}. \tag{56}$$

Condition A1 therefore implies rank $\begin{bmatrix} M_{c2} & A_{c4} & Q_{c2} \end{bmatrix} = n + \bar{n} + q + h$. To show the necessity of A1, suppose it is not met: rank $\begin{bmatrix} M_{c2} & A_{c4} & Q_{c2} \end{bmatrix} = \tilde{q} < q + n - \bar{n} + h$. Hence, performing SVD on T_{e1} yields $T_{e1} \begin{bmatrix} M_{c2} & A_{c4} & Q_{c2} \end{bmatrix} T_{e2} = \begin{bmatrix} I_{\tilde{q}} & 0 \\ 0 & 0 \end{bmatrix}$, where T_{e1} and T_{e2} are non-singular. This implies that the structure in (8) cannot be obtained, i.e., B1 is not satisfied, showing the necessity of A1.

On the other hand, A1 being satisfied i.e., rank $\begin{bmatrix} M_{c2} & A_{c4} & Q_{c2} \end{bmatrix} = n + \bar{n} + q + h$, implies $\begin{bmatrix} M_{c2} & A_{c4} & Q_{c2} \end{bmatrix}$ is full-column rank. Hence T_{d1} can be obtained by applying QR decomposition on $\begin{bmatrix} M_{c2} & A_{c4} & Q_{c2} \end{bmatrix}$: $T_{d1} \begin{bmatrix} M_{c2} & A_{c4} & Q_{c2} \end{bmatrix} = \begin{bmatrix} I_{q+n-\bar{n}+h} \\ 0 \end{bmatrix}$, satisfying B1. Therefore, the sufficiency of A1 is shown, and the proof completed. □

Proposition 4 *Assumption B2 is satisfied if and only if A2 holds.* ♯

Proof Matrix T_{44} has the dimensions $T_{44} \in \mathbb{R}^{(p-q-n+\bar{n}-h)\times(p-q-h)}$. Therefore, if A2 is not satisfied (i.e., $p < q + n - \bar{n} + h$), then T_{44} would not exist, and B2 not be satisfied, proving the necessity of A2.

On the other hand, if A2 is satisfied (i.e., $p \geq q + n - \bar{n} + h$), since $n - \bar{n} > 0$ then T_{44} will have its assigned dimensions, satisfying B2 and proving the sufficiency of A2. Thus, the proof is complete. □

Proposition 5 *Condition A3 is necessary and sufficient for B3 to hold.* ♯

Proof From [23], B3 implies that the vector space spanning the columns of $\begin{bmatrix} M & Q \end{bmatrix}$ is a subset of the vector space spanning the columns of B, i.e.,

$$\operatorname{span}\begin{bmatrix} M & Q \end{bmatrix} \subseteq \operatorname{span}(B). \tag{57}$$

Suppose A3 is not satisfied, i.e., $\operatorname{rank}\begin{bmatrix} B & M & Q \end{bmatrix} > \operatorname{rank}(B)$; this would imply that $\operatorname{span}\begin{bmatrix} M & Q \end{bmatrix} \nsubseteq \operatorname{span}(B)$, contradicting (57), resulting in B3 not holding and proving the necessity of A3.

On the other hand, if A3 is satisfied, there exists a matrix $B^{\dagger}$ satisfying $\left(I_n - BB^{\dagger}\right)\begin{bmatrix} M & Q \end{bmatrix} = 0$ [27], which in turn satisfies $BB^{\dagger}\begin{bmatrix} M & Q \end{bmatrix} = \begin{bmatrix} M & Q \end{bmatrix}$, showing the sufficiency of A3. Thus the proof is complete. □

Proposition 6 *For B4 to be satisfied, A4 is necessary. If $p \geq q + n - \bar{n} + h$, then A4 is also sufficient; otherwise (for $p = q + n - \bar{n} + h$), A5 is sufficient.* ♯

Proof Substituting for the structures of (E, A, M, C) in (9)–(11) into the left-hand side of A4 gives

$$\begin{aligned} \operatorname{rank}\begin{bmatrix} sE - A & M & Q \\ C & 0 & 0 \end{bmatrix} &= \operatorname{rank}\begin{bmatrix} 0 & sI_{\bar{n}-p} - A_2 & sE_1 - A_3 & 0 & 0 \\ -A_4 & -A_5 & sE_2 - A_6 & M_2 & Q_2 \\ 0 & 0 & I_p & 0 & 0 \end{bmatrix} \\ &= \underbrace{\operatorname{rank}\begin{bmatrix} sI_{\bar{n}-p} - A_2 \\ A_{54} \end{bmatrix}}_{K_0(s)} + p + q + n - \bar{n} + h. \end{aligned} \tag{58}$$

Then, by applying the Popov-Hautus-Rosenbrock (PHR) rank test [30], if the values of s that make $K_0(s)$ lose rank (the unobservable modes of A_2) are stable, then the pair (A_2, A_{54}) is said to be *detectable*. Hence, recast A4 as: (A_2, A_{54}) is detectable.

The necessity of A4 is shown as follows: B4 implies $A_2 + R_{24}T_{44}A_{54}$ is stable, i.e., $\lambda\left(A_2 + R_{24}T_{44}A_{54}\right) < 0$, which implies $(A_2, T_{44}A_{54})$ is also detectable. Hence, B4 can be recast as: there exists T_{44} such that $(A_2, T_{44}A_{54})$ is detectable.

Using the PHR test, the values that make $K_1(s)$ are stable, where

$$K_1(s) = \begin{bmatrix} sI_{\bar{n}-p} - A_2 \\ T_{44}A_{54} \end{bmatrix} = \begin{bmatrix} I_{\bar{n}-p} & 0 \\ 0 & T_{44} \end{bmatrix} \underbrace{\begin{bmatrix} sI_{\bar{n}-p} - A_2 \\ A_{54} \end{bmatrix}}_{K_0(s)}, \tag{59}$$

which implies rank $(K_1(s)) \leq$ rank $(K_0(s))$, i.e., a value of s making $K_0(s)$ lose rank would also make $K_1(s)$ lose rank, showing the necessity of A4 for the detectability of $(A_2, T_{44}A_{54})$.

To show the sufficiency of A4 when $p \geq q + n - \bar{n} + h$, let N be a matrix containing the right-eigenvectors of A_2; then $N^{-1}A_2N$ is a diagonal matrix containing the eigenvalues of A_2.

Pre-multiply and post-multiply $K_1(s)$ with $N_0 = \begin{bmatrix} N^{-1} & 0 \\ 0 & I_{p-q-n+\bar{n}-h} \end{bmatrix}$ and N, respectively, to obtain

$$N_0K_1(s)N = \begin{bmatrix} I_{\bar{n}-p} & 0 \\ 0 & T_{44} \end{bmatrix} \begin{bmatrix} N^{-1}A_2N \\ A_{54}N \end{bmatrix}. \tag{60}$$

Unobservable modes of $(A_2, T_{44}A_{54})$ would also be unobservable modes of $\left(N^{-1}A_2N, A_{54}N\right)$, and would appear as elements of $N^{-1}A_2N$ with their corresponding columns in A_{54} (and therefore $T_{44}A_{54}$) are zero.

If A4 is satisfied however, T_{44} can always be chosen such that $T_{44}A_{54}$ corresponding to elements within $N^{-1}A_2N$ are non-zero, satisfying B4 and showing the sufficiency of A4 when $p > q + n - \bar{n} + h$.

If $p = q + n - \bar{n} + h$, then T_p in (12) becomes $T_p = \begin{bmatrix} I_{q+n-\bar{n}+h} & 0 \end{bmatrix}$, and R_{24} does not exist. Therefore it is sufficient that A_2 has stable eigenvalues (i.e., A5 is satisfied) for error system (31)–(32) to be stable, completing the proof. □

Remark 2 The presented method is also applicable to regular state-space systems (where $E = I_n$) and infinitely observable descriptor systems $\left(\text{where rank} \begin{bmatrix} E \\ C \end{bmatrix} = n\right)$. This can be shown by setting $E = I_n$, and E and C such that rank $\begin{bmatrix} E \\ C \end{bmatrix} = n$, respectively, and applying the analysis as shown. The proposed scheme is therefore even more general than those applicable only to regular state-space systems [1, 6, 19–23, 27–29], or to infinitely observable descriptor systems [7, 8, 10, 15, 17, 18]. ♯

5.2 Design Algorithm

A summarized design procedure for the FTC scheme is given as follows:

Step 1. Check that A1–A4 (and A5 if necessary) hold. If these conditions are not satisfied, do not continue as the scheme is not applicable.

Step 2. Calculate T_a from (5), T_b and T_c from before (6), and T_d from (8). Then, apply the state equation transformation $T_d T_b$ and state transformation $T_c T_a$ onto system (1)–(2) to obtain the structures in (9)–(11).

Step 3. Choose T_{44} such that it is full-row rank, and $(A_2, T_{44}A_{54})$ is detectable. Apply the state equation transformation T given in (12).

Step 4. Construct the reduced-order system (13)–(14).

Step 5. Select a value for γ, and use a LMI solver to determine P_2 and $\bar{P}_2$ from (24). Calculate R_{24}, and then R and V from (19).

Step 6. Pick values for A_o and G_{n2} such that (19) is satisfied.

Step 7. Determine A_{51}, A_{52}, A_{53}, and $T_{44}A_{54}$ and choose L_{11}, L_{f1}, and L_{x1} to satisfy (37), and set ρ in (18) to satisfy (25).

Step 8. Estimate x_2 and y from (17), and f, x_1, and ξ from (33).

Step 9. Obtain H, J, and S from LMI (45) using a LMI solver, and then calculate K_x from (46), and K_f and K_ξ from (40).

6 Simulation Example

A modified version of the model describing the movement of the endocrine disruptor diethylstilbestrol (DES) through the human body in [12] is used to show the effectiveness of the proposed scheme. The actuator signals u_1 and u_3 are considered separately in this example. The input u_1 is generated by a faulty first-order device

$$\dot{u}_1 = -4u_1 + 4u_{1,e} + f, \tag{61}$$

where $u_{1,e}$ is the reference signal, and f is the fault.

Incorporating the dynamics of u_1 into the system yields the system matrices

$$E = \begin{bmatrix} I_4 & 0 & 0 \\ 0 & 0 & 0 \\ 0 & 0 & 1 \end{bmatrix}, \quad A = \begin{bmatrix} -1.6 & 0.2 & 0.5 & 0.2 & 0 & 1 \\ 0.5 & -0.2 & 0 & 0 & 0 & 0 \\ 0.6 & 0 & -0.5 & 0 & 0 & 0 \\ 0.5 & 0 & 0 & -0.5 & 0 & 0 \\ 1 & 1 & 1 & 1 & 1 & -1 \\ 0 & 0 & 0 & 0 & 0 & -4 \end{bmatrix}, \quad B = \begin{bmatrix} 0 & 0 \\ 0 & 0 \\ 0 & 3 \\ 0 & 0 \\ 0 & -3 \\ 4 & 0 \end{bmatrix}, \tag{62}$$
$$M = \begin{bmatrix} 0 & 0 & 0 & 0 & 0 & 1 \end{bmatrix}^T,$$

for the system variables

$$x_0 = \begin{bmatrix} \text{quantity of DES in the circulation system, } x_a \\ \text{quantity of DES in the reproductive system, } x_b \\ \text{quantity of DES in the digestive system, } x_c \\ \text{quantity of DES in the excretion system, } x_d \\ \text{quantity of DES outside the body, } x_e \\ \text{quantity of drug one injected into blood, } u_1 \end{bmatrix}, \tag{63}$$
$$u = \begin{bmatrix} \text{reference for } u_1,\ u_{1,e} \\ \text{quantity of drug two injected into liver, } u_3 \end{bmatrix}.$$

The drug intake into the liver causes DES quantities to fluctuate by ξ.

Additionally, assume that measurements are only available for x_a, x_c, x_d, and u_1. Hence, Q and C have the structures

$$Q = \begin{bmatrix} 0 \; 0 \; 1 \; 0 \; -1 \; 0 \end{bmatrix}^T, \; C = \begin{bmatrix} 1 \; 0 \; 0 \; 0 \; 0 \\ 0 \; 0 \; I_2 \; 0 \; 0 \\ 0 \; 0 \; 0 \; 0 \; 1 \end{bmatrix}. \tag{64}$$

Since $\bar{n}$ (5) $< n$ (6), system (62)–(64) is *non-infinitely observable*. The FTC scheme would now be designed according to the steps outlined in Sect. 5.2.

Step 1. It can be seen that

$$\begin{aligned} &\text{rank} \begin{bmatrix} E & A & M & Q \\ C & 0 & 0 & 0 \\ 0 & E & 0 & 0 \\ 0 & C & 0 & 0 \end{bmatrix} = n + \bar{n}\,, \; p > q + n - \bar{n} + h, \\ &\text{rank} \begin{bmatrix} sE - A & M \\ C & 0 \end{bmatrix} = n + q \forall\, s \in \mathbb{C}^+, \\ &\text{rank} \begin{bmatrix} B & M & Q \end{bmatrix} = \text{rank}(B) = 2, \end{aligned} \tag{65}$$

that is, A1–A4 are satisfied, guaranteeing the existence of the proposed FTC scheme.

Step 2. The transformations $T_d T_b$ and $T_c T_a$ are calculated to be

$$T_d T_b = \begin{bmatrix} 0 \; 1 \; 0 \; 0 \; 0 \; 0 \\ 0 \; 0 \; 0 \; 0 \; 0 \; 1 \\ 0 \; 0 \; 1 \; 0 \; 1 \; 0 \\ 0 \; 0 \; 1 \; 0 \; 0 \; 0 \\ 1 \; 0 \; 0 \; 0 \; 0 \; 0 \\ 0 \; 0 \; 0 \; 1 \; 0 \; 0 \end{bmatrix}, \; T_c T_a = \begin{bmatrix} 0 \; 0 \; 0 \; 1 \; 0 \\ 0 \; 1 \; 0 \; 0 \; 0 \\ 1 \; 0 \; 0 \; 0 \; 0 \\ 0 \; 0 \; I_2 \; 0 \; 0 \\ 0 \; 0 \; 0 \; 0 \; 1 \end{bmatrix}. \tag{66}$$

Step 3. The following partitions are obtained: $A_2 = -0.2,\ A_{54} = \begin{bmatrix} 0.2 \\ 0 \end{bmatrix}$.

Hence, choose $T_{44} = \begin{bmatrix} -1 & 1 \end{bmatrix}$ to get $T = \begin{bmatrix} I_4 & 0 & 0 \\ 0 & -1 & 1 \\ 0 & 0 & 1 \end{bmatrix}$.

Step 4. The reduced-order system (13)–(14) therefore has the matrices

$$\bar{E} = \begin{bmatrix} 1 & 0 & 0 & 0 & 0 \\ 0 & 0 & 0 & 0 & 1 \\ 0 & 0 & 1 & 0 & 0 \\ 0 & 0 & 1 & 0 & 0 \\ 0 & -1 & 0 & 1 & 0 \end{bmatrix}, \ \bar{A} = \begin{bmatrix} -0.2 & 0.5 & 0 & 0 & 0 \\ 0 & 0 & 0 & 0 & -4 \\ 1 & 1.6 & 0.5 & 1 & -1 \\ 0 & 0.6 & -0.5 & 0 & 0 \\ -0.2 & 2.1 & -0.5 & -0.7 & -1 \end{bmatrix}, \ \bar{B} = \begin{bmatrix} 0 & 0 \\ 4 & 0 \\ 0 & 0 \\ 0 & 3 \\ 0 & 0 \end{bmatrix}, \tag{67}$$
$$\bar{M} = \begin{bmatrix} 0 & I_3 & 0 \end{bmatrix}^T, \ \bar{C} = \begin{bmatrix} 0 & I_4 \end{bmatrix}.$$

Step 5. The parameter γ is chosen as $\gamma = 0.001$. Using the SeDuMi solver for YALMIP within MATLAB on LMI (24), the following are obtained:

$$P_2 = 1,\ \bar{P}_2 = 1.502,\ R_{24} = P_2^{-1}\bar{P}_2 = 1.502.$$
$$R = \begin{bmatrix} 1 & 0 & 1.502 \\ 0 & I_3 & 0 \\ 0 & 0 & 1 \end{bmatrix}, \ V = \begin{bmatrix} 1.502 & 0 & -1.502 & 0 \\ 1 & 0 & 0 & -1 \\ 0 & 0 & 0 & 0 \\ 0 & -1 & 1 & 0 \\ 1 & 0 & -1 & 1 \end{bmatrix}. \tag{68}$$

Steps 6–8. The following partitions are obtained: $A_{51} = 0,\ A_{52} = 1,\ A_{53} = 0,\ T_{44}A_{54} = -0.2$. Hence $A_o,\ G_{n2},\ L_f,\ L_1,\ L_x$, and ρ are chosen as $A_o = 2I_4,\ G_{n2} = I_4,\ L_f = L_x = 0,\ L_1 = 5$, and $\rho = 6$.

Step 9. Using the SeDuMi solver for YALMIP within MATLAB on LMI (45), the following are obtained:

$$H = \begin{bmatrix} 537.6 & 491.6 & 447.7 & 435.8 & 457.9 & 491.4 \\ 626.1 & 1930 & 436.4 & 549.9 & 457.9 & 409.1 \\ 277.0 & 182.4 & 505.2 & 283.6 & 457.9 & 553.3 \\ 572.2 & 738.5 & 403.4 & 994.5 & 457.9 & 357.7 \\ 457.9 & 457.9 & 457.9 & 457.9 & 457.9 & 457.9 \\ 76.56 & 47.41 & 134.7 & 83.60 & 457.9 & 186.8 \end{bmatrix}, \ J = \begin{bmatrix} -3003 \\ -4246 \\ -1547 \\ -3230 \\ -1652 \\ -687.8 \end{bmatrix},$$
$$S = \begin{bmatrix} 99.48 & 58.47 & 142.1 & 101.9 & 457.9 & 146.4 \\ 203.0 & 164.3 & 268.2 & 197.6 & 457.9 & 145.6 \end{bmatrix},$$
$$K_f = \begin{bmatrix} 0.25 \\ 0 \end{bmatrix}, \ K_\xi = \begin{bmatrix} 0 \\ 0.3333 \end{bmatrix},$$
$$K_x = \begin{bmatrix} -0.9983 & -0.1708 & -0.2798 & -0.4975 & -0.2771 & 5.224 \\ -0.9017 & -0.1724 & 0.7694 & -0.4520 & -0.2771 & 1.774 \end{bmatrix}. \tag{69}$$

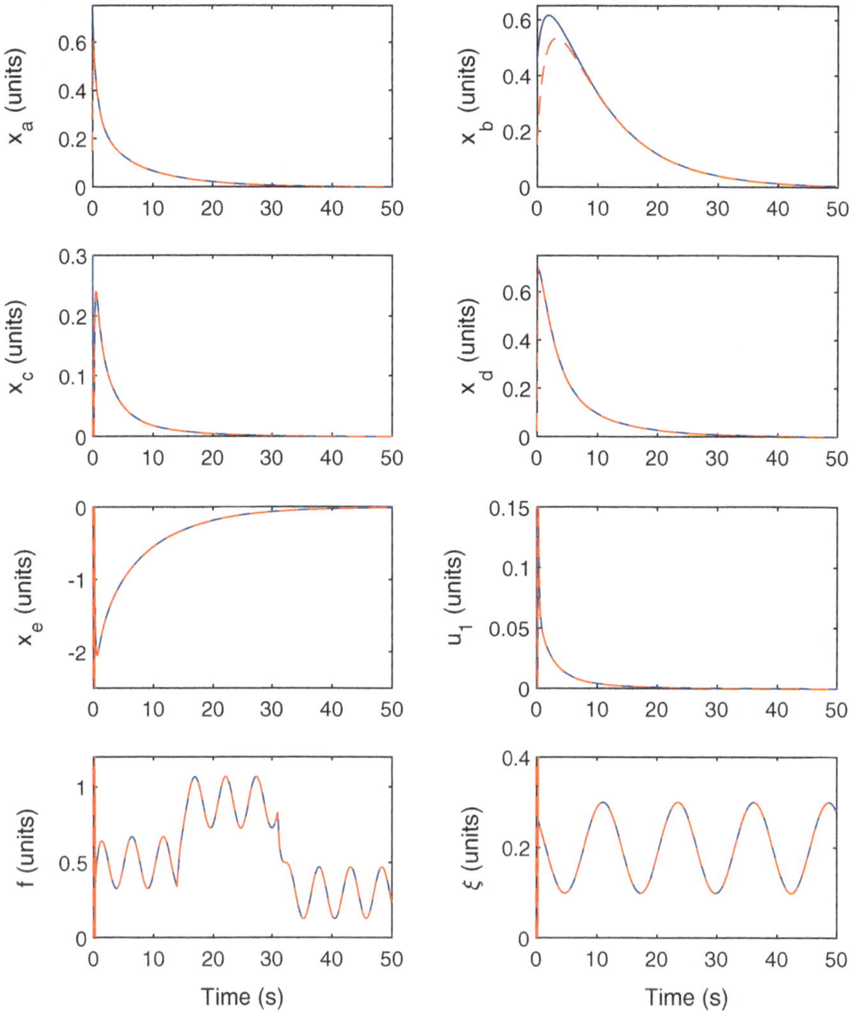

Fig. 1 States and unknown inputs (solid) and their estimates (dashed)

The design of the FTC scheme is now complete. To demonstrate the effectiveness of the proposed scheme, the disturbance signal is set as $\xi = 0.1 \sin(0.5t + 3\pi/4) + 0.2$. The initial condition of the system is set to be $\{0.8, 0.45, 0.68, 0.7, -20.10, 0.65\}$, while the observer is set to have zero initial conditions.

The fault signal is $\dot{f} = -2f + 0.4 \sin(1.2t + \pi/5) + 1 + 0.8h(t - 14) - 1.2h(t - 31)$, where $h(\cdot)$ is the Heaviside unit step function.

Figure 1 shows the states and unknown inputs, and their estimates. It can be seen that sliding motion takes place close to the start of the simulation. The estimates

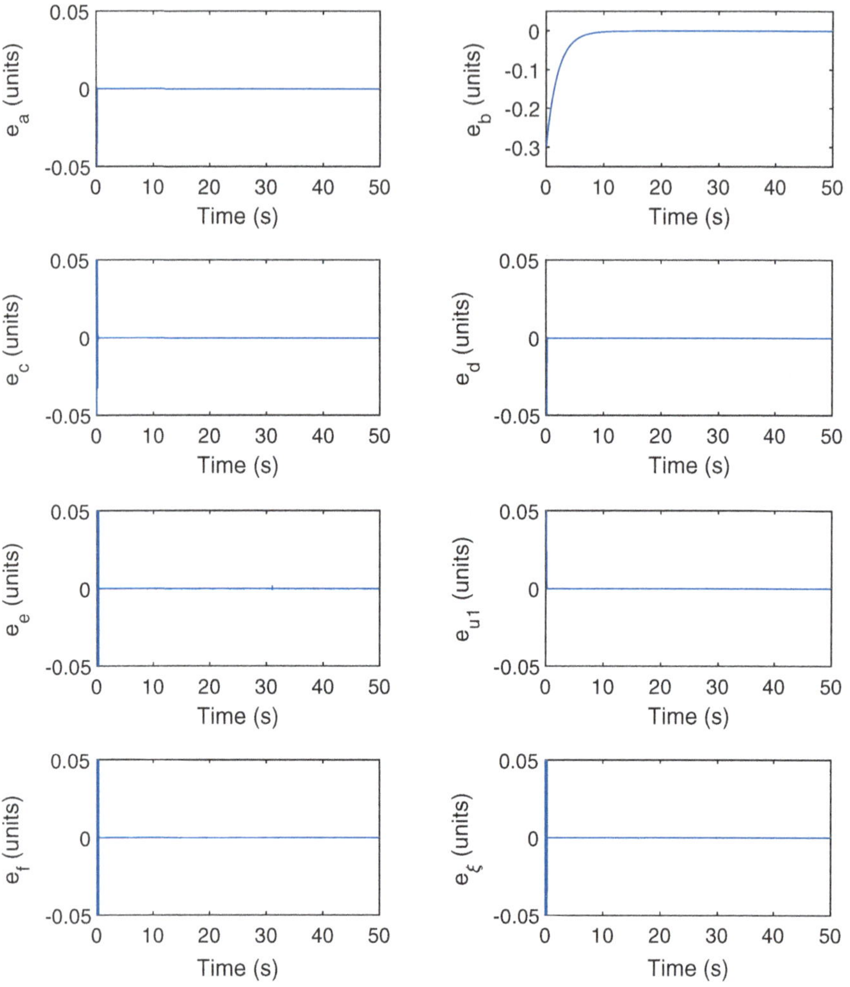

Fig. 2 Estimation errors for the states and unknown inputs

for the non-infinitely observable state x_e, output states x_a, x_c, x_d, and u_1, fault f, and disturbance ξ converge onto the actual values very quickly, as per the gain design in **Step 7**. This convergence is also reflected in Fig. 2, which shows the estimation errors for the observer. The estimate for x_b (which is a non-output, infinitely observable state), however, experiences an initial mismatch due to the system and observer having different initial conditions. Figure 2 however shows that e_b goes to zero asymptotically, in accordance with Proposition 2. Hence, the effectiveness of the SMO is shown.

Finally, Fig. 3 shows the control inputs fed into the system. As can be seen from

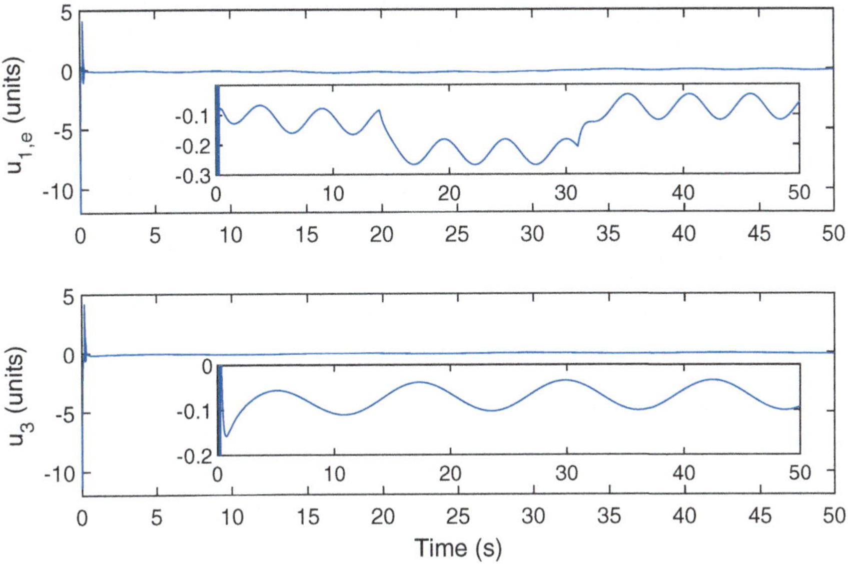

Fig. 3 Inputs into system from the fault-tolerant controller

Fig. 1, all states of the system converge to the reference of zero, demonstrating the efficacy of the proposed FTC scheme.

7 Conclusion

An FTC scheme for NIODS was presented in this chapter. The FTC scheme consists of an SMO to estimate the states and unknown inputs in the system, and a controller to negate the effects of the unknown inputs. The NIODS is first re-expressed as an infinitely-reduced order system by treating some of its states as unknown inputs. The SMO is then designed based on this reduced-order system to estimate the states and unknown inputs. The controller is then utilized to generate the control input signal such that it counters the unknown inputs while driving the system states to the reference. LMIs were used to design the SMO and controller such that the SMO estimates and system states converge asymptotically. Necessary and sufficient conditions for the feasibility of the scheme were presented in terms of the original system matrices. A simulation example was then used to verify the efficacy of the FTC scheme.

Acknowledgements This work was supported by the National Research Foundation of Korea (NRF) grant funded by the Korea government (MSIT) (No. 2020R1C1C1012707). The authors would like to thank Prof. H. Trinh for his invaluable input on the development of the observer utilized in this scheme.

References

1. Chua, W.S., Chan, J.C.L., Tan, C.P., Chong, E.K.P., Saha, S.: Robust fault reconstruction for a class of nonlinear systems. Automatica **113**, Art. no. 108718 (2020)
2. Shen, H., Park, J.H., Wu, Z.G.: Finite-time reliable $\mathcal{L}_2$-$\mathcal{L}_\infty/\mathcal{H}_\infty$ control for Takagi-Sugeno fuzzy systems with actuator faults. IET Control Theory Appl. **8**(9), 688–696 (2014)
3. Chan, J.C.L., Tan, C.P., Trinh, H.: Robust fault reconstruction for a class of infinitely unobservable descriptor system. Int. J. Syst. Sci. **48**(8), 1646–1655 (2017)
4. Liu, G., Park, J.H., Xu, X., Zhuang, G.: Robust non-fragile $\mathcal{H}_\infty$ fault detection filter design for delayed singular Markovian jump systems with linear fractional parametric uncertainties. Nonlinear Anal. Hybrid Syst. **32**, 65–78 (2019)
5. Hao, L.Y., Park, J.H., Ye, D.: Integral sliding mode fault-tolerant control for uncertain linear systems over networks with signals quantization. IEEE Trans. Neural Netw. Learn. Syst. **28**(9), 2088–2100 (2017)
6. Lee, T.H., Lim, C.P., Nahavandi, S., Park, J.H.: Network-based synchronization of T-S fuzzy chaotic systems with asynchronous samplings. J. Frankl. Inst. **355**(13), 5736–5758 (2018)
7. Gao, Z., Ding, S.X.: Actuator fault robust estimation and fault-tolerant control for a class of nonlinear descriptor systems. Automatica **43**, 912–920 (2007)
8. Yao, L., Cocquempot, V.: Fault diagnosis and fault tolerant control for a class of nonlinear singular systems. In: Paper presented at: IEEE International Conference on Control and Automation, Xiamen, China, pp. 1536–1541 (2010)
9. Chan, J.C.L., Tan, C.P., Ooi, J.H.T., Trinh, H.: New results in robust fault reconstruction for a class of non-infinitely observable descriptor systems. In: Paper Presented at: American Control Conference; Philadelphia, PA, USA, pp. 5058–5064 (2019)
10. Mu, Y., Zhang, H., Su, H., Zhang, K.: Observer-based actuator fault estimation and proportional derivative fault tolerant control for continuous-time singular systems. Optimal Control Appl. Methods **40**(6), 979–997 (2019)
11. Xiao, X., Park, J.H., Zhou, L., Lu, G.: New results on stability analysis of Markovian switching singular systems. IEEE Trans. Autom. Control **64**(5), 2084–2091 (2019)
12. Liu, P., Zhang, Q., Yang, X., Yang, L.: Passivity and optimal control of descriptor biological complex systems. IEEE Trans. Autom. Control **53**, 122–125 (2008)
13. Chan, J.C.L., Tan, C.P., Trinh, H., Kamal, M.A.S.: State and fault estimation for a class of non-infinitely observable descriptor systems using two sliding mode observers in cascade. J. Frankl. Inst. **356**(5), 3010–3029 (2019)
14. Chan, J.C.L., Tan, C.P., Kamal, M.A.S., Chiew, Y.S.: Robust fault reconstruction for a class of non-infinitely observable descriptor systems using two sliding mode observers in cascade. Appl. Math. Comput. **350**, 78–92 (2019)
15. Duan, G.R.: Analysis and design of descriptor linear systems. Advances in Mechanics and Mathematics. Springer Science & Business Media (2010)
16. Fahmy, M.M., O'Reilly, J.: Observers for descriptor systems. Int. J. Control **49**(6), 2013–2028 (1989)
17. Yeu, T.K., Kim, H.S., Kawaji, S.: Fault detection, isolation and reconstruction for descriptor systems. Asian J. Control **7**(4), 356–367 (2005)
18. Zhang, J., Zhu, F., Karimi, H.R., Wang, F.: Observer-based sliding mode control for T-S fuzzy descriptor systems with time delay. IEEE Trans. Fuzzy Syst. **27**(10), 2009–2023 (2019)
19. Lan, J., Patton, R.J.: A decoupling approach to integrated fault-tolerant control for linear systems with unmatched non-differentiable faults. Automatica **89**, 290–299 (2018)
20. Lee, T.H., Lim, C.P., Nahavandi, S., Roberts, R.G.: Observer-based $\mathcal{H}_\infty$ fault-tolerant control for linear systems with sensor and actuator faults. IEEE Syst. J. **13**(2), 1981–1990 (2019)
21. Li, X., Ahn, C.K., Lu, D., Guo, S.: Robust simultaneous fault estimation and nonfragile output feedback fault-tolerant control for Markovian jump systems. IEEE Trans. Syst. Man Cybern. Syst. **49**(9), 1769–1776 (2019)

22. Wang, Z., Rodrigues, M., Theilliol, D., Shen, Y.: Fault-tolerant control for discrete linear systems with consideration of actuator saturation and performance degradation. IFAC-PapersOnLine **48**(21), 499–504 (2015)
23. Zhang, K., Jiang, B., Chen, M., Yan, X.G.: Distributed fault estimation and fault-tolerant control of interconnected systems. IEEE Trans. Cybern. (2019). in press
24. Ooi, J.H.T., Tan, C.P., Ng, K.Y.: State and fault estimation for infinitely unobservable descriptor systems using sliding mode observers. Asian J. Control **17**(4), 1458–1461 (2015)
25. Ooi, J.H.T., Tan, C.P., Nurzaman, S.G., Ng, K.Y.: A sliding mode observer for infinitely unobservable descriptor systems. IEEE Trans. Autom. Control **62**(7), 3580–3587 (2017)
26. Ooi, J.H.T., Tan, C.P., Chua, W.S., Wang, X.: State and unknown input estimation for a class of infinitely unobservable descriptor systems using two observers in cascade. J. Frankl. Inst. **354**(18), 8374–8397 (2017)
27. Rastegari, A., Arefi, M.M., Asemani, M.H.: Robust H_∞ sliding mode observer-based fault-tolerant control for one-sided Lipschitz nonlinear systems. Asian J. Control **21**(1), 114–129 (2018)
28. Tan, C.P., Edwards, C.: Sliding mode observers for robust detection and reconstruction of actuator and sensor faults. Int. J. Robust Nonlinear Control **13**(5), 443–463 (2003)
29. Yan, X.G., Edwards, C.: Nonlinear robust fault reconstruction and estimation using a sliding mode observer. Automatica **43**(9), 1605–1614 (2007)
30. Hautus, M.L.J.: Controllability and observability conditions of linear autonomous systems. Proceedings of the Koninklijke Nederlandse Akademie van Wetenschappen Series A, Mathematical Sciences **72**, 443–448 (1969)

Fault Estimator for Parabolic Systems with Distributed Inputs and Outputs

K. Mathiyalagan

Abstract This chapter is concerned with the $\mathcal{H}_\infty$ fault estimation of a class of parabolic systems with distributed inputs and outputs. The fault in the considered parabolic system is characterized by a minimal state-space description of a low-pass filter, which incorporates fault's prior knowledge. The fault estimator is designed by using the Lyapunov theory and linear matrix inequalities (LMIs) approach. The sufficient conditions in the form of LMIs are developed to ensure the resulting error system is asymptotically stable with an optimized $\mathcal{H}_\infty$ disturbance attenuation level. The designed estimator gain matrices can be easily determined by using the standard optimization toolboxes. Finally, two numerical examples and simulation results are provided to illustrate the effectiveness and advantages of the obtained results.

Keywords Fault detection · Low pass filter · $\mathcal{H}_\infty$ analysis · Exponential stability

1 Introduction

Generally, partial differential equations (PDEs) exhibits its special features, which usually mirror the physical phenomena, many of the fundamental theories of physics and engineering are expressed by means of systems of PDEs.

Analyzing the existence and stability of PDEs are far more difficult and form a key topic of contemporary mathematical research. There is no common hypothesis known concerning the solvability of all PDEs. Such a hypothesis is greatly impossible to exist, given the wealthy variety of physical, probabilistic and geometric phenomena that can be modeled by PDEs. Moreover, many research concentrates on different specific PDEs that are essential for applications in the real world scenario, for example see [1–4].

In PDEs, the typical additional constraint is so-called boundary condition, in which specified values are imposed at points on the boundary of the domain where

K. Mathiyalagan (✉)
Department of Applied Mathematics, Bharathiar University, Coimbatore 641 046, India
e-mail: kmathimath@gmail.com

J. H. Park (ed.), *Recent Advances in Control Problems of Dynamical Systems and Networks*, Studies in Systems, Decision and Control 301,
https://doi.org/10.1007/978-3-030-49123-9_7

the solution is supposed to be defined. PDEs are usually built around the equations of classical mathematical physics. They are used to describe the evolution of gases in fluid dynamics, the formation of galaxies, describing the nature of quantum mechanics and so on [5–7].

PDEs are used to mathematically express, and thus aid the solution of physical and other problems involving functions of several variables. Parabolic PDEs which describe diffusion processes such as heat equation, chemical concentration, etc. These equations play a substantial role in understanding problems related to flow and heat motion of free fall body. Solutions are smooth in space but may have singularities. Dynamical frameworks may be a massive field, diverse sorts of tributes and PDEs that emerge as models in material science, physics, etc [8–12] and references therein. Boskovic et al., [13] discussed about the problem of nonlinear parabolic PDEs have been discussed and constructed certain Lyapunov structures to prove stability.

Stability is the fundamental requirement in all dynamical systems it is mainly inspired by differential equations of motion and flows. Stability in motion means, the system is evolving in a direction such that it will reach a minimum energy state. Stability of solutions is important in physical problems if slight deviations from the mathematical model caused by unavoidable errors in measurement do not have a correspondingly slight effect on the solution, so the mathematical equations describing the problem will not accurately predict the future outcome [14].

In real-world, situations uncertainties and disturbances like noise in sensors and actuators can unstabilize the system [15–18]. Since most PDEs are express the real control systems, so they may encounter different kinds of faults due to malfunctions or imperfect performances exist. These faults are caused because of the unexpected variations, sudden changes in signals, environments, external disturbances or parameter shifting and so on [19, 20]. So, the fault detection problem becomes an important topic among the research community [21, 22].

The objective in the fault detection problem is to detect a fault value when it happens to take protection measures, or reconfiguring the running control scheme. Ferdowsi and Jagannathan [23] discussed the fault diagnosis problem for a class of distributed parameter systems modeled by parabolic partial differential equations. Demetriou [24] has discussed about model-based fault detection and diagnosis scheme for distributed parameter system. Ferdowsi et al., [23] designed fault detection schemes for distributed parameter parabolic system has been investigated.

It is noted that an optimization problem is one of the most commonly implemented ways to deal the fault detection. In literature, many results related to fault estimation/detection can be found for varies types of practical systems including parabolic type ones [25, 26]. Baruh [27] studied the actuator failure detection in distributed systems.

Among the many approaches in the literature, the common way is the observer-based approach [28]. This approach is based on generating residuals by designing proper state observers or filters and using these residuals to set a threshold to detect the fault. The problem of $\mathcal{H}_\infty$ fault estimation for the discrete-time switched nonlinear system is considered by using the multiple-Krasovskii functional and average dwell-time approach in [29]. Yuan et al., [30] derived some sufficient conditions for $\mathcal{H}_\infty$

fault estimator for switched linear systems to be asymptotically stable and determined $\mathcal{L}_2$ gain bound. Zhang et al., [31] discussed about observer-based fault diagnosis techniques for faulty systems.

It is well known that, an observer or filter should be designed to guarantee the improved robustness against unknown inputs and sensitivity to faults [32, 33]. There are increasing awareness to ensure robustness in performance requires simpler and stable adaptive observer schemes.

The main objective of the optimization problem is to introduce a performance index on the residual signal between the robustness to disturbance and the sensitivity to faults and formulate the fault detection and isolation as an optimization problem. Moreover, it is well known that the design problem of $\mathcal{H}_\infty$-based observer or filter expresses the design procedure as a mathematical optimization problem for finding some required solutions [34]. The $\mathcal{H}_\infty$-norm measures robustness against the unknown inputs to achieving results of guaranteed performance $\mathcal{H}_\infty$ methods are utilized to achieve stabilization with ensured performance, and to constrict the external disturbance with the reason for minimizing the effect of the disturbance of the system performance [35]. Baniamerian et al., [36] discussed about dissipative parabolic PDE and designed geometric fault detection obtained from the Galerkin method. These techniques are receiving considerable attention in research fields. Lien et al., discussed about discrete switched system then the results are extended for exponential stability and robust $\mathcal{H}_\infty$ performance. Fridman et al., [37] discussed about $\mathcal{H}_\infty$ boundary control for semilinear parabolic and hyperbolic systems in 2009. Fischer et al., [38] discussed about fault detection approach for parabolic systems using modulation function.

However, to the best of the authors' knowledge, the $\mathcal{H}_\infty$ fault estimation of a class of parabolic systems with distributed inputs and outputs has not been fully investigated, which motivates this present investigation. Motivated by the above, in this chapter, the observer design for a class of distributed parameter system of parabolic type with fault inputs is discussed. The sufficient conditions are derived by using the Lyapunov functional method together with the LMI approach under which the error system is robustly stable with a guaranteed $\mathcal{H}_\infty$ performance index. Moreover, the sufficient conditions to design the parameters of the observer gains are established in terms of LMIs, which can be easily checked by using the standard numerical toolbox. Also, the results are expressed in the form of an optimization problem that can guarantee the minimum $\mathcal{H}_\infty$ performance index. Finally, numerical examples are presented to validate the proposed results.
Now, the problem under investigation can be summarized as

- Study the exponential stability conditions for the distributed parameter system of a parabolic type system.
- Design a suitable $\mathcal{H}_\infty$ fault estimator that can guarantee the minimum $\mathcal{H}_\infty$ performance index.

2 Problem Description

2.1 System Description

Consider the parabolic partial differential equation

$$\partial_t w(z,t) = \partial_z^2 w(z,t) + \beta(z)w(z,t) + a(z)f(t) + b(z)d(t), \tag{1}$$

where $(z,t) \in (0,1) \times \mathbb{R}^+$ with $\beta(z) \in C[0,1], a(z), b(z)$ are piecewise continuous, $f(t)$ is the unknown fault signal the unknown disturbance $d(t) \in \mathbb{R}^{p_d}$.

The initial condition for the above system is

$$w(z,0) = w_0(z), \ \forall\, w_0 \in L_2(0,1).$$

The distributed output $y(t) \in \mathbb{R}^m$ is assumed as

$$y(t) = \int_0^1 w(z,t)\mathrm{d}z. \tag{2}$$

2.2 Low Pass Filter

Most of the dynamical systems typically need to deal with noise, in such cases, low pass filters (LPFs) are used to remove noise from the measured signals.

In order to achieve a better fault estimation results, incorporating the faults prior knowledge into the design process is very much useful, which is much more suitable and reasonable in many practical situations.

So, using an LPF, we define a minimal state-space description of the fault $f(t)$ as

$$\left.\begin{aligned} &\partial_t w_f(z,t) = \partial_z^2 w_f(z,t) + a_f(z)w_f(z,t) + b_f(z)f_0(t), \\ &w_f(z_0,t_0) = 0, \end{aligned}\right\} \tag{3}$$

where $w_f(z,t) \in \mathbb{R}^{n_f}$ is the LPF state, $f_0(t)$ is a fictitious signal in $L_2(0,1)$ and $a_f(z)$, $b_f(z)$ and c_f are weight functions obtained from the prior knowledge of the fault.

In this work, the fault is considered as the output of the above model and described as

$$f(t) = c_f \int_0^1 w_f(z,t)\mathrm{d}z, \tag{4}$$

2.3 Observer Description

Consider the following observer to estimate the fault in (1)

$$\partial_t \hat{w}(z,t) = \partial_z^2 \hat{w}(z,t) + \beta(z)\hat{w}(z,t) + a(z)\hat{f}(t) + k_1(y(t) - \hat{y}(t)), \tag{5}$$

with an output of the form

$$\hat{y}(t) = \int_0^1 \hat{w}(z,t)\mathrm{d}z,$$

where $\hat{w}(z,t)$, $\hat{y}(t)$ are the estimates of $w(z,t)$, $y(t)$.

The goal is to find $\hat{f}(t)$, such that the following ratio is satisfied under zero initial condition

$$\left\| f(t) - \hat{f}(t) \right\|_2 \leq \gamma \left\| w(t) \right\|_2,$$

where $\gamma > 0$, $w(t) = \left[d^T(t) \ f_0^T(t) \right]^T$.

Now, we assume the fault estimation $\hat{f}(t)$ as

$$\hat{f}(t) = c_f \int_0^1 \hat{w}_f(z,t)\mathrm{d}z,$$

where $\hat{w}_f(z,t)$ is the state vector of the system having the dynamics

$$\partial_t \hat{w}_f(z,t) = \partial_z^2 \hat{w}_f(z,t) + a_f(z)\hat{w}_f(z,t) + k_2(y(t) - \hat{y}(t)), \tag{6}$$

in which k_1 and k_2 are observer gains to be designed.

Let the error signals be

$$e(z,t) = w(z,t) - \hat{w}(z,t),$$

$$e_f(z,t) = w_f(z,t) - \hat{w}_f(z,t)$$

and

$$r_f(t) = f(t) - \hat{f}(t).$$

Then, the error systems can be represented as

$$\partial_t e(z,t) = \partial_z^2 e(z,t) + \beta(z)e(z,t) + a(z)c_f \int_0^1 e_f(z,t) + b(z)d(t)$$
$$-k_1 \int_0^1 e(z,t)\mathrm{d}z, \tag{7}$$

and

$$\partial_t e_f(z,t) = \partial_z^2 e_f(z,t) + a_f(z)e_f(z,t) + b_f(z)f_0(t) - k_2 \int_0^1 e(z,t)\mathrm{d}z, \tag{8}$$

and the residual function is given by

$$r_f(t) = c_f \int_0^1 e_f(z,t)\mathrm{d}z. \tag{9}$$

Next, denote $\xi(z,t) = \begin{bmatrix} e(z,t) \\ e_f(z,t) \end{bmatrix}$, then the augmented system can be rewritten as

$$\partial_t \xi(z,t) = A\partial_z^2 \xi(z,t) + C(z)\xi(z,t) + B(z)w(t) - K \textstyle\int_0^1 \xi(z,t)\mathrm{d}z, \tag{10}$$

where

$$A = \begin{bmatrix} 1 & 0 \\ 0 & 1 \end{bmatrix}, \quad C(z) = \begin{bmatrix} \beta(z) & 0 \\ 0 & a_f(z) \end{bmatrix},$$
$$B(z) = \begin{bmatrix} b(z) & 0 \\ 0 & b_f(z) \end{bmatrix}, \quad K = \begin{bmatrix} k_1 & -a(z)c_f \\ k_2 & 0 \end{bmatrix},$$

and the residual function turns of the form

$$r_f(t) = C_f \int_0^1 \xi(z,t)\mathrm{d}z \tag{11}$$

with

$$C_f = \begin{bmatrix} 0 \; c_f \end{bmatrix},$$

and the periodic boundary conditions for (10) are taken as

$$\xi(0,t) = 0,$$
$$\xi(1,t) = 0.$$

Now, denote that

$$g(z,t) = \int_0^1 \xi(z,t)\mathrm{d}z - \xi(z,t),$$

which implies that

$$\int_0^1 \xi(z,t)\mathrm{d}z = g(z,t) + \xi(z,t).$$

Now, the problem under investigation can be summarized as

- to obtain the sufficient conditions for the exponential stability of the system (10) when $w(t) = 0$.
- to obtain the sufficient conditions for the existence of a suitable $\mathcal{H}_\infty$ fault estimator that can guarantee the $\mathcal{H}_\infty$ performance for all nonzero $w(t) \in l_2[0,\infty)$.

3 Preliminaries

In this section, we are introducing some basic definitions and lemma.

Lemma 1 *Given constant matrices G_{11}, G_{12}, G_{22} with appropriate dimensions, where $G_{11}^T = G_{11}$ and $G_{22}^T = G_{22}$, then*

$$G_{11} + G_{12}^T G_{22}^{-1} G_{12} < 0$$

if and only if

$$\begin{bmatrix} G_{11} & G_{12}^T \\ * & -G_{22} \end{bmatrix} < 0.$$

Lemma 2 (Vector Poincare-Wirtinger inequality) *Let $\xi(z,t) \in \mathcal{H}$ be a vector function. Then, the following inequality holds for any matrix $0 \leq M \in \mathbb{R}^{n\times n}$, we have*

$$\int_\Omega g^T(z,t)Mg(z,t)dz \leq 4\phi_s\pi^{-2}\int_\Omega \xi_z^T(z,t)M\xi_z(z,t)dz \tag{12}$$

where

$$g(z,t) = (l_2 - l_1)^{-1}\int_{l_1}^{l_2} \xi(z,t)dz - \xi(z,t),$$

the limits satisfying $\alpha_1 \leq l_1 < l_2 \leq \alpha_2$ and $\phi_s = max\{(l_2-\alpha_1)^2, (\alpha_2 - l_1)^2\}$.

Definition 1 System (10) with $w(t) = 0$ is said to be exponentially stable, if there exist scalars $\rho > 0$, $\alpha > 0$ such that the solution $\xi(z,t)$ of the system (10) satisfies the following condition

$$\|\xi(\cdot,t_0)\| \leq \rho e^{-\alpha(t-t_0)}\|\theta(\cdot)\|, \forall t \geq t_0$$

where $\xi(z,t_0) = \theta(z)$, $\alpha > 0$ is called the decay rate.

Definition 2 The distributed parameter system (10) is said to be exponentially stable with given disturbance attenuation level $\gamma > 0$, if it is robustly stable and the response $r_f(t)$ under zero initial condition satisfies,

$$||r_f(t)||_2 \leq \gamma||w(t)||_2$$

for every non-zero $w(t) \in L_2[0, \infty)$.

The following specified notations are used throughout this chapter.

- $\mathbb{R}^n$ denotes the n-dimensional Euclidean space.
- $\mathbb{R}^{n\times m}$ is the set of all $n \times m$ matrices with real entries.
- The superscript $'T'$ stands for matrix transposition.
- L_2 denotes the space of square integrable functions on (0, 1) with Euclidean norm $\|\cdot\| = \left(\int_0^\infty \|\cdot\|^2 \mathrm{d}t\right)^{\frac{1}{2}}$.
- I and 0 are identity matrix and zero matrix respectively.
- The spatial detection domain $\Omega = [z_0, z_1] \subset [0, 1]$ where $z_0 < z_1$.
- For a real matrix P, $P > 0$ means P is Positive definite.
- The symmetric term in a symmetric matrix is denoted by asterisk $(*)$.
- diag$\{\cdot\}$ denotes the diagonal matrix.

4 Exponential Stability and $\mathcal{H}_\infty$ Performance

For known k_1 and k_2, first we prove exponential stability for the system (10) and the result extended to ensure the $\mathcal{H}_\infty$ performance.

Theorem 1 *For given scalars $\gamma > 0$, $\alpha > 0$ and known k_1, k_2, there exist a positive-definite symmetric matrix P such that the following LMI holds*

$$\begin{bmatrix} \Omega_{11} & -PK + C_f^T C_f & PB(z) \\ * & \Omega_{22} & 0 \\ * & * & -\gamma^2 \end{bmatrix} < 0, \tag{13}$$

where

$$\Omega_{11} = PC(z) + C^T(z)P^T + P\alpha - PK - K^T P^T + C_f^T C_f,$$

$$\Omega_{22} = -\frac{\Pi^2}{4\phi_s}(PA + A^T P^T) + C_f^T C_f,$$

then the system (10) *is exponentially stable with $\mathcal{H}_\infty$ performance level γ.*

Proof Now, we choose a Lyapunov function as follows

$$V(z, t, \xi(z, t)) = \int_\Omega \xi^T(z, t)P\xi(z, t)\mathrm{d}z, \tag{14}$$

and taking the time derivative of $V(z, t, \xi(z, t))$ along the trajectory of the system (10) yields

$$\begin{aligned}\partial_t V(z, t, \xi(z, t)) &= 2\int_\Omega \xi^T(z, t) P \partial_t \xi(z, t) \mathrm{d}z, \\ &= 2\int_\Omega \xi^T(z, t) P\{A\partial_z^2 \xi(z, t) + C(z)\xi(z, t) + B(z)w(t) \\ &\quad - K[g(z, t) + \xi(z, t)]\}\mathrm{d}z, \\ &= 2\int_\Omega \xi^T(z, t)(PA)\partial_z^2 \xi(z, t)\mathrm{d}z \\ &\quad + 2\int_\Omega \xi^T(z, t) P\{C(z)\xi(z, t) \\ &\quad + B(z)w(t) - K[g(z, t) + \xi(z, t)]\}\mathrm{d}z.\end{aligned}$$

Here, using integration by parts that leads to

$$\begin{aligned}\partial_t V(z, t, \xi(z, t)) &= 2(\xi^T(z, t)(PA)\partial_z \xi(z, t))_\Omega \\ &\quad - 2\int_\Omega \partial_z \xi(z, t)(PA)\partial_z \xi^T(z, t)\mathrm{d}z \\ &\quad + 2\int_\Omega \xi^T(z, t) P\{C(z)\xi(z, t) + B(z)w(t) \\ &\quad - K[g(z, t) + \xi(z, t)]\}\mathrm{d}z\end{aligned}$$

and applying boundary conditions, we get

$$\begin{aligned}\partial_t V(z, t, \xi(z, t)) &= -2\int_\Omega \partial_z \xi^T(z, t)(PA)\partial_z \xi(z, t)\mathrm{d}z \\ &\quad + 2\int_\Omega \xi^T(z, t) P\{C(z)\xi(z, t) \\ &\quad + B(z)w(t) - K[g(z, t) + \xi(z, t)]\}\mathrm{d}z.\end{aligned}$$

By using vector Poincare-Wirtinger inequality, above equation takes the form,

$$\begin{aligned}\partial_t V(z, t, \xi(z, t)) &\leq -2\frac{\Pi^2}{4\phi_s}\int_\Omega g^T(z, t)(PA)g(z, t)\mathrm{d}z \\ &\quad + 2\int_\Omega \xi^T(z, t) P\{C(z)\xi(z, t) \\ &\quad + B(z)w(t) - K[g(z, t) + \xi(z, t)]\}\mathrm{d}z.\end{aligned}$$

First, we consider the stability of the system (10), for this case if $w(t) = 0$, we obtained the following inequality.

$$\partial_t V(z,t,\xi(z,t)) + \alpha V(z,t,\xi(z,t)) \le \int_0^1 \eta^T(z,t)\Phi\eta(z,t)\mathrm{d}z, \tag{15}$$

where $\eta^T(z,t) = \left[\, \xi^T(z,t)\; g^T(z,t) \,\right]$ and

$$\Phi = \begin{bmatrix} PC(z) + C^T(z)P^T + P\alpha - PK - K^TP^T & -PK \\ * & -\frac{\Pi^2}{4\phi_s}(PA + A^TP^T) \end{bmatrix},$$

if the $\Phi < 0$ holds, then it implies that LMI (15) holds.
Then, we have

$$\partial_t V(z,t,\xi(z,t)) + \alpha V(z,t,\xi(z,t)) < 0$$

for all $\eta(z,t) \neq 0$.
Integrating this inequality over $[t_0, t]$ gives

$$V(z,t,\xi(z,t)) \le e^{-\alpha(t-t_0)} V(z,t_0,\xi(z,t_0)),$$

for arbitrary initial condition $\xi(z,t_0) = \theta(z)$ and

$$\begin{aligned} \int_\Omega \xi^T(z,t)P\xi(z,t)\mathrm{d}z &= V(z,t,\xi(z,t)), \\ &\le e^{-\alpha(t-t_0)} V(z,t_0,\xi(z,t_0)), \\ &\le e^{-\alpha(t-t_0)} \int_0^1 \theta^T(z)P\theta(z)\mathrm{d}z. \end{aligned}$$

Then, it is easy to show that

$$\|\xi(.,t)\| \le e^{-\alpha(t-t_0)}\|\theta(\cdot)\|, \forall t \ge t_0. \tag{16}$$

Therefore, by Definition 1, we conclude that the system (10) with $w(t) = 0$ is exponentially stable.

Now, we build $\mathcal{H}_\infty$ performance for the augmented system (10),

$$\begin{aligned} \partial_t V(z,t,\xi(z,t)) + \alpha V(z,t,\xi(z,t)) + \int_0^1 J(t)\mathrm{d}z \\ \le \int_0^1 \eta_1^T(z,t)\Psi\eta_1(z,t)\mathrm{d}z, \end{aligned}$$

by setting $\eta_1^T(z,t) = \left[\, \xi^T(z,t)\; g^T(z,t)\; w^T(t) \,\right]$, then we easily obtain (13).
By ensuring (13)< 0, for all $\eta_1(z,t) \neq 0$, we get

$$\partial_t V(z,t,\xi(z,t)) + \alpha V(z,t,\xi(z,t)) + J(t) < 0. \tag{17}$$

follows from by denoting

$$J(\cdot) = r_f^T(t)r_f(t) - w^T(t)\gamma^2 I w(t).$$

Integrating (17) over $[t_0, t]$ gives

$$V(z, t, \xi(z, t)) \le e^{-\alpha(t-t_0)} V(z, t_0, \xi(z, t_0)) + \int_{t_0}^{t} e^{-\alpha(t-s)} J(s)\mathrm{d}s. \tag{18}$$

In the same way to prove the exponential stability, (18) produces

$$V(z, t, \xi(z, t)) \le e^{-\lambda(t-t_0)} V(z, t_0, \xi(z, t_0)) + \int_{t_0}^{t} e^{-\lambda(t-s)} J(s)\mathrm{d}s. \tag{19}$$

Taking $t_0 = 0$, (19) implies

$$\int_0^t e^{-\lambda(t-s)} J(s)\mathrm{d}s \le 0. \tag{20}$$

Integrating (20) over $[0, \infty)$ to obtain

$$\int_0^\infty \int_0^t e^{-\lambda(t-s)} J(s)\mathrm{d}s\mathrm{d}t \le 0.$$

changing the order of integration which yields that

$$\int_0^\infty J(s)\left(\int_s^\infty e^{-\lambda(t-s)}\mathrm{d}t\right)\mathrm{d}s = \frac{1}{\lambda}\int_0^\infty J(s)\mathrm{d}s \le 0.$$

Therefore, from the above it is easy to verify

$$||r_f(t)||_2 \le \gamma||w(t)||_2,$$

then by Definition 2, we conclude that the system (10) is exponentially stable with $\mathcal{H}_\infty$ performance level γ. □

Suppose that the observer gains k_1 and k_2 are not known, now it can be designed using the results obtained in the above theorem.

Theorem 2 *For given scalars $\gamma > 0$, $\alpha > 0$, the system (7) is exponentially stable with $\mathcal{H}_\infty$ performance level γ, there exist a positive-definite symmetric matrix diag{p_1, p_2}$>$ 0 such that the following LMI holds*

$$\begin{bmatrix} \Sigma_{11} & \Sigma_{12} & -\kappa_1 & p_1 a(z) c_f & p_1 b(z) & 0 \\ * & \Sigma_{22} & -\kappa_2 & c_f^T c_f & 0 & p_2 b_f(z) \\ * & * & \Sigma_{33} & 0 & 0 & 0 \\ * & * & * & -\frac{\Pi^2}{4\phi_s}(p_2 + p_2^T) + c_f^T c_f & 0 & 0 \\ * & * & * & * & -\gamma^2 & 0 \\ * & * & * & * & * & -\gamma^2 \end{bmatrix} < 0 \tag{21}$$

where

$$\begin{aligned} \Sigma_{11} &= p_1\beta(z) + \beta^T(z)p_1^T + p_1\alpha - \kappa_1 - \kappa_1^T \\ \Sigma_{12} &= p_1 a(z) c_f - \kappa_2^T, \\ \Sigma_{22} &= p_2 a_f(z) + a_f^T(z) p_2^T + p_2\alpha + c_f^T c_f, \\ \Sigma_{33} &= -\frac{\Pi^2}{4\phi_s}(p_1 + p_1^T), \end{aligned}$$

and moreover the observer gains are given by

$$k_1 = p_1^{-1}\kappa_1, \quad k_2 = p_2^{-1}\kappa_2.$$

Proof By letting $\kappa_1 = p_1 k_1$ and $\kappa_2 = p_2 k_2$, in the proof of the Theorem 1 and using Schur complement lemma, it is easy that the LMI (21) holds.

Therefore, by Definitions 1 and 2, we conclude that the system (7) is exponentially stable with $\mathcal{H}_\infty$ performance level γ. □

5 Mixed $\mathcal{H}_\infty$ and Passivity Performance

It is worth pointing that the passivity framework has proved to be useful one as the main idea behind the passivity theory is that the passive properties can make sure the internal stability of the system. It is noted that with the different supply functions, the passive theory can be established for various approaches including the $\mathcal{H}_\infty$ analysis [35, 39].

The distributed parameter system (10) is said to be exponentially stable with a mixed $\mathcal{H}_\infty$ and passivity performance index $\gamma > 0$, if it is robustly stable and the response $r_f(t)$ under zero initial condition satisfies,

$$\int_0^\infty (\gamma^{-1}\delta r_f^T(t) r_f(t) - 2(1-\delta) r_f^T(t) w(t) \leq \gamma w^T(t) w(t)) \mathrm{d}t,$$

for every non-zero $w(t) \in L_2[0, \infty)$.

The result can be deduced by letting in (17) as

$$J(\cdot) = \gamma^{-1}\delta r_f^T(t) r_f(t) - 2(1-\delta) r_f^T(t) w(t) - \gamma w^T(t) w(t),$$

we can easily deduce the the following theorem based on mixed $\mathcal{H}_\infty$ and passivity performance index.

Theorem 3 *For given scalars $\gamma > 0$, $\alpha > 0$ if there exist a positive-definite symmetric matrix $diag\{p_1, p_2\} > 0$ such that*

$$\begin{bmatrix} \hat{\Sigma}_{11} & \hat{\Sigma}_{12} & -\kappa_1 & p_1 a(z) c_f & \hat{\Sigma}_{15} & 0 & \sqrt{\delta} c_f^T \\ * & \hat{\Sigma}_{22} & -\kappa_2 & 0 & 0 & p_2 b_f(z) - (1-\delta) c_f^T & 0 \\ * & * & \hat{\Sigma}_{33} & 0 & 0 & 0 & \sqrt{\delta} c_f^T \\ * & * & * & -\frac{\Pi^2}{4\phi_s}(p_2 + p_2^T) & 0 & -(1-\delta) c_f^T & 0 \\ * & * & * & * & -\gamma & 0 & 0 \\ * & * & * & * & * & -\gamma & 0 \\ * & * & * & * & * & * & -\gamma \end{bmatrix} < 0 \quad (22)$$

where

$$\begin{aligned} \hat{\Sigma}_{11} &= p_1 \beta(z) + \beta^T(z) p_1^T + p_1 \alpha - \kappa_1 - \kappa_1^T, \\ \hat{\Sigma}_{12} &= p_1 a(z) c_f(z) - \kappa_2^T, \\ \hat{\Sigma}_{15} &= p_1 b(z) \\ \hat{\Sigma}_{22} &= p_2 a_f(z) + a_f^T(z) p_2^T + p_2 \alpha, \\ \hat{\Sigma}_{33} &= -\frac{\Pi^2}{4\phi_s}(p_1 + p_1^T), \end{aligned}$$

then system (7) is exponentially stable with mixed $\mathcal{H}_\infty$ and passivity performance level γ and observer gains are given by

$$k_1 = p_1^{-1}\kappa_1, \quad k_2 = p_2^{-1}\kappa_2.$$

Proof The proof of the above theorem is immediately follows from (17) that

$$\begin{bmatrix} \check{\Sigma}_{11} & -PK + \gamma^{-1}\delta C_f^T C_f & PB(z) - (1-\delta) C_f^T \\ * & -\frac{\Pi^2}{4\phi_s}(PA + A^T P^T) + \gamma^{-1}\delta C_f^T C_f & (1-\delta) C_f^T \\ * & * & -\gamma \end{bmatrix} < 0$$

where $\check{\Sigma}_{11} = PC(z) + C^T(z)P^T + P\alpha(z) - PK + \gamma^{-1}\delta C_f^T C_f$.
Then, by applying Schur complement it is easy to obtain (22). □

Remark 1 It is noted that the mixed $\mathcal{H}_\infty$ and passivity performance index presented above turns to be a $\mathcal{H}_\infty$ performance index when $\delta = 1$.

As mentioned earlier, the main objective in the $\mathcal{H}_\infty$ analysis is to minimize the performance index on the residual signal and fault signal to increase the robustness, To achieve this next the optimization problem to obtain a minimum $\mathcal{H}_\infty$ performance index value will be presented by using the results shown in Theorems 1 and 3.

Corollary 1 *Let the scalar $\alpha > 0$, gains k_1 and k_2 are known. The system* (10) *is exponentially stable with $\mathcal{H}_\infty$ performance level γ, there exist a positive-definite symmetric matrix P such that the optimization problem:*

$$\min \gamma \tag{23}$$

subject to the LMI

$$\begin{bmatrix} \bar{\Omega}_{11} & -PK + C_f^T C_f & PB(z) \\ * & \bar{\Omega}_{22} & 0 \\ * & * & -\gamma^2 \end{bmatrix} < 0,$$

is solvable,

$$\bar{\Omega}_{11} = PC(z) + C^T(z)P^T + P\alpha - PK + C_f^T C_f,$$
$$\bar{\Omega}_{22} = -\frac{\Pi^2}{4\phi_s}(PA + A^T P^T) + C_f^T C_f.$$

Corollary 2 *Let the scalar $\alpha > 0$ is known. The system* (10) *is exponentially stable with $\mathcal{H}_\infty$ performance level γ, there exist a positive-definite symmetric matrix diag$\{p_1, p_2\}$ such that the optimization problem:*

$$\min \gamma \tag{24}$$

subject to the LMI

$$\begin{bmatrix} \bar{\Sigma}_{11} & \bar{\Sigma}_{12} & -\kappa_1 & p_1 a(z) c_f & p_1 b(z) & 0 \\ * & \bar{\Sigma}_{22} & -\kappa_2 & c_f^T c_f & 0 & p_2 b_f(z) \\ * & * & \bar{\Sigma}_{33} & 0 & 0 & 0 \\ * & * & * & -\frac{\Pi^2}{4\phi_s}(p_2 + p_2^T) + c_f^T c_f & 0 & 0 \\ * & * & * & * & -\gamma^2 & 0 \\ * & * & * & * & * & -\gamma^2 \end{bmatrix} < 0 \tag{25}$$

is solvable, where

$$\bar{\Sigma}_{11} = p_1 \beta(z) + \beta^T(z) p_1^T + p_1 \alpha - \kappa_1 - \kappa_1^T$$
$$\bar{\Sigma}_{12} = p_1 a(z) c_f(z) - \kappa_2^T,$$
$$\bar{\Sigma}_{22} = p_2 a_f(z) + a_f^T(z) p_2^T + p_2 \alpha + c_f^T c_f,$$
$$\bar{\Sigma}_{33} = -\frac{\Pi^2}{4\phi_s}(p_1 + p_1^T)$$

and further the observer gains are given by

$$k_1 = p_1^{-1}\kappa_1, \quad k_2 = p_2^{-1}\kappa_2.$$

6 Numerical Example

In this section, two numerical examples are presented to show the effectiveness of the proposed method.

Example 1 Consider the system (1) with distributed parameter $\beta(z)$ has the values bounded by 0.01 and the fault value has the shape function

$$a(z) = \begin{cases} 0.5 & 0.7 \le z \le 1, \\ 0 & \text{otherwise} \end{cases}$$

and the disturbance input has the shape function

$$b(z) = \begin{cases} 0.7 & 0 \le z \le 0.3 \\ 0, & \text{otherwise.} \end{cases}$$

Also, the weight functions of the observer system (3) is defined as

$$a_f(z) = -0.5, \quad b_f(z) = 0.05, \quad c_f(z) = 0.05.$$

Then, for the given values of exponential decay rate $\alpha = 0.01$ and the $\mathcal{H}_\infty$ attenuation level $\gamma = 0.5$, by using the LMI conditions in Theorem 2, the observer gains are attained as

$$k_1 = 0.5988, \quad k_2 = 0.0035.$$

Further, it is easy to verify that the exponential stability condition in (16) will also be satisfied, so it can be concluded that the considered PDE is exponentially stable with a given prescribed performance level γ.

Moreover, to reflect the effectiveness of the developed fault design scheme, simulation results are presented in Figs. 1, 2, 3 and 4. For this, it is assumed the disturbance signal $w(t)$ be a arbitrary function in $L_2[0, \infty)$.

The simulation results of the considered system state and observer state are plotted in Figs. 1 and 2. It is clear from shown figures that the obtained gains for the PDEs with fault have been tracked well and the observer design scheme proposed in this chapter produces a good performance. To confirm this, the simulation results for fault and its estimation are shown in Figs. 3 and 4. It is clear that the fault signal given in Fig. 3 has been estimated through the proposed results and the estimated one is

Fig. 1 Response of $w(x, t)$

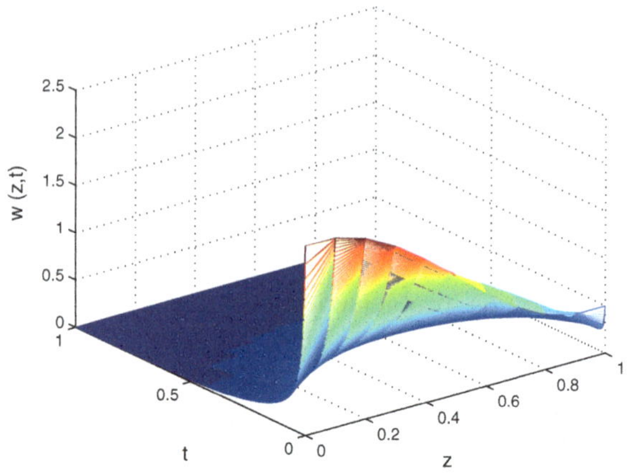

Fig. 2 Response of $\hat{w}(x, t)$

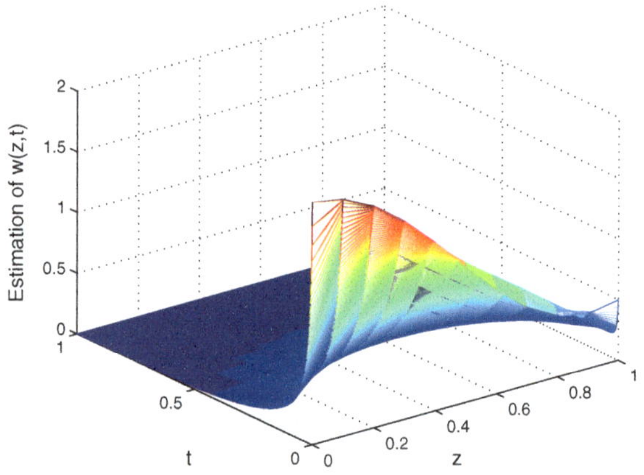

shown in Fig. 4. So, it can be concluded that using the proposed results a suitable fault estimator can be found with prescribed $\mathcal{H}_\infty$ performance attenuation level γ.

Example 2 Consider the faulty heat conducting rod

$$\partial_t w(z, t) = \partial_z^2 w(z, t) + a(z) f(t) + b(z) d(t),$$

with the shape function of the fault value given by

$$a(z) = \begin{cases} 1 : 0.4 \leq z \leq 0.6 \\ 0 : \quad \text{otherwise}, \end{cases}$$

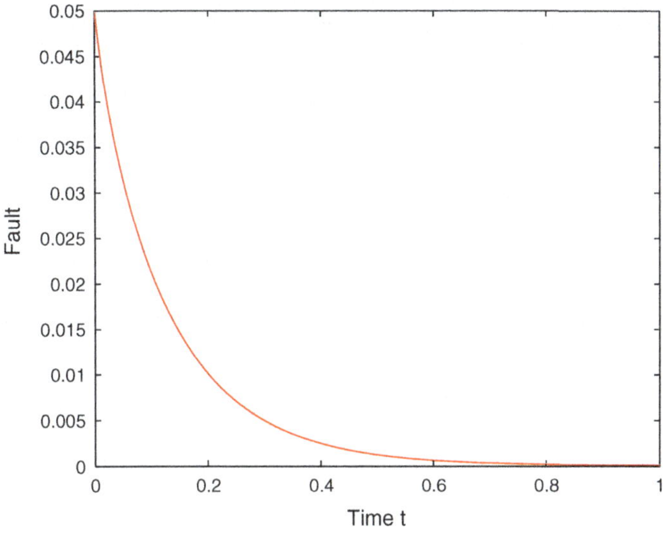

Fig. 3 Fault signal $f(t)$

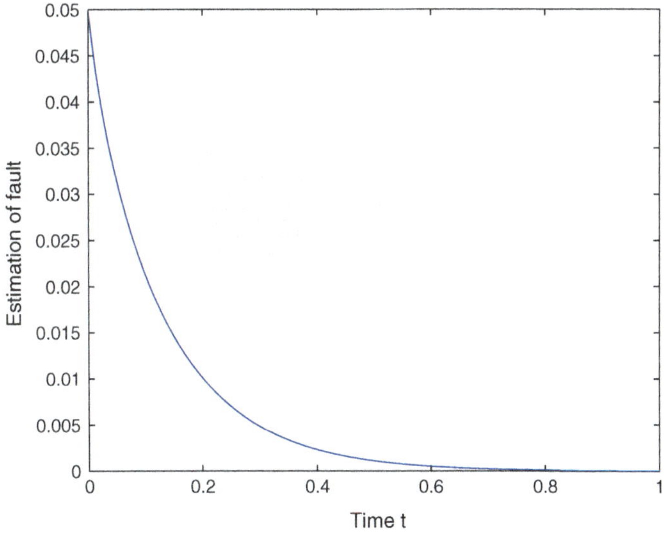

Fig. 4 Estimation of fault signal $\hat{f}(t)$

and the external disturbance input has the shape function

$$b(z) = \begin{cases} 0.5 : 0 \leq z \leq 0.6 \\ 0 : \quad \text{otherwise.} \end{cases}$$

Also, the weight functions of the observer system (3) is defined as

$$a_f(z) = -1.5, \quad b_f(z) = 0.05, \quad c_f(z) = 0.05.$$

Now, for $\alpha = 0.05$, the LMI conditions in Corollary 2 has feasible solution with the observer gains

$$k_1 = 0.6094, \quad k_2 = 0.0094,$$

and optimum $\mathcal{H}_\infty$ attenuation level γ is obtained as 0.093.

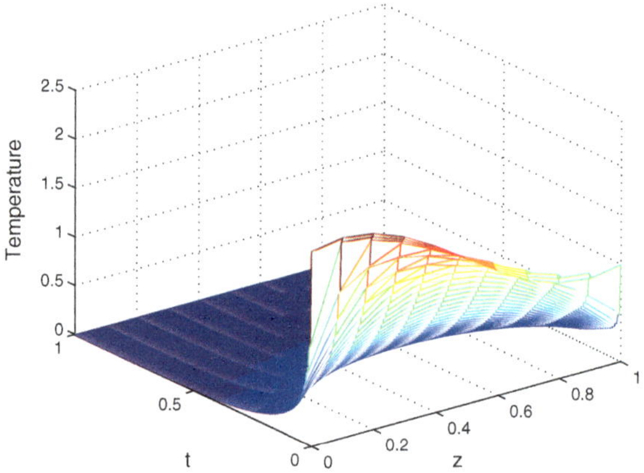

Fig. 5 Temperature profile $w(x, t)$

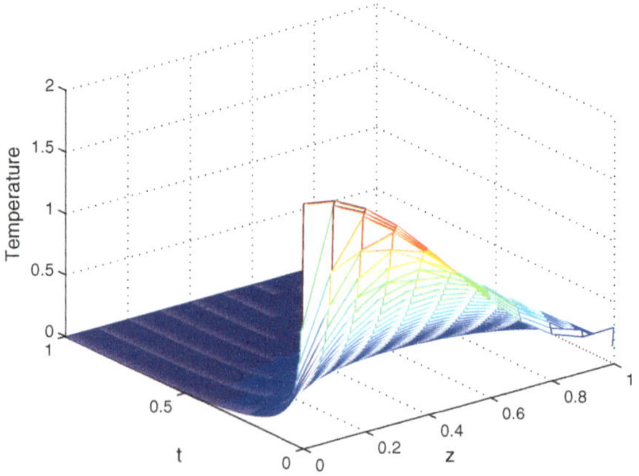

Fig. 6 Estimation of temperature profile $\hat{w}(x, t)$

Finally, in order to show the effectiveness of the developed fault design scheme, simulation results are presented in Figs. 5 and 6. The simulation results of the considered system state and observer state are shown and it is clear that the obtained gains for the heat conducting rod have been observed well and the proposed design scheme produces a good performance.

7 Conclusion

$\mathcal{H}_\infty$ fault estimation problem and exponential stability for a parabolic system has been discussed. Most of the dynamical systems can be modelled by PDEs, sometimes such systems affected by the fault. In such cases, the prior knowledge of the fault is incorporated by a filter. Then, an observer designed to estimate both the fault and the states of the considered system. Exponential stability was discussed for the system without external disturbance. Then, the results are extended to guarantee the improved robustness ensure that the system is the $\mathcal{H}_\infty$ performance in terms of an optimization problem. Also, the results for mixed $\mathcal{H}_\infty$ and passivity performance based fault estimation results have presented. Finally, numerical examples are presented to validate the results, the simulation results in the example show that the performance and effectiveness of the developed fault design scheme.

Acknowledgements The work was supported by University Grants Commission (UGC), New Delhi, India through BSR Start-up Research Grant under the grant No. F.30-410/2018(BSR).

References

1. Boskovic, D.M., Krstic, M.: Stabilization of a solid propellent rocket instability by state feedback. Int. J. Robust Nonlinear **13**, 483–495 (2003)
2. El-Farra, N.H., Armaou, A., Christofides, P.D.: Analysis and control of parabolic PDE systems with input constraints. Automatica **39**, 715–725 (2003)
3. Mazenc, F., Prieur, C.: Strict Lyapunov functions for semilinear parabolic partial differential equations. Math. Control Relat. Fields AIMS **1**, 231–250 (2011)
4. Moura, J., Chaturvedi, N.A., Krstic, M.: PDE estimation techniques for advanced battery management systems - Part I: SOC estimation. In: ACC, pp. 559–565 (2012)
5. Papachristodoulou, A., Peet, M.M.: On the analysis of systems described by classes of partial differential equations. In: IEEE Decision and Control (2006)
6. Friedman, A.: Partial Differential Equations. Dover Publications, Incorporated (2008)
7. Wang, Z.P., Wu, H.N., Li, H.X.: Fuzzy control under spatially local averaged measurements for nonlinear distributed parameter systems with time-varying delay. IEEE Trans. Cybern. **99**, 1–11 (2019)
8. Shearer, M., Levy, R.: Partial Differential Equations: An Introduction to Theory and Applications. Princeton University Press (2015)
9. Christofides, P.D.: Nonlinear and Robust Control of PDE Systems: Methods and Applications to Transport Reaction Processes. Springer (2001)
10. Garg, S.: Controls and health management technologies for intelligent aerospace propulsion systems. In: 42nd AIAA Aerospace Sciences Meeting and Exhibit (2004)

11. Liu, X., Yuan, S.: On designing $\mathcal{H}_\infty$ fault estimator for switched nonlinear systems of neutral type. Commun. Nonlinear Sci. Numer. Simul. **16**, 4379–4389 (2011)
12. Omatu, S., Seinfeld, J.H.: Distributed Parameter Systems: Theory and Applications. Clarendon Press (1989)
13. Boskovic, D.M., Krstic, M., Weijiu, L.: Boundary control of an unstable heat equation via measurement of domain-averaged temperature. IEEE Trans. Autom. Control **46**, 2022–2028 (2001)
14. Troltzsch, F.: Optimal control of partial differential equations: theory, methods and applications. AMS **112**, 49–01 (2000)
15. Cheng, M.B., Radisavljevic, V., Su, W.C.: Sliding mode boundary control of a parabolic pde system with parameter variations and boundary uncertainties. Automatic **47**, 381–387 (2011)
16. Glashoff, W.N.: Boundary control of parabolic differential equations in arbitrary dimensions: supremum-norm problems. SIAM J. Control Optim. **14**, 662–681 (1976)
17. Gordji, M.E., Ghaemi, M.B., Alizadeh, B.: Stability of the second order partial differential equations. J. Inequal. Appl. **81**, 1–10 (2011)
18. Krstic, M.: Lyapunov adaptive stabilization of parabolic PDEs-part I: a benchmark for boundary control. In: IEEE Decision and Control (2005)
19. Marzat, J., et al.: Model-based fault diagnosis for aerospace systems: a survey proceedings of the institution of mechanical engineers, Part-G. J. Aerosp. Eng. **226**, 1329–1360 (2012)
20. Park, J.H., Shen, H., Chang, X.H., Lee, T.H.: Recent Advances in Control and Filtering of Dynamic Systems with Constrained Signals. Springer, Cham, Switzerland (2018). https://doi.org/10.1007/978-3-319-96202-3
21. Shen, H., Park, J.H., Wu, Z.G.: Finite-time reliable $\mathcal{L}_2 - \mathcal{L}_\infty/\mathcal{H}_\infty$ control for Takagi-Sugeno fuzzy systems with actuator faults. IET Control Theory Appl. **8**(9), 688–696 (2014)
22. Liu, G., Park, J.H., Xu, X., Zhuang, G.: Robust non-fragile $\mathcal{H}_\infty$ fault detection filter design for delayed singular Markovian jump systems with linear fractional parametric uncertainties. Nonlinear Anal. Hybrid Syst. **32**, 65–78 (2019)
23. Ferdowsi, H., Jagannathan, S.: Fault diagnosis of a class of distributed parameter systems modeled by parabolic partial differential equations. In: ACC, pp. 5434–5439 (2014)
24. Demetriou, M.A.: A model-based fault detection and diagnosis scheme for distributed parameter system: a learning systems approach. ESAIM Control Optim. Calc. Var. **7**, 43–67 (2002)
25. Ghantasala, S., El-Flarra, N.H.: Active fault-tolerant control of sampled-data nonlinear distributed parameter systems. Int. J. Robust Nonlinear **22**, 24–42 (2011)
26. Ghantasala, S., El-Flarra, N.H.: Robust actuator fault isolation and management in constrained uncertain parabolic PDE systems. Automatica **45**, 2368–2373 (2009)
27. Baruh, H.: Actuator failure detection in the control of distributed systems. J. Guid Control Dyn. **9**, 181–189 (1986)
28. Fridman, E., Terushkin, M.: New stability and exact observability conditions for semilinear wave equations. Automatica **63**, 1–10 (2016)
29. Park, J.H., Mathiyalagan, K., Sakthivel, R.: Fault estimation for discrete-time switched nonlinear systems with discrete and distributed delays. Int. J. Robust Nonlinear **26**(17), 3755–3771 (2016)
30. Yuan, S., Liu, X.: Fault estimator design for a class of switched systems with time-varying delay. Int. J. Syst. Sci. **42**, 2125–2135 (2011)
31. Zhang, K., et al.: Observer-based fault estimation techniques. In: Studies in Systems, Decision and Control, p. 127. Springer (2017)
32. Ferdowsi, H., Jagannathan, S.: Fault diagnosis of distributed parameter systems modeled by linear parabolic partial differential equations with state faults. J. Dyn. Syst-T ASME **140**, 011–010 (2017)
33. Mathiyalagan, K., Park, J.H., Sakthivel, R.: Observer-based dissipative control for networked control systems: a switched system approach. Complexity **21**(2), 297–308 (2016)
34. James, M.R.: A partial differential inequality for dissipative nonlinear systems. Syst. Control Lett. **21**, 315–320 (1993)

35. Ku, C.C., Chen, R.W.: Mixed $\mathcal{H}_\infty$/passivity performances control of discrete time linear stochastic systems. J. Mar. Sci. Technol. **25**, 424–434 (2017)
36. Baniamerian, A., Khorasani, K.: Fault detection and isolation of dissipative parabolic partial differential equations: finite dimensional geometric approach. In: ACC, pp. 5894–5899 (2012)
37. Fridman, E., Orlov, Y.: An LMI approach to $\mathcal{H}_\infty$ boundary control of semilinear parabolic and hyperbolic systems. Automatica **45**, 2060–2066 (2009)
38. Fischer, F., Deutscher, J.: Fault detection for parabolic systems with distributed inputs and outputs using the modulation function approach. IFAC **50**, 6774–6779 (2017)
39. Mathiyalagan, K., Park, J.H., Sakthivel, R., Marshal Anthoin, S.: Robust mixed $\mathcal{H}_\infty$ and passive filtering for networked Markov jump systems with impulses. Signal Process. **101**, 162–173 (2014)

Switched Systems and Hybrid Control

Event-Triggered Control of Discrete-Time Switched Linear Systems via Communication Channels with Limited Capacity

Xiaoqing Xiao, Lei Zhou, and Ju H. Park

Abstract This chapter considers the state feedback control problem for discrete-time switched linear systems via communication channels with limited capacity. It is assumed that the controller only accesses the successfully event-triggered transmitted signal of switching mode and system state. Considering the signal transmission delays and packet dropout respectively, switching mode based event-triggered transmission conditions are proposed to mitigate the unnecessary data transmission. Based on the new switching signal generated by merging the original switching signal with event-triggering transmitted signal, the closed-loop systems become switched systems with time delay. Then, utilizing the average dwell time technical and multiple Lyapunov functional methods, exponential stability conditions for the closed-loop systems are derived. Furthermore, linear matrix inequalities based co-design methods of event-trigger parameters and state feedback controller gains are proposed. Numerical examples illustrate the effectiveness of obtained results.

Keywords Discrete-time switched linear systems · State feedback stabilization · Event-triggered mechanisms · Packet dropout · Transmission delay · Exponential stability

---

X. Xiao · L. Zhou (✉)
School of Information Science and Technology, Nantong University, Nantong, 226019 Jiangsu, People's Republic of China
e-mail: zlxxq80@ntu.edu.cn

X. Xiao
e-mail: xiaoxqzl@126.com

J. H. Park
Department of Electrical Engineering, Yeungnam University, 280 Daehak-Ro, Gyeongsan 38541, Republic of Korea
e-mail: jessie@ynu.ac.kr

J. H. Park (ed.), *Recent Advances in Control Problems of Dynamical Systems and Networks*, Studies in Systems, Decision and Control 301,
https://doi.org/10.1007/978-3-030-49123-9_8

1 Introduction

Switched systems, which consist of a finite number of subsystems and a switching law that determines the operation order between them, are a special type of hybrid systems. Many applications for switched systems have been found in chemical processes, power electronics and other practical fields. Furthermore, in order to improve the trade-off of conflicting design objectives in some complex situations, the multiple controllers switching technique is applied, which also leads to switched systems with special switching signals. Therefore, considerable attentions have been received for the study of switched systems and many efficient methods have been proposed recently, please refer to the books [1, 2], survey papers [3, 4], and some recent published papers [5–8].

In the traditional control of switched systems, the assumption that the controller could acquire the exact knowledge of both the switching signal and system measurement simultaneously was made, then the synchronization of the switching among controllers and systems could be guaranteed. On the other hand, with the rapid development of communication networks, microelectronics technology and high-speed computation, the system control via communication networks has become a hot research topic in recent decades. However, the limited capacity of the communication channel and unreliability of the communication network links lead to network induced delay and packet losses inevitably, which might cause performance degradation or even instability of system [9–13].

Furthermore, when the controllers are connected with systems via communication channels with limited capacity, they can only receive the transmitted signal of the switching mode and system measurement. In this case, the mode of controllers might lag behind that of the systems, and then the switching between systems and controllers becomes asynchronous, which will be even more serious when the transmission delays or packet losses exist. Therefore, the introduction of the network into the control of switching systems will lead to the complexity of mode switching, which makes the control analysis and design of switched systems in the networked environment more complicated and challenging. Based on the estimation of ultimate bounds or reachable sets of the system states, the authors in [14, 15] considered the quantized stabilization problem of switched systems. When there exists network-induced transmission delay, networked stabilization of switched linear systems is considered in [16], where it is assumed that the data sampling and system switching are with the same fixed period simultaneously. However, the controller has no information on the system switching signal in many practical applications. For example, in the vehicle suspension control design problem, the switched control strategy is determined by the calculation of motion-mode energy [17]; the switching law of the voltage controller in [18] is dependent on the measurement of the dc-bus voltage and load current. The control problem for switched systems with sampled data is studied in [19] with consideration of mismatch phenomenon. Sufficient conditions for exponential stabilization are presented by multiple Lyapunov functions method.

In networked control systems, one main concern is how to reduce the unnecessary data transmission since the communication network has constrained bandwidth. In the case of time-triggered communication, the control signal is transmitted at prescribed instants without regard to the system operation state, which might cause network congestion and waste of communication resources. Therefore, it is significant to investigate other better schemes to reduce redundant data transmission and inefficient utilization of the communication network. In the late nineties, the event-triggered mechanism for networked control of non-switched systems was proposed [20–23], which may keep the desired system performance and decrease the data transmission efficiently. Recently, wide applications of the event-triggered scheme have been found in many areas, for example, networked control [24, 25], networked filtering [26–29], consensus and synchronization [30–32], and so on.

Recently, the event-triggered scheme was extended to networked control for switched systems [33–35]. For example, based on the assumption that system only switches at sampling instants, the sampled-data control of the continuous switched system with event-triggered scheme was considered in [36]. Taking an asynchronization between system and controller into consideration, a mode based event-triggered transmission scheme and switched controllers were co-designed in [37] for a switched linear system, which was recently extended to the investigation of switched affine systems in [38] by event-triggered control approach with periodic sampling. By using the common Lyapunov function method, an event-triggered algorithm for the feedback stabilization of switched linear systems was proposed in [39]. The optimal finite-time event-triggered controller for switched systems was designed in [40]. However, the results above-mentioned are mainly about switched systems in the continuous-time domain. To the best of our knowledge, few results are reported for the discrete-time case when considering the event-triggered transmission.

The objective of this chapter is to address the event-triggered control problem of discrete-time switched linear systems via communication channels with limited capacity. It is assumed that the controller and switched system are connected by a limited capacity communication channel, and the controller can only receive the successfully transmitted signal of switching mode and system state. Considering the signal transmission delays and packet dropout respectively, mode based event-triggered transmission mechanisms are developed, and the switching signal merging technique is applied to generate a new switching signal of the augmented type. Then, considering the multi-mode state feedback controller, the closed-loop system becomes a discrete-time switched system with the delayed state and augmented switching signal. Utilizing the average dwell time technology and multiple Lyapunov functional approaches, stability conditions are obtained such that the closed-loop systems is exponentially stable. Furthermore, event-trigger parameters and feedback controller gains are designed by the feasible solution of a series of linear matrix inequalities. Numerical examples are given to illustrate the effectiveness of our proposed event-triggered design approach. The obtained results extend the input delay approach and event-triggered control mechanism of non-switched systems to discrete-time switched systems, which will be useful for further study of event-triggering analysis and synthesis for discrete-time switched systems.

2 System Description and Some Preliminaries

The discrete-time switched linear systems to be controlled over limited capacity channel in this chapter is as follows.

$$x(k+1) = A_{\sigma(k)}x(k) + B_{\sigma(k)}u(k), \tag{1}$$

where $x(k) \in \mathcal{R}^n$ is the system state and $u(k) \in \mathcal{R}^m$ is control input. The piecewise constant function $\sigma(k) : \mathbf{N}^+ \to \mathcal{M} = \{1, 2, \ldots, p\}$ denotes the switching law, p is the number of switching modes. A_i and $B_i, \forall i \in \mathcal{M}$ are constant matrices of proper dimension. For simplicity, we denotes the i-th subsystem as (A_i, B_i) and make the following assumption.

Assumption 1 Assume that each subsystem (A_i, B_i) is stabilizable, $\forall i \in \mathcal{M}$. □

For any time interval (k_1, k_2), let $N_\sigma(k_1, k_2)$ be the number of switches for the switching signal $\sigma(k)$ and we give the following definition.

Definition 1 [1] Given the switching signal $\sigma(k)$, two constants τ_d and τ_a.

(1) τ_d is the dwell time of $\sigma(k)$ if $N_\sigma(k_1, k_2) \leq 1$ when $k_2 - k_1 \leq \tau_d$ for some constant $\tau_d > 0$.
(2) τ_a is the average dwell time of $\sigma(k)$ if $N_\sigma(k_1, k_2) \leq N_0 + \frac{k_2-k_1}{\tau_a}$, $\forall\, k_2 \geq k_1 \geq 0$ for a positive number N_0, and N_0 is called as the chatter bound of $\sigma(k)$.

For simplicity, let $\mathcal{S}[\tau_a, N_0]$ be the set of the switching signals whose average dwell time and chatter bound are τ_a and N_0 respectively. The following assumption for the $\sigma(k)$ in the system (1) is necessary.

Assumption 2 The dwell time of $\sigma(k)$ is τ_d and $\sigma(k)$ belongs to the set $\mathcal{S}[\tau_a, N_0]$. □

Then, based on Assumptions 1 and 2 and the classical results for switched system, exponential stability of the closed-loop system of (1) under a proper feedback controller $u(k) = K_{\sigma(k)}x(k)$ is guaranteed if τ_d or τ_a of the switching signal is large enough.

In this chapter, the primary objective is how to maintain the stability when system (1) is controlled over limited capacity channel, where the switching signal and system state need to be detected and transmitted remotely. To this end, we firstly present some preliminary results of the merging technique on switching signal, which might be viewed as the counterpart of that in [41] for the discrete-time case.

Suppose that $\hat{\sigma}(k) = \sigma(k - \tau(k))$ is the delayed signal of $\sigma(k)$, where the time-varying delay $\tau(k) \in [\tau_1, \tau_2)$. Then, for the switching signal $\sigma(k)$ satisfying Assumption 2, we define the augmented signal $\theta(k) = (\sigma(k), \hat{\sigma}(k))$ by merging $\sigma(k)$ with its delayed signal $\hat{\sigma}(k)$. Based on the similar analysis as that in Lemmas 1 and 2 in [41], we can get the following conclusion.

Lemma 1 *Suppose that the switching signal $\sigma(k)$ belongs to the set $\mathcal{S}[\tau_a, N_0]$. If the delay $\tau(k) \in [\tau_1, \tau_2)$, then $\hat{\sigma}(k) = \sigma(k - \tau(k)) \in \mathcal{S}\left[\tau_a, N_0 + \frac{\tau_2-\tau_1}{\tau_a}\right]$ and $\theta(k) = (\sigma(k), \hat{\sigma}(k)) \in \mathcal{S}\left[\frac{\tau_a}{2}, 2N_0 + \frac{\tau_2-\tau_1}{\tau_a}\right]$.*

Suppose that $T_s(k_1, k_2)$ denotes the total synchronous time in any given time interval $[k_1, k_2)$ for $\sigma(k)$ and $\hat{\sigma}(k)$. Then, $T_{as}(k_1, k_2) = k_2 - k_1 - T_s(k_1, k_2)$ denotes the asynchronous time for $\sigma(k)$ and $\hat{\sigma}(k)$ during $[k_1, k_2)$. In addition, $[\bar{k}_i, \bar{k}_{i+1})$ denotes any switching interval of $\sigma(k)$ and $\bar{\tau}_a$ be the upper bound of the asynchronous time length for $\sigma(k)$ and $\hat{\sigma}(k)$ during $[\bar{k}_i, \bar{k}_{i+1})$. Similar to Lemma 3 of [41], we have the conclusion as follows.

Lemma 2 *Given scalar constants $\eta > 0$, $\rho \in (0, 1)$ and $\bar{\beta} \in (1 - \rho, 1)$. Suppose that $(1 + \eta)^{\bar{\tau}_a}(1 - \rho)^{\tau_a - \bar{\tau}_a} \leq \bar{\beta}^{\tau_a}$, then the inequality $(1 - \rho)^{T_s(k_1,k_2)}(1 + \eta)^{T_{as}(k_1,k_2)} \leq \bar{\alpha}\bar{\beta}^{(k_2-k_1)}$ holds, where $\bar{\alpha} = \left(\frac{1+\eta}{1-\rho}\right)^{N_0\bar{\tau}_a}$.*

3 Event-Triggered Control of Discrete-Time Switched Linear Systems with Package Dropout

The event-triggered control framework for the considered switched systems (1) is illustrated in Fig. 1, in which the event generator is proposed to determine transmission instants of the switching signal and system state. The corresponding event-triggered transmission time sequence is denoted by $\{k_s\}_{s\geq 0}$ with $k_0 = 0$.

The event-triggered transmission scheme in this section is designed as follows:

$$\begin{aligned} k_{s+1} &= \min\left\{k_s + \tau_d, \hat{k}_{s+1}\right\}, s \geq 1, \\ \hat{k}_{s+1} &= \min_{k>k_s}\{k|(x(k) - x(k_s))^T \Phi_{\sigma(k_s)}(x(k) - x(k_s)) \\ &\geq \nu x^T(k_s)\Phi_{\sigma(k_s)}x(k_s)\}, \end{aligned} \tag{2}$$

where the event-trigger parameters, to be determined, $\nu \in (0, 1)$ is a positive scalar and Φ_i is a positive defined matrix.

Let $\Omega_s = [k_s, k_{s+1})$ denote the holding interval generated by the proposed event-triggered scheme (2). Now, we propose the following state feedback controller generated by the zero-order holder (ZOH):

$$u(k) = K_{\sigma(k_s)}x(k_s), \ k \in \Omega_s. \tag{3}$$

It is obvious that the mode of controllers in (3) remains constant within each holding interval.

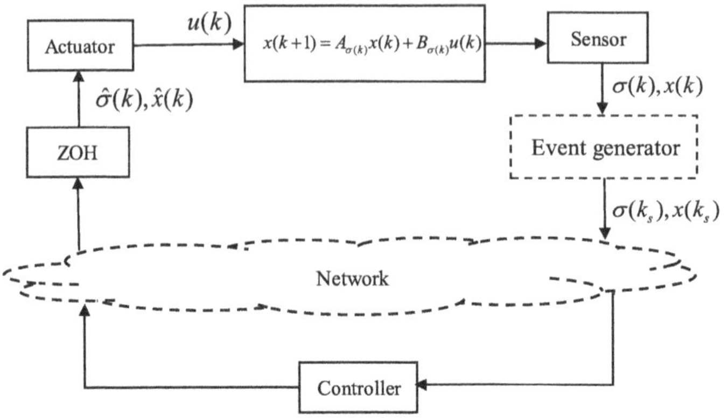

Fig. 1 The event-triggered control framework for switched system (1)

In addition, the event-triggering scheme (2) indicates that $k_{s+1} - k_s \leq \tau_d$, which combining Assumption 2 implies that the system might switch once within the holding interval Ω_s. Therefore, the switching modes of the controller and the system become mismatch after the system switches.

For $k \in \Omega_s$, let $e(k) = x(k) - x(k_s)$ and $\tau_1(k) = k - k_s \in [0, \tau_d)$, then $\hat{\sigma}(k) = \sigma(k - \tau_1(k))$. In addition, denote $\theta(k) = (\sigma(k), \hat{\sigma}(k))$ and $\widetilde{B}_{\theta(k)} = B_{\sigma(k)} K_{\hat{\sigma}(k)}$, then the closed-loop system of (1) and (3) can be represented as follows:

$$x(k+1) = A_{\sigma(k)} x(k) + \widetilde{B}_{\theta(k)} \left(x(k) - e(k) \right). \tag{4}$$

Furthermore, the event-triggering condition in (2) implies that

$$\nu(x(k) - e(k))^T \Phi_{\sigma(k_s)} (x(k) - e(k)) - e^T(k) \Phi_{\sigma(k_s)} e(k) > 0, \ k \in \Omega_s. \tag{5}$$

3.1 Stability Analysis Without Pack Dropout

The first main result is obtained as follows.

Theorem 1 *Assume that there exist positive definite matrices P_{ij} and Φ_j such that the matrix inequalities (6) and (7) hold for the given scalars $\mu \geq 1$, $\eta > 0$, $\rho \in (0, 1)$, $\nu \in (0, 1)$ and state feedback gains K_j. Then, under event-triggered scheme (2), the closed-loop system (4) is exponentially stable if the switching signal satisfies the average dwell time condition (8).*

$$\mu^{-1} P_{ii} \leq P_{ij} \leq \mu P_{jj}, \ P_{ii} \leq \mu P_{jj}, \ \forall i, j \in \mathcal{M}, \tag{6}$$

$$\begin{bmatrix} -\lambda_{ij}^2 P_{ij} + \nu\Phi_j & -\nu\Phi_j & (A_i^T + \tilde{B}_{ij}^T)P_{ij} \\ * & (\nu - 1)\Phi_j & -\tilde{B}_{ij}^T P_{ij} \\ * & * & -P_{ij} \end{bmatrix} < 0,\ \forall i, j \in \mathcal{M}, \tag{7}$$

$$\lambda_{ij} = \delta_{ij}(1-\rho) + (1-\delta_{ij})(1+\eta),$$

$$-\tau_a \ln(1-\rho) > \ln\mu + \tau_d \ln\frac{1+\eta}{1-\rho}. \tag{8}$$

Proof To obtain the conclusion in Theorem 1, we construct a Lyapunov function as follows:

$$V_{\theta(k)}(k) = x^T(k)P_{\theta(k)}x(k). \tag{9}$$

Let $\bar{k}$ be an any switching instant of $\theta(k)$. In other words, $\bar{k}$ is some transmission instant or switching instant of $\sigma(k)$. From (6) we have

$$V_{\theta(\bar{k})}(\bar{k}) \le \mu V_{\theta(\bar{k}-1)}(\bar{k}). \tag{10}$$

Then, for $k \in \Omega_s$, considering the event-triggering condition (5), we compute the difference $V_{\theta(k)}(k+1) - \lambda_{\theta(k)}^2 V_{\theta(k)}(k)$ along the closed-loop system (4) as:

$$\begin{aligned} & V_{\theta(k)}(k+1) - \lambda_{\theta(k)}^2 V_{\theta(k)}(k) \\ & = x^T(k+1)P_{\theta(k)}x(k+1) - \lambda_{\theta(k)}^2 x^T(k)P_{\theta(k)}x(k) \\ & \le (A_{\sigma(k)}x(k) + \widetilde{B}_{\theta(k)}(x(k)-e(k)))^T P_{\theta(k)}(A_{\sigma(k)}x(k) + \widetilde{B}_{\theta(k)}(x(k)-e(k))) \\ & \quad - \lambda_{\theta(k)}^2 x^T(k)P_{\theta(k)}x(k) - e^T(k)\Phi_{\hat{\sigma}(k)}e(k) \\ & \quad + \nu(x(k)-e(k))^T\Phi_{\hat{\sigma}(k)}(x(k)-e(k)) \\ & = \zeta^T(k)\Omega_{\theta(k)}\zeta(k), \end{aligned} \tag{11}$$

where

$$\zeta(k) = \left[x^T(k)\ e^T(k)\right]^T,$$

$$\lambda_{\theta(k)} = \begin{cases} 1+\eta, & k \in T_{as}(\Omega_s); \\ 1-\rho, & k \in T_s(\Omega_s), \end{cases}$$

$$\Omega_{\theta(k)} = \begin{bmatrix} -\lambda_{\theta(k)}^2 P_{\theta(k)} & 0 \\ * & -\Phi_{\hat{\sigma}(k)} \end{bmatrix} + \nu\left[I\ -I\right]^T \Phi_j \left[I\ -I\right] + \tilde{A}_{\theta(k)}^T P_{\theta(k)} \tilde{A}_{\theta(k)},$$

$$\tilde{A}_{\theta(k)} = \left[A_{\sigma(k)} + \tilde{B}_{\theta(k)}\ -\tilde{B}_{\theta(k)}\right].$$

Note that at most one switching exists in $\Omega_s = [k_s, k_{s+1})$ for $\theta(k)$, therefore, we consider the following two cases respectively.

Firstly, one switching exists for $\theta(k)$ in $[k_s, k_{s+1})$. In general, let the switching instant be $\bar{k}$ and $\sigma(\bar{k}) = i$, $\sigma(k) = j$ for $k \in [k_s, \bar{k})$. Therefore, $\theta(k) = (j, j)$ and

$\Omega_{\theta(k)} = \Omega_{jj}$ when $k_s \le k < \bar{k}$; $\theta(k) = (i, j)$ and $\Omega_{\theta(k)} = \Omega_{ij}$ when $\bar{k} \le k < k_{s+1}$. Secondly, if $\theta(k)$ has no switch in the time interval $[k_s, k_{s+1})$, we have $\theta(k) = (j, j)$ and $\Omega_{\theta(k)} = \Omega_{jj}$ for $k_s \le k < k_{s+1}$.

Now, applying Schur complement lemma to the matrix inequality (7), we obtain $\Omega_{ij} < 0$ for $i, j \in \mathcal{M}$. Therefore, from (11), we have

$$V_{\theta(k)}(k+1) - \lambda_{\theta(k)}^2 V_{\theta(k)}(k) < 0, \ k \in \Omega_s. \tag{12}$$

To prove that the closed-loop system (4) is exponentially stable, we should analyze the jump behavior of the Lyapunov function (9) after each switching instant of the signal $\theta(k)$. In general, assume that $\bar{k}_1, \ldots, \bar{k}_{N_\theta(0,k)}$ are the switching instants of $\theta(k)$ during the arbitrary interval $(0, k)$ and denote $\bar{k}_0 = 0$ and $\bar{k}_{N_\theta(0,k)+1} = k$. Then, it follows from (10) and (12) that

$$\begin{aligned}
V_{\theta(k)}(k) &\le \lambda_{\theta(k)}^2 V_{\theta(k)}(k-1) \\
&\le \lambda_{\theta(k)}^4 V_{\theta(k)}(k-2) \\
&\le \cdots \\
&\le \lambda_{\theta(k)}^{2(k-\bar{k}_{N_\theta(0,k)})} V_{\theta(k)}(\bar{k}_{N_\theta(0,k)}) \\
&\le \lambda_{\theta(k)}^{2(k-\bar{k}_{N_\theta(0,k)})} \mu V_{\theta(\bar{k}_{N_\theta(0,k)}-1)}(\bar{k}_{N_\theta(0,k)}) \\
&\le \mu\lambda_{\theta(k)}^{2(k-\bar{k}_{N_\theta(0,k)})} \lambda_{\theta(\bar{k}_{N_\theta(0,k)}-1)}^2 V_{\theta(\bar{k}_{N_\theta(0,k)}-1)}(\bar{k}_{N_\theta(0,k)}-1) \\
&\le \cdots \\
&\le \mu^{N_\theta(0,k)} \lambda_{\theta(k)}^{2(k-\bar{k}_{N_\theta(0,k)})} \lambda_{\theta(\bar{k}_{N_\theta(0,k)-1})}^{2(\bar{k}_{N_\theta(0,k)}-\bar{k}_{N_\theta(0,k)-1})} \cdots \lambda_{\theta(\bar{k}_1)}^{2(\bar{k}_2-\bar{k}_1)} V_{\theta(\bar{k}_0)}(\bar{k}_1) \\
&\le \mu^{N_\theta(0,k)} \lambda_{\theta(k)}^{2(k-\bar{k}_{N_\theta(0,k)})} \lambda_{\theta(\bar{k}_{N_\theta(0,k)-1})}^{2(\bar{k}_{N_\theta(0,k)}-\bar{k}_{N_\theta(0,k)-1})} \cdots \lambda_{\theta(\bar{k}_0)}^{2(\bar{k}_1-\bar{k}_0)} V_{\theta(\bar{k}_0)}(\bar{k}_0) \\
&\le (1-\rho)^{2T_s(0,k)}(1+\eta)^{2T_{as}(0,k)} \mu^{N_\theta(0,k)} V_{\theta(0)}(0).
\end{aligned} \tag{13}$$

In addition, note that $\tau_1(k) \in [0, \tau_d)$, then it follows from Lemma 1 that $\hat{\sigma}(k) \in \mathcal{S}[\tau_a, N_0 + \frac{\tau_d}{\tau_a}]$ and $\theta(k) \in \mathcal{S}[\frac{\tau_a}{2}, 2N_0 + \frac{\tau_d}{\tau_a}]$. Therefore, $N_\theta(0,k) \le 2N_0 + \frac{\tau_d}{\tau_a} + \frac{2k}{\tau_a}$. Furthermore, the inequality (8) yields $\mu^{\frac{1}{\tau_a}} < \frac{1}{1-\rho}\left(\frac{1-\rho}{1+\eta}\right)^{\frac{\tau_d}{\tau_a}}$.

Then, for any $\bar{\beta} \in \left((1-\rho)\left(\frac{1+\eta}{1-\rho}\right)^{\frac{\tau_d}{\tau_a}}, \mu^{-\frac{1}{\tau_a}}\right)$, we have $\mu^{\frac{1}{\tau_a}}\bar{\beta} \in (0, 1)$ with $\bar{\beta} \in (1-\rho, 1)$. Now combining with Lemma 2, it can be obtained from (13) that

$$\begin{aligned}
V_{\theta(k)}(k) &\le (1-\rho)^{2T_s(0,k)}(1+\eta)^{2T_{as}(0,k)} \mu^{N_\theta(0,k)} V_{\theta(0)}(0) \\
&\le \mu^{\left(2N_0+\frac{\tau_d}{\tau_a}+\frac{2k}{\tau_a}\right)} \bar{\alpha}^2 \bar{\beta}^{2k} V_{\theta(0)}(0) \\
&= \tilde{\alpha}\tilde{\beta}^{2k} V_{\theta(0)}(0),
\end{aligned}$$

where $\tilde{\alpha} = \mu^{2N_0 + \frac{\tau_d}{\tau_a}} \cdot \bar{\alpha}^2$, $\tilde{\beta} = \mu^{\frac{1}{\tau_a}} \cdot \bar{\beta} \in (0, 1)$. Thus, the exponential stability of closed-loop system (4) when there is no packet dropout is obtained. □

3.2 Stability Analysis with Packet Dropout

The influence of packet dropout on the stability of the considered switched system during the network transmission is considered in this subsection. It is assumed that the transmitted signal $x(k_s)$ and $\sigma(k_s)$ might be lost because of the limited network capacity in the event-triggered control framework shown in Fig. 1. Suppose that the maximum number of allowable successive packet dropouts is τ_D. To this end, we propose a new event-triggered scheme based on (2) as follows:

$$\begin{aligned} k_{s+1} &= \min\left\{k_s + \tau_d, \hat{k}_{s+1}\right\}, s \geq 1, \\ \hat{k}_{s+1} &= \min_{k > k_s}\{k | (x(k) - x(k_s))^T \Phi_{\sigma(k_s)}(x(k) - x(k_s)) \geq \tilde{\nu} x^T(k_s) \Phi_{\sigma(k_s)} x(k_s)\}, \end{aligned} \tag{14}$$

where Φ_i is the event-trigger parameter defined as (2) and $\tilde{\nu} > 0$ is an unknown scalar.

Considering the event-triggered mechanism (14), we will give a criterion for the exponential stability of the closed-loop system (4) with packet dropout based on a set of matrix inequalities.

Theorem 2 *Assume that there exist positive definite matrices P_{ij} and Φ_j such that the matrix inequalities (6) and (7) hold for the given scalars $\mu \geq 1$, $\eta > 0$, $\rho \in (0, 1)$, $\nu \in (0, 1)$, $\tilde{\nu} > 0$ and state feedback gain K_j. Then, the closed-loop system (4) is exponentially stable under event-triggered scheme (14) when the average dwell time τ_a satisfies (8) and τ_D satisfies*

$$\frac{\lambda_m}{\lambda_n} \psi(\tilde{\nu}, \tau_D) \leq \nu, \tag{15}$$

where

$$\begin{aligned} \psi(\tilde{\nu}, \tau_D) &= \left(\left(\alpha - \left(\sqrt{\tilde{\nu}} + 1\right)\right) \phi(\tau_D) + \sqrt{\tilde{\nu}} \phi(\tau_D + 1) + \tau_D \beta\right)^2, \\ \phi(\tau_D) &= \frac{1 - \alpha^{\tau_D} + \beta}{1 - \alpha} - \frac{\beta(1 - \alpha^{\tau_D})}{(1 - \alpha)^2}, \\ \alpha &= (1 + \sqrt{\tilde{\nu}})(1 + \alpha_1), \\ \alpha_1 &= \max_i \|A_i - I\|, \ \beta = \max_{i,j} \left\|B_i K_j\right\|, \\ \lambda_m &= \max_i \lambda(\Phi_i), \ \lambda_n = \min_i \lambda(\Phi_i). \end{aligned}$$

Proof In the case of packet dropout, we only need to show that the event-triggered mechanism (14) implies the event-triggered mechanism (2). Then, it follows from Theorem 1 that the closed-loop system (4) is exponentially stable.

Suppose that $[k_s, k_{s+1})$ is a triggered interval. Without loss of generality, we assume that there are τ_D unsuccessfully transmitted packets during $[k_s, k_{s+1})$ and denote

$$k_s = p_0 < p_1 < p_2 < \cdots < p_{\tau_D} < p_{\tau_D+1} = k_{s+1}.$$

For any $i = 1, \cdots, \tau_D$, let $k_s + l \in (p_i,\ p_{i+1})$, then

$$\begin{aligned}\|x(k_s+l) - x(k_s)\| &\le \|x(k_s+l) - x(p_i)\| + \|x(k_s) - x(p_i)\| \\ &\le \|x(p_i) - x(k_s)\| + \sqrt{\tilde{\nu}}\|x(p_i)\|.\end{aligned}$$

Furthermore, one has

$$\begin{aligned}\|x(p_i)\| &\le \|x(p_{i-1})\| + \|x(p_i - 1) - x(p_{i-1})\| + \|x(p_i) - x(p_i - 1)\| \\ &\le \|x(p_i) - x(p_i - 1)\| + (1 + \sqrt{\tilde{\nu}})\|x(p_{i-1})\|.\end{aligned} \tag{16}$$

It is assumed that no switching occurs in the interval $[p_i - 1,\ p_i]$, then

$$x(p_i) = A_i x(p_i - 1) + B_i K_j x(k_s).$$

Then, we have

$$\|x(p_i) - x(p_i - 1)\| \le \|B_i K_j\| \cdot \|x(k_s)\| + \|A_i - I\| \cdot \|x(p_i - 1)\|. \tag{17}$$

In addition, it follows that

$$\|x(p_i - 1) - x(p_{i-1})\| \le \sqrt{\tilde{\nu}}\|x(p_{i-1})\|.$$

Therefore,

$$\begin{aligned}\|x(p_i - 1)\| &\le \|x(p_{i-1})\| + \|x(p_i - 1) - x(p_{i-1})\| \\ &\le (1 + \sqrt{\tilde{\nu}})\|x(p_{i-1})\|.\end{aligned} \tag{18}$$

It follows from (17) and (18) that

$$\begin{aligned}\|x(p_i) - x(p_i - 1)\| &\le \|A_i - I\|(1 + \sqrt{\tilde{\nu}})\|x(p_{i-1})\| + \|B_i K_j\|\|x(k_s)\| \\ &= \alpha_1(1 + \sqrt{\tilde{\nu}})\|x(p_{i-1})\| + \beta\|x(k_s)\|.\end{aligned} \tag{19}$$

Substitute (19) into (16), and we obtain

$$\begin{aligned}\|x(p_i)\| &\le (1+\sqrt{\tilde{\nu}})(1+\alpha_1)\|x(p_{i-1})\| + \beta\|x(k_s)\| \\ &= \alpha\|x(p_{i-1})\| + \beta\|x(k_s)\|.\end{aligned} \tag{20}$$

Then, we can obtain from (19) by recursion process that

$$\begin{aligned}|x(p_i)| &\le \alpha(\alpha\|x(p_{i-2})\| + \beta\|x(k_s)\|) + \beta\|x(k_s)\| \\ &= \alpha^2\|x(p_{i-2})\| + (1+\alpha)\beta\|x(k_s)\| \\ &\dots \\ &\le \alpha^i\|x(p_0)\| + (1+\alpha+\cdots+\alpha^{i-1})\beta\|x(k_s)\| \\ &= \left(\alpha^i + \frac{(1-\alpha^i)\beta}{1-\alpha}\right)\|x(k_s)\|.\end{aligned}$$

Therefore, we have

$$\begin{aligned}\|x(p_i) - x(k_s)\| &\le \|x(p_1-1) - x(p_0)\| + \|x(p_1) - x(p_1-1)\| + \cdots \\ &\quad + \|x(p_i-1) - x(p_{i-1})\| + \|x(p_i) - x(p_i-1)\| \\ &= \sum_{j=1}^{i}\|x(p_j-1) - x(p_{j-1})\| + \sum_{j=1}^{i}\|x(p_j) - x(p_j-1)\| \\ &\le \sum_{j=1}^{i}\sqrt{\tilde{\nu}}\|x(p_{j-1})\| + \beta\|x(k_s)\|) \\ &\quad + \sum_{j=1}^{i}(\alpha_1(1+\sqrt{\tilde{\nu}})\|x(p_{j-1})\|.\end{aligned} \tag{21}$$

Furthermore, from (21) we can get

$$\begin{aligned}\|x(k) - x(k_s)\| &\le \|x(p_{\tau_D}) - x(k_s)\| + \|x(k) - x(p_{\tau_D})\| \\ &= \sum_{j=1}^{\tau_D+1}\|x(p_j-1) - x(p_{j-1})\| + \sum_{j=1}^{\tau_D}\|x(p_j) - x(p_j-1)\| \\ &\le \sum_{j=1}^{\tau_D+1}\sqrt{\tilde{\nu}}\|x(p_{j-1})\| + \beta\|x(k_s)\|) \\ &\quad + \sum_{j=1}^{\tau_D}(\alpha_1(1+\sqrt{\tilde{\nu}})\|x(p_{j-1})\| \\ &\le \sqrt{\psi(\tilde{\nu},\tau_D)}\|x(k_s)\|.\end{aligned}$$

Finally, we obtain

$$
\begin{aligned}
(x(k)-x(k_s))^T \Phi_{\sigma(k_s)}(x(k)-x(k_s)) &\le \lambda_m (x(k)-x(k_s))^T (x(k)-x(k_s)) \\
&\le \lambda_m \psi(\tilde{\nu}, \tau_D) x^T(k_s) x(k_s) \\
&\le \frac{\lambda_m}{\lambda_n} \psi(\tilde{\nu}, \tau_D) x^T(k_s) \Phi_{\sigma(k_s)} x(k_s) \\
&\le \nu x^T(k_s) \Phi_{\sigma(k_s)} x(k_s).
\end{aligned} \tag{22}
$$

Inequality (22) indicates that event-triggered mechanism (14) can guarantee that the event-triggering condition (2) holds. Therefore, by Theorem 1, we can obtain that based on new event-triggered mechanism (14), the closed-loop system (4) is exponentially stable in the case of packet dropouts. □

3.3 Controller Design

Based on the stability analysis result in Theorem 2, a co-design approach for the controller gain matrices in (3) and event-trigger parameters in (14) will be given in this subsection.

Let $P_{ij} = P_j$ and $\mathcal{P}_j = P_j^{-1}$, then pre-and post-multiply both sides of (7) with $\mathrm{diag}(\mathcal{P}_j, \mathcal{P}_j, \mathcal{P}_j)$.

Let

$$
\Theta_j = \mathcal{P}_j \Phi_j \mathcal{P}_j, \; Y_j = K_j \mathcal{P}_j,
$$

then by applying Schur complement lemma to (6) and (7) in Theorem 1, we can obtain the following linear matrix inequality based co-design method for controller gains and event-trigger parameters.

Theorem 3 *Assume that there exist positive definite matrices $\mathcal{P}_j$, Θ_j such that the following linear matrix inequalities hold for the given scalars $\mu \ge 1$, $\eta > 0$, $\rho \in (0, 1)$, $\nu \in (0, 1)$ and $\tilde{\nu} > 0$.*

$$
\begin{aligned}
&\mathcal{P}_j \le \mu \mathcal{P}_i, \\
&\begin{bmatrix} -\lambda_{ij}^2 \mathcal{P}_j + \nu \Theta_j & -\nu \Theta_j & \mathcal{P}_j A_i^T + Y_j^T B_i^T \\ * & -(1-\nu)\Theta_j & -Y_j^T B_i^T \\ * & * & -\mathcal{P}_j \end{bmatrix} < 0, \; \forall i, j \in \mathcal{M}.
\end{aligned} \tag{23}
$$

Then, exponential stability of the closed-loop system (4) under the event-triggered mechanism (14) is obtained when the average dwell time τ_a *satisfies (8), and* τ_D *satisfies (15). The corresponding controller gains* $K_j = Y_j \mathcal{P}_j^{-1}$ *and mode dependent event-trigger parameters* $\Phi_j = \mathcal{P}_j^{-1} \Theta_j \mathcal{P}_j^{-1}$.

3.4 Numerical Example

In this section, a numerical example is given to show the effectiveness of the above approach. A switched system with two subsystems is considered, and the corresponding system parameters are chosen as

$$A_1 = \begin{bmatrix} 1.01 & 0 \\ 0.03 & 0.95 \end{bmatrix}, \quad B_1 = \begin{bmatrix} 0.5 & 0.3 \\ -0.6 & 0.8 \end{bmatrix},$$
$$A_2 = \begin{bmatrix} 0.95 & 0.03 \\ 0 & 1.01 \end{bmatrix}, \quad B_2 = \begin{bmatrix} 0.5 & -0.2 \\ -0.3 & 0.9 \end{bmatrix}.$$

Assume the dwell time $\tau_d = 2$. Then, for the given constants $\eta = 0.02$, $\rho = 0.01$, $\nu = 0.85$ and $\mu = 1.01$, we obtain the following results by solving the linear matrix inequalities (23) for $i, j = 1, 2$.

$$K_1 = \begin{bmatrix} -0.1239 & 0.0289 \\ -0.0596 & -0.0685 \end{bmatrix}, \quad K_2 = \begin{bmatrix} -0.1180 & 0.0139 \\ -0.0478 & -0.0777 \end{bmatrix},$$

$$\Phi_1 = \begin{bmatrix} 0.0126 & 0 \\ 0 & 0.0136 \end{bmatrix}, \quad \Phi_2 = \begin{bmatrix} 0.0117 & -0.0001 \\ -0.0001 & 0.0127 \end{bmatrix}.$$

Suppose that the maximum number of allowable successive packet dropout is $\tau_D = 2$ and $\tilde{\nu} = 0.01$, then we get $\frac{\lambda_m}{\lambda_n}\psi(\tilde{\nu}, \tau_D) = 0.8265 < \nu = 0.85$ from (15). In addition, according to (8), if average dwell time $\tau_a > 6.9307$, then the exponential stability of closed-loop system can be guaranteed.

The simulation of the closed-loop system state trajectories is illustrated in Fig. 2, which shows that the both system states converge to zero asymptotically. The switching signal $\sigma(k)$ with $\tau_a = 7$ and $\hat{\sigma}(k)$ are simulated in Fig. 3. Figure 4 illustrates the interval length of two successive event-triggered instants, where packet dropout instants are also labelled. It shows that there are 47 event-triggered transmission and 10 packet losses within the time interval [0, 60], which implies that the event-triggered mechanism proposed in this paper is effective.

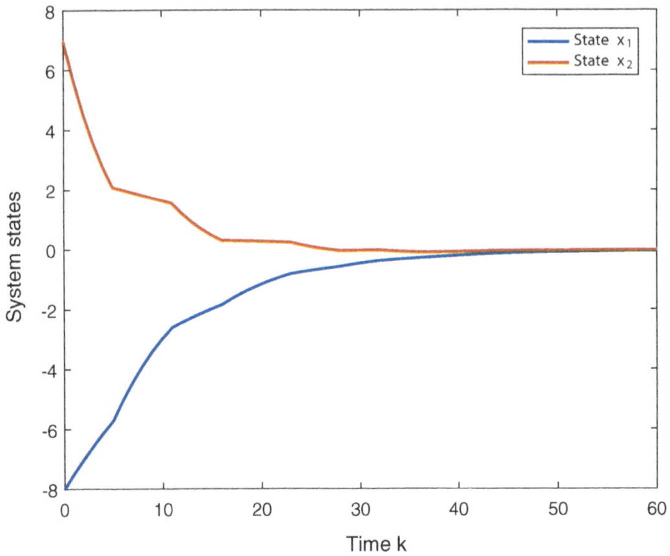

Fig. 2 The simulation of the closed-loop system state with $\tau_a = 7$

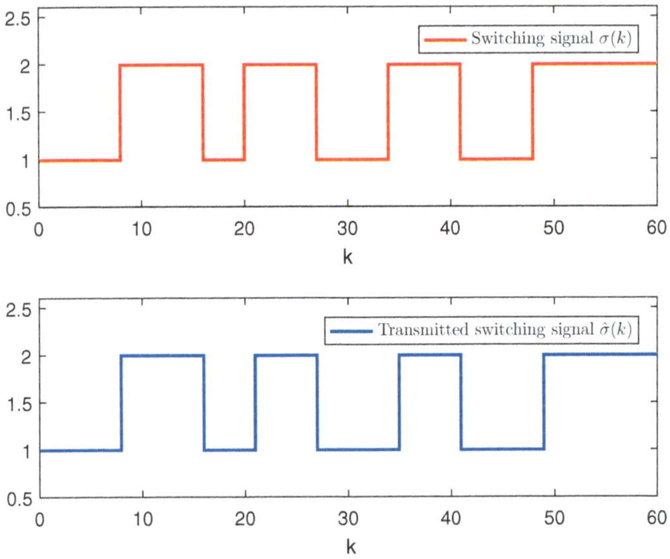

Fig. 3 The switching signals $\sigma(k)$ of the system and $\hat{\sigma}(k)$ of the controller

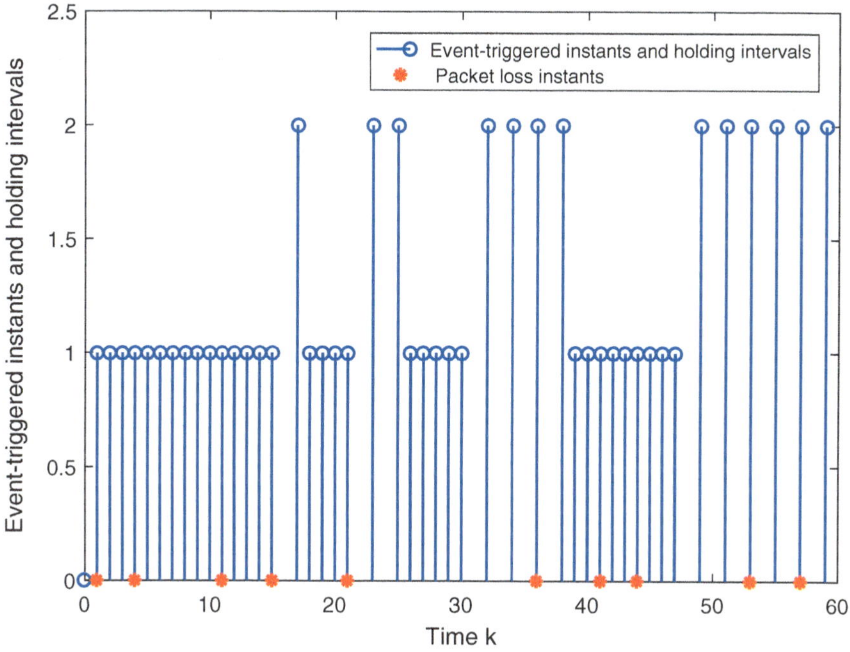

Fig. 4 Length of event-triggered intervals and packet dropout instants

4 Event-Triggered Control of Discrete-Time Switched Linear Systems with Network-Induced Delays

4.1 Problem Formulation

We will focus on the event-triggered control analysis and design of system (1) when there exist transmission delays in this section.

The primary objective is to propose a proper event-triggered scheme for the considered system (1) when only delayed signal of the system state and the system mode is accessible to the remote controller. More specially, the proposed framework is shown in Fig. 5, where k_s is the event-triggered instant generated by the event generator, which should be properly designed. Moreover, both the system state $x(k_s)$ and the system mode $\sigma(k_s)$ are transmitted via the communication channel, and the transmitted information is subjected to transmission delay $d(k_s) \in [d_1, d_2]$, where d_1 and d_2 are the lower and upper bound of $d(k_s)$ respectively. The upper bound d_2 of $d(k_s)$ satisfies the following assumption.

Assumption 3 Assume that the maximum value d_2 of the transmission delay $d(k_s)$ satisfies $d_2 \leq \tau_d$. □

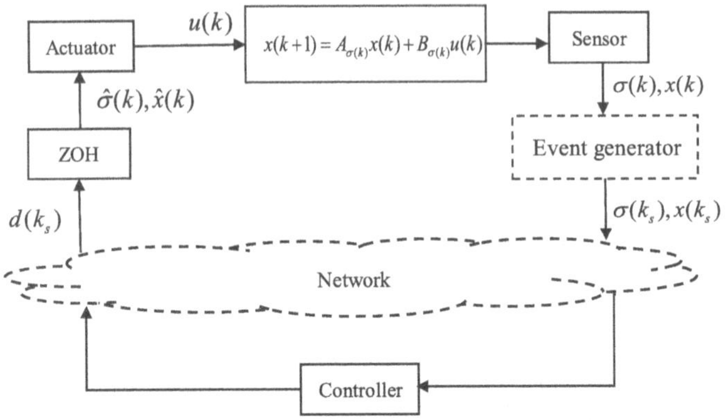

Fig. 5 Event-triggered control framework for switched system (1) with network-induced delays

We propose the following event-triggered mechanism to generate the data transmitting instants $\{k_s\}_{s\geq 1}$:

$$
\begin{aligned}
k_0 &= 0,\\
k_{s+1} &= \hat{k}_{s+1} I_{\{\hat{k}_{s+1}<\tilde{k}_{s+1}\}} + \min\{\hat{k}_{s+1}, \tilde{k}_{s+1} + \tau_d - d_2\} I_{\{\hat{k}_{s+1}\geq\tilde{k}_{s+1}\}}, \ s > 0,\\
\hat{k}_{s+1} &= \min_{k>k_s}\big\{k | (x(k) - x(k_s))^T \Phi_{\sigma(k_s)} (x(k) - x(k_s))\\
&\quad \geq x^T(k_s)\Psi_{\sigma(k_s)} x(k_s)\big\},\\
\tilde{k}_{s+1} &= \min_{k>k_s}\{k | \sigma(k) \neq \sigma(k_s)\},
\end{aligned}
\tag{24}
$$

where $I_{\{\cdot\}} \in \{0, 1\}$ is the indicator function, whose value is 1 if and only if the condition $\{\cdot\}$ holds; positive definite matrices Φ_i and Ψ_i are event-trigger parameters to be designed.

Obviously, it follows from (24) that the transmission will only be triggered when the system switches and the current state is comparatively larger than the previously transmitted one. In addition, we have

$$
k_{s+1} + d(k_{s+1}) \leq \tilde{k}_{s+1} + \tau_d. \tag{25}
$$

In fact, when $\hat{k}_{s+1} < \tilde{k}_{s+1}$, we can get $k_{s+1} + d(k_{s+1}) = \hat{k}_{s+1} + d(k_{s+1}) \leq \hat{k}_{s+1} + \tau_d < \tilde{k}_{s+1} + \tau_d$. Similarly, $k_{s+1} + d(k_{s+1}) \leq \tilde{k}_{s+1} + \tau_d - d_2 + d(k_{s+1}) \leq \tilde{k}_{s+1} + \tau_d$ when $\hat{k}_{s+1} \geq \tilde{k}_{s+1}$. Therefore, it follows from (25) and Assumption 3 that at most one switching occurs during every holding interval $\Omega_s = [k_s + d(k_s), k_{s+1} + d(k_{s+1}))$.

Based on the transmitted system state $x(k_s)$ and switching signal $\sigma(k_s)$, the state feedback controller can be generated in each Ω_s by ZOH as

$$
u(k) = K_{\sigma(k_s)} x(k_s), \ k \in \Omega_s. \tag{26}
$$

For $k \in \Omega_s = [k_s + d(k_s), k_{s+1} + d(k_{s+1}))$, define $e(k) = x(k - \tau_1(k)) - x(k_s)$, where $\tau_1(k) \in [d_1, d_2]$ is given as follows.

When $d(k_{s+1}) - d(k_s) \in (k_s - k_{s+1}, 0]$, define

$$\tau_1(k) = d(k_s). \tag{27}$$

When $d(k_{s+1}) - d(k_s) \in (0, k_{s+1} - k_s]$, define

$$\tau_1(k) = \begin{cases} d(k_s), & k \in [k_s + d(k_s), k_s + d(k_{s+1})), \\ d(k_{s+1}), & k \in [k_s + d(k_{s+1}), k_{s+1} + d(k_{s+1})). \end{cases} \tag{28}$$

When $k_{s+1} - k_s < d(k_{s+1}) - d(k_s)$, we assume κ be the nonnegative integer such that $2^{\kappa+1}(k_{s+1} - k_s) \geq d(k_{s+1}) - d(k_s) > 2^{\kappa}(k_{s+1} - k_s)$.

In this case, we define

$$\tau_1(k) = \begin{cases} \frac{2i}{2^{\kappa+1}} d(k_{s+1}) + \frac{2^{\kappa+1}-2i}{2^{\kappa+1}} d(k_s), & k \in \Lambda_{1i}, \\ \frac{2i-1}{2^{\kappa+1}} d(k_{s+1}) + \frac{2^{\kappa+1}-(2i-1)}{2^{\kappa+1}} d(k_s), & k \in \Lambda_{2i}, \end{cases} \tag{29}$$

where

$$\begin{aligned} \Lambda_{1i} &= \left[k_s + \frac{i}{2^{\kappa}} d(k_{s+1}) + \frac{2^{\kappa} - i}{2^{\kappa}} d(k_s), k_{s+1} + \frac{i}{2^{\kappa}} d(k_{s+1}) + \frac{2^{\kappa} - i}{2^{\kappa}} d(k_s)\right], \\ \Lambda_{2i} &= \left[k_{s+1} + \frac{i-1}{2^{\kappa}} d(k_{s+1}) + \frac{2^{\kappa} - (i-1)}{2^{\kappa}} d(k_s),\right. \\ &\qquad \left. k_s + \frac{i}{2^{\kappa}} d(k_{s+1}) + \frac{2^{\kappa} - i}{2^{\kappa}} d(k_s)\right], \\ \Omega_s &= \left(\bigcup_{i=0}^{2^{\kappa}} \Lambda_{1i}\right) \bigcup \left(\bigcup_{i=1}^{2^{\kappa}} \Lambda_{2i}\right), \quad i = 0, \ldots, 2^{\kappa}. \end{aligned}$$

In summary, when $k \in \Omega_s$, it follows from (27), (28) and (29) that $\tau_1(k) \in [d_1, d_2]$ and $k - \tau_1(k) \in (k_s, k_{s+1})$.

Furthermore, for $k \in \Omega_s$, denote $\tau_2(k) = k - k_s$, then we get $\hat{\sigma}(k) = \sigma(k - \tau_2(k))$. Then we can rewrite the feedback controller (26) as follows

$$u(k) = K_{\hat{\sigma}(k)} \left(x(k - \tau_1(k)) - e(k)\right), \ k \in \Omega_s. \tag{30}$$

Consequently, under the state feedback controller (26), the closed-loop system of (1) is formulated as

$$x(k+1) = A_{\sigma(k)} x(k) + \widetilde{B}_{\theta(k)} \left(x(k - \tau_1(k)) - e(k)\right), \tag{31}$$

with $\widetilde{B}_{\theta(k)} = B_{\sigma(k)} K_{\hat{\sigma}(k)}$, $\tau_1(k) \in [d_1, d_2]$, and $\tau_2(k) \in [d_1, 2\tau_d)$.

Remark 1 To define $e(k)$ according to the event-triggered transmission scheme (24), we present a explicit mathematical definition for $\tau_1(k)$, $k \in \Omega_s$ in (27), (28) and (29), which is simpler and easier to understand than that in [24].

In addition, combining $\hat{\sigma}(k) = \sigma(k - \tau_2(k))$ and $\tau_2(k) \in [d_1, 2\tau_d)$, it follows from Lemma 1 that $\hat{\sigma}(k)$ belongs to the set $\mathcal{S}\left[\tau_a, N_0 + \frac{2\tau_d - d_1}{\tau_a}\right]$ and $\theta(k)$ belongs to the set $\mathcal{S}\left[\frac{\tau_a}{2}, 2N_0 + \frac{2\tau_d - d_1}{\tau_a}\right]$.

Remark 2 Since only the transmitted information of switching signal is available to the controllers, then there exists a time lag between the controller mode $\sigma(k_s)$ and system mode $\sigma(k)$, therefore, the controllers might switch asynchronously with the system. It can be derived from the transmission scheme (24) that the asynchronous time can be estimated by τ_d.

4.2 Stability Analysis

The exponential stability of the closed-loop system (31) will be investigated in this subsection. A proper quadratic Lyapunov functional will be constructed to obtain a sufficient stability condition in the form of matrix inequality.

Theorem 4 *If there exist positive definite matrices P_{ij}, Q_{1ij}, Q_{2ij}, Q_{3ij}, Z_{1ij}, Z_{2ij}, Ψ_j, Φ_j, and matrices N_{1ij}, N_{2ij}, N_{3ij} such that matrix inequalities (32), (33) (34) and (35) hold for the scalars $\mu \geq 1$, $\eta > 0$, $\rho \in (0, 1)$ and state feedback gain matrices K_j, and the switching signal satisfies (36), then the closed-loop system (31) is exponentially stable.*

$$\mu^{-1} P_{ii} \leq P_{ij} \leq \mu P_{jj}, \ \forall i, j \in \mathcal{M}, \tag{32}$$

$$Q_{\iota ii} \leq \mu Q_{\iota ij}, \ \bar{\lambda}_{ij}^{2\bar{d}_\iota} Q_{\iota ij} \leq \mu Q_{\iota jj}, \ \iota = 1, 2, 3, \ \forall i, j \in \mathcal{M}, \tag{33}$$

$$Z_{\nu ii} \leq \mu Z_{\nu ij}, \ \bar{\lambda}_{ij}^{2(d_\nu - 1)} Z_{\nu ij} \leq \mu Z_{\nu jj}, \ \nu = 1, 2, \ \forall i, j \in \mathcal{M}, \tag{34}$$

$$\tilde{\Omega}_{ij} + (d_2 - d_1) N_{lij}^T Z_{2ij}^{-1} N_{lij} < 0, \ l = 2, 3, \ \forall i, j \in \mathcal{M}, \tag{35}$$

$$-\tau_a \ln(1 - \rho) > \left(\ln \mu + \tau_d \ln \frac{1 + \eta}{1 - \rho}\right), \tag{36}$$

where

$$\begin{aligned}
\tilde{\Omega}_{ij} = & \begin{bmatrix} \Omega_{ij} & \tilde{\Psi}_j \\ * & \Psi_j - \Phi_j \end{bmatrix} + \varrho_{1ij}(N_{1ij}^T E_1 + E_1^T N_{1ij}) + \varrho_{2ij}(N_{2ij}^T E_2 + E_2^T N_{2ij}) \\
& + \varrho_{2ij}(N_{3ij}^T E_3 + E_3^T N_{3ij}) + d_1 N_{1ij}^T Z_{1ij}^{-1} N_{1ij} + \mathcal{A}_{ij}^T P_{ij} \mathcal{A}_{ij} \\
& + d_1 \tilde{\mathcal{A}}_{ij}^T Z_{1ij} \tilde{\mathcal{A}}_{ij} + (d_2 - d_1) \tilde{\mathcal{A}}_{ij}^T Z_{2ij} \tilde{\mathcal{A}}_{ij},
\end{aligned}$$

$$\begin{aligned}
\Omega_{ij} &= \mathrm{diag}\{\Omega_{11ij}, \Omega_{22ij}, \Omega_{33ij}, \Omega_{44ij}\}, \\
\Omega_{11ij} &= -\lambda_{ij}^2 P_{ij} + Q_{1ij} + (d_2 - d_1 + 1)Q_{3ij}, \\
\Omega_{22ij} &= -\lambda_{ij}^{2(\delta_{ij}d_2+(1-\delta_{ij})d_1)} Q_{3ij} + \Psi_j, \\
\Omega_{33ij} &= -\lambda_{ij}^{2d_1} Q_{1ij} + Q_{2ij}, \\
\Omega_{44ij} &= -\lambda_{ij}^{2(d_2-d_1)} Q_{2ij}, \ \tilde{\Psi}_j = \begin{bmatrix} 0 & -\Psi_j & 0 & 0 \end{bmatrix}^T, \\
\mathcal{A}_{ij} &= \begin{bmatrix} A_i & B_i K_j & 0 & 0 & -B_i K_j \end{bmatrix}, \ \tilde{\mathcal{A}}_{ij} = \begin{bmatrix} A_i - I & B_i K_j & 0 & 0 & -B_i K_j \end{bmatrix}, \\
E_1 &= \begin{bmatrix} I & 0 & -I & 0 & 0 \end{bmatrix}, \ E_2 = \begin{bmatrix} 0 & I & 0 & -I & 0 \end{bmatrix}, \ E_3 = \begin{bmatrix} 0 & -I & I & 0 & 0 \end{bmatrix}, \\
\bar{d}_1 &= d_1 - 1, \ \bar{d}_2 = d_2 - d_1 - 1, \ \bar{d}_3 = d_2 - 1, \\
\lambda_{ij} &= \delta_{ij}(1-\rho) + (1-\delta_{ij})(1+\eta), \\
\bar{\lambda}_{ij} &= \lambda_{ij}\lambda_{jj}^{-1}, \ \varrho_{1ij} = \lambda_{ij}^{\delta_{ij}+(1-\delta_{ij})d_1}, \\
\varrho_{2ij} &= \lambda_{ij}^{\delta_{ij}(d_1+1)+(1-\delta_{ij})d_2}.
\end{aligned}$$

Proof A quadratic Lyapunov functional $V_{\theta(k)}(k)$ is constructed as follows

$$V_{\theta(k)}(k) = V_{1\theta(k)}(k) + V_{2\theta(k)}(k) + V_{3\theta(k)}(k), \tag{37}$$

where

$$\begin{aligned}
V_{1\theta(k)}(k) &= x^T(k) P_{\theta(k)} x(k), \\
V_{2\theta(k)}(k) &= \sum_{m=k-d_1}^{k-1} \lambda_{\theta(k)}^{2(k-1-m)} x^T(m) Q_{1\theta(k)} x(m) \\
&\quad + \sum_{m=k-d_2}^{k-1-d_1} \lambda_{\theta(k)}^{2(k-1-d_1-m)} x^T(m) Q_{2\theta(k)} x(m) \\
&\quad + \sum_{m=k-\tau_1(k)}^{k-1} \lambda_{\theta(k)}^{2(k-1-m)} x^T(m) Q_{3\theta(k)} x(m) \\
&\quad + \sum_{n=-d_2+1}^{-d_1} \sum_{m=k+n}^{k-1} \lambda_{\theta(k)}^{2(k-1-m)} x^T(m) Q_{3\theta(k)} x(m), \\
V_{3\theta(k)}(k) &= \sum_{n=-d_1}^{-1} \sum_{m=k+n}^{k-1} \lambda_{\theta(k)}^{2(k-1-m)} \Delta x^T Z_{1\theta(k)} \Delta x \\
&\quad + \sum_{n=-d_2}^{-1-d_1} \sum_{m=k+n}^{k-1} \lambda_{\theta(k)}^{2(k-1-m)} \Delta x^T Z_{2\theta(k)} \Delta x,
\end{aligned} \tag{38}$$

with $\Delta x = x(m+1) - x(m)$.

Let $\bar{k} \in \Omega_s$ be any switching instant of switching signal $\theta(k)$. Since $\theta(k)$ is the augment of $\sigma(k)$ and $\hat{\sigma}(k)$, then $\bar{k}$ must be either a transmitting instant or some switching time of $\sigma(k)$.

Case 1. Assume that $\bar{k}$ be some switching instant of $\sigma(k)$ in Ω_s. In general, denote $\sigma(\bar{k}-1) = j$ and $\sigma(\bar{k}) = i$.
Since there exists at most one switch in Ω_s, then both $\hat{\sigma}(\bar{k})$ and $\hat{\sigma}(\bar{k}-1)$ are equal to j, therefore, $\theta(\bar{k}) = (i, j)$ and $\theta(\bar{k}-1) = (j, j)$. Then we have $P_{\theta(\bar{k})} = P_{ij}$, $P_{\theta(\bar{k}-1)} = P_{jj}$, $Q_{\iota\theta(\bar{k})} = Q_{\iota ij}$, $Q_{\iota\theta(\bar{k}-1)} = Q_{\iota jj}$, $Z_{\nu\theta(\bar{k})} = Z_{\nu ij}$, $Z_{\nu\theta(\bar{k}-1)} = Z_{\nu jj}$, and $\lambda_{\theta(\bar{k})} = 1+\eta$, $\lambda_{\theta(\bar{k}-1)} = 1-\rho$.

According to (32), (33) and (34), we can obtain the following estimation of $V_{\theta(\bar{k})}$ at the switching instant $\bar{k}$:

$$V_{1\theta(\bar{k})}(\bar{k}) = x^T(\bar{k})P_{ij}x(\bar{k}) \le \mu x^T(\bar{k})P_{jj}x(\bar{k}) = \mu V_{1\theta(\bar{k}-1)}(\bar{k}),$$

and

$$\begin{aligned}
V_{2\theta(\bar{k})}(\bar{k}) = & \sum_{m=\bar{k}-d_1}^{\bar{k}-1} \lambda_{ij}^{2(\bar{k}-1-m)} x^T(m)Q_{1ij}x(m) \\
& + \sum_{m=\bar{k}-d_2}^{\bar{k}-1-d_1} \lambda_{ij}^{2(\bar{k}-1-d_1-m)} x^T(m)Q_{2ij}x(m) \\
& + \sum_{m=\bar{k}-\tau_1(\bar{k})}^{\bar{k}-1} \lambda_{ij}^{2(\bar{k}-1-m)} x^T(m)Q_{3ij}x(m) \\
& + \sum_{n=-d_2+1}^{-d_1} \sum_{m=\bar{k}+n}^{\bar{k}-1} \lambda_{ij}^{2(\bar{k}-1-m)} x^T(m)Q_{3ij}x(m) \\
\le & \sum_{m=\bar{k}-d_1}^{\bar{k}-1} \lambda_{jj}^{2(\bar{k}-1-m)} x^T(m)\mu Q_{1jj}x(m) \\
& + \sum_{m=\bar{k}-d_2}^{\bar{k}-1-d_1} \lambda_{jj}^{2(\bar{k}-1-d_1-m)} x^T(m)\mu Q_{2jj}x(m) \\
& + \sum_{m=\bar{k}-\tau_1(\bar{k})}^{\bar{k}-1} \lambda_{jj}^{2(\bar{k}-1-m)} x^T(m)\mu Q_{3jj}x(m) \\
& + \sum_{n=-d_2+1}^{-d_1} \sum_{m=\bar{k}+n}^{\bar{k}-1} \lambda_{jj}^{2(\bar{k}-1-m)} x^T(m)\mu Q_{3jj}x(m) \\
= & \mu V_{2\theta(\bar{k}-1)}(\bar{k}),
\end{aligned}$$

and finally,

$$
\begin{aligned}
&V_{3\theta(\bar{k})}(\bar{k})\\
&=\sum_{n=-d_1}^{-1}\sum_{m=\bar{k}+n}^{\bar{k}-1}\lambda_{ij}^{2(\bar{k}-1-m)}(x(m+1)-x(m))^T Z_{1ij}(x(m+1)-x(m))\\
&\quad+\sum_{n=-d_2}^{-1-d_1}\sum_{m=\bar{k}+n}^{\bar{k}-1}\lambda_{ij}^{2(\bar{k}-1-m)}(x(m+1)-x(m))^T Z_{2ij}(x(m+1)-x(m))\\
&\le\sum_{n=-d_1}^{-1}\sum_{m=\bar{k}+n}^{\bar{k}-1}\lambda_{jj}^{2(\bar{k}-1-m)}(x(m+1)-x(m))^T \mu Z_{1jj}(x(m+1)-x(m))\\
&\quad+\sum_{n=-d_2}^{-1-d_1}\sum_{m=\bar{k}+n}^{\bar{k}-1}\lambda_{jj}^{2(\bar{k}-1-m)}(x(m+1)-x(m))^T \mu Z_{2jj}(x(m+1)-x(m))\\
&=\mu V_{3\theta(\bar{k}-1)}(\bar{k}).
\end{aligned}
$$

Therefore, we have $V_{\theta(\bar{k})}(\bar{k})\le\mu V_{\theta(\bar{k}-1)}(\bar{k})$.

Case 2. Assume that $\bar{k}=k_{s+1}+d(k_{s+1})$, while $\bar{k}$ is not a switching instant of $\sigma(k)$.

In general, suppose $\hat{\sigma}(\bar{k}-1)=\sigma(k_s)=j$, $\hat{\sigma}(\bar{k})=\sigma(k_{s+1})=i$, then, $\sigma(\bar{k})=\sigma(\bar{k}-1)=i$ and one switch occurs in Ω_s. Therefore, we have $\theta(\bar{k})=(i,i)$, $\theta(\bar{k}-1)=(i,j)$. Consequently, $P_{\theta(\bar{k})}=P_{ii}$, $P_{\theta(\bar{k}-1)}=P_{ij}$, $Q_{\iota\theta(\bar{k}-1)}=Q_{\iota ij}$, $Q_{\iota\theta(\bar{k})}=Q_{\iota ii}$, $Z_{\nu\theta(\bar{k}-1)}=Z_{\nu ij}$, $Z_{\nu\theta(\bar{k})}=Z_{\nu ii}$, $\lambda_{\theta(\bar{k}-1)}=1+\eta$, $\lambda_{\theta(\bar{k})}=1-\rho$.

Then, similar to the analysis as Case 1, we can obtain from (32), (33) and (34) that

$$V_{\theta(\bar{k})}(\bar{k})\le\mu V_{\theta(\bar{k}-1)}(\bar{k}). \tag{39}$$

The following estimation of Lyapunov functional (37) along the closed-loop system (31) can be computed for $k\in\Omega_s$:

$$
\begin{aligned}
&V_{1\theta(k)}(k+1)-\lambda_{\theta(k)}^2 V_{1\theta(k)}(k)\\
&=x^T(k+1)P_{\theta(k)}x(k+1)-\lambda_{\theta(k)}^2 x^T(k)P_{\theta(k)}x(k),
\end{aligned}
\tag{40}
$$

and

$$
\begin{aligned}
&V_{2\theta(k)}(k+1)-\lambda_{\theta(k)}^2 V_{2\theta(k)}(k)\\
&\le x^T(k)(Q_{1\theta(k)}+(d_2-d_1+1)Q_{3\theta(k)})x(k)\\
&\quad-\lambda_{\theta(k)}^{2(\delta_{\theta(k)}d_2+(1-\delta_{\theta(k)})d_1)}x^T(k-\tau_1(k))Q_{3\theta(k)}x(k-\tau_1(k))\\
&\quad+x^T(k-d_1)(-\lambda_{\theta(k)}^{2d_1}Q_{1\theta(k)}+Q_{2\theta(k)})x(k-d_1)\\
&\quad-\lambda_{\theta(k)}^{2(d_2-d_1)}x^T(k-d_2)Q_{2\theta(k)}x(k-d_2),
\end{aligned}
\tag{41}
$$

and finally,

$$\begin{aligned}
&V_{3\theta(k)}(k+1) - \lambda_{\theta(k)}^2 V_{3\theta(k)}(k)\\
&= d_1(x(k+1)-x(k))^T Z_{1\theta(k)}(x(k+1)-x(k))\\
&\quad - \sum_{m=k-d_1}^{k-1} \lambda_{\theta(k)}^{2(k-m)}(x(m+1)-x(m))^T Z_{1\theta(k)}(x(m+1)-x(m))\\
&\quad + (d_2-d_1)(x(k+1)-x(k))^T Z_{2\theta(k)}(x(k+1)-x(k))\\
&\quad - \sum_{m=k-d_2}^{k-1-\tau_1(k)} \lambda_{\theta(k)}^{2(k-m)}(x(m+1)-x(m))^T Z_{2\theta(k)}(x(m+1)-x(m))\\
&\quad - \sum_{m=k-\tau_1(k)}^{k-1-d_1} \lambda_{\theta(k)}^{2(k-m)}(x(m+1)-x(m))^T Z_{2\theta(k)}(x(m+1)-x(m)).
\end{aligned} \tag{42}$$

Furthermore, it follows from event-triggering condition (24) that

$$\begin{aligned}
e^T(k)\Phi_{\hat{\sigma}(k)}e(k) < {} & [x(k-\tau_1(k))-e(k)]^T \Psi_{\hat{\sigma}(k)}\\
&\times[x(k-\tau_1(k))-e(k)],\ k\in\Omega_s.
\end{aligned} \tag{43}$$

Combining (31), (40), (41), (42) and (43), we obtain

$$V_{\theta(k)}(k+1) - \lambda_{\theta(k)}^2 V_{\theta(k)}(k) \le \xi^T(k)\Upsilon_{\theta(k)}(\tau_1(k))\xi(k), \tag{44}$$

where $\xi(k) = \left[x^T(k)\ x^T(k-\tau_1(k))\ x^T(k-d_1)\ x^T(k-d_2)\ e^T(k)\right]^T$ and

$$\begin{aligned}
\Upsilon_{\theta(k)}(\tau_1(k)) = {} & \tilde{\Omega}_{\theta(k)} + (d_2-\tau_1(k))N_{2\theta(k)}^T Z_{2\theta(k)}^{-1} N_{2\theta(k)}\\
&+ (\tau_1(k)-d_1)N_{3\theta(k)}^T Z_{2\theta(k)}^{-1} N_{3\theta(k)}.
\end{aligned}$$

For given scalars $\eta > 0$ and $0 < \rho < 1$, define $\hat{\lambda}_{\theta(k)} = 1 - \lambda_{\theta(k)}^2$ and

$$\lambda_{\theta(k)} = \begin{cases} 1+\eta, & k \in T_{as}(\Omega_s);\\ 1-\rho, & k\in T_s(\Omega_s). \end{cases}$$

Then, we can obtain from (44) that $\Delta V_{\theta(k)}(k)$ along the switched system (31) satisfies the following estimation if $\Upsilon_{\theta(k)}(\tau_1(k)) < 0$.

$$\Delta V_{\theta(k)}(k) + \hat{\lambda}_{\theta(k)} V_{\theta(k)}(k) \le 0,\ k\in\Omega_s. \tag{45}$$

Now, we show that $\Upsilon_{\theta(k)}(\tau_1(k)) < 0$ can be guaranteed by (35). Note that $\Upsilon_{\theta(k)}(\tau_1(k))$ is linearly dependent on $\tau_1(k)$ and $\tau_1(k) \in [d_1, d_2]$. Thus, $\Upsilon_{\theta(k)}(\tau_1(k)) < 0$ if and only if $\Upsilon_{\theta(k)}(d_1) < 0$ and $\Upsilon_{\theta(k)}(d_2) < 0$ hold simultaneously.

Case 1. The $\theta(k)$ has one switch in Ω_s, then $\sigma(k)$ switches one time in Ω_s. The corresponding switching instant is supposed to be $\bar{k}$.

Without loss of generality, denote $\sigma(\bar{k}) = i$, then we have $\Upsilon_{\theta(k)}(\tau_1(k)) = \Upsilon_{jj}(\tau_1(k))$ when $k \in [k_s + d(k_s), \bar{k})$. For $k \in [\bar{k}, k_{s+1} + d(k_{s+1}))$, we have $\Upsilon_{\theta(k)}(\tau_1(k)) = \Upsilon_{ij}(\tau_1(k))$.

Case 2. The $\theta(k)$ has no switch in Ω_s, then we have $\Upsilon_{\theta(k)}(\tau_1(k)) = \Upsilon_{jj}(\tau_1(k))$, for $k \in \Omega_s$.

Therefore, $\Upsilon_{\theta(k)}(d_1) < 0$ and $\Upsilon_{\theta(k)}(d_2) < 0$ are equivalent to $\Upsilon_{ij}(d_1) < 0$ and $\Upsilon_{ij}(d_2) < 0$ for all $i, j \in \mathcal{M}$, which can be guaranteed by (35).

Finally, if the switching signal satisfies (36), then the exponential stability of the closed-loop system (31) is formulated as follows.

During any time interval $(0, k)$, the switching signal $\theta(k)$ has $N_\theta(0, k)$ switching instants $\tilde{k}_1, \ldots, \tilde{k}_{N_\theta(0,k)}$. In general, we suppose $\tilde{k}_{N_\theta(0,k)+1} = k$ and $\tilde{k}_0 = 0$. From (39) and (45), we can obtain that

$$\begin{aligned}
V_{\theta(k)}(k) &\leq \lambda_{\theta(k)}^2 V_{\theta(k)}(k-1) \\
&\leq \cdots \\
&\leq \lambda_{\theta(k)}^{2(k-\tilde{k}_{N_\theta(0,k)})} \mu V_{\theta(\tilde{k}_{N_\theta(0,k)}-1)}(\tilde{k}_{N_\theta(0,k)}) \\
&\leq \cdots \\
&\leq \mu^{N_\theta(0,k)}(1-\rho)^{2T_s(0,k)}(1+\eta)^{2T_{as}(0,k)} V_{\theta(0)}(0).
\end{aligned}$$

Note that $\sigma(k) \in \mathcal{S}[\tau_a, N_0]$, then, according to Lemma 1, we have $\theta(k) \in \mathcal{S}[\frac{\tau_a}{2}, 2N_0 + \frac{\tau_d - d_1}{\tau_a}]$, therefore, $N_\theta(0, k) \leq 2N_0 + \frac{2\tau_d - d_1}{\tau_a} + \frac{2k}{\tau_a}$.

Furthermore, from (36) we obtain that $\mu^{\frac{1}{\tau_a}} < \frac{1}{1-\rho}\left(\frac{1-\rho}{1+\eta}\right)^{\frac{\tau_d - d_2}{\tau_a}}$.

Then for any $\bar{\beta} \in \left((1-\rho)\left(\frac{1+\eta}{1-\rho}\right)^{\frac{\tau_d}{\tau_a}}, \mu^{-\frac{1}{\tau_a}}\right)$, we have $0 < \mu^{\frac{1}{\tau_a}} \cdot \bar{\beta} < 1$.

By Lemma 2, we get

$$\begin{aligned}
V_{\theta(k)}(k) &\leq (1-\rho)^{2T_s(0,k)}(1+\eta)^{2T_{as}(0,k)} \mu^{N_\theta(0,k)} V_{\theta(0)}(0) \\
&\leq \mu^{(2N_0 + \frac{2\tau_d - d_1}{\tau_a} + \frac{2k}{\tau_a})} \bar{\alpha}^2 \bar{\beta}^{2k} V_{\theta(0)}(0) \\
&= \tilde{\alpha} \beta^{2k} V_{\theta(0)}(0),
\end{aligned}$$

where $\tilde{\alpha} = \mu^{2N_0 + \frac{2\tau_d - d_1}{\tau_a}} \cdot \bar{\alpha}^2$, $\beta = \mu^{\frac{1}{\tau_a}} \cdot \bar{\beta}$. Therefore, exponential stability of the closed-loop system (31) is obtained. □

Remark 3 The average dwell time condition (36) indicates that switched systems can maintain stability when the overall activation time of the unstable subsystem (with $\lambda_{\theta(k)} = 1 + \eta$, $k \in T_{as}(\Omega_s)$) is comparatively smaller than that of the stable subsystem (with $\lambda_{\theta(k)} = 1 - \rho,\ k \in T_s(\Omega_s)$). We mention that condition (36) can be viewed as a counterpart of the following stability condition obtained in [5]

$$\tau_a > -[\mathcal{T}_{\max}(\ln(1+\eta) - \ln(1-\rho)) + \ln\mu]/\ln(1-\rho), \tag{46}$$

where $\mathcal{T}_{\max}$ represents the maximal delay of asynchronous switching. $\mathcal{T}_{\max}$ (corresponding to τ_d in (36)) is assumed to be a priori in [5]. However, the dwell time τ_d in our proposed condition (36) can be measured from $\sigma(k)$ directly.

Remark 4 It should be mentioned that the obtained condition (36) has some conservativeness, since we estimate the mismatch time τ_d between two switching signals $\sigma(k)$ and $\hat{\sigma}(k)$ in the worst-case scenario.

On the other hand, when consider the time-triggered transmission and the delay $d(k) = 0$, then $\sigma(k)$ and $\hat{\sigma}(k)$ always switch synchronously. In this case, the asynchronous time is zero. Then, the average dwell time condition (36) reduces to $\tau_a > -\ln\mu/\ln(1-\rho)$, which recovers to the classic stability condition of discrete-time switched systems obtained in [6].

4.3 Controller Design

In the following, we give a co-design approach for the stabilizing controller according to the obtained stability analysis conclusion in previous subsection.

To this end, we assume P_{ij} is invariant with respect to i in Theorem 4, that is $P_{ij} = P_j, \forall i \in \mathcal{M}$.

Let us denote

$$\begin{aligned}
&\mathcal{P}_j = P_j^{-1},\ \mathcal{Q}_{1ij} = P_j^{-1}Q_{1ij}P_j^{-1},\ \mathcal{Q}_{2ij} = P_j^{-1}Q_{2ij}P_j^{-1},\\
&\mathcal{Q}_{3ij} = P_j^{-1}Q_{3ij}P_j^{-1},\ \mathcal{Z}_{1ij} = Z_{1ij}^{-1},\ \mathcal{Z}_{2ij} = Z_{2ij}^{-1},\\
&\Xi_j = P_j^{-1}\Psi_jP_j^{-1}, \Theta_j = P_j^{-1}\Phi_jP_j^{-1},\ Y_j = K_jP_j^{-1},\\
&\mathbb{P}_j = \operatorname{diag}\left(\mathcal{P}_j, \mathcal{P}_j, \mathcal{P}_j, \mathcal{P}_j, \mathcal{P}_j\right).
\end{aligned}$$

After pre-and post-multiplying both sides of matrix inequality (35) with $\mathbb{P}_j$, a co-design approach for event-trigger parameters and state feedback controllers is formulated as follows by utilizing Schur complement lemma.

Theorem 5 *If there exist $\mu \geq 1$, $\eta > 0$, $\rho \in (0, 1)$, constants α_{ij}, β_{ij}, positive definite matrices $\mathcal{P}_j$, $\mathcal{Q}_{1ij}$, $\mathcal{Q}_{2ij}$, $\mathcal{Q}_{3ij}$, $\mathcal{Z}_{1ij}$, $\mathcal{Z}_{2ij}$, Θ_j, Ξ_j, and matrices M_{1ij}, M_{2ij}, M_{3ij}, Y_j satisfying linear matrix inequalities as follows.*

$$\mathcal{P}_j \leq \mu\mathcal{P}_i,\ \mathcal{Q}_{\kappa ii} \leq \mu\mathcal{Q}_{\kappa ij},\ \bar{\lambda}_{ij}^{2\bar{d}_\kappa}\mathcal{Q}_{\kappa ij} \leq \mu\mathcal{Q}_{\kappa jj},\ \kappa = 1, 2, 3, \forall i, j \in \mathcal{M}, \tag{47}$$

$$\mathcal{Z}_{\nu ii} \leq \mu\mathcal{Z}_{\nu ij},\ \bar{\lambda}_{ij}^{2(d_\nu - 1)}\mathcal{Z}_{\nu ij} \leq \mu\mathcal{Z}_{\nu jj},\ \nu = 1, 2, \forall i, j \in \mathcal{M}, \tag{48}$$

$$\begin{bmatrix} \bar{\Omega}_{ij} & \bar{\mathcal{A}}_{ij}^T & d_1^{\frac{1}{2}}\hat{\mathcal{A}}_{ij}^T & \tilde{\Omega}_{1ij} & d_1^{\frac{1}{2}}M_{1ij}^T & \tilde{\Omega}_{2ij} \\ * & -\mathcal{P}_j & 0 & 0 & 0 & 0 \\ * & * & -\mathcal{Z}_{1ij} & 0 & 0 & 0 \\ * & * & * & -\mathcal{Z}_{2ij} & 0 & 0 \\ * & * & * & * & \tilde{\Upsilon}_{55} & 0 \\ * & * & * & * & * & \tilde{\Upsilon}_{66} \end{bmatrix} < 0,\ l = 2, 3, \forall i, j \in \mathcal{M}, \tag{49}$$

where

$$\bar{\Omega}_{ij} = \begin{bmatrix} \bar{\Omega}_{11ij} & 0 & 0 & 0 & 0 \\ * & \bar{\Omega}_{22ij} & 0 & 0 & -\Xi_j \\ * & * & \bar{\Omega}_{33ij} & 0 & 0 \\ * & * & * & \bar{\Omega}_{44ij} & 0 \\ * & * & * & * & \Xi_j - \Theta_j \end{bmatrix}$$

$$+ \varrho_{1ij}(M_{1ij}^T E_1 + E_1^T M_{1ij}) + \varrho_{2ij} \sum_{l=2}^{3} (M_{lij}^T E_l + E_l^T M_{lij}),$$

$$\bar{\Omega}_{11ij} = -\lambda_{ij}^2 \mathcal{P}_j + \mathcal{Q}_{1ij} + (d_2 - d_1 + 1)\mathcal{Q}_{3ij},$$

$$\bar{\Omega}_{22ij} = -\lambda_{ij}^{2(\delta_{ij} d_2 + (1-\delta_{ij})d_1)} \mathcal{Q}_{3ij} + \Xi_j,$$

$$\bar{\Omega}_{33ij} = -\lambda_{ij}^{2d_1} \mathcal{Q}_{1ij} + \mathcal{Q}_{2ij},$$

$$\bar{\Omega}_{44ij} = -\lambda_{ij}^{2(d_2-d_1)} \mathcal{Q}_{2ij},$$

$$E_1 = \begin{bmatrix} I & 0 & -I & 0 & 0 \end{bmatrix},\ E_2 = \begin{bmatrix} 0 & I & 0 & -I & 0 \end{bmatrix},\ E_3 = \begin{bmatrix} 0 & -I & I & 0 & 0 \end{bmatrix},$$

$$\bar{\mathcal{A}}_{ij} = \begin{bmatrix} A_i \mathcal{P}_j & B_i Y_j & 0 & 0 & -B_i Y_j \end{bmatrix},$$

$$\hat{\mathcal{A}}_{ij} = \begin{bmatrix} A_i \mathcal{P}_j - \mathcal{P}_j & B_i Y_j & 0 & 0 & -B_i Y_j \end{bmatrix},$$

$$\tilde{\Omega}_{1ij} = (d_2 - d_1)^{\frac{1}{2}} \hat{\mathcal{A}}_{ij}^T,\ \tilde{\Omega}_{2ij} = (d_2 - d_1)^{\frac{1}{2}} M_{lij}^T,$$

$$\tilde{\Upsilon}_{55} = -2\alpha_{ij}\mathcal{P}_j + \alpha_{ij}^2 \mathcal{Z}_{1ij},\ \tilde{\Upsilon}_{66} = -2\beta_{ij}\mathcal{P}_j + \beta_{ij}^2 \mathcal{Z}_{2ij},$$

$$\lambda_{ij} = (1-\rho)\delta_{ij} + (1+\eta)(1-\delta_{ij}),\ \bar{\lambda}_{ij} = \lambda_{ij}\lambda_{jj}^{-1},$$

$$\bar{d}_1 = d_1 - 1,\ \bar{d}_2 = d_2 - d_1 - 1,\ \bar{d}_3 = d_2 - 1,$$

$$\varrho_{1ij} = \lambda_{ij}^{\delta_{ij} + (1-\delta_{ij})d_1},\ \varrho_{2ij} = \lambda_{ij}^{\delta_{ij}(d_1+1) + (1-\delta_{ij})d_2}.$$

Then, exponential stability of the closed-loop system (31) can be guaranteed for the switching signal $\sigma(k)$ whose average dwell time τ_a satisfies the constraint (36). Furthermore, the feedback gain matrices are computed as $K_j = Y_j \mathcal{P}_j^{-1}$ and event-trigger parameters are $\Psi_j = \mathcal{P}_j^{-1} \Xi_j \mathcal{P}_j^{-1}$, $\Phi_j = \mathcal{P}_j^{-1} \Theta_j \mathcal{P}_j^{-1}$.

4.4 Numerical Example

To show the effectiveness of the design method, the switched system with two subsystems will be considered and system parameters are given as follows:

$$A_1 = \begin{bmatrix} 1.05 & 0.1 \\ 0 & 0.1 \end{bmatrix}, \; B_1 = \begin{bmatrix} 0.6 & 0 \\ 0.2 & 0.4 \end{bmatrix},$$
$$A_2 = \begin{bmatrix} 0.6 & 0.1 \\ 0.5 & 0.9 \end{bmatrix}, \; B_2 = \begin{bmatrix} 0.1 & 0.3 \\ 0.2 & 0.2 \end{bmatrix}.$$

For the given dwell time $\tau_d = 4$ and the transmission delay $d(k) \in [0, 2]$, let $\mu = 1.2$, $\rho = 0.1$, $\eta = 0.05$, $\alpha_{ij} = \beta_{ij} = 0.5$, $i, j = 1, 2$. It can be verified that LMIs (47), (48) and (49) are feasible, the corresponding controller gain matrices are computed as

$$K_1 = \begin{bmatrix} -0.4268 & -0.0498 \\ 0.2068 & -0.0104 \end{bmatrix}, \; K_2 = \begin{bmatrix} -0.3877 & 0.0666 \\ -0.3027 & -0.3921 \end{bmatrix}.$$

The event-trigger parameters are

$$\Psi_1 = \begin{bmatrix} 1.1965 & -0.1422 \\ -0.1422 & 0.3701 \end{bmatrix}, \; \Psi_2 = \begin{bmatrix} 0.4391 & -0.0843 \\ -0.0843 & 0.0350 \end{bmatrix},$$

$$\Phi_1 = \begin{bmatrix} 18.7865 & 2.3630 \\ 2.3630 & 0.8000 \end{bmatrix}, \; \Phi_2 = \begin{bmatrix} 13.7775 & 5.9591 \\ 5.9591 & 3.7998 \end{bmatrix}.$$

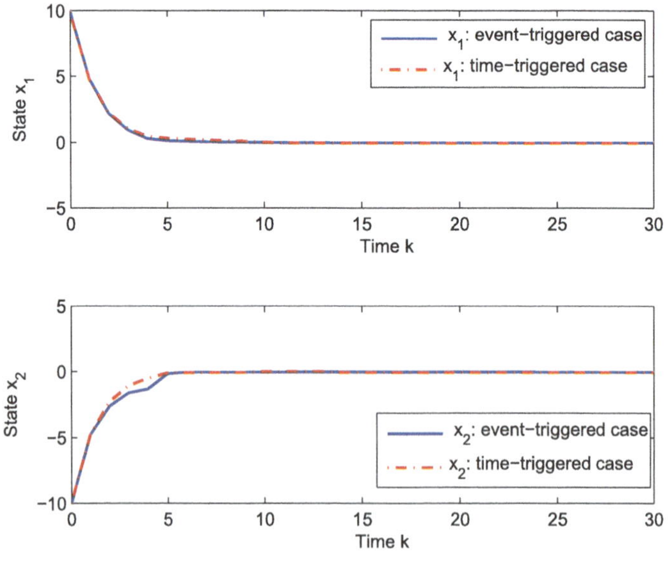

Fig. 6 The simulation of the closed-loop system state with $\tau_a = 4.7$

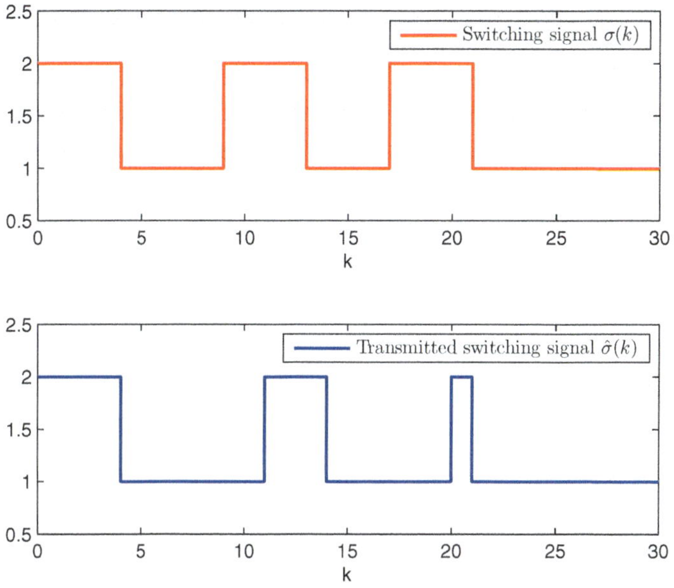

Fig. 7 The switching signals $\sigma(k)$ of the system and $\hat{\sigma}(k)$ of the controller

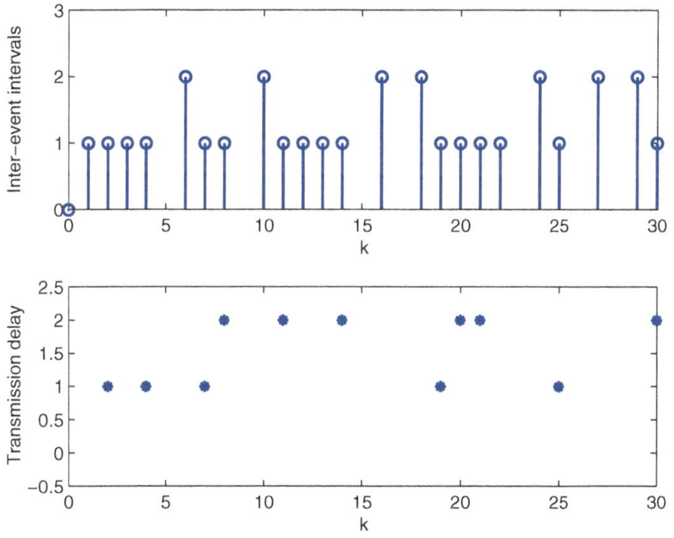

Fig. 8 Length of event-triggered intervals and network-induced delays

Then, it follows from (36) that exponential stability of the closed-loop system can be guaranteed if the average dwell time $\tau_a > 4.6566$.

The state responses of the closed-loop system in both the event-triggered case and time-triggered case when $\tau_a = 4.7$ are shown in Fig. 6, which converge to zero asymptotically. The switching signal $\sigma(k)$ of the system and $\hat{\sigma}(k)$ of the controller are simulated in Fig. 7. Event-triggered intervals and network-induced delays are illustrated in Fig. 8. In the simulation, there are 23 transmissions during the whole time interval $[0, 30]$, which means that the proposed event-triggered scheme can reduce about $7/30 = 23.33\%$ of the data transmission.

5 Conclusion

In this chapter, considering network induced packet dropouts and transmission delays, new mode dependent event-triggered schemes with dwell time constraints are given to solve the networked control problem for discrete-time switched linear systems. Utilizing multiple Lyapunov functional approach and switching signal merging technique, exponential stability conditions for the closed-loop systems are established, co-design methods for event-trigger parameters and feedback stabilizing controllers are obtained. The proposed method extends the event-triggered transmission scheme of non-switched systems, which can also be adopted to solve event-triggered $\mathcal{H}_\infty$ control/filtering and other related event-triggered design problems of switched linear systems.

Acknowledgements The authors would be very grateful for the support of Jiangsu Overseas Visiting Scholar Program for University Prominent Young & Middle-aged Teachers and Presidents, Six Talent Peaks Project of Jiangsu (No. DZXX-044) and Nantong 226 High-level Talents Project. The work of Xiaoqing Xiao was also supported by China Scholarship Council (CSC NO. 201908320096), National Natural Science Foundation of China(Nos. 61374061, 61403216 and 61573201). Also, the work of J.H. Park was supported by the National Research Foundation of Korea (NRF) grant funded by the Korea government (Ministry of Science and ICT) (No. 2019R1A5A8080290l1).

References

1. Liberzon, D.: Switching in Systems and Control. Birkhäuser, Boston (2003)
2. Zhao, X.D., Kao, Y.G., Niu, B., Wu, T.T.: Control synthesis of switched systems. Studies in Systems, Decision and Control, vol. 80. Springer (2017)
3. Lin, H., Antsaklis, P.J.: Stability and stabilizability of switched linear systems: a survey of recent results. IEEE Trans. Autom. Control **54**, 308–322 (2009)
4. Zhu, F., Antsaklis, P.J.: Optimal control of hybrid switched systems: a brief survey. Discret. Event Dyn. Syst. **25**(3), 345–364 (2015)
5. Zhang, L.X., Gao, H.J.: Asynchronously switched control of switched linear systems with average dwell time. Automatica **46**(5), 953–958 (2010)

6. Zhang, L.X., Shi, P.: $l_2 - l_\infty$ model reduction for switched LPV systems with average dwell time. IEEE Trans. Autom. Control **53**, 2443–2448 (2008)
7. Yu, Q., Zhao, X.D.: Stability analysis of discrete-time switched linear systems with unstable subsystems. Appl. Math. Comput. **273**, 718–725 (2016)
8. Zhang, D., Xu, Z., Karimi, H.R., Wang, Q.: Distributed filtering for switched linear systems with sensor networks in presence of packet dropouts and quantization. IEEE Trans. Circuits Syst. I: Regul. Pap. **64**(10), 2783–2796 (2013)
9. Zhang, W.A., Yu, L.: Modelling and control of networked control systems with both network-induced delay and packet-dropout. Automatica **44**, 3206–3210 (2008)
10. Tan, C., Li, L., Zhang, H.S.: Stabilization of networked control systems with both network-induced delay and packet dropout. Automatica **59**, 194–199 (2015)
11. Tang, Z., Park, J.H., Lee, T.H.: Dynamic output-feedback-based $\mathcal{H}_\infty$ design for networked control systems with multipath packet dropouts. Appl. Math. Comput. **275**, 121–133 (2016)
12. Wen, S.P., Zeng, Z.G., Huang, T.W.: Observer-based $\mathcal{H}_\infty$ control of discrete time-delay systems with random communication packet losses and multiplicative noises. Appl. Math. Comput. **219**, 6484–6493 (2013)
13. Wang, J.F., Yang, H.Z.: Exponential stability of a class of networked control systems with time delays and packet dropouts. Appl. Math. Comput. **218**, 8887–8894 (2012)
14. Liberzon, D.: Finite data-rate feedback stabilization of switched and hybrid linear systems. Automatica **50**, 409–420 (2014)
15. Wakaiki, M., Yamamoto, Y.: Stability analysis of sampled-data switched systems with quantization. Automatica **69**, 157–168 (2016)
16. Ma, D., Zhao, J.: Stabilization of networked switched linear systems: an asynchronous switching delay system approach. Syst. Control Lett. **77**, 46–54 (2015)
17. Du, H.P., Zhang, N., Wang, L.F.: Switched control of vehicle suspension based on motion-mode detection. Veh. Syst. Dyn. **52**(1), 142–164 (2014)
18. Zhang, W., Hou, Y.Z., Liu, X.B., Zhou, Y.J.: Switched control of three-phase voltage source PWM rectifier under a wide-range rapidly varying active load. IEEE Trans. Power Electron. **27**(2), 881–890 (2012)
19. Lian, J., Li, C., Xia, B.: Sampled-data control of switched linear systems with application to an F-18 aircraft. IEEE Trans. Ind. Electron. **64**(2), 1332–1340 (2017)
20. Åström, K., Bernhardsson, B.: Comparison of periodic and event based sampling for first order stochastic systems. In: Proceedings of IFAC World Conference, pp. 301–306 (1999)
21. Hendricks, E., Jensen, M., Chevalier, A., Vesterholm, T.: Problems in event based engine control. In: Proceedings of the American Control Conference, pp. 1585–1587 (2010)
22. Garcia, E., Antsaklis, P.J.: Model-based event-triggered control for systems with quantization and time-varying network delays. IEEE Trans. Autom. Control **58**(2), 422–434 (2013)
23. Heemels, W.P.M.H., Donkers, M.C.F., Teel, A.R.: Periodic event-triggered control for linear systems. IEEE Trans. Autom. Control **58**(4), 847–861 (2013)
24. Hu, S.L., Yin, X., Zhang, Y., Tian, E.G.: Event-triggered guaranteed cost control for uncertain discrete-time networked control systems with time-varying transmission delays. IET Control Theory Appl. **6**(18), 2793–2804 (2012)
25. Peng, C., Yang, T.C.: Event-triggered communication and $\mathcal{H}_\infty$ control co-design for networked control systems. Automatica **49**, 1326–1332 (2013)
26. Meng, X.Y., Chen, T.W.: Event triggered robust filter design for discrete-time systems. IET Control Theory Appl. **8**(2), 104–113 (2014)
27. Peng, C., Fei, M.R.: Networked $\mathcal{H}_\infty$ filtering for discrete linear systems with a periodic event-triggering communication scheme. IET Signal Process. **7**(8), 754–765 (2013)
28. Zhang, Y., Wang, Z.D., Zou, L., Fang, H.J.: Event-based finite-time filtering for multi-rate systems with fading measurements. IEEE Trans. Aerosp. Electron. Syst. **53**(3), 1431–1441 (2017)
29. Zhou, L., Pan, R., Xiao, X.Q., Sun, T.T.: Event-triggered $\mathcal{H}_\infty$ filtering for discrete-time systems over unreliable networks with package dropouts. Neurocomputing **218**, 346–353 (2016)

30. Wu, Z.G., Xu, Y., Pan, Y.J., Su, H.S., Tang, Y.: Event-triggered control for consensus problem in multi-agent systems with quantized relative state measurement and external disturbance. IEEE Trans. Circuit Syst. I: Regul. Pap. **65**(7), 2232–2242 (2018)
31. Wu, Z.G., Xu, Y., Lu, R.Q., Wu, Y.Q., Huang, T.W.: Event-triggered control for consensus of multi-agent systems with fixed/switching topologies. IEEE Trans. Syst. Man Cybern. Syst. **48**(10), 1736–1746 (2018)
32. Liu, S.J., Zhou, L.: Network synchronization and application of chaotic Lur'e systems based on event-triggered mechanism. Nonlinear Dyn. **83**(4), 2497–2507 (2016)
33. Park, J.H., Shen, H., Chang, X.H., Lee, T.H.: Recent Advances in Control and Filtering of Dynamic Systems with Constrained Signals. Springer, Cham, Switzerland (2018). https://doi.org/10.1007/978-3-319-96202-3
34. Liu, J., Gu, Y., Xie, X., Yue, D., Park, J.H.: Hybrid-driven-based $\mathcal{H}_\infty$ control for networked cascade control systems with actuator saturations and stochastic cyber attacks. IEEE Trans. Syst. Man Cybern. Syst. **49**(12), 2452–2463 (2019)
35. Wang, J., Chen, M., Shen, H., Park, J.H., Wu, Z.G.: A Markov jump model approach to reliable event-triggered retarded dynamic output feedback $\mathcal{H}_\infty$ control for networked systems. Nonlinear Anal. Hybrid Syst. **26**, 137–150 (2017)
36. Xiang, W., Johnson, T.: Event-triggered control for continuous-time switched linear systems. IET Control Theory Appl. **11**(11), 1694–1703 (2017)
37. Xiao, X.Q., Zhou, L., Ho, D.W.C., Lu, G.P.: Event-triggered control of continuous-time switched linear systems. IEEE Trans. Autom. Control **64**(4), 1710–1717 (2019)
38. Li, Z., Ma, D., Zhao, J.: Periodic event-triggered control for switched affine systems. In: IEEE 15th International Conference on Control and Automation (ICCA), Edinburgh, United Kingdom, pp. 567–572 (2019)
39. Zobiri, F., Meslem, N., Bidegaray-Fesquet, B.: Event-triggered stabilizing controllers for switched linear systems. Nonlinear Anal. Hybrid Syst. **36**, 100831 (2020)
40. Yang, F., Gu, Z., Cheng, J., Liu, J.L.: Event-driven finite-time control for continuous-time networked switched systems under cyber attacks. J. Frankl. Inst. (2019). https://doi.org/10.1016/j.jfranklin.2019.09.037
41. Linh, V., Morgansen, K.A.: Stability of time-delay feedback switched linear systems. IEEE Trans. Autom. Control **55**(10), 2385–2390 (2010)

Fault Alarm-Based Hybrid Control Design for Periodic Piecewise Time-Delay Systems

R. Sakthivel and S. Harshavarthini

Abstract In this chapter, fault alarm-based hybrid control scheme is developed for periodic piecewise systems with time delay and disturbances. Precisely, the hybrid controller is proposed to handle both the normal and faulty systems. Moreover, a switching among the robust and reliable controller is accomplished with respect to the alarm signal. Based on the threshold value of the alarm signal, the timely alert will be encountered. Further, the linear matrix inequality based stability conditions are derived by considering suitable Lyapunov–Krasovskii functional and employing some advanced inequalities. With the aid of delay-dependent sufficient conditions, the asymptotic stability of the addressed system is ensured with a minimum disturbance attenuation level. In particular, the gain matrices of the control scheme is obtained by solving the developed stability conditions. At last, the impact and significance of the proposed hybrid controller is validated by presenting simulation results of a numerical example.

Keywords Periodic piecewise time-invariant systems · Time-varying delay · Uncertainties and disturbances · Hybrid control · Lyapunov stability method

1 Introduction

In recent years, the system with periodic characteristics receives considerable research attention because of its potential applications in various engineering and science fields. To mention a few, DC-DC converters, vehicle suspension, transmission lines with distinct loads, rotor-bearing systems, helicopter rotors and power electronic equipments with multi-function converters. More specifically, the periodic systems are a special class of switched systems with fixed dwell time. In particular,

R. Sakthivel (✉) · S. Harshavarthini
Department of Applied Mathematics, Bharathiar University, Coimbatore 641046, India
e-mail: krsakthivel@yahoo.com

S. Harshavarthini
e-mail: varthiniksp@gmail.com

J. H. Park (ed.), *Recent Advances in Control Problems of Dynamical Systems and Networks*, Studies in Systems, Decision and Control 301,
https://doi.org/10.1007/978-3-030-49123-9_9

the periodic systems are subdivided into a finite number of subsystems to accurately analyze the spectrum of the considered periodic systems in each instant of time, which is referred to as periodic piecewise systems (PPSs). Thus, the study on stability and stabilization of PPSs become an active research topic in which various control protocol is designed to stabilize the system dynamics [1–5]. Xie et al. [1] examined the stabilization problem of uncertain PPSs by Lyapunov stability theory. Notably, in [2], stability and $\mathcal{L}_1$ gain analysis of positive PPSs with time-invariant subsystems is studied by constructing continuous Lyapunov function. With the aid of polynomial Lyapunov function, a linear inequality based on sufficient conditions are obtained in [3] for the exponential stability of PPSs.

On the other hand, the time-delays [6] are often encountered in many practical systems that are inexorable and it may lead to oscillation, poor performance and even instability of the control systems [7]. Besides, the external disturbances also may have a consequence in the performance of dynamical control systems [8–13]. However, the exact modeling of physical systems is not always guaranteed due to variation in system parameters, modeling errors and neglected nonlinearities. The differences or errors between the mathematical model and the actual physical systems are commonly referred to as model uncertainties [14–17]. Thus, the significance of system performance motivates the researchers in the study on stability analysis of uncertain control systems with time-delay and external disturbances. As a result, a countable number of work is reported in the stability and stabilization of PPSs with uncertainties, time-delay, and disturbances. Precisely, the stabilization of delayed PPSs subject to disturbances is retained with the aid of various control schemes such as $\mathcal{H}_\infty$ control [8], $\mathcal{L}_2-$ gain analysis [9] and guaranteed cost control [10] addressed in the existing literature.

On another research frontier, the occurrence of faults in system components is unavoidable and will happen at any instant of time. In such circumstances, the system trajectories may attain unanticipated performance degradation and substantial damage to the plant. Thus, it is necessary to tolerate the effects of actuator faults and also to retain the robust performance of the system dynamics. The reliable control strategy is an effective technique to tackle the impact of faults in system components and it is widely reported for various dynamical systems [18–22].

Moreover, the hybrid control design comprised of both the robust and reliable control, where the fault alarm is designed to timely alert the controller once the fault is occurred in system components. Based on the alarm-signal, the controller automatically switched to the reliable control from conventional control scheme. Precisely to switch between the controllers we design a switching method by selecting a suitable threshold value of faults. However, implementing the reliable controller, where the system does not undergo failures will cause conservative in such cases the robust controller is activated [23–27].

Despite the advantages of above-mentioned studies and based on fault-alarm approach, the hybrid control is designed for delayed PPSs with uncertainties and disturbances.

In this chapter, we mainly focused on the alarm design to timely alert the controller with the aid of the appropriate threshold value. To be precise, the fault is detected with the aid of state residual observer and by simple manipulation, the threshold value is calculated. Consequently, the delay-dependent conditions are derived by constructing periodic piecewise Lyapunov–Krasovskii functional with time-varying matrices. Further, the periodically time-varying controller gain matrices are obtained by solving the developed linear matrix inequality based conditions. More specifically, the asymptotic stability of the considered PPSs is ensured by the aid of the Lyapunov stability theory.

Notations: The superscripts (-1) and T stand for matrix inverse and matrix transposition, respectively; $\mathcal{R}^n$ denotes the set of all n-dimensional Euclidean space; $L_2[0, \infty)$ denotes the space of all square integrable function on $[0, \infty)$; We use an asterisk "$*$" to represent the term induced by symmetry and we denote $Sym(A) = A^T + A$; $P > 0$ represents P is real, symmetric and positive definite; $\mathbb{N}$ denotes the set of all natural numbers; I represents unit matrix with compatible dimension; Moreover, the matrices are assumed to be compatible for algebraic operations if their dimensions are not explicitly stated.

2 Problem Formulation and Preliminaries

Consider a class of periodic systems with uncertainties, time-varying delay and exogenous disturbances in the following form;

$$\begin{cases} \dot{x}(t) & = (A(t) + \Delta A(t))x(t) + (A_\tau(t) + \Delta A_\tau(t))x(t - \tau(t)) \\ & \quad + B(t)u^G(t) + C(t)v(t), \\ y(t) & = D(t)x(t), \\ x(t_0) & = \phi(t_0),\ \forall t_0 \in [-\tau, 0], \end{cases} \tag{1}$$

where $x(t) \in \mathcal{R}^x$ is the state vector; $u^G(t) \in \mathcal{R}^u$ is the control input vector; $v(t) \in L_2[0, \infty)$ is the exogenous disturbance input; $\tau(t)$ is the time-varying delay which satisfies $0 \le \tau(t) \le \tau$ and $\dot{\tau}(t) \le \upsilon < 1$; $y(t) \in \mathcal{R}^y$ represents output vector. $A(t) = A(t + \nu F_p)$, $A_\tau(t) = A_\tau(t + \nu F_p)$, $B(t) = B(t + \nu F_p)$, $C(t) = C(t + \nu F_p)$ and $D(t) = D(t + \nu F_p)$ are known matrices with fundamental period F_p, $\nu = 0, 1, \ldots$. Further, the time interval $[\nu F_p, (\nu + 1)F_p)$ is partitioned into Q subintervals $[\nu F_p + t_{i-1}, \nu F_p + t_i)$, $i \in \mathbb{Q} = [1, 2, \ldots, Q]$, $\sum_{i=1}^{Q} F_i = F_p$, $t_0 = 0$ and $t_Q = F_p$.

In this connection, the system (1) is rewritten as periodic piecewise systems (PPSs) in the following form;

$$\begin{cases} \dot{x}(t) & = (A_i + \Delta A_i(t))x(t) + (A_{\tau i} + \Delta A_{\tau i}(t))x(t - \tau(t)) \\ & \quad + B_i u^G(t) + C_i v(t), \\ y(t) & = D_i x(t), \\ x(t_0) & = \phi(t_0),\ \forall t_0 \in [-\tau, 0], \end{cases} \tag{2}$$

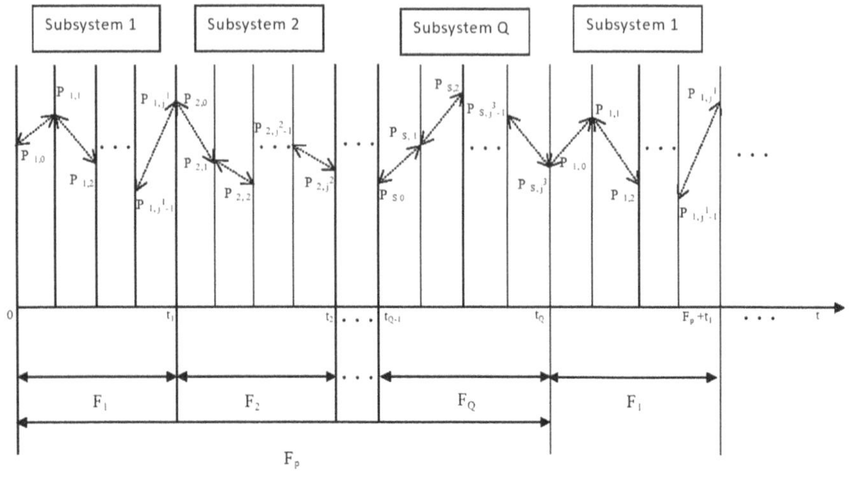

Fig. 1 Evolution of PPTV positive-definite matrix $P(t)$

where A_i, $A_{\tau i}$, B_i, C_i and D_i, $i \in \mathbb{M}$ are the appropriate dimensioned constant matrices of ith subsystem; $\Delta A_i(t)$ and $\Delta A_{\tau i}(t)$ represent the time-varying parameter uncertainties of the form $[\, \Delta A_i(t) \;\; \Delta A_{\tau i}(t) \,] = U_i \Delta_i(t)[\, V_{Ai} \;\; V_{A\tau i} \,]$, in which U_i and V_i are known constant matrices and $\Delta_i(t)$ is an unknown time-varying matrix with $\Delta_i^T(t)\Delta_i(t) < I$.

On the other hand, the periodic piecewise control input in (2) is considered as follows;

$$u^G(t) = \tilde{\hat{K}}_i(t)x(t) = (K_i + \Delta K_i(t))x(t), \tag{3}$$

where K_i is the controller gain matrix of ith subsystem and $\Delta K_i(t)$ represents the controller gain variation of the form $\Delta K_i(t) = U_{Ki}\Delta_i(t)V_{Ki}$, here U_{Ki} and V_{Ki} are known constant matrices with appropriate dimensions and $\Delta_i(t)$ is a time-varying matrix with $\Delta_i^T(t)\Delta_i(t) < I$.

Now, by substituting the designed control input (3) in system (4), we obtain the following closed-loop system;

$$\begin{cases} \dot{x}(t) & = (A_i + \Delta A_i(t))x(t) + (A_{\tau i} + \Delta A_{\tau i}(t))x(t - \tau(t)) \\ & \quad + B_i(K_i + \Delta K_i(t))x(t) + C_i v(t), \\ y(t) & = D_i x(t), \\ x(t_0) & = \phi(t_0), \; \forall t_0 \in [-\tau, 0]. \end{cases} \tag{4}$$

Moreover, the following lemma is more useful for deriving our main results.

Lemma 1 ([15]) *For given symmetric matrices* $\mathbf{Q} > 0$ *and* $\mathbf{D} > 0$, *scalars* a *and* b *satisfying* $a < b$, *the following inequalities hold for all continuously differentiable function* $\kappa : [a, b] \to \mathbb{R}^{\kappa}$, *such that*

$$(b-a)\int_a^b \kappa^T(s)\mathbf{Q}\kappa(s)ds \geq \pi^T \begin{bmatrix} 9\mathbf{Q} & -\frac{36}{b-a}\mathbf{Q} & \frac{60}{(b-a)^2}\mathbf{Q} \\ * & \frac{192}{(b-a)^2}\mathbf{Q} & -\frac{360}{(b-a)^3}\mathbf{Q} \\ * & * & \frac{720}{(b-a)^4}\mathbf{Q} \end{bmatrix} \pi,$$

$$\frac{(b-a)^2}{2}\int_a^b ds \int_a^s \kappa^T(u)D\kappa(u)du \geq \hat{\pi}^T \begin{bmatrix} 6\mathbf{D} & \frac{30}{(b-a)}\mathbf{D} & \frac{60}{(b-a)^2}\mathbf{D} \\ * & \frac{720}{(b-a)^4}\mathbf{D} & -\frac{480}{(b-a)^3}\mathbf{D} \\ * & * & \frac{1200}{(b-a)^4}\mathbf{D} \end{bmatrix} \hat{\pi},$$

where

$$\pi^T = \left[\int_a^b \kappa^T(s)ds \ \int_a^b ds \int_a^s \kappa^T(u)du \ \ \int_a^b ds \int_a^s du \int_a^u \kappa^T(v)dv \right]^T$$

and

$$\hat{\pi} = \left[\int_a^b ds \int_a^s \kappa^T(u)du \quad \int_a^b ds \int_a^s du \int_a^u \kappa^T(v)dv \right.$$
$$\left. \int_a^b ds \int_a^s du \int_a^u dv \int_a^v \kappa^T(w)dw \right]^T.$$

3 Main Results

In this section, $\mathcal{H}_\infty$ based hybrid control scheme is designed for PPSs with time-varying delay, uncertainties and exogenous disturbances. To be precise, this section consists of the following three theorems.

First, linear matrix inequality based sufficient stability conditions are derived by constructing Lyapunov–Krasovskii functional with periodic piecewise time-varying (PPTV) positive-definite matrix for known gain matrix which is presented in Theorem 1. Further, the result will be extended to the case of an unknown gain matrix with gain fluctuations. Finally, Theorem 3 provides a design of a non-fragile reliable control with unknown gain and unknown fault matrix.

Before proceeding, we construct a time-varying piecewise periodic positive-definite matrix $P(t), \forall t > 0$ with fundamental period F_p and it satisfies $\lim_{t \to \nu F_p + t_i} P(t) = P(t + \nu F_p)$. Further, the sub-interval $[t_{i-1}, t_i)$ of each piecewise subsystems in each period F_p is divided into J_i segments with length $\sigma_i = F_i/J_i$, where $J_i \in \mathbb{N}$.

Let $\omega_{i,j} = \nu F_p + t_{i-1} + j\sigma_i$, where $j \in \mathbb{J} = 0, 1, \ldots, J_i - 1$. Now, the periodic piecewise positive-definite matrix $P_i(t)$ is defined as

$$P_i(t) = P_{i,j} + \rho_{i,j}(t)(P_{i,j+1} - P_{i,j}), \ \ i \in Q, t \in [\omega_{i,j}, \omega_{i,j+1}) \in F_i, \tag{5}$$

where $\rho_{i,j}(t) = \frac{J_i(t-(\nu F_p + t_{i-1} + j\sigma_i))}{F_i} \in [0, 1)$ and $P_{i,J_i} = P_{i+1,0}$ $P_{Q,J_i} = P_{i,0}$.

For brevity, the continuous time-varying matrix $P_i(t)$ is rewritten as $P_i(t) = (1-\rho_{i,j}(t))P_{i,j} + \rho_{i,j}(t)P_{i,j+1}$. Moreover, the schematic representation of PPTV positive-definite matrix $P(t)$ is given in Fig. 1.

3.1 $\mathcal{H}_\infty$ Performance Analysis

In this subsection, a delay-dependent stability condition is derived by considering Lyapunov–Krasovskii functional with the PPTV matrix for the design of a robust controller with known gain matrix which can ensure the asymptotic stabilization of the considered PPS (1) with prescribed $\mathcal{H}_\infty$ performance index.

Theorem 1 *For given control gain K and scalars α_i, μ, τ, the closed-loop PPS* (4) *with $\mathcal{H}_\infty$ performance index $\gamma > 0$ is asymptotically stable if there exist matrices $P_{i,j} > 0$, $Q_1 > 0$, $Q_2 > 0$, $Q_3 > 0$, $Q_4 > 0$, $j \in \mathbb{J}$, $i \in S$ and positive scalars $\varepsilon_1 > 0$, $\varepsilon_2 > 0$, such that the following constraints hold:*

$$\Psi = \begin{bmatrix} [\Psi^1]_{8\times 8} & [\Psi^2]_{8\times 5} \\ * & [\Psi^3]_{5\times 5} \end{bmatrix} < 0, \tag{6}$$

$$\Phi = \begin{bmatrix} [\Phi^1]_{8\times 8} & [\Phi^2]_{8\times 5} \\ * & [\Phi^3]_{5\times 5} \end{bmatrix} < 0, \tag{7}$$

$$P_{i,J_i} = P_{i+1,0}, \quad P_{S,J_i} = P_{1,0}, \tag{8}$$

where

$$\Psi^1_{1,1} = Sym(P_{i,j}A_i + P_{i,j}B_iK_i) + \alpha_i P_{i,j} + \frac{J_i(P_{i,j+1} - Pi, j)}{F_i}$$
$$+Q_1 + Q_2 + \tau^2 Q_3 + \frac{\tau^4}{4}Q_4,$$
$$\Phi^1_{1,1} = Sym(P_{i,j+1}A_i + P_{i,j+1}B_iK_i) + \alpha_i P_{i,j+1} - \frac{J_i(P_{i,j+1} - P_{i,j})}{F_i}$$
$$+Q_1 + Q_2 + \tau^2 Q_3 + \frac{\tau^4}{4}Q_4,$$
$$\Psi^1_{1,2} = P_{i,j}A_{\tau i}, \ \Phi^1_{1,2} = P_{i,j+1}A_{\tau i},$$
$$\Psi^1_{1,8} = P_{i,j}C_i, \ \Phi^1_{1,8} = P_{i,j+1}C_i,$$
$$\Psi^1_{2,2} = \Phi^1_{2,2} = -(1-\mu)Q_1, \ \Psi^1_{3,3} = \Phi^1_{3,3} = -Q_2,$$
$$\Psi^1_{4,4} = \Phi^1_{4,4} = -9Q_3, \ \Psi^1_{4,5} = \Phi^1_{4,5} = \frac{36}{\tau}Q_3,$$
$$\Psi^1_{4,6} = \Phi^1_{4,6} = \frac{-60}{\tau^2}Q_3,$$
$$\Psi^1_{5,5} = \Phi^1_{5,5} = \frac{-192}{\tau^2}Q_3 - 6Q_4,$$

$$
\begin{aligned}
&\Psi^1_{5,6} = \Phi^1_{5,6} = \frac{360}{\tau^3} Q_3 + \frac{30}{\tau} Q_4,\\
&\Psi^1_{5,7} = \Phi^1_{5,7} = -\frac{60}{\tau^2} Q_4,\\
&\Psi^1_{6,6} = \Phi^1_{6,6} = -\frac{210}{\tau^2} Q_4 - \frac{720}{\tau^4} Q_3,\\
&\Psi^1_{6,7} = \Phi^1_{6,7} = \frac{480}{\tau^3} Q_4,\\
&\Psi^1_{7,7} = \Phi^1_{7,7} = -\frac{1200}{\tau^4} Q_4,\ \Psi^1_{8,8} = \Phi^1_{8,8} = -\gamma I,\\
&\Psi^2_{1,1} = \Phi^2_{1,1} = D_i^T,\ \Psi^2_{1,2} = \varepsilon_1 P_{i,j} U_i,\\
&\Phi^2_{1,2} = \varepsilon_1 P_{i,j+1} U_i,\ \Psi^2_{1,3} = \Phi^2_{1,2} = V^T_{Ai},\\
&\Psi^2_{1,4} = \varepsilon_2 P_{i,j} U_i,\ \Psi^2_{1,4} = \varepsilon_2 P_{i,j+1} U_i,\\
&\Psi^2_{2,5} = \Phi^2_{2,4} = V^T_{A\tau i},\\
&\Psi^3 = \Phi^3 = diag\{-\gamma I, -\varepsilon_1 I, -\varepsilon_1 I, -\varepsilon_2 I, -\varepsilon_2 I\}.
\end{aligned}
$$

Proof To establish the required stability constraints, the piecewise periodic Lyapunov–Krasovskii functional with time-varying positive-definite matrices for the close-loop system (4) is considered as follows;

$$
V(x,t) = \sum_{i=1}^{3} V_i(x,t), \tag{9}
$$

where

$$
\begin{aligned}
V_1(x,t) &= x^T(t) P_i(t) x(t),\\
V_2(x,t) &= \int_{t-\tau(t)}^{t} x^T(s) Q_1 x(s) ds + \int_{t-\tau}^{t} x^T(s) Q_2 x(s) ds,\\
V_3(x,t) &= \tau \int_{-\tau}^{0} ds \int_{t+s}^{t} x^T(u) Q_3 x(u) du\\
&+ \frac{\tau^2}{2} \int_{-\tau}^{0} ds \int_{-\tau}^{s} du \int_{t+u}^{t} x^T(v) Q_4 x(v) dv.
\end{aligned}
$$

Now, the time derivative of the above defined Lyapunov–Krasovskii functional $V(x,t)$ together with the trajectories of closed-loop system (4) is calculated as follows:

$$
\begin{aligned}
\dot{V}_1(x,t) &= 2x^T(t) P_i(t) \dot{x}(t) + x^T(t) \dot{P}_i(t) x(t)\\
&= 2x^T(t) P_i(t)((A_i + \Delta A_i(t)) x(t) + (A_{\tau i} + \Delta A_{\tau i}(t)) x(t-\tau(t))\\
&+ B_i K_i(t) x(t) + C_i w(t) + x^T(t) \frac{P_{i,j+1} - P_{i,j}}{\sigma_i} x(t),
\end{aligned} \tag{10}
$$

$$\dot{V}_2(x,t) \leq x^T(t)(Q_1+Q_2)(t)x(t) - (1-\mu)x^T(t-\tau(t))Q_1x(t-\tau(t)) - x^T(t-\tau)Q_2x(t-\tau), \tag{11}$$

$$\dot{V}_3(x,t) = x^T(t)\left[\tau^2 Q_3 + \frac{\tau^4}{4}Q_4\right]x(t) - \tau\int_{t-\tau}^{t} x^T(s)Q_3x(s)ds - \frac{\tau^2}{2}\int_{t-\tau}^{t} ds\int_{t-\tau}^{t+s} x^T(u)Q_4x(u)du. \tag{12}$$

By applying Lemma 1 to the single and double integral terms in Eq. (12), we get

$$-\tau\int_{-\tau}^{0} x^T(t+s)Q_3x(t+s)ds \leq \hat{\varphi}^T(t)\begin{bmatrix} -9Q_3 & \frac{36}{\tau}Q_3 & \frac{-60}{\tau^2}Q_3 \\ * & \frac{-192}{\tau^2}Q_3 & \frac{360}{\tau^3}Q_3 \\ * & * & \frac{-720}{\tau^4}Q_3 \end{bmatrix}\hat{\varphi}(t), \tag{13}$$

$$-\frac{\tau^2}{2}\int_{-\tau}^{0} ds\int_{-\tau}^{s} x^T(t+u)Q_4x(t+u)du \leq \tilde{\varphi}^T(t)\begin{bmatrix} -6Q_4 & \frac{30}{\tau}Q_4 & \frac{-60}{\tau^2}Q_4 \\ * & \frac{-720}{\tau^4}Q_4 & \frac{480}{\tau^3}Q_4 \\ * & * & \frac{-1200}{\tau^4}Q_4 \end{bmatrix}\tilde{\varphi}(t), \tag{14}$$

where

$$\hat{\varphi}^T(t) = \left[\int_{t-\tau}^{t} x^T(s)ds \;\; \int_{-\tau}^{0} ds\int_{t-\tau}^{t+s} x^T(u)du \;\; \int_{-\tau}^{0} ds\int_{t-\tau}^{t+s} du\int_{t-\tau}^{t+u} x^T(v)dv\right]^T$$

and

$$\tilde{\varphi}^T(t) = \left[\hat{\varphi}^T(t) \;\; \int_{-\tau}^{0} ds\int_{t-\tau}^{t+s} du\int_{t-\tau}^{t+u} dv\int_{t-\tau}^{t+v} x^T(w)dw\right]^T.$$

Further, by combining the inequalities (10)–(14), we obtain

$$\dot{V}(x,t) + \alpha V(x,t) \leq \bar{\varphi}^T(t)[\Theta]_{8\times 8}\bar{\varphi}(t) \tag{15}$$

where

$$\bar{\varphi}^T(t) = \left[x^T(t)\; x^T(t-\tau(t))\; x^T(t-\tau)\; \tilde{\varphi}^T(t)\; w^T(t)\right]^T,$$

$$\Theta_{1,1} = Sym(P_i(t)(A_i + B_iK_i + \Delta A_i(t)) + \alpha_i P_{i,j} + \frac{J_i(P_{i,j+1} - Pi,j)}{F_i} + Q_1 + Q_2 + \tau^2 Q_3 + \frac{\tau^4}{4}Q_4,$$

$$\Theta_{1,2} = P_i(t)(A_{\tau i} + \Delta A_{\tau i}(t)),\;\; \Theta_{1,2} = P_iC_i,$$

$$\Theta_{2,2} = -(1-\mu)Q_1,\;\; \Theta_{3,3} = -Q_2,\;\; \Theta_{4,4} = -9Q_3,$$

$$\Theta_{4,5} = \frac{36}{\tau} Q_3, \ \Theta_{4,6} = \frac{-60}{\tau^2} Q_3,$$
$$\Theta_{5,5} = \frac{-192}{\tau^2} Q_3 - 6Q_4, \ \Theta_{5,6} = \frac{360}{\tau^3} Q_3 + \frac{30}{\tau} Q_4,$$
$$\Theta_{5,7} = -\frac{60}{\tau^2} Q_4, \ \Theta_{6,6} = -\frac{210}{\tau^2} Q_4 - \frac{720}{\tau^4} Q_3,$$
$$\Theta_{6,7} = \frac{480}{\tau^3} Q_4, \ \Theta_{7,7} = -\frac{1200}{\tau^4} Q_4, \ \Theta_{8,8} = -\gamma I.$$

Now, by using the Eq. (5) and by simple manipulation, the matrix Θ in above inequality (15) can be expressed in the following form

$$\Theta = (1 - \sigma_{i,j}(t))\tilde{\Psi} + \sigma_{i,j}(t)\tilde{\Phi}, \tag{16}$$

where

$$\tilde{\Psi} = [\Psi^1]_{8\times 8} + (\upsilon_1 \Delta_i(t)\upsilon_2) + (\upsilon_1 \Delta_i(t)\upsilon_2)^T + (\upsilon_1 \Delta_i(t)\upsilon_4) + (\upsilon_1 \Delta_i(t)\upsilon_4)^T,$$
$$\tilde{\Phi} = [\Phi^1]_{8\times 8} + (\tilde{\upsilon}_1 \Delta_i(t)\upsilon_2) + (\tilde{\upsilon}_1 \Delta_i(t)\upsilon_2)^T + (\tilde{\upsilon}_1 \Delta_i(t)\upsilon_4) + (\tilde{\upsilon}_1 \Delta_i(t)\upsilon_4)^T,$$

with

$$\upsilon_1 = \left[(P_{i,j}U_i)^T \ 0_{1\times 7}\right]^T, \ \tilde{\upsilon}_1 = \left[(P_{i,j+1}U_i)^T \ 0_{1\times 7}\right]^T,$$
$$\upsilon_2 = \left[V_{Ai}^T \ 0_{1\times 7}\right], \ \upsilon_4 = \left[0 \ V_{A_{\tau i}}^T \ 0_{1\times 6}\right],$$

in which $\Psi_{1,1}^1 = \Psi_{1,1}^1 + \gamma^{-1} D^T i D_i$, and $\Phi_{1,1}^1 = \Phi_{1,1}^1 + \gamma^{-1} D^T i D_i$.

Here, by employing S-procedure lemma and Schur complement in the above mentioned inequality (16), we obtain the LMIs stated in theorem statement. If the constraints (6)–(8) hold, we get $\Theta < 0$. Hence, by Lyapunov stability theory, the closed-loop form of PPS (4) is asymptotically stable. Thus, the proof is completed. □

3.2 *Hybrid Control Design*

This subsection will be focused on deriving the sufficient condition for ensuring the asymptotic stability of PPSs (1) through the robust and reliable control design. More specifically, a robust control law is designed with an unknown gain matrix and without any actuator faults. Further, the controller gain and actuator faults are considered as unknown in reliable control design.

Case (i) (Robust controller design): In this case, we aim to design robust controller with gain fluctuations and without actuator fault. Thus, the control gain matrix in (3) takes the form $\tilde{\hat{K}}_i(t) = \hat{K}_i(t) + \Delta\hat{K}_i(t)$ in which $\Delta\hat{K}_i(t) = \hat{U}_{Ki}\Delta_i(t)\hat{V}_{Ki}$

where $\hat{U}_{Ki}$ and $\hat{V}_{Ki}$ are known constant matrices with appropriate dimensions and $\Delta_i(t)$ is the time-varying matrix satisfying $\Delta_i^T(t)\Delta_i(t) < I$.

In this connection, the robust control scheme with gain fluctuation is considered as;

$$u^G(t) = (\hat{K}_i(t) + \Delta\hat{K}_i(t))x(t). \tag{17}$$

Then, we have the following theorem.

Theorem 2 *For given scalars α_i, μ, τ, if there exist matrices $\hat{M}_{i,j} > 0$, $N_{i,j}$, $\tilde{Q}_1 > 0$, $\tilde{Q}_2 > 0$, $\tilde{Q}_3 > 0$, $\tilde{Q}_4 > 0$, $j \in \mathbb{J}$, $i \in S$ and positive scalars $\varepsilon_1 > 0$, $\varepsilon_2 > 0$, $\varepsilon_3 > 0$, such that the following LMIs hold:*

$$[\Upsilon]_{15\times 15} < 0, \tag{18}$$

$$[\Omega]_{15\times 15} < 0, \tag{19}$$

$$\hat{M}_{i,J_i} = \hat{M}_{i+1,0}, \hat{M}_{S,J_i} = \hat{M}_{1,0}, \tag{20}$$

$$\hat{N}_{i,J_i} = \hat{N}_{i+1,0}, \hat{N}_{S,J_i} = \hat{N}_{1,0}, \tag{21}$$

where

$$\Upsilon_{1,1} = Sym(A_i\hat{M}_{i,j} + B_i\hat{N}_{i,j}) + \alpha_i\hat{M}_{i,j} - \frac{J_i(\hat{M}_{i,j+1} - \hat{M}_{i,j})}{F_i}$$
$$+\tilde{Q}_1 + \tilde{Q}_2 + \tau^2\tilde{Q}_3 + \frac{\tau^4}{4}\tilde{Q}_4,$$
$$\Omega_{1,1} = Sym(A_i\hat{M}_{i,j+1} + B_i\hat{N}_{i,j+1}) + \alpha_i\hat{M}_{i,j+1} - \frac{J_i(\hat{M}_{i,j+1} - \hat{M}_{i,j})}{F_i}$$
$$+\tilde{Q}_1 + \tilde{Q}_2 + \tau^2\tilde{Q}_3 + \frac{\tau^4}{4}\tilde{Q}_4,$$
$$\Upsilon_{1,2} = A_{\tau i}\hat{M}_{i,j},\ \Omega_{1,2} = A_{\tau i}\hat{M}_{i,j+1},\ \Upsilon_{1,8} = C_i,$$
$$\Omega_{1,8} = C_i,\ \Upsilon_{1,9} = \hat{M}_{i,j}D_i^T,\ \Omega_{1,9} = \hat{M}_{i,j+1}D_i^T,$$
$$\Upsilon_{1,10} = \Omega_{1,10} = \varepsilon_1 U_i,\ \Upsilon_{1,11} = \hat{M}_{i,j}V_{Ai}^T,$$
$$\Omega_{1,11} = \hat{M}_{i,j+1}V_{Ai}^T,\ \Upsilon_{1,12} = \Omega_{1,12} = \varepsilon_2 U_i,$$
$$\Upsilon_{1,14} = \Omega_{1,14} = \varepsilon_3 B_i\mathbb{G}_0\hat{U}_{Ki},\ \Upsilon_{1,15} = \hat{M}_{i,j}\hat{V}_{Ki}^T,$$
$$\Omega_{1,15} = \hat{M}_{i,j+1}\hat{V}_{Ki}^T,\ \Upsilon_{2,2} = \Omega_{2,2} = -(1-\mu)\tilde{Q}_1,$$
$$\Upsilon_{2,13} = \hat{M}_{i,j}V_{Bi}^T,\ \Omega_{2,13} = M_{i,j+1}V_{Bi}^T,\ \Upsilon_{3,3} = \Omega_{3,3} = -\tilde{Q}_2,$$
$$\Upsilon_{4,4} = \Omega_{4,4} = -9\tilde{Q}_3,\ \Upsilon_{4,5} = \Omega_{4,5} = \frac{36}{\tau}\tilde{Q}_3,$$
$$\Upsilon_{4,6} = \Omega_{4,6} = \frac{-60}{\tau^2}\tilde{Q}_3,\ \Upsilon_{5,5} = \Omega_{5,5} = \frac{-192}{\tau^2}\tilde{Q}_3 - 6\tilde{Q}_4,$$
$$\Upsilon_{5,6} = \Omega_{5,6} = \frac{360}{\tau^3}\tilde{Q}_3 + \frac{30}{\tau}\tilde{Q}_4,\ \Upsilon_{5,7} = \Omega_{5,7} = -\frac{60}{\tau^2}\tilde{Q}_4,$$

$$\Upsilon_{6,6} = \Omega_{6,6} = -\frac{210}{\tau^2}\tilde{Q}_4 - \frac{720}{\tau^4}\tilde{Q}_3,\ \Upsilon_{6,7} = \Omega_{6,7} = \frac{480}{\tau^3}\tilde{Q}_4,$$
$$\Upsilon_{7,7} = \Omega_{7,7} = -\frac{1200}{\tau^4}\tilde{Q}_4,$$
$$\Upsilon_{8,8} = \Omega_{8,8} = -\gamma I,\ \Upsilon_{9,9} = \Omega_{9,9} = -\gamma I,$$
$$\Upsilon_{10,10} = \Omega_{10,10} = \Upsilon_{11,11} = \Omega_{11,11} = -\varepsilon_1 I,$$
$$\Upsilon_{12,12} = \Omega_{12,12} = \Upsilon_{13,13} = \Omega_{13,13} = -\varepsilon_2 I,$$
$$\Upsilon_{14,14} = \Omega_{14,14} = \Upsilon_{15,15} = \Omega_{15,15} = -\varepsilon_3 I.$$

Thus, the considered PPSs is asymptotically stabilized through the controller (17) *with* $\mathcal{H}_\infty$ *performance index* $\gamma > 0$*. Moreover, the periodic controller gain matrix is computed by the relation*

$$\hat{K}_i(t) = \hat{N}_i(t)\hat{M}_i^{-1}(t), \tag{22}$$

where the time-varying matrix function M_i *and* N_i *are determined by*

$$\hat{M}_i(t) = \hat{M}_{i,j} + \rho_{i,j}(t)(\hat{M}_{i,j+1} - \hat{M}_{i,j}), \tag{23}$$
$$\hat{N}_i(t) = \hat{N}_{i,j} + \rho_{i,j}(t)(\hat{N}_{i,j+1} - \hat{N}_{i,j}), \tag{24}$$

where $\rho_{i,j}(t) = J_i \frac{(t-(lT_p+t_{i-1}+j\sigma_i))}{F_i}$.

Proof In order to obtain the stabilization condition of PPS (1) through the controller (17), the similar derivations of Theorem 1 is followed by substituting $K_i = \hat{K}_i$ in inequalities (6) and (7), then pre and post multiplying the obtained inequality by $diag\{\underbrace{P^{-1}, \ldots, P^{-1}}_{7}, \underbrace{I, \ldots, I}_{8}\}$ and letting $P^{-1} = \hat{M}$, $\hat{K}\hat{M} = \hat{N}$, $\tilde{Q}_q = MQ_qM$, $q = 1, \ldots, 4$, we obtain the inequalities (18)–(21), which completes the proof. □

Case (ii) (Reliable controller design): It should be noted that the faults in actuators are inevitable in control systems due to many practical reasons, which results in performance deterioration or even instability.

In this connection, the considered PPS (1) is formulated with actuator fault in control input.

In order to overcome the performance degradation of the system, the reliable control is designed in the following form

$$u^G(t) = G\tilde{K}_i(t)x(t), \tag{25}$$

where $\tilde{K}_i(t) = \bar{K}_i(t) + \Delta\bar{K}_i(t)$ in which $\Delta\bar{K}_i(t) = \bar{U}_{Ki}\Delta_i(t)\bar{V}_{Ki}$ in which $\bar{U}_{Ki}$ and $\bar{V}_{Ki}$ are known constant matrices with appropriate dimensions and $\Delta_i(t)$ is the time-varying matrix satisfying $\Delta_i^T(t)\Delta_i(t) < I$ and $G = diag\{g_1, g_2, \ldots, g_k\}$ is the actuator fault matrix with $0 \le g_v^l \le g_v \le g_v^u \le 1$, here g_v^l and g_v^u are the lower and upper bound of g_v and assumed to be known positive constants.

It is noted that if $g_v = 0$ the vth actuator fails, $g_v = 1$ means that vth actuator is normal. If $0 \le g_v \le 1$ means the vth actuator has partial failure.

Subsequently, the fault matrix G is reformulated as $G = \mathbb{G}_0 + \mathbb{G}_1\mathbb{M}$, $|\mathbb{M}| \leq I$, where $\mathbb{G}_0 = diag\{g_{10}, g_{20}, \ldots, g_{k0}\}$, $\mathbb{G}_1 = diag\{g_{11}, g_{21}, \ldots, g_{k1}\}$, $g_{v0} = \frac{g_v^l + g_v^u}{2}$ and $g_{v1} = \frac{g_v^u - g_v^l}{2}$.

Theorem 3 *For given scalars α_i, μ and τ, the PPS* (1) *is asymptotically stabilized through the controller* (25) *with $\mathcal{H}_\infty$ performance index $\gamma > 0$, if there exist matrices $M_{i,j} > 0$, $N_{i,j}$, $\tilde{Q}_1 > 0$, $\tilde{Q}_2 > 0$, $\tilde{Q}_3 > 0$, $\tilde{Q}_4 > 0$, $j \in \mathbb{J}$, $i \in S$ and positive scalars $\varepsilon_1 > 0$, $\varepsilon_2 > 0$, $\varepsilon_3 > 0$, $\varepsilon_4 > 0$, $\varepsilon_5 > 0$, such that the following LMIs hold:*

$$[\tilde{\Upsilon}]_{19\times 19} < 0, \tag{26}$$

$$[\tilde{\Omega}]_{19\times 19} < 0, \tag{27}$$

$$M_{i,J_i} = M_{i+1,0},\ M_{S,J_i} = M_{1,0}, \tag{28}$$

$$N_{i,J_i} = N_{i+1,0},\ N_{S,J_i} = N_{1,0}, \tag{29}$$

where

$$\Upsilon_{1,1} = Sym(A_i M_{i,j} + B_i \mathbb{G}_0 N_{i,j}) + \alpha_i M_{i,j} - \frac{J_i(M_{i,j+1} - M_{i,j})}{F_i} + \tilde{Q}_1 + \tilde{Q}_2 + \tau^2 \tilde{Q}_3 + \frac{\tau^4}{4}\tilde{Q}_4,$$

$$\tilde{\Omega}_{1,1} = Sym(A_i M_{i,j+1} + B_i \mathbb{G}_0 N_{i,j+1}) + \alpha_i M_{i,j+1} - \frac{J_i(M_{i,j+1} - M_{i,j})}{F_i} + \tilde{Q}_1 + \tilde{Q}_2 + \tau^2 \tilde{Q}_3 + \frac{\tau^4}{4}\tilde{Q}_4,$$

$$\tilde{\Upsilon}_{1,2} = A_{\tau i} M_{i,j},\ \tilde{\Omega}_{1,2} = A_{\tau i} M_{i,j+1},\ \tilde{\Upsilon}_{1,8} = C_i,\ \tilde{\Omega}_{1,8} = C_i,$$
$$\tilde{\Upsilon}_{1,9} = M_{i,j} D_i^T,\ \tilde{\Omega}_{1,9} = M_{i,j+1} D_i^T,$$
$$\tilde{\Upsilon}_{1,10} = \tilde{\Omega}_{1,10} = \varepsilon_1 U_i,\ \tilde{\Upsilon}_{1,11} = M_{i,j} V_{Ai}^T,$$
$$\tilde{\Omega}_{1,11} = M_{i,j+1} V_{Ai}^T,\ \tilde{\Upsilon}_{1,12} = \tilde{\Omega}_{1,12} = \varepsilon_2 U_i,$$
$$\tilde{\Upsilon}_{1,14} = \tilde{\Omega}_{1,14} = \varepsilon_3 B_i \mathbb{G}_0 \bar{U}_{Ki},\ \tilde{\Upsilon}_{1,15} = M_{i,j} \bar{V}_{Ki}^T,$$
$$\tilde{\Omega}_{1,15} = M_{i,j+1} \bar{V}_{Ki}^T,\ \tilde{\Upsilon}_{1,16} = \tilde{\Omega}_{1,16} = \varepsilon_4 B_i \mathbb{G}_1,$$
$$\tilde{\Upsilon}_{1,17} = N_{i,j}^T,\ \tilde{\Omega}_{1,17} = N_{i,j+1}^T,\ \tilde{\Upsilon}_{1,18} = \tilde{\Omega}_{1,18} = \varepsilon_5 B_i \mathbb{G}_1,$$
$$\tilde{\Upsilon}_{2,2} = \tilde{\Omega}_{2,2} = -(1-\mu)\tilde{Q}_1,\ \tilde{\Upsilon}_{2,13} = M_{i,j} V_{Bi}^T,$$
$$\tilde{\Omega}_{2,13} = M_{i,j+1} V_{Bi}^T,\ \tilde{\Upsilon}_{3,3} = \tilde{\Omega}_{3,3} = -\tilde{Q}_2,$$
$$\tilde{\Upsilon}_{4,4} = \tilde{\Omega}_{4,4} = -9\tilde{Q}_3,\ \tilde{\Upsilon}_{4,5} = \tilde{\Omega}_{4,5} = \frac{36}{\tau}\tilde{Q}_3,$$
$$\tilde{\Upsilon}_{4,6} = \tilde{\Omega}_{4,6} = \frac{-60}{\tau^2}\tilde{Q}_3,\ \tilde{\Upsilon}_{5,5} = \tilde{\Omega}_{5,5} = \frac{-192}{\tau^2}\tilde{Q}_3 - 6\tilde{Q}_4,$$
$$\tilde{\Upsilon}_{5,6} = \tilde{\Omega}_{5,6} = \frac{360}{\tau^3}\tilde{Q}_3 + \frac{30}{\tau}\tilde{Q}_4,\ \tilde{\Upsilon}_{5,7} = \tilde{\Omega}_{5,7} = -\frac{60}{\tau^2}\tilde{Q}_4,$$

$$\tilde{\Upsilon}_{6,6} = \tilde{\Omega}_{6,6} = -\frac{210}{\tau^2}\tilde{Q}_4 - \frac{720}{\tau^4}\tilde{Q}_3, \ \tilde{\Upsilon}_{6,7} = \tilde{\Omega}_{6,7} = \frac{480}{\tau^3}\tilde{Q}_4,$$
$$\tilde{\Upsilon}_{7,7} = \tilde{\Omega}_{7,7} = -\frac{1200}{\tau^4}\tilde{Q}_4, \ \tilde{\Upsilon}_{8,8} = \tilde{\Omega}_{8,8} = -\gamma I,$$
$$\tilde{\Upsilon}_{9,9} = \tilde{\Omega}_{9,9} = -\gamma I, \ \tilde{\Upsilon}_{14,19} = \tilde{\Omega}_{14,19} = \varepsilon_3 U_{Ki}^T,$$
$$\tilde{\Upsilon}_{10,10} = \tilde{\Omega}_{10,10} = \tilde{\Upsilon}_{11,11} = \tilde{\Omega}_{11,11} = -\varepsilon_1 I,$$
$$\tilde{\Upsilon}_{12,12} = \tilde{\Omega}_{12,12} = \tilde{\Upsilon}_{13,13} = \tilde{\Omega}_{13,13} = -\varepsilon_2 I,$$
$$\tilde{\Upsilon}_{14,14} = \tilde{\Omega}_{14,14} = \tilde{\Upsilon}_{15,15} = \tilde{\Omega}_{15,15} = -\varepsilon_3 I,$$
$$\tilde{\Upsilon}_{16,16} = \tilde{\Omega}_{16,16} = \tilde{\Upsilon}_{17,17} = \tilde{\Omega}_{17,17} = -\varepsilon_4 I,$$
$$\tilde{\Upsilon}_{18,18} = \tilde{\Omega}_{18,18} = \tilde{\Upsilon}_{19,19} = \tilde{\Omega}_{19,19} = -\varepsilon_5 I.$$

Moreover, the corresponding gain matrices of the designed controller (25) *is obtained by using the relations* (22)–(24) *with* $\hat{K}_i(t) = \bar{K}_i(t)$, $\hat{M}_i(t) = M_i(t)$ *and* $\hat{N}_i(t) = N_i(t)$.

Proof The sufficient stability conditions in Theorem 3 can be easily obtained by following the similar lines of Theorems 1–2 and Schur complement. Thus, the proof is completed. □

4 Design of Fault Alarm

In this section, the alarm system is designed to timely alert the controller to switch over from robust controller to the reliable controller, which tolerates the effect of the faults occurred in system dynamics. In particular, the threshold value of the fault is estimated with the aid state observer. Thus, the robust residual observer system is considered in the following form;

$$\begin{cases} \dot{\tilde{x}}(t) &= (A_i + \Delta A_i(t))\tilde{x}(t) + (A_{\tau i} + \Delta A_{\tau i}(t))\tilde{x}(t-\tau(t)) \\ & \quad + B_i u^g(t) + L_i(t)(y(t) - \tilde{y}(t)), \\ \tilde{y}(t) &= D_i \tilde{x}(t), \end{cases} \tag{30}$$

where $\tilde{x}(t) \in \mathcal{R}^x$ is the observer state; $u^g(t) \in \mathcal{R}^u$ is the control input; $L_i(t)$ is the time-varying periodic observer gain and $\tilde{y}(t) \in \mathbb{R}^y$ is the observer output.

Now, the error residue is defined as $\check{x}(t) = x(t) - \tilde{x}(t)$.

Based on the above relation, the error system is obtained in the following form:

$$\begin{aligned} \dot{\check{x}}(t) =& (A_i + \Delta A_i(t))\check{x}(t) + (A_{\tau i} + \Delta A_{\tau i}(t))\check{x}(t-\tau(t)) \\ &+ B_i(u^G(t) - u^g(t)) + C_i v(t) - LD_i\check{x}(t). \end{aligned} \tag{31}$$

It should be noted that the relation between the control input of the state and observer is varying with respect to the presence of actuator faults.

In this connection, the control input is selected as $u^G(t) = u^g(t)$ for fault free case and $u^G(t) = Gu^g(t)$ when the fault occurs.

Consequently, the critical relationship between the $\check{x}(t)$ and $w(t)$ is defined as follows:

$$(L_i D_i - A_i - A_{\tau i})\check{x}(t) = C_i \bar{v}. \tag{32}$$

where $\bar{v}$ is the upper bound of $v(t)$.

Then, by the simple manipulation, we obtain

$$||e(t)|| \leq ||\zeta_i \bar{v}|| \leq ||\zeta_i|| ||\bar{v}||,$$

where $\zeta_i = D_i(L_i D_i - A_i - A_{\tau i})^+ C_i, e(t) = y(t) - \tilde{y}(t)$.

Now, the threshold value of fault is defined as $T = \sup ||\zeta_i|| ||\bar{v}||$. Moreover, the algorithm used to detect the fault is designed in the following form;

$$S_w(t) = \begin{cases} 0, & normal, \quad ||e(t)|| \leq T, \\ 1, & alarm, \quad ||e(t)|| > T. \end{cases} \tag{33}$$

Finally, the hybrid control scheme based on the fault-alarm approach is given as follows;

$$u^G(t) = (1 - S_w(t))\hat{K}_i(t)x(t) + S_w(t)G\tilde{K}_i(t)x(t). \tag{34}$$

5 Numerical Example

Consider a uncertain periodic piecewise system (2) with three subsystems and fundamental period $F_p = 3s$ and $F_1 = 0.9s$, $F_2 = 0.6s$, $F_3 = 1.5s$.

Subsystem 1:

$$A_1 = \begin{bmatrix} -2.51 & 1.21 \\ 0.49 & -0.9 \end{bmatrix}, A_{\tau 1} = \begin{bmatrix} 2.53 & 0.6 \\ 1.49 & 0.7 \end{bmatrix}, B_1 = \begin{bmatrix} 1 & 1 \\ 1.41 & 0.52 \end{bmatrix},$$

$$C_1 = \begin{bmatrix} 0.21 \\ 0.1 \end{bmatrix}, D_1 = \begin{bmatrix} 1 \\ 0.8 \end{bmatrix}^T, U_1 = \begin{bmatrix} 0.51 \\ 0.1 \end{bmatrix}, V_{A1} = \begin{bmatrix} -0.1 \\ -0.51 \end{bmatrix}^T,$$

$$V_{A\tau 1} = \begin{bmatrix} -0.1 \\ -0.5 \end{bmatrix}, \hat{U}_{K1} = \bar{U}_{K1} = \begin{bmatrix} 0.12 \\ 0.5 \end{bmatrix}, \hat{V}_{K1} = \bar{V}_{K1} = \begin{bmatrix} 0.61 \\ 0.51 \end{bmatrix}^T.$$

Subsystem 2:

$$A_2 = \begin{bmatrix} 1 & 1.79 \\ 2.51 & -1 \end{bmatrix}, A_{\tau 2} = \begin{bmatrix} 0.59 & 0.51 \\ 1.5 & 0.61 \end{bmatrix}, B_2 = \begin{bmatrix} 1.51 & 1 \\ 0.95 & 1 \end{bmatrix},$$

$$C_2 = \begin{bmatrix} 0.20 \\ 0.15 \end{bmatrix}, D_2 = \begin{bmatrix} 1 \\ 0.42 \end{bmatrix}^T, U_2 = \begin{bmatrix} 0.19 \\ 0.2 \end{bmatrix}, V_{A2} = \begin{bmatrix} -0.49 \\ 0.9 \end{bmatrix}^T,$$
$$V_{A\tau 2} = \begin{bmatrix} 0.61 \\ -0.8 \end{bmatrix}^T, \hat{U}_{K2} = \bar{U}_{K2} = \begin{bmatrix} 0.22 \\ 0.4 \end{bmatrix}, \hat{V}_{K2} = \bar{V}_{K2} = \begin{bmatrix} 0.89 \\ 0.11 \end{bmatrix}^T.$$

Subsystem 3:

$$A_3 = \begin{bmatrix} 1.22 & 3 \\ 0.99 & -1.1 \end{bmatrix}, A_{\tau 3} = \begin{bmatrix} 0.39 & 0.8 \\ 1.33 & 0.6 \end{bmatrix}, B_3 = \begin{bmatrix} 1 & 1 \\ 1.82 & 1 \end{bmatrix},$$
$$C_3 = \begin{bmatrix} 0.12 \\ 0.31 \end{bmatrix}, D_3 = \begin{bmatrix} 0.9 \\ 1.1 \end{bmatrix}^T, U_3 = \begin{bmatrix} 0.49 \\ 0.1 \end{bmatrix}, V_{A3} = \begin{bmatrix} 0.49 \\ 0.15 \end{bmatrix}^T,$$
$$V_{A\tau 3} = \begin{bmatrix} 0.21 \\ 0.8 \end{bmatrix}, \hat{U}_{K3} = \bar{U}_{K3} = \begin{bmatrix} 0.52 \\ 0.1 \end{bmatrix}, \hat{V}_{K3} = \bar{V}_{K3} = \begin{bmatrix} 0.49 \\ -0.1 \end{bmatrix}^T.$$

Further, the external disturbance signal $w(t)$ is chosen as

$$w(t) = \begin{cases} 4 + 4\cos(8\pi t), & \text{t} < 1.5, \\ 0.01e^{(-0.07t)} \sin t, & \textit{otherwise.} \end{cases}$$

Moreover, the upper and lower bound of g_k is considered as $g_1^u = 0.3$, $g_2^u = 0.3$, $g_1^l = 0.1$ and $g_2^l = 0.1$ with respect to the initial condition $x(0) = [0.1 \ \ -0.1]^T$.

Now, by solving the conditions developed in Theorems 2 and 3 with the considered parameters together with the values of $\alpha = 0.7$, $\tau = 1.5$, $J_i = 1$ and $\mu = 0.3$, we

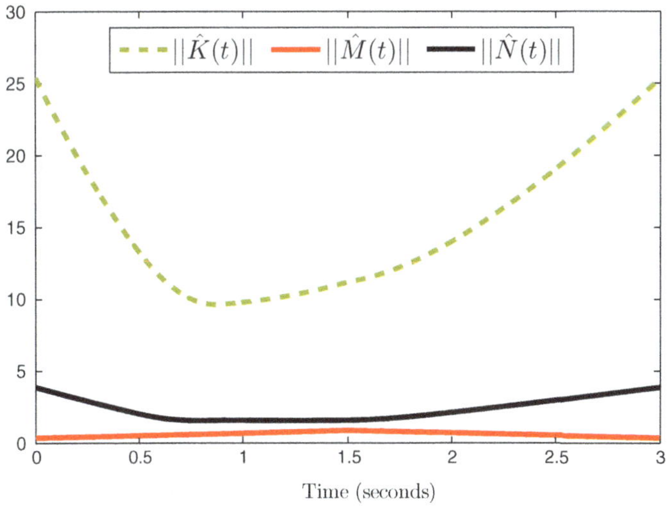

Fig. 2 Variations of $||\hat{K}(t)||$, $||\hat{M}(t)||$ and $||\hat{N}(t)||$ over one period based on Theorem 2

obtain the gain matrices of the corresponding control and observer with satisfactory $\mathcal{H}_\infty$ performance index $\gamma = 0.17$.

Further, the computed gain matrices of the corresponding robust and reliable controller with respect to the relation (22)–(24) are plotted in Figs. 2 and 3, respectively. Further, the observer gain matrix is plotted in Fig. 4. Consequently, in the presence of unknown actuator faults, the trajectories of the system (4) with respect to the robust controller (17) is plotted in Fig. 5. From Fig. 5, it can be seen that the robust

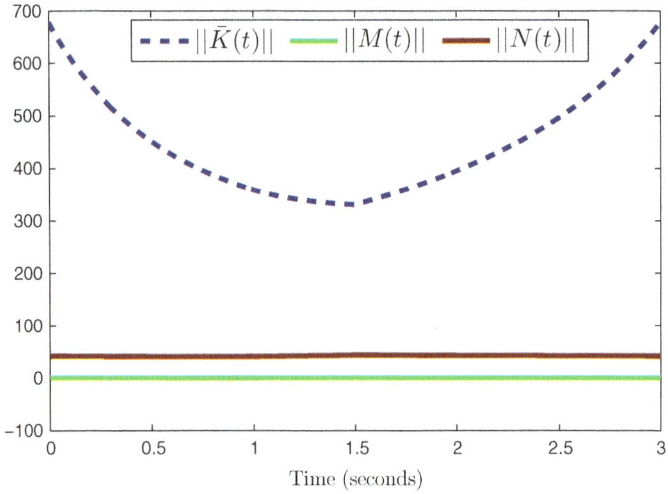

Fig. 3 Variations of $||\bar{K}(t)||$, $||M(t)||$ and $||N(t)||$ over one period based on Theorem 3

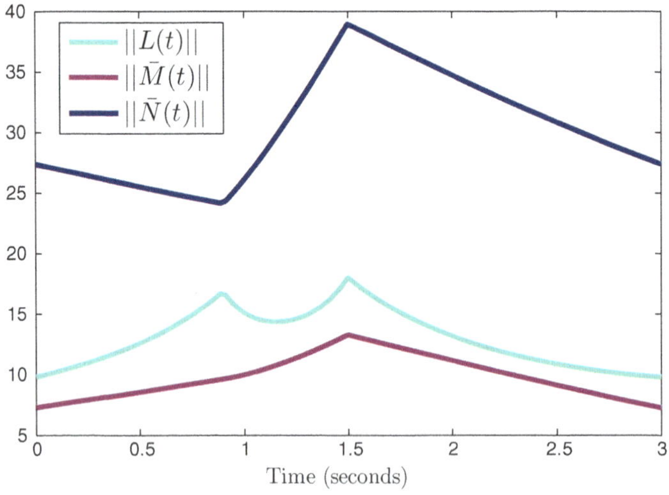

Fig. 4 Variations of observer gain matrix over one period

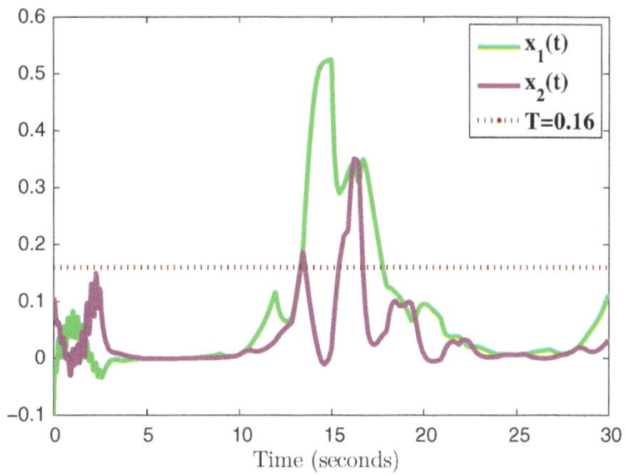

Fig. 5 State responses of the closed-loop system (4) with fault under robust control (17)

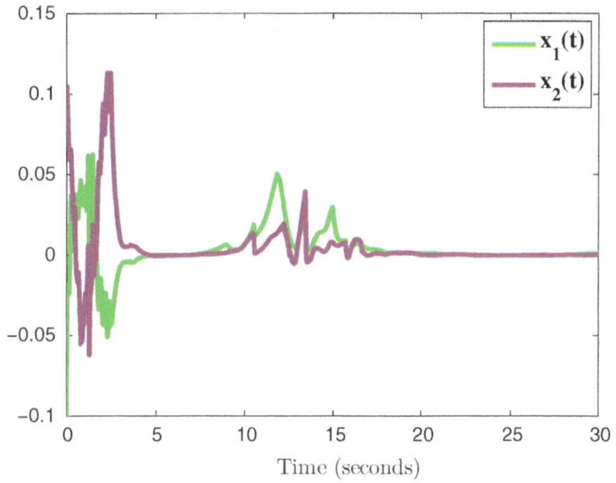

Fig. 6 State responses of the closed-loop system (4) with fault under hybrid control (34)

controller exhibits poor performance with respect to the actuator faults in system (1). Moreover, in Fig. 5, the calculated threshold value $T = 0.16$ is also plotted.

Furthermore, the state response of the system (2) concerning the fault-alarm based switching controller (hybrid controller) is provided in Fig. 6. Moreover, the proposed hybrid controller via fault alarm-based scheme ensures the asymptotic stability of the system even in the presence of actuator faults and also the automatic switching among the controllers is achieved with the aid of timely alert of the alarm signal. Finally, the simulation results reveal the applicability and effectiveness of the fault-alarm based hybrid control scheme.

6 Conclusion

In this chapter, fault-alarm based hybrid control design for a class of uncertain periodic piecewise systems subject to time-varying delay, external disturbances, and actuator faults is discussed. Specifically, fault-alarm is designed to automatically switch over from robust control to reliable control once if the faults occur. By constructing periodically time-varying Lyapunov–Krasovskii functional, the sufficient conditions are derived in the form of LMIs to guarantee the asymptotic stability of the addressed system together with satisfactory $\mathcal{H}_\infty$ performance level. Finally, the effectiveness of the theoretical result is illustrated via numerical simulations.

References

1. Xie, X., Lam, J., Fan, C.: Robust time-weighted guaranteed cost control of uncertain periodic piecewise linear systems. Inf. Sci. **460**, 238–253 (2018)
2. Zhu, B., Lam, J., Song, X.: Stability and $\mathcal{L}_1$-gain analysis of linear periodic piecewise positive systems. Automatica **101**, 232–240 (2019)
3. Li, P., Li, P., Liu, Y., Bao, H., Lu, R.: $\mathcal{H}_\infty$ control of periodic piecewise polynomial time-varying systems with polynomial Lyapunov function. J. Frankl. Inst. **356**, 6968–6988 (2019)
4. Li, P., Lam, J., Cheung, K.C.: $l\mathcal{H}_\infty$ control of periodic piecewise vibration systems with actuator saturation. J. Vib. Control **23**, 3377–3391 (2017)
5. Li, P., Lam, J., Kwok, K.W., Lu, R.: Stability and stabilization of periodic piecewise linear systems: a matrix polynomial approach. Automatica **94**, 1–8 (2018)
6. Park, J.H., Lee, T.H., Liu, Y., Chen, J.: Dynamic Systems with Time Delays: Stability and Control. Springer-Nature, Singapore (2019). https://doi.org/10.1007/978-981-13-9254-2
7. Lee, T.H., Park, J.H.: A novel Lyapunov functional for stability of time-varying delay systems via matrix-refined-function. Automatica **80**, 239–242 (2017)
8. Xie, X., Lam, J., Li, P.: Finite-time $\mathcal{H}_\infty$ control of periodic piecewise linear systems. Int. J. Syst. Sci. **48**, 2333–2344 (2017)
9. Li, P., Lam, J., Cheung, K.C.: Stability, stabilization and $\mathcal{L}_2-$gain analysis of periodic piecewise linear systems. Automatica **61**, 218–226 (2015)
10. Xie, X., Lam, J.: Guaranteed cost control of periodic piecewise linear time-delay systems. Automatica **94**, 274–282 (2018)
11. Xie, X., Lam, J., Li, P.: $\mathcal{H}_\infty$ control problem of linear periodic piecewise time-delay systems. Int. J. Syst. Sci. **49**, 997–1011 (2018)
12. Park, M.J., Kwon, O.M., Ryu, J.H.: Advanced stability criteria for linear systems with time-varying delays. J. Frankl. Inst. **355**, 520–543 (2018)
13. Liu, Q., Zhou, B.: Extended observer based feedback control of linear systems with both state and input delays. J. Frankl. Inst. **354**, 8232–8255 (2017)
14. Yu, J., Yang, C., Tang, X., Wang, P.: $\mathcal{H}_\infty$ control for uncertain linear system over networks with Bernoulli data dropout and actuator saturation. ISA Trans. **74**, 1–13 (2018)
15. Samidurai, R., Sriraman, R., Cao, J., Tu, Z.: Non-fragile stabilization for uncertain system with interval time-varying delays via a new double integral inequality. Math. Methods Appl. Sci. **41**, 6272–6287 (2018)
16. Xiao, B., Yang, X., Karimi, H.R., Qiu, J.: Asymptotic tracking control for a more representative class of uncertain nonlinear systems with mismatched uncertainties. IEEE Trans. Ind. Electron. **66**, 9417–9427 (2019)
17. Xiang, W.: Necessary and sufficient condition for stability of switched uncertain linear systems under dwell-time constraint. IEEE Trans. Autom. Control **61**, 3619–3624 (2016)

18. Sakthivel, R., Wang, C., Santra, S., Kaviarasan, B.: Non-fragile reliable sampled-data controller for nonlinear switched time-varying systems. Nonlinear Anal.: Hybrid Syst. **27**, 62–76 (2018)
19. Shen, M., Lim, C.C., Shi, P.: Reliable $\mathcal{H}_\infty$ static output control of linear time-varying delay systems against sensor failures. Int. J. Robust Nonlinear Control **27**, 3109–3123 (2017)
20. Zhao, D., Karimi, H.R., Sakthivel, R., Li, Y.: Non-fragile fault-tolerant control for nonlinear Markovian jump systems with intermittent actuator fault. Nonlinear Anal.: Hybrid Syst. **32**, 337–350 (2019)
21. Shen, H., Wu, Z.W., Park, J.H.: Reliable mixed passive and $\mathcal{H}_\infty$ filtering for semi-Markov jump systems with randomly occurring uncertainties and sensor failures. Int. J. Robust Nonlinear Control **25**, 3231–3251 (2015)
22. Sakthivel, R., Saravanakumar, T., Ma, Y.K., Anthoni, S.M.: Finite-time resilient reliable sampled-data control for fuzzy systems with randomly occurring uncertainties. Fuzzy Sets Syst. **329**, 1–18 (2017)
23. Sakthivel, R., Divya, H., Sakthivel, R., Liu, Y.: Robust hybrid control design for stochastic Markovian jump system via fault alarm approach. IEEE Trans. Circuits Syst. II: Express Briefs (2019). https://doi.org/10.1109/TCSII.2019.2941008
24. Li, J.N., Xu, Y.F., Bao, W.D., Xu, X.B.: Passivity and fault alarm-based hybrid control for a Markovian jump delayed system with actuator failures. IEEE Access **6**, 797–805 (2017)
25. Harshavarthini, S., Selvi, S., Sakthivel, R., Almakhles, D.J.: Non-fragile fault alarm-based hybrid control for the attitude quadrotor model with actuator saturation. IEEE Trans. Circuits Syst. II: Express Briefs (2020). https://doi.org/10.1109/TCSII.2020.2964906
26. Shen, Q., Jiang, B., Shi, P., Lim, C.C.: Novel neural networks-based fault tolerant control scheme with fault alarm. IEEE Trans. Cybern. **44**, 2190–2201 (2014)
27. Zheng, Q., Zhang, H., Ling, Y., Guo, X.: Mixed $\mathcal{H}_\infty$ and passive control for a class of nonlinear switched systems with average dwell time via hybrid control approach. J. Frankl. Inst. **355**, 1156–1175 (2018)

Event-Triggered Sliding Mode Control for Stochastic Markov Jump Systems

Wenhai Qi

Abstract This chapter considers the problem of sliding mode control (SMC) design for stochastic Markov jump systems (MJSs) via a suitable event-triggered communication scheme. Event-triggered strategy, which can save computational resources, is designed to determine whether the current data should be transmitted or not in the design process. First, by the key points of stochastic Markov Lyapunov functional and observer design theory, a desired SMC law is constructed to guarantee that the system trajectories can arrive at the specified sliding surface. Then, sojourn-time-dependent sufficient conditions are established to ensure the required stochastic stability performance of the whole closed-loop system. Finally, a boost converter circuit model is given to illustrate the proposed method.

Keywords Sliding mode control · Stochastic Markov jump systems · Event-triggered communication scheme · Stochastic stability

1 Introduction

It is well known that Markov jump systems (MJSs) are of great value in practical applications, such as networked control systems, manufacturing systems, fault-detection systems, economic systems, power systems, and aircraft systems [1–8]. Many actual control processes subject to random abrupt variations in their structures due to random failures or repairs of components can be described by MJSs. As a special kind of hybrid systems, MJSs consist of some dynamics and mode transitions governed by a Markov process (or a Markov chain) taking values in a finite set.

Over the past decades, most leading research mainly focuses on MJSs with stochastic stability [9, 10], stochastic stabilization [11, 12], filter design [13], actuator saturation [14], passivity [15, 16], finite-time control [17, 18], and positivity analysis [19].

W. Qi (✉)
School of Engineering, Qufu Normal University, Rizhao 276826, People's Republic of China
e-mail: qiwhtanedu@163.com

J. H. Park (ed.), *Recent Advances in Control Problems of Dynamical Systems and Networks*, Studies in Systems, Decision and Control 301,
https://doi.org/10.1007/978-3-030-49123-9_10

On another research front, sliding mode control (SMC) has been active in a wide variety of complex systems for its insensitivity to parameter variations, its complete rejection of external disturbances, and order reduction. The purpose of SMC is to drive the state signals onto an artificially specified hyper-surface (i.e. the sliding mode surface), on which the requested stability and disturbance inhibition can be achieved [20–22]. In a limited time, the convergence of the sliding mode surface can be guaranteed when an appropriate control law is designed to suppress the adverse effects of the uncertainty and nonlinearity.

Nowadays, with the help of MJSs describing complex dynamical systems, various SMC approaches have been immersed with stochastic MJSs; for details, see [23–33].

With the rapid development of computer and networked technologies, data transmission via communication networks has received considerable research attention. Compared with traditional control systems, communication network schemes have incomparable advantages in realizing remote control and detection, increasing flexibility and reliability of control system [34, 35]. However, the data transmission process inevitably suffers from limited bandwidth, network-induced delay [36], communication packet losses, and network congestion. The traditional approach is to update the control signal in a time-triggered manner, which will cause unnecessary signal transmission. For the time-triggered strategy, it will result in resource waste since some redundant sampled signals are sent even when those signals keep the same values all the time.

In order to overcome this shortcoming, the event-triggered communication scheme has been put forward [37, 38], in which the sampled signals are transmitted when the updated sampled signal satisfies the predefined triggered condition. Its advantage is to avoid the transmission of redundant data and save precious communication resources. Thus, plenty of noticeable results have been devoted to the event-triggered strategy [39–43].

To name a few, a novel event-triggered scheme has been proposed for a class of fuzzy Markov jump systems with general switching policies to improve the transmission efficiency at each sampling instance [42]. By constructing a novel common sliding surface, an event-triggered asynchronous SMC law for networked Markov jump Lur'e systems has been designed to depend on the hidden mode information [43]. However, the SMC design for stochastic MJSs via a suitable event-triggered communication scheme has received little research attention, since many complicated factors, such as uncertainty, time-triggered communication scheme, and Markov process, exist in the design of SMC law.

These stimulate our interests to address the event-triggered SMC design for a class of stochastic MJSs.

The main contributions can be concluded as follows:

(i) Considering the effects of sensor information constraints and limited network resource, the observer design theory and suitable event-triggered communication scheme is adopted to remove the constraint of unmeasurable system states and save network resource, respectively.
(ii) By utilizing Lyapunov functional, sufficient conditions for stochastic stability are proposed for the resulting closed-loop system.

(iii) An appropriate observer-based SMC law via the event-triggered strategy is proposed to ensure the finite-time reachability of the sliding mode surface.
(iv) Representing a boost converter circuit model as MJSs shows the validity of the SMC algorithm.

Notations: $\mathcal{P}^T$ represents the transpose of $\mathcal{P}$. Symbol $\mathbb{R}^n$ stands for the n dimensional Euclidean space, $\mathbb{R}^{n\times m}$ is the set of $n \times m$ real matrices, and $\mathcal{S} = \{1, 2, \ldots, \mathcal{N}\}$ means a set of positive numbers. The superscript $\mathcal{I}$ denotes identity matrix with appropriate dimensions. Given a probability space $(\varXi, \varUpsilon, \varTheta)$, $\varXi$ is the sample space, $\varUpsilon$ is the σ-algebra of subsets of the sample space and $\varTheta$ is the probability measure on $\varUpsilon$. Symbol $\mathcal{E}\{\cdot\}$ represents the mathematical expectation, $\| * \|$ means the Euclidean norm of vectors. $\mathcal{P} > 0$ (≥ 0) means $\mathcal{P}$ is real symmetric positive (semi-positive) definite. For simplicity, symbol $*$ is represented as an ellipsis for symmetry. Matrices, if not explicitly stated, are assumed to have compatible dimensions.

2 Preliminaries

Consider a class of MJSs on the probability space $(\varXi, \varUpsilon, \varTheta)$ as

$$\begin{aligned}\dot{x}(t) =&(\mathcal{A}(r_t) + \Delta\mathcal{A}(r_t, t))x(t) + \mathcal{B}(r_t)u(t),\\ y(t) =&\mathcal{C}(r_t)x(t),\end{aligned} \tag{1}$$

where $x(t) \in \mathbb{R}^n$, $u(t) \in \mathbb{R}^m$, and $y(t) \in \mathbb{R}^l$ are the state, control input, and output, respectively. $\Delta\mathcal{A}(r_t, t)$ is represented by $\Delta\mathcal{A}(r_t, t) = \mathcal{M}(r_t)\mathcal{H}(r_t, t)\mathcal{N}(r_t)$, in which $\mathcal{H}(r_t, t)$ satisfies $\mathcal{H}^T(r_t, t)\mathcal{H}(r_t, t) \leq \mathcal{I}$. $\{r_t, t \geq 0\}$ denotes the Markov process in $\mathcal{S} = \{1, 2, \ldots, \mathcal{N}\}$ with probability transitions:

$$Pr\{r_{t+\bar{\Delta}} = \beta | r_t = \alpha\} = \begin{cases} \pi_{\alpha\beta}\bar{\Delta} + o(\bar{\Delta}), & \alpha \neq \beta,\\ 1 + \pi_{\alpha\alpha}\bar{\Delta} + o(\bar{\Delta}), & \alpha = \beta,\end{cases} \tag{2}$$

where $\pi_{\alpha\beta} \geq 0$ denotes the transition rate from α to β for $\alpha \neq \beta$, $\sum_{\beta=1,\beta\neq\alpha}^{\mathcal{N}} \pi_{\alpha\beta} = -\pi_{\alpha\alpha}$.

For $r_t = \alpha \in \mathcal{S}$, $\mathcal{A}(r_t)$, $\Delta\mathcal{A}(r_t)$, $\mathcal{B}(r_t)$, $\mathcal{C}(r_t)$, $\mathcal{M}(r_t)$, $\mathcal{H}(r_t, t)$, and $\mathcal{N}(r_t)$ are, respectively, denoted as $\mathcal{A}_\alpha$, $\Delta\mathcal{A}_\alpha(t)$, $\mathcal{B}_\alpha$, $\mathcal{C}_\alpha$, $\mathcal{M}_\alpha$, $\mathcal{H}_\alpha(t)$, and $\mathcal{N}_\alpha$.

Consider the output signal $y(t)$ sampled at a fixed period h. Denote the current sampled signal as $y(ih)$ ($i \in \mathbb{N}$) and the latest transmitted one as $y(i_r h)$ ($i_r \in \mathbb{N}$, $r = 0, 1, 2, \ldots, \infty$, $i_0 = 0$, r means the number of triggering).

Whether the sampled signal $y(ih)$ is sent or not is determined by the event trigger. Then, the event-triggered condition is formulated as

$$\begin{aligned}i_{r+1}h =&i_r h + \min_{j\in\mathbb{N}}\{jh | e^T((i_r + j)h)\varPi_\alpha e((i_r + j)h)\\ &> \kappa_\alpha y^T((i_r + j)h)\varPi_\alpha y((i_r + j)h)\},\end{aligned} \tag{3}$$

where $e((i_r + j)h) = y((i_r + j)h) - y(i_r h)$. κ_α and Π_α, $\alpha \in \mathcal{S}$, mean the predefined thresholds and weighting matrices. $\{(i_r + j)h\}$, $j \in \mathbb{N}$, denotes the set of sampling instants during $[i_r h, i_{r+1} h)$.

Remark 1 It is noted that if there is no latest signal, the zero-order holder keeps the previous input signal of observer and SMC law. Once new signals are updated by the event-triggered communication scheme, the zero-order holder transmits the data to observer and SMC law.

In the network, we denote τ_{i_r} as the network-induced transmission delay at the sampling instants $i_r h$. The zero-order holder successfully receives the signals at instants $i_r h + \tau_{i_r}$. It is assumed that τ_{i_r} satisfies $\tau_m \leq \tau_{i_r} \leq \tau_M \leq h$, where d_m and d_M are two positive constants.

It is noted that the output signal $\bar{y}(t)$ keeps the same data until new one reaches the zero-order holder, that is

$$\bar{y}(t) = y(i_r h), t \in [i_r h + \tau_{i_r}, i_{r+1} h + \tau_{i_{r+1}}). \tag{4}$$

Due to the existence of the network-induced transmission delay, the signal holding interval $[i_r h + \tau_{i_r}, i_{r+1} h + \tau_{i_{r+1}})$ can be divided into the following subintervals:

$$[i_r h + \tau_{i_r}, i_{r+1} h + \tau_{i_{r+1}}) = \cup_{j=0}^{k_r} \Omega_{r,j}, \tag{5}$$

where $\Omega_{r,j} = [q_{r,j} h + \tau_{q_{r,j}}, q_{r,j+1} h + \tau_{q_{r,j+1}})$, $q_{r,j} h = i_r + jh$ are the sampling instants during $[i_r h, i_{r+1} h)$, and $k_r = i_{r+1} - i_r$.

The invisible time delays $\tau_{q_{r,j}}$ satisfy $\tau_m \leq \tau_{q_{r,j}} \leq \tau_M \leq h$ and $q_{r,j} h + \tau_{q_{r,j}} < q_{r,j+1} h + \tau_{q_{r,j+1}}$.

Define the piecewise function $\eta(t) = t - q_{r,j} h$, $t \in \Omega_{r,j}$. Then, we have $\eta_1 \leq \eta(t) \leq \eta_2$, where $\eta_1 = \tau_m$ and $\eta_2 = \tau_M + h$.

Furthermore, one has

$$\bar{y}(t) = y(i_r h) = y(q_{r,j} h) - e(q_{r,j} h) = y(t - \eta(t)) - e(q_{r,j} h), t \in \Omega_{r,j}. \tag{6}$$

According to (3), it is obtained that

$$e^T(q_{r,j} h) \Pi_\alpha e(q_{r,j} h) \leq \kappa_\alpha y^T(t - \eta(t)) \Pi_\alpha y(t - \eta(t)), \; t \in \Omega_{r,j}. \tag{7}$$

In many practical industrial control systems, due to the limitation of sensor, special working environment, and other factors, it is difficult to measure the signal variables directly. In such case, we need to use other methods to obtain the numerical values of these physical quantities. The observer-design approach is an important method in solving these problems. Then, it is necessary and important to take the observer-design approach into account when studying the characteristics of those signal variables. Based on the event-triggered communication scheme, the observer is constructed as

$$\dot{\check{x}}(t) =\mathcal{A}_\alpha \check{x}(t) + \mathcal{B}_\alpha u(t) + \mathcal{L}_\alpha(\bar{y}(t) - \mathcal{C}_\alpha \check{x}(t-\eta(t))),\ t \in \Omega_{r,j}, \tag{8}$$

where $\check{x}(t) \in \mathbb{R}^n$ and $\mathcal{L}_\alpha$ mean the estimated state and the observer gains.

Then, it follows from (6) and (8) that

$$\begin{aligned}\dot{\check{x}}(t) =&\mathcal{A}_\alpha \check{x}(t) + \mathcal{B}_\alpha u(t) + \mathcal{L}_\alpha(y(t-\eta(t)) \\ &- \mathcal{C}_\alpha \check{x}(t-\eta(t)) - e(q_{r,j}h)),\ t \in \Omega_{r,j}.\end{aligned} \tag{9}$$

Define error signal $e(t) = x(t) - \check{x}(t)$.

Combining (1) and (9) leads to the error dynamics as

$$\begin{aligned}\dot{e}(t) =&(\mathcal{A}_\alpha + \Delta\mathcal{A}_\alpha(t))e(t) + \Delta\mathcal{A}_\alpha(t)\check{x}(t) \\ &- \mathcal{L}_\alpha\mathcal{C}_\alpha e(t-\eta(t)) + \mathcal{L}_\alpha e(q_{r,j}h).\end{aligned} \tag{10}$$

Therefore, according to (9) and (10), one has

$$\begin{aligned}\dot{\xi}(t) &= (\bar{\mathcal{A}}_\alpha + \Delta\bar{\mathcal{A}}_\alpha(t))\xi(t) + \bar{\mathcal{A}}_{d\alpha}\xi(t-\eta(t)) + \bar{\mathcal{B}}_\alpha u(t) \\ &\quad + \bar{\mathcal{L}}_\alpha e(q_{r,j}h), \\ \xi(t_0+\theta) &= \phi(\theta), \theta \in [-\eta_2, 0],\end{aligned} \tag{11}$$

where

$$\begin{aligned}&\xi(t) = \left[\check{x}^T(t)\ e^T(t)\right]^T,\ \ \bar{\mathcal{A}}_\alpha = \begin{bmatrix}\mathcal{A}_\alpha & 0\\ 0 & \mathcal{A}_\alpha\end{bmatrix},\\ &\Delta\bar{\mathcal{A}}_\alpha(t) = \begin{bmatrix}0 & 0\\ \Delta\mathcal{A}_\alpha(t) & \Delta\mathcal{A}_\alpha(t)\end{bmatrix} = \bar{\mathcal{M}}_\alpha\mathcal{H}_\alpha(t)\bar{\mathcal{N}}_\alpha,\\ &\bar{\mathcal{A}}_{d\alpha} = \begin{bmatrix}0 & \mathcal{L}_\alpha\mathcal{C}_\alpha\\ 0 & -\mathcal{L}_\alpha\mathcal{C}_\alpha\end{bmatrix},\ \ \bar{\mathcal{B}}_\alpha = \left[\mathcal{B}_\alpha^T\ 0\right]^T,\\ &\bar{\mathcal{L}}_\alpha = \left[-\mathcal{L}_\alpha^T\ \mathcal{L}_\alpha^T\right]^T,\ \ \bar{\mathcal{M}}_\alpha = \left[0\ \mathcal{M}_\alpha^T\right]^T,\ \ \bar{\mathcal{N}}_\alpha = \left[0\ \mathcal{N}_\alpha\right].\end{aligned}$$

Definition 1 ([9]) System (11) is said to be stochastically stable if, for $r_0 \in \mathcal{S}$ and $\phi(\theta) \in \mathbb{R}^n, \mathcal{E}\left\{\int_0^\infty ||\xi(t)||^2 dt|(\phi(\theta), \delta_0)\right\} < \mathcal{T}(\phi(\theta), r_0)$, where $\mathcal{T}(\phi(\theta), r_0)$ is a positive constant.

Lemma 1 ([44]) *For given matrix $\mathcal{P} > 0$ and all continuously $\vartheta(\cdot)$ in $[a, b] \to \mathcal{R}^{2n}$, one has*

$$\begin{aligned}&-(b-a)\int_a^b \dot{\vartheta}^T(s)\mathcal{P}\dot{\vartheta}(s)ds \\ &\quad\le -(\vartheta(b)-\vartheta(a))^T\mathcal{P}(\vartheta(b)-\vartheta(a)) - 3\Pi^T\mathcal{P}\Pi,\end{aligned} \tag{12}$$

where $\Pi = \vartheta(b) + \vartheta(a) - \frac{2}{b-a}\int_a^b \vartheta(s)ds$.

Lemma 2 ([45]) *Consider a continuous function as* $0 < \tau_1 \le \tau(t) \le \tau_2$. *For positive-definite real matrix* $\mathcal{X} \in \mathcal{R}^{2n\times 2n}$, *a vector* $\dot{x}$: $[-\tau_2, 0] \rightarrow \mathcal{X}^{2n}$, *and real matrix* $\mathcal{Y} \in \mathcal{R}^{4n\times 4n}$ *satisfying* $\begin{bmatrix} \tilde{\mathcal{X}} & \mathcal{Y} \\ \mathcal{Y}^T & \tilde{\mathcal{X}} \end{bmatrix} \ge 0$, *one has*

$$-(\tau_2 - \tau_1)\int_{t-\tau_2}^{t-\tau_1} \dot{x}^T(s)\mathcal{X}\dot{x}(s)ds \le 2\psi_1^T\mathcal{Y}\psi_2^T - \psi_1^T\tilde{\mathcal{X}}\psi_1^T - \psi_2^T\tilde{\mathcal{X}}\psi_2^T, \tag{13}$$

where

$$\begin{aligned}
\psi_1 &= \begin{bmatrix} x(t-\tau(t)) - x(t-\tau_2) \\ x(t-\tau(t)) + x(t-\tau_2) - 2\varrho_2(t) \end{bmatrix}, \\
\psi_2 &= \begin{bmatrix} x(t-\tau_1) - x(t-\tau(t)) \\ x(t-\tau_1) + x(t-\tau(t)) - 2\varrho_1(t) \end{bmatrix}, \\
\tilde{\mathcal{X}} &= diag\{\mathcal{X}, 3\mathcal{X}\}, \\
\varrho_1(t) &= \frac{1}{\tau(t)-\tau_1}\int_{t-\tau(t)}^{t-\tau_1} x(s)ds, \\
\varrho_2(t) &= \frac{1}{\tau_2-\tau(t)}\int_{t-\tau_2}^{t-\tau(t)} x(s)ds.
\end{aligned}$$

3 Stochastic Stability

The sliding switching surface is chosen as

$$s(t) = \mathcal{D}_\sigma\check{x}(t) - \int_0^t \mathcal{D}_\sigma[(\mathcal{A}_\sigma + \mathcal{B}_\sigma\mathcal{F}_\sigma)\check{x}(s) - \mathcal{L}_\sigma\mathcal{C}_\sigma\check{x}(s-\eta(s))]ds, \tag{14}$$

where $\mathcal{D}_\sigma \in \mathbb{R}^{m\times n}$ are chosen such that $\mathcal{D}_\sigma\mathcal{B}_\sigma$ are nonsingular and $\mathcal{F}_\sigma \in \mathbb{R}^{m\times n}$ will be designed later.

When the system states arrive at the sliding switching surface, we get

$$u_{eq}(t) = \mathcal{F}_\sigma\check{x}(t) - (\mathcal{D}_\sigma\mathcal{B}_\sigma)^{-1}\mathcal{D}_\sigma\mathcal{L}_\alpha(y(t-\eta(t)) - e(q_{r,j}h)). \tag{15}$$

Substituting (15) into (11) yields

$$\dot{\xi}(t) = (\bar{\bar{\mathcal{A}}}_\alpha + \Delta\bar{\bar{\mathcal{A}}}_\alpha(t))\xi(t) + \bar{\bar{\mathcal{A}}}_{d\alpha}\xi(t-\eta(t)) + \bar{\bar{\mathcal{L}}}_\alpha e(q_{r,j}h), \tag{16}$$

where

$$\xi(t) = \left[\check{x}^T(t)\; e^T(t)\right]^T, \quad \bar{\bar{\mathcal{A}}}_\alpha = \begin{bmatrix} \mathcal{A}_\alpha + \mathcal{B}_\alpha \mathcal{F}_\alpha & 0 \\ & \mathcal{A}_\alpha \end{bmatrix},$$
$$\Delta\bar{\bar{\mathcal{A}}}_\alpha(t) = \begin{bmatrix} 0 & 0 \\ \Delta\mathcal{A}_\alpha(t) & \Delta\mathcal{A}_\alpha(t) \end{bmatrix} = \bar{\bar{\mathcal{M}}}_\alpha \mathcal{H}_\alpha(t) \bar{\bar{\mathcal{N}}}_\alpha,$$
$$\bar{\bar{\mathcal{A}}}_{d\alpha} = \begin{bmatrix} -\mathcal{L}_\alpha\mathcal{C}_\alpha & 0 \\ 0 & -\mathcal{L}_\alpha\mathcal{C}_\alpha \end{bmatrix}, \quad \bar{\bar{\mathcal{L}}}_\alpha = \left[0\; \mathcal{L}_\alpha^T\right]^T,$$
$$\bar{\bar{\mathcal{M}}}_\alpha = \left[0\; \mathcal{M}_\alpha^T\right]^T, \quad \bar{\bar{\mathcal{N}}}_\alpha = \left[0\; \mathcal{N}_\alpha\right].$$

Next, we will construct novel sufficient conditions to make sure that the closed-loop system (16) can realize stochastic stability.

Theorem 1 *If we find symmetric matrices $\Pi_\alpha > 0$, $\mathcal{P}_\alpha > 0$, $\mathcal{Q}_{1\alpha} > 0$, $\mathcal{Q}_{2\alpha} > 0$, $\mathcal{R}_1 > 0, \mathcal{R}_2 > 0, \mathcal{Z}_1 > 0, \mathcal{Z}_2 > 0$, matrices $\mathcal{X}_\sigma$ and $\mathcal{Y}$, and scalars $\varepsilon_{1\alpha} > 0, \varepsilon_{2\alpha} > 0$, $\forall\alpha \in \mathcal{S}$, such that*

$$\begin{bmatrix} \tilde{\mathcal{Z}}_2 & \mathcal{Y} \\ \mathcal{Y}^T & \tilde{\mathcal{Z}}_2 \end{bmatrix} \geq 0, \tag{17}$$
$$\Omega_1 < 0, \tag{18}$$
$$\sum\nolimits_{\beta=1}^{\mathcal{N}} \pi_{\alpha\beta}\mathcal{Q}_{1\beta} \leq \mathcal{R}_1, \sum\nolimits_{\beta=1}^{\mathcal{N}} \pi_{\alpha\beta}\mathcal{Q}_{2\beta} \leq \mathcal{R}_2, \tag{19}$$

where

$$\begin{aligned}
\Omega_1 = {} & a_1^T\big(\mathcal{P}_\alpha\bar{\bar{\mathcal{A}}}_\alpha + \bar{\bar{\mathcal{A}}}_\alpha^T\mathcal{P}_\alpha + \sum\nolimits_{\beta=1}^{\mathcal{N}} \pi_{\alpha\beta}\mathcal{P}_\beta + \mathcal{Q}_{1\alpha} + \mathcal{Q}_{2\alpha} \\
& + \eta_1\mathcal{R}_1 + \eta_2\mathcal{R}_2 + \varepsilon_{1\alpha}\mathcal{P}_\alpha\bar{\bar{\mathcal{M}}}_\alpha\bar{\bar{\mathcal{M}}}_\alpha^T\mathcal{P}_\alpha + \varepsilon_{1\alpha}^{-1}\bar{\bar{\mathcal{N}}}_\alpha^T\bar{\bar{\mathcal{N}}}_\alpha \\
& + \varepsilon_{2\alpha}^{-1}\bar{\bar{\mathcal{N}}}_\alpha^T\bar{\bar{\mathcal{N}}}_\alpha\big)a_1 + a_1^T\mathcal{P}_\alpha\bar{\bar{\mathcal{A}}}_{d\alpha}a_2 + a_2^T\bar{\bar{\mathcal{A}}}_{d\alpha}^T\mathcal{P}_\alpha a_1 \\
& + a_1^T\mathcal{X}_\alpha\bar{\bar{\mathcal{A}}}_\sigma a_8 + a_8^T\bar{\bar{\mathcal{A}}}_\sigma^T\mathcal{X}_\alpha a_1 + a_1^T\mathcal{P}_\alpha\bar{\bar{\mathcal{L}}}_\sigma a_9 + a_9^T\bar{\bar{\mathcal{L}}}_\sigma^T\mathcal{P}_\alpha a_1 \\
& + a_1^T\mathcal{P}_\alpha\bar{\bar{\mathcal{G}}}_\sigma a_{10} + a_{10}^T\bar{\bar{\mathcal{G}}}_\sigma^T\mathcal{P}_\alpha a_1 - a_1^T\mathcal{P}_\alpha\bar{\bar{\mathcal{B}}}_\sigma a_{11} - a_{11}^T\bar{\bar{\mathcal{B}}}_\sigma^T\mathcal{P}_\alpha a_1 \\
& + \kappa_\alpha a_2^T[\mathcal{I}\;\; \mathcal{I}]^T\mathcal{C}_\alpha^T\Pi_\alpha\mathcal{C}_\alpha[\mathcal{I}\;\; \mathcal{I}]a_2 + a_2^T\mathcal{X}_\alpha\bar{\bar{\mathcal{A}}}_{d\alpha}a_8 \\
& + a_8^T\bar{\bar{\mathcal{A}}}_{d\alpha}^T\mathcal{X}_\alpha a_2 - a_3^T\mathcal{Q}_{1\alpha}a_3 - a_4^T\mathcal{Q}_{2\alpha}a_4 + a_8^T\big((\eta_2-\eta_1)^2\mathcal{Z}_2 \\
& - 2\mathcal{X}_\sigma + \varepsilon_{2\alpha}\mathcal{X}_\alpha\bar{\bar{\mathcal{M}}}_\alpha\bar{\bar{\mathcal{M}}}_\alpha^T\mathcal{X}_\alpha\big)a_8 + a_8^T\mathcal{X}_\alpha\bar{\bar{\mathcal{L}}}_\sigma a_9 + a_9^T\bar{\bar{\mathcal{L}}}_\sigma^T\mathcal{X}_\alpha a_8 \\
& + a_8^T\mathcal{X}_\alpha\bar{\bar{\mathcal{G}}}_\sigma a_{10} + a_{10}^T\bar{\bar{\mathcal{G}}}_\sigma^T\mathcal{X}_\alpha a_8 - a_8^T\mathcal{X}_\alpha\bar{\bar{\mathcal{B}}}_\sigma a_{11} - a_{11}^T\bar{\bar{\mathcal{B}}}_\sigma^T\mathcal{X}_\alpha a_8 \\
& - a_9^T\Pi_\alpha a_9 - a_{10}^T\gamma\mathcal{I}a_{10} - a_{11}^T\gamma\mathcal{I}a_{11} - b_1^T\tilde{\mathcal{Z}}_1 b_1 + 2b_2^T\mathcal{Y}b_3 \\
& - b_2^T\tilde{\mathcal{Z}}_2 b_2 - b_3^T\tilde{\mathcal{Z}}_2 b_3,
\end{aligned}$$

with

$$b_1 = \left[a_1^T - a_3^T \ a_1^T + a_3^T - a_7^T\right]^T, \quad \tilde{\mathcal{Z}}_1 = diag\{\mathcal{Z}_1, 3\mathcal{Z}_1\},$$
$$b_2 = \left[a_2^T - a_4^T \ a_2^T + a_4^T - a_5^T\right]^T, \quad \tilde{\mathcal{Z}}_2 = diag\{\mathcal{Z}_2, 3\mathcal{Z}_2\},$$
$$b_3 = \left[a_3^T - a_2^T \ a_3^T + a_2^T - a_6^T\right]^T, \quad \mathcal{Y} = \begin{bmatrix} \mathcal{Y}_1 & \mathcal{Y}_2 \\ \mathcal{Y}_3 & \mathcal{Y}_4 \end{bmatrix},$$

in which $a_{s_1} \in \mathcal{R}^{(16n+l)\times 2n}$ ($s_1 = 1, 2, \ldots, 8$), and $a_9 \in \mathcal{R}^{(16n+l)\times l}$ mean the block entry matrices, e.g. $a_2\zeta(t) = \xi^T(t - \eta(t))$, then system (16) achieves stochastic stability.

Proof Choose Lyapunov functional as

$$\mathcal{V}(\xi(t), r_t, t) = \mathcal{V}_1(\xi(t), r_t, t) + \mathcal{V}_2(\xi(t), r_t, t) + \mathcal{V}_3(\xi(t), t) + \mathcal{V}_4(\xi(t), t), \tag{20}$$

where

$$\mathcal{V}_1(\xi(t), r_t, t) = \xi^T(t)\mathcal{P}(r_t)\xi(t),$$
$$\mathcal{V}_2(\xi(t), r_t, t) = \int_{t-\eta_1}^{t} \xi^T(s)\mathcal{Q}_1(r_t)\xi(s)ds + \int_{t-\eta_2}^{t} \xi^T(s)\mathcal{Q}_2(r_t)\xi(s)ds,$$
$$\mathcal{V}_3(\xi(t), t) = \int_{-\eta_1}^{0}\int_{t+\theta}^{t} \xi^T(s)\mathcal{R}_1\xi(s)dsd\theta + \int_{-\eta_2}^{0}\int_{t+\theta}^{t} \xi^T(s)\mathcal{R}_2\xi(s)dsd\theta,$$
$$\mathcal{V}_4(\xi(t), t) = \eta_1 \int_{-\eta_1}^{0}\int_{t+\theta}^{t} \dot{\xi}^T(s)\mathcal{Z}_1\dot{\xi}(s)dsd\theta + (\eta_2 - \eta_1)\int_{-\eta_2}^{-\eta_1}\int_{t+\theta}^{t} \dot{\xi}^T(s)\mathcal{Z}_2\dot{\xi}(s)dsd\theta.$$

At time t, $r_t = \alpha$, for $\alpha \in \mathcal{S}$, one has the infinitesimal operator as

$$\Gamma\mathcal{V}_1(\xi(t), r_t, t) = 2\xi^T(t)\mathcal{P}_\alpha[(\bar{\bar{\mathcal{A}}}_\alpha + \Delta\bar{\bar{\mathcal{A}}}_\alpha(t))\xi(t) + \bar{\bar{\mathcal{A}}}_{d\alpha}\xi(t - \eta(t)) + \bar{\bar{\mathcal{L}}}_\alpha e(q_{r,j}h)] + \xi^T(t)\sum_{\beta=1}^{\mathcal{N}} \pi_{\alpha\beta}\mathcal{P}_\beta\xi(t). \tag{21}$$

Similarly, one has

$$\begin{aligned}
&\Gamma\mathcal{V}_2(\xi(t), r_t, t)\\
&\quad \le \xi^T(t)(\mathcal{Q}_{1\alpha} + \mathcal{Q}_{2\alpha})\xi(t) - \xi^T(t-\eta_1)\mathcal{Q}_{1\alpha}\xi(t-\eta_1)\\
&\quad - \xi^T(t-\eta_2)\mathcal{Q}_{2\alpha}\xi(t-\eta_2) + \sum_{\beta=1}^{\mathcal{N}} \pi_{\alpha\beta}\int_{t-\eta_1}^{t} e^T(s)\mathcal{Q}_{1\beta}e(s)ds\\
&\quad + \sum_{\rho=1}^{\mathcal{N}} \pi_{\alpha\beta}\int_{t-\eta_2}^{t} e^T(s)\mathcal{Q}_{2\beta}e(s)ds,
\end{aligned} \tag{22}$$

$$\begin{aligned}
&\Gamma\mathcal{V}_3(\xi(t), t)\\
&\quad = \xi^T(t)(\eta_1\mathcal{R}_1 + \eta_2\mathcal{R}_2)\xi(t) - \int_{t-\eta_1}^{t}\xi^T(s)\mathcal{R}_1\xi(s)ds\\
&\quad - \int_{t-\eta_2}^{t}\xi^T(s)\mathcal{R}_2\xi(s)ds,
\end{aligned} \tag{23}$$

$$\begin{aligned}
&\Gamma\mathcal{V}_4(\xi(t), t)\\
&\quad = \dot{\xi}^T(t)(\eta_1^2\mathcal{Z}_1 + (\eta_2-\eta_1)^2\mathcal{Z}_2)\dot{\xi}(t) - \eta_1\int_{t-\eta_1}^{t}\dot{\xi}^T(s)\mathcal{Z}_1\dot{\xi}(s)ds\\
&\quad - (\eta_2-\eta_1)\int_{t-\eta_2}^{t-\eta_1}\dot{\xi}^T(s)\mathcal{Z}_2\dot{\xi}(s)ds.
\end{aligned} \tag{24}$$

For the uncertain term $2\xi^T(t)\mathcal{P}_\alpha\Delta\bar{\bar{\mathcal{A}}}_\alpha(t)\xi(t)$, there exist scalars $\varepsilon_{1\alpha}$, such that

$$\begin{aligned}
&2\xi^T(t)\mathcal{P}_\alpha\Delta\bar{\bar{\mathcal{A}}}_\alpha(t)\xi(t)\\
&\quad \le \varepsilon_{1\alpha}\xi^T(t)\mathcal{P}_\alpha\bar{\bar{\mathcal{M}}}_\alpha\bar{\bar{\mathcal{M}}}_\alpha^T\mathcal{P}_\alpha\xi(t) + \varepsilon_{1\alpha}^{-1}\xi^T(t)\bar{\bar{\mathcal{N}}}_\alpha^T\bar{\bar{\mathcal{N}}}_\alpha\xi(t).
\end{aligned} \tag{25}$$

Applying Lemma 1 gives rise to

$$\begin{aligned}
&-\eta_1\int_{t-\eta_1}^{t}\dot{\xi}^T(s)\mathcal{Z}_1\dot{\xi}(s)ds\\
&\quad \le -\begin{bmatrix}\xi(t)-\xi(t-\eta_1)\\ \xi(t)+\xi(t-\eta_1)-\frac{2}{\eta_1}\int_{t-\eta_1}^{t}\xi(s)ds\end{bmatrix}^T\begin{bmatrix}\mathcal{Z}_1 & 0\\ 0 & 3\mathcal{Z}_1\end{bmatrix}\\
&\qquad \times\begin{bmatrix}\xi(t)-\xi(t-\eta_1)\\ \xi(t)+\xi(t-\eta_1)-\frac{2}{\eta_1}\int_{t-\eta_1}^{t}\xi(s)ds\end{bmatrix}.
\end{aligned} \tag{26}$$

By Lemma 2 and (17), one has

$$-(\eta_2-\eta_1)\int_{t-\eta_2}^{t-\eta_1}\dot{\xi}^T(s)\mathcal{Z}_2\dot{\xi}(s)ds \le 2\kappa_1^T\mathcal{Y}\kappa_2 - \kappa_1^T\tilde{\mathcal{Z}}_2\kappa_1 - \kappa_2^T\tilde{\mathcal{Z}}_1\kappa_2, \tag{27}$$

where

$$\kappa_1 = \begin{bmatrix} \xi(t-\eta(t)) - \xi(t-\eta_2) \\ \xi(t-\eta(t)) + \xi(t-\eta_2) - \frac{2}{\eta_2-\eta(t)} \int_{t-\eta_2}^{t-\eta(t)} \xi(s)ds \end{bmatrix},$$
$$\kappa_2 = \begin{bmatrix} \xi(t-\eta_1) - \xi(t-\eta(t)) \\ \xi(t-\eta_1) + \xi(t-\eta(t)) - \frac{2}{\eta(t)-\eta_1} \int_{t-\eta(t)}^{t-\eta_1} \xi(s)ds \end{bmatrix}.$$

For any appropriately matrices $\mathcal{X}_\sigma$, it is shown that

$$2\dot{\xi}^T(t)\mathcal{X}_\sigma\big[-\dot{\xi}(t) + (\bar{\bar{\mathcal{A}}}_\alpha + \Delta\bar{\bar{\mathcal{A}}}_\alpha(t))\xi(t) + \bar{\bar{\mathcal{A}}}_{d\alpha}\xi(t-\eta(t)) + \bar{\bar{\mathcal{L}}}_\alpha e(q_{r,j}h)\big] = 0. \tag{28}$$

For the uncertain term $2\dot{\xi}^T(t)\mathcal{X}_\sigma\Delta\bar{\bar{\mathcal{A}}}_\alpha(t)\xi(t)$ in (28), there exist scalars $\varepsilon_{2\alpha}$ such that

$$\begin{aligned} &2\dot{\xi}^T(t)\mathcal{X}_\alpha\Delta\bar{\bar{\mathcal{A}}}_\alpha(t)\xi(t) \\ &\le \varepsilon_{2\alpha}\dot{\xi}^T(t)\mathcal{X}_\alpha\bar{\bar{\mathcal{M}}}_\alpha\bar{\bar{\mathcal{M}}}_\alpha^T\mathcal{X}_\alpha\dot{\xi}(t) + \varepsilon_{2\alpha}^{-1}\xi^T(t)\bar{\bar{\mathcal{N}}}_\alpha^T\bar{\bar{\mathcal{N}}}_\alpha\xi(t). \end{aligned} \tag{29}$$

Define

$$\begin{aligned} \zeta(t) = \Big[&\xi^T(t) \;\; \xi^T(t-\eta(t)) \;\; \xi^T(t-\eta_1) \;\; \xi^T(t-\eta_2) \\ &\frac{2}{\eta_2-\eta(t)}\int_{t-\eta_2}^{t-\eta(t)} \xi^T(s)ds \quad \frac{2}{\eta(t)-\eta_1}\int_{t-\eta(t)}^{t-\eta_1} \xi^T(s)ds \\ &\frac{2}{\eta_1}\int_{t-\eta_1}^{t} \xi^T(s)ds \quad \dot{\xi}^T(t) \quad e^T(q_{r,j}h)\Big]^T. \end{aligned} \tag{30}$$

Applying the condition (7) leads to

$$\Gamma\mathcal{V}(\xi(t), r_t, t) \le \zeta^T(t)\Omega_1\zeta(t), \tag{31}$$

with $a_{s_1} \in \mathcal{R}^{(16n+l)\times 2n}$ $(s_1 = 1, 2, \ldots, 8)$ and $a_9 \in \mathcal{R}^{(16n+l)\times l}$ described in Theorem 1.

According to the condition (18), we have

$$\Gamma\mathcal{V}(\xi(t), r_t, t) < 0, \tag{32}$$

which means that system (16) achieves stochastic stability. This completes proof. □

Remark 2 Here, an appropriate mode-dependent Lyapunov function (20) is constructed, and parameter in integral term $\mathcal{V}(\xi(t), r_t, t)$ is mode-dependent, which may reduce some conservativeness.

Next, in order to deal with some nonlinear coupling terms in (18), we will design the controller gains and observer gains for satisfying the aforementioned conditions Theorem 1 under linear matrix framework.

Theorem 2 *Consider $\mathcal{D}_\alpha$ in (14) given as $\mathcal{D}_\alpha = \mathcal{B}_\alpha^T \mathcal{P}_{1\alpha}$. If there exist symmetric matrices $\Pi_\alpha > 0$, $\mathcal{P}_\alpha > 0$, $\mathcal{Q}_{1\alpha} > 0$, $\mathcal{Q}_{2\alpha} > 0$, $\mathcal{R}_1 > 0$, $\mathcal{R}_2 > 0$, $\mathcal{Z}_1 > 0$, $\mathcal{Z}_2 > 0$, matrices $\mathcal{Y}$ and $\tilde{\mathcal{L}}$, and scalars $\varepsilon_{1\alpha} > 0$, $\varepsilon_{2\alpha} > 0$, $\epsilon_\alpha > 0$, $\forall \alpha \in \mathcal{S}$, such that (17), (19) and*

$$\begin{bmatrix} \Omega_2 & a_1^T \varepsilon_{1\alpha} \mathcal{P}_\alpha \bar{\bar{\mathcal{M}}}_\alpha & a_8^T \varepsilon_{2\alpha} \epsilon_\alpha \mathcal{P}_\alpha \bar{\bar{\mathcal{M}}}_\alpha \\ * & -\varepsilon_{1\alpha}\mathcal{I} & 0 \\ * & * & -\varepsilon_{2\alpha}\mathcal{I} \end{bmatrix} < 0, \tag{33}$$

where

$$\begin{aligned}
\Omega_2 = {} & a_1^T \sum_{\beta=1}^{\mathcal{N}} \pi_{\alpha\beta} \mathcal{P}_\beta a_1 + a_1^T (\mathcal{P}_\alpha \bar{\bar{\mathcal{A}}}_\alpha + \bar{\bar{\mathcal{A}}}_\alpha^T \mathcal{P}_\alpha + \mathcal{Q}_{1\alpha} + \mathcal{Q}_{2\alpha} \\
& + \eta_1 \mathcal{R}_1 + \eta_2 \mathcal{R}_2 + \varepsilon_{1\alpha}^{-1} \bar{\bar{\mathcal{N}}}_\alpha^T \bar{\bar{\mathcal{N}}}_\alpha + \varepsilon_{2\alpha}^{-1} \bar{\bar{\mathcal{N}}}_\alpha^T \bar{\bar{\mathcal{N}}}_\alpha) a_1 \\
& + a_1^T diag\{-\tilde{\mathcal{L}}_\alpha \mathcal{C}_\alpha, -\tilde{\mathcal{L}}_\alpha \mathcal{C}_\alpha\} a_2 + a_2^T diag\{-\tilde{\mathcal{L}}_\alpha \mathcal{C}_\alpha, -\tilde{\mathcal{L}}_\alpha \mathcal{C}_\alpha\}^T a_1 \\
& + \epsilon_\alpha a_1^T \mathcal{P}_\alpha \bar{\bar{\mathcal{A}}}_\alpha a_8 + \epsilon_\alpha a_8^T \bar{\bar{\mathcal{A}}}_\alpha^T \mathcal{P}_\alpha a_1 + a_1^T \left[0 \ \tilde{\mathcal{L}}_\alpha^T\right]^T a_9 \\
& + a_9^T \left[0 \ \tilde{\mathcal{L}}_\alpha^T\right] a_1 + a_1^T \mathcal{P}_\alpha \bar{\bar{\mathcal{G}}}_\alpha a_{10} + a_{10}^T \bar{\bar{\mathcal{G}}}_\alpha^T \mathcal{P}_\alpha a_1 - a_1^T \mathcal{P}_\alpha \bar{\bar{\mathcal{B}}}_\alpha a_{11} \\
& - a_{11}^T \bar{\bar{\mathcal{B}}}_\alpha^T \mathcal{P}_\alpha a_1 + \kappa_\alpha a_2^T [\mathcal{I} \ \mathcal{I}]^T \mathcal{C}_\alpha^T \Pi_\alpha \mathcal{C}_\alpha [\mathcal{I} \ \mathcal{I}] a_2 \\
& + \epsilon_\alpha a_2^T diag\{-\tilde{\mathcal{L}}_\alpha \mathcal{C}_\alpha, -\tilde{\mathcal{L}}_\alpha \mathcal{C}_\alpha\} a_8 \\
& + \epsilon_\alpha a_8^T diag\{-\tilde{\mathcal{L}}_\alpha \mathcal{C}_\alpha, -\tilde{\mathcal{L}}_\alpha \mathcal{C}_\alpha\}^T a_2 \\
& - a_3^T \mathcal{Q}_{1\alpha} a_3 - a_4^T \mathcal{Q}_{2\alpha} a_4 + a_8^T ((\eta_2 - \eta_1)^2 \mathcal{Z}_2 - 2\epsilon_\alpha \mathcal{P}_\alpha) a_8 \\
& + \epsilon_\alpha a_8^T \left[0 \ \tilde{\mathcal{L}}_\alpha^T\right]^T a_9 + \epsilon_\alpha a_9^T \left[0 \ \tilde{\mathcal{L}}_\alpha^T\right] a_8 + \epsilon_\alpha a_8^T \mathcal{P}_\alpha \bar{\bar{\mathcal{G}}}_\alpha a_{10} \\
& + \epsilon_\alpha a_{10}^T \bar{\bar{\mathcal{G}}}_\alpha^T \mathcal{P}_\alpha a_8 - \epsilon_\alpha a_8^T \mathcal{P}_\alpha \bar{\bar{\mathcal{B}}}_\alpha a_{11} - \epsilon_\alpha a_{11}^T \bar{\bar{\mathcal{B}}}_\alpha^T \mathcal{P}_\alpha a_8 \\
& - a_9^T \Pi_\alpha a_9 - a_{10}^T \gamma \mathcal{I} a_{10} - a_{11}^T \gamma \mathcal{I} a_{11} - b_1^T \tilde{\mathcal{Z}}_1 b_1 + 2 b_2^T \mathcal{Y} b_3 \\
& - b_2^T \tilde{\mathcal{Z}}_2 b_2 - b_3^T \tilde{\mathcal{Z}}_2 b_3,
\end{aligned}$$

with

$$\mathcal{P}_\alpha = diag\{\mathcal{P}_{1\alpha}, \mathcal{P}_{1\alpha}\},$$

in which $\mathcal{Y}$, $\tilde{\mathcal{Z}}_1$, $\tilde{\mathcal{Z}}_2$, b_1, b_2, and b_3 are described in Theorem 1, then system (16) achieves stochastic stability, in which $\mathcal{F}_\alpha$ are selected such that $\mathcal{A}_\alpha + \mathcal{B}_\alpha \mathcal{F}_\alpha$ are Hurwitz. Furthermore, the observer gains are given as $\mathcal{L}_\alpha = \mathcal{P}_{1\alpha}^{-1} \tilde{\mathcal{L}}_\alpha$.

Proof First, let us define the followings

$$\mathcal{P}_\alpha = diag\{\mathcal{P}_{1\alpha}, \mathcal{P}_{1\alpha}\}, \ \ \tilde{\mathcal{L}}_\alpha = \mathcal{P}_{1\alpha}\mathcal{L}_\alpha, \ \ \mathcal{X}_\alpha = \epsilon_\alpha \mathcal{P}_\alpha. \tag{34}$$

Then, applying Schur complement lemma to (17), we can easily get (33) without detail procedure. This completes proof. □

4 Reachability

The observer-based SMC law will be constructed to ensure the reachability of sliding switching surface, which guarantees that the state trajectories can be driven onto the sliding switching surface in a finite time.

Theorem 3 *Consider feasible solutions in Theorem 2. Then, the finite-time attractiveness of the sliding switching surface can be realized by*

$$\begin{aligned} u(t) = &\mathcal{F}_\alpha \check{x}(t) - (\|(\mathcal{D}_\alpha\mathcal{B}_\alpha)^{-1}\mathcal{D}_\alpha\mathcal{L}_\alpha\|\|\bar{y}(t)\| \\ &+ \frac{1}{2}\|\sum\nolimits_{\beta=1,\beta\neq\alpha}^{\mathcal{N}} \pi_{\alpha\beta}(\mathcal{D}_\beta\mathcal{B}_\beta)^{-1}\|\|s(t)\| + \omega_\alpha)\text{sign}(s(t)), \end{aligned} \tag{35}$$

with the positive constant ω_α.

Proof Choose Lyapunov function:

$$\mathcal{V}(\xi(t), r_t) = \frac{1}{2}s^T(t)(\mathcal{D}(r_t)\mathcal{B}(r_t))^{-1}s(t). \tag{36}$$

When $r_t = \alpha$, we have

$$\begin{aligned} &\Gamma\mathcal{V}(\xi(t), r_t) \\ &= s^T(t)(\mathcal{D}_\alpha\mathcal{B}_\alpha)^{-1}\dot{s}(t) + \frac{1}{2}s^T(t)\sum\nolimits_{\beta=1}^{\mathcal{N}} \pi_{\alpha\beta}(\mathcal{D}_\beta\mathcal{B}_\beta)^{-1}s(t) \\ &= s^T(t)[u(t) + (\mathcal{D}_\alpha\mathcal{B}_\alpha)^{-1}\mathcal{D}_\alpha\mathcal{L}_\alpha\bar{y}(t) + \mathcal{F}_\alpha\check{x}(t)] \\ &\quad + \frac{1}{2}s^T(t)\sum\nolimits_{\beta=1}^{\mathcal{N}} \pi_{\alpha\beta}(\mathcal{D}_\beta\mathcal{B}_\beta)^{-1}s(t) \\ &= s^T(t)[-(\|(\mathcal{D}_\alpha\mathcal{B}_\alpha)^{-1}\mathcal{D}_\alpha\mathcal{L}_\alpha\|\|\bar{y}(t)\| \\ &\quad + \frac{1}{2}\|\sum\nolimits_{\beta=1,\beta\neq\alpha}^{\mathcal{N}} \pi_{\alpha\beta}(\mathcal{D}_\beta\mathcal{B}_\beta)^{-1}\|\|s(t)\| + \omega_\alpha]\text{sign}(s(t)) \\ &\quad + (\mathcal{D}_\alpha\mathcal{B}_\alpha)^{-1}\mathcal{D}_\alpha\mathcal{L}_\alpha\bar{y}(t) + \frac{1}{2}s^T(t)\sum\nolimits_{\beta=1}^{\mathcal{N}} \pi_{\alpha\beta}(\mathcal{D}_\beta\mathcal{B}_\beta)^{-1}s(t). \end{aligned}$$

Further, it can be led as

$$
\begin{aligned}
&\Gamma\mathcal{V}(\xi(t), r_t) \\
&\le s^T(t)[-(\|\mathcal{D}_\alpha\mathcal{B}_\alpha)^{-1}\mathcal{D}_\alpha\mathcal{L}_\alpha\|\|\bar{y}(t)\| \\
&\quad + \frac{1}{2}\|\sum\nolimits_{\beta=1,\beta\neq\alpha}^{\mathcal{N}} \pi_{\sigma\beta}(\mathcal{D}_\beta\mathcal{B}_\beta)^{-1}\|\|s(t)\| + \omega_\alpha)]\mathrm{sign}(s(t)) \\
&\quad + \|s(t)\|\|(\mathcal{D}_\alpha\mathcal{B}_\alpha)^{-1}\mathcal{D}_\alpha\mathcal{L}_\alpha\|\|\bar{y}(t)\| \\
&\quad + \frac{1}{2}\|s(t)\|\|\sum\nolimits_{\beta=1,\beta\neq\alpha}^{\mathcal{N}} \pi_{\alpha\beta}(\mathcal{D}_\beta\mathcal{B}_\beta)^{-1}\|\|s(t)\| \\
&= |s(t)|\big[-(\|\mathcal{D}_\alpha\mathcal{B}_\alpha)^{-1}\mathcal{D}_\alpha\mathcal{L}_\alpha\|\|\bar{y}(t)\| \\
&\quad + \frac{1}{2}\|\sum\nolimits_{\beta=1,\beta\neq\alpha}^{\mathcal{N}} \pi_{\alpha\beta}(\mathcal{D}_\beta\mathcal{B}_\beta)^{-1}\|\|s(t)\| + \omega_\alpha)\big] \\
&\quad + \|s(t)\|\|(\mathcal{D}_\alpha\mathcal{B}_\alpha)^{-1}\mathcal{D}_\alpha\mathcal{L}_\alpha\|\|\bar{y}(t)\| \\
&\quad + \frac{1}{2}\|s(t)\|\|\sum\nolimits_{\beta=1,\beta\neq\alpha}^{\mathcal{N}} \pi_{\alpha\beta}(\mathcal{D}_\beta\mathcal{B}_\beta)^{-1}\|\|s(t)\|.
\end{aligned}
$$

Noting $\|s^T(t)\| \le |s^T(t)|$, one has

$$\Gamma\mathcal{V}(\xi(t), r_t) \le -\omega_\alpha\|s(t)\|. \tag{37}$$

Therefore, the finite-time reachability can be realized. This completes proof. □

Remark 3 Considering the multi-mode characteristic of MJSs, it is necessary to choose a mode-dependent Lyapunov function (36). In addition, the proposed SMC law (35) depends on the upper bound of stochastic Markov switching transition rate $\pi_{\alpha\beta}$, $\beta \neq \alpha$, the feedback controller gains $\mathcal{F}_\alpha$, and the event-triggered transmit data $\bar{y}(t)$. Hence, under the effects of event-triggered communication scheme, the SMC law (35) can guarantee the finite-time attractiveness and improve the system performance.

5 Case Study

Consider the boost converter circuit model. As shown in Fig. 1, there exist two modes that obey the SMP $\{r_t, t \ge 0\}$ taking values in $\mathcal{S} = \{1, 2\}$.
Applying Kirchoff voltage and current law gives rise to

$$L\frac{di_L}{dt} = u(t) - V_c,\ C\frac{dV_c}{dt} = i_L - \frac{V_c}{R},\ \delta_t = 1, \tag{38}$$

$$L\frac{di_L}{dt} = u(t),\ RC\frac{dV_c}{dt} = -V_c,\ \delta_t = 2. \tag{39}$$

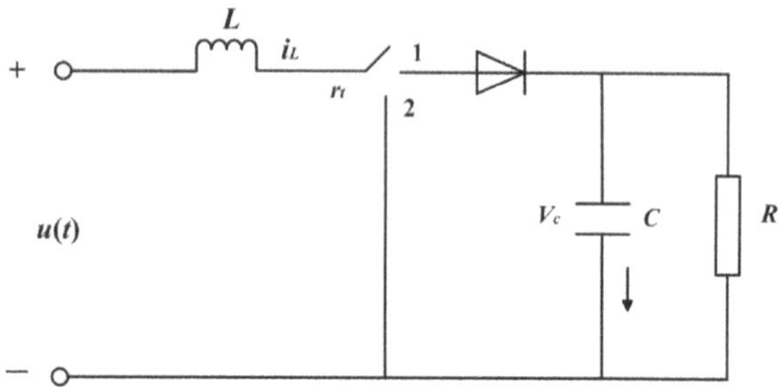

Fig. 1 Boost converter circuit

Define the state vector $z(t) = \begin{bmatrix} i_L & V_c \end{bmatrix}^T$. Then, we can get the system matrices as

$$\mathcal{A}_1 = \begin{bmatrix} 0 & -\frac{1}{L} \\ \frac{1}{C} & -\frac{1}{RC} \end{bmatrix}, \quad \mathcal{A}_2 = \begin{bmatrix} 0 & 0 \\ 0 & -\frac{1}{RC} \end{bmatrix}, \quad \mathcal{B}_1 = \mathcal{B}_2 = \begin{bmatrix} \frac{1}{L} \\ 0 \end{bmatrix}.$$

Suppose the other system matrices given as

$$\mathcal{C}_\alpha = \begin{bmatrix} 0 & 1 \end{bmatrix}, \quad \mathcal{M}_\alpha = \begin{bmatrix} 0 & 0.3 \end{bmatrix}^T,$$
$$\mathcal{N}_\alpha = \begin{bmatrix} 0.1 & -0.1 \end{bmatrix}, \quad \mathcal{H}_\alpha(t) = \sin(t), \quad (\alpha = 1, 2).$$

The values of system parameters R, L, and C are given as $R = 200\ \Omega$, $L = 2\ H$, and $C = 200\ mF$, respectively.

Also, the TR matrix is listed as

$$\begin{bmatrix} -0.8 & 0.8 \\ 0.6 & -0.6 \end{bmatrix}.$$

In this model, an appropriate SMC law is designed to ensure that the equilibrium point zero of the state space is stochastically stable and it dispels the adverse effects of parametric uncertainty.

The controller gains are chosen as $\mathcal{F}_1 = \begin{bmatrix} -1.9789 & -9.1881 \end{bmatrix}$ and $\mathcal{F}_2 = \begin{bmatrix} 1 & 0 \end{bmatrix}$, such that $\mathcal{A}_\alpha + \mathcal{B}_\alpha \mathcal{F}_\alpha$, $\alpha = 1, 2$, are Hurwitz.

For given $\tau_m = 0.01$, $\tau_M = 0.1$, $h = 0.02$, $\kappa_\alpha = 0.1$, $\varepsilon_{1\alpha} = 0.1$, $\varepsilon_{2\alpha} = 0.1$, $\epsilon_\alpha = 1.2$, and $\alpha = 1, 2$, solving Theorem 2 results in

$$\Pi_1 = 114.4548, \ \Pi_2 = 114.9490,$$
$$\mathcal{P}_{11} = \begin{bmatrix} 0.6962 & 0.0101 \\ 0.0101 & 0.2976 \end{bmatrix}, \quad \mathcal{P}_{12} = \begin{bmatrix} 2.8600 & -0.3184 \\ -0.3184 & 2.1869 \end{bmatrix},$$
$$\mathcal{L}_1 = \begin{bmatrix} -0.2274 \\ 1.2440 \end{bmatrix}, \quad \mathcal{L}_2 = \begin{bmatrix} 0.0141 \\ 0.8914 \end{bmatrix}.$$

Then, $\mathcal{B}_\alpha^T \mathcal{P}_{1\alpha} \mathcal{B}_\alpha$ is nonsingular and the integral sliding variable is given in the form of (14), for $\alpha = 1, 2$.

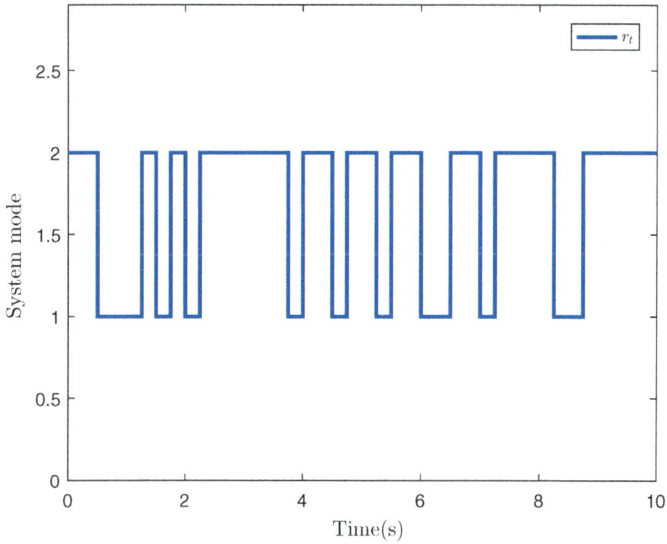

Fig. 2 System mode

Fig. 3 Release instants

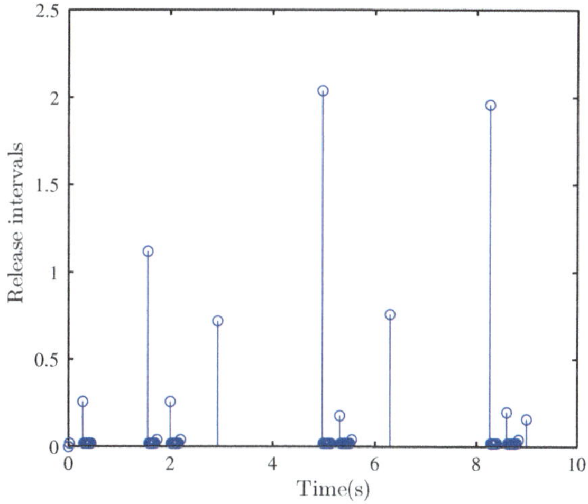

For given $\omega_\sigma = 0.01$, $\alpha = 1, 2$, the SMC law based on the event-triggered communication scheme can be computed by (35).
For given $r_0 = 2$, $x_0 = \left[-0.8\ 1.2\right]^T$ and $\check{x}_0 = \left[-0.3\ 0.8\right]^T$, the simulation results with the event-triggered SMC law (35) are given in Figs. 2, 3, 4, 5, 6 and 7.

Fig. 4 System state $x(t)$

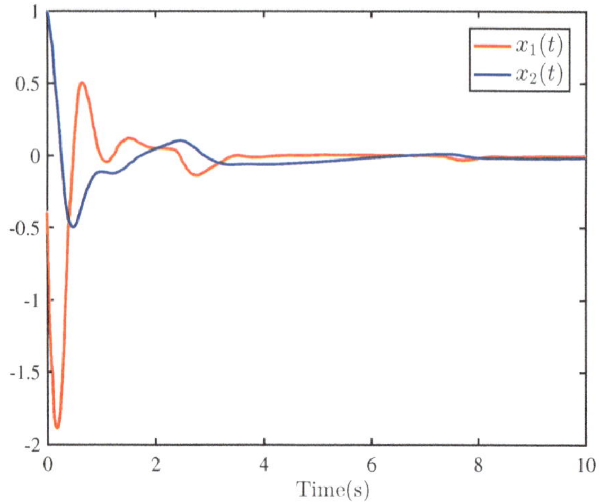

Fig. 5 Observer state $\check{x}(t)$

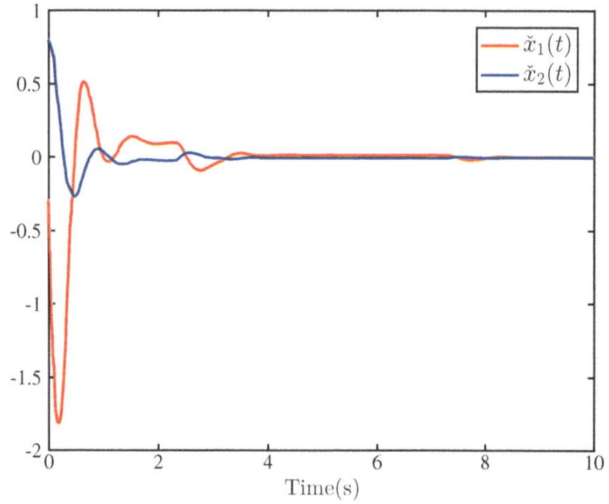

Figure 2 describes the system mode. Figure 3 plots the event-triggering release instants and intervals. In Figs. 4 and 5, the state responses $x(t)$ and $\check{x}(t)$ converge to a small region of the origin, which explain the availability of the proposed SMC approach.

Figure 6 stands for the sliding variable $s(t)$. Figure 7 represents the SMC law $u(t)$. Hence, the dynamic performance of the closed-loop system is satisfied under the event-triggered communication scheme.

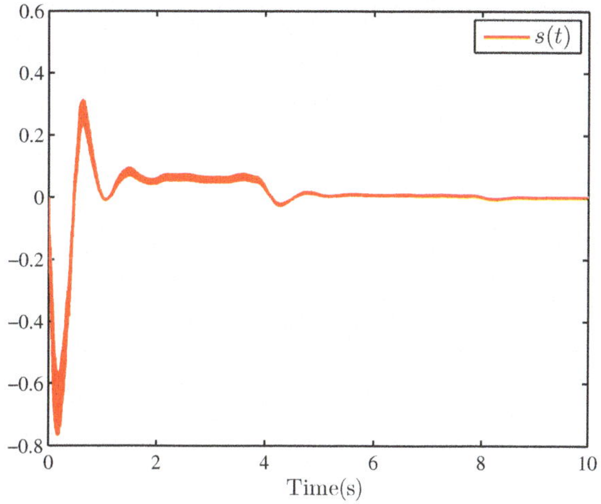

Fig. 6 Sliding variable $s(t)$

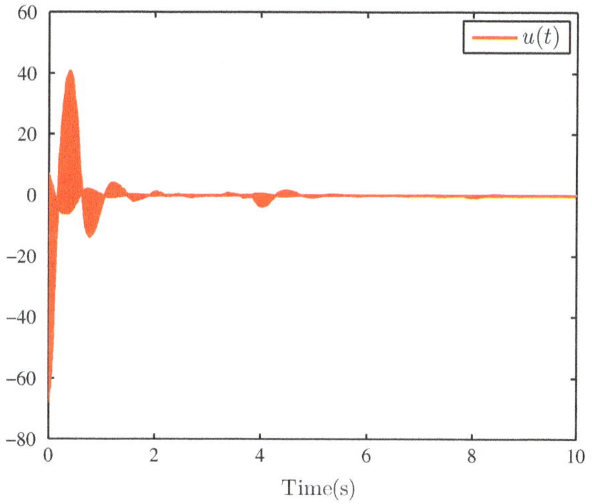

Fig. 7 SMC law $u(t)$

6 Conclusion

In this chapter, the SMC for S-MSSs has been developed by applying an observer design. In order to reduce the occupancy of network bandwidth resources, the event-triggered communication scheme has been considered in the design of the SMC law. Novel sufficient conditions have been proposed to guarantee that the system is stochastically stable. Based on these criteria, a new event-triggered SMC law is constructed to realize the finite-time attractiveness of the sliding surface.

Acknowledgements The author would be very grateful to thank National Natural Science Foundation of China (61703231, 61773236, and 61873331), Natural Science Foundation of Shandong (ZR2017QF001, ZR2017MF063, and ZR2019YQ29), Natural Science Foundation of Hubei Province (2017CFB501), Chinese Postdoctoral Science Foundation (2018T110670), and Excellent Experiment Project of Qufu Normal University (jp201728).

References

1. Ugrinovskii, V., Pota, H.: Decentralized control of power systems via robust control of uncertain Markov jump parameter systems. Int. J. Control **78**, 662–677 (2005)
2. Shen, L., Buscher, U.: Solving the serial batching problem in job shop manufacturing systems. Eur. J. Oper. Res. **221**, 14–26 (2012)
3. Dong, J., Kim, W.: Markov-chain-based output feedback control for stabilization of networked control systems with random time delays and packet losses. Int. J. Control Autom. Syst. **10**, 1013–1022 (2012)
4. Ge, X., Han, Q.: Distributed fault detection over sensor networks with Markovian switching topologies. Int. J. Syst. **43**, 305–318 (2014)
5. Park, J.H., Shen, H., Chang, X.H., Lee, T.H.: Recent Advances in Control and Filtering of Dynamic Systems with Constrained Signals. Springer, Cham (2018). https://doi.org/10.1007/978-3-319-96202-3
6. Shen, H., Zhu, Y., Zhang, L., Park, J.H.: Extended dissipative state estimation for Markov jump neural networks with unreliable links. IEEE Trans. Neural Netw. Learn. Syst. **28**(2), 346–358 (2017)
7. Wei, Y., Park, J.H., Qiu, J., Wu, L., Jung, H.Y.: Sliding mode control for semi-Markovian jump systems via output feedback. Automatica **81**, 133–141 (2017)
8. Gray, W., González, O., Doğan, M.: Stability analysis of digital linear flight controllers subject to electromagnetic disturbances. IEEE Trans. Aerosp. Electr. Syst. **36**, 1204–1218 (2000)
9. Mao, X.: Stability of stochastic differential equations with Markovian switching. Stoch. Process Appl. **79**(1), 45–67 (1999)
10. Guo, Y., Wang, Z.: Stability of Markovian jump systems with generally uncertain transition rates. J. Frankl. Inst. **350**(9), 2826–2836 (2013)
11. Xiong, J., Lam, J., Gao, H., et al.: On robust stabilization of Markovian jump systems with uncertain switching probabilities. Automatica **41**(5), 897–903 (2005)
12. Zhang, L., Boukas, E.: Stability and stabilization for Markovian jump linear systems with partly unknown transition probabilities. Automatica **45**(2), 463–468 (2009)
13. Huang, H., Huang, T., Cao, Y.: Reduced-order filtering of delayed static neural networks with Markovian jumping parameters. IEEE Trans. Neural Netw. Learn. Syst. **29**(11), 5606–5618 (2018)
14. Liu, H., Boukas, E., Sun, F., et al.: Controller design for Markov jump systems subject to actuator saturation. Automatica **42**(3), 459–465 (2006)
15. Ali, M., Arik, S., Rani, M.: Passivity analysis of stochastic neural networks with leakage delay and Markovian jumping parameters. Neurocomputing **218**, 139–145 (2016)
16. Mathiyalagan, K., Park, J.H., Sakthivel, R., Anthoni, S.M.: Robust mixed $\mathcal{H}_\infty$ and passive filtering for networked Markov jump systems with impulses. Signal Process. **101**, 162–173 (2014)
17. Shen, M., Yan, S., Zhang, G.M., Park, J.: Finite-time $\mathcal{H}_\infty$ static output control of Markov jump systems with an auxiliary approach. Appl. Math. Comput. **273**, 553–561 (2016)
18. Shen, H., Park, J.H., Wu, Z.W., Zhang, Z.: Finite-time $\mathcal{H}_\infty$ synchronization for complex networks with semi-Markov jump topology. Commun. Nonlinear Sci. Numer. Simul. **24**(1–3), 40–51 (2015)

19. Lian, J., Li, S., Liu, J.: T-S fuzzy control of positive Markov jump nonlinenar systems. IEEE Trans. Fuzzy Syst. **26**(4), 2374–2383 (2018)
20. Emelyanov, S.: Variable Structure Control Systems. Nauka, Moscow (1967)
21. Zhang, Q., Li, L., Yan, X., Spurgeon, S.: Sliding mode control for singular stochastic Markovian jump systems with uncertainties. Automatica **79**, 27–34 (2017)
22. Wu, L., Su, X., Shi, P.: Sliding mode control with bounded $\mathcal{L}_2$ gain performance of Markovian jump singular time-delay systems. Automatica **82**(8), 1929–1933 (2012)
23. Niu, Y., Ho, D., Wang, X.: Sliding mode control for It$\hat{o}$ stochastic systems with Markovian switching. Automatica **43**, 1784–1790 (2007)
24. Li, H., Shi, P., Yao, D., et al.: Observer-based adaptive sliding mode control of nonlinear Markovian jump systems. Automatica **64**, 133–142 (2016)
25. Feng, Z., Shi, P.: Sliding mode control of singular stochastic Markov jump systems. IEEE Trans. Autom. Control **62**(8), 4266–4273 (2017)
26. Zhang, Q., Zhang, J., Wang, Y.: Sliding-mode control for singular Markovian jump systems with Brownian motion based on stochastic sliding mode surface. IEEE Trans. Syst. Man Cybern. Syst. **49**(3), 494–505 (2019)
27. Zhang, J., Shi, P., Lin, W.: Extended sliding mode observer based control for Markovian jump linear systems with disturbances. Automatica **70**, 140–147 (2016)
28. Song, J., Niu, Y., Zou, Y.: Asynchronous sliding mode control of Markovian jump systems with time-varying delays and partly accessible mode detection probabilities. Automatica **93**, 33–41 (2018)
29. Liu, M., Zhang, L., Shi, P., et al.: Sliding mode control of continuous-time Markovian jump systems with digital data transmission. Automatica **80**, 200–209 (2017)
30. Li, H., Wang, Y., Yao, D., et al.: A sliding mode approach to stabilization of nonlinear Markovian jump singularly perturbed systems. Automatica **97**, 404–413 (2018)
31. Wu, L., Shi, P., Gao, H.: State estimation and sliding-mode control of Markovian jump singular systems. IEEE Trans. Autom. Control **55**(5), 1213–1219 (2010)
32. Qi, W., Park, J.H., Cheng, J., Kao, Y.: Robust stabilisation for nonlinear time-delay semi-Markovian jump systems via sliding mode control. IET Control Theory Appl. **11**(10), 1504–1513 (2017)
33. Wang, Y., Xia, Y., Shen, H., et al.: SMC design for robust stabilization of nonlinear Markovian jump singular systems. IEEE Trans. Autom. Control **63**(1), 219–224 (2018)
34. Shi, Y., Yu, B.: Output feedback stabilization of networked control systems with random delays modeled by Markov chains. IEEE Trans. Autom. Control **54**(7), 1668–1674 (2009)
35. Wu, Y., Karimi, H., Lu, R.: Sampled-data control of network systems in industrial manufacturing. IEEE Trans. Ind. Electr. **65**(11), 9016–9024 (2018)
36. Park, J.H., Lee, T.H., Liu, Y., Chen, J.: Dynamic Systems with Time Delays: Stability and Control. Springer-Nature, Singapore (2019). https://doi.org/10.1007/978-981-13-9254-2
37. Arzen, K.: A simple event-based PID controller. In: Proceedings of 14th IFAC World Congress, vol. 18, pp. 423–428 (1999)
38. Astrom, K., Bernhardsson, B.: Comparison of periodic and event based sampling for first order stochastic systems. In: Proceedings of 14th IFAC World Congress, vol. 11, pp. 301–306 (1999)
39. Yue, D., Tian, E.G., Han, Q.L.: A delay system method for designing event-triggered controllers of networked control systems. IEEE Trans. Autom. Control **58**(2), 475–481 (2013)
40. Peng, C., Yue, D., Han, Q.: Communication and Control for Networked Complex Systems. Springer, Berlin (2015)
41. Wu, L., Gao, Y., Liu, J., Li, H.: Event-triggered sliding mode control of stochastic systems via output feedback. Automatica **82**, 79–92 (2017)
42. Cheng, J., Park, J., Zhang, L., et al.: An asynchronous operation approach to event-triggered control for fuzzy Markovian jump systems with general switching policies. IEEE Trans. Fuzzy Syst. **26**(1), 6–18 (2018)

43. Song, J., Niu, Y., Xu, J.: An event-triggered approach to sliding mode control of Markovian jump Lur'e systems under hidden mode detections. IEEE Trans. Syst. Man Cybern. Syst. (2018). https://doi.org/10.1109/TSMC.2018.2847315
44. Seuret, A., Gouaisbaut, F.: Wirtinger-based integral inequality: application to time-delay systems. Automatica **49**(9), 2860–2866 (2013)
45. Zhang, X., Han, Q.: Global asymptotic stability analysis for delayed neural networks using a matrix-based quadratic convex approach. Neural Netw. **54**, 57–69 (2014)

Asynchronous Quantized Control for Markov Switching Systems with Channel Fading

Jun Cheng, Yaonan Shan, and Ju H. Park

Abstract This chapter considers the asynchronous control for Markov switching systems with channel fading, where a more general class of discrete-time Markov switching system is developed by exploring the channel fading. The original systems and controller are characterized by a variety of stochastic variables. Following this, the relationship between stochastic variables is considered. In order to reduce energy consumption, a mode-dependent event-triggered communication mechanism is employed. By establishing the mode-dependent Lyapunov function, sufficient conditions are derived to guarantee the stochastic stability of closed-loop systems. Furthermore, by solving achieved results, the gains of an asynchronous controller are obtained. In the end, a simulation example is borrowed to illustrate the effectiveness of the theoretical results.

Keywords Markov switching systems · Asynchronous control · Channel fading · Lyapunov function

1 Introduction

Over the past few decades, as a special class of hybrid systems, Markov switching systems (MSSs) have been received increasing attention due to its powerful modeling physical applications with sudden environmental changes [1–4]. Benefit from its

J. Cheng (✉)
College of Mathematics and Statistics, Guangxi Normal University, Guilin 541006, People's Republic of China
e-mail: jcheng6819@126.com

Y. Shan
University of Electronic Science and Technology of China, Chengdu 611731, Sichuan, People's Republic of China
e-mail: ynanshan@126.com

J. H. Park
Department of Electrical Engineering, Yeungnam University, Gyeongsan 38541, Republic of Korea
e-mail: jessie@ynu.ac.kr

J. H. Park (ed.), *Recent Advances in Control Problems of Dynamical Systems and Networks*, Studies in Systems, Decision and Control 301,
https://doi.org/10.1007/978-3-030-49123-9_11

advantage, MSSs have been applied to in many fields, such as robotics, networked systems, economic systems, et al. Accordingly, many scholars draw their attention to MSSs and many valuable results are reported, including stability and stabilization [5–8], control [9–11], filtering [12–15], sliding mode control [16, 17], et al. For instance, a class of MSS has been studied by taking unreliable communication links into consideration in [8]. The issue of dissipativity-based control for MSS with unreliable measurements is investigated in [9].

In addition, due to the scenario that limitation of bandwidth and unmeasurable state in the networked systems, the system's performance is degrading, even instable, which may lead to the fading phenomenon. Such limited communication is not acceptable in reality. To tackle with the shortage of unreliable channels, fading channels are adopted. Note that some important results are reported on fading channels, including control [18–20], state estimation [21–23], filtering [24, 25]. Despite the successfully utilizing of fading channels in networked control systems, a few results are expressed on the topic of MSSs.

On the other hand, most of signals is transmitted to others by communication networks, and the sampling data control strategy (SDCS) was the most used technology. As stated in [26–31], SDCS has been extended to many fields, including MSSs. Recently, as an improvement of conventional SDCS, an event-triggered mechanism (ETM) has been proposed. Compared with conventional SDCS, ETM based control is aperiodic control, which plays an important role in reducing useless data. Following this, many MSSs have been investigated by employing ETM [30–33]. It is remarkable that, most of the reported ETMs is not consider the systems mode information, which left room for further improvement.

In most existing works, synchronization issues of MSSs are studied by assuming controllers/filters keep the same modes with plants [34, 35]. In practical, it is difficult to achieve the information of original systems, and the existing assumption is hard to be satisfied. Recently, the asynchronous controllers/filters are developed, where the controller/filter modes are different from the system modes. By introducing a hidden Markov model (HMM), the original system modes are hidden to the controllers/filters. Accordingly, many asynchronous topics of MSSs are addressed [36–40]. However, there is little literature on asynchronous control for MSSs subject to ETM and CF, which remains open and challenging for MSSs.

Motivated by aforementioned discussions, in this chapter, an asynchronous controller is developed for SMMs by utilizing HMM strategy, which covers synchronous ore mode-dependent one as a special case. By a set of random variables, a mode-dependent CF model is expressed. An improved ETM is established, where original system information is involved in ETM. Together with mode-dependent Lyapunov functional, sufficient conditions are derived and the closed-loop MSSs are stochastically stable.

Notation $\mathfrak{A} > 0(\mathfrak{A} \geq 0)$ means the symmetric positive (semi-positive) definite matrix; $\mathscr{E}\{\cdot\}$ denotes the mathematical expectation; $\mathfrak{A}^{-1}$ and $\mathfrak{A}^{\top}$ represent the inverse and transpose of $\mathfrak{A}$, respectively. $\|\cdot\|$ stands for the Euclidean vector norm; $\otimes$ indicates the Kronecker product.

2 Preliminaries

Let's concerning on the following MSS:

$$\begin{aligned} x(k+1) &= A_{\varphi(k)}x(k) + B_{1\varphi(k)}u(k) + D_{1\varphi(k)}\omega(k), \\ z(k) &= C_{\varphi(k)}x(k) + B_{2\varphi(k)}u(k) + D_{2\varphi(k)}\omega(k), \end{aligned} \tag{1}$$

where $x(k) \in \mathbb{R}^n$ means the system state, $u(k) \in \mathbb{R}^u$ denotes the control input, $z(k) \in \mathbb{R}^z$ represents the controlled output, $\omega(k) \in \mathbb{R}^w$ stands for the external disturbance belonging to $l_2[0, \infty)$, $A_{\varphi(k)}$, $B_{1\varphi(k)}$, $B_{2\varphi(k)}$, $C_{\theta\varphi(k)}$, $D_{1\varphi(k)}$, and $D_{2\varphi(k)}$ are the given matrices subject to proper dimensions, and $\varphi(k) \in \mathcal{M} \triangleq \{1, 2, \ldots, M\}$ is a homogenous Markovian chain subject to the following transition probability matrix (TPM) $\Pi = \{\pi_{ab}\}$:

$$\Pr\{\varphi(k+1) = b \mid \varphi(k) = a\} = \pi_{ab} \tag{2}$$

where $0 \leq \pi_{ab} \leq 1$ and $\sum_{b\in\mathcal{M}} \pi_{ab} = 1$.

As is well known that, in the point-to-point systems, the signal transition between sensor and actuator is reliable. In reality, signal transmission is always experienced with bandwidth and limitation, which may lead to the instability of conventional data transmission. Following this trend, the network induced time delays have been taken into consideration in the networked system channels, for instance, the sensor to the controller, the controller to the actuator, et al. Therefore, it is important to deal with the consumption of network resources. Event-triggered mechanism (ETM) has been introduced in [31–33], which is an efficient way to tackle with issue of limited bandwidth. With the aid of ETM, the signals are transmitted at triggered time $\{k_s\}_{s=0}^{\infty}$. Inspired by above discussion, in this chapter, we consider the mode-dependent ETM (MDETM) described by

$$(x(k) - x(k_s))^\top \Omega_{1\varphi(k)}(x(k) - x(k_s)) \leq x^\top(k_s)\Omega_{2\varphi(k)}x(k_s), \tag{3}$$

where $\Omega_{1\varphi(k)}$ and $\Omega_{2\varphi(k)}$ are two positive weighting matrices to be determined. Defining $e(k)$ as the error during current instant $x(k)$ and latest triggered instant $x(k_s)$, one has $e(k) = x(k) - x(k_s)$. The MDETM (3) can be changed as follows:

$$e^\top(k)\Omega_{1\varphi(k)}e(k) \leq x^\top(k_s)\Omega_{2\varphi(k)}x(k_s). \tag{4}$$

In the networked systems the phenomenon of unreliability often inevitably occurs during data transmission. Channel fading (CF) is widely exists in the networked systems, which has been attracted increasing attention in the past years. Combing with MDETM, the improved CF is given by

$$\bar{x}(k) = \sum_{t=0}^{l_k} \alpha_t(k)x(k_{s-t}), \tag{5}$$

where $l_k = \min\{l, k\}$ with k represents the current sampled instant and l means the given number of paths. $\alpha_t(k) \in [0, 1]$, $(t = 0, \ldots, l_k)$ are the mutually independent stochastic variables, note that $\alpha_t(k)$ satisfies the probability density functions $g(\alpha_t(k))$ with

$$\mathscr{E}\{\alpha_t(k)\} = \alpha_t, \quad \mathscr{E}\{(\alpha_t(k) - \alpha_t)^2\} = \overline{\alpha}_t^2, \tag{6}$$

where $\overline{\alpha}_t = \sqrt{\alpha_t(1 - \alpha_t)}$.

Recalling $e(k) = x(k) - x(k_s)$, it is easy to achieve that $e(k - t) = x(k_{s-t+1}) - x(k_{s-t}), t = 1, 2, \ldots, l_k$. Following this trend, one has

$$x(k_{s-t}) = x(k) - e(k) - \cdots - e(k - t). \tag{7}$$

It follows from (5) and (7) that

$$\begin{aligned}\bar{x}(k) &= \sum_{t=0}^{l_k} \alpha_t(k)[x(k) - e(k) - \cdots - e(k - t)] \\ &= (\alpha_0(k) + D_{l_k})x(k) - (\alpha_0(k) + D_{l_k})e(k) - De(l_k),\end{aligned}$$

where

$$\begin{aligned}e(l_k) &= [e^T(k - 1), \ldots, e^T(k - l_k)]^T, \\ D_{l_k-i+1} &= \sum_{t=i}^{l_k} \alpha_t(k), \, D = [D_{l_k} I, \ldots, D_1 I].\end{aligned}$$

In the networked systems, the asynchronous phenomena is encountered because of the time lags, data dropout, et al. By taking MDETM and FC into consideration, the asynchronous controller can be designed as below:

$$\overline{u}(k) = K_{\phi(k)}\overline{x}(k), \tag{8}$$

where $K_{\phi(k)}$ is the asynchronous controller to be given later. $\phi(k)$ decides the controller mode and takes finite values in $\mathcal{N} = \{1, 2, \ldots, N\}$. The discrete-time MC $\phi(k)$ satisfying the conditional probability matrix $\Gamma = \{\lambda_{ac}\}$

$$\Pr\{\phi(k) = c \mid \varphi(k) = a\} = r_{ac}, \tag{9}$$

where $r_{ac} \in [0, 1]$, for any $c \in \mathcal{N}$, and $\sum_{c=1}^{N} r_{ac} = 1$.

Under the unreliable communication environment, it is difficult to get the raw measurement signals completely and continuously. Therefore, control $u(k)$ should be quantized before transmission:

$$u(k) = q(\overline{u}(k)), \tag{10}$$

where

$$q(\overline{u}(k)) = [q_1(\overline{u}_1(k)), q_2(\overline{u}_2(k)), \ldots, q_t(\overline{u}_t(k))]^\top,$$

with $\overline{u}_s(k), s \in \{1, 2, \ldots, t\}$ is the sth component of $\overline{u}(k)$, and $q(-\overline{u}(k)) = -q(\overline{u}(k))$.

$\forall q_s(\overline{u}_s(k))$ $(s \in \{1, 2, \ldots, t\})$, the logarithmic quantizer has the following quantization levels

$$\mathcal{T}_s = \{\pm\zeta_s^l : \zeta_s^l = \rho_s^l \zeta_{s0}, l = \pm 1, \pm 2, \ldots\} \cup \{0\},$$

where $\rho_s \in (0, 1)$ and $\zeta_{s0} > 0$.

To be continued, the quantizer $q_{s\psi(k)}(\overline{u}_s(k))$ is expressed as below

$$q_s(\overline{u}_s(k)) = \begin{cases} \zeta_s^l, & \frac{\zeta_s^l}{1+\delta_s} < \overline{u}_s(k) < \frac{\zeta_s^l}{1-\delta_s} \\ 0, & \overline{u}_s(k) = 0 \\ -q_s(-\overline{u}_s(k)), & \overline{u}_s(k) < 0 \end{cases}$$

where

$$\delta_s = \frac{1 - \rho_s}{1 + \rho_s}.$$

Similar to [15], one has

$$q_s(\overline{u}_s(k)) = (I + \Delta_s(k))\overline{u}_s(k), \tag{11}$$

where $\mid \Delta_s(k) \mid \leq \nabla_s$.

Defining

$$\Delta(k) = \text{diag}\{\Delta_1(k), \Delta_2(k), \ldots, \Delta_t(k)\},$$
$$\nabla = \text{diag}\{\nabla_1, \nabla_2, \ldots, \nabla_t\}.$$

Equation (11) indicates that

$$q(\overline{u}_s(k)) = (I + \Delta(k))\overline{u}(k). \tag{12}$$

The relationship between $\Delta(k)$ and ∇ is achieved as

$$(\Delta(k)\nabla^{-1})^\top(\Delta(k)\nabla^{-1}) \leq I. \tag{13}$$

Letting $\varphi(k) = a$, $\phi(k) = c$, and $\psi(k) = d$, the closed-loop MSS is derived by combining original system (1), asynchronous controller (8) and quantized controller (12):

$$\begin{aligned} x(k+1) =& [A_a + B_{1a}(I + \Delta(k))K_c(\alpha_0(k) + D_{l_k})]x(k) \\ & - B_{1a}(I + \Delta(k))K_c De(l_k) - B_{1a}(I + \Delta(k))K_c(\alpha_0(k) + D_{l_k})e(k) \\ & + D_{1a}\omega(k), \\ z(k) =& [C_a + B_{2a}(I + \Delta(k))K_c(\alpha_0(k) + D_{l_k})]x(k) \\ & - B_{2a}(I + \Delta(k))K_c De(l_k) - B_{2a}(I + \Delta(k))K_c(\alpha_0(k) + D_{l_k})e(k) \\ & + D_{2a}\omega(k) \end{aligned} \tag{14}$$

By taking $\alpha_s(k)$ $(s = 0, 1, \ldots, l_k)$ into consideration, one has

$$\begin{aligned} x(k+1) &= \mathscr{F}_1\eta(k) + (\alpha_0(k) - \alpha_0 + D_{l_k} - \overline{D}_{l_k})\mathscr{F}_2\eta(k) \\ &\quad - (D - \overline{D})\mathscr{F}_3\eta(k), \\ z(k) &= \mathscr{F}_4\eta(k) + (\alpha_0(k) - \alpha_0 + D_{l_k} - \overline{D}_{l_k})\mathscr{F}_5\eta(k) \\ &\quad - (D - \overline{D})\mathscr{F}_6\eta(k), \end{aligned} \tag{15}$$

where

$$\begin{aligned} \mathscr{F}_1 =& [A_a + B_{1a}(I + \Delta(k))K_c(\alpha_0 + \overline{D}_{l_k}) \;\; D_{1a} \\ & - B_{1a}(I + \Delta(k))K_c(\alpha_0 + \overline{D}_{l_k}) \;\; B_{1a}(I + \Delta(k))K_c \\ & B_{1a}(I + \Delta(k))K_c \;\cdots\; B_{1a}(I + \Delta(k))K_c], \\ \mathscr{F}_2 =& [B_{1a}(I + \Delta(k))K_c \;\; 0 \;\; B_{1a}(I + \Delta(k))K_c \;\; 0], \\ \mathscr{F}_3 =& [0\;0\;0\; B_{1a}(I + \Delta(k))K_c, B_{1a}(I + \Delta(k))K_c \;\cdots\; B_{1a}(I + \Delta(k))K_c], \\ \mathscr{F}_4 =& [C_a + B_{2a}(I + \Delta(k))K_c(\alpha_0 + \overline{D}_{l_k}) \;\; D_{2a} \\ & - B_{2a}(I + \Delta(k))K_c(\alpha_0 + \overline{D}_{l_k}) \;\; B_{2a}(I + \Delta(k))K_c \\ & B_{2a}(I + \Delta(k))K_c \;\cdots\; B_{2a}(I + \Delta(k))K_c], \\ \mathscr{F}_5 =& [B_{2a}(I + \Delta(k))K_c \;\; 0 \;\; B_{2a}(I + \Delta(k))K_c \;\; 0], \\ \mathscr{F}_6 =& [0\;0\;0\; B_{2a}(I + \Delta(k))K_c, B_{2a}(I + \Delta(k))K_c \;\cdots\; B_{2a}(I + \Delta(k))K_c], \\ D - \overline{D} =& \operatorname{diag}\left\{0, 0, 0, \sum_{t=1}^{l_k}(\alpha_t(k) - \alpha_t), \sum_{t=2}^{l_k}(\alpha_t(k) - \alpha_t), \ldots, (\alpha_{l_k}(k) - \alpha_{l_k})\right\}, \\ \eta(k) =& [x^\top(k) \;\; \omega^\top(k) \;\; e^\top(k) \;\; e^\top(l_k)]. \end{aligned}$$

Definition 1 ([12]) The MSS (15) is stochastically stable (SS) when $\omega(k) = 0$, that is

$$\mathscr{E}\left\{\sum_{k=0}^{\infty} \| x(k) \|^2\right\} < \infty.$$

The purpose of this chapter is designing asynchronous controller such that closed-loop MSS (15) is SS and meeting the following conditions.

(1) The MSS (15) is SS with $\omega(k) = 0$.

(2) $\forall \omega(k) \neq 0$ and zero initial condition, the MSS (15) achieves an $\mathcal{H}_\infty$ performance index γ, such that

$$\sum_{k=0}^{\infty} z^\top(k)z(k) < \gamma^2 \sum_{k=0}^{\infty} \omega^\top(k)\omega(k).$$

3 Main Results

In the following section, the $\mathcal{H}_\infty$ performance analysis of MSS (15) will be presented and the HMM based asynchronous control strategy will be expressed.

Theorem 1 *Given scalars $\gamma > 0$, the MSS (15) is called SS and achieves the prescribed $\mathcal{H}_\infty$ performance γ, if there exists matrices $P_a > 0$, $W_{ac} > 0$, $Q > 0$, such that for $\forall a, b \in \mathcal{M}$ and $c \in \mathcal{N}$,*

$$\sum_{c=1}^{N} r_{ac} W_{ac} - P_a < 0, \tag{16}$$

and

$$\begin{bmatrix} \Xi_a & \Psi_{1a} & \Psi_{2a} \\ * & \Psi_{3a} & 0 \\ * & * & \Psi_4 \end{bmatrix} < 0, \tag{17}$$

where

$$X_a = \sum_{b=1}^{M} \pi_{ab} P_b,$$

$$\Xi_a = l_k \chi_3^\top Q \chi_3 - \chi_4^\top \overline{Q} \chi_4 - \chi_3^\top \Omega_{1a} \chi_3 - \chi_1^\top W_{ac} \chi_1 + (\chi_1 - \chi_3)^\top \Omega_{2a} (\chi_1 - \chi_3) - \gamma^2 \chi_2^\top \chi_2,$$

$$\Psi_{1a} = \left[\mathscr{F}_1^\top \ \sqrt{1 + \overline{\alpha}_0^2 + \sum_{t=1}^{l_k} \overline{\alpha}_t^2} \mathscr{F}_2^\top \ (\mathscr{F}_3 \hat{D})^\top \ \mathscr{F}_7^\top \right],$$

$$\Psi_{2a} = \left[\mathscr{F}_4^\top \ \sqrt{1 + \overline{\alpha}_0^2 + \sum_{t=1}^{l_k} \overline{\alpha}_t^2} \mathscr{F}_5^\top \ (\mathscr{F}_6 \hat{D})^\top \ \mathscr{F}_8^\top \right],$$

$$\mathscr{F}_7 = \left[\overline{\alpha}_1 \Phi_1^{(1)\top} \ \overline{\alpha}_2 \Phi_2^{(1)\top} \ \cdots \ \overline{\alpha}_{l_k} \Phi_{l_k}^{(1)\top} \right],$$

$$\mathscr{F}_8 = \left[\overline{\alpha}_1 \Phi_1^{(2)\top} \ \overline{\alpha}_2 \Phi_2^{(2)\top} \ \cdots \ \overline{\alpha}_{l_k} \Phi_{l_k}^{(2)\top} \right],$$

$$\Psi_{3a} = \mathrm{diag} \left\{ -X_a^{-1}, -X_a^{-1}, -X_a^{-1}, \underbrace{-X_a^{-1}, \ldots, -X_a^{-1}}_{l_k} \right\},$$

$$\Psi_4 = \mathrm{diag} \left\{ -I, -I, -I, \underbrace{-I, \ldots, -I}_{l_k} \right\},$$

$$\Phi_t^{(1)} = \left[0, 0, 0, \sum_{s=1}^{t} \Upsilon_s \otimes B_{1a} (I + \Delta(k)) K_c \right],$$

$$\Phi_t^{(2)} = \left[0, 0, 0, \sum_{s=1}^{t} \Upsilon_s \otimes B_{2a}(I + \triangle(k))K_c\right],$$
$$\Upsilon_s = [0_{1\times(s-1)}, 1, 0_{1\times(l_k-s)}], \ (s = 1, 2, \ldots, l_k),$$
$$\hat{D} = \text{diag}\left\{0, 0, 0, \sum_{t=1}^{l_k} \overline{\alpha}_t^2, \sum_{t=2}^{l_k} \overline{\alpha}_t^2, \ldots, \overline{\alpha}_{l_k}^2\right\},$$
$$\overline{Q} = \text{diag}\{Q, Q, \ldots, Q\}.$$

Proof Utilizing the Schur complement to (17), such that

$$\begin{aligned}
&\overline{\mathscr{F}}_1^\top X_a \overline{\mathscr{F}}_1 + (1 + \overline{\alpha}_0^2 + \sum_{t=1}^{l_k} \overline{\alpha}_t^2) \overline{\mathscr{F}}_2^\top X_a \overline{\mathscr{F}}_2 \\
&+ \overline{\chi}_3^\top \sum_{t=1}^{l_k} \overline{\alpha}_t^2 \left(\sum_{s=1}^{t} \Upsilon_s \otimes K_c\right)^\top X_a \left(\sum_{s=1}^{t} \Upsilon_s \otimes K_c\right) \overline{\chi}_3 \\
&+ \hat{D} \overline{\mathscr{F}}_3^\top X_a \overline{\mathscr{F}}_3 + l_k \overline{\chi}_2^\top Q \overline{\chi}_2 - \overline{\chi}_3^\top \overline{Q} \overline{\chi}_3 - \overline{\chi}_2^\top \Omega_{1a} \overline{\chi}_2 \\
&+ (\overline{\chi}_1 - \overline{\chi}_2)^\top \Omega_{2a} (\overline{\chi}_1 - \overline{\chi}_2) - \overline{\chi}_1^\top W_{ac} \overline{\chi}_1 < 0,
\end{aligned} \tag{18}$$

and

$$\begin{aligned}
&\mathscr{F}_1^\top X_a \mathscr{F}_1 + (1 + \overline{\alpha}_0^2 + \sum_{t=1}^{l_k} \overline{\alpha}_t^2) \mathscr{F}_2^\top X_a \mathscr{F}_2 \\
&+ \chi_4^\top \sum_{t=1}^{l_k} \overline{\alpha}_t^2 \left(\sum_{s=1}^{t} \Upsilon_s \otimes K_c\right)^\top X_a \left(\sum_{s=1}^{t} \Upsilon_s \otimes K_c\right) \chi_4 \\
&+ \hat{D} \mathscr{F}_3^\top X_a \mathscr{F}_3 + l_k \chi_3^\top Q \chi_3 - \chi_4^\top \overline{Q} \chi_4 - \chi_3^\top \Omega_{1a} \chi_3 \\
&+ (\chi_1 - \chi_3)^\top \Omega_{2a} (\chi_1 - \chi_3) \\
&+ \mathscr{F}_4^\top \mathscr{F}_4 + (1 + \overline{\alpha}_0^2 + \sum_{t=1}^{l_k} \overline{\alpha}_t^2) \mathscr{F}_5^\top \mathscr{F}_5 \\
&+ \chi_4^\top \sum_{t=1}^{l_k} \overline{\alpha}_t^2 \left(\sum_{s=1}^{t} \Upsilon_s \otimes K_c\right)^\top \left(\sum_{s=1}^{t} \Upsilon_s \otimes K_c\right) \chi_4 \\
&+ \hat{D} \mathscr{F}_6^\top \mathscr{F}_6 - \gamma^2 \chi_2^\top \chi_2 - \chi_1^\top W_{ac} \chi_1 < 0.
\end{aligned} \tag{19}$$

According to the condition that $\sum_{c=1}^{N} r_{ac} W_{ac} < P_a$ in (16), it is clear to observe that

$$\sum_{c=1}^{N} r_{ac}\Bigg\{\overline{\mathscr{F}}_1^{\top} X_a \overline{\mathscr{F}}_1 + (1+\overline{\alpha}_0^2+\sum_{t=1}^{l_k}\overline{\alpha}_t^2)\overline{\mathscr{F}}_2^{\top} X_a \overline{\mathscr{F}}_2$$
$$+\overline{\chi}_3^{\top}\sum_{t=1}^{l_k}\overline{\alpha}_t^2\left(\sum_{s=1}^{t}\Upsilon_s\otimes K_c\right)^{\top} X_a\left(\sum_{s=1}^{t}\Upsilon_s\otimes K_c\right)\overline{\chi}_3$$
$$+\hat{D}\overline{\mathscr{F}}_3^{\top} X_a \overline{\mathscr{F}}_3 + l_k\overline{\chi}_2^{\top} Q\overline{\chi}_2 - \overline{\chi}_3^{\top}\overline{Q}\overline{\chi}_3 - \overline{\chi}_2^{\top}\Omega_{1a}\overline{\chi}_2$$
$$+(\overline{\chi}_1-\overline{\chi}_2)^{\top}\Omega_{2a}(\overline{\chi}_1-\overline{\chi}_2)\Bigg\} - \overline{\chi}_1^{\top} P_a\overline{\chi}_1 < 0, \tag{20}$$

and

$$\sum_{c=1}^{N} r_{ac}\Bigg\{\mathscr{F}_1^{\top} X_a \mathscr{F}_1 + (1+\overline{\alpha}_0^2+\sum_{t=1}^{l_k}\overline{\alpha}_t^2)\mathscr{F}_2^{\top} X_a \mathscr{F}_2$$
$$+\chi_4^{\top}\sum_{t=1}^{l_k}\overline{\alpha}_t^2\left(\sum_{s=1}^{t}\Upsilon_s\otimes K_c\right)^{\top} X_a\left(\sum_{s=1}^{t}\Upsilon_s\otimes K_c\right)\chi_4$$
$$+\hat{D}\mathscr{F}_3^{\top} X_a \mathscr{F}_3 + l_k\chi_3^{\top} Q\chi_3 - \chi_4^{\top}\overline{Q}\chi_4 - \chi_3^{\top}\Omega_{1a}\chi_3$$
$$+(\chi_1-\chi_3)^{\top}\Omega_{2a}(\chi_1-\chi_3)$$
$$+\mathscr{F}_4^{\top}\mathscr{F}_4 + (1+\overline{\alpha}_0^2+\sum_{t=1}^{l_k}\overline{\alpha}_t^2)\mathscr{F}_5^{\top}\mathscr{F}_5 \tag{21}$$
$$+\chi_4^{\top}\sum_{t=1}^{l_k}\overline{\alpha}_t^2\left(\sum_{s=1}^{t}\Upsilon_s\otimes K_c\right)^{\top}\left(\sum_{s=1}^{t}\Upsilon_s\otimes K_c\right)\chi_4$$
$$+\hat{D}\mathscr{F}_6^{\top}\mathscr{F}_6 - \gamma^2\chi_2^{\top}\chi_2\Bigg\} - \chi_1^{\top} P_a\chi_1 < 0.$$

We select the following mode-dependent Lyapunov functional for MSS (15)

$$V(k, x_k, \varphi_k) = V_1(k, x_k, \varphi_k) + V_2(k, x_k, \varphi_k), \tag{22}$$

where

$$V_1(k, x_k, \varphi_k) = x^{\top}(k)P_{\varphi(k)}x(k),$$
$$V_2(k, x_k, \varphi_k) = \sum_{p=1}^{l_k}\sum_{q=k-p}^{k-1} e^{\top}(q)e(q).$$

Define $\mathscr{E}\{\Delta V(k)\} = \mathscr{E}\{V(k+1, x_{k+1}, \varphi_{k+1}) \mid -V(k, x_k, \varphi_k)\}$. Here, denote $\varphi(k) = a$, $\varphi(k+1) = b$, $\phi(k) = c$, we get

$$
\begin{aligned}
\mathscr{E}\{\Delta V_1(k)\} &= x^\top(k+1)P_{\varphi(k+1)}x(k+1) - x^\top(k)P_{\varphi(k)}x(k) \\
&= \sum_{c=1}^{N} r_{ac}\eta^\top(k) f_1^\top X_a \mathscr{F}_1 \eta(k) - x^\top(k)P_{\varphi(k)}x(k) \\
&\quad + \sum_{c=1}^{N} r_{ac}\mathscr{E}\left\{(\alpha_0(k) - \alpha_0 + D_{l_k} - \overline{D}_{l_k})^2\right\}\eta^\top(k) f_2^\top X_a \mathscr{F}_2\eta(k) \\
&\quad + \sum_{c=1}^{N} r_{ac}\mathrm{He}\{\mathscr{E}\left\{(D_{l_k} - \overline{D}_{l_k})\right\}\}\eta^\top(k)\mathscr{F}_2^\top X_a \mathscr{F}_3\eta(k) \\
&\quad + \sum_{c=1}^{N} r_{ac}\mathscr{E}\left\{(D - \overline{D})^2\right\}\eta^\top(k)\mathscr{F}_3^\top X_a \mathscr{F}_3\eta(k) \\
&= \sum_{c=1}^{N} r_{ac}\eta^\top(k)\mathscr{F}_1^\top X_a \mathscr{F}_1\eta(k) - x^\top(k)P_{\varphi(k)}x(k) \\
&\quad + \sum_{c=1}^{N} r_{ac}(\mathscr{E}\left\{(\alpha_0(k) - \alpha_0)^2\right\}\eta^\top(k)\mathscr{F}_2^\top X_a \mathscr{F}_2\eta(k) \\
&\quad + \sum_{c=1}^{N} r_{ac}\mathscr{E}\left\{(D_{l_k} - \overline{D}_{l_k})^2\right\})\eta^\top(k)\mathscr{F}_2^\top X_a \mathscr{F}_2\eta(k) \\
&\quad + \sum_{c=1}^{N} r_{ac}\mathrm{He}\{\mathscr{E}\left\{(D_{l_k} - \overline{D}_{l_k})\right\}\}\eta^\top(k)\mathscr{F}_2^\top X_a \mathscr{F}_3\eta(k) \\
&\quad + \sum_{c=1}^{N} r_{ac}\mathscr{E}\left\{(D - \overline{D})^2\right\}\eta^\top(k)\mathscr{F}_3^\top X_a \mathscr{F}_3\eta(k).
\end{aligned} \tag{23}
$$

Then, it is easy to achieve that

$$
\begin{aligned}
\mathscr{E}\{\Delta V_1(k)\} = &\sum_{c=1}^{N} r_{ac}\eta^\top(k)\mathscr{F}_1^\top X_a \mathscr{F}_1\eta(k) \\
&+ \sum_{c=1}^{N} r_{ac}(\overline{\alpha}_0^2 + \sum_{t=1}^{l_k}\overline{\alpha}_t^2)\eta^\top(k)\mathscr{F}_2^\top X_a \mathscr{F}_2\eta(k) \\
&+ 2\sum_{c=1}^{N} r_{ac}\sum_{t=1}^{l_k}\overline{\alpha}_t^2\eta^\top(k)\mathscr{F}_2^\top X_a \mathscr{F}_3\eta(k) \\
&+ \sum_{c=1}^{N} r_{ac}\hat{D}\eta^\top(k)\mathscr{F}_3^\top X_a \mathscr{F}_3\eta(k) - x^\top(k)P_{\varphi(k)}x(k).
\end{aligned} \tag{24}
$$

By using Jensen's inequality to (24), it yields that

$$
\begin{aligned}
\mathscr{E}\{\Delta V_1(k)\} =& x^\top(k+1)P_{\varphi(k+1)}x(k+1) - x^\top(k)P_{\varphi(k)}x(k) \\
\le& \sum_{c=1}^{N} r_{ac}\eta^\top(k)\mathscr{F}_1^\top X_a \mathscr{F}_1\eta(k) - x^\top(k)P_{\varphi(k)}x(k) \\
&+ \sum_{c=1}^{N} r_{ac}(1+\overline{\alpha}_0^2 + \sum_{t=1}^{l_k}\overline{\alpha}_t^2)\eta^\top(k)\mathscr{F}_2^\top X_a\mathscr{F}_2\eta(k) \\
&+ \sum_{c=1}^{N} r_{ac}e^\top(l_k)\sum_{t=1}^{l_k}\overline{\alpha}_t^2\left(\sum_{s=1}^{t}\Upsilon_s\otimes B_{1a}(I+\Delta(k)\right)K_c)^\top \\
&\times X_a\left(\sum_{s=1}^{t}\Upsilon_s\otimes B_{1a}(I+\Delta(k))K_c\right)e(l_k) \\
&+ \sum_{c=1}^{N} r_{ac}\hat{D}\eta^\top(k)\mathscr{F}_3^\top X_a\mathscr{F}_3\eta(k).
\end{aligned}
\tag{25}
$$

On the other hand, it follows from (22) that

$$
\begin{aligned}
\mathscr{E}\{\Delta V_2(k)\} =& \mathscr{E}\{l_k e^\top(k)Qe(k) - \sum_{p=1}^{l_k} e^\top(k-p)Qe(k-p)\} \\
=& \mathscr{E}\{l_k e^\top(k)Qe(k) - e^\top(l_k)\overline{Q}e(l_k)\}.
\end{aligned}
\tag{26}
$$

Recalling (4) and using the term $x(k_s) = x(k) - e(k)$, we have

$$
0 \le -e^\top(k)\Omega_{1a}e(k) + (x(k)-e(k))^\top\Omega_{2a}(x(k)-e(k)). \tag{27}
$$

Firstly, when $\omega(k) = 0$, one has

$$
\begin{aligned}
\mathscr{E}\{\Delta V(k)\} =& \sum_{c=1}^{N} r_{ac}\overline{\eta}^\top(k)\overline{\mathscr{F}}_1^\top X_a\overline{\mathscr{F}}_1\overline{\eta}(k) - \overline{\eta}^\top(k)\chi_1^\top P_a\chi_1\overline{\eta}(k) \\
&+ \sum_{c=1}^{N} r_{ac}(1+\overline{\alpha}_0^2 + \sum_{t=1}^{l_k}\overline{\alpha}_t^2)\overline{\eta}^\top(k)\overline{\mathscr{F}}_2^\top X_a\overline{\mathscr{F}}_2\overline{\eta}(k) \\
&+ \sum_{c=1}^{N} r_{ac}\overline{\eta}^\top(k)\chi_3^\top\sum_{t=1}^{l_k}\overline{\alpha}_t^2\left(\sum_{s=1}^{t}\Upsilon_s\otimes K_c\right)^\top X_a\left(\sum_{s=1}^{t}\Upsilon_s\otimes K_c\right)\chi_3\overline{\eta}(k) \\
&+ \sum_{c=1}^{N} r_{ac}\hat{D}\overline{\eta}^\top(k)\overline{\mathscr{F}}_3^\top X_a\overline{\mathscr{F}}_3\overline{\eta}(k) + l_k\overline{\eta}^\top(k)\chi_1^\top Q\chi_1\overline{\eta}(k) \\
&- \overline{\eta}^\top\chi_3^\top\overline{Q}\chi_3\overline{\eta}(k) - \overline{\eta}^\top(k)\chi_2^\top\Omega_{1a}\chi_2\overline{\eta}(k) \\
&+ \overline{\eta}^\top(k)(\chi_1-\chi_2)^\top\Omega_{2a}(\chi_1-\chi_2)\overline{\eta}(k) \\
=& \overline{\eta}^\top(k)\overline{\Xi}_a\overline{\eta}(k),
\end{aligned}
\tag{28}
$$

where

$$
\begin{aligned}
\overline{\Xi}_a &= \sum_{c=1}^{N} r_{ac}\overline{\mathscr{F}}_1^{\top} X_a \overline{\mathscr{F}}_1 - \chi_1^{\top} P_{\varphi(k)} \chi_1 \\
&+ \sum_{c=1}^{N} r_{ac}(1+\overline{\alpha}_0^2 + \sum_{t=1}^{l_k} \overline{\alpha}_t^2)\overline{\mathscr{F}}_2^{\top} X_a \overline{\mathscr{F}}_2 \\
&+ \sum_{c=1}^{N} r_{ac}\chi_3^{\top} \sum_{t=1}^{l_k} \overline{\alpha}_t^2 \left(\sum_{s=1}^{t} \Upsilon_s \otimes K_c\right)^{\top} X_a \left(\sum_{s=1}^{t} \Upsilon_s \otimes K_c\right) \chi_3 \\
&+ \sum_{c=1}^{N} r_{ac}\hat{D}\overline{\mathscr{F}}_3^{\top} X_a \overline{\mathscr{F}}_3 + l_k \chi_1^{\top} Q \chi_1 - \chi_3^{\top} \overline{Q} \chi_3 - \chi_2^{\top} \Omega_{1a} \chi_2 \\
&+ (\chi_1 - \chi_2)^{\top} \Omega_{2a} (\chi_1 - \chi_2), \\
\overline{\mathscr{F}}_1 &= [A_a + B_{1a}(I + \Delta(k))K_c(\alpha_0 + \overline{D}_{l_k}) \;\; - B_{1a}(I+\Delta(k))K_c(\alpha_0 + \overline{D}_{l_k}) \\
&\quad B_{1a}(I+\Delta(k))K_c \;\; B_{1a}(I+\Delta(k))K_c \; \cdots \; B_{1a}(I+\Delta(k))K_c], \\
\overline{\mathscr{F}}_2 &= [B_{1a}(I+\Delta(k))K_c(\alpha_0(k) - \alpha_0 + D_{l_k} - \overline{D}_{l_k}) \\
&\quad B_{1a}(I+\Delta(k))K_c(\alpha_0(k) - \alpha_0 + D_{l_k} - \overline{D}_{l_k}) \;\; 0], \\
\overline{\mathscr{F}}_3 &= [0\; 0\; B_{1a}(I+\Delta(k))K_c \;\; B_{1a}(I+\Delta(k))K_c \; \cdots \; B_{1a}(I+\Delta(k))K_c], \\
\overline{\eta}(k) &= [x^{\top}(k) \;\; e^{\top}(k) \;\; e^{\top}(l_k)], \overline{\chi}_1 = [I_{n\times n}, 0_{n\times(1+l_k)n}], \\
\overline{\chi}_2 &= [0_{n\times n}, I_{n\times n}, 0_{n\times(l_k)n}],\; \overline{\chi}_3 = [0_{n\times 2n}, I_{n\times n}, \ldots, I_{n\times n}].
\end{aligned}
$$

From (22), we get

$$
\mathscr{E}\{\Delta V(k)\} < -\inf\{\lambda_{\min}(-\overline{\Xi}_a)\} x^{\top}(k) x(k) < 0. \tag{29}
$$

where

$$
\beta = \inf\{\lambda_{\min}(-\overline{\Xi}_a)\}.
$$

From $k = 0$ to ∞, summing up the both sides of (29), one can derive that

$$
\mathscr{E}\left\{\sum_{k=0}^{\infty} \| x^{\top}(k) \|^2\right\} < \frac{1}{\beta}(\mathscr{E}\{V(0)\} - \mathscr{E}\{V(\infty)\}) < \infty. \tag{30}
$$

It follows from Definition 1 that the MSS (15) is SS.

Next, when $\omega(k) \neq 0$, one has

$$
\begin{aligned}
\mathscr{E}\{\Delta V(k)\} = & \sum_{c=1}^{N} r_{ac}\eta^\top(k)\mathscr{F}_1^\top X_a \mathscr{F}_1\eta(k) - \eta^\top(k)\chi_1^\top P_a \chi_1 \eta(k) \\
& + \sum_{c=1}^{N} r_{ac}(1+\overline{\alpha}_0^2 + \sum_{t=1}^{l_k}\overline{\alpha}_t^2)\eta^\top(k)\mathscr{F}_2^\top X_a \mathscr{F}_2\eta(k) \\
& + \sum_{c=1}^{N} r_{ac}\eta^\top(k)\chi_4^\top \sum_{t=1}^{l_k}\overline{\alpha}_t^2 \left(\sum_{s=1}^{t}\Upsilon_s \otimes K_c\right)^\top X_a \left(\sum_{s=1}^{t}\Upsilon_s \otimes K_c\right)\chi_4\eta(k) \\
& + \sum_{c=1}^{N} r_{ac}\hat{D}\eta^\top(k)\mathscr{F}_3^\top X_a \mathscr{F}_3\eta(k) + l_k\eta^\top(k)\chi_3^\top Q\chi_3\eta(k) \\
& - \eta^\top(k)\chi_4^\top \overline{Q}\chi_4\eta(k) - \eta^\top(k)\chi_3^\top \Omega_{1a}\chi_3\eta(k) \\
& + \eta^\top(k)(\chi_1-\chi_3)^\top \Omega_{2a}(\chi_1-\chi_3)\eta(k) \\
& + \eta^\top(k)\mathscr{F}_4^\top \mathscr{F}_4\eta(k) + (1+\overline{\alpha}_0^2+\sum_{t=1}^{l_k}\overline{\alpha}_t^2)\eta^\top(k)\mathscr{F}_5^\top \mathscr{F}_5\eta(k) \\
& + \eta^\top(k)\chi_4^\top \sum_{t=1}^{l_k}\overline{\alpha}_t^2 \left(\sum_{s=1}^{t}\Upsilon_s \otimes K_c\right)^\top \left(\sum_{s=1}^{t}\Upsilon_s \otimes K_c\right)\chi_4\eta(k) \\
& + \hat{D}\eta^\top(k)\mathscr{F}_6^\top \mathscr{F}_6\eta(k) - \gamma^2\eta^\top(k)\chi_2^\top \chi_2\eta(k) \\
& - (z^\top(k)z(k) - \gamma^2\omega^\top(k)\omega(k)).
\end{aligned} \tag{31}
$$

It follows from (12) that

$$
\mathscr{E}\{\Delta V(k)\} < -(z^\top(k)z(k) - \gamma^2\omega^\top(k)\omega(k)). \tag{32}
$$

Summing above inequality from 0 to h, one has

$$
\sum_{k=0}^{h}\mathscr{E}\{\Delta V(k)\} < -\sum_{k=0}^{h}(z^\top(k)z(k) - \gamma^2\omega^\top(k)\omega(k)). \tag{33}
$$

Under the condition of zero initial value, we obtain that

$$0 < V(h) < -\sum_{k=0}^{h}(z^{\top}(k)z(k) - \gamma^{2}\omega^{\top}(k)\omega(k)). \tag{34}$$

Thus, one has

$$\sum_{k=0}^{h} z^{\top}(k)z(k) < \gamma^{2}\sum_{k=0}^{h}\omega^{\top}(k)\omega(k). \tag{35}$$

For $h \to \infty$, we conclude that MSS (15) is SS with $\mathcal{H}_{\infty}$ performance. This completes proof. □

Now, we are ready to design the asynchronous controller in Theorem 2.

Theorem 2 *Given scalars $\gamma > 0$, the MSS (15) is called SS and achieves the prescribed $\mathcal{H}_{\infty}$ performance γ, if there exists matrices $\widehat{P}_a > 0$, $\widehat{W}_{ac} > 0$, $\widehat{Q} > 0$, and matrices Y_c, X, such that for $\forall a, b \in \mathcal{M}$ and $c \in \mathcal{N}$,*

$$\sum_{c=1}^{N} r_{ac}\widehat{W}_{ac} - \widehat{P}_a < 0, \tag{36}$$

and

$$\begin{bmatrix} \widehat{\Xi}_a & \Pi^{(1)} & \Pi^{(2)} \\ * & \Pi^{(3)} & 0 \\ * & * & \Pi^{(3)} \end{bmatrix} < 0, \tag{37}$$

where

$$\Pi_a = \begin{bmatrix} \widehat{\Xi}_a & \widehat{\Psi}_{1a} & \widehat{\Psi}_{2a} \\ * & \widehat{\Psi}_{3a} & 0 \\ * & * & \Psi_4 \end{bmatrix}, \quad \Pi^{(1)} = [\Lambda_{1c}\nabla \;\; \Lambda_{2c}\nabla \;\; \Lambda_{3c}\nabla],$$

$$\Pi^{(2)} = [\varepsilon_1\Theta_{1a} \;\; \varepsilon_2\Theta_{2a} \;\; \varepsilon_3\Theta_{1a}], \; \Pi^{(3)} = diag\{-\varepsilon_1 I, -\varepsilon_2 I, -\varepsilon_3 I\},$$

$$\widehat{\Psi}_{1a} = \left[\widehat{\mathscr{F}}_1^{\top} Z_a \;\; \sqrt{1 + \overline{\alpha}_0^2 + \sum_{t=1}^{l_k}\overline{\alpha}_t^2}\,\widehat{\mathscr{F}}_2^{\top} Z_a \;\; (\widehat{\mathscr{F}}_3\hat{D})^{\top} Z_a \;\; \widehat{\mathscr{F}}_7^{\top} Z_a \right],$$

$$
\begin{aligned}
\widehat{\Psi}_{2a} &= \left[\widehat{\mathscr{F}}_4^\top \ \sqrt{1+\overline{\alpha}_0^2+\sum_{t=1}^{l_k}\overline{\alpha}_t^2}\,\widehat{\mathscr{F}}_5^\top \ (\widehat{\mathscr{F}}_6\hat{D})^\top \ \widehat{\mathscr{F}}_8^\top \right],\\
\widehat{\mathscr{F}}_1 &= [A_a X + B_{1a}Y_c(\alpha_0+\overline{D}_{l_k}) \ \ D_{1a} \ \ -B_{1a}Y_c(\alpha_0+\overline{D}_{l_k})\\
&\quad\ B_{1a}Y_c \ \ B_{1a}Y_c \ \cdots \ B_{1a}Y_c],\\
\widehat{\mathscr{F}}_2 &= [B_{1a}Y_c \ \ 0 \ \ B_{1a}Y_c \ \ 0],\\
\widehat{\mathscr{F}}_3 &= [0\,0\,0\ B_{1a}Y_cX,\ B_{1a}Y_c \ \cdots \ B_{1a}Y_c],\\
\widehat{\mathscr{F}}_4 &= [C_a X + B_{2a}Y_c(\alpha_0+\overline{D}_{l_k}) \ \ D_{2a} \ \ -B_{2a}Y_c(\alpha_0+\overline{D}_{l_k})\\
&\quad\ B_{2a}Y_c \ \ B_{2a}Y_c \ \cdots \ B_{2a}Y_c],\\
\widehat{\mathscr{F}}_5 &= [B_{2a}Y_c \ \ 0 \ \ B_{2a}Y_c \ \ 0],\\
\widehat{\mathscr{F}}_6 &= [0\,0\,0\ B_{2a}Y_c,\ B_{2a}Y_c \ \cdots \ B_{2a}Y_c],\\
\widehat{\mathscr{F}}_7 &= \left[\overline{\alpha}_1\widehat{\Phi}_1^{(1)\top} \ \overline{\alpha}_2\widehat{\Phi}_2^{(1)\top} \ \cdots \ \overline{\alpha}_{l_k}\widehat{\Phi}_{l_k}^{(1)\top}\right],\\
\widehat{\mathscr{F}}_8 &= \left[\overline{\alpha}_1\widehat{\Phi}_1^{(2)\top} \ \overline{\alpha}_2\widehat{\Phi}_2^{(2)\top} \ \cdots \ \overline{\alpha}_{l_k}\widehat{\Phi}_{l_k}^{(2)\top}\right],\\
\Lambda_{1c} &= [0\,0\,0\ Y_c \ \ Y_c \ \cdots \ Y_c\ 0 \ 0 \ \cdots \ 0]^\top,\\
\Lambda_{2c} &= [Y_c\ 0\ 0 \ \cdots \ 0]^\top,\ \Lambda_{3c} = [0\ 0\ \ Y_c \ \ 0 \ \ 0 \ \cdots \ 0]^\top,\\
\Theta_{1a} &= [0 \ \ 0 \ \cdots \ 0\ \mathcal{B}_a^\top\ 0 \cdots 0\ \mathcal{B}_a^\top \ \ \digamma_{1a} \ \ B_{2a}^\top\ 0\ B_{2a}^\top\ \digamma_{2a}],\\
\Theta_{2a} &= [0\,0 \cdots 0\ B_{1a}^\top \cdots B_{1a}^\top \ \ 0 \cdots 0\ B_{2a}^\top \ \ B_{2a}^\top \ \ 0 \cdots 0],\\
\Theta_{3a} &= [0\,0 \cdots 0 - B_{1a}^\top \ \cdots - B_{1a}^\top\ B_{1a}^\top \ \cdots \ B_{1a}^\top \ 0 \cdots 0 - B_{2a}^\top \ \ B_{2a}^\top \ \ 0 \cdots 0],\\
\digamma_{1a} &= [\overline{\alpha}_1\widehat{\Phi}_1^{(3)\top} \ \cdots \ \overline{\alpha}_1\widehat{\Phi}_1^{(3)\top} \ \overline{\alpha}_2\widehat{\Phi}_2^{(3)\top} \ \cdots \ \overline{\alpha}_{l_k}\widehat{\Phi}_{l_k}^{(3)\top}],\\
\digamma_{2a} &= [\overline{\alpha}_1\widehat{\Phi}_1^{(4)\top} \ \overline{\alpha}_2\widehat{\Phi}_2^{(4)\top} \ \cdots \ \overline{\alpha}_{l_k}\widehat{\Phi}_{l_k}^{(4)\top}],\\
\widehat{\Phi}_t^{(1)} &= \left[0,0,0,\sum_{s=1}^{t}\Upsilon_s\otimes B_{1a}Y_c\right],\ \widehat{\Phi}_t^{(2)} = \left[0,0,0,\sum_{s=1}^{t}\Upsilon_s\otimes B_{2a}Y_c\right],\\
\widehat{\Phi}_t^{(3)} &= \left[0,0,0,\sum_{s=1}^{t}\Upsilon_s\otimes B_{1a}\right],\ \widehat{\Phi}_t^{(4)} = \left[0,0,0,\sum_{s=1}^{t}\Upsilon_s\otimes B_{2a}\right].
\end{aligned}
$$

Here, the control gain matrices are given by

$$K_c = Y_c X^{-1}. \tag{38}$$

Proof The inequality (17) can be rewritten as follows:

$$\begin{bmatrix} \Xi_a & \overline{\Psi}_{1a} & \Psi_{2a} \\ * & \overline{\Psi}_{3a} & 0 \\ * & * & \Psi_4 \end{bmatrix} < 0, \tag{39}$$

where

$$\overline{\Psi}_{1a} = \left[\mathscr{F}_1^\top Z_a \; \sqrt{1+\overline{\alpha}_0^2+\sum_{t=1}^{l_k}\overline{\alpha}_t^2}\mathscr{F}_2^\top Z_a \; (\mathscr{F}_3\hat{D})^\top Z_a \; \mathscr{F}_7^\top Z_a \right],$$

$$\overline{\Psi}_{3a} = \text{diag}\left\{ -\mathscr{P}^{-1}, -\mathscr{P}^{-1}, -\mathscr{P}^{-1}, \underbrace{-\mathscr{P}^{-1}, \ldots, -\mathscr{P}^{-1}}_{l_k} \right\},$$

$$Z_a = [\sqrt{\pi_{a1}}I \; \sqrt{\pi_{a2}}I \; \cdots \; \sqrt{\pi_{aN}}I],$$

$$\mathscr{P} = \text{diag}\{P_1, P_2, \ldots, P_N\}.$$

For each $a, b \in \mathcal{M}$ and $c \in \mathcal{N}$, we denote

$$\begin{aligned} P_a &= X^{-\top}\widehat{P}_a X^{-1}, \; W_{ac} = X^{-\top}\widehat{W}_{ac}X^{-1}, \\ Q &= X^{-\top}\widehat{Q}X^{-1}, \; \Omega_{la} = X^{-\top}\widehat{\Omega}_{la}X^{-1}, l = 1, 2. \end{aligned} \tag{40}$$

Then, one has

$$-P_a^{-1} = \widehat{P}_a - \text{He}(X). \tag{41}$$

According to (40) and (41), (39) can be rewritten as

$$\begin{bmatrix} \widehat{\Xi}_a & \overline{\Psi}_{1a} & \Psi_{2a} \\ * & \widehat{\Psi}_{3a} & 0 \\ * & * & \Psi_4 \end{bmatrix} < 0, \tag{42}$$

where

$$\widehat{\Psi}_{3a} = \text{diag}\left\{ \widehat{\mathscr{P}}, \widehat{\mathscr{P}}, \widehat{\mathscr{P}}, \underbrace{\widehat{\mathscr{P}}, \ldots, \widehat{\mathscr{P}}}_{l_k} \right\},$$

$$\widehat{\mathscr{P}} = \text{diag}\{\widehat{P}_1 - \text{He}(X), \widehat{P}_2 - \text{He}(X), \ldots, \widehat{P}_N - \text{He}(X)\}.$$

Pre- and post-multiple (42) by $\{X, I, X, \underbrace{X, \ldots, X}_{l_k}, I, \ldots, I\}$, then, (42) can be rewritten as

$$\begin{bmatrix} \widehat{\Xi}_a & \grave{\Psi}_{1a} & \grave{\Psi}_{2a} \\ * & \widehat{\Psi}_{3a} & 0 \\ * & * & \Psi_4 \end{bmatrix} < 0, \tag{43}$$

where

$$\begin{aligned} \widehat{\Xi}_a =& l_k \chi_3^\top \widehat{Q}\chi_3 - \chi_4^\top \widehat{\widehat{Q}}\chi_4 - \chi_3^\top \widehat{\Omega}_{1a}\chi_3 - \chi_1^\top \widehat{W}_{ac}\chi_1 \\ &+ (\chi_1 - \chi_3)^\top \widehat{\Omega}_{2a}(\chi_1 - \chi_3) - \gamma^2 \chi_2^\top \chi_2, \end{aligned}$$

$$\grave{\Psi}_{1a} = \left[\grave{\mathscr{F}}_1^\top Z_a \ \sqrt{1+\overline{\alpha}_0^2+\sum_{t=1}^{l_k}\overline{\alpha}_t^2}\grave{\mathscr{F}}_2^\top Z_a \ (\grave{\mathscr{F}}_3\hat{D})^\top Z_a \ \grave{\mathscr{F}}_7^\top Z_a \right],$$

$$\grave{\Psi}_{2a} = \left[\grave{\mathscr{F}}_4^\top \ \sqrt{1+\overline{\alpha}_0^2+\sum_{t=1}^{l_k}\overline{\alpha}_t^2}\grave{\mathscr{F}}_5^\top \ (\grave{\mathscr{F}}_6\hat{D})^\top \ \grave{\mathscr{F}}_8^\top \right],$$

$$\begin{aligned}
\grave{\mathscr{F}}_1 =& [A_a X + B_{1a}(I+\Delta(k))K_c X(\alpha_0+\overline{D}_{l_k}) \ D_{1a} \\
& - B_{1a}(I+\Delta(k))K_c X(\alpha_0+\overline{D}_{l_k}) \ B_{1a}(I+\Delta(k))K_c X \\
& B_{1a}(I+\Delta(k))K_c X \ \cdots \ B_{1a}(I+\Delta(k))K_c X], \\
\grave{\mathscr{F}}_2 =& [B_{1a}(I+\Delta(k))K_c X \ 0 \ B_{1a}(I+\Delta(k))K_c X \ 0], \\
\grave{\mathscr{F}}_3 =& [0\ 0\ 0\ B_{1a}(I+\Delta(k))K_c X, B_{1a}(I+\Delta(k))K_c X \ \cdots \\
& B_{1a}(I+\Delta(k))K_c X], \\
\grave{\mathscr{F}}_4 =& [C_a X + B_{2a}(I+\Delta(k))K_c X(\alpha_0+\overline{D}_{l_k}) \ D_{2a} \\
& - B_{2a}(I+\Delta(k))K_c X(\alpha_0+\overline{D}_{l_k}) \ B_{2a}(I+\Delta(k))K_c X \\
& B_{2a}(I+\Delta(k))K_c X \ \cdots \ B_{2a}(I+\Delta(k))K_c X], \\
\grave{\mathscr{F}}_5 =& [B_{2a}(I+\Delta(k))K_c X \ 0 \ B_{2a}(I+\Delta(k))K_c X \ 0], \\
\grave{\mathscr{F}}_6 =& [0\ 0\ 0\ B_{2a}(I+\Delta(k))K_c X, B_{2a}(I+\Delta(k))K_c X \ \cdots \\
& B_{2a}(I+\Delta(k))K_c X].
\end{aligned}$$

Adding (13) to (43), one has

$$\begin{bmatrix} \widehat{\Xi}_a & \widehat{\Psi}_{1a} & \widehat{\Psi}_{2a} \\ * & \widehat{\Psi}_{3a} & 0 \\ * & * & \Psi_4 \end{bmatrix} + \sum_{s=1}^{3} \mathrm{He}(\Lambda_{sc}\Delta(k)\Theta_{sa}) < 0, \tag{44}$$

Applying Schur complement to (44), one can obtain (37). □

4 Numerical Example

A single link robot arm model [39], which is described by the differential equation as below

$$\begin{aligned}
\dot{x}_1(t) =& x_2(t) \\
\dot{x}_2(t) =& -\frac{glM(\varphi(t))}{J(\varphi(t))}\sin(x_1(t)) - \frac{R}{J(\varphi(t))} - \frac{R}{J(\varphi(t))}x_2(t) \\
& + \frac{1}{J(\varphi(t))}u(t) + \omega(t)
\end{aligned} \tag{45}$$

Table 1 The meaning of g, R, l, $M(\varphi(t))$ and $J(\varphi(t))$

	Description
g	The gravity acceleration
R	The coefficient of the viscous friction
l	The arm length
$M(\varphi(t))$	The payload mass
$J(\varphi(t))$	The inertia moment

Table 2 The values of $M(\varphi(t))$ and $J(\varphi(t))$

$\varphi(t)$	1	2	3
$M(\varphi(t))$	1	5	10
$J(\varphi(t))$	1	5	10

where the description of g, R, l, $M(\varphi(t))$ and $J(\varphi(t))$ are given in Table 1.

Accordingly, the values of $M(\varphi(t))$ and $J(\varphi(t))$ are given in Table 2.

The TPM is assumed to be

$$\Pi = \begin{bmatrix} 0.3 & 0.4 & 0.3 \\ 0.4 & 0.2 & 0.4 \\ 0.5 & 0.3 & 0.2 \end{bmatrix}.$$

Applying the first-order Euler approximation method to discretize the MSS (1) with $T = 0.1$. Then, the parameters of MSS (1) is given as follows

$$\begin{aligned} A_{\varphi(k)} &= \begin{bmatrix} 1 & T \\ -\frac{TglM(\varphi(k))}{J(\varphi(k))} & 1 - \frac{TR}{J(\varphi(k))} \end{bmatrix}, \\ B_{1\varphi(k)} &= \begin{bmatrix} 0 \\ \frac{T}{J(\varphi(k))} \end{bmatrix}, \; D_{1\varphi(k)} = \begin{bmatrix} 0 \\ T \end{bmatrix}, \\ C_{\varphi(k)} &= [1 \;\; 0], \; B_{2\varphi(k)} = 0, \; D_{2\varphi(k)} = 0.1. \end{aligned}$$

On the other hand, the conditional probability matrix of Makov jump $\phi(k)$, is shown as

$$\Pi = \begin{bmatrix} 0.2 & 0.55 & 0.25 \\ 0.3 & 0.6 & 0.1 \\ 0.4 & 0.2 & 0.4 \end{bmatrix}.$$

Additionally, one gives the fading channel $l_k = 1$, and the probability density functions of $\alpha_\iota(k)$ with the following forms:

$$\begin{cases} f(\alpha_0(k)) = 0.0005(\exp(9.89\alpha_0(k)) - 1), \; 0 \le \alpha_0(k) \le 1, \\ f(\alpha_1(k)) = 8.5017(\exp(8.5\alpha_1(k)) - 1), \;\; 0 \le \alpha_0(k) \le 1. \end{cases}$$

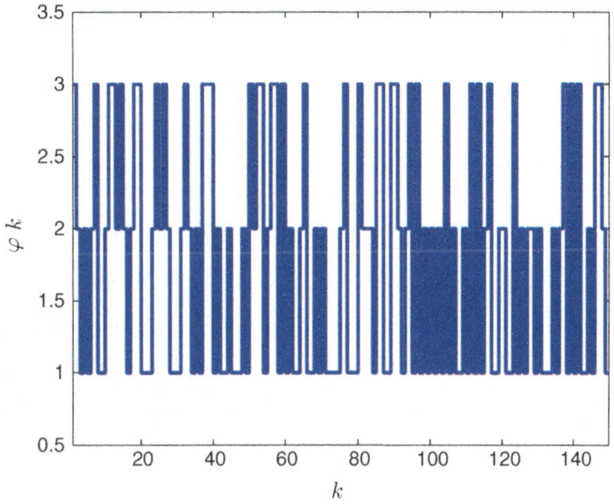

Fig. 1 System mode $\varphi(k)$

Thus, it is easy to obtain that $\alpha_0 = 0.8989$, $\alpha_1 = 0.1174$, $\overline{\alpha}_0^2 = 0.0105$ and $\overline{\alpha}_0^2 = 0.5870$.

For given initial condition $x(0) = [0.2\pi \quad -0.2]^\top$, and the external disturbance

$$\omega(k) = \begin{cases} \exp(-0.1(k-15))\sin(0.3(k-15)), & 1 \leq k \leq 15, \\ 0, & 16 \leq k \leq 80. \end{cases}$$

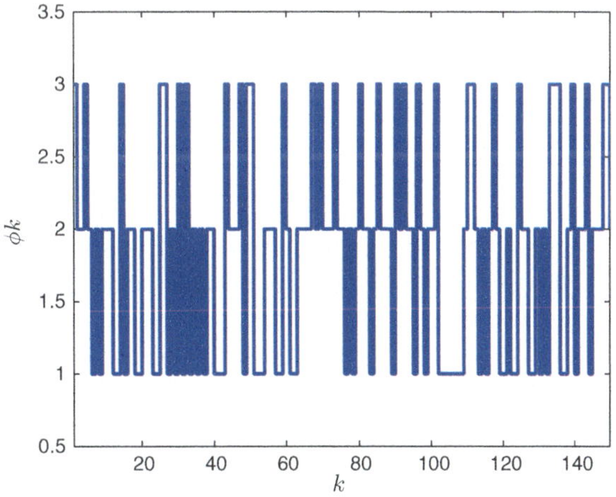

Fig. 2 Control mode $\phi(k)$

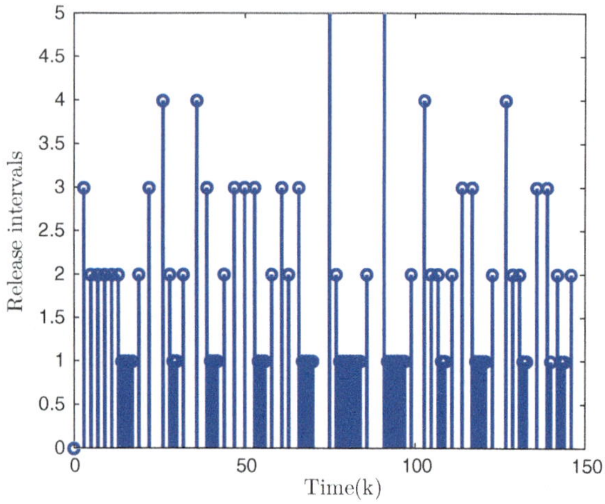

Fig. 3 Triggered instants and release intervals

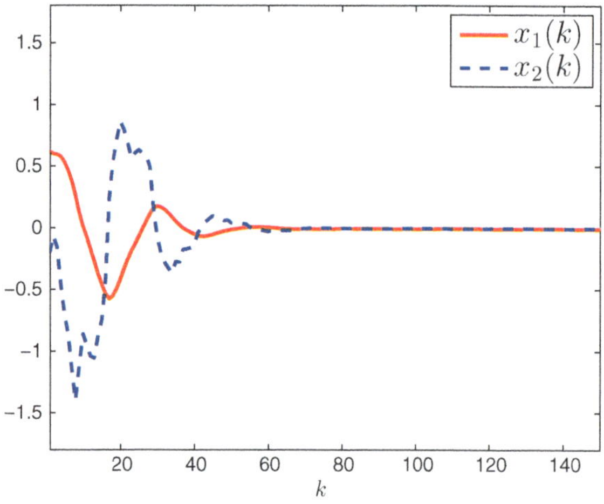

Fig. 4 State response

According to Theorem 2, the network is SS under the designed control strategy. The modes evolution of the system and the controller are given in Figs. 1 and 2. By using ETM, the triggered instants and release intervals are presented in Fig. 3. With the designed controller gain, the state response and control input are expressed in Figs. 4 and 5, respectively. As shown in Figs. 4 and 5, the simulation result shows the feasibility of the theory.

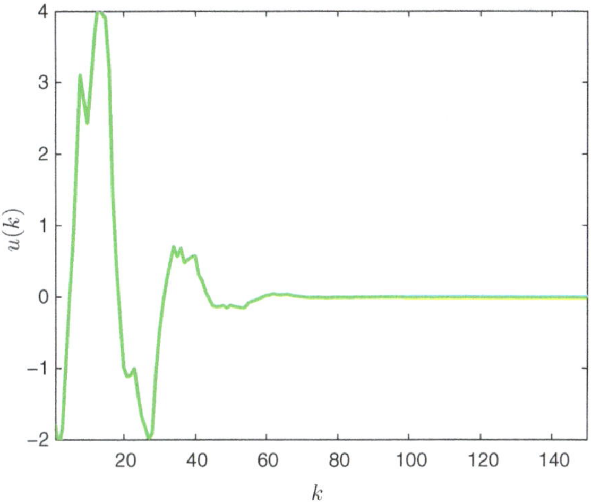

Fig. 5 Control input

5 Conclusion

In this chapter, by utilizing an HMM technique, the asynchronous control for Markov switching systems with channel fading, where a more general class of discrete-time MSS is developed. The original systems and controller are characterized by two stochastic variables. Following this, the relationship between stochastic variables is considered. In order to reduce energy consumption, a mode-dependent event-triggered communication mechanism is employed. A simulation example is employed to demonstrate the correctness and effectiveness of the derived results.

Acknowledgements This work of J. Cheng was supported by the National Natural Science Foundation of China (No. 61703150), and the Training Program for 1,000 Young and Middle-aged Cadre Teachers in Universities of Guangxi Province. Also, the work of J. H. Park was supported by the National Research Foundation of Korea (NRF) grant funded by the Korea government (Ministry of Science and ICT) (No. 2019R1A5A8080290l1).

References

1. Park, J.H., Shen, H., Chang, X.H., Lee, T.H.: Recent Advances in Control and Filtering of Dynamic Systems with Constrained Signals. Springer, Cham (2018). https://doi.org/10.1007/978-3-319-96202-3
2. Shen, H., Park, J.H., Zhang, L., Wu, Z.G.: Robust extended dissipative control for sampled-data Markov jump systems. Int. J. Control **87**(8), 1549–1564 (2014)
3. Wei, Y., Park, J.H., Qiu, J., Wu, L., Jung, H.Y.: Sliding mode control for semi-Markovian jump systems via output feedback. Automatica **81**, 133–141 (2017)
4. Mathiyalagan, K., Park, J.H., Sakthivel, R., Anthoni, S.M.: Robust mixed $\mathcal{H}_\infty$ and passive filtering for networked Markov jump systems with impulses. Signal Process. **101**, 162–173 (2014)

5. Hou, T., Ma, H.: Exponential stability for discrete-time infinite Markov jump systems. IEEE Trans. Autom. Control **61**(12), 4241–4246 (2016)
6. Zhang, L., Boukas, E.: Stability and stabilization of Markovian jump linear systems with partly unknown transition probabilities. Automatica **45**(2), 463–468 (2019)
7. Zhang, L., Leng, Y., Colaneri, P.: Stability and stabilization of discrete-time semi-Markov jump linear systems via semi-Markov kernel approach. IEEE Trans. Autom. Control **61**(2), 503–508 (2016)
8. Zhang, L., Ning, Z., Shi, P.: Input-output approach to control for fuzzy Markov jump systems with time-varying delays and uncertain packet dropout rate. IEEE Trans. Cybern. **45**(11), 2449–2460 (2015)
9. Dong, S., Fang, M., Shi, P., Wu, Z., Zhang, D.: Dissipativity-based control for suzzy systems with asynchronous modes and intermittent measurements. IEEE Trans. Cybern. (2019). https://doi.org/10.1109/TCYB.2018.2887060
10. Xu, S., Lam, J., Mao, X.: Delay-dependent $\mathcal{H}_\infty$ control and filtering for uncertain markovian jump systems with time-varying delays. IEEE Trans. Circuits Syst. Regul. Pap. **54**(9), 2070–2077 (2007)
11. Shen, H., Li, F., Wu, Z., Park, J.H., Sreeram, V.: Fuzzy-model-based nonfragile control for nonlinear singularly perturbed systems with semi-Markov jump parameters. IEEE Trans. Fuzzy Syst. **26**(6), 3428–3439 (2018)
12. Dai, M., Xia, J., Park, J.H., et al.: Asynchronous dissipative filtering for Markov jump discrete-time systems subject to randomly occurring distributed delays. J. Frankl. Inst. **356**(4), 2395–2420 (2019)
13. Wu, Z., Shi, P., Su, H., Chu, J.: Asynchronous $l_2 - l_\infty$ filtering for discrete-time stochastic Markov jump systems with randomly occurred sensor nonlinearities. Automatica **50**(1), 180–186 (2014)
14. Goncalves, A., Fioravanti, A., Geromel, J.: $\mathcal{H}_\infty$ filtering of discrete-time Markov jump linear systems through linear matrix inequalities. IEEE Trans. Autom. Control **54**(6), 1347–1351 (2008)
15. Cheng, J., Park, J.H., Zhao, X., Karimi, H., Cao, J.: Quantized nonstationary filtering of network-based Markov switching RSNSs: a multiple hierarchical structure strategy. IEEE Trans. Autom. Control (2019). https://doi.org/10.1109/TAC.2019.2958824
16. Feng, Z., Shi, P.: Sliding mode control of singular stochastic Markov jump systems. IEEE Trans. Autom. Control **62**(8), 4266–4273 (2017)
17. Yang, H., Yin, S.: Reduced order sliding mode observer-based fault estimation for Markov jump systems. IEEE Trans. Autom. Control (2019). https://doi.org/10.1109/TAC.2019.2904435
18. Yan, H., Zhang, H., Yang, F., et al.: Event-triggered asynchronous guaranteed cost control for Markov jump discrete-time neural networks with distributed delay and channel fading. IEEE Trans. Neural Netw. Learn. Syst. **29**(8), 3588–3598 (2018)
19. Xiao, N., Xie, L., Qiu, L.: Feedback stabilization of discrete-time networked systems over fading channels. IEEE Trans. Autom. Control **57**(9), 2176–2189 (2012)
20. Shan, Y., She, K., Zhong, S., et al.: Event-triggered passive control for Markovian jump discrete-time systems with incomplete transition probability and unreliable channels. J. Frankl. Inst. **356**(15), 8093–8117
21. Quevedo, D., Ahlen, A., Johansson, K.: State estimation over sensor networks with correlated wireless fading channels. IEEE Trans. Autom. Control **58**(3), 581–593 (2013)
22. Dong, H., Wang, Z., Ding, S., Gao, H.: On $\mathcal{H}_\infty$ estimation of randomly occurring faults for a class of nonlinear time-varying systems with fading channels. IEEE Trans. Autom. Control **61**(2), 479–484 (2016)
23. Xu, Y., Lu, R., Shi, P., et al.: Finite-time distributed state estimation over sensor networks with round-robin protocol and fading channels. IEEE Trans. Cybern. **48**(1), 336–345
24. Quevedo, D., Ahlen, A., Leong, A., Dey, A.: On Kalman filtering over fading wireless channels with controlled transmission powers. Automatica **48**(7), 1306–1316 (2012)
25. Ding, D., Wang, Z., Shen, B., Dong, H.: Envelope-constrained $\mathcal{H}_\infty$ filtering with fading measurements and randomly occurring nonlinearities: the finite horizon case. Automatica **55**, 37–45 (2015)

26. Xu, W., Ho, D., Li, L., Cao, J.: Event-triggered schemes on leader-following consensus of general linear multiagent systems under different topologies. IEEE Trans. Cybern. **47**(1), 212–223 (2017)
27. Zhang, X., Han, Q.: Event-based $\mathcal{H}_\infty$ filtering for sampled-data systems. Automatica **51**, 55–69 (2015)
28. Zhang, H., Feng, G., Yan, H., Chen, Q.: Observer-based output feedback event-triggered control for consensus of multi-agent systems. IEEE Trans. Ind. Electr. **61**(9), 4885–4894 (2014)
29. Peng, C., Zhang, J.: Event-triggered output-feedback $\mathcal{H}_\infty$ control for networked control systems with time-varying sampling. IET Control Theory Appl. **9**(9), 1384–1391 (2015)
30. Zhang, L., Liang, H., Sun, Y., Ahn, C.: Adaptive event-triggered fault detection for semi-Markovian jump systems with output quantization. IEEE Trans. Syst. Man Cybern.: Syst. (2019). https://doi.org/10.1109/TSMC.2019.2912846
31. Li, H., Zuo, Z., Wang, Y.: Event triggered control for Markovian jump systems with partially unknown transition probabilities and actuator saturation. Neurocomputing **353**(8), 1848–1861 (2016)
32. Zha, L., Fang, J., Li, X., Liu, J.: Event-triggered output feedback $\mathcal{H}_\infty$ control for networked Markovian jump systems with quantizations. Nonlinear Anal. Hybrid Syst. **24**, 146–158 (2017)
33. Cheng, J., Ahn, C., Karimi, H., Cao, J., Qi, W.: An event-based asynchronous approach to Markov jump systems with hidden mode detections and missing measurements. IEEE Trans Syst. Man Cybern. Syst. (2018). https://doi.org/10.1109/TSMC.2018.2866906
34. Xia, J., Park, J.H., Zhang, B., Shen, H.: Robust $\mathcal{H}_\infty$ tracking control for uncertain Markovian jumping systems with interval time-varying delay. Complexity **21**(2), 355–366 (2014)
35. Zhuang, G., Xia, J., Zhao, J., Zhang, H.: Nonfragile $\mathcal{H}_\infty$ output tracking control for uncertain singular Markovian jump delay systems with network-induced delays and data packet dropouts. Complexity **21**(6), 396–411 (2015)
36. Dong, S., Wu, Z., Shi, P., et al.: Networked fault detection for Markov jump nonlinear systems. IEEE Trans. Fuzzy Syst. **26**(6), 3368–3378 (2018)
37. Shen, Y., Wu, Z., Shi, P., Shu, et al.: $\mathcal{H}_\infty$ control of Markov jump time-delay systems under asynchronous controller and quantizer. Automatica **99**, 352–360 (2019)
38. Wu, Z., Dong, S., Shi, P., Zhang, D., Huang, T.: Reliable filter design of Takagi-Sugeno fuzzy switched systems with imprecise modes. IEEE Trans. Cybern. (2018). https://doi.org/10.1109/TCYB.2018.2885505
39. Wu, Z., Dong, S., Su, H., Li, C.: Asynchronous dissipative control for fuzzy Markov jump systems. IEEE Trans. Cybern. **48**(8), 2426–2436 (2018)
40. Dong, S., Wu, Z., Pan, Y., et al.: Hidden-Markovmodel- based asynchronous filter design of nonlinear Markov jump systems in continuous-time domain. IEEE Trans. Cybern. (2018). https://doi.org/10.1109/TCYB.2018.2824799

Fractional-Order Systems

Observer-Based Controller Design for Fractional-Order Neutral-Type Systems

Yongxia Qu, Youggui Kao, and Cunchen Gao

Abstract This article is devoted to investigating how to design the observer-based controller for a type of fractional-order neutral-type systems (FONTSs) by applying the Lyapunov direct method. The feedback control system based on the observer is designed, and by solving the linear matrix inequality (LMI) feasible problems, the gains of the controller and the observer are given. Then, an observer-based guaranteed cost control of fractional-order systems is studied. At last, simulation examples are provided to demonstrate that our method is effective.

Keywords Fractional-order systems · Neutral-type · Time-delay · Observer-based control · LMI approach · Guaranteed cost control

1 Introduction

With the continuous improvement of modern industry and advanced technology, it is necessary to analyze and synthesize the related mathematical model, so as to achieve the higher performances of the systems to a greater extent. Recently, scholars have realized most of the systems are non-integer differential systems [1], especially for

Y. Qu · C. Gao
School of Mathematical Science, Ocean University of China,
Qingdao 266100, People's Republic of China
e-mail: 664704120@qq.com

C. Gao
e-mail: ccgao@ouc.edu.cn

Y. Kao (✉)
Department of Mathematics, Harbin Institute of Technology (Weihai),
Weihai 264209, People's Republic of China
e-mail: ygkao2008@gmail.com

J. H. Park (ed.), *Recent Advances in Control Problems of Dynamical Systems and Networks*, Studies in Systems, Decision and Control 301,
https://doi.org/10.1007/978-3-030-49123-9_12

dynamic systems with non-rigid motion, such as thermal systems, flexible systems, and viscous systems [2–5]. Fractional-order differential equations (FODEs) are more accurate and appropriate to describe them. The widespread applications of FODEs in describing the nature of things have become the hot topic [6–16]. Stability is the basic character of all control systems, and, of course, includes fractional-order control systems. Li et al. [17, 18] discussed the generalized Mittag-Leffler (M-L) stability (asymptotic stability) by the characteristics of fractional order differential operators. Further, the literature [19] studied the M-L stability of FODEs with impulses. The authors of [20] gave the Lyapunov method to study the FODEs with uncertainties and time-delay. And there are more and more researches about the fractional-order discrete-time systems, such as the literature [21].

In practice, the time delay phenomenon [22] often occurs in the dynamical systems. It restricts the structure, nature, and performance of the systems to some extent. The neutral system [20, 23, 24] is a kind of important time-delay system, which exists in many engineering practices such as ship stability, population ecology and protein distribution in blood. Naturally, as a dynamical system, the first control task for the neutral systems is to ensure the stability of the controlled system. However, until now, there have been few research results on the fractional-order neutral systems, and their control problems are of theoretical importance and practical significance [25–29]. Liu et al. [25] studied the Riemann–Liouville FONTSs, and discussed the asymptotical stability of it by using a new method to calculate the Lyapunov functions. In the literature [26], the Riemann–Liouville fractional-order neutral delay neural network is studied and its sufficient conditions for delay-independent stability are given. Also, the literatures [27–29] studied the stability of some kinds of FONTSs.

In many practical systems, the feedback states are usually not directly available, so it is necessary to create a controller based on the observer for such a system. There are many reports about the design approach of the controller based on state observers. In [30], a neutral linear delay differential system is studied, and a method of constructing an observer-based controller is presented. Further, there are many other researches on the observer design, for instance, the guaranteed cost control based on the observer and the $\mathcal{H}_\infty$ control based on the observer. In [31], the author considered a new control problem for uncertain neutral time-delay systems (UNTDSs), and considered how to design a guaranteed cost controller based on the observer. Then, in [32], the authors investigated how to create a $\mathcal{H}_\infty$ controller based on the observer for UNTDSs. In [33], the authors focused on a class of stochastic systems, which is more complicated. They gave a sliding mode method to study how to design an observer-based $\mathcal{H}_\infty$ controller, which is non-fragile at the same time. Then, in the literature [34], the author gave an integral sliding mode method to study the design of the $\mathcal{H}_\infty$ controllers. However, to the best knowledge of ours, the similar reports on FONTSs are few.

Base on the aforementioned concerns, how to design the observer-based controller for FONTSs by the knowledge of Lyapunov stability and fractional calculus is discussed in this chapter. Section 2 presents problem preliminaries and formulation. Then, in Sect. 3, an observer-based fractional order controller for FONTSs is designed to guarantee the closed-loop system asymptotically stable. By using the

LMI technology, the gains of the observer and the controller are calculated respectively. And in this section, an observer-based guaranteed cost controller is designed for the fractional-order systems. In Sect. 4, the effectiveness of our findings is verified by numerical examples. At last, the full chapter is summarized in Sect. 5.

Notations: In this chapter, matrix $\boldsymbol{M} > \boldsymbol{N}$ means that $\boldsymbol{M} - \boldsymbol{N}$ is a positive definite matrix. A unit matrix with appropriate dimension is represented as $\boldsymbol{I}$. $*$ represents the symmetric term in a matrix.

2 Problem Formulation and Preliminaries

In this section, we present some basic definitions, lemmas, and basic calculation rules of fractional-order systems, such as Γ functions, Riemann–Liouville fractional derivative and integral, and some properties of them.

Definition 1 ([3]) The Riemann–Liouville fractional-order integral with $a > 0$ for a function $g(t)$ is defined as

$$ {}_0\mathrm{D}_t^{-a} g(t) = \frac{1}{\Gamma(a)} \int_0^t (t-s)^{a-1} g(s)\mathrm{d}s, $$

where $\Gamma(\cdot)$ is the Γ function, $\Gamma(s) = \int_0^{+\infty} e^{-t} t^{s-1} \mathrm{d}t, s > 0$.

Definition 2 ([3]) The Riemann–Liouville fractional-order derivative with $a > 0$ for a function $g(t)$ is defined as

$$ {}_0\mathrm{D}_t^{a} g(t) = \frac{1}{\Gamma(n-a)} \frac{\mathrm{d}^n}{\mathrm{d}t^n} \int_0^t \frac{g(s)}{(t-s)^{a+1-n}} \mathrm{d}s, n-1 \leq a < n, t \geq 0, $$

n is a positive integer. And when $0 < a < 1$, the upper form changed to

$$ {}_0\mathrm{D}_t^{a} g(t) = \frac{1}{\Gamma(1-a)} \frac{\mathrm{d}}{\mathrm{d}t} \int_0^t \frac{g(s)}{(t-s)^{a}} \mathrm{d}s, t \geq 0. $$

Property 1 ${}_0\mathrm{D}_t^{a} C = \frac{C(t-t_0)^{-a}}{\Gamma(1-a)}$ *holds, where* C *is a constant.*

Property 2 *When* $n-1 \leq a < n, b > 0$, *first calculate the integrals and then derivative, and the following equality*

$$ {}_0\mathrm{D}_t^{a}\,{}_0\mathrm{D}_t^{-b} g(t) = {}_0\mathrm{D}_t^{a-b} g(t) $$

holds for function $g(t)$ *with good property. In particular, if* $g(t)$ *is integrable, the upper relation still holds.*

Now, we consider the following FONTS:

$$\begin{cases} {}_0\mathrm{D}_t^{\alpha}\boldsymbol{x}(t) = \boldsymbol{A}_0\boldsymbol{x}(t) + \boldsymbol{A}_1\boldsymbol{x}(t-\omega) + \mathbf{B}_0\mathrm{D}_t^{\alpha}\boldsymbol{x}(t-\omega) + \boldsymbol{G}\boldsymbol{u}(t), & t \geq 0 \\ \boldsymbol{x}(t) = \boldsymbol{\varphi}(t), \quad t \in [-\omega, 0], \\ \boldsymbol{y}(t) = \boldsymbol{C}\boldsymbol{x}(t) + \boldsymbol{D}\boldsymbol{u}(t), \end{cases} \tag{1}$$

where $0 < \alpha < 1, \boldsymbol{x}(t) \in \mathbf{R}^n$ is the state vector, $\boldsymbol{y}(t) \in \mathbf{R}^q$ is the output vector, $\boldsymbol{u}(t) \in \mathbf{R}^m$ is the control input vector, $\boldsymbol{\varphi}(t) \in \mathbf{C}([-\omega, 0], \mathbf{R}^n)$ is the initial function, and $\boldsymbol{A}_0, \boldsymbol{A}_1, \mathbf{B} \in \mathbf{R}^{n\times n}, \boldsymbol{C} \in \mathbf{R}^{q\times n}, \boldsymbol{G} \in \mathbf{R}^{n\times m}, \boldsymbol{D} \in \mathbf{R}^{q\times m}$ are known constant real-valued matrices.

Lemma 1 (Schur Complement lemma) ([35]) *For a given symmetric matrix*

$$\boldsymbol{M} = \boldsymbol{M}^{\mathrm{T}} = \begin{bmatrix} \boldsymbol{M}_{11} & \boldsymbol{M}_{12} \\ \boldsymbol{M}_{12}^{\mathrm{T}} & \boldsymbol{M}_{22} \end{bmatrix} \tag{2}$$

with $\boldsymbol{M}_{11} \in \mathbf{R}^{r\times r}$, the following conditions are equivalent:

(1) $\boldsymbol{M} < 0$,
(2) $\boldsymbol{M}_{11} < 0, \quad \boldsymbol{M}_{22} - \boldsymbol{M}_{12}^{\mathrm{T}}\boldsymbol{M}_{11}^{-1}\boldsymbol{M}_{12} < 0$,
(3) $\boldsymbol{M}_{22} < 0, \quad \boldsymbol{M}_{11} - \boldsymbol{M}_{12}\boldsymbol{M}_{22}^{-1}\boldsymbol{M}_{12}^{\mathrm{T}} < 0$.

The observer-based control of system (1) is designed by

$$\begin{cases} {}_0\mathrm{D}_t^{\alpha}\hat{\boldsymbol{x}}(t) = \boldsymbol{A}_0\hat{\boldsymbol{x}}(t) + \boldsymbol{A}_1\hat{\boldsymbol{x}}(t-\omega) + \mathbf{B}_0\mathrm{D}_t^{\alpha}\hat{\boldsymbol{x}}(t-\omega) \\ \qquad + \boldsymbol{G}\boldsymbol{u}(t) + \boldsymbol{L}[\boldsymbol{y}(t) - \hat{\boldsymbol{y}}(t)], \quad t \geq 0, \\ \hat{\boldsymbol{x}}(t) = \hat{\boldsymbol{\varphi}}(t), \quad t \in [-\omega, 0], \\ \hat{\boldsymbol{y}}(t) = \boldsymbol{C}\hat{\boldsymbol{x}}(t) + \boldsymbol{D}\boldsymbol{u}(t), \end{cases} \tag{3}$$

and

$$\boldsymbol{u}(t) = -\boldsymbol{K}\hat{\boldsymbol{x}}(t) \tag{4}$$

where $\hat{\boldsymbol{x}}(t) \in \mathbf{R}^n$ is the estimation of the state vector $\boldsymbol{x}(t)$, $\hat{\boldsymbol{y}}(t) \in \mathbf{R}^q$ is the observer output, $\boldsymbol{L} \in \mathbf{R}^{n\times q}$ is the gain of the observer, $\boldsymbol{K} \in \mathbf{R}^{m\times n}$ is the gain of the controller, and $\hat{\boldsymbol{\varphi}}(t) \in \mathbf{C}([-\omega, 0], \mathbf{R}^n)$ is the initial function of $\hat{\boldsymbol{x}}(t)$.

Here, the vector $\boldsymbol{e}(t) = \boldsymbol{x}(t) - \hat{\boldsymbol{x}}(t)$ is the error of the estimated system.

Lemma 2 ([3]) *Let $\boldsymbol{x}(t) \in \mathbf{R}^n$ be a derivable and continuous function, then, the following inequalities holds*

$$\frac{1}{2}{}_0\mathrm{D}_t{}^{\alpha}(\boldsymbol{x}^T(t)\boldsymbol{P}\boldsymbol{x}(t)) \leq \boldsymbol{x}^T(t)\boldsymbol{P}_0\mathrm{D}_t{}^{\alpha}\boldsymbol{x}(t) \quad \forall\alpha \in (0, 1), \forall t \geq 0$$

where $\boldsymbol{P} \in \mathbf{R}^{n\times n}$ is a symmetric positive definite matrix.

By (2)–(3), (1) and (2) can be sorted and written as

$$\begin{cases} {}_0\mathrm{D}_t^{\alpha}\hat{\boldsymbol{x}}(t) = (\boldsymbol{A}_0 - \boldsymbol{GK})\hat{\boldsymbol{x}}(t) + \boldsymbol{A}_1\hat{\boldsymbol{x}}(t-\omega) + \mathbf{B}_0{}_0\mathrm{D}_t^{\alpha}\hat{\boldsymbol{x}}(t-\omega) + \boldsymbol{LCe}(t), \quad t \geq 0 \\ {}_0\mathrm{D}_t^{\alpha}\boldsymbol{e}(t) = (\boldsymbol{A}_0 - \boldsymbol{LC})\boldsymbol{e}(t) + \boldsymbol{A}_1\boldsymbol{e}(t-\omega) + \mathbf{B}_0{}_0\mathrm{D}_t^{\alpha}\boldsymbol{e}(t-\omega), \quad t \geq 0. \end{cases} \tag{5}$$

Then, write a matrix as

$$\begin{aligned} {}_0\mathrm{D}_t^{\alpha}\boldsymbol{Z}(t) = {}_0\mathrm{D}_t^{\alpha}\begin{bmatrix} \hat{\boldsymbol{x}}(t) \\ \boldsymbol{e}(t) \end{bmatrix} &= \begin{bmatrix} \boldsymbol{A}_0 - \boldsymbol{GK} & \boldsymbol{LC} \\ \boldsymbol{0} & \boldsymbol{A}_0 - \boldsymbol{LC} \end{bmatrix}\begin{bmatrix} \hat{\boldsymbol{x}}(t) \\ \boldsymbol{e}(t) \end{bmatrix} \\ &+ \begin{bmatrix} \boldsymbol{A}_1 & \boldsymbol{0} \\ \boldsymbol{0} & \boldsymbol{A}_1 \end{bmatrix}\begin{bmatrix} \hat{\boldsymbol{x}}(t-\omega) \\ \boldsymbol{e}(t-\omega) \end{bmatrix} + \begin{bmatrix} \boldsymbol{B}_0 & \boldsymbol{0} \\ \boldsymbol{0} & \boldsymbol{B}_0 \end{bmatrix}\begin{bmatrix} {}_0\mathrm{D}_t^{\alpha}\hat{\boldsymbol{x}}(t-\omega) \\ {}_0\mathrm{D}_t^{\alpha}\boldsymbol{e}(t-\omega) \end{bmatrix}. \end{aligned} \tag{6}$$

The problem considered here is to find the controller based on the observer (2)–(3) to ensure the system (4) above asymptotically stable, then calculate the gains of the controller and the observer.

3 Main Results

In this section, we study the design of the controller based on the observer for FONTSs in this section. First, we give the main theorem as follows:

Theorem 1 *The system (1) with (2)–(3) is asymptotically stable if there exist some symmetric positive definite matrices* $\bar{\boldsymbol{H}}_i \in \mathbf{R}^{n\times n}$, $\bar{\boldsymbol{J}}_i \in \mathbf{R}^{n\times n}$, $(i = 1, 2, 3)$ *and matrices* $\boldsymbol{X} \in \mathbf{R}^{n\times q}$, $\boldsymbol{Y} \in \mathbf{R}^{m\times n}$, $\hat{\boldsymbol{R}}_1 \in \mathbf{R}^{q\times q}$, *such that*

$$\begin{bmatrix} \Omega_{11} & \Omega_{12} & \Omega_{13} & \mathbf{0} & \Omega_{15} & \mathbf{0} & \Omega_{17} & \mathbf{0} & \bar{\boldsymbol{H}}_1 & \mathbf{0} \\ * & \Omega_{22} & \mathbf{0} & \Omega_{24} & \mathbf{0} & \Omega_{26} & \Omega_{27} & \Omega_{28} & \mathbf{0} & \bar{\boldsymbol{J}}_1 \\ * & * & -\bar{\boldsymbol{H}}_2 & \mathbf{0} & \mathbf{0} & \mathbf{0} & \Omega_{37} & \mathbf{0} & \mathbf{0} & \mathbf{0} \\ * & * & * & -\bar{\boldsymbol{J}}_2 & \mathbf{0} & \mathbf{0} & \mathbf{0} & \Omega_{48} & \mathbf{0} & \mathbf{0} \\ * & * & * & * & -\bar{\boldsymbol{H}}_3 & \mathbf{0} & \Omega_{57} & \mathbf{0} & \mathbf{0} & \mathbf{0} \\ * & * & * & * & * & -\bar{\boldsymbol{J}}_3 & \mathbf{0} & \Omega_{68} & \mathbf{0} & \mathbf{0} \\ * & * & * & * & * & * & -\bar{\boldsymbol{H}}_3 & \mathbf{0} & \mathbf{0} & \mathbf{0} \\ * & * & * & * & * & * & * & -\bar{\boldsymbol{J}}_3 & \mathbf{0} & \mathbf{0} \\ * & * & * & * & * & * & * & * & -\bar{\boldsymbol{H}}_2 & \mathbf{0} \\ * & * & * & * & * & * & * & * & * & -\bar{\boldsymbol{J}}_2 \end{bmatrix} < 0, \tag{7}$$

$$\boldsymbol{C}\bar{\boldsymbol{J}}_1 = \hat{\boldsymbol{R}}_1\boldsymbol{C}, \tag{8}$$

where

$$\begin{aligned} &\Omega_{11} = \bar{\boldsymbol{H}}_1\boldsymbol{A}_0^{\mathrm{T}} + \boldsymbol{A}_0\bar{\boldsymbol{H}}_1 - \boldsymbol{Y}^{\mathrm{T}}\boldsymbol{G}^{\mathrm{T}} - \boldsymbol{GY}, \\ &\Omega_{12} = \boldsymbol{XC}, \ \Omega_{13} = \boldsymbol{A}_1\bar{\boldsymbol{H}}_2, \\ &\Omega_{15} = \boldsymbol{B}_0\bar{\boldsymbol{H}}_3, \ \ \Omega_{17} = \bar{\boldsymbol{H}}_1\boldsymbol{A}_0^{\mathrm{T}} - \boldsymbol{Y}^{\mathrm{T}}\boldsymbol{G}^{\mathrm{T}}, \\ &\Omega_{22} = \bar{\boldsymbol{J}}_1\boldsymbol{A}_0^{\mathrm{T}} + \boldsymbol{A}_0\bar{\boldsymbol{J}}_1 - \boldsymbol{C}^{\mathrm{T}}\boldsymbol{X}^{\mathrm{T}} - \boldsymbol{XC}, \end{aligned}$$

$$\begin{aligned}
&\Omega_{24} = A_1\bar{J}_2,\ \Omega_{26} = B_0\bar{J}_3,\ \Omega_{27} = C^{\mathrm{T}}X^{\mathrm{T}},\\
&\Omega_{28} = \bar{J}_1A_0^{\mathrm{T}} - C^{\mathrm{T}}X^{\mathrm{T}},\ \Omega_{37} = \bar{H}_2A_1^{\mathrm{T}},\\
&\Omega_{48} = \bar{J}_2A_1^{\mathrm{T}},\ \Omega_{57} = \bar{H}_3B_0^{\mathrm{T}},\ \Omega_{68} = \bar{J}_3B_0^{\mathrm{T}}.
\end{aligned}$$

The observer gain and the control gain are given by

$$K = Y\bar{H}_1^{-1}$$

and

$$L = X\hat{R}_1^{-1}$$

respectively, where $H_i = \bar{H}_i^{-1}, J_i = \bar{J}_i^{-1}, i = 1, 2, 3.$

Proof Defining the Lyapunov function

$$\begin{aligned}
V(Z(t)) = {}_0\mathrm{D}_t^{\alpha-1}Z^{\mathrm{T}}(t)\begin{bmatrix} H_1 & 0 \\ 0 & J_1 \end{bmatrix}Z(t) + \int_{t-\omega}^{t} Z^{\mathrm{T}}(s)\begin{bmatrix} H_2 & 0 \\ 0 & J_2 \end{bmatrix}Z(s)\mathrm{d}s \\
+ \int_{t-\omega}^{t} {}_0\mathrm{D}_t^{\alpha}Z^{\mathrm{T}}(s)\begin{bmatrix} H_3 & 0 \\ 0 & J_3 \end{bmatrix}{}_0\mathrm{D}_t^{\alpha}Z(s)\mathrm{d}s.
\end{aligned} \tag{9}$$

From Property 2 and Lemma 2, the integer derivative of $V(Z(t))$ with respect to t along the trajectories of (4) is calculated as

$$\begin{aligned}
\dot{V}(Z(t)) &= {}_0\mathrm{D}_t^{\alpha}Z^{\mathrm{T}}(t)\begin{bmatrix} H_1 & 0 \\ 0 & J_1 \end{bmatrix}Z(t) + Z^{\mathrm{T}}(t)\begin{bmatrix} H_2 & 0 \\ 0 & J_2 \end{bmatrix}Z(t) \\
&\quad - Z^{\mathrm{T}}(t-\omega)\begin{bmatrix} H_2 & 0 \\ 0 & J_2 \end{bmatrix}Z(t-\omega) \\
&\quad + {}_0\mathrm{D}_t^{\alpha}Z^{\mathrm{T}}(t)\begin{bmatrix} H_3 & 0 \\ 0 & J_3 \end{bmatrix}{}_0\mathrm{D}_t^{\alpha}Z(t) \\
&\quad - {}_0\mathrm{D}_t^{\alpha}Z^{\mathrm{T}}(t-\omega)\begin{bmatrix} H_3 & 0 \\ 0 & J_3 \end{bmatrix}{}_0\mathrm{D}_t^{\alpha}Z(t-\omega) \\
&\le Z^{\mathrm{T}}(t)\left(\begin{bmatrix} A_0 - GK & LC \\ 0 & A_0 - LC \end{bmatrix}^{\mathrm{T}}\begin{bmatrix} H_1 & 0 \\ 0 & J_1 \end{bmatrix}\right. \\
&\quad \left. + \begin{bmatrix} H_1 & 0 \\ 0 & J_1 \end{bmatrix}\begin{bmatrix} A_0 - GK & LC \\ 0 & A_0 - LC \end{bmatrix} + \begin{bmatrix} H_2 & 0 \\ 0 & J_2 \end{bmatrix}\right)Z(t) \\
&\quad + 2Z^{\mathrm{T}}(t)\begin{bmatrix} H_1 & 0 \\ 0 & J_1 \end{bmatrix}\begin{bmatrix} A_1 & 0 \\ 0 & A_1 \end{bmatrix}Z(t-\omega) \\
&\quad + 2Z^{\mathrm{T}}(t)\begin{bmatrix} H_1 & 0 \\ 0 & J_1 \end{bmatrix}\begin{bmatrix} B_0 & 0 \\ 0 & B_0 \end{bmatrix}{}_0\mathrm{D}_t^{\alpha}Z(t-\omega) \\
&\quad + {}_0\mathrm{D}_t^{\alpha}Z^{\mathrm{T}}(t)\begin{bmatrix} H_3 & 0 \\ 0 & J_3 \end{bmatrix}{}_0\mathrm{D}_t^{\alpha}Z(t)
\end{aligned} \tag{10}$$

$$
\begin{aligned}
&- \boldsymbol{Z}^{\mathrm{T}}(t-\omega)\begin{bmatrix} \boldsymbol{H}_2 & \boldsymbol{0} \\ \boldsymbol{0} & \boldsymbol{J}_2 \end{bmatrix}\boldsymbol{Z}(t-\omega) \\
&- {}_0\mathrm{D}_t^{\alpha}\boldsymbol{Z}^{\mathrm{T}}(t-\omega)\begin{bmatrix} \boldsymbol{H}_3 & \boldsymbol{0} \\ \boldsymbol{0} & \boldsymbol{J}_3 \end{bmatrix}{}_0\mathrm{D}_t^{\alpha}\boldsymbol{Z}(t-\omega) \\
= {}& \boldsymbol{X}^{\mathrm{T}}(t)\Phi\boldsymbol{X}(t)
\end{aligned}
$$

where

$$
\boldsymbol{X}(t) = \begin{bmatrix} \boldsymbol{Z}(t) \\ \boldsymbol{Z}(t-\omega) \\ {}_0\mathrm{D}_t^{\alpha}\boldsymbol{Z}(t-\omega) \end{bmatrix},
$$

$$
\Phi = \begin{bmatrix}
\Phi_{11} & \Phi_{12} & \Phi_{13} & \boldsymbol{0} & \Phi_{15} & \boldsymbol{0} \\
* & \Phi_{22} & \boldsymbol{0} & \Phi_{24} & \boldsymbol{0} & \Phi_{26} \\
* & * & -\boldsymbol{H}_2 & \boldsymbol{0} & \boldsymbol{0} & \boldsymbol{0} \\
* & * & * & -\boldsymbol{J}_2 & \boldsymbol{0} & \boldsymbol{0} \\
* & * & * & * & -\boldsymbol{H}_3 & \boldsymbol{0} \\
* & * & * & * & * & -\boldsymbol{J}_3
\end{bmatrix} + \boldsymbol{M}^{\mathrm{T}}\begin{bmatrix} \boldsymbol{H}_3 & \boldsymbol{0} \\ \boldsymbol{0} & \boldsymbol{J}_3 \end{bmatrix}\boldsymbol{M},
$$

and

$$
\boldsymbol{M} = \begin{bmatrix} \boldsymbol{A}_0 - \boldsymbol{GK} & \boldsymbol{LC} & \boldsymbol{A}_1 & \boldsymbol{0} & \boldsymbol{B}_0 & \boldsymbol{0} \\ \boldsymbol{0} & \boldsymbol{A}_0 - \boldsymbol{LC} & \boldsymbol{0} & \boldsymbol{A}_1 & \boldsymbol{0} & \boldsymbol{B}_0 \end{bmatrix},
$$

with

$$
\begin{aligned}
&\Phi_{11} = \boldsymbol{A}_0^{\mathrm{T}}\boldsymbol{H}_1 + \boldsymbol{H}_1\boldsymbol{A}_0 - \boldsymbol{K}^{\mathrm{T}}\boldsymbol{G}^{\mathrm{T}}\boldsymbol{H}_1 - \boldsymbol{H}_1\boldsymbol{GK} + \boldsymbol{H}_2, \\
&\Phi_{12} = \boldsymbol{H}_1\boldsymbol{LC},\ \Phi_{13} = \boldsymbol{H}_1\boldsymbol{A}_1,\ \Phi_{15} = \boldsymbol{H}_1\boldsymbol{B}_0, \\
&\Phi_{22} = \boldsymbol{A}_0^{\mathrm{T}}\boldsymbol{J}_1 + \boldsymbol{J}_1\boldsymbol{A}_0 - \boldsymbol{C}^{\mathrm{T}}\boldsymbol{L}^{\mathrm{T}}\boldsymbol{J}_1 - \boldsymbol{J}_1\boldsymbol{LC} + \boldsymbol{J}_2, \\
&\Phi_{24} = \boldsymbol{J}_1\boldsymbol{A}_1,\ \Phi_{26} = \boldsymbol{J}_1\boldsymbol{B}_0.
\end{aligned}
$$

Define a real symmetric matrix

$$
\Sigma = \begin{bmatrix}
\Sigma_{11} & \Sigma_{12} & \Sigma_{13} & \boldsymbol{0} & \Sigma_{15} & \boldsymbol{0} & \Sigma_{17} & \boldsymbol{0} & \bar{\boldsymbol{H}}_1 & \boldsymbol{0} \\
* & \Sigma_{22} & \boldsymbol{0} & \Sigma_{24} & \boldsymbol{0} & \Sigma_{26} & \Sigma_{27} & \Sigma_{28} & \boldsymbol{0} & \bar{\boldsymbol{J}}_1 \\
* & * & -\bar{\boldsymbol{H}}_2 & \boldsymbol{0} & \boldsymbol{0} & \boldsymbol{0} & \Sigma_{37} & \boldsymbol{0} & \boldsymbol{0} & \boldsymbol{0} \\
* & * & * & -\bar{\boldsymbol{J}}_2 & \boldsymbol{0} & \boldsymbol{0} & \boldsymbol{0} & \Sigma_{48} & \boldsymbol{0} & \boldsymbol{0} \\
* & * & * & * & -\bar{\boldsymbol{H}}_3 & \boldsymbol{0} & \Sigma_{57} & \boldsymbol{0} & \boldsymbol{0} & \boldsymbol{0} \\
* & * & * & * & * & -\bar{\boldsymbol{J}}_3 & \boldsymbol{0} & \Sigma_{68} & \boldsymbol{0} & \boldsymbol{0} \\
* & * & * & * & * & * & -\bar{\boldsymbol{H}}_3 & \boldsymbol{0} & \boldsymbol{0} & \boldsymbol{0} \\
* & * & * & * & * & * & * & -\bar{\boldsymbol{J}}_3 & \boldsymbol{0} & \boldsymbol{0} \\
* & * & * & * & * & * & * & * & -\bar{\boldsymbol{H}}_2 & \boldsymbol{0} \\
* & * & * & * & * & * & * & * & * & -\bar{\boldsymbol{J}}_2
\end{bmatrix} \tag{11}
$$

where

$$\begin{aligned}
&\Sigma_{11}=\bar{H}_1A_0^{\mathrm{T}}+A_0\bar{H}_1-Y^{\mathrm{T}}G^{\mathrm{T}}-GY,\\
&\Sigma_{12}=XC,\ \ \Sigma_{13}=A_1\bar{H}_2,\\
&\Sigma_{15}=B_0\bar{H}_3,\ \Sigma_{17}=\bar{H}_1A_0^{\mathrm{T}}-Y^{\mathrm{T}}G^{\mathrm{T}},\\
&\Sigma_{22}=\bar{J}_1A_0^{\mathrm{T}}+A_0\bar{J}_1-C^{\mathrm{T}}X^{\mathrm{T}}-XC,\\
&\Sigma_{24}=A_1\bar{J}_2,\ \ \Sigma_{26}=B_0\bar{J}_3,\ \ \Sigma_{27}=C^{\mathrm{T}}X^{\mathrm{T}},\\
&\Sigma_{28}=\bar{J}_1A_0^{\mathrm{T}}-C^{\mathrm{T}}X^{\mathrm{T}},\ \ \Sigma_{37}=\bar{H}_2A_1^{\mathrm{T}},\\
&\Sigma_{48}=\bar{J}_2A_1^{\mathrm{T}},\ \ \Sigma_{57}=\bar{J}_3B_0^{\mathrm{T}},\ \ \Sigma_{68}=\bar{J}_3B_0^{\mathrm{T}}.
\end{aligned}$$

Here, let $K=Y\bar{H}_1^{-1}$, $LC=XC\bar{H}_2^{-1}$. Then, the LMI condition (6) equals to (10) < 0, then it implies the following result:

$$\begin{aligned}
N=&\begin{bmatrix}
\Sigma_{11}+\bar{H}_1\bar{H}_2^{-1}\bar{H}_1 & \Sigma_{12} & \Sigma_{13} & 0 & \Sigma_{15} & 0\\
* & \Sigma_{22}+\bar{J}_1\bar{J}_2^{-1}\bar{J}_1 & 0 & \Sigma_{24} & 0 & \Sigma_{26}\\
* & * & -\bar{H}_2 & 0 & 0 & 0\\
* & * & * & -\bar{J}_2 & 0 & 0\\
* & * & * & * & -\bar{H}_3 & 0\\
* & * & * & * & * & -\bar{J}_3
\end{bmatrix}\\
&+N_1^{\mathrm{T}}\begin{bmatrix}\bar{H}_3 & 0\\ 0 & \bar{J}_3\end{bmatrix}^{-1}N_1\\
&<0
\end{aligned}\tag{12}$$

where

$$N_1=\begin{bmatrix}\Sigma_{17}^{\mathrm{T}} & \Sigma_{27}^{\mathrm{T}} & \Sigma_{37}^{\mathrm{T}} & 0 & \Sigma_{57}^{\mathrm{T}} & 0\\ 0 & \Sigma_{28}^{\mathrm{T}} & 0 & \Sigma_{48}^{\mathrm{T}} & 0 & \Sigma_{68}^{\mathrm{T}}\end{bmatrix}.$$

Pre and post-multiply N by Ψ^{T} and Ψ, where

$$\Psi=\begin{bmatrix}
\bar{H}_1^{-1} & 0 & 0 & 0 & 0 & 0\\
* & \bar{J}_1^{-1} & 0 & 0 & 0 & 0\\
* & * & \bar{H}_2^{-1} & 0 & 0 & 0\\
* & * & * & \bar{J}_2^{-1} & 0 & 0\\
* & * & * & * & \bar{H}_3^{-1} & 0\\
* & * & * & * & * & \bar{J}_3^{-1}
\end{bmatrix}=\begin{bmatrix}
H_1 & 0 & 0 & 0 & 0 & 0\\
* & J_1 & 0 & 0 & 0 & 0\\
* & * & H_2 & 0 & 0 & 0\\
* & * & * & J_2 & 0 & 0\\
* & * & * & * & H_3 & 0\\
* & * & * & * & * & J_3
\end{bmatrix},$$

we can obtain the following results

$$\Phi<0,\tag{13}$$

which leads

$$\dot{V}(Z(t)) = X^{\mathrm{T}}(t)\Phi X(t) < 0, \quad t \geq 0.$$

So, the observer-based closed-loop fractional-order system above is proven to be asymptotically stable. The proof is complete. □

Remark 1 The content of Theorem 1 is also applicable to the integer-order system, and the Lyapunov function can be chosen similar to the form in this chapter.

Remark 2 Theorem 1 provides sufficient conditions for the FONTSs based on the LMI technique. Through Matlab's LMI toolbox, the LMI condition can be calculated effectively. So it can be used more conveniently.

Remark 3 We calculate the integer-order derivatives of the Lyapunov function in this chapter, which is more convenient and easier to deal with the neutral system.

Corollary 1 *When the neutral items and the time delay items of system (1) are removed, the condition in Theorem 1 becomes to*

$$\begin{bmatrix} \Sigma_1 & XC \\ * & \Sigma_2 \end{bmatrix} < 0 \tag{14}$$

where

$$\Sigma_1 = \bar{H}_1 A_0^{\mathrm{T}} + A_0 \bar{H}_1 - Y^{\mathrm{T}} G^{\mathrm{T}} - GY,$$

$$\Sigma_2 = \bar{J}_1 A_0^{\mathrm{T}} + A_0 \bar{J}_1 - C^{\mathrm{T}} X^{\mathrm{T}} - XC.$$

That is, if there exit some symmetric positive definite matrices $\bar{H}_1 \in \mathbf{R}^{n\times n}, \bar{J}_1 \in \mathbf{R}^{n\times n}$, *and matrices* $X \in \mathbf{R}^{n\times q}$, $Y \in \mathbf{R}^{m\times n}$, $\hat{R}_1 \in \mathbf{R}^{n\times q}$, *such that conditions (13) and (7) hold, then the fractional-order system without time-delay and neutral items is asymptotically stable. The gains are the same as Theorem 1.*

Proof Proof process is the same as Theorem 1, so the proof process is omitted here. □

When the neutral items and the time delay items of system (1) are removed, we can further study the guaranteed cost control problem of the system. Firstly, let us introduce some definitions and lemmas.

Lemma 3 ([3]) *If* $h(t) \in \mathbf{C}^n([0, +\infty), \mathbf{R})$, *and the order* a *satisfies* $a \in (n-1, n]$, *where* $n \geq 1, n \in \mathbf{Z}^+$, *then*

$${}_0\mathrm{D}_t^{-a}({}_0\mathrm{D}_t^{a} h(t)) = h(t) - \sum_{k=0}^{n-1} \frac{t^k}{k!} h^{(k)}(0).$$

In particular, when $0 < a < 1$, *the upper form changed to*

$${}_0\mathrm{D}_t^{-a}({}_0\mathrm{D}_t^{a} h(t)) = h(t) - h(0).$$

Lemma 4 ([20, 35]) *If $\boldsymbol{U}, \boldsymbol{V}$ are matrices with appropriate dimensions, then the following inequality is established for any positive number α*

$$\boldsymbol{U}^{\mathrm{T}}\boldsymbol{V} + \boldsymbol{V}^{\mathrm{T}}\boldsymbol{U} \leq \alpha \boldsymbol{U}^{\mathrm{T}}\boldsymbol{U} + \alpha^{-1}\boldsymbol{V}^{\mathrm{T}}\boldsymbol{V}.$$

Definition 3 ([15, 36]) For the system (1), define the cost function indicator as

$$J = \frac{1}{\Gamma(a)}\int_0^t (t-s)^{a-1}[\hat{\boldsymbol{x}}^{\mathrm{T}}(s)\boldsymbol{Q}_1\hat{\boldsymbol{x}}(s) + \boldsymbol{e}^{\mathrm{T}}(s)\boldsymbol{Q}_2\boldsymbol{e}(s) + \boldsymbol{u}^{\mathrm{T}}(s)\boldsymbol{Q}_3\boldsymbol{u}(s)]\mathrm{d}s \quad (15)$$

where $\boldsymbol{Q}_1, \boldsymbol{Q}_2 \in \mathbf{R}^{n\times n}$, $\boldsymbol{Q}_3 \in \mathbf{R}^{m\times m}$ are the given positive-definite symmetric matrices, then the vector $\boldsymbol{e}(t) = \boldsymbol{x}(t) - \hat{\boldsymbol{x}}(t)$ is the error of the estimated system.

According to the observer and the controller above, when the neutral items and the time delay items are removed, we can rewrite the system as

$$\begin{cases} {}_0\mathrm{D}_t^{\alpha}\hat{\boldsymbol{x}}(t) = (\boldsymbol{A}_0 - \boldsymbol{G}\boldsymbol{K})\hat{\boldsymbol{x}}(t) + \boldsymbol{L}\boldsymbol{C}\boldsymbol{e}(t), & t \geq 0 \\ {}_0\mathrm{D}_t^{\alpha}\boldsymbol{e}(t) = (\boldsymbol{A}_0 - \boldsymbol{L}\boldsymbol{C})\boldsymbol{e}(t), & t \geq 0 \end{cases} \quad (16)$$

The system (15) is written as a form of the vector and matrix:

$${}_0\mathrm{D}_t^{\alpha}\mathbf{X}(t) = {}_0\mathrm{D}_t^{\alpha}\begin{bmatrix}\hat{\boldsymbol{x}}(t)\\ \boldsymbol{e}(t)\end{bmatrix} = \begin{bmatrix}\boldsymbol{A}_0 - \boldsymbol{G}\boldsymbol{K} & \boldsymbol{L}\boldsymbol{C}\\ \mathbf{0} & \boldsymbol{A}_0 - \boldsymbol{L}\boldsymbol{C}\end{bmatrix}\begin{bmatrix}\hat{\boldsymbol{x}}(t)\\ \boldsymbol{e}(t)\end{bmatrix}. \quad (17)$$

Then, we will have the following theorem

Theorem 2 *The system (16) is asymptotically stable by the controller and the observer above if there exist some symmetric positive-definite matrices $\bar{\boldsymbol{H}}_1, \bar{\boldsymbol{J}}_1 \in \boldsymbol{R}^{n\times n}$, and matrices $\boldsymbol{X} \in \boldsymbol{R}^{n\times q}$, $\boldsymbol{Y} \in \boldsymbol{R}^{m\times n}$, $\hat{\boldsymbol{R}}_1 \in \boldsymbol{R}^{q\times q}$, such that*

$$\begin{bmatrix} \boldsymbol{\Sigma}_1 & \boldsymbol{X}\boldsymbol{C} & \bar{\boldsymbol{H}}_1\boldsymbol{Q}_1 & \mathbf{0} & \boldsymbol{Y}^{\mathrm{T}}\boldsymbol{Q}_3 \\ * & \boldsymbol{\Sigma}_2 & \mathbf{0} & \bar{\boldsymbol{J}}_1\boldsymbol{Q}_2 & \mathbf{0} \\ * & * & -\boldsymbol{Q}_1 & \mathbf{0} & \mathbf{0} \\ * & * & * & -\boldsymbol{Q}_2 & \mathbf{0} \\ * & * & * & * & -\boldsymbol{Q}_3 \end{bmatrix} < \boldsymbol{0}, \quad (18)$$

and

$$\boldsymbol{C}\bar{\boldsymbol{J}}_1 = \hat{\boldsymbol{R}}_1\boldsymbol{C}. \quad (19)$$

where

$$\boldsymbol{\Sigma}_1 = \boldsymbol{A}_0\bar{\boldsymbol{H}}_1 + \bar{\boldsymbol{H}}_1\boldsymbol{A}_0^{\mathrm{T}} - \boldsymbol{G}\boldsymbol{Y} - \boldsymbol{Y}^{\mathrm{T}}\boldsymbol{G}^{\mathrm{T}},$$

$$\boldsymbol{\Sigma}_2 = \boldsymbol{A}_0\bar{\boldsymbol{J}}_1 + \bar{\boldsymbol{J}}_1\boldsymbol{A}_0^{\mathrm{T}} - \boldsymbol{X}\boldsymbol{C} - \boldsymbol{C}^{\mathrm{T}}\boldsymbol{X}^{\mathrm{T}}.$$

The gains are given by, respectively,

$$K = Y\bar{H}_1^{-1}$$

and

$$L = X\hat{R}_1^{-1}.$$

The guaranteed cost value is given by

$$J^* = \hat{x}^{\mathrm{T}}(0)H_1\hat{x}(0) + e^{\mathrm{T}}(0)J_1e(0) \tag{20}$$

where $H_1 = \bar{H}_1^{-1}$ *and* $J_1 = \bar{J}_2^{-1}$.

Proof Defining the Lyapunov functional

$$V(X(t)) = \hat{x}^{\mathrm{T}}(t)H_1\hat{x}(t) + e^{\mathrm{T}}(t)J_1e(t) \tag{21}$$

by Lemma 2, the Caputo derivative of $V(X(t))$ with respect t along the trajectories of (15) is defined by

$$\begin{aligned}
{}_0\mathrm{D}_t^{\alpha}V(X(t)) &\le 2\hat{x}^{\mathrm{T}}(t)H_1({}_0\mathrm{D}_t^{\alpha}\hat{x}(t)) + 2e^{\mathrm{T}}(t)J_1({}_0\mathrm{D}_t^{\alpha}e(t)) \\
&= 2\hat{x}^{\mathrm{T}}(t)H_1[(A_0 - GK)\hat{x}(t) + LCe(t)] \\
&\quad + 2e^{\mathrm{T}}(t)J_1[(A_0 - LC)e(t)] \\
&= \hat{x}^{\mathrm{T}}(t)(H_1A_0 + A_0^{\mathrm{T}}H_1 - H_1GK - K^{\mathrm{T}}G^{\mathrm{T}}H_1)\hat{x}(t) \\
&\quad + e^{\mathrm{T}}(t)(J_1A_0 + A_0^{\mathrm{T}}J_1 - J_1LC - C^{\mathrm{T}}L^{\mathrm{T}}J_1)e(t) \\
&\quad + 2\hat{x}^{\mathrm{T}}(t)H_1LCe(t) \\
&= X^{\mathrm{T}}(t)\bar{\Sigma}X(t)
\end{aligned} \tag{22}$$

where

$$\begin{aligned}
X(t) &= \begin{bmatrix} \hat{x}(t) \\ e(t) \end{bmatrix}, \\
\bar{\Sigma} &= \begin{bmatrix} \bar{\Sigma}_1 & \bar{\Sigma}_2 \\ * & \bar{\Sigma}_3 \end{bmatrix}, \\
\bar{\Sigma}_1 &= A_0^{\mathrm{T}}H_1 + H_1A_0 - K^{\mathrm{T}}G^{\mathrm{T}}H_1 - H_1GK, \\
\bar{\Sigma}_2 &= H_1LC,\ \bar{\Sigma}_3 = J_1A_0 + A_0^{\mathrm{T}}J_1 - J_1LC - C^{\mathrm{T}}L^{\mathrm{T}}J_1.
\end{aligned}$$

The above LMI condition (17) equals to

$${}_0\mathrm{D}_t^{\alpha}V(X(t)) + \hat{x}^{\mathrm{T}}(t)Q_1\hat{x}(t) + e^{\mathrm{T}}(t)Q_2e(t) + u^{\mathrm{T}}(t)Q_3u(t) < 0.$$

Define a real symmetric matrix

$$\Sigma' = \begin{bmatrix} \Sigma'_{11} & \Sigma'_{12} & \bar{H}_1 & 0 & Y^{\mathrm{T}} \\ * & \Sigma'_{22} & 0 & \bar{J}_1 & 0 \\ * & * & -Q_1^{-1} & 0 & 0 \\ * & * & * & -Q_2^{-1} & 0 \\ * & * & * & * & -Q_3^{-1} \end{bmatrix} \tag{23}$$

where

$$\Sigma'_{11} = \bar{H}_1 A_0^{\mathrm{T}} + A_0 \bar{H}_1 - Y^{\mathrm{T}} G^{\mathrm{T}} - GY, \ \Sigma'_{12} = XC,$$

and

$$\Sigma'_{22} = \bar{J}_1 A_0^{\mathrm{T}} + A_0 \bar{J}_1 - C^{\mathrm{T}} X^{\mathrm{T}} - XC.$$

The LMI condition (17) equals to (22) < 0, so we have

$$N' = \begin{bmatrix} \Sigma'_{11} + \bar{H}_1 Q_1 \bar{H}_1 + Y^{\mathrm{T}} Q_3 Y & \Sigma'_{12} \\ * & \Sigma'_{22} + \bar{J}_1 Q_2 \bar{J}_1 \end{bmatrix} < 0. \tag{24}$$

Pre and post-multiply N' by H^{T} and H, where

$$H = \begin{bmatrix} \bar{H}_1^{-1} & 0 \\ * & \bar{J}_1^{-1} \end{bmatrix} = \begin{bmatrix} H_1 & 0 \\ * & J_1 \end{bmatrix},$$

we can obtain the following results

$$\bar{\Sigma} < \begin{bmatrix} -Q_1 - K^{\mathrm{T}} Q_3 K & 0 \\ * & -Q_2 \end{bmatrix}. \tag{25}$$

So, we have

$$\begin{aligned} {}_0\mathrm{D}_t^{\alpha} V(X(t)) &\leq X^{\mathrm{T}}(t) \bar{\Sigma} X(t) \\ &< X^{\mathrm{T}}(t) \begin{bmatrix} -Q_1 - K^{\mathrm{T}} Q_3 K & 0 \\ * & -Q_2 \end{bmatrix} \\ &= \begin{bmatrix} \hat{x}^{\mathrm{T}}(t) & e^{\mathrm{T}}(t) \end{bmatrix} \begin{bmatrix} -Q_1 - K^{\mathrm{T}} Q_3 K & 0 \\ * & -Q_2 \end{bmatrix} \begin{bmatrix} \hat{x}(t) \\ e(t) \end{bmatrix} \\ &= -\hat{x}^{\mathrm{T}}(t)(Q_1 + K^{\mathrm{T}} Q_3)\hat{x}(t) - e^{\mathrm{T}}(t) Q_2 e(t) \\ &= -(\hat{x}^{\mathrm{T}}(t) Q_1 \hat{x}(t) + e^{\mathrm{T}}(t) Q_2 e(t) + u^{\mathrm{T}}(t) Q_3 u(t)). \end{aligned} \tag{26}$$

Then,

$${}_0\mathrm{D}_t^{\alpha} V(X(t)) + \hat{x}^{\mathrm{T}}(t) Q_1 \hat{x}(t) + e^{\mathrm{T}}(t) Q_2 e(t) + u^{\mathrm{T}}(t) Q_3 u(t) < 0.$$

According to the conditions (17), (18), (25), and

$$\hat{x}^{\mathrm{T}}(t) Q_1 \hat{x}(t) + e^{\mathrm{T}}(t) Q_2 e(t) + u^{\mathrm{T}}(t) Q_3 u(t) \geq 0$$

we have

$$ {}_0\mathrm{D}_t^{\alpha} \boldsymbol{V}(\boldsymbol{X}(t)) \le 2\boldsymbol{X}^{\mathrm{T}}(t) \begin{bmatrix} \boldsymbol{H}_1 & \boldsymbol{0} \\ * & \boldsymbol{J}_1 \end{bmatrix} ({}_0\mathrm{D}_t^{\alpha} \boldsymbol{X}(t)) < 0. \tag{27} $$

We conclude that system (15) is asymptotically stable by the observer and the controller. Then, we can calculate the guaranteed cost value, integrating both sides of the system (25) with order α from 0 to t and using Lemma 3, the following result is established

$$ {}_0\mathrm{D}_t^{-\alpha}({}_0\mathrm{D}_t^{\alpha} \boldsymbol{V}(\boldsymbol{X}(t))) = \boldsymbol{V}(\boldsymbol{X}(t)) - \boldsymbol{V}(\boldsymbol{X}(0)) \le -\mathbf{J}, \tag{28} $$

that is

$$ 0 < \boldsymbol{V}(\boldsymbol{X}(t)) \le \boldsymbol{V}(\boldsymbol{X}(0)) - \mathbf{J}. $$

So, we have the following result

$$ \begin{aligned} \mathbf{J}^* &= \boldsymbol{V}(\boldsymbol{X}(0)) \\ &= \boldsymbol{X}^{\mathrm{T}}(0) \begin{bmatrix} \boldsymbol{H}_1 & \boldsymbol{0} \\ * & \boldsymbol{J}_1 \end{bmatrix} \boldsymbol{X}(0) \\ &= \begin{bmatrix} \hat{\boldsymbol{x}}^{\mathrm{T}}(0) & \boldsymbol{e}^{\mathrm{T}}(0) \end{bmatrix} \begin{bmatrix} \boldsymbol{J}_1 & \boldsymbol{0} \\ * & \boldsymbol{H}_1 \end{bmatrix} \begin{bmatrix} \hat{\boldsymbol{x}}(0) \\ \boldsymbol{e}(0) \end{bmatrix} \\ &= \hat{\boldsymbol{x}}^{\mathrm{T}}(0) \boldsymbol{H}_1 \hat{\boldsymbol{x}}(0) + \boldsymbol{e}^{\mathrm{T}}(0) \boldsymbol{J}_1 \boldsymbol{e}(0). \end{aligned} \tag{29} $$

The proof is now completed. □

4 Numerical Example

In this section, numerical examples are provided below to verify the effectiveness of Theorems 1 and 2 in this chapter.

Example 1 Considering the state equation with the following parameters:

$$ \boldsymbol{A}_0 = \begin{bmatrix} -8 & 0 \\ -3 & -4 \end{bmatrix}, \ \boldsymbol{A}_1 = \begin{bmatrix} -0.5 & 0.9 \\ 0.1 & -0.6 \end{bmatrix}, \ \boldsymbol{G} = \begin{bmatrix} 4 \\ 3 \end{bmatrix}, $$

$$ \boldsymbol{B}_0 = \begin{bmatrix} 0.5 & 0 \\ 0 & 0.5 \end{bmatrix}, \ \boldsymbol{C} = \begin{bmatrix} 0.5 & 1 \end{bmatrix}, \ \boldsymbol{D} = \boldsymbol{0}. $$

First, let $\tau = 2$. Then, by applying Theorem 1, we can get the following results through Matlab's LMI toolbox:

$$ \bar{\boldsymbol{H}}_1 = \begin{bmatrix} 0.0882 & -0.0963 \\ -0.0963 & 0.1086 \end{bmatrix}, \ \bar{\boldsymbol{H}}_2 = \begin{bmatrix} 1.0425 & 0.2915 \\ 0.2915 & 0.5101 \end{bmatrix}, $$

$$\bar{H}_3 = \begin{bmatrix} 1.3518 & -0.0429 \\ -0.0429 & 1.2789 \end{bmatrix}, \ \bar{J}_1 = \begin{bmatrix} 0.0879 & -0.0306 \\ -0.0306 & 0.1968 \end{bmatrix},$$

$$\bar{J}_2 = \begin{bmatrix} 1.0381 & 0.2997 \\ 0.2997 & 0.4992 \end{bmatrix}, \ \bar{J}_3 = \begin{bmatrix} 1.3262 & -0.0215 \\ -0.0215 & 1.2838 \end{bmatrix},$$

$$Y = \begin{bmatrix} 0.0189 & 0.1768 \end{bmatrix}, \ X = \begin{bmatrix} 0.0659 \\ -0.0221 \end{bmatrix}, \ \hat{R}_1 = 0.8498.$$

The gains are in the following forms

$$K = \begin{bmatrix} 5.0630 & 56.2758 \end{bmatrix}, \ L = \begin{bmatrix} 0.0776 \\ -0.0260 \end{bmatrix}.$$

For the simulation results, when $\alpha = 0.3$, the state trajectories and the state estimation trajectories are presented in Figs. 1, 2 and 3.

Also, when $\alpha = 0.6$, the state trajectories and the state estimation trajectories are presented in Figs. 4, 5 and 6.

Finally, when $\alpha = 0.7$, the state trajectories and the state estimation trajectories are given in Figs. 7, 8 and 9.

Remark 4 We can see from the figures that the stability of the FONTSs is associated with the order number α. The increase of α, the later the system gets stable. So, it's necessary to study the various properties of the FONTSs.

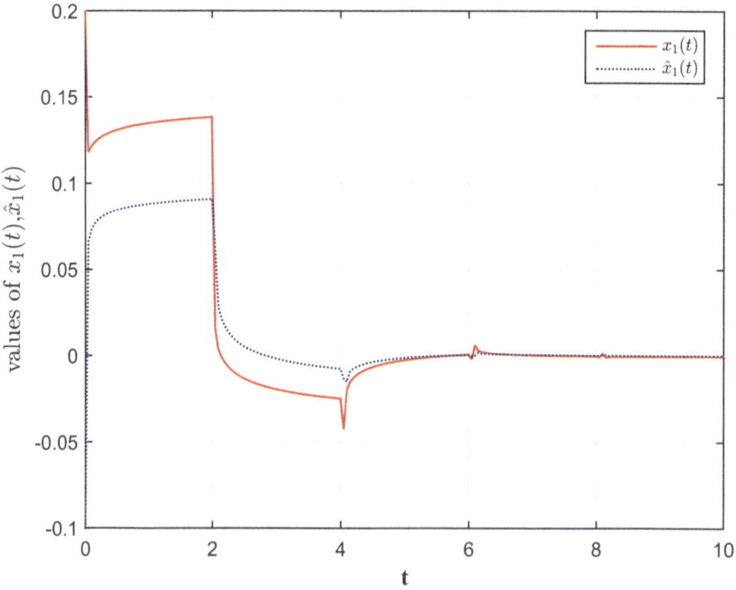

Fig. 1 The trajectory of state x_1 and trajectory of state estimation $\hat{x}_1$

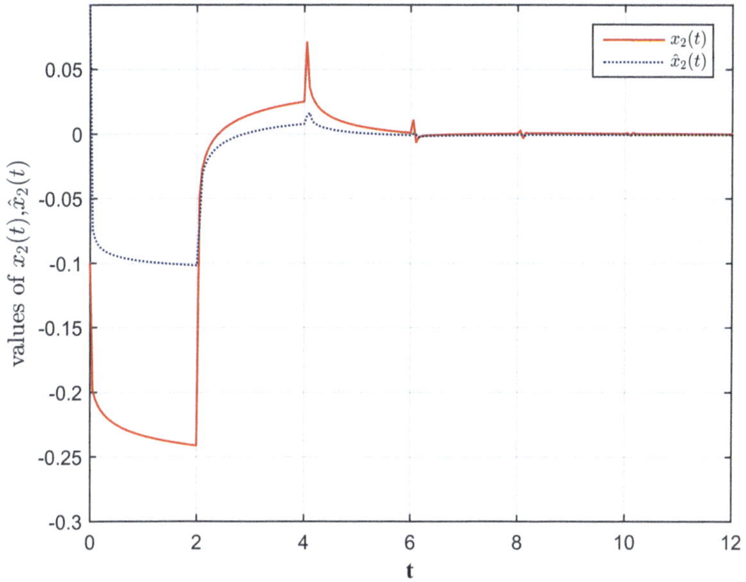

Fig. 2 The trajectory of state $\boldsymbol{x}_2$ and trajectory of state estimation $\hat{\boldsymbol{x}}_2$

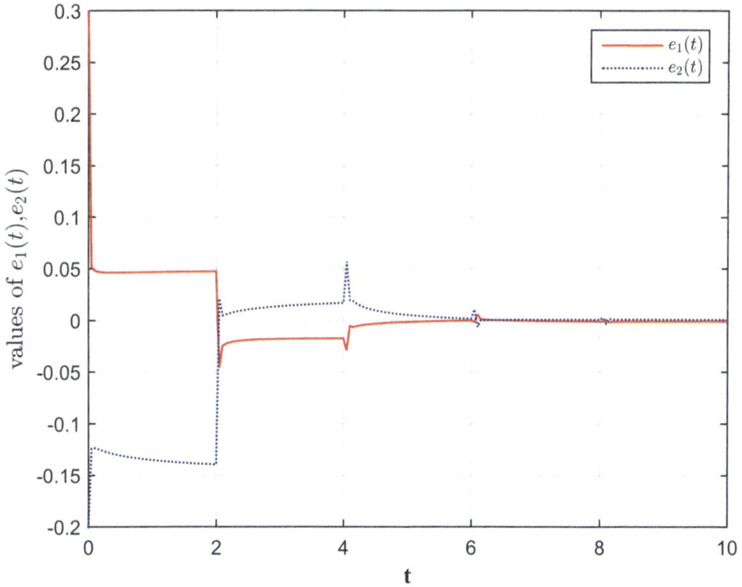

Fig. 3 The trajectory of errors

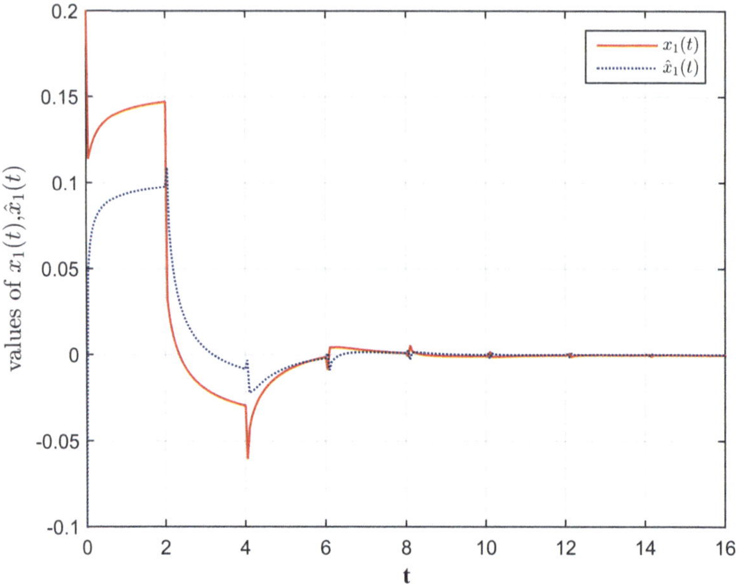

Fig. 4 The trajectory of state $\boldsymbol{x}_1$ and trajectory of state estimation $\hat{\boldsymbol{x}}_1$

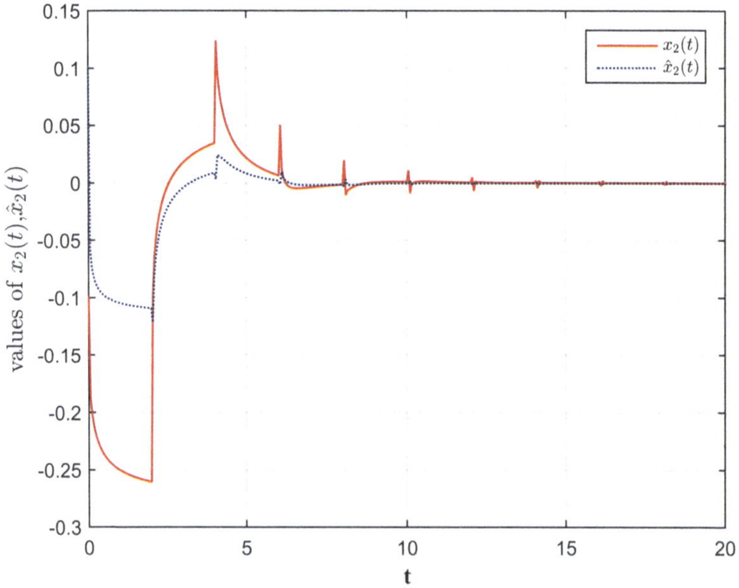

Fig. 5 The trajectory of state $\boldsymbol{x}_2$ and trajectory of state estimation $\hat{\boldsymbol{x}}_2$

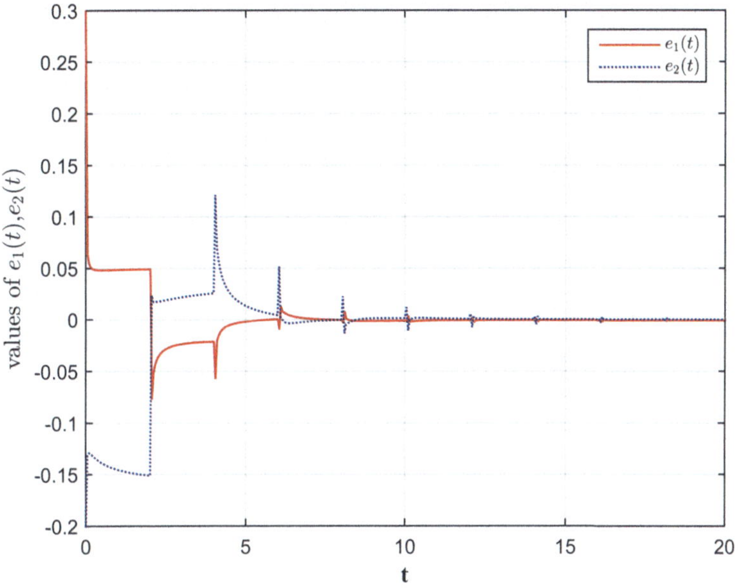

Fig. 6 The trajectory of errors

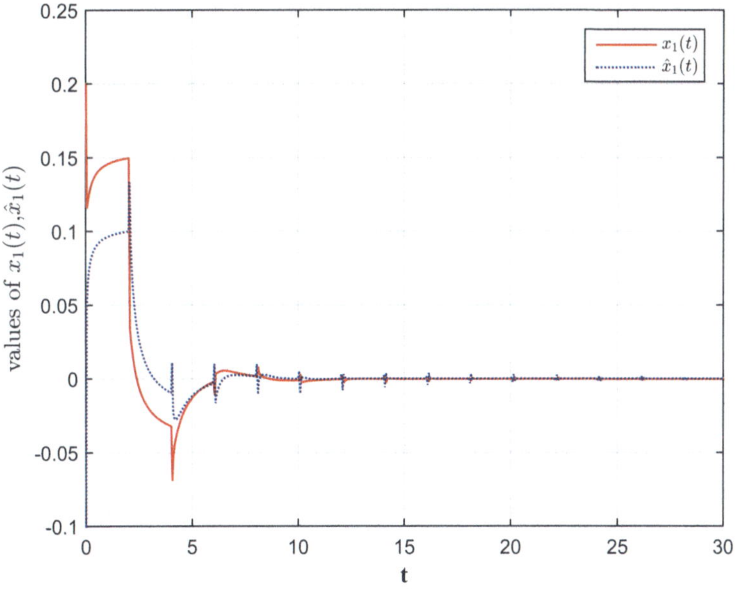

Fig. 7 The trajectory of state $\boldsymbol{x}_1$ and trajectory of state estimation $\hat{\boldsymbol{x}}_1$

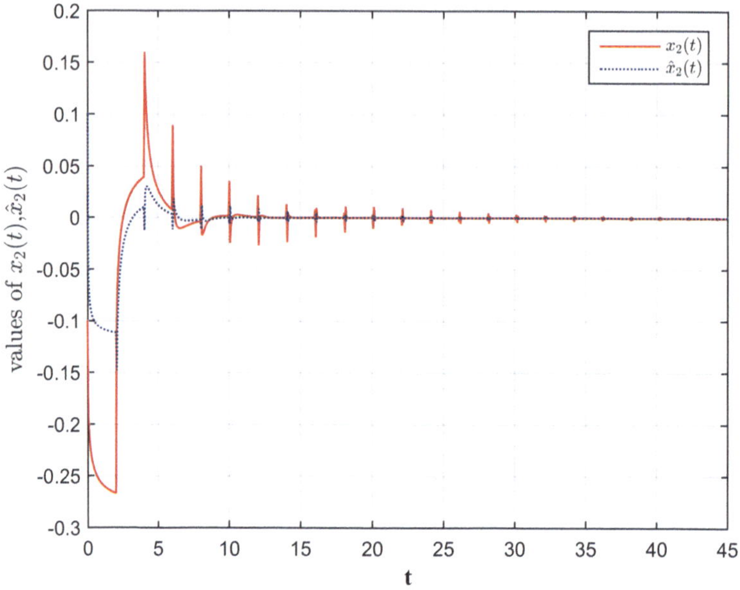

Fig. 8 The trajectory of state $\boldsymbol{x}_2$ and trajectory of state estimation $\hat{\boldsymbol{x}}_2$

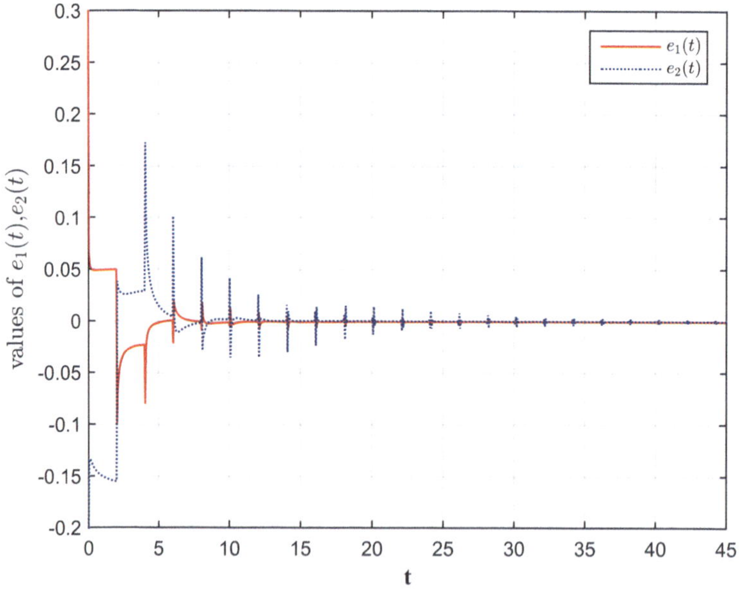

Fig. 9 The trajectory of errors

Example 2 Consider the fractional-order system with the following parameters

$$\boldsymbol{A}_0 = \begin{bmatrix} -6 & 0 \\ 3 & -4 \end{bmatrix}, \ \boldsymbol{G} = \begin{bmatrix} 4 \\ 3 \end{bmatrix},$$

$$\boldsymbol{C} = \begin{bmatrix} 0.5 & 1 \end{bmatrix}, \ \boldsymbol{D} = \boldsymbol{0},$$

for the cost function (14), where

$$\boldsymbol{Q}_1 = \begin{bmatrix} 1 & 0 \\ 0 & 1 \end{bmatrix}, \ \boldsymbol{Q}_2 = \begin{bmatrix} 2 & 0 \\ 0 & 2 \end{bmatrix}, \ \boldsymbol{Q}_3 = \boldsymbol{I}.$$

Let $\alpha = 0.85$, from Theorem 2, we can get the following results through Matlab's LMI toolbox:

$$\bar{\boldsymbol{H}}_1 = \begin{bmatrix} 0.1089 & -0.0378 \\ -0.0378 & 0.1350 \end{bmatrix}, \ \bar{\boldsymbol{J}}_1 = \begin{bmatrix} 0.1471 & 0.0167 \\ 0.0167 & 0.1936 \end{bmatrix},$$

$$\boldsymbol{Y} = \begin{bmatrix} 0.0854 & 0.1115 \end{bmatrix}, \ \boldsymbol{X} = \begin{bmatrix} 0.1107 \\ 0.1835 \end{bmatrix}, \ \hat{\boldsymbol{R}}_1 = 0.1977.$$

The gains are in the following forms

$$\boldsymbol{K} = \begin{bmatrix} 1.1865 & 1.1582 \end{bmatrix}, \ \boldsymbol{L} = \begin{bmatrix} 0.5599 \\ 0.9281 \end{bmatrix}.$$

Also, the guaranteed cost value is obtained as $J^* = 0.0222$.

The state trajectories and state estimation trajectories are shown in Figs. 10, 11 and 12.

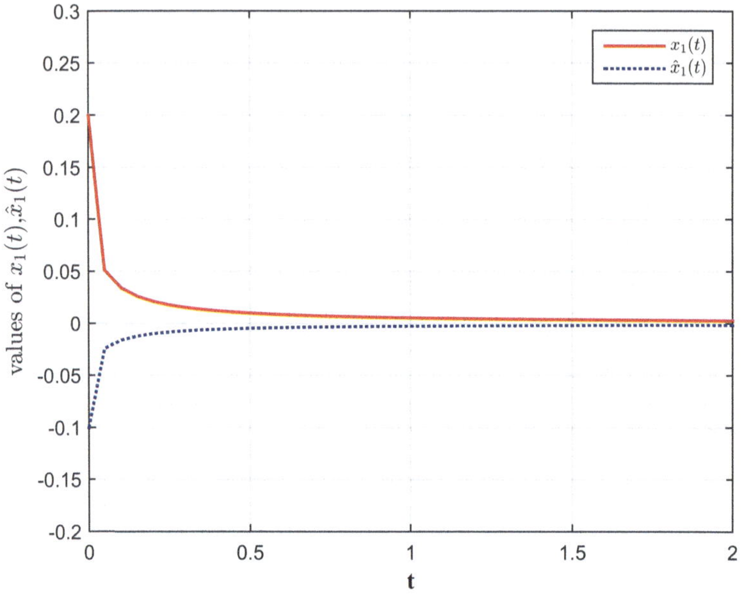

Fig. 10 The trajectory of state $\boldsymbol{x}_1$ and trajectory of state estimation $\hat{\boldsymbol{x}}_1$

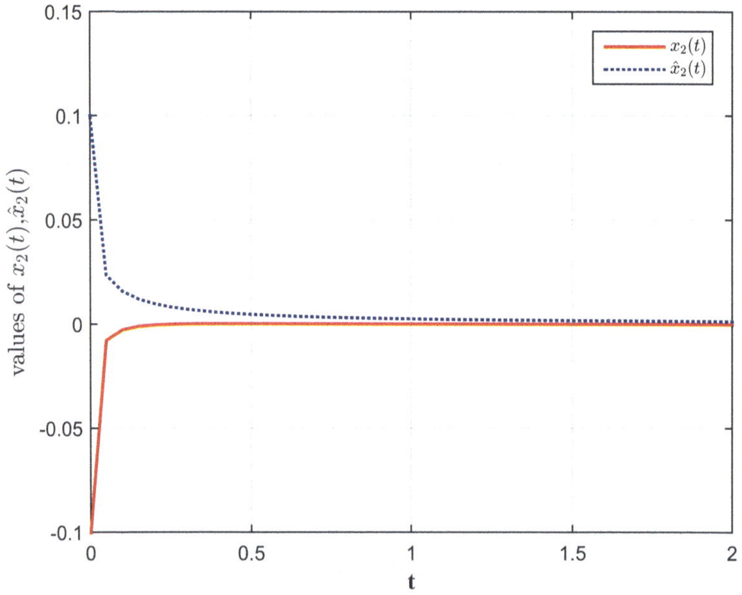

Fig. 11 The trajectory of state $\boldsymbol{x}_2$ and trajectory of state estimation $\hat{\boldsymbol{x}}_2$

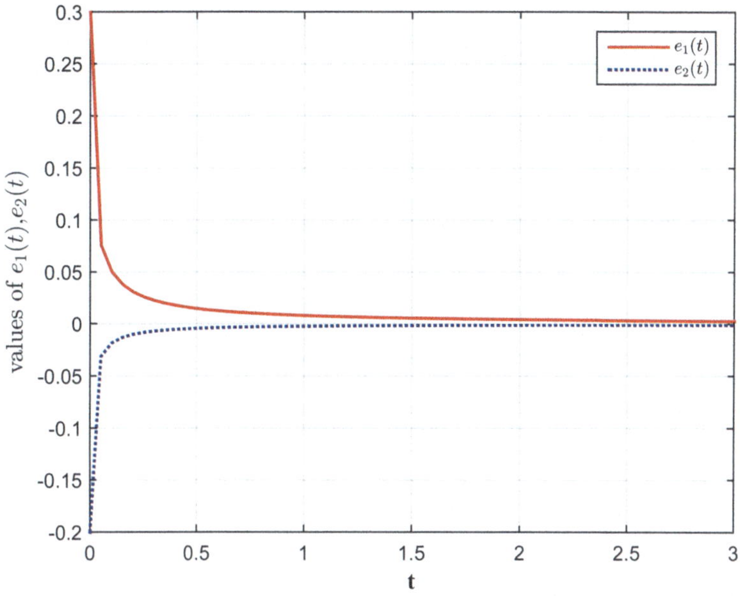

Fig. 12 The trajectory of errors

5 Conclusion

In this chapter, we discussed how to design a state observer for the FONTSs and how to design an observer-based feedback controller. A sufficient condition for judging whether the closed-loop system is asymptotically stable is given, and the design of the observer and the design of the feedback controller are attributed to the feasible solution problem of a class of LMIs. Then, the observer-based guaranteed cost controller is designed for the fractional-order systems. The work in this chapter is more practice-oriented. At last, the feasibility of the control scheme is illustrated by simulation examples. The more complex control problems of this system can be further discussed.

Acknowledgements The authors would be very grateful to the editor of this volume for their helpful comments and constructive suggestions. This research is supported by Shandong Provincial Natural Science Foundations (Grant No. ZR2019MF006 and ZR2019MF027).

References

1. Mandelbort, B.B.: The Fractal Geometry of Nature. Freeman, New York (1983)
2. Kaczorek, T.: Selected Problems of Fractional Systems Theory. Lecture Notes in Control and Information Sciences, vol. 411. Springer, Berlin (2011)

3. Kilbas, A.A., Srivastava, H.M., Trujillo, J.J.: Theory and Application of Fractional Differential Equations. Elsevier, New York (2006)
4. Miller, K.S., Ross, B.: An Introduction to the Fractional Calculus and Fractional Differential Equations. Wiley, New York (1993)
5. Li, C., Deng, W.: Remarks on fractional derivatives. Appl. Math. Comput. **187**, 777–784 (2007)
6. Podlubny, I.: Fractional Differential Equations. Academic, New York (1999)
7. Hilfer, R.: Application of Fractional Calculus in Physics. World Scientific, Singapore (2000)
8. Oldham, K.B., Spanier, J.: The Fractional Calculus. Academic, New York (1974)
9. Bao, H.B., Park, J.H., Cao, J.: Adaptive synchronization of fractional-order memristor-based neural networks with time delay. Nonlinear Dyn. **82**, 1343–1354 (2015)
10. Zhan, T., Liu, X., Ma, S.: A new singular system approach to output feedback sliding mode control for fractional order nonlinear systems. J. Frankl. Inst. **355**, 6746–6762 (2018)
11. Li, P., Chen, L., Wu, R., Machado, J.A.T., Lopes, A.M., Yuan, L.: Robust asymptotic stability of interval fractional-order nonlinear systems with time-delay. J. Frankl. Inst. **355**, 7749–7763 (2018)
12. Kong, S., Saif, M., Liu, B.: Observer design for a class of nonlinear fractional-order systems with unknown input. J. Frankl. Inst. **354**, 5503–5518 (2017)
13. Zhang, X.: Some results of linear fractional order time-delay system. Appl. Math. Comput. **197**(1), 407–411 (2008)
14. Bao, H.B., Park, J.H., Cao, J.: Non-fragile state estimation for fractional-order delayed memristive BAM neural networks. Neural Netw. **119**, 190–199 (2019)
15. Song, S., Park, J.H., Zhang, B., Song, X.: Adaptive hybrid fuzzy output feedback control for fractional-order nonlinear systems with time-varying delays and input saturation. Appl. Math. Comput. **365**, Article 124662 (2020). https://doi.org/10.1016/j.amc.2019.124662
16. Ye, H., Gao, J., Ding, Y.: A generalized Gronwall inequality and its application to a fractional differential equation. J. Math. Anal. Appl. **328**(2), 1075–1081 (2007)
17. Li, Y., Chen, Y.Q., Podlubny, I.: Mittag-Leffler stability of fractional order nonlinear dynamic systems. Automatica **45**(8), 1965–1969 (2009)
18. Li, Y., Chen, Y.Q., Podlubny, I.: Stability of fractional-order nonlinear dynamic systems: Lyapunov direct method and generalized Mittag-Leffler stability. Comput. Math. Appl. **59**, 1810–1821 (2010)
19. Li, H., Kao, Y.G.: Mittag-Leffler stability for a new coupled system of fractional-order differential equations with impulses. Appl. Math. Comput. **361**, 22–31 (2019)
20. Wen, Y., Zhou, X., Zhang, Z., Liu, S.: Lyapunov method for nonlinear fractional differential systems with delay. Nonlinear Dyn. **82**, 1015–1025 (2015)
21. Li, H., Kao, Y.G.: Synchronous stability of the fractional-order discrete-time dynamical network system model with impulsive couplings. Neurocomputing. https://doi.org/10.1016/j.neucom. Accessed 09 June 2019
22. Park, J.H., Lee, T.H., Liu, Y., Chen, J.: Dynamic Systems with Time Delays: Stability and Control. Springer-Nature, Singapore (2019). https://doi.org/10.1007/978-981-13-9254-2
23. Park, J.H., Won, S.: Asymptotic stability of neutral systems with multiple delays. J. Optim. Theory Appl. **103**, 183–200 (1999)
24. Park, J.H., Won, S.: Stability analysis for neutral delay-differential systems. J. Frankl. Inst. **337**, 1–9 (2000)
25. Liu, S., Wu, X., Zhang, Y.J., Yang, R.: Asymptotical stability of Riemann-Liouville fractional neutral systems. Appl. Math. Lett. **69**, 168–173 (2007)
26. Zhang, H., Ye, R., Cao, J., Alsaedi, A.: Delay-independent stability of Riemann-Liouville fractional neutral-type delayed neural networks. Neural Process. Lett. **47**, 427–442 (2018)
27. Moornani, K.A., Haeri, M.: On robust stability of LTI fractional-order delay systems of retarded and neutral type. Automatica **46**(2), 362–368 (2010)
28. Pakzad, M.A.: On the stability of fractional-order systems of neutral type. J. Comput. Nonlinear Dyn. **10**(5), 1415–1555 (2015)
29. Vadivoo, B.S., Ramachandran, R., Cao, J., et al.: Controllability analysis of nonlinear neutral-type fractional-order differential systems with state delay and impulsive effects. Int. J. Control Autom. Syst. **16**(2), 659–669 (2018)

30. Park, J.H.: On the design of observer-based controller of linear neutral delay-differential systems. Appl. Math. Comput. **150**, 195–202 (2004)
31. Lien, C.H.: Guaranteed cost observer-based controls for a class of uncertain neutral time-delay systems. J. Optim. Theory Appl. **126**(1), 137–156 (2005)
32. Lien, C.H.: $\mathcal{H}_\infty$ observer-based control for a class of uncertain neutral time-delay systems via LMI optimization approach. J. Optim. Theory Appl. **127**(1), 129–144 (2005)
33. Kao, Y.G., Xie, J., Wang, C., Karimi, H.R.: A sliding mode approach to H_∞ non-fragile observer-based control design for uncertain Markovian neutral-type stochastic systems. Automatica **52**, 218–226 (2015)
34. Liu, Z., Gao, C., Kao, Y.G.: Robust $\mathcal{H}_\infty$ control for a class of neutral-type systems via sliding mode observer. Appl. Math. Comput. **271**, 669–681 (2015)
35. Boyd, S., El Ghaoui, L., Feron, E., Balakrishnan, V.: Linear Matrix Inequalities in System and Control Theory. SIAM, Philadelphia (1994)
36. Phat, V.N., Thuan, M.V., Tuan, T.N.: New criteria for guaranteed cost control of nonlinear fractional-order delay systems: a Razumikhin approach. Vietnam J. Math. 1–13 (2018)

Adaptive Synchronization of Fractional-Order Delayed Memristive Neural Networks

Haibo Bao, Ju H. Park, and Jinde Cao

Abstract In this chapter, an adaptive synchronization problem is considered for fractional-order memristor-based neural networks (FMNNs) with time delay. Adaptive delay feedback control and a fractional-order inequality are adopted to get synchronization between the master system and the slave system. Novel synchronization criteria are constructed. Ultimately, numerical simulations are carried out to testify the main results.

Keywords Memristor-based neural networks · Adaptive synchronization · Fractional-order · Time-delay

1 Introduction

As an important phenomenon in nature and human society, synchronization has been found an increasingly wide utilization in secure communication, biological systems, signal processing, combinatorial optimization, see [1–10] and the references therein. Pecora and Carroll's work on chaos synchronization has set off a trend to study networks synchronization [11]. With the help of control theory, many methods for realizing and controlling synchronization have come into being, such as adaptive

---

H. Bao
School of Mathematics and Statistics, Southwest University, Chongqing 400715, People's Republic of China
e-mail: hbbao@swu.edu.cn

J. H. Park (✉)
Department of Electrical Engineering, Yeungnam University, 280 Daehak-Ro, Gyeongsan 38541, Republic of Korea
e-mail: jessie@ynu.ac.kr

J. Cao
School of Mathematics, Southeast University, Nanjing 210096, People's Republic of China
e-mail: jdcao@seu.edu.cn

J. H. Park (ed.), *Recent Advances in Control Problems of Dynamical Systems and Networks*, Studies in Systems, Decision and Control 301,
https://doi.org/10.1007/978-3-030-49123-9_13

control [12, 13], linear and nonlinear feedback control [14, 15], pinning control [16–18], intermittent control [19, 20], impulsive control [21–23], and so on.

During recent decades, much more attention has been paid to fractional-order models due to their memory and hereditary merits in a lot of materials and courses. Some phenomena in science and engineering, were well simulated or analyzed, which benefits from the advantage of fractional-order calculus and fractional-order differential equations [24–27].

Recently, many researchers focus on the problem of synchronization and dynamical behaviors for fractional-order networks and systems [28–36]. Compared with integer order calculus and integer-order differential systems, models and analysis by fractional-order equations and calculus in science and engineering fields have higher accuracy. The issue of projective synchronization was addressed by means of open-loop control [32]. Some results showed that it is a good way to stabilize fractional-order networks through decreasing the order of the fractional-order networks [33]. Numerical simulations are adopted by Huang et al. to investigate chaos and hyperchaos for a class of fractional-order neural networks (FNNs) [35].

Time delays often result in undesirable behaviors, such as chaos, oscillation and bifurcations, which do exist in artificial neural networks [37]. Hence, time delays play an unavoidable role in investigating the dynamical behaviors of FNNs [38–40]. From the fact that most existing synchronization works for FNNs are the ones without time delay, the synchronization of FNNs with time delay still needs further investigation.

An outstanding feature of memristor, proposed by Prof. Chua [41], is its memory quality, which was verified by Hewlett Packard lab 40 years later [42, 43]. In novel analog circuits, memristors succeeded in emulating the basic associative memory functions of the human brain, which cannot be realized by resistors. Thus, investigations on dynamics analysis and synchronization for memristor-based neural networks (MNNs) have become a hotspot [44–50]. Due to the fact that no papers reported synchronization for FMNNs with time delays, it makes urgent sense to design proper controllers to reach the synchronization of FMNNs, from the aspect of time delays.

To summarized the above discussions, it is of great significance to look into the synchronization of FMNNs with time delay. The main contributions of this chapter are listed as below.

(1) First attempt to study the synchronization of FMNNs is realized by designing a novel adaptive delay feedback controller;
(2) Different from other existing papers, two important factors, discontinuous right-hand system and time delay are both considered when establishing fractional-order models;
(3) Some existing results can be seen as the special cases of our results.

The chapter content is as follows. Section 2 introduces some preliminaries, lemmas and problem formulation. In Sect. 3, we give new adaptive synchronization criteria. In Sect. 4, two examples are utilized to examine the proposed synchronization criteria. At last, the concluding statements are made in Sect. 5.

2 Preliminaries

Some preparations about definitions, lemmas and fractional calculus are made to establish the models in this chapter.

Definition 1 ([24, 25]) The q order fractional integral for the function h is defined as

$$I^q h(x) = \frac{1}{\Gamma(q)} \int_{x_0}^{x} (x-t)^{q-1} h(t)dt,$$

where $x \geq x_0$ and $q > 0$.

Definition 2 ([24, 25]) For function $h \in \mathcal{C}^k([x_0, +\infty), \mathcal{R})$, its q order Caputo's derivative has the following form:

$$D^q h(x) = \frac{1}{\Gamma(k-q)} \int_{x_0}^{x} (x-t)^{k-q-1} h^{(k)}(t)dt,$$

where $x \geq x_0$ and k is a positive integer, $k-1 < q < k$.
Furthermore, $D^q h(x) = \frac{1}{\Gamma(1-q)} \int_{x_0}^{x} (x-t)^{-q} h'(t)dt$, $0 < q < 1$.

The model of FMNN with time delay as the master system is depicted by

$$D^q u_i(t) = -a_i u_i(t) + \sum_{j=1}^{m} w_{ij}(u_j(t)) h_j(u_j(t)) + \sum_{j=1}^{m} v_{ij}(u_j(t)) h_j(u_j(t-\delta)) + l_i, \tag{1}$$

where $0 < q < 1$, $i = 1, 2, \ldots, m$, $t \geq 0$, $u(t) = (u_1(t), \ldots, u_m(t))^T$, $u(t)$ stands for neurons' states, $a_i > 0$, l_i represents a constant external input, $h_i(\cdot)$ ($i = 1, 2, \ldots, m$) and $\delta > 0$ represent nonlinear activation functions and time delay, respectively. Memristive connection weights $w_{ij}(u_j(t))$ and $v_{ij}(u_j(t))$ have the presentations like

$$w_{ij}(u_j(t)) = \begin{cases} w_{ij}^*, & |u_j(t)| > X_j, \\ w_{ij}^{**}, & |u_j(t)| < X_j, \end{cases} \quad v_{ij}(u_j(t)) = \begin{cases} v_{ij}^*, & |u_j(t)| > X_j, \\ v_{ij}^{**}, & |u_j(t)| < X_j, \end{cases}$$

$w_{ij}(\pm X_j) = w_{ij}^*$ or w_{ij}^{**}, $v_{ij}(\pm X_j) = v_{ij}^*$ or v_{ij}^{**}, for $i,\ j = 1, 2, \ldots, m$. Switching jumps $X_j > 0$, weights w_{ij}^*, w_{ij}^{**}, v_{ij}^* and v_{ij}^{**} are all invariant scalars.

The slave system is built as follows:

$$D^q \tilde{u}_i(t) = -a_i \tilde{u}_i(t) + \sum_{j=1}^{m} w_{ij}(\tilde{u}_j(t)) h_j(\tilde{u}_j(t))$$
$$+ \sum_{j=1}^{m} v_{ij}(\tilde{u}_j(t)) h_j(\tilde{u}_j(t-\delta)) + l_i + r_i(t), \tag{2}$$

where $r_i(t)$ is a controller , $i = 1, 2, \ldots, m$, $t \geq 0$.

Since the discontinuity of $w_{ij}(u_j(t))$ and $v_{ij}(u_j(t))$, all solutions of master-slave systems are Filippov's ones.

With the aid of the theories of differential inclusions and set-valued maps [51], from (1) and (2), we get

$$D^q u_i(t) \in -a_i u_i(t) + \sum_{j=1}^{m} co[w_{ij}(u_j(t))] h_j(u_j(t))$$
$$+ \sum_{j=1}^{m} co[v_{ij}(u_j(t))] h_j(u_j(t-\delta)) + l_i, \tag{3}$$

and

$$D^q \tilde{u}_i(t) \in -a_i \tilde{u}_i(t) + \sum_{j=1}^{m} co[w_{ij}(\tilde{u}_j(t))] h_j(\tilde{u}_j(t))$$
$$+ \sum_{j=1}^{m} co[v_{ij}(\tilde{u}_j(t))] h_j(\tilde{u}_j(t-\delta)) + l_i + r_i(t), \tag{4}$$

for a.e. $t \geq 0$, $i = 1, 2, \ldots, m$, where $0 < q < 1$,

$$co[w_{ij}(u_j(t))] = \begin{cases} w_{ij}^*, & |u_j(t)| > X_j, \\ co\{w_{ij}^*, w_{ij}^{**}\}, & |u_j(t)| = X_j, \\ w_{ij}^{**}, & |u_j(t)| < X_j. \end{cases}$$

$$co[v_{ij}(u_j(t))] = \begin{cases} v_{ij}^*, & |u_j(t)| > X_j, \\ co\{v_{ij}^*, v_{ij}^{**}\}, & |u_j(t)| = X_j, \\ v_{ij}^{**}, & |u_j(t)| < X_j. \end{cases}$$

$$co[w_{ij}(\tilde{u}_j(t))] = \begin{cases} w_{ij}^*, & |\tilde{u}_j(t)| > X_j, \\ co\{w_{ij}^*, w_{ij}^{**}\}, & |\tilde{u}_j(t)| = X_j, \\ w_{ij}^{**}, & |\tilde{u}_j(t)| < X_j. \end{cases}$$

$$co[v_{ij}(\tilde{u}_j(t))] = \begin{cases} v_{ij}^*, & |\tilde{u}_j(t)| > X_j, \\ co\{v_{ij}^*, v_{ij}^{**}\}, & |\tilde{u}_j(t)| = X_j, \\ v_{ij}^{**}, & |\tilde{u}_j(t)| < X_j, \end{cases}$$

with $co\{w_{ij}^*, w_{ij}^{**}\} = [\underline{w}_{ij}, \overline{w}_{ij}]$, $co\{v_{ij}^*, v_{ij}^{**}\} = [\underline{v}_{ij}, \overline{v}_{ij}]$, $\underline{w}_{ij} = \min\{w_{ij}^*, w_{ij}^{**}\}$, $\overline{w}_{ij} = \max\{w_{ij}^*, w_{ij}^{**}\}$, $\underline{v}_{ij} = \min\{v_{ij}^*, v_{ij}^{**}\}$, $\overline{v}_{ij} = \max\{v_{ij}^*, v_{ij}^{**}\}$.

Or, there exist $\xi_{ij}(u_j(t)) \in co[w_{ij}(u_j(t))]$, $\eta_{ij}(\tilde{u}_j(t)) \in co[w_{ij}(\tilde{u}_j(t))]$, $\zeta_{ij}(u_j(t)) \in co[v_{ij}(u_j(t))]$ and $\chi_{ij}(\tilde{u}_j(t)) \in co[v_{ij}(\tilde{u}_j(t))]$ satisfying

$$\begin{aligned} D^q u_i(t) &= -a_i u_i(t) + \sum_{j=1}^{m} \xi_{ij}(u_j(t)) h_j(u_j(t)) \\ &\quad + \sum_{j=1}^{m} \zeta_{ij}(u_j(t)) h_j(u_j(t-\delta)) + l_i, \end{aligned} \tag{5}$$

and

$$\begin{aligned} D^q \tilde{u}_i(t) &= -a_i \tilde{u}_i(t) + \sum_{j=1}^{m} \eta_{ij}(\tilde{u}_j(t)) h_j(\tilde{u}_j(t)) \\ &\quad + \sum_{j=1}^{m} \chi_{ij}(\tilde{u}_j(t)) h_j(\tilde{u}_j(t-\delta)) + l_i + r_i(t), \end{aligned} \tag{6}$$

for a.e. $t \geq 0$, $i = 1, 2, \ldots, m$.

It is assumed that the initial conditions of master-slave systems (1) and (2) are $u_i(s) = \phi_i(s)$ and $\tilde{u}_i(s) = \psi_i(s)$, respectively, for $s \in [-\delta, 0]$, $\phi_i(s)$, $\psi_i(s) \in \mathcal{C}([-\delta, 0], \mathcal{R})$, $i = 1, 2, \ldots, m$.

Definition 3 ([52]) $u(t) = (u_1(t), u_2(t), \ldots, u_m(t))^T$ is called a Filippov solution of system (1) with initial condition $\phi_i(s) = (\phi_1(s), \phi_2(s), \ldots, \phi_m(s))^T \in \mathcal{C}([-\delta, 0], \mathcal{R}^m)$, if $u(t)$ is absolutely continuous on any compact interval of $[0, +\infty)$ and satisfies the differential inclusions (4) or (6).

The following assumption determines that the solution of system (1) or fractional differential inclusion (4), is existent and unique.

Assumption 1 The functions h_j are Lipschitz-continuous on $\mathcal{R}$ i.e.,

$$|h_j(x) - h_j(y)| \leq L_j |x - y|$$

for all x, $y \in \mathcal{R}$ and $j = 1, 2, \ldots, m$. □

Lemma 1 ([32, 53]) *The following inequality holds if function $h(t)$ belongs to the nondecreasing and differential functions set,*

$$D^q (h(t) - g)^2 \leq 2(h(t) - g) D^q h(t),$$

where g is any constant, $t \in [0, \infty)$, $0 < q < 1$.

Lemma 2 ([50]) *If Assumption 1 holds and* $h_j(\pm X_j) = 0$ $(j = 1, 2, \ldots, m)$, *then*

$$|co(w_{ij}(u_j(t)))h_j(u_j(t)) - co(w_{ij}(\tilde{u}_j(t)))h_j(\tilde{u}_j(t))| \\ \leq w_{ij}^z L_j|u_j - \tilde{u}_j|, \ for\ i, j = 1, 2, \ldots, m,$$

that is, for any $\xi_{ij}(u_j(t)) \in co[w_{ij}(u_j(t))]$, $\eta_{ij}(\tilde{u}_j(t)) \in co[w_{ij}(\tilde{u}_j(t))]$,

$$|\xi_{ij}(u_j(t)) - \eta_{ij}(\tilde{u}_j(t))| \leq w_{ij}^z L_j|u_j - \tilde{u}_j|,$$

for $i, j = 1, 2, \ldots, m$, $w_{ij}^z = \max\{|w_{ij}^*|, |w_{ij}^{**}|\}$.

3 Main Results

In this section, adaptive synchronization schemes are taken for FMNNs to get effective and simple conditions.

Let $\bar{u}_i(t) = \tilde{u}_i(t) - u_i(t)$, for $i = 1, 2, \ldots, m$ and

$$\begin{cases} r_i(t) = -\alpha_i(t)\bar{u}_i(t) - sgn(\bar{u}_i(t))\beta_i(t)|\bar{u}_i(t-\delta)|, \\ D^q\alpha_i(t) = \rho_i|\bar{u}_i(t)|, \\ D^q\beta_i(t) = \theta_i|\bar{u}_i(t-\delta)|. \end{cases} \tag{7}$$

where $\rho_i > 0$ and $\theta_i > 0$ are any scalars. $\bar{u}_i(t) \to 0$ as $t \to \infty$ $(i = 1, 2, \ldots, m)$ implies that the synchronization between master system (1) and the slave system (2) is reached.

The error dynamical system can be got from (5) and (6):

$$\begin{aligned} D^q\bar{u}_i(t) = & -a_i\bar{u}_i(t) + \sum_{j=1}^{m}[\eta_{ij}(\tilde{u}_j(t))h_j(\tilde{u}_j(t)) - \xi_{ij}(u_j(t))h_j(u_j(t))] \\ & + \sum_{j=1}^{m}[\chi_{ij}(\tilde{u}_j(t))h_j(\tilde{u}_j(t-\delta)) - \zeta_{ij}(u_j(t))h_j(u_j(t-\delta)] \\ & -\alpha_i(t)\bar{u}_i(t) - sgn(\bar{u}_i(t))\beta_i(t)|\bar{u}_i(t-\delta)|, \end{aligned} \tag{8}$$

for a.e. $t \geq 0$, $i = 1, 2, \ldots, m$.

Now, we have the following main result.

Theorem 1 *If Assumption 1 and* $h_j(\pm X_j) = 0$ $(j = 1, 2, \ldots, m)$ *hold, the globally asymptotical synchronization will be derived for master-slvae systems (1) and (2) based on the controller (7).*

Proof Let $u(t) = (u_1(t), \ldots, u_m(t))^T$ and $\tilde{u}(t) = (\tilde{u}_1(t), \ldots, \tilde{u}_m(t))^T$ denote two arbitrary solutions of systems (1) and (2) with initial conditions $u(t_0) = (u_1(t_0), \ldots, u_m(t_0))^T$ and $\tilde{u}(t_0) = (\tilde{u}_1(t_0), \ldots, \tilde{u}_m(t_0))^T$ meeting $\bar{u}_i(t_0) \neq 0$ for $i = 1, \ldots, m$.

Apparently, $\bar{u}_i(t) = 0$ is a solution of the error system (8).

Since the solution of fractional order differential functions is existent and unique [24], $\bar{u}_i(t_0)\bar{u}_i(t) > 0$ for $t > t_0$.

If $\bar{u}_i(t_0) > 0$, then $\bar{u}_i(t) > 0$ and

$$D^q|\bar{u}_i(t)| = \frac{1}{\Gamma(1-q)}\int_{t_0}^{t}\frac{\bar{u}_i'(s)}{(t-s)^q}ds = D^q\bar{u}_i(t),$$

and if $\bar{u}_i(t_0) < 0$, then $\bar{u}_i(t) < 0$ and

$$D^q|\bar{u}_i(t)| = -\frac{1}{\Gamma(1-q)}\int_{t_0}^{t}\frac{\bar{u}_i'(s)}{(t-s)^q}ds = -D^q\bar{u}_i(t),$$

So,

$$D^q|\bar{u}_i(t)| = sgn(\bar{u}_i(t))D^q\bar{u}_i(t).$$

Building a Lyapunov function:

$$V(t) = \sum_{i=1}^{m}|\bar{u}_i(t)| + \sum_{i=1}^{m}\frac{1}{2\rho_i}(\alpha_i(t)-\alpha_i)^2 + \sum_{i=1}^{m}\frac{1}{2\theta_i}(\beta_i(t)-\beta_i)^2, \tag{9}$$

where adaptive constants α_i and β_i will be decided later.

Combing $D^q\alpha_i(t) = \rho_i|\bar{u}_i(t)|$, $D^q\beta_i(t) = \theta_i|\bar{u}_i(t-\delta)|$, with Lemmas 1 and 2, the following inequality holds

$$\begin{aligned}
D^qV(t) &= \sum_{i=1}^{m}D^q|\bar{u}_i(t)| + \sum_{i=1}^{m}\frac{1}{2\rho_i}D^q(\alpha_i(t)-\alpha_i)^2 + \sum_{i=1}^{m}\frac{1}{2\theta_i}D^q(\beta_i(t)-\beta_i)^2\\
&\le \sum_{i=1}^{m}sgn(\bar{u}_i(t))\Big\{-a_i\bar{u}_i(t)\\
&\quad+\sum_{j=1}^{m}[\eta_{ij}(\tilde{u}_j(t))h_j(\tilde{u}_j(t)) - \xi_{ij}(u_j(t))h_j(u_j(t))]\\
&\quad+\sum_{j=1}^{m}[\chi_{ij}(\tilde{u}_j(t))h_j(\tilde{u}_j(t-\delta)) - \zeta_{ij}(u_j(t))h_j(u_j(t-\delta))] - \alpha_i(t)\bar{u}_i(t)\\
&\quad-sgn(\bar{u}_i(t))\beta_i(t)|\bar{u}_i(t-\delta)|\Big\} + \sum_{i=1}^{m}\frac{1}{\rho_i}(\alpha_i(t)-\alpha_i)\rho_i|\bar{u}_i(t)|\\
&\quad+\sum_{i=1}^{m}\frac{1}{\theta_i}(\beta_i(t)-\beta_i)\theta_i|\bar{u}_i(t-\delta)|
\end{aligned}$$

$$\le \sum_{i=1}^{m}\Big\{-a_i|\bar{u}_i(t)| + \sum_{j=1}^{m} w_{ij}^z L_j|\bar{u}_j(t)| + \sum_{j=1}^{m} v_{ij}^z L_j|\bar{u}_j(t-\delta)| - \alpha_i|\bar{u}_i(t)|$$
$$-\beta_i|\bar{u}_i(t-\delta)|\Big\}$$
$$= -\sum_{i=1}^{m}\Big\{a_i+\alpha_i-\sum_{j=1}^{m} w_{ji}^z L_i\Big\}|\bar{u}_i(t)| - \sum_{i=1}^{m}\Big\{\beta_i - \sum_{j=1}^{m} v_{ji}^z L_i\Big\}|\bar{u}_i(t-\delta)|.$$

Suitably selecting α_i and β_i such that: $a_i+\alpha_i-\sum_{j=1}^{m} w_{ji}^z L_i > 0$ and $\beta_i - \sum_{j=1}^{m} v_{ji}^z L_i > 0$, for $i = 1, 2, \ldots, m$.

Let

$$\pi_1 = \min\Big\{a_i+\alpha_i-\sum_{j=1}^{m} w_{ji}^z L_i\Big\} > 0,$$

and

$$\pi_2 = \min\Big\{\beta_i - \sum_{j=1}^{m} v_{ji}^u L_i\Big\} > 0,$$

then we have

$$D^q V(t) \le -\pi_1 \sum_{i=1}^{m} |\bar{u}_i(t)| - \pi_2 \sum_{i=1}^{m} |\bar{u}_i(t-\delta)|,\ t \ge t_0.$$

Let $t > t_0$, and

$$\begin{aligned} D^q V(t) &= h(t, \bar{u}(t), \bar{u}(t-\delta)) \\ &\le -\pi_1 \sum_{i=1}^{m} |\bar{u}_i(t)| - \pi_2 \sum_{i=1}^{m} |\bar{u}_i(t-\delta)| \\ &\le -\pi_1 \sum_{i=1}^{m} |\bar{u}_i(t)| = -\pi_1 Y(t) \le 0, \end{aligned} \tag{10}$$

where $Y(t) = \sum_{i=1}^{m} |\bar{u}_i(t)|$.
We can get

$$V(t) - V(t_0) = \frac{1}{\Gamma(q)} \int_{t_0}^{t} (t-s)^{q-1} h(s, \bar{u}(s), \bar{u}(s-\delta))ds \le 0,$$

from Definition 1, which implied that $V(t) \le V(t_0),\ t \ge t_0$.

And, the boundness of $\bar{u}_i(t)$, $\alpha_i(t)$ and $\beta_i(t)$ are direct results of (9) on $t \ge t_0$. Thus, there exists a positive constant $H > 0$ satisfying

$$|D^q Y(t)| \le H,\ \ t \ge t_0. \tag{11}$$

Next, we prove $\lim_{t\to\infty} Y(t) = 0$ is true. And the method of contradiction is utilized.

If (11) is not true, the inequality

$$Y(t_n) \geq \epsilon,\ n = 1, 2, \ldots, \tag{12}$$

holds for some constant $\epsilon > 0$ and the time series $\{t_n\}$ meeting $t_0 < t_1 < t_2 < \ldots < t_n < \ldots$.

Let $Q = (\frac{\Gamma(q+1)\epsilon}{4H})^{\frac{1}{q}} > 0$. If $t_n \leq t \leq t_n + Q$, $n = 1, 2, \ldots$, we get

$$\begin{aligned}
|Y(t) - Y(t_n)| &= \frac{1}{\Gamma(q)}|\int_{t_0}^{t}(t-s)^{q-1}D^qY(s)ds \\
&\quad - \int_{t_0}^{t_n}(t_n-s)^{q-1}D^qY(s)ds| \\
&= \frac{1}{\Gamma(q)}|\int_{t_0}^{t_n}[(t-s)^{q-1}-(t_n-s)^{q-1}]D^qY(s)ds \\
&\quad + \int_{t_n}^{t}(t-s)^{q-1}D^qY(s)ds| \\
&\leq \frac{1}{\Gamma(q)}[\int_{t_0}^{t_n}|[(t-s)^{q-1}-(t_n-s)^{q-1}]D^qY(s)|ds \\
&\quad + \int_{t_n}^{t}|(t-s)^{q-1}D^qY(s)|ds] \\
&\leq \frac{H}{\Gamma(q)}[\int_{t_0}^{t_n}[(t_n-s)^{q-1}-(t-s)^{q-1}]ds + \int_{t_n}^{t}(t-s)^{q-1}ds] \\
&= \frac{H}{\Gamma(q+1)}[(t_n-t_0)^q-(t-t_0)^q+2(t-t_n)^q] \\
&\leq \frac{2H}{\Gamma(q+1)}(t-t_n)^q \leq \frac{\epsilon}{2}.
\end{aligned}$$

From (11) and (12), this indicates $Y(t) \geq \frac{\epsilon}{2}$, $t_n \leq t \leq t_n + Q$, $n = 1, 2, \ldots$.

Analogously, for $t_n - Q \leq t \leq t_n$ and $n = 1, 2 \ldots$, we can get

$$\begin{aligned}
|Y(t) - Y(t_n)| &= \frac{1}{\Gamma(q)}|\int_{t_0}^{t_n}(t_n-s)^{q-1}D^qY(s)ds \\
&\quad - \int_{t_0}^{t}(t-s)^{q-1}D^qY(s)ds| \\
&= \frac{1}{\Gamma(q)}|\int_{t_0}^{t}[(t_n-s)^{q-1}-(t-s)^{q-1}]D^qY(s)ds \\
&\quad + \int_{t}^{t_n}(t_n-s)^{q-1}D^qY(s)ds|
\end{aligned}$$

$$\le \frac{1}{\Gamma(q)}[\int_{t_0}^{t} |[(t_n-s)^{q-1}-(t-s)^{q-1}]D^qY(s)|ds$$
$$+\int_{t}^{t_n} |(t_n-s)^{q-1}D^qY(s)|ds]$$
$$\le \frac{H}{\Gamma(q)}[\int_{t_0}^{t}[(t-s)^{q-1}-(t_n-s)^{q-1}]ds+\int_{t}^{t_n}(t_n-s)^{q-1}ds]$$
$$=\frac{H}{\Gamma(q+1)}[(t-t_0)^q-(t_n-t_0)^q+2(t_n-t)^q]$$
$$\le \frac{2H}{\Gamma(q+1)}(t_n-t)^q\le\frac{\epsilon}{2}.$$

From (11) and (12) , $Y(t)\ge\frac{\epsilon}{2}$, $t_n-Q\le t\le t_n+Q$, $n=1,2,\ldots$.
Hence,

$$Y(t)\ge\frac{\epsilon}{2},\quad t_n-Q\le t\le t_n+Q, n=1,2,\ldots. \tag{13}$$

Under the assumption of these intervals are disjoint and $t_1-Q>t_0$, for any $n=1,2,\ldots$, it holds

$$t_{n-1}+Q<t_n-Q<t_n+Q<t_{n+1}-Q. \tag{14}$$

Combing with (10) and (13), when $t_n-Q\le t\le t_n+Q$,

$$D^qV(t)\le-\frac{\epsilon}{2}\pi_1$$

holds.

So, for any $n=1,2,\ldots$, we get

$$V(t_n+Q)-V(t_0)=\frac{1}{\Gamma(q)}\int_{t_0}^{t_n+Q}(t_n+Q-s)^{q-1}D^qV(s)ds$$
$$=\frac{1}{\Gamma(q)}\Bigg[\int_{t_0}^{t_1-Q}+\int_{t_1-Q}^{t_1+Q}+\int_{t_1+Q}^{t_2-Q}+\int_{t_2-Q}^{t_2+Q}+\int_{t_2+Q}^{t_3-Q}+\cdots$$
$$+\int_{t_n-Q}^{t_n+Q}\Bigg](t_n+Q-s)^{q-1}D^qV(s)ds$$
$$\le\frac{1}{\Gamma(q)}\sum_{i=1}^{n}\int_{t_i-Q}^{t_i+Q}(t_n+Q-s)^{q-1}D^qV(s)ds$$
$$\le-\frac{\epsilon}{2}\pi_1\frac{1}{\Gamma(q)}\sum_{i=1}^{n}\int_{t_i-Q}^{t_i+Q}(t_n+Q-s)^{q-1}ds$$
$$=\frac{\epsilon}{2}\pi_1\frac{1}{\Gamma(q+1)}\sum_{i=1}^{n}\big[(t_n-t_i)^q-(t_n-t_i+2Q)^q\big]. \tag{15}$$

From (15), we get $V(t_n+Q) \leq V(t_0) + \frac{\epsilon}{2}\pi_1 \frac{1}{\Gamma(q+1)} \sum_{i=1}^{n} [(t_n - t_i)^q - (t_n - t_i + 2Q)^q]$. This shows that $V(t_n + Q) \to -\infty$ as $n \to +\infty$, and this contradicts with $V(t) \geq 0$.

Hence, $\lim_{t\to\infty} Y(t) = 0$. Furthermore, globally asymptotical synchronization is acquired between the master-slave systems (1) and (2) under the controller (7). This completes proof. □

Remark 1 Apparently, the adaptive control gains $\alpha_i(t)$ and $\beta_i(t)$ converge to some positive constants when the master system (1) and the slave system (2) are synchronous.

Remark 2 If we don't consider the value changes of w_{ij}^*, w_{ij}^{**}, v_{ij}^*, v_{ij}^{**} for (1) and (2), i.e. we consider an FNN built by resistors, then we can obtain the adaptive synchronization conditions for the FNNs.

Suppose the fractional-order delayed master-slave systems have the following form:

$$D^q u_i(t) = -a_i u_i(t) + \sum_{j=1}^{m} w_{ij} h_j(u_j(t)) + \sum_{j=1}^{m} v_{ij} h_j(u_j(t-\delta)) + l_i, \quad (16)$$

$$\begin{aligned} D^q \tilde{u}_i(t) = -a_i \tilde{u}_i(t) + \sum_{j=1}^{m} w_{ij} h_j(\tilde{u}_j(t)) + \sum_{j=1}^{m} v_{ij} h_j(\tilde{u}_j(t-\delta)) \\ + l_i + r_i(t). \end{aligned} \quad (17)$$

Then, the error system can be got under the controller (7):

$$\begin{aligned} D^q \bar{u}_i(t) = -a_i \bar{u}_i(t) + \sum_{j=1}^{m} w_{ij}[h_j(\tilde{u}_j(t)) - h_j(u_j(t))] + \sum_{j=1}^{m} v_{ij}[h_j(\tilde{u}_j(t-\delta)) \\ - h_j(u_j(t-\delta))] - \alpha_i(t)\bar{u}_i(t) - sgn(\bar{u}_i(t))\beta_i(t)|\bar{u}_i(t-\delta)|. \end{aligned} \quad (18)$$

Now, we are in the position to give synchronization for FNNs.

Theorem 2 *Under Assumption 1, the master system (16) and the slave system (17) with the controller (7) come to globally asymptotical synchronization.*

Proof Making use of the Lyapunov function (9), from Assumption 1, let us carry out the fractional-order derivative computation for $V(t)$,

$$
\begin{aligned}
D^q V(t) &= \sum_{i=1}^{m} D^q |\bar{u}_i(t)| + \sum_{i=1}^{m} \frac{1}{2\rho_i} D^q (\alpha_i(t) - \alpha_i)^2 + \sum_{i=1}^{m} \frac{1}{2\theta_i} D^q (\beta_i(t) - \beta_i)^2 \\
&\leq \sum_{i=1}^{m} sgn(\bar{u}_i(t)) \Big\{ -a_i \bar{u}_i(t) + \sum_{j=1}^{m} w_{ij} [h_j(\tilde{u}_j(t)) - h_j(u_j(t))] \\
&\quad + \sum_{j=1}^{m} v_{ij} [h_j(\tilde{u}_j(t-\delta)) - h_j(u_j(t-\delta))] - \alpha_i(t) \bar{u}_i(t) \\
&\quad - sgn(\bar{u}_i(t)) \beta_i(t) |\bar{u}_i(t-\delta)| \Big\} + \sum_{i=1}^{m} (\alpha_i(t) - \alpha_i) |\bar{u}_i(t)| \\
&\quad + \sum_{i=1}^{m} (\beta_i(t) - \beta_i) |\bar{u}_i(t-\delta)| \\
&\leq \sum_{i=1}^{m} \Big\{ -a_i |\bar{u}_i(t)| + \sum_{j=1}^{m} |w_{ij}| L_j |\bar{u}_j(t)| + \sum_{j=1}^{m} |v_{ij}| L_j |\bar{u}_j(t-\delta)| \\
&\quad - \alpha_i |\bar{u}_i(t)| - \beta_i |\bar{u}_i(t-\delta)| \Big\} \\
&= - \sum_{i=1}^{m} \Big\{ a_i + \alpha_i - \sum_{j=1}^{m} |w_{ji}| L_i \Big\} |\bar{u}_i(t)| \\
&\quad - \sum_{i=1}^{m} \Big\{ \beta_i - \sum_{j=1}^{m} |v_{ji}| L_i \Big\} |\bar{u}_i(t-\delta)|.
\end{aligned}
$$

The selection of α_i and β_i meets the conditions:

$$
a_i + \alpha_i - \sum_{j=1}^{m} |w_{ji}| L_i > 0,
$$

and

$$
\beta_i - \sum_{j=1}^{m} |v_{ji}| L_i > 0,
$$

for $i = 1, 2, \ldots, m$.
Let

$$
\pi_1 = \min \Big\{ a_i + \alpha_i - \sum_{j=1}^{m} |w_{ji}| L_i \Big\} > 0,
$$

and

$$
\pi_2 = \min \Big\{ \beta_i - \sum_{j=1}^{m} |v_{ji}| L_i \Big\} > 0,
$$

then

$$D^q V(t) \le -\pi_1 \sum_{i=1}^{m} |\bar{u}_i(t)| - \pi_2 \sum_{i=1}^{m} |\bar{u}_i(t-\delta)|,\ t \ge t_0.$$

The other proof is similar to that in the proof of Theorem 1. This completes proof. □

Remark 3 As far as we know, there is no investigation about adaptive synchronization of FMNNs or FNNs with time delay. Compared with the conclusions in [28, 31, 32], the results and models in this chapter have more generality and less conservativeness for two reasons: The systems under investigation are discontinuous right-hand ones and they are FNNs with time delay. It should be emphasized that methods used in [28, 31] do not work for the systems in this chapter.

Remark 4 An important thing we should emphasize is that Theorem 1 and 2 are new and still hold for the case $q = 1$. Moreover, results in [12, 45] are exceptional circumstances of our results.

Remark 5 The control cost is obviously cut down and easy to be accomplished by designing a simple delay feedback controller (7). Clearly, the order of the fractional-order systems affect the synchronization of FMNNs.

4 Numerical Examples

We will give two examples to testify the validity of the proposed synchronization results.

Example 1 Suppose the master system is described by the following FMNN:

$$D^q u(t) = -Au(t) + W(u(t))h(u(t)) + V(u(t))h(u(t-\delta)) + l, \tag{19}$$

where $q = 0.98$, $u = (u_1, u_2)^T$, $h(u) = (h_1(u_1), h_2(u_2))^T$, $h_j(u_j) = \tanh(|u_j| - 1)$, $L_j = 1$ $(j = 1, 2)$, $\delta = 1$, $l = (0, 0)^T$, and

$$A = 2I_{2\times 2},\quad W(u(t)) = \begin{bmatrix} w_{11}(u_1) & -0.2 \\ -10 & w_{22}(u_2) \end{bmatrix},$$

$$V(u(t)) = \begin{bmatrix} v_{11}(u_1) & -0.2 \\ -0.4 & v_{22}(u_2) \end{bmatrix},$$

$$w_{11}(u_1) = \begin{cases} -4, & |u_1| < 1 \\ -3.5, & |u_1| > 1 \end{cases},\quad w_{22}(u_2) = \begin{cases} 8.8, & |u_2| < 1 \\ 7.8, & |u_2| > 1 \end{cases},$$

$$v_{11}(u_1) = \begin{cases} -2.9, & |u_1| < 1 \\ -2.8, & |u_1| > 1 \end{cases},\quad v_{22}(u_2) = \begin{cases} -7.9, & |u_2| < 1 \\ -7.4, & |u_2| > 1 \end{cases}.$$

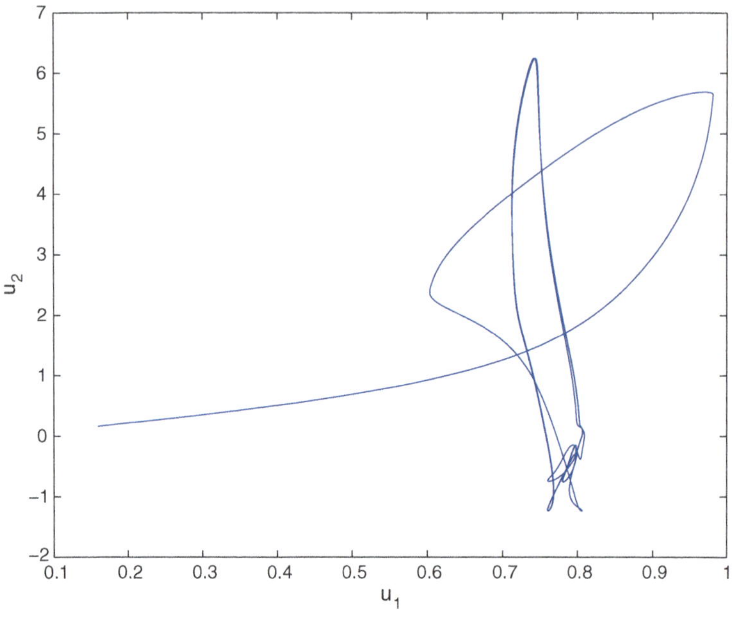

Fig. 1 Chaotic attractor of FMNN (19)

The salve system is given by:

$$D^q\tilde{u}(t) = -A\tilde{u}(t) + W(\tilde{u}(t))h(\tilde{u}(t)) + V(\tilde{u}(t))h(\tilde{u}(t-\delta)) \\ +l + r(t), \tag{20}$$

where $r(t) = (r_1(t), r_2(t))^T$.

The FMNN (19) is a chaotic system as indicated in Fig. 1.

Next, The coupling strength update law (7) will be proved to work. All the initial values are $(u_1(s), u_2(s))^T = (0.16, 0.17)^T$, $(\tilde{u}_1(s), \tilde{u}_2(s))^T = (-0.28, -0.28)^T$, $\forall s \in [-1, 0]$, $\beta_i(0) = 0.55$, $\alpha_i(0) = 0.51$ $(i = 1, 2)$.

Let $\rho_i = 2, \theta_i = 1$, and $\alpha_1 = 13.2, \alpha_2 = 8.1, \beta_1 = 4.1, \beta_2 = 8.5$, by simple computing, we can get $a_i + \alpha_i - \sum_{j=1}^m w_{ji}^z L_i > 0$, $\beta_i - \sum_{j=1}^m v_{ji}^z L_i > 0$ $(i = 1, 2)$, therefore, from Theorem 1, synchronization occurs between the master system (19) and the slave system (20).

Figure 2 gives the synchronization error. Also, Fig. 3 reveals the variable evolutions for FMNNs $u_1(t), u_2(t), \tilde{u}_1(t), \tilde{u}_2(t)$.

We can see the master-slave systems (19) and (20) are synchronized in Figs. 2 and 3. Figures 4 and 5 demonstrate the convergence of the adaptive gains $\alpha_i(t), \beta_i(t)$ $(i = 1, 2)$.

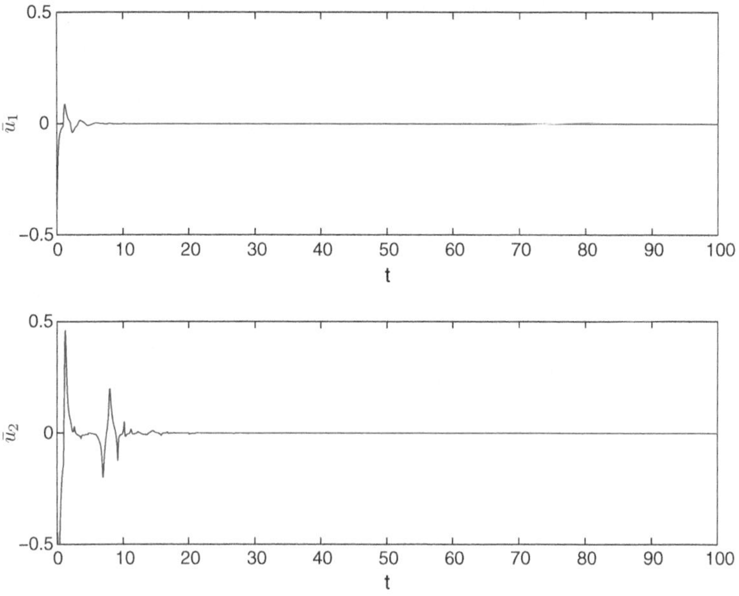

Fig. 2 The synchronization error of $\bar{u}_i(t) = \tilde{u}_i(t) - u_i(t),\ i = 1, 2$

Example 2 Suppose the master system is described by the following FMNN:

$$D^q u(t) = -Au(t) + W(u(t))h(u(t)) + V(u(t))h(u(t-\delta)) + l, \tag{21}$$

where $q = 0.98$, $u = (u_1, u_2)^T$, $h(u) = (h_1(u_1), h_2(u_2))^T$, $h_1(u_1) = \sin(|u_1| - 1)$, $h_2(u_2) = \tanh(|u_2| - 1)$, $L_j = 1\ (j = 1, 2)$, $\delta = 1$, $l = (0, 0)^T$, and

$$A = I_{2\times 2},\quad W(u(t)) = \begin{bmatrix} 1 & w_{12}(u_2) \\ w_{21}(u_1) & 1.8 \end{bmatrix},$$

$$V(u(t)) = \begin{bmatrix} v_{11}(u_1) & v_{12}(u_2) \\ v_{21}(u_1) & v_{22}(u_2) \end{bmatrix},$$

$$w_{12}(u_2) = \begin{cases} 7.1, & |u_2| < 1 \\ 5.1, & |u_2| > 1 \end{cases},\quad w_{21}(u_1) = \begin{cases} 0.85, & |u_1| < 1 \\ 1.05, & |u_1| > 1 \end{cases},$$

$$v_{11}(u_1) = \begin{cases} -1.4, & |u_1| < 1 \\ -1.1, & |u_1| > 1 \end{cases},\quad v_{12}(u_2) = \begin{cases} 1.1, & |u_2| < 1 \\ 0.9, & |u_2| > 1 \end{cases},$$

$$v_{21}(u_1) = \begin{cases} 0.9, & |u_1| < 1 \\ 1.1, & |u_1| > 1 \end{cases},\quad v_{22}(u_2) = \begin{cases} -1.45, & |u_2| < 1 \\ -1.65, & |u_2| > 1 \end{cases}.$$

The salve system is:

$$\begin{aligned} D^q \tilde{u}(t) = & -A\tilde{u}(t) + W(\tilde{u}(t))h(\tilde{u}(t)) + V(\tilde{u}(t))h(\tilde{u}(t-\delta)) \\ & + l + r(t), \end{aligned} \tag{22}$$

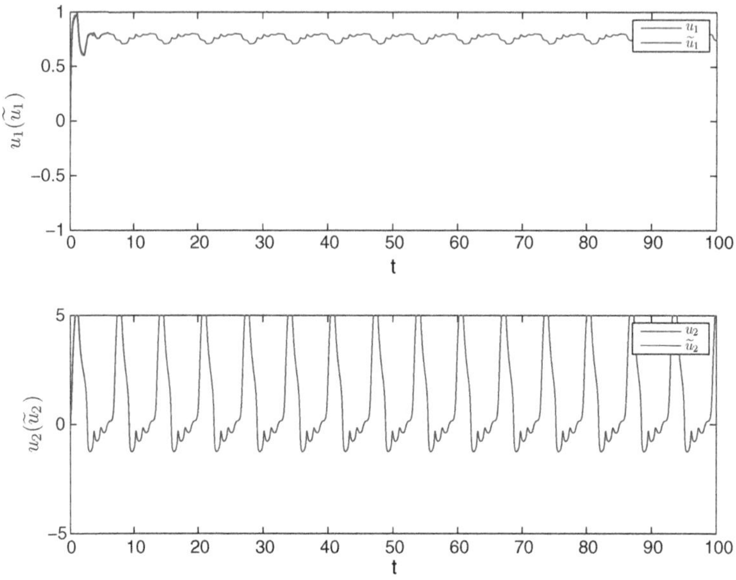

Fig. 3 Time evolution of $u_1(t)$, $u_2(t)$, $\tilde{u}_1(t)$, $\tilde{u}_2(t)$

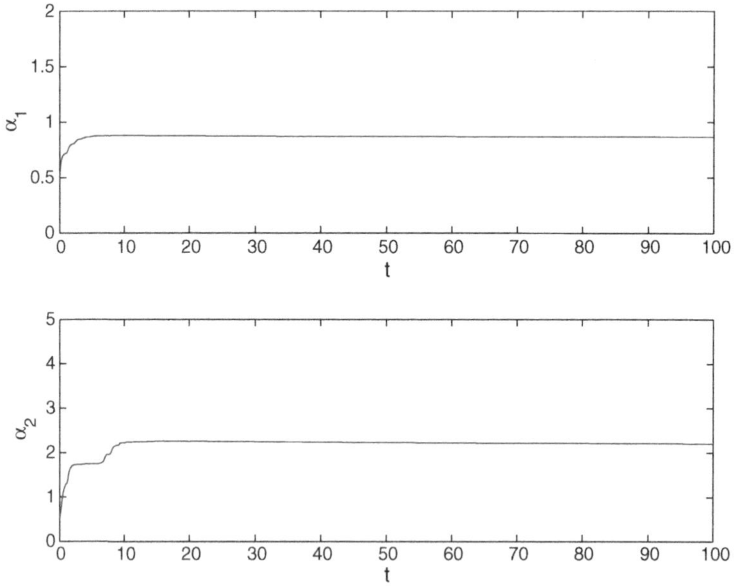

Fig. 4 Trajectories of $\alpha_i(t)$, $i = 1, 2$

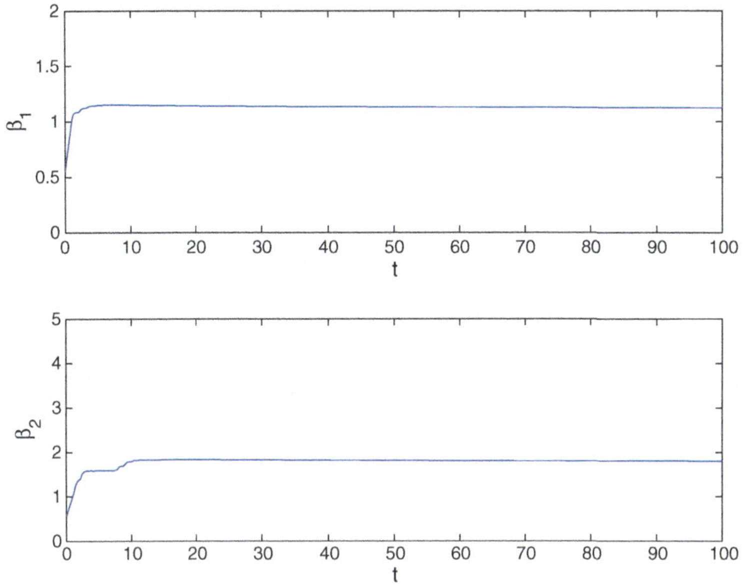

Fig. 5 Trajectories of $\beta_i(t),\ i=1,2$

where $r(t)=(r_1(t),r_2(t))^T$.

System (21) is demonstrated to be chaotic in Fig. 6. The initial values are $(u_1(s),u_2(s))^T=(0.46,0.59)^T$, $(\tilde{u}_1(s),\tilde{u}_2(s))^T=(0.2,0.3)^T$, $\forall s\in[-1,0]$, $\beta_1(0)=0.1$, $\beta_2(0)=-0.1$, $\alpha_1(0)=-0.1$, $\alpha_2(0)=0.1$.

Let $\rho_i=3.1$, $\theta_i=2.1$, and $\alpha_1=1.5$, $\alpha_2=8$, $\beta_1=3$, $\beta_2=3$.

Through calculation, we can get $a_i+\alpha_i-\sum_{j=1}^{m}w_{ji}^{z}L_i>0$, $\beta_i-\sum_{j=1}^{m}v_{ji}^{z}L_i>0$ $(i=1,2)$.

From the conditions of Theorem 1, the slave system (22) synchronizes with the master system (21). Figures 7 and 8 give the synchronization errors and the states' trajectories of FMNN, respectively, and reveals the fact that synchronization takes place between the master-slave systems (21) and (22).

Figures 9 and 10 demonstrate the convergence of the adaptive gains $\alpha_i(t)$ and $\beta_i(t)$, $(i=1,2)$.

5 Conclusion

In this chapter, the issue of adaptive synchronization has been concerned for FMNNs subject to time delay. Adaptive feedback control and fractional-order inequality are employed to obtain sufficient synchronization conditions. The proposed approach is

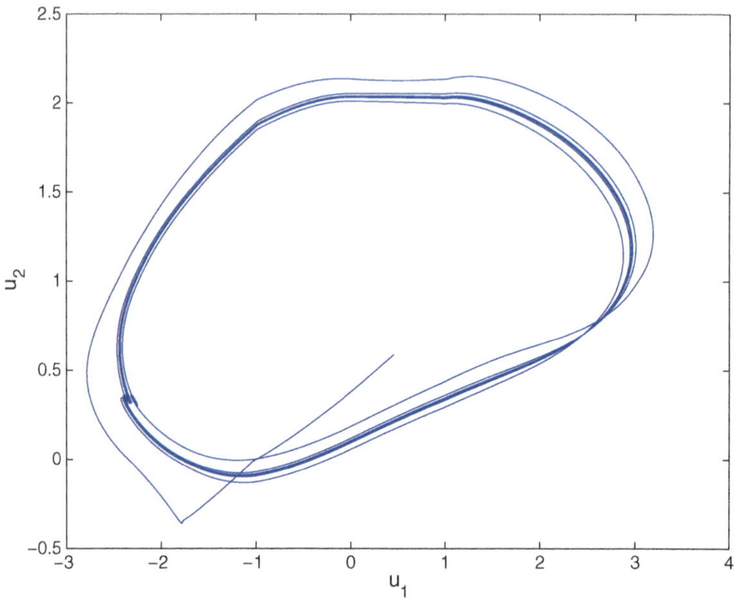

Fig. 6 Chaotic attractor of FMNN (21)

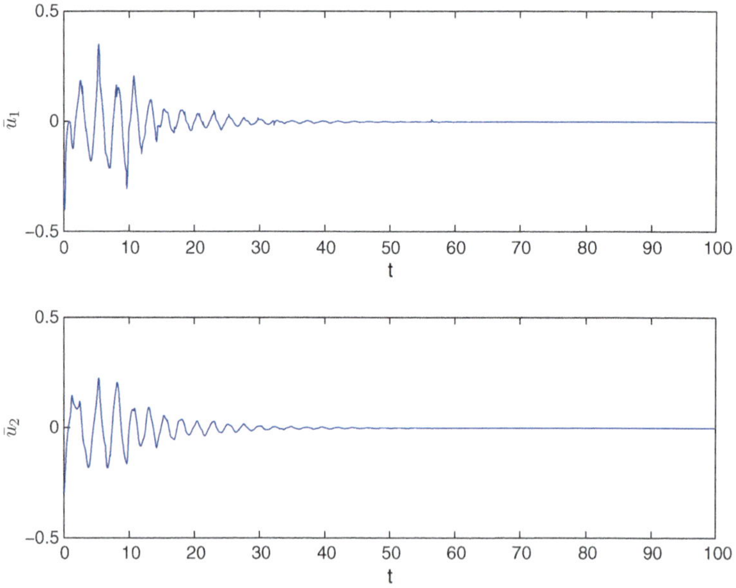

Fig. 7 The synchronization error of $\bar{u}_i(t) = \tilde{u}_i(t) - u_i(t),\ i = 1, 2$

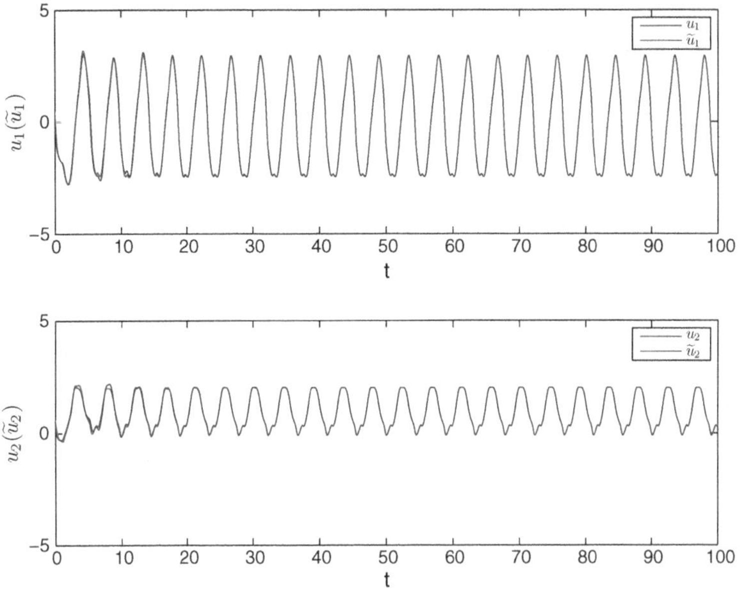

Fig. 8 Trajectories of $u_1(t), u_2(t), \tilde{u}_1(t), \tilde{u}_2(t)$

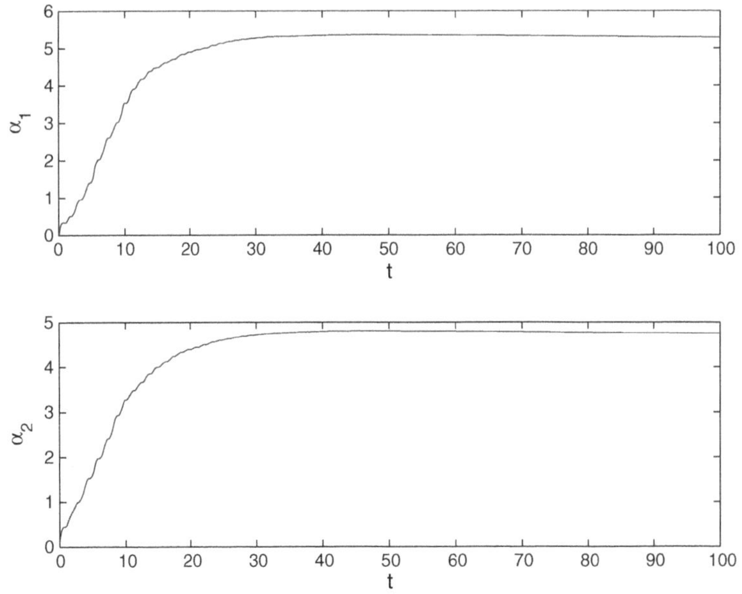

Fig. 9 Trajectories of $\alpha_i(t),\ i = 1, 2$

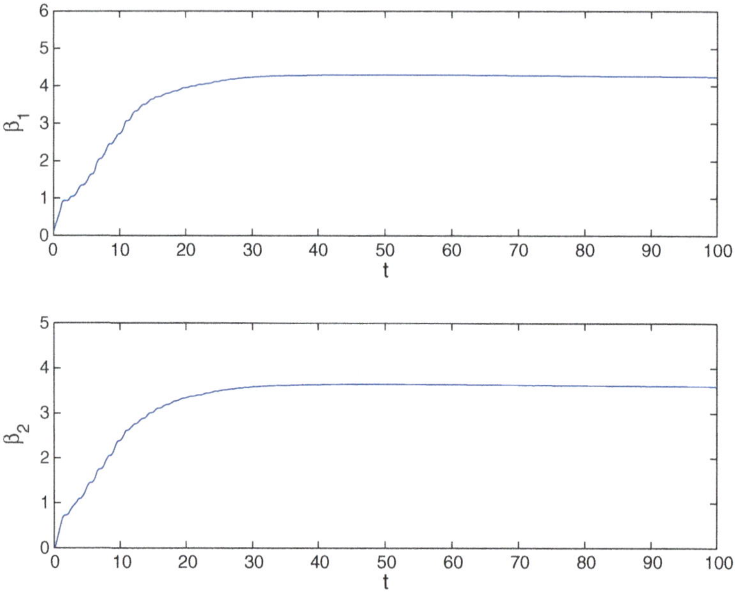

Fig. 10 Trajectories of $\beta_i(t), \ i = 1, 2$

easy to be implemented. Ultimately, we gave two examples to illustrate the correctness of the newly designed synchronization method.

Acknowledgements The work of H. Bao was jointly supported by the National Natural Science Foundation of China under Grant No. 61203096, the Chinese Postdoctoral Science Foundation under Grant 2013M513924, the Fundamental Research Funds for Central Universities XDJK2013C001 and the scientific research support project for teachers with doctor's degree, Southwest University under Grant No. SWU112024. The work of J.H. Park was supported by the National Research Foundation of Korea (NRF) grant funded by the Korea government (MSIT) (No. 2020R1A2B5B02002002).

References

1. Liu, X., Cao, J., Yu, W.: Filippov systems and quasi-synchronization control for switched networks. Chaos **22**(3), 033110 (2012)
2. Liang, J., Wang, Z., Liu, X.: Exponential synchronization of stochastic delayed discrete-time complex networks. Nonlinear Dyn. **53**(1–2), 153–165 (2008)
3. Park, J.H.: Synchronization of Genesio chaotic system via backstepping approach. Chaos, Solitons Fractals **27**(5), 1369–1375 (2006)
4. Park, J.H.: Robust stability of bidirectional associative memory neural networks with time delays. Phys. Lett. A **349**(6), 494–499 (2006)
5. Cheng, J., Park, J.H., Karimi, H.R., Shen, H.: A flexible terminal approach to sampled-data exponentially synchronization of Markovian neural networks with time-varying delayed signals. IEEE Trans. Cybern. **48**(8), 2232–2244 (2018)

6. Lu, J., Ho, D.W.: Globally exponential synchronization and synchronizability for general dynamical networks. IEEE Trans. Syst. Man Cybern. Part B: Cybern. **40**(2), 350–361 (2010)
7. Yu, W., Cao, J., Lü, J.: Global synchronization of linearly hybrid coupled networks with time-varying delay. SIAM J. Appl. Dyn. Syst. **7**(1), 108–133 (2008)
8. Cao, J., Chen, G., Li, P.: Global synchronization in an array of delayed neural networks with hybrid coupling. IEEE Trans. Syst. Man Cybern. Part B: Cybern. **38**(2), 488–498 (2008)
9. Park, J.H., Shen, H., Chang, X.H., Lee, T.H.: Recent Advances in Control and Filtering of Dynamic Systems with Constrained Signals. Springer, Cham, Switzerland (2018). https://doi.org/10.1007/978-3-319-96202-3
10. Lee, T.H., Wu, Z.G., Park, J.H.: Synchronization of a complex dynamical network with coupling time-varying delays via sampled-data control. Appl. Math. Comput. **219**(3), 1354–1366 (2012)
11. Pecora, L.M., Carroll, T.L.: Synchronization in chaotic systems. Phys. Rev. Lett. **64**(8), 821–825 (1990)
12. He, W., Cao, J.: Adaptive synchronization of a class of chaotic neural networks with known or unknown parameters. Phys. Lett. A **372**(4), 408–416 (2008)
13. Yu, W., Cao, J.: Adaptive synchronization and lag synchronization of uncertain dynamical system with time delay based on parameter identification. Phys. A: Stat. Mech. Appl. **375**(2), 467–482 (2007)
14. Chen, J., Jiao, L., Wu, J., Wang, X.: Projective synchronization with different scale factors in a driven-response complex network and its application in image encryption. Nonlinear Anal. Real World Appl. **11**(4), 3045–3058 (2010)
15. Song, Q.: Design of controller on synchronization of chaotic neural networks with mixed time-varying delays. Neurocomputing **72**(13), 3288–3295 (2009)
16. Li, L., Cao, J.: Cluster synchronization in an array of coupled stochastic delayed neural networks via pinning control. Neurocomputing **74**(5), 846–856 (2011)
17. Lee, T.H., Ma, Q., Xu, S., Park, J.H.: Pinning control for cluster synchronisation of complex dynamical networks with semi-Markovian jump topology. Int. J. Control **88**(6), 1223–1235 (2015)
18. Xia, W., Cao, J.: Pinning synchronization of delayed dynamical networks via periodically intermittent control. Chaos **19**(1), 013120 (2009)
19. Yang, X., Cao, J.: Stochastic synchronization of coupled neural networks with intermittent control. Phys. Lett. A **373**(36), 3259–3272 (2009)
20. Zheng, C., Cao, J.: Robust synchronization of coupled neural networks with mixed delays and uncertain parameters by intermittent pinning control. Neurocomputing **141**, 153–159 (2014)
21. Lu, J., Ho, D.W., Cao, J.: A unified synchronization criterion for impulsive dynamical networks. Automatica **46**(7), 1215–1221 (2010)
22. Bao, H., Park, J.H., Cao, J.: Exponential synchronization of coupled stochastic memristor-based neural networks with time-varying probabilistic delay coupling and impulsive delay. IEEE Trans. Neural Netw. Learn. Syst. **27**(1), 190–201 (2016)
23. Yang, Y., Cao, J.: Exponential synchronization of the complex dynamical networks with a coupling delay and impulsive effects. Nonlinear Anal. Real World Appl. **11**(3), 1650–1659 (2010)
24. Kilbas, A.A.A., Srivastava, H.M., Trujillo, J.J.: Theory and Applications of Fractional Differential Equations. Elsevier, New York (2006)
25. Podlubny, I.: Fractional Differential Equations. Academic Press, New York (1999)
26. Bao, H., Park, J.H., Cao, J.: Adaptive synchronization of fractional-order memristor-based neural networks with time delay. Nonlinear Dyn. **82**(3), 1343–1354 (2015)
27. Huang, L.L., Park, J.H., Wu, G.C., Mo, Z.W.: Variable-order fractional discrete-time recurrent neural networks. J. Comput. Appl. Math. **370**, Article ID 112633 (2020)
28. Chai, Y., Chen, L., Wu, R., Sun, J.: Adaptive pinning synchronization in fractional-order complex dynamical networks. Physica A **391**(22), 5746–5758 (2012)
29. Rakkiyappan, R., Sivasamy, R., Park, J.H.: Synchronization of fractional-order different memristor based chaotic systems using active control. Can. J. Phys. **92**(12), 1688–1695 (2014)

30. Mathiyalagan, K., Park, J.H., Sakthivel, R.: Exponential synchronization for fractional-order chaotic systems with mixed uncertainties. Complexity **21**(1), 114–125 (2015)
31. Yang, L., Jiang, J.: Adaptive synchronization of drive-response fractional-order complex dynamical networks with uncertain parameters. Commun. Nonlinear Sci. Numer. Simul. **19**(5), 1496–1506 (2014)
32. Yu, J., Hu, C., Jiang, H., Fan, X.: Projective synchronization for fractional neural networks. Neural Netw. **49**, 87–95 (2014)
33. Tang, Y., Wang, Z., Fang, J.A.: Pinning control of fractional-order weighted complex networks. Chaos **19**(1), 013112 (2009)
34. Shen, J., Lam, J.: State feedback H-infty control of commensurate fractional-order systems. Int. J. Syst. Sci. **45**(3), 363–372 (2014)
35. Huang, X., Zhao, Z., Wang, Z., Li, Y.: Chaos and hyperchaos in fractional-order cellular neural networks. Neurocomputing **94**, 13–21 (2012)
36. Bao, H., Park, J.H., Cao, J.: Synchronization of fractional-order complex-valued neural networks with time delay. Neural Netw. **81**, 16–28 (2016)
37. Park, J.H., Lee, T.H., Liu, Y., Chen, J.: Dynamic Systems with Time Delays: Stability and Control. Springer-Nature, Singapore (2019). https://doi.org/10.1007/978-981-13-9254-2
38. Rakkiyappan, R., Velmurugan, G., Cao, J.: Finite-time stability analysis of fractional-order complex-valued memristor-based neural networks with time delays. Nonlinear Dyn. **78**(4), 2823–2836 (2014)
39. Wu, R., Lu, Y., Chen, L.: Finite-time stability of fractional delayed neural networks. Neurocomputing **149**, 700–707 (2015)
40. Stamova, I.: Global Mittag-Leffler stability and synchronization of impulsive fractional-order neural networks with time-varying delays. Nonlinear Dyn. **77**(4), 1–10 (2014)
41. Chua, L.O.: Memristor-the missing circuit element. IEEE Trans. Circuit Theory **18**(5), 507–519 (1971)
42. Strukov, D.B., Snider, G.S., Stewart, D.R., Williams, R.S.: The missing memristor found. Nature **453**(7191), 80–83 (2008)
43. Tour, J.M., He, T.: Electronics: the fourth element. Nature **453**(7191), 42–43 (2008)
44. Guo, Z., Wang, J., Yan, Z.: Attractivity analysis of memristor-based cellular neural networks with time-varying delays. IEEE Trans. Neural Netw. Learn. Syst. **25**(4), 704–717 (2014)
45. Li, N., Cao, J.: New synchronization criteria for memristor-based networks: adaptive control and feedback control schemes. Neural Netw. **61**, 1–9 (2015)
46. Wen, S., Bao, G., Zeng, Z., Chen, Y., Huang, T.: Global exponential synchronization of memristor-based recurrent neural networks with time-varying delays. Neural Netw. **48**, 195–203 (2013)
47. Wu, A., Wen, S., Zeng, Z.: Synchronization control of a class of memristor-based recurrent neural networks. Inf. Sci. **183**(1), 106–116 (2012)
48. Zhang, G., Shen, Y.: New algebraic criteria for synchronization stability of chaotic memristive neural networks with time-varying delays. IEEE Trans. Neural Netw. Learn. Syst. **24**(10), 1701–1707 (2013)
49. Zhang, G., Shen, Y.: Exponential synchronization of delayed memristor-based chaotic neural networks via periodically intermittent control. Neural Netw. **55**, 1–10 (2014)
50. Chen, J., Zeng, Z., Jiang, P.: Global Mittag-Leffler stability and synchronization of memristor-based fractional-order neural networks. Neural Netw. **51**, 1–8 (2014)
51. Henderson, J., Ouahab, A.: Fractional functional differential inclusions with finite delay. Nonlinear Anal. Theory Methods Appl. **70**(5), 2091–2105 (2009)
52. Filippov, A.F.: Differential Equations with Discontinuous Righthand Sides. Soviet Series, Mathematics and Its Applications. Kluwer Academic Publishers, Boston (1988)
53. Yu, J., Hu, C., Jiang, H.: Corrigendum to "Projective synchronization for fractional neural networks". Neural Netw. **67**, 152–154 (2015)

New Results on Stability of Coupled Impulsive Fractional-Order Systems on Networks

Li Zhang, Youggui Kao, and Cunchen Gao

Abstract The chapter is devoted to researching the stability issue of fractional-order impulsive coupled systems over networks (FICSNs). Five new results on stability, uniform stability, uniform asymptotic stability, finite-time stability and Mittag-Leffler stability of FICSNs are derived respectively by the graph theory and Lyapunov second technique. These stability principles have a close relation to the topology property of the network. A numerical example is put forward to verify the validity of the obtained findings.

Keywords Fractional-order · Impulsive coupled systems · Lyapunov second method · Graph theory

1 Introduction

Fractional calculus is an old mathematical concept and dates back to 17 century. Regards to the complexity and lack of background, not enough attention has been paid by researchers for quite a long time. Recently, fractional differential equations have been extensively applied in a variety of fields, such as neural networks, complex

L. Zhang · C. Gao
School of Mathematical Science, Ocean University of China,
Qingdao 266100, People's Republic of China
e-mail: 13153108596@163.com

C. Gao
e-mail: ccgao123@126.com

Y. Kao (✉)
Department of Mathematics, Harbin Institute of Technology (Weihai),
Weihai 264209, People's Republic of China
e-mail: ygkao2008@gmail.com

J. H. Park (ed.), *Recent Advances in Control Problems of Dynamical Systems and Networks*, Studies in Systems, Decision and Control 301,
https://doi.org/10.1007/978-3-030-49123-9_14

switched networks, stochastic systems, hybrid systems, chaotic systems, and discrete systems, etc., see [1, 2]. On the other hand, impulsive differential equations were widely employed to simulate some real circumstances, such as population medicine, population ecology, radio engineering, neural networks, economics, dynamics, communication security, etc., see [3–10]. Simultaneously, plenty of important conclusions about the stability issue of impulsive systems have been reported [11–17]. Li [11–15] investigated the stability issue of nonlinear systems with the influence of delayed impulses on input-to-state, distributed-delayed impulses, state-independent impulses, and impulsive control. In the research works [16, 17], the finite-time stability of impulsive reaction-diffusion systems and impulsive switched delay systems with nonlinear disturbances was investigated. Particularly, there are a large number of research results about the stability issue of various impulsive fractional systems [18–23]. Hei [18] studied the finite-time stability of impulsive fractional-order systems with time-delay. Yang [19] studied the Mittag-Leffler stability for fractional nonlinear fractional-order systems with impulses. Stamova [20, 21] researched the Mittag-Leffler stability and synchronization of fractional neural networks with time-varying delays and reaction-diffusion terms using impulsive and linear controllers. Li [22] and Xi [23] investigated the impulsive synchronization and adaptive impulsive synchronization of fractional-order chaotic systems with time-delay, respectively.

Besides, more and more researchers have spent more effort to study coupled nonlinear differential systems, such as social science systems, biological systems, engineering systems, and physical systems. Strenuous results about stability analysis on networks have been provided in literatures [24–31]. In particular, Chen [24] gave some criteria on uniform stability of the coupled system with constant delay over networks. Kao et al. [25–29] resolved the stability analysis of coupled reaction-diffusion systems (SCRDS). Suo et al. [30] discussed a stability analysis issue for impulsive coupled systems on networks (ICSNs). Wang et al. [31] provided networked synchronization controller design method for coupled delay dynamic networks. In addition, there are some stability analysis results of the fractional coupled systems (FCSNs) [32–39]. Li [32, 33] investigated the Mittage-Leffler stability analysis of FCSNs and global Mittag-Leffler stability for FCSNs with feedback controls. Zhang [34, 35] proposed the synchronization issues of fractional coupled delayed systems. In the chapter [36–38], hybrid synchronization, Q-S synchronization, periodically intermittent discrete observation control for synchronization of coupled fractional-order systems, respectively. Especially, the exponential stability related to the integer-order systems is generalized to the fractional-order case by means of the Mittag-Leffler stability. Li [39] studied the Mittag-Leffler stability for a new coupled system of fractional-order differential equations with impulses, which have a great significance to the investigation of FICSNs.

As we all know, in order to investigate the stability analysis for integer-order nonlinear systems, the direct Lyapunov method is derived. The direct Lyapunov method that it is not necessary to find the specific solution of the equation then we can analyze the equilibrium solution stability. When we considered stability analysis for fractional order nonlinear systems, constructed systematically a global Lyapunov function for the system on networks was a significant point in the research. Li et al. [40] settled the global stability issue for coupled systems by utilizing outcomes of the graph theory and the direct Lyapunov method. And Li [41] utilizes the same technique for some stochastic coupled systems on networks (SCSNs). To the best knowledge of ours, there are no research findings about the finite-time stability, uniform stability, asymptotic stability as well as uniform asymptotic stability analysis of FICSNs, which inspires us to research further.

Motivated by the above analysis, in this chapter, we mainly study the various stability analysis of FICSNs. Firstly, based on the results of graph theory, we provide a systematic method that constructs global Lyapunov functions for large-scale impulsive coupled systems from building blocks of individual vertex systems. Secondly, five sufficient criterions on stability, uniform stability, uniform asymptotic stability, finite-time stability, and Mittag-Leffler stability of FICSNs are derived respectively. It should be noted that the study in this aspect is meaningful and challenging.

At the rest of the chapter, Sect. 2 presents a few essential definitions and lemmas. In Sect. 3, we derived five criteria on various stability for FICSNs. In Sect. 4, an example is given to verify the validity and feasibility of the outcomes. The conclusion is given in Sect. 5.

2 Preliminaries

Let's introduce a few essential lemmas and definitions. Denote $\mathbf{R}^n$ as the n-dimensional Euclidean space, Ξ is an open set in $\mathbf{R}^n$, and $\mathbf{R}_+ = [0, \infty)$.

Definition 1 ([42]) Caputo fractional-order derivative for a function $h(t) \in C^1([t_o, +\infty), R)$ denote as

$$
{}_{t_o}^{C}D_t^{\beta} h(t) = \frac{1}{\Gamma(1-\beta)} \int_{t_o}^{t} \frac{h^{'}(\xi)}{(t-\xi)^{\beta}} d\xi \tag{1}
$$

$\Gamma(\cdot)$ denotes the Gamma function, $t \geq t_o, n > 0, 0 < \beta < 1$.

Definition 2 ([30]) The function γ is defined as the Kamke function if $\gamma \in C[\mathbf{R}_+, \mathbf{R}_+]$, when $\nu = 0$, $\gamma(\nu) = 0$, $\gamma(\nu)$ is monotonically increasing with ν.

Definition 3 ([30]) $\lambda_{max}(\mathbf{C})$ represents the maximum eigenvalue of matrix $\mathbf{C}$.

Let us consider a network can be denoted as the digraph $\mathcal{G}$ accompany n vertices ($n > 1$), we will assume that the l-th vertex dynamic is described by a system of impulsive differential equations:

$$
\begin{cases}
{}_{t_o}^{C}D_t^{\beta} x_l = h_l(t, x_l), & t \neq t_w, \\
\Delta x_l = I_w(x_l), & t = t_w, \quad w = 1, 2, \ldots, \\
x_l(t_o^+) = x_{lo},
\end{cases} \tag{2}
$$

where $x_l \in \mathbf{R}^{n_l}$, $h_l : \mathbf{R}^{n_l} \to \mathbf{R}^{n_l}$.

Let $g_{lk}: \mathbf{R}_{+} \times \mathbf{R}^{n_l} \times \mathbf{R}^{n_k} \rightarrow \mathbf{R}^{n_l}$ reflect the effect from the vertex k to l and $g_{lk} \equiv 0$ represents there have no arc between l and k.

The following system can be get:

$$\begin{cases} {}_{t_o}^{C}D_t^{\beta} x_l = h_l(t, x_l) + \sum_{k=1}^{n} g_{lk}(t, x_l, x_k), & t \neq t_w, \\ \Delta x_l = I_w(x_l), & t = t_w, \quad w = 1, 2, \ldots, \\ x_l(t_o^+) = x_{lo}, \end{cases} \tag{3}$$

$l = 1, 2, \ldots, n$, assuming these conditions hold:

(i) $0 < t_1 < t_2 < \cdots < t_w < \cdots$, and $t_w \rightarrow \infty$ as $w \rightarrow \infty$,

(ii) h_l: $(t_{w-1}, t_w] \times \Xi \rightarrow \mathbf{R}^{n_l}$. g_{lk}: $(t_{w-1}, t_w] \times \mathbf{R}^{n_l} \times \mathbf{R}^{n_k} \rightarrow \mathbf{R}^{n_l}$ is continuous,

(iii) $I_w : \mathbf{R}^{n_l} \rightarrow \mathbf{R}^{n_l}$.

When $t = t_w$, $w = 1, 2, \ldots$, for each x_l, x_k the functions h_l, g_{lk} and $I_w : \Xi \rightarrow \mathbf{R}^{n_k}$ are right continuous and satisfy the global Lipschitz condition. Respectively, to guarantee existence, uniqueness and continuability of the solution $(x_1, x_2, \ldots, x_n) = 0$ for $t \geq t_o$ of system (3). The solution $(x_1, x_2, \ldots, x_n) = 0$ is, in general, piecewise continuous functions with points of discontinuity of the first type at with they are left continuous. At the moments $t = t_w$, $w = 1, 2, \ldots$, the following relationship can be get: $x(t_w^-) = x(t_w)$ and $x(t_w^+) = x(t_w) + I_w(x(t_w))$.

Definition 4 ([42]) $(x_1, x_2, \ldots, x_n) = 0$ for $t \geq t_o$ is the solution of system (3):

(a) Mittag-Leffler stable if

$$\|x(t; t_o, x_o)\| \leq \{n[x(t_o)]E_{\beta}(-\nu(t - t_o)^r)\}^s \tag{4}$$

where $\beta \in (0, 1)$, $\nu > 0$, $s > 0$, $n(x) \geq 0$, $n(0) = 0$, $n(x)$ contents the locally Lipschitz, and n_0 is the Lipschitz constant;

$$E_{\beta}(t) = \sum_{k=0}^{\infty} \frac{t^k}{\Gamma(\beta k + 1)} \tag{5}$$

represents Mittag-Leffler function, and then follow two parameters β and γ:

$$E_{\beta,\gamma}(t) = \sum_{k=0}^{\infty} \frac{t^k}{\Gamma(\beta k + \gamma)} \tag{6}$$

where $\beta > 0$ and $\gamma > 0$. When $\gamma = 1$, then obtain $E_{\beta}(t) = E_{\beta,1}(t)$.

(b) Globally Mittag-Leffler stable if (a) satisfies that $\Xi \in \mathbf{R}^n$.

Definition 5 ([43]) $(x_1, x_2, \ldots, x_n) = 0, \forall t \geq t_o^{'}$ is the solution of system (3):

(a) stable supposing that $\forall t_o^{'} \in \mathbf{R}_+, \forall \epsilon > 0, \exists \eta = \eta(t_o^{'}, \epsilon) > 0, \forall x_o^{'} \in \Xi : \|x_o^{'}\| < \eta, \forall t \geq t_o^{'}$, then $\|x(t; t_o^{'}, x_o^{'})\| < \epsilon$;
(b) uniformly stable supposing that η in (a) is not depend on $t_o^{'} \in \mathbf{R}_+$;
(c) attractive supposing that $\forall t^{'} \in \mathbf{R}_+, \exists \chi = \chi(t_o^{'}) > 0$, and

$$\forall x_o^{'} \in \Xi : \|x_o^{'}\| < \chi, \lim_{t_o \to \infty} x(t; t_o^{'}, x_o^{'}) = 0;$$

(d) equi-attractive supposing that $\forall t_o^{'} \in \mathbf{R}_+, \exists \chi = \chi(t_o^{'}) > 0, \quad \forall \epsilon > 0, \exists$ $T_o^* = T_o^*(t_o^{'}, \epsilon) > 0$, and

$$\forall x_o^{'} \in \Xi : \|x_o^{'}\| < \chi, \ \forall t \geq t_o^{'} + T_o : \|x(t; t_o^{'}, x_o^{'}\| < \epsilon;$$

(e) uniformly attractive supposing that the χ and T_o^* in (d) are not depend on $t^{'} \in \mathbf{R}_+$;
(f) asymptotically stable supposing that (a) and (c) hold;
(g) uniformly asymptotically stable supposing that (b) and (e) hold.

Definition 6 ([44]) There are three constants $\omega_o^{'} > 0, \ \omega_o^{''} > 0, \ T_o, \ \omega_o^{'} < \omega_o^{''}$, the system (3) is defined as finite-time stable, $\|x_o(t_o^{'})\| \leq \omega_o^{'} \Rightarrow \|x_o(t)\| \leq \omega_o^{''}, \ \forall t_o \in (t_o^{'}, t_o^{'} + T_o]$.

Remark 1 Assume that the equilibrium solution of system (3) is Mittag-Leffler stable, then must be asymptotic stable.

There are some fundamental notations about the graph theory [19] as follows: The digraph $\mathcal{G} = (\mathcal{U}, \mathcal{F})$ consist of a set $\mathcal{U} = \{1, 2, \ldots, n\}$ vertices and a set $\mathcal{F}$ of arcs (l, k), which represents the vertex from l to k, $\boldsymbol{B} = (b_{lk})_{n \times n}$ is the weight matrix. The elements b_{lk} have consistent with the weight of arcs (l, k).

A weighted digraph $(\mathcal{G}, \boldsymbol{B})$ is called balanced assume that the $W(\Pi) = W(-\Pi)$ for every directed cycle Π. The $-\Pi$ represents the reverse of Π. Ψ represents a unicyclic graph. $L = (q_{lk})_{n \times n}$ is the Laplacian matrix of $(\mathcal{G}, \boldsymbol{B})$ is defined as

$$L = \begin{bmatrix} \sum_{k \neq 1} q_{1k} & -q_{12} & \cdots & -q_{1n} \\ -q_{21} & \sum_{k \neq 2} q_{2k} & \cdots & -q_{2n} \\ \cdot & \cdot & \cdot & \cdot \\ \cdot & \cdot & \cdot & \cdot \\ \cdot & \cdot & \cdot & \cdot \\ -q_{n1} & -q_{n2} & \cdots & \sum_{k \neq n} q_{nk} \end{bmatrix}.$$

Lemma 1 ([32]) *Supporting that $n \geq 2$. Denote e_l represent the cofactor of the l-th diagonal element of $L^{'}$.*

$$\sum_{(l,k)=1}^{n} e_l a_{lk} F_{lk}(x_l, x_k) = \sum_{(\Psi \in \Theta)} W(\Psi) \sum_{(u,v) \in E(\Pi_\Psi)} F_{uv}(x_u, x_v). \tag{7}$$

The $F_{lk}(x_l, x_k)$, $l, k \in \mathbf{L}$ represent any functions, Θ denotes the set that whole spanning unicyclic graphs of $(\mathcal{G}, \mathbf{B})$ in. $W(\Psi)$ represents the weight of Ψ, Π_Ψ is the directional cycle of Ψ. The $(\mathcal{G}, \mathbf{B})$ is strongly connected, $e_l > 0$, $l \in \mathbf{L}$. If $(\mathcal{G}, \mathbf{B})$ is balanced

$$\sum_{(l,k)=1}^{n} e_l a_{lk} F_{lk}(t, x_l, x_k) = \frac{1}{2} \sum_{(\Psi \in \Theta)} W(\Psi) \sum_{(k,l)\in E(\Pi_\Psi)} \{F_{lk}(t, x_l, x_k) + F_{kl}(t, x_k, x_l)\}. \tag{8}$$

Lemma 2 ([43]) *Assume the inequality*

$${}_{t_o}^{C}D_t^{\alpha} x(t) \leq {}_{t_o}^{C} D_t^{\alpha} y_o \tag{9}$$

is effective when $t \geq t_o$, $t \neq t_w$, $w = 1, 2, \ldots$, then

$$x(t; t_o, x_o) \leq y_o, \quad t \in [t_o, \infty). \tag{10}$$

Lemma 3 (Generalized Gronwall's inequality) ([44]) *Assume that $x(\zeta) > 0$, $k(\zeta) > 0$ are local integrable for $0 \leq \zeta < T^*$, $T^* \leq \infty$, $h(\zeta) > 0$ is the nondecreasing continuous function with $0 \leq \zeta < T^*$; $\beta > 0$, content that*

$$x(\zeta) \leq k(\zeta) + h(\zeta) \int_0^{\zeta} (\zeta - s)^{\beta-1} x(s) ds. \tag{11}$$

Then

$$x(\zeta) \leq k(\zeta) + h(\zeta) \int_0^{\zeta} \left[\sum_{i=1}^{\infty} \frac{(h(\zeta)\Gamma(\beta))^n}{\Gamma(n\beta)} (\zeta - s)^{n\beta-1} k(s) \right] ds. \tag{12}$$

Let $k(\zeta)$ is a nondecreasing function for $[0, T^*)$, then

$$x(\zeta) \leq k(\zeta) E_\beta(h(\zeta)\Gamma(\beta)\zeta^\beta). \tag{13}$$

The E_β represents the Mittag-Leffler function.

Lemma 4 ([45]) *Let $z(\zeta) \in \mathbf{R}$ is a continuous differentiable function. $\forall \zeta \geq \zeta_0$,*

$$\frac{1}{2}({}_{t_0}^{C}D_t^{\beta} z(\zeta)^2) \leq z(\zeta) {}_{\zeta_0}^{C} D_\zeta^{\beta} z(\zeta), \qquad \forall \beta \in (0, 1). \tag{14}$$

3 Stability Analysis for Fractional-Order Impulsive Coupled Systems on Networks

Let $D_w = (t_{w-1}, t_w) \times \Xi$, $w = 1, 2, \ldots$, $D = \bigcup_{w=1} D_w$.

We shall study the stability analysis the system (3) with

$A^0: \ f_l(t, 0) = 0,\ t \geq 0,$

$A^1: \ I_w(0) = 0,\ w = 1, 2, \ldots.$

To proof the mainly theorems of this work, the piecewise continues auxiliary Lyapunov functions [42] is used: $V \in C[D, \mathbf{R}_+] : \mathbf{R}_+ \times \Xi \to \mathbf{R}_+$.

The V is the continuous function and content the locally Lipschitz with respect to its second argument on each D_w, $V(t_w^-, x) = V(t_w, x)$, and $V(t_w^-, x) = V(t_w, x)(t \to t_w, t > t_w)$ exists.

The following condition contents that

$A^2: \ V_l(t, 0) = 0,\ t \geq 0.$

Theorem 1 *Assume theses conditions hold:*

(A1) $\exists\ V_l(t, x_l)$, $F_{lk}(x_l, x_k)$, *a matrix* $\mathbf{B} = (b_{lk})_{m\times n}$ *where* $b_{lk} \geq 0$ *satisfy:*

$$ {}_{t_o}^{C}D_t^{\beta} V_l(t, x_l) \leq \sum_{k=1}^{n} b_{lk} F_{lk}(x_l, x_k), \quad t \neq t_w,\ t \geq t_o,\ l = 1, 2, \ldots, n. \tag{15}$$

(A2) Towards every directional cycle Π *of the weight digraph* $(\mathcal{G}, \mathbf{B})$

$$\sum_{(u,v)\in E(\Pi)} F_{uv}(x_u, x_v) \leq 0, \quad t \geq t_o,\ x_u \in D_u,\ x_v \in D_v. \tag{16}$$

(A3)

$$V_l(t^+, x_l + I_w(x_l)) \leq V_l(t, x_l) \qquad x_l \in \Xi,\ t = t_w,\ w = 1, 2, \ldots, \tag{17}$$

(A4) $\exists\ \alpha_1^l \in K,\ l = 1, 2, \ldots, n$ *satisfy:*

$$\alpha_1^l(\|x_l\|) \leq V_l(t, x_l), \quad (t, x_l) \in D_w. \tag{18}$$

(A5) The functions $I_w \in [\Xi, \mathbf{R}^{n^l}]$, $w = 1, 2, \ldots$, *are nondecreasing in* Ξ *and* $(E + I_w) : \Xi \to \Xi$, $w = 1, 2, \ldots$, *in which E is the identity of* Ξ.

The $V(t, x) = \sum_{l=1}^{n} e_l V_l(t, x_l)$ *is a Lyapunov function and contents* ${}_{t_o}^{C}D_t^{\beta} V(t, x) \leq 0$, $t \geq t_o, t \neq t_w$. *The* $(x_1, x_2, \ldots, x_n) = 0$ *for system (3) is stable.*

Proof Denote $\alpha_1(\|x(t)\|) = \sum_{l=1}^{n} c_l \alpha_1^l(\|x_l(t)\|)$ and $\delta = \min\{\delta_0^1, \delta_0^2, \ldots, \delta_0^n\}$.

Let $\varepsilon > 0. Then$, $\exists \delta_0 = \delta_0(t_o, \varepsilon) > 0$ if $\|x\| \leq \delta_0 < \delta$, later

$$\sum_{\|x\|\leq\delta_0} V(t_o^+, x) < \alpha_1(\varepsilon). \tag{19}$$

Next, to proof: $\forall t_o \in \mathbf{R}_+, \forall\varepsilon > 0, \exists\delta_0 = \delta(t_o, \varepsilon) > 0, \forall x_o \in \varXi : \|x_o\| < \delta_0, \forall t \geq t_o$, then $\|x(t; t_o, x_o)\| < \varepsilon$.

Suppose that is not true. There exists $x(t; t_o, x_o)$ for system (3), then $\|x_o\| < \delta_0, \exists t^\star > t_o, t_w < t^\star \leq t_{w+1}$, then $\|x(t^\star)\| \geq \varepsilon$, and $\|x(t; t_o, x_o\| < \varepsilon, t \in [t_o, t_w]$.

In view of the conditions (A1) and (A2), when $t \in D_w$, we will have

$$\begin{aligned}{}_{t_o}^{C}D_t^{\beta} V(t, x) &={}_{t_o}^{C} D_t^{\beta} \sum_{l=1}^{n} e_l V_l(t, x_l)\\ &\leq \sum_{l=1}^{n} e_l({}_{t_0}^{C} D_t^{\beta} V_l(t, x_l)) \leq \sum_{l,k=1}^{n} e_l a_{lk} F_{lk}(x_l, x_k).\end{aligned} \tag{20}$$

Regards of the conditions (A2), (7), $W(\varPsi) > 0$, then

$${}_{t_o}^{C}D_t^{\beta} V(t, x) \leq \sum_{\varPsi\in\Theta} W(\varPsi) \sum_{(u,v)\in E(\varPi_\varPsi)} F_{uv}(x_u, x_v) \leq 0. \tag{21}$$

When $t = t_w$, make use of the condition (A3), then

$$\begin{aligned}{}_{t_o}^{C}D_t^{\beta} V(t^+, x + I_w(x)) &={}_{t_o}^{C} D_t^{\beta} \sum_{l=1}^{n} e_l V_l(t^+, x_l + I_w(x_l))\\ \leq{}_{t_o}^{C} D_t^{\beta} \sum_{l=1}^{n} e_l V_l(t, x_l)) &\leq \sum_{l=1}^{n} e_l({}_{t_o}^{C} D_t^{\beta} V_l(t, x_l)) \leq 0.\end{aligned} \tag{22}$$

Considered condition (A5) and I_w is nondecreasing, $\exists t^o$, satisfy $t_w < t^o \leq t^\star$, then $\|x(t^o)\| > \varepsilon$ and $x(t^o; t_o, x_o) \in \varXi$, when $t \in [t_o, t^o]$, then we can obtain

$$\begin{aligned}V(t, x(t; t_o, x_o)) &= \sum_{l=1}^{n} e_l V_l(t, x_l(t; t_o, x_{lo}))\\ &\leq \sum_{l=1}^{n} e_l V_l(t_o^+, x_{lo})) = V(t_o^+, x_o).\end{aligned} \tag{23}$$

From above discussion, the inequalities hold

$$\begin{aligned}\alpha_1(\varepsilon) < \alpha_1(\|x(t^o)\|) &= \sum_{l=1}^{n} e_l\alpha_1^l(\|(x_{lt^o})\|) \leq \sum_{l=1}^{n} e_l V_l(t^o; t_o, x_{lo}))\\ &= V(t^o; t_o, x_o) \leq V(t_o^+, x_o) < \alpha_1(\varepsilon).\end{aligned} \tag{24}$$

This is contradictory. Then, we can obtain $\forall\varepsilon > 0, \exists\delta_0 = \delta_0(t_o, \varepsilon) > 0, \forall x_o \in \Xi : \|x_o\| < \delta_0, \forall t \geq t_o, \forall t_o \in \mathbf{R}_+$, therefore $\|x(t; t_o, x_o\| < \varepsilon$. □

Corollary 1 *Assume the* $(\mathcal{G}, \mathbf{B})$ *is balanced, later*

$$\sum_{l,k=1}^{n} e_l a_{lk} F_{lk}(x_l, x_k) = \frac{1}{2}\sum_{\Psi\in\Theta} W(\Psi) \sum_{(u,v)\in E(\Pi_\Psi)} [F_{uv}(x_u, x_v) + F_{vu}(x_v, x_u)],$$

the result for Theorem 1 is established when the condition (A2) is substitute for (A2′)

$$\sum_{(u,v)\in E(\Pi_\Psi)} [F_{uv}(x_u, x_v) + F_{vu}(x_v, x_u)] \leq 0, \quad t \geq t_o, x_u \in D_u, x_v \in D_v. \tag{25}$$

Theorem 2 *Assume that the requirements (A1)–(A6) set up:*
(A6) Existing $\delta_0^{(l)} > 0$ *and functions* $b_1^l \in K$ *satisfy:*

$$V_l(t, x_l) \leq b_l(\|x_l\|), \quad \text{provided } \|x_l\| < \delta_0^{(l)}, \quad (t, x_l) \in (t_o, \infty) \times \Xi, \tag{26}$$

the solution $(x_1, x_2, \ldots, x_n) = 0$ *for system (3) is uniformly stable.*

Proof For all $\varepsilon > 0$, there exists $\delta_0 = \delta_0(\varepsilon) < \delta$ such that $b_1^l(\delta_0) < \alpha_1^l(\varepsilon)$.
Let $x \in \Xi : \|x_o\| < \delta_0$.
From Lemma 2, we can get:

$$\begin{aligned}
\alpha_1(\|x(t; t_o, x_o)\|) &= \sum_{l=1}^{n} \alpha_1^l(\|x_l(t; t_o, x_o)\|) \leq \sum_{l=1}^{n} e_l V_l(t; x_l(t; t_o, x_o)) \\
&= V(t, x(t; t_o x_o)) < V(t_o^+, x_o) = \sum_{l=1}^{n} e_l V_l(t_o^+, x_{lo}) \\
&\leq \sum_{l=1}^{n} e_l b_1^l(\|x_{lt_o^+}\|) < \sum_{l=1}^{n} e_l \alpha_1^l(\varepsilon) = \alpha_1(\varepsilon),
\end{aligned}$$

thus there exists $\delta_0 = \delta_0(\varepsilon) < \delta$ such that $\|x(t; t_o, x_o)\| < \varepsilon, t \geq t_o$. Then, the proof process is ended. □

Theorem 3 *Assume the conditions (A2)–(A7) hold:*
(A7) Existing functions $V_l \in C[\mathbf{R}_+ \times D_l, \mathbf{R}_+]$, $F_{lk}(t, x_l, x_k)$, *a matrix* $B = (b_{lk})_{n\times n}$, *where* $b_{lk} > 0$ *and* $q_l > 0$, $l = 1, 2, \ldots, n$. *satisfy:*

$$ {}_{t_0}^{C}D_t^{\beta} V_l(t, x_l) \leq -q_l V_l(t, x_l) + \sum_{l=1}^{n} b_{lk} F_{lk}(x_l, x_k) \quad t \neq t_w, \quad t \geq t_o. \tag{27}$$

The solution $(x_1, x_2, \ldots, x_n) = 0$ for system (3) is uniformly asymptotically stable.

Proof Denote $q = \min\{q_1, q_2, \ldots, q_n\}$ and $\eta = \max\{\eta_1, \eta_2 \cdots, \eta_n\}$. Let $H_0 > 0$ be a constant satisfying $\{x \in \mathbf{R}^n : \|x\| \leq H_0\} \subset \Xi, \forall t \in [t_o, \infty)$.

We shall denote

$$W_{t,H_0} = \{x \in \Xi : V(t^+, x) \leq \alpha_1(H_0)\}.$$

From conditions (A4) and (A6): $\exists\, \alpha_1^l, b_1^l \in K, (t, x) \in D_w$, we have

$$\alpha_1^l \|x_l\| \leq V_l(t, x_l) \leq b_1^l \|x_l\|, \tag{28}$$

then we reduce

$$W_{t,H_0} \subset \{x \in \mathbf{R}^n : \|x\| \leq H_0\} \subset \Xi.$$

Let $\lambda = constant > 0$, which satisfies $\alpha_1^l(\lambda) < b_1^l(H_0)$. If $x_o \in \Xi : \|x_o\| < \lambda$, from (28) implies

$$\begin{aligned} V(t_o^+, x_o) &= \sum_{l=1}^{n} e_l V_l(t_o^+, x_{lo}) \leq \sum_{l=1}^{n} e_l b_1^l(\|x_{lt_o^+}\|) = b_1(\|x_{lt_o^+}\|) \\ &< \sum_{l=1}^{n} e_l b_1^l(\lambda) = b_1(\lambda) < \sum_{l=1}^{n} e_l b_1^l(H_0) = b_1(H_0) \end{aligned}$$

which shows that $x_o \in W_{t,H_0}$.

From conditions (A7) and (28), we can get $\forall t_o \in \mathbf{R}_+$, $\forall x_o \in W_{t,H_0}, x(t; t_o, x_o) \in W_{t,H_0}$ for $t \geq t_o$.

Next, we prove the equilibrium solution $(x_1, x_2, \ldots, x_n) = 0$ for the system (3) is uniformly attractive. Let $\forall \varepsilon > 0, \exists \eta = \eta(\varepsilon)$ content $b_1^l(\eta) < \alpha_1^l(\lambda)$,

$$T > \left[\frac{\alpha_1(\eta) q}{b_1(H_0)\Gamma(\beta + 1)}\right]^{\frac{1}{\beta}}.$$

Assume $\forall t \in [t_o, t_o + T]$, then $\|x(t; t_o, x_o)\| \geq \eta$ is valid, in view of condition (A7)

$$\begin{aligned} {}_{t_o}^{C}D_t^{\beta} V(t, x) &= {}_{t_0}^{C}D_t^{\beta} \sum_{l=1}^{n} e_l V_l(t, x) \leq \sum_{l=1}^{n} e_l ({}_{t_o}^{C}D_t^{\beta} V_l(t, x_l)) \\ &\leq \sum_{l=1}^{n} e_l \left[-q_l V_l(t, x_l) + \sum_{l=1}^{n} b_{lk} F_{lk}(x_l, x_k) \right] \\ &\leq \sum_{l=1}^{n} e_l q_l V_l(t, x_l)) = -q V(t, x) \qquad t \neq t_w \end{aligned}$$

and from condition (A3), we can have

$$ {}_{t_o}^{C}D_t^{\beta} V(t,x) \le -qV(t,x) \qquad t \in [t_o, \infty). $$

Integral of degree β on both sides of the equation gives

$$ \begin{aligned} V(t,x) - V(t_o^+, x(t_o^+)) &\le \frac{-q}{\Gamma(\beta)} \int_{t_o^+}^{t} V(s,x)(t-s)^{\beta-1} ds \\ &= \frac{-q}{\Gamma(\beta)} \int_{t_o^+}^{t} \sum_{l=1}^{n} e_l V_l(s, x_l)(t-s)^{\beta-1} ds \\ &\le \frac{-q}{\Gamma(\beta)} \int_{t_o^+}^{t} \sum_{l=1}^{n} e_l \alpha_1^l \|x_l(\tau; t_o, x_{lo})\| (t-s)^{\beta-1} ds \\ &= \frac{-q}{\Gamma(\beta)} \int_{t_o^+}^{t} \alpha_1 \|x_l(\tau; t_o, x_o)\| (t-s)^{\beta-1} ds \end{aligned} $$

and

$$ \frac{-q}{\Gamma(\beta)} \int_{t_o^+}^{t} \alpha_1(\eta)(t-s)^{\beta-1} ds \le \frac{-q\alpha_1(\eta)}{\Gamma(\beta+1)T^{\beta}}, $$

which leads

$$ V(t,x) \le V(t_o^+, x(t_o^+)) - \frac{-q\alpha_1(\eta)}{\Gamma(\beta+1)T^{\beta}} \le b_1(H_0) - \frac{-q\alpha_1(\eta)}{\Gamma(\beta+1)T^{\beta}} < 0. $$

This is contradictory with (28). So, the assumption is not valid. This contradiction indicates $\exists\, t^\star \in [t_o, t_o + T]$ satisfies $\|x(t^\star; t_o, x_o)\| < \eta$.

Regard to (A3), (A7), and (28), we can obtain $\forall\, t \ge t^\star$ (thus for any $t \ge t_o + T$), the inequalities set up:

$$ \begin{aligned} \alpha_1(\|x(t; t_o, x_o)\|) &= \sum_{l=1}^{n} e_l \alpha_1^l (\|x_l(t; t_o, x_{lo})\|) \le \sum_{l=1}^{n} e_l V_l(t, x(t; t_o, x_{lo})) \\ &\le \sum_{l=1}^{n} e_l V_l(t^\star; x(t; t^\star, x_{lo})) \le \sum_{l=1}^{n} e_l b_1^l (\|x_l(t^\star; t_o, x_{lo})\|) \\ &= b_1 \|x(t^\star; t_o, x_o)\| < b_1(\eta) < \alpha_1(\varepsilon). \end{aligned} $$

Then, $\|x(t; t_o, x_o)\| < \varepsilon$, $\forall t \ge t_o + T$. Combining with Theorem 2, the results can be obtained. □

Corollary 2 *Assume* $(\mathcal{G}, \mathbf{B})$ *is balanced, the results for Theorem 3 established if condition (A2) is substitute for (A2′).*

Remark 2 When $\beta = 1$, the stability analysis of equilibrium solution for FICSNs shall degenerate the stability analysis of ICSNs, Theorems 1–3 in this chapter in accordance with Theorems 3.1 and 3.2 in the literature [32].

Remark 3 When $\sum_{k=1}^{n} g_{lk}(t, x_l, x_k) = 0$, the system will transform into:

$$\begin{cases} {}_{t_o}^{C}D_t^{\beta} x_l = h_l(t, x_l), \quad t \neq t_w, \\ \Delta x_l = I_w(x_l), \quad t = t_w, \quad w = 1, 2, \ldots, \\ x_l(t_o^+) = x_{lo}. \end{cases} \tag{29}$$

The stability analysis of equilibrium solution for FICSNs will degenerate the stability of fractional impulsive nonlinear system, Theorem 1, 2, and 3 of this chapter consistent with Theorems 5.1, 5.2, and 5.3 in the paper [43].

Theorem 4 *Assume the conditions (A2), (A3), (A5), (A7) and (A8), (A9) set up:*

(A8) There have $r_1^l, r_2^l \in k$ satisfying:

$$r_1^l \|x_l\| \leq V_l(t, x_l) \leq r_2^l \|x_l\| \quad t \geq t_o, \tag{30}$$

(A9) when $\|x_{lt_o}\| < \varrho$,

$$\frac{r_2}{r_1} E_\alpha(-qT^\alpha) < \frac{\epsilon}{\varrho}, \tag{31}$$

the equilibrium solution is finite-time stable.

Proof In view of conditions (A7) and (A2) :

$${}_{t_o}^{C}D_t^{\beta} V_l(t, x_l) \leq -q_l V_l(t, x_l) + \sum_{l=1}^{n} b_{lk} F_{lk}(x_l, x_k) \quad t \neq t_k, t \geq t_0$$

$$\sum_{(u,v)\in E(\Pi)} F_{uv}(x_u, x_v) \leq 0, \quad t \geq t_o, \ x_u \in D_u, \ x_v \in D_v,$$

we can obtain:

$${}_{t_o}^{C}D_t^{\beta} V_l(t, x_l) \leq -q_l V_l(t, x_l).$$

From condition (A3), we will get:

$${}_{t_o}^{C}D_t^{\beta} V_l(t, x_l) \leq -q_l V_l(t, x_l), \quad t \in [t_o, \infty).$$

Integral of degree β on both sides of the equation gives

$$V_l(t, x_l) - V(t_o^+, x_l(t_o) + I_w x_l(t_o)) \leq \frac{-q}{\Gamma(\beta)} \int_{t_o^+}^{t} V_l(s, x_l)(t-s)^{\beta-1} ds,$$

then,

$$V_l(t, x_l) \leq V(t_o^+, x_l(t_o) + I_w x_l(t_o)) + \frac{-q}{\Gamma(\beta)} \int_{t_o^+}^{t} V_l(s, x_l)(t-s)^{\beta-1} ds,$$

which finally leads

$$V_l(t, x_l) \leq V_l(t_o, x_{lt_o}) + \frac{-q}{\Gamma(\beta)} \int_{t_o}^{t} V_l(s, x_l)(t-s)^{\beta-1} ds.$$

By using Lemma 3, we can get:

$$V_l(t, x_l) \leq V_l(t_o, x_{lt_o}) \cdot E_\beta(-qT^\beta).$$

When considered the condition (A8), we can see:

$$\begin{aligned} r_1 \|x\| &= \sum_{l=1}^{n} e_l r_1^l \|x_l\| \leq \sum_{l=1}^{n} e_l V_l(t, x_l) \\ &= V(t, x) \leq \sum_{l=1}^{n} e_l V_l(t_o, x_{lt_o}) \cdot E_\beta(-qT^\beta) \\ &\leq \sum_{l=1}^{n} e_l r_2^l \|x_{lt_o}\| \cdot E_\beta(-qT^\beta) = r_2 \|x_{lt_o}\| \cdot E_\beta(-qT^\beta), \end{aligned}$$

which leads

$$\|x\| \leq \frac{r_2}{r_1} \|x_{lt_o}\| E_\beta(-qT^\beta), \quad t \geq t_o.$$

Let $\frac{r_2}{r_1} E_\beta(-qT^\beta) < \frac{\epsilon}{\varrho}$, when $\|x_{lt_o}\| < \varrho$, we can get

$$\|x\| \leq \frac{r_2}{r_1} \|x_{lt_o}\| E_\beta(-qT^\beta) < \varrho \frac{r_2}{r_1} E_\beta(-qT^\beta) < \epsilon.$$

Then, the proof process is ended. □

Remark 4 When $\beta = 1$, the Mittag-Leffler stable of equilibrium solution for FIC-SNs shall translate into exponentially stable of Theorem 3.3 in [30].

Theorem 5 *Let $t_o = 0$, Assume the conditions (A2), (A3), (A5), (A7) and (A10) satisfying:*

(A10) There have functions $r_1^l, r_2^l \in k$, such that

$$r_1^l \|x_l\|^a \leq V_l(t, x_l) \leq r_2^l \|x_l\|^{ab} \quad t \geq 0. \tag{32}$$

The $r_1^l > 0,\ r_2^l > 0,\ a > 0$ *and* $b > 0$ *are arbitrary numbers. The equilibrium solution* $(x_1, x_2, \ldots, x_n) = 0$ *for system (3) is Mittag-Leffler stable.*

Proof The proof process has the similarity with Theorem 4. When $t_o = 0$

$$V_l(t, x_l) \le V_l(0, x_l(0)) + \frac{-q}{\Gamma(\beta)} \int_o^t V_l(s, x_l)(t-s)^{\beta-1} ds.$$

By using Lemma 3, we can get:

$$V_l(t, x_l) \le V_l(0, x_l(0)) \cdot E_\beta(-qT^\beta).$$

When considered the condition (A10), we can see:

$$\begin{aligned} r_1 \|x\|^a &= \sum_{l=1}^n e_l r_1^l \|x_l\|^a \le \sum_{l=1}^n e_l V_l(t, x_l) \\ &= V(t, x) \le \sum_{l=1}^n e_l V_l(0, x_l(0)) \cdot E_\beta(-qT^\beta) \\ &= V(0, x(0)) \cdot E_\beta(-qT^\beta), \end{aligned}$$

which leads

$$\|x\| \le \left\{ \frac{V(0, x(0))}{r_1} E_\beta(-qT^\beta) \right\}^{\frac{1}{a}}, \quad t \ge 0.$$

Let $m = \frac{V(0,x(0))}{r_1} \ge 0$. The we obtain

$$\|x\| \le \{mE_\beta(-qT^\beta)\}^{\frac{1}{a}}, \quad t \ge 0,$$

then $m = 0$ satisfies only if $x(0) = 0$. m is a Lipschitz constant, $m(0) = 0$, therefore the equilibrium solution $(x_1, x_2, \ldots, x_n) = 0$ for system (3) is Mittag-Leffler stable. □

Corollary 3 *From Theorem 5, the equilibrium solution* $(x_1, x_2, \ldots, x_n) = 0$ *for system (3) is Mittag-Leffler stable and must be asymptotically stable.*

Remark 5 When $I_w(x_l) \equiv 0$ for all l, k and the constant $x^\star \neq 0$, then the Mittag-Leffler stable of equilibrium solution for FICSNs shall transform into the Mittag-Leffler stable of FCSNs in the paper [43] according with Theorem 3.1.

Remark 6 Above five theorems are our main works. It is different that the results in this chapter compared with the previous in [30, 32, 43]. We consider the stability analysis for FICSNs. This is the highlight of this work.

4 Numerical Example

Let's consider the following example:

$$\begin{cases} {}_{t_o}^{C}D_t^{\alpha} x_l = \alpha_l x_l + h_l(x_l) - \sum\limits_{k=1}^{n} \beta_{lk}(x_l - x_k), \quad t \neq t_w, \\ {}_{t_o}^{C}D_t^{\alpha} y_l = \alpha_l y_l + h_l(y_l) - \sum\limits_{k=1}^{n} \beta_{lk}(y_l - y_k), \quad t \neq t_w, \\ {}_{t_o}^{C}D_t^{\alpha} z_l = \alpha_l z_l + h_l(z_l) - \sum\limits_{k=1}^{n} \beta_{lk}(z_l - z_k), \quad t \neq t_w, \\ \Delta x_l = I_w(x_l(t_w)) = \mathbf{C}_w x_l(t_w) - x_l(t_w), \quad t = t_w, \\ \Delta y_l = I_w(y_l(t_w)) = \mathbf{C}_w y_l(t_w) - y_l(t_w), \quad t = t_w, \\ \Delta z_l = I_w(z_l(t_w)) = \mathbf{C}_w z_l(t_w) - z_l(t_w), \quad t = t_w, \\ x_l(t_o^+) = x_{lo}, \\ y_l(t_o^+) = y_{lo}, \\ z_l(t_o^+) = z_{lo}, \end{cases} \tag{33}$$

where x_l, y_l, z_l are n-dimensional column vectors, $h_l : \mathbf{R}^n \to \mathbf{R}^n$ are continuous and satisfy that $\|h_l(x_l) - h_l(y_l)\| \leq L_l \|x_l - y_l\|$ for any $x_l \neq y_l$. $\beta_{lk} \leq 0, \beta_{lk} = -\beta_{kl}$ and $\beta_{lk} \neq 0$ for $l \neq k$. $\mathbf{C}_w$ is an $n \times n$ constant matrix and $\lambda_{\max}(\mathbf{C}_w) \leq 1, k = 1, 2, \ldots$.

Let $\mathcal{G}$ be a digraph with n vertices, $b_{lk} = |\beta_{lk}|, (l, k = 1, 2, \ldots, n),\ \boldsymbol{B} = (b_{lk})_{n \times n}$, $(\mathcal{G}, \boldsymbol{B})$ is strongly connected and balanced; $\exists p_l, \quad p_l = -\alpha_l + \sum\limits_{k=1}^{n} \beta_{lk} - L_l > 0$.

Here, let $X_l = (x_l, y_l, z_l), l = 1, 2, \ldots, n$, then constitute the Lyapunov functions

$$V_l(t, X_l) = \frac{1}{2}x_l^2 + \frac{1}{2}y_l^2 + \frac{1}{2}z_l^2.$$

From then, we have

$${}_{t_o}^{C}D_t^{\alpha} V_l(t, X_l) = {}_{t_o}^{C} D_t^{\alpha} \left(\frac{1}{2}x_l^2 + \frac{1}{2}y_l^2 + \frac{1}{2}z_l^2\right) \leq x_l({}_{t_o}^{C}D_t^{\alpha} x_l) + y_l({}_{t_o}^{C}D_t^{\alpha} y_l) + z_l({}_{t_o}^{C}D_t^{\alpha} z_l)$$

$$= x_l \left(\alpha_l x_l + h_l(x_l) - \sum_{k=1}^{n} \beta_{lk}(x_l - x_k)\right) + y_l \left(\alpha_l y_l + h_l(y_l) - \sum_{k=1}^{n} \beta_{lk}(y_l - y_k)\right)$$

$$+ z_l \left(\alpha_l z_l + h_l(z_l) - \sum_{k=1}^{n} \beta_{lk}(z_l - z_k)\right)$$

$$\leq \left(\alpha_l + L_l - \sum_{k=1}^{n} \beta_{lk}\right)(x_l^2 + y_l^2 + z_l^2) + \sum_{k=1}^{n} \beta_{lk} x_l x_k + \sum_{k=1}^{n} \beta_{lk} y_l y_k + \sum_{k=1}^{n} \beta_{lk} z_l z_k$$

$$\leq \left(\alpha_l + L_l - \sum_{k=1}^{n} \beta_{lk}\right) V_l(t, X_l) + \sum_{k=1}^{n} \beta_{lk}(x_l x_k + y_l y_k + z_l z_k)$$

$$= -p_l V_l(t, X_l) + \sum_{k=1}^{n} b_{lk} F_{lk}(X_l, X_k)$$

where $p_l = -\alpha_l - L_l + \sum_{k=1}^{n} \beta_{lk} > 0$, $F_{lk}(X_l, X_k) = sgn(\beta_{lk})(x_l x_k + y_l y_k + z_l z_k)$ for $l \neq k$, and

$$F_{lk}(X_l, X_k) = sgn(\beta_{lk})(x_l x_k + y_l y_k + z_l z_k) = -sgn(\beta_{kl})(x_k x_l + y_k y_l + z_k z_l).$$

Towards every directional cycle Π of the weighted digraph $(\mathcal{G}, \boldsymbol{B})$ satisfies the following relationship:

$$\sum_{(l,k)\in E(\Pi_\psi)} F_{lk}(X_l, X_k) + F_{kl}(X_k, X_l) = 0.$$

Then, we can obtain:

$$V_l(t_w^+, X_l t_w + I_w(X_l t_w)) = \frac{1}{2}(\mathbf{C}_w x_l(t_w))^2 + (\mathbf{C}_w y_l(t_w))^2 + (\mathbf{C}_w z_l(t_w))^2$$

$$\leq \frac{1}{2}\lambda^2(\mathbf{C}_w)(x_l(t_w))^2 + (y_l(t_w))^2 + (z_l(t_w))^2) = \lambda^2(\mathbf{C}_w) V_l(t_w, X_l(t_w)).$$

From the condition $\lambda_{\max}(\mathbf{C}_w) \leq 1$, $w = 1, 2, \ldots$, we can get:

$$V_l(t_w^+, X_l t_w + I_w(X_l t_w)) \leq V_l(t_w, X_l(t_w)).$$

Let $\alpha_1^l \|X_l\| = \frac{1}{4}\|X_l\|$, $b_1^l \|X_l\| = \|X_l\|$, then $\alpha_1^l \|X_l\| \leq V_l(t, X_l) = \frac{1}{2}\|X_l\| \leq b_1^l \|X_l\|$. Considering Theorem 3, the $(x_1, x_2, \ldots, x_n) = 0$ for system (33) is uniformly asymptotically stable.

Let $h_l(x_l) = \sin(l x_l/(n-1))$, $\alpha_l = -2n/(n-1)$, $\beta_{ll} = 0$, $\beta_{lk} = -1/(n-1)$ if $l < k$ and $\beta_{lk} = 1/(n-1)$ if $l > k$.

Let $n = 3$, $l = 1, 2, 3$, $\mathbf{C}_w = 1/2\ I_3$, where I_3 is 3×3 identity matrix.

Since $b_{lk} = |\beta_{lk}|$, we can obtain that

$$\boldsymbol{A} = \begin{bmatrix} 0 & \frac{1}{2} & \frac{1}{2} \\ \frac{1}{2} & 0 & \frac{1}{2} \\ \frac{1}{2} & \frac{1}{2} & 0 \end{bmatrix}, \quad \boldsymbol{L} = \begin{bmatrix} 1 & -\frac{1}{2} & -\frac{1}{2} \\ -\frac{1}{2} & 1 & -\frac{1}{2} \\ -\frac{1}{2} & -\frac{1}{2} & 1 \end{bmatrix}.$$

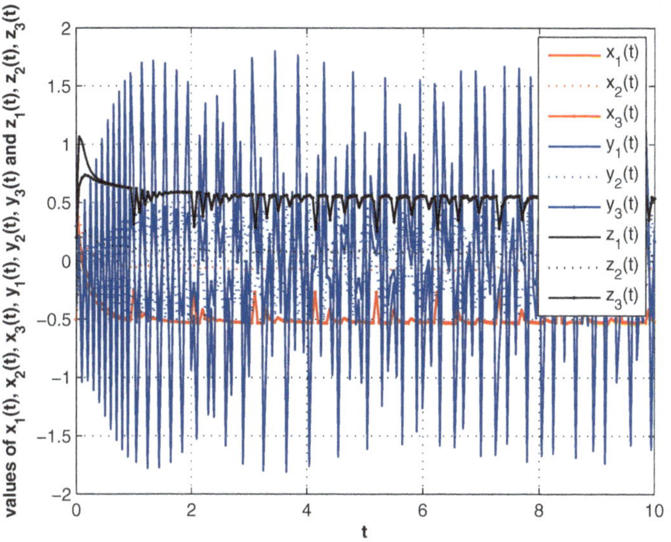

Fig. 1 Times-series of three-dimension system with initial value $X_1 = [0.6,\ 0.2,\ -0.5]$, $Y_1 = [-0.4,\ 0.6,\ 0.3]$, $Z_1 = [0.4,\ -0.5,\ 0.2]$

By simple calculation, $e_1 = e_2 = e_3 = \frac{3}{4}$, $L_o^1 = 0.5$, $L_o^2 = 1$, $L_o^3 = 1.5$.

Therefore, $(\mathcal{G}, \boldsymbol{B})$ is balanced and strongly connected. For $i = 1, 2, 3$, we can compute

$$p_l = -\alpha_l + \sum_{k=1}^{n} \beta_{lk} - L_l = 1 > 0, \quad \lambda^2(\mathbf{C}_w) = \frac{1}{4} < 1$$

which satisfy all conditions of the system (33).

Firstly, we consider the system with the parameters that not content with the condition $p_l = -\alpha_l + \sum_{k=1}^{n} \beta_{lk} - L_l < 0$, then from the Fig. 1, we can get the equilibrium point is instability.

Secondly, we adjust the parameters of the system (33) so that content all the conditions above analysis. From Fig. 2, the system is uniformly asymptotically stable. Here, the initial values are taken as $X_1 = [0.6,\ 0.2,\ -0.5]$, $Y_1 = [-0.4,\ 0.6,\ 0.3]$, and $Z_1 = [0.4,\ -0.5,\ 0.2]$.

Remark 7 Distinguish from the numerical example in [30], they gave an impulsive coupled system with integer derivative. In this chapter, we study the FICSNs (33) with the dimension extended to three.

Next, we discuss the influence of fractional-order α of system (33).

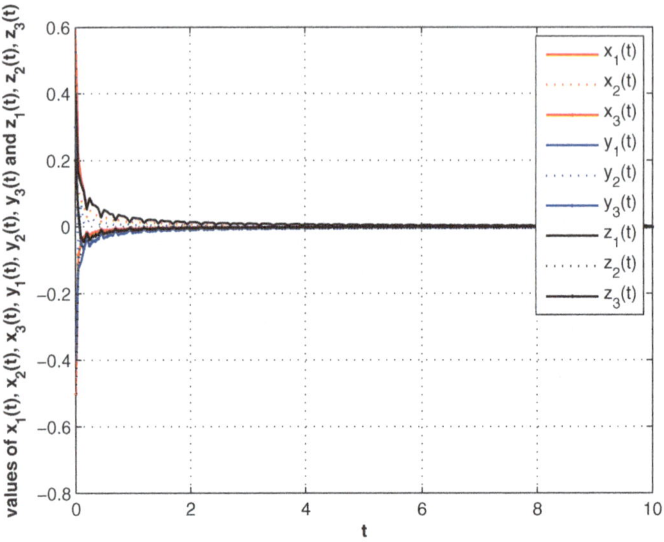

Fig. 2 Times-series of three-dimension system with initial value $X_1 = [0.6,\ 0.2,\ -0.5]$, $Y_1 = [-0.4,\ 0.6,\ 0.3]$, $Z_1 = [0.4,\ -0.5,\ 0.2]$

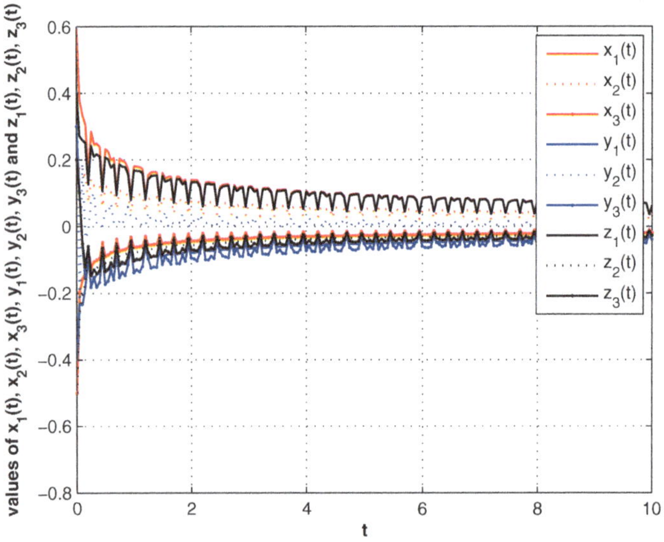

Fig. 3 $\alpha = 0.5$

Keeping the other parameters fixed and altering the parameter α with $\alpha_1 = 0.98$, $\alpha_2 = 0.75$, $\alpha_3 = 0.5$. Figures 3, 4 and 5 show that the differences in times-series of three-dimension system with initial value $X_1 = [0.6,\ 0.2,\ -0.5]$, $Y_1 = [-0.4,\ 0.6,\ 0.3]$, and $Z_1 = [0.4,\ -0.5,\ 0.2]$.

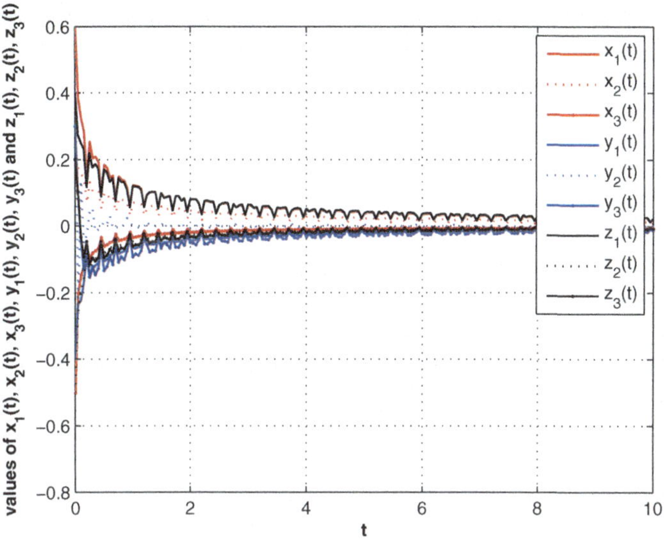

Fig. 4 $\alpha = 0.75$

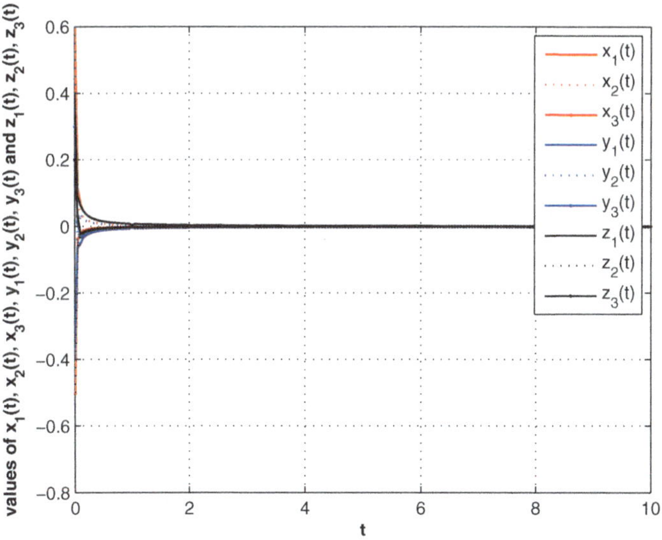

Fig. 5 $\alpha = 0.98$

Remark 8 From Figs. 3, 4 and 5, α is an essential element that can affect the convergence velocities. With the ascending of the fractional-order $\alpha \in (0, 1)$, accelerates the convergence velocities.

5 Conclusion

In this chapter, we studied the stability, uniform stability, asymptotic stability, uniform asymptotic stability, finite-time stability and the Mittag-Leffler stability for FICSNs via utilizing the graph theory and Lyapunov second technique. We first build Lyapunov functions for individual vertex and build global Lyapunov functions for FICSNs. Then, five stability criteria of FICSNs have been derived. Moreover, our method and results can be expanded to large scale complex networks and agents.

Acknowledgements The authors would be very grateful to the editor of this edited book for their helpful comments and constructive suggestions. This research is supported by the National Natural Science Foundations of China (61873071) and the Shandong Provincial Natural Science Foundations (ZR2019MF027).

References

1. Podlubny, I.: Fractional Differential Equations: an Introduction to Fractional Derivatives, Fractional Differential Equations, to Methods of Their Solution and Some of Their Applications. Elsevier, Amsterdam (1998)
2. Carpinteri, A., Mainardi, F.: Fractional calculus: some basic problems in continuum and statistical mechanics. Fractals and Fractional Calculus in Continuum Mechanics, pp. 291–348. Springer, Berlin (1997)
3. Stamova, I.: Stability Analysis of Impulsive Functional Differential Equations. Walter de Gruyter, New York (2009)
4. Yang, Y., He, Y., Wang, Y., et al.: Stability analysis for impulsive fractional hybrid systems via variational Lyapunov method. Commun. Nonlinear Sci. Numer. Simul. **45**, 140–157 (2016)
5. Luo, R., Su, H.: The stability of impulsive incommensurate fractional order chaotic systems with Caputo derivative. Chin. J. Phys. **56**(4), 1599–1608 (2018)
6. Jiao, T., Park, J.H., Zong, G., Zhao, Y., Du, Q.: On stability analysis of random impulsive and switching neural networks. Neurocomputing **350**, 146–154 (2019)
7. Mathiyalagan, K., Park, J.H., Sakthivel, R.: Synchronization for delayed memristive BAM neural networks using impulsive control with random nonlinearities. Appl. Math. Comput. **259**, 967–979 (2015)
8. Zhang, Y., Sun, J., Gang, F.: Impulsive control of discrete systems with time delay. IEEE Trans. Autom. Control **54**(4), 830–834 (2009)
9. Tang, Z., Park, J.H., Wang, Y., Feng, J.: Distributed impulsive quasi-synchronization of Lur'e networks with proportional delay. IEEE Trans. Cybern. **49**(8), 3105–3115 (2019)
10. Zhang, Y.: Robust exponential stability of uncertain impulsive neural networks with time-varying delays and delayed impulses. Neurocomputing **74**(17), 3268–3276 (2011)
11. Li, X., Zhang, X., Song, S.: Effect of delayed impulses on input-to-state stability of nonlinear systems. Automatica **76**, 378–382 (2017)

12. Li, X., Wu, J.: Stability of nonlinear differential systems with state-dependent delayed impulses. Automatica **64**, 63–69 (2016)
13. Zhang, X., Li, X.: Input-to-state stability of non-linear systems with distributed-delayed impulses. IET Control Theory Appl. **11**(1), 81–89 (2017)
14. Li, X., Song, S.: Stabilization of delay systems: delay-dependent impulsive control. IEEE Trans. Autom. Control **62**(1), 406–411 (2017)
15. Lin, D., Li, X., O'Regan, D.: Stability analysis of generalized impulsive functional differential equations. Math. Comput. Model. **55**(5–6), 1682–1690 (2012)
16. Wu, K., Na, M., Wang, L., et al.: Finite-time stability of impulsive reaction-diffusion systems with and without time delay. Appl. Math. Comput. **363**, 124591 (2019)
17. Tian, Y., Cai, Y., Sun, Y., et al.: Finite-time stability for impulsive switched delay systems with nonlinear disturbances. J. Frankl. Inst. **353**(14), 3578–3594 (2016)
18. Hei, X., Wu, R.: Finite-time stability of impulsive fractional-order systems with time-delay. Appl. Math. Model. **40**(7–8), 4285–4290 (2016)
19. Yang, X., Li, C., Huang, T., et al.: Mittag-Leffler stability analysis of nonlinear fractional-order systems with impulses. Appl. Math. Comput. **293**, 416–422 (2017)
20. Stamova, I.: Global Mittag-Leffler stability and synchronization of impulsive fractional-order neural networks with time-varying delays. Nonlinear Dyn. **77**(4), 1251–1260 (2014)
21. Stamova, I., Stamov, G.: Mittag-Leffler synchronization of fractional neural networks with time-varying delays and reaction-diffusion terms using impulsive and linear controllers. Neural Netw. **96**, 22–32 (2017)
22. Li, D., Zhang, X.: Impulsive synchronization of fractional order chaotic systems with time-delay. Neurocomputing **216**, 39–44 (2016)
23. Xi, H., Yu, S., Zhang, R., et al.: Adaptive impulsive synchronization for a class of fractional-order chaotic and hyperchaotic systems. Optik-Int. J. Light Electron Opt. **125**(9), 2036–2040 (2014)
24. Chen, H., Sun, J.: Stability analysis for coupled systems with time delay on networks. Phys. A: Stat. Mech. Appl. **391**(3), 528–534 (2012)
25. Li, Y., Kao, K.: Stability of coupled impulsive Markovian jump reaction-diffusion systems on networks. J. Syst. Sci. Complex. **29**(5), 1269–1280 (2016)
26. Kao, Y., Sun, H., Cao, H.: Stability analysis for coupled stochastic systems with time delay on networks. Math. Appl. **26**, 67–75 (2013)
27. Kao, Y., Wang, C., Karimi, H.R., et al.: Global stability of coupled Markovian switching reaction-diffusion systems on networks. Nonlinear Anal.: Hybrid Syst. **13**, 61–73 (2014)
28. Kao, Y., Zhu, Q., Qi, W.: Exponential stability and instability of impulsive stochastic functional differential equations with Markovian switching. Appl. Math. Comput. **271**, 795–804 (2015)
29. Kao, Y., Wang, C.: Global stability analysis for stochastic coupled reaction-diffusion systems on networks. Nonlinear Anal.: Real World Appl. **14**(3), 1457–1465 (2013)
30. Suo, J., Sun, J., Zhang, Y.: Stability analysis for impulsive coupled systems on networks. Neurocomputing **99**, 172–177 (2013)
31. Wang, Y., Zhang, H., Wang, X., et al.: Networked synchronization control of coupled dynamic networks with time-varying delay. IEEE Trans. Syst. Man Cybern. Part B (Cybern.) **40**(6), 1468–1479 (2010)
32. Li, H., Jiang, Y., Wang, Z., et al.: Global Mittag-Leffler stability of coupled system of fractional-order differential equations on network. Appl. Math. Comput. **270**, 269–277 (2015)
33. Li, H., Hu, C., Jiang, Y., et al.: Global Mittag-Leffler stability for a coupled system of fractional-order differential equations on network with feedback controls. Neurocomputing **214**, 233–241 (2016)
34. Zhang, Y., Liu, S., Yang, R., et al.: Global synchronization of fractional coupled networks with discrete and distributed delays. Phys. A: Stat. Mech. Appl. **514**, 830–837 (2019)
35. Zhang, H., Ye, M., Ye, R., et al.: Synchronization stability of Riemann-Liouville fractional delay-coupled complex neural networks (2018)
36. Ma, T., Zhang, J.: Hybrid synchronization of coupled fractional-order complex networks. Neurocomputing **157**, 166–172 (2015)

37. Ouannas, A., Odibat, Z., Alsaedi, A., et al.: Investigation of QS synchronization in coupled chaotic incommensurate fractional order systems. Chin. J. Phys. **56**(5), 1940–1948 (2018)
38. Xu, Y., Qiang, L., Li, W.: Periodically intermittent discrete observation control for synchronization of fractional-order coupled systems. Commun. Nonlinear Sci. Numer. Simul. **74**, 219–235 (2019)
39. Li, H., Kao, Y.: Mittag-Leffler stability for a new coupled system of fractional-order differential equations with impulses. Appl. Math. Comput. **361**, 22–31 (2019)
40. Li, M., Shuai, Z.: Global-stability problem for coupled systems of differential equations on networks. J. Differ. Equ. **248**(1), 1–20 (2010)
41. Li, W., Su, H., Wang, K.: Global stability analysis for stochastic coupled systems on networks. Automatica **47**(1), 215–220 (2011)
42. Li, Y., Chen, Y., Podlubny, I.: Mittag-Leffler stability of fractional order nonlinear dynamic systems. Automatica **45**(8), 1965–1969 (2009)
43. Stamova, I.: Mittag-Leffler stability of impulsive differential equations of fractional order. Q. Appl. Math. **73**(3), 525–535 (2015)
44. Phat, V., Thanh, N.: New criteria for finite-time stability of nonlinear fractional-order delay systems: a Gronwall inequality approach. Appl. Math. Lett. **83**, 169–175 (2018)
45. Aguila-Camacho, N., Duarte-Mermoud, M., Gallegos, J.: Lyapunov functions for fractional order systems. Commun. Nonlinear Sci. Numer. Simul. **19**(9), 2951–2957 (2014)

Dynamical Networks

Positivity and Stability of Nonlinear Time-Delay Systems in Neural Networks

Le Van Hien

Abstract This chapter is concerned with the problems of positivity and stability of nonlinear time-delay systems in the area of neural networks. The concept of positivity is first developed and characterized for various models of neural networks with delays including Hopfield neural networks, bidirectional associative memory neural networks and inertial neural networks. It will be shown by utilizing the property of order-preserving of neuron activation functions that for nonnegative connection weights of neurons, state trajectories of the networks initializing from a positive cone called the admissible set of initial conditions are always nonnegative subject to nonnegative inputs. Then, the exponential stability of positive equilibriums of the underlying models is investigated by a systematic approach involving extended comparison techniques via differential and integral inequalities. Unified conditions for the existence and exponential stability of positive equilibriums are derived in the form of linear programming (LP) conditions with M-matrix, which can be effectively solved by various convex algorithms. Numerical examples and simulations are provided to demonstrate the applicability and effectiveness of the derived theoretical results.

Keywords Positive neural networks · Positive equilibrium · Exponential stability · Time-varying delay · M-matrix.

1 Introduction

Neural networks (NNs), including artificial neural networks (ANNs) and biological neural networks, can be found in a wide range of practical applications. An instance example can be found in the area of time series forecasting where ANNs are used for the test of ill-defined temporal problems like predictions in financial markets. In the

L. V. Hien (✉)
Faculty of Mathematics and Informatics, Hanoi National University of Education, 136 Xuan Thuy, Hanoi, Vietnam
e-mail: hienlv@hnue.edu.vn

J. H. Park (ed.), *Recent Advances in Control Problems of Dynamical Systems and Networks*, Studies in Systems, Decision and Control 301,
https://doi.org/10.1007/978-3-030-49123-9_15

field of pattern recognition for medical visualization aids and image processing, due to their capabilities in pattern-matching and learning, ANNs are used to solve many problems in image realization, speech recognition, parallel computation or natural language processing, which are normally difficult to solve by conventional methods [1–4]. For modelling of system dynamics and adaptivity features, ANNs can also be used in various state estimation problems as system identification and fault detection [5, 6] resorting to the approximation capability of nonlinear functions. Many other applications of neural networks (NNs) can be found in transportation, health care, industrial robotics, human-machine interface, process control, associative memory or computer security [7, 8]. In practical applications of NNs, it is important to ensure the existence and stability normally in the sense of asymptotic behavior of a unique equilibrium [9]. In addition, the implementation process of NNs in practical is frequently encountered with time delays according to many technical reasons such as the limit of switching speed of amplifiers or the signal processing transmission through layers in the network. The presence of time-delay usually makes the system behaviors more complicated and unpredictable [10–12]. Thus, in the past few decades, the problems of performance analysis and synthesis of NNs with delays have been extensively studied and, as a consequence, many results concerning the stability of NNs with delays have been reported. To mention a few, we refer the reader to [13–25] and the references therein.

Positive systems are dynamical systems whose states and outputs initiating from nonnegative inputs and initial conditions are always nonnegative [26]. This type of systems can be used to model various practical systems where the associated state variables are subject to positivity constraints according to the nature of phenomena [27]. It has been well recognized in the area of systems and control that the problem of stability and/or performance analysis of dynamical time-delay systems plays a key role in the design and implement of controller/observer/filter in practical applications. Like general dynamical systems, the problem of stability analysis and control of positive systems with or without delays has received significant research attention from researchers in the past few decades [28–35]. While many important problems in the systems and control theory have been well-studied for various classes of positive time-delay systems and, in particular, linear positive time-delay systems, this area is still considerably less well-developed for positive nonlinear systems describing neural network models. It is noted that when a neural network is designed for application purposes of positive systems, for example, in identification [36], control [37], monotone-regular behavior implement [38] and various disciplines in the academia and industries including computer vision, pattern recognition, alignment and detection [39], states of the designed networks are expected to inherit positivity constraints of the practical model. Besides, for neural network systems, the nonlinearity of neuron activation functions makes the study of positive neural networks with delays more complicated and challenging. Thus, as a key and primary issue to apply in control of positive neural networks, the problem of stability analysis of positive nonlinear time-delay systems in network structures is very relevant and of great importance to study. In particular, a systematic approach and effective tools used in the analysis of positivity and stability of such models are obviously important to develop.

Typically, the dynamics of a conventional neural network is described by first-order differential equations. Recently, many authors have devoted attention to studies a type of network models called inertial neural networks (INNs) [40]. In the state-space model, INNs are described by second-order differential equations with inertial terms represented by the first-order derivative. The modeling of inertial terms in neural systems has strong biological and engineering backgrounds especially for those containing an inductance [41, 42]. In addition, in the presence of inertial terms, it is much more difficult and challenging to analyze the dynamic behaviors of INNs. Thus, due to both theoretical and practical reasons, the investigation of INNs has attracted considerable attention in recent years. For example, based on extended Halanay inequalities and a Lyapunov-like functional method, the problems of dissipativity and exponential convergence were studied for various types of INNs with time-varying delays in [43–46]. Lagrange stability and stabilization via sampled-data control were also investigated for Hopfield INNs in [47–49] and [50] using characteristic function and Lyapunov–Krasovskii functional methods. However, it should be mentioned here that the above papers only deal with traditional neural networks. Furthermore, the methodologies proposed in the existing literature cannot be directly applied or extended to positive nonlinear systems which describe INNs models with delays. Extending the theory of positive systems to nonlinear time-delay systems that describe INNs proves to be an interesting issue. In particular, the existence, uniqueness, and global exponential stability of the positive equilibrium of positive INNs with delays need further investigation, which will be addressed in this chapter.

Another important model of neural networks which appear in artificial intelligence is referred to as bidirectional associative memory (BAM) neural networks. First introduced in the pioneering work [51] when studied stability and encoding properties of two-layer nonlinear feedback neural networks, the terminology of bidirectionality, referred to forward and backward information flows, was adapted to neural nets to produce a two-way associative search for stored paired-data associations. This model was later called BAM neural network. Typically, a BAM neural network is constructed of two neuron layers namely X-layer and Y-layer. Neurons in each layer are completely integrated into neurons in the other layer, whereas there is no interconnection among neurons in the same layer. This structure performs a two-way associative search for stored bipolar vector pairs and generalizes the single-layer autoassociative Hebbian correlation to a two-layer pattern-matched heteroassociative circuits [52]. Thus, the BAM model possesses many application prospects in the areas of pattern recognition, signal and image processing. As discussed in the preceding paragraphs, practical applications of neural networks in many areas require that the designed networks admit only one equilibrium point which is globally asymptotically stable thereby avoiding the risks of having spurious equilibrium point and being trapped in local minima [53]. Thus, it is important to study the problems of stability and performance analysis of a unique equilibrium of BAM neural networks with delays [54–57]. The problems of synchronization [58, 59], stabilization [60, 61] or impulsive control [62, 63] have also been investigated for various types of BAM neural networks with delays. However, the theory for positive BAM neural networks with delays has not been well investigated despite their elegant properties and potential applications. It should be noted that the methods developed for Hopfield-type neural networks cannot be simply extended to BAM neural networks with delays according to the nature of the structure themselves.

Motivated from the above discussion, in this chapter, we study the stability problem of nonlinear time-delay systems in three types of neural networks. Specifically, we develop the concept of positivity for popular neural networks models with delays including Hopfield neural networks, BAM neural networks and inertial neural networks. A systematic approach involving extended comparison techniques via differential and integral inequalities is developed for studying exponential stability of positive equilibriums of considering models. By utilizing the proposed approach, unified conditions for the existence and exponential stability of positive equilibriums are derived in the form of linear programming conditions with M-matrix, which can be effectively solved by various convex algorithms. Numerical examples and simulations are provided to demonstrate the applicability and effectiveness of the derived theoretical results.

2 Preliminaries

2.1 Notation

$\mathbb{R}^n$ the n-dimensional Euclidean space with the vector norm $\|x\| = \max_{1\le i\le n} |x_i|$, $\mathbf{1}_n \in \mathbb{R}^n$ is the vector with all entries equal one. $\mathbb{R}^{m\times n}$ is the set of $m \times n$-matrices. $\sigma(A)$ denotes the set of eigenvalues of $A \in \mathbb{R}^{n\times n}$ and $\rho(A) = \max\{|\lambda| : \lambda \in \sigma(A)\}$ is its spectral radius. $vec(v_1, v_2, \ldots, v_k)$ denotes the augmented vector formulated by stacking components of $v_1, v_2, \ldots, v_k$. For vectors $x = (x_i) \in \mathbb{R}^n$ and $y = (y_i) \in \mathbb{R}^n$, we write $x \preceq y$ if $x_i \le y_i$ and $x \prec y$ if $x_i < y_i$ for all $i \in [n] \triangleq \{1, 2, \ldots, n\}$. $\mathbb{R}^n_+ = \{x \in \mathbb{R}^n : x \succeq 0\}$ and $|x| = (|x_i|) \in \mathbb{R}^n_+$ for $x = (x_i) \in \mathbb{R}^n$. A matrix $A = (a_{ij}) \in \mathbb{R}^{m\times n}$ is nonnegative, $A \succeq 0$, if $a_{ij} \ge 0$ for all i, j, A is a Metzler matrix if its off-diagonal entries are nonnegative and A is an M-matrix if $a_{ij} \le 0$ for all $i \ne j$. $C([a, b], \mathbb{R}^n)$ denotes the set of $\mathbb{R}^n$-valued continuous functions on $[a, b]$ endowed with the *supremum* norm $\|\phi\|_C = \sup_{a\le t\le b} \|\phi(t)\|$ for a $\phi \in C([a, b], \mathbb{R}^n)$.

2.2 Auxiliary Results

We recall here some concepts in nonlinear analysis and the theory of monotone dynamical systems which will be used in the derivation of our results. A vector field $\varphi : \mathbb{R}^n \to \mathbb{R}^n$ is said to be *order-preserving* on $\mathbb{R}^n_+$ if $\varphi(x) \preceq \varphi(y)$ for any $x, y \in \mathbb{R}^n_+$ satisfying $x \preceq y$ [27]. A mapping $\Psi : \mathbb{R}^n \to \mathbb{R}^n$ is *proper* if $\Psi^{-1}(K)$ is compact for any compact subset $K \subset \mathbb{R}^n$. It is well-known that a continuous mapping $\Psi : \mathbb{R}^n \to \mathbb{R}^n$ is proper if and only if Ψ has the property that for any sequence $\{p_k\} \subset \mathbb{R}^n$, $\|p_k\| \to \infty$ then $\|\Psi(p_k)\| \to \infty$ as $k \to \infty$.

Lemma 1 ([64]) *A locally invertible continuous mapping $\Psi : \mathbb{R}^n \to \mathbb{R}^n$ is a homeomorphism of $\mathbb{R}^n$ onto itself if and only if it is proper.*

Next, we introduce the following Brouwer's fixed point theorem which will be used to establish the existence of equilibrium points of neural networks models. For more details, we refer the reader to Zeidler E (1986) [65, Sect. 2.3]. Let $f : X \to X$ be a mapping from a metric or topology space X into itself. An $x \in X$ is said to be a fixed point of f if it is unchanged under the effect of f, that is, $f(x) = x$. The following result is a simple form of Brouwer fixed point theorem.

Proposition 1 ([65]) *Suppose that M is a nonempty, convex, compact subset of $\mathbb{R}^n$ ($n \geq 1$) and that $f : M \to M$ is a continuous mapping. Then f possesses at least a fixed point in M.*

Finally, we recall some properties of nonsingular M-matrix [66]. A matrix $A = (a_{ij}) \in \mathbb{R}^{n\times n}$ is an M-matrix if it can be expressed in the form $A = sI_n - B$, where $B = (b_{ij}) \succeq 0$ and $s \geq \rho(B)$, the maximum of the moduli of eigenvalues of B (also known as the spectral radius of B). In the aforementioned expression, A is a nonsingular M-matrix if and only if $s > \rho(B)$. The following proposition summarizes widely used properties of nonsingular M-matrix.

Proposition 2 ([66]) *Let $A = (a_{ij}) \in \mathbb{R}^{n\times n}$ be an M-matrix. The following statements are equivalent.*

(i) A is a nonsingular M-matrix.
(ii) All the principal minors of A are positive.
(iii) $A + D$ is nonsingular for any nonnegative diagonal matrix D.
(iv) A is inverse-positive, that is, the inverse matrix A^{-1} exists and $A^{-1} \succeq 0$.
(v) A has a convergent regular splitting, that is, A has a representation of the form $A = M - N$ with $M^{-1} \succeq 0$, $N \succeq 0$ and $\rho(M^{-1}N) < 1$ ($M^{-1}N$ convergent).
(vi) Every regular splitting of A is convergent.
(vii) There exists a vector $\zeta \in \mathbb{R}^n$, $\zeta \succ 0$, such that $A\zeta \succ 0$.

It follows from Proposition 2 that if $K = (k_{ij}) \in \mathbb{R}^{n\times n}$ is a nonnegative matrix whose spectral radius $\rho(K) < 1$ then $(I_n - K)^{-1} \succeq 0$ and there exists a positive vector $\zeta = (\zeta_i)$ such that $(I_n - K)\zeta \succ 0$. Therefore,

$$\sum_{j=1}^{n} k_{ij}\zeta_j < \zeta_i, \; i \in [n].$$

This fact will be useful for our later derivation.

3 Exponential Stability of Positive Hopfield Neural Networks with Time-Varying Delay

This section investigates the problem of exponential stability of a unique positive equilibrium of positive nonlinear systems with multiple time-varying delays which describe Hopfield-type neural networks with nonlinear self-inhibition rates. Based on a novel comparison technique via a differential and integral inequalities, testable conditions are derived to ensure system state trajectories converge exponentially to a unique positive equilibrium.

3.1 Model Description and Preliminaries

Consider the following nonlinear system with heterogeneous delays

$$\begin{aligned} x_i'(t) = & -d_i\varphi_i(x_i(t)) + \sum_{j=1}^{n} a_{ij} f_j(x_j(t)) \\ & + \sum_{j=1}^{n} b_{ij} g_j(x_j(t-\tau_j(t))) + J_i, \ \ i \in [n], \ \ t \geq 0. \end{aligned} \tag{1}$$

System (1) describes a model of Hopfield neural networks, where n is the number of neurons in the network, $x(t) = (x_i(t)) \in \mathbb{R}^n$ and $J = (J_i) \in \mathbb{R}^n$ are the state vector and the external input vector, respectively; $f(x(t)) = (f_i(x_i(t))) \in \mathbb{R}^n$, $g(x(t)) = (g_i(x_i(t))) \in \mathbb{R}^n$ are neuron activation functions, $\varphi_i(x_i(t))$ and $d_i > 0$, $i \in [n]$, are nonlinear self-excitation rates and self-inhibition coefficients; $A = (a_{ij}) \in \mathbb{R}^{n\times n}$ and $B = (b_{ij}) \in \mathbb{R}^{n\times n}$ are neuron connection weight matrices; $\tau_j(t)$, $j \in [n]$, represent time-varying delays satisfying $0 \leq \tau_j(t) \leq \tau_j^+$ for all $t \geq 0$, where τ_j^+ is a known scalar. Initial condition of system (1) is specified as

$$x(\theta) = \phi(\theta), \ \ \theta \in [-\tau^+, 0]$$

where $\tau^+ = \max_{1\leq j\leq n} \tau_j^+$ and $\phi \in C([-\tau^+, 0], \mathbb{R}^n$ is a given function.

Let $\mathcal{F}$ be the set of continuous functions $\varphi : \mathbb{R} \to \mathbb{R}$ satisfying $\varphi(0) = 0$ and there exist positive scalars l_φ^-, l_φ^+ such that

$$l_\varphi^- \leq \frac{\varphi(u) - \varphi(v)}{u - v} \leq l_\varphi^+ \tag{2}$$

for all $u, v \in \mathbb{R}$, $u \neq v$. It is clear that the function class $\mathcal{F}$ includes all linear functions $\varphi(u) = \gamma_\varphi u$ where γ_φ is some positive scalar.

Assumptions

(A1) The decay rate functions φ_i, $i \in [n]$, belong to the function class $\mathcal{F}$.

(A2) The activation functions $f_j(.)$ and $g_j(.)$ are continuous and satisfy the following conditions

$$0 \leq \frac{f_j(u) - f_j(v)}{u - v} \leq l_j^f, \quad 0 \leq \frac{g_j(u) - g_j(v)}{u - v} \leq l_j^g, \ \forall u \neq v, \tag{3}$$

where l_j^f and l_j^g, $j \in [n]$, are positive constants.

Remark 1 It follows from Assumption (A2) that the functions $f(x) = (f_i(x_i))$ and $g(x) = (g_i(x_i))$, $x = (x_i) \in \mathbb{R}^n$, are globally Lipschitz continuous on $\mathbb{R}^n$. Thus, by utilizing fundamental results in the theory of functional differential equations [67], it can be verified that for any initial function $\phi \in C([-\tau^+, 0], \mathbb{R}^n)$, there exists a unique solution $x(t) = x(t, \phi)$ of (1) on the interval $[0, \infty)$, which is absolutely continuous in t. In the sequel, each solution of (1) will be denoted simply as $x(t)$, $t \in [0, \infty)$, if it does not make any confusion.

Definition 1 System (1) is said to be *positive* if for any initial function $\phi \in C([-\tau^+, 0], \mathbb{R}^n_+)$ and nonnegative input vector $J \in \mathbb{R}^n_+$, the corresponding state trajectory is nonnegative, that is, $x(t) \in \mathbb{R}^n_+$ for all $t \geq 0$.

Definition 2 Given an input vector $J \in \mathbb{R}^n_+$. A vector $x_e \in \mathbb{R}^n_+$ is said to be a *positive equilibrium* of system (1) if it satisfies the following algebraic equation

$$-D\Phi(x_e) + Af(x_e) + Bg(x_e) + J = 0, \tag{4}$$

where the vector function $\Phi : \mathbb{R}^n \to \mathbb{R}^n$ is defined as $\Phi(x) = (\varphi_i(x_i))$.

Definition 3 A positive equilibrium x_e of (1) is said to be globally exponentially stable if there exist positive scalars β, η such that any solution $x(t)$ of (1) satisfies the following inequality

$$\|x(t) - x_e\|_\infty \leq \beta\|\phi - x_e\|_C e^{-\eta t}, \quad t \geq 0.$$

3.2 *Positivity of Model* (1)

In this section, sufficient conditions are derived to ensure that for any initial function $\phi \in C^+ \triangleq C([-\tau^+, 0], \mathbb{R}^n_+)$ and nonnegative input vector $J \in \mathbb{R}^n_+$, the corresponding state trajectory of the system is nonnegative for all time.

Proposition 3 *Let Assumptions* (A1)–(A2) *hold and assume that the connection weight matrices A, B are nonnegative (i.e. $A \succeq 0$ and $B \succeq 0$). Then, system* (1) *is positive for any bounded delays.*

Proof Let $x(t)$ be a solution of (1) with initial function $\phi \in C^+$ and input vector $J \in \mathbb{R}^n_+$. For a given $\epsilon > 0$, let $x_\epsilon(t)$ denote the solution (1) with initial condition $\phi_\epsilon(.) = \phi(.) + \epsilon\mathbf{1}_n$. Note that $x_\epsilon(t) \to x(t)$ as $\epsilon \to 0$. Thus, it suffices to show that $x_\epsilon(t) \succ 0$ for all $t \geq 0$. Suppose in contrary that there exists an index $i \in [n]$ and a $t_f > 0$ such that

$$x_{i\epsilon}(t_f) = 0, \;\; x_{i\epsilon}(t) > 0 \text{ for all } t \in [0, t_f)$$

and $x_{j\epsilon}(t) \geq 0$ for all $j \in [n]$. Then,

$$\psi_i(t) = \sum_{j=1}^{n} a_{ij} f_j(x_{j\epsilon}(t)) + \sum_{j=1}^{n} b_{ij} g_j(x_{j\epsilon}(t - \tau_j(t))) + J_i \geq 0$$

for all $t \in [0, t_f]$. On the other hand, by condition (4), we have

$$l_{\varphi_i}^- \leq \frac{\varphi_i(x_{i\epsilon}(t))}{x_{i\epsilon}(t)} \leq l_{\varphi_i}^+, \quad t \in [0, t_f).$$

This, together with (1), leads to

$$x'_{i\epsilon}(t) \geq -l_{\varphi_i}^+ x_{i\epsilon}(t) + \psi_i(t), \;\; t \in [0, t_f). \tag{5}$$

By integrating both sides of inequality (5) we then obtain

$$\begin{aligned} x_{i\epsilon}(t) &\geq e^{-l_{\varphi_i}^+ t}\left(\phi_i(0) + \epsilon + \int_0^t e^{l_{\varphi_i}^+ s}\psi_i(s)ds\right) \\ &\geq e^{-l_{\varphi_i}^+ t}(\phi_i(0) + \epsilon), \quad t \in [0, t_f). \end{aligned} \tag{6}$$

Let $t \uparrow t_f$, inequality (6) gives

$$0 < (\phi_i(0) + \epsilon)e^{-l_{\varphi_i}^+ t_f} \leq x_{i\epsilon}(t_f) = 0$$

which clearly raises a contradiction. This shows that $x_\epsilon(t) \succ 0$ for $t \in [0, \infty)$. Finally, let $\epsilon \to 0$ we obtain $x(t) \succeq 0$ for all $t \geq 0$. The proof is completed. □

3.3 Exponential Stability of a Unique Positive Equilibrium

According to Eq. (4), for a given input vector $J \in \mathbb{R}^n$, an equilibrium of (1) exists if and only if the equation $\Psi(x) = 0$ has a solution $x_e \in \mathbb{R}^n$, where the mapping $\Psi : \mathbb{R}^n \to \mathbb{R}^n$ is defined as $\Psi(x) = -D\Phi(x) + Af(x) + Bg(x) + J$. Clearly, Ψ is continuous on $\mathbb{R}^n$. Based on Lemma 1, we have the following result.

Proposition 4 *Let Assumptions* (A1)–(A2) *hold. Assume that the connection weight matrices A, B are nonnegative and there exists a vector $v \in \mathbb{R}^n$, $v \succ 0$, such that*

$$\sum_{i=1}^{n} \left(a_{ij} l_j^f + b_{ij} l_j^g\right) v_i < d_j l_{\varphi_j}^- v_j, \ \ j \in [n], \tag{7}$$

that is,

$$v^\top \left(-DL_\varphi^- + AL_f + BL_g\right) \prec 0,$$

where

$$\begin{aligned} L_\varphi^- &= \mathtt{diag}\{l_{\varphi_1}^-, \ldots, l_{\varphi_n}^-\}, \\ L_f &= \mathtt{diag}\{l_1^f, \ldots, l_n^f\}, \\ L_g &= \mathtt{diag}\{l_1^g, \ldots, l_n^g\}. \end{aligned}$$

Then, for a given vector $J \in \mathbb{R}^n$, system (1) *has a unique equilibrium $x_e \in \mathbb{R}^n$.*

Proof For any vectors $x, y \in \mathbb{R}^n$, we have

$$\Psi(x) - \Psi(y) = -D\left(\Phi(x) - \Phi(y)\right) + A[f(x) - f(y)] + B[g(x) - g(y)]. \tag{8}$$

According to Eqs. (4) and (3), we have

$$\begin{aligned} \mathtt{sgn}(x_i - y_i)\left(\varphi_i(x_i) - \varphi_i(y_i)\right) &\geq l_{\varphi_i}^- |x_i - y_i|, \\ \mathtt{sgn}(x_j - y_j)(f_j(x_j) - f_j(y_j)) &\leq l_j^f |x_j - y_j|. \end{aligned}$$

Thus, by multiplying both sides of (8) with $\mathcal{S}(x-y) \triangleq \mathtt{diag}\{\mathtt{sgn}(x_i - y_i)\}$, we obtain

$$\mathcal{S}(x-y)\left(\Psi(x) - \Psi(y)\right) \preceq \left(-DL_\varphi^- + AL_f + BL_g\right)|x-y|$$

which leads to

$$|\Psi(x) - \Psi(y)| \succeq \left(DL_\varphi^- - AL_f - BL_g\right)|x-y|. \tag{9}$$

For a vector $v \in \mathbb{R}^n$, $v \succ 0$, it follows from (9) that

$$v^\top |\Psi(x) - \Psi(y)| \succeq v^\top \left(DL_\varphi^- - AL_f - BL_g\right)|x-y|.$$

Therefore, if $\Psi(x) = \Psi(y)$ then, by condition (7), $v^\top \left(D - AL_f - BL_g\right)|x-y| = 0$ and $|x-y| = 0$ which clearly gives $x = y$. This shows that the mapping Ψ is an injective mapping in $\mathbb{R}^n$. On the other hand, (9) also gives

$$\|\Psi(x)\|_\infty \geq \frac{1}{\|v\|_\infty} v^\top \left(DL_\varphi^- - AL_f - BL_g\right)|x| - \|\Psi(0)\|_\infty.$$

The above estimate implies that $\|\Psi(x_k)\|_\infty \to \infty$ for any sequence $\{x_k\} \subset \mathbb{R}^n$ satisfying $\|x_k\|_\infty \to \infty$. By Lemma 1, Ψ is a homeomorphism onto $\mathbb{R}^n$, and thus, the equation $\Psi(x) = 0$ has a unique solution $x_e \in \mathbb{R}^n$ which is an equilibrium of (1). The proof is completed. □

Remark 2 It should be pointed out that the approach based on homeomorphism as presented in the proof of Proposition 4 does not make sure x_e is a positive equilibrium even when J is a positive input vector. The positivity of x_e will be shown simultaneously with exponential stability of x_e.

Remark 3 Clearly, $\mathcal{M} = -D + AL_f + BL_g$ and $\mathcal{M}^\top$ are Metzler matrices. In addition, (7) holds if and only if $\mathcal{M}^\top v < 0$. This condition is feasible for a vector $v \succ 0$ if and only if $\mathcal{M}^\top$ (and thus $\mathcal{M}$) is a Metzler–Hurwitz matrix [68]. In the following, we will show that the derived conditions in Propositions 3 and 4 ensure that system (1) is positive and the unique equilibrium point x_e is positive for any given input vector $J \in \mathbb{R}^n_+$ which is globally exponentially stable.

Theorem 1 *Let Assumptions* (A1)–(A2) *hold. Assume that* $A \succeq 0$, $B \succeq 0$ *and there exists a vector* $\chi \in \mathbb{R}^n$, $\chi \succ 0$, *such that*

$$\mathcal{M}\chi = \left(-DL_\varphi^- + AL_f + BL_g\right)\chi \prec 0. \tag{10}$$

Then, for any $J \in \mathbb{R}^n_+$, *system* (1) *has a unique positive equilibrium* $x_e \in \mathbb{R}^n_+$ *which is globally exponentially stable for any delays* $\tau_j(t) \in [0, \tau_j^+]$.

Proof By Proposition 4, there exists a unique equilibrium $x_e \in \mathbb{R}^n$ of (1). We first prove that x_e is globally exponentially stable.

Indeed, let $x(t) = (x_i(t))$ be a solution of (1). Then, by (1) and (4), we have

$$\begin{aligned}(x_i(t) - x_{*i})' = &- d_i\left(\varphi_i(x_i(t)) - \varphi_i(x_{*i})\right) + \sum_{j=1}^n a_{ij}[f_j(x_j(t)) - f_j(x_{*j})] \\ &+ \sum_{j=1}^n b_{ij}[g_j(x_j(t - \tau_j(t))) - g_j(x_{*j})].\end{aligned} \tag{11}$$

We define $z(t) = |x(t) - x_e|$ and $\mathcal{S}(x(t) - x_e) = \mathtt{diag}\{\mathtt{sgn}(x_i(t) - x_{ei})\}$. It follows from (11) that

$$\begin{aligned} D^{-}z_i(t) &= \operatorname{sign}(x_i(t) - x_{*i})(x_i(t) - x_{*i})' \\ &\leq -d_i l_{\varphi_i}^{-} |x_i(t) - x_{*i}| + \sum_{j=1}^{n} a_{ij} l_j^f |x_j(t) - x_{*j}| \\ &\quad + \sum_{j=1}^{n} b_{ij} l_j^g |x_j(t - \tau_j(t)) - x_{*j}|, \end{aligned} \tag{12}$$

where $D^{-}z_i(t)$ denotes the upper left Dini derivative of $z_i(t)$. Therefore, it follows from (12) that

$$D^{-}z_i(t) \leq -d_i l_{\varphi_i}^{-} z_i(t) + \sum_{j=1}^{n} a_{ij} l_j^f z_j(t) + \sum_{j=1}^{n} b_{ij} l_j^g z_j(t - \tau_j(t)). \tag{13}$$

We now establish an exponential estimate for $z(t)$. From (10) we have

$$-d_i l_{\varphi_i}^{-} \chi_i + \sum_{j=1}^{n} (a_{ij} l_j^f + b_{ij} l_j^g) \chi_j < 0, \quad i \in [n]. \tag{14}$$

Consider the following scalar functions

$$H_i(\eta) = (\eta - d_i l_{\varphi_i}^{-}) \chi_i + \sum_{j=1}^{n} a_{ij} l_j^f \chi_j + (\sum_{j=1}^{n} b_{ij} l_j^g \chi_j) e^{\eta \tau^+}, \ \eta \geq 0, \quad i \in [n].$$

Clearly, the function $H_i(\eta)$ is continuous and strictly increasing on $[0, \infty)$, $H_i(0) < 0$ and $H_i(\eta) \to \infty$ as $\eta \to \infty$. Thus, there exists a unique positive scalar η_i such that $H_i(\eta_i) = 0$.

Let $\eta_0 = \min_{1 \leq i \leq n} \eta_i$ and define the following functions for $i \in [n]$

$$\rho_i(t) = \frac{\chi_i}{\chi_+} \|\phi - x_e\|_{\mathcal{C}} e^{-\eta_0 t}, \ t \geq 0,$$

and $\rho_i(t) = \rho_i(0)$, $t \in [-\tau^+, 0]$, where $\chi_+ = \min_{1 \leq i \leq n} \chi_i$.

Note also that

$$\rho_j(t - \tau_j(t)) = e^{\eta_0 \tau_j(t)} \rho_j(t) \leq e^{\eta \tau^+} \rho_j(t)$$

for any $t \geq 0$. Therefore,

$$
\begin{aligned}
-d_il^-_{\varphi_i}\rho_i(t) &+ \sum_{j=1}^n a_{ij}l^f_j\rho_j(t) + \sum_{j=1}^n b_{ij}l^g_j\rho_j(t-\tau_j(t))\\
&\le \Big[-d_il^-_{\varphi_i}\chi_i + \sum_{j=1}^n a_{ij}l^f_j\chi_j + \big(\sum_{j=1}^n b_{ij}l^g_j\chi_j\big)e^{\eta_0\tau^+}\Big]\\
&\quad\times \frac{1}{\chi_+}\|\phi - x_e\|_C e^{-\eta_0 t}\\
&\le \frac{H_i(\eta_0)-\eta_0\chi_i}{\chi_+}\|\phi - x_e\|_C e^{-\eta_0 t}.
\end{aligned}
\tag{15}
$$

Since $H_i(\eta)$ is increasing in η, $H_i(\eta_0) \le 0$ for all $i \in [n]$. Thus, (15) gives

$$
\rho_i'(t) \ge -d_il^-_{\varphi_i}\rho_i(t) + \sum_{j=1}^n a_{ij}l^f_j\rho_j(t) + \sum_{j=1}^n b_{ij}l^g_j\rho_j(t-\tau_j(t)) \tag{16}
$$

for all $t \ge 0$ and $i \in [n]$.

Combining (13) with (16) we obtain

$$
D^-\zeta_i(t) \le -d_il^-_{\varphi_i}\zeta_i(t) + \sum_{j=1}^n a_{ij}l^f_j\zeta_j(t) + \sum_{j=1}^n b_{ij}l^g_j\zeta_j(t-\tau_j(t)), \tag{17}
$$

where $\zeta_i(t) = z_i(t) - \rho_i(t)$.

It follows from (17) that

$$
\begin{aligned}
\zeta_i(t) \le e^{-d_il^-_{\varphi_i}t}\zeta_i(0) &+ \sum_{j=1}^n a_{ij}l^f_j \int_0^t e^{d_il^-_{\varphi_i}(s-t)}\zeta_j(s)ds\\
&+ \sum_{j=1}^n b_{ij}l^g_j \int_0^t e^{d_il^-_{\varphi_i}(s-t)}\zeta_j(s-\tau_j(s))ds,\ t \ge 0.
\end{aligned}
\tag{18}
$$

Obviously, $\zeta(0) \preceq 0$. For any $t_f > 0$, if $\zeta(t) \preceq 0$ for all $t \in [0, t_f)$ then from (18), $\zeta_i(t_f) \le 0$ for all $i \in [n]$, that is, $\zeta(t_f) \preceq 0$. This shows that $\zeta(t) \preceq 0$ for all $t \ge 0$. Consequently,

$$
\|x(t) - x_e\|_\infty \le (\max_{1\le i\le n} \chi_i/\chi_+)\|\phi - x_e\|_C e^{-\eta_0 t}
$$

by which we can conclude the exponential stability of the equilibrium x_e.

Finally, for any initial function $\phi \in C^+$, by Proposition 3, the corresponding trajectory $x(t) \succeq 0$ for all $t \ge 0$. Thus, $x_e = \lim_{t\to\infty} x(t) \succeq 0$. This shows that x_e is the unique positive equilibrium of system (1). The proof is completed. □

Remark 4 The result of Theorem 1 can be extended to a general class of neural networks with delays. More precisely, if the neuron activation functions $f_j(.)$ and

$g_j(.)$ satisfy Lipschitz condition

$$l_{jf}^- \leq \frac{f_j(u) - f_j(v)}{u - v} \leq l_{jf}^+, \quad l_{jg}^- \leq \frac{g_j(u) - g_j(v)}{u - v} \leq l_{jg}^+, \ u \neq v,$$

for real constants $l_{jf}^-, l_{jf}^+, l_{jg}^-$ and l_{jg}^+ and assume that there exists a vector $\chi \in \mathbb{R}^n$, $\chi \succ 0$, such that

$$\left(-DL_\varphi^- + |A|L_f + |B|L_g\right)\chi \prec 0$$

where $|A| = (|a_{ij}|)$ and $|B| = (|b_{ij}|)$, then system (1) has a unique equilibrium x_e which is globally exponentially stable.

3.4 Simulations

Consider a nonlinear system in the form of (1), which describes a Hopfield neural networks with linear decay rates (i.e. $\varphi_i(x_i) = x_i$) and Bolzmann sigmoid activation functions [69]

$$f_j(x_j) = g_j(x_j) = \frac{1 - \exp(-\frac{x_j}{\theta_j})}{1 + \exp(-\frac{x_j}{\theta_j})}, \ \theta_j > 0, \ j = 1, 2, 3. \tag{19}$$

The connection weight matrices are given by

$$A = \begin{bmatrix} 0.5 & 0.6 & 0.25 \\ 0.7 & 0.4 & 0 \\ 0.2 & 0.15 & 0.1 \end{bmatrix}, \quad B = \begin{bmatrix} 0.15 & 0 & 0.35 \\ 0.2 & 0.3 & 0.3 \\ 0.1 & 0.4 & 0.25 \end{bmatrix}, \quad D = \mathtt{diag}\{1.0, 1.2, 1.5\}.$$

Clearly, condition (3) is satisfied with $l_j^f = l_j^g = \frac{1}{2\theta_j}$.

For illustrative purpose, let $\theta_1 = 2.0, \theta_2 = 1.0$ and $\theta_3 = 1.0$, we have $L_f = L_g = \mathtt{diag}\{0.25, 0.5, 0.5\}$ and hence

$$(-D + AL_f + BL_g)\begin{bmatrix} 1 \\ 1 \\ 1 \end{bmatrix} = \begin{bmatrix} -0.2375 \\ -0.475 \\ -0.975 \end{bmatrix} \prec 0.$$

By Theorem 1, for a given input vector $J \in \mathbb{R}_+^3$, system (1) has a unique positive equilibrium $x_e \in \mathbb{R}_+^3$ which is globally exponentially stable.

Let $J = (1.0, 1.5, 0.5)^\top \in \mathbb{R}_+^3$. By solving Eq. (4) using Matlab Toolbox, the equilibrium x_e is obtained as

$$x_e = (2.0059, 2.1552, 0.8057)^\top.$$

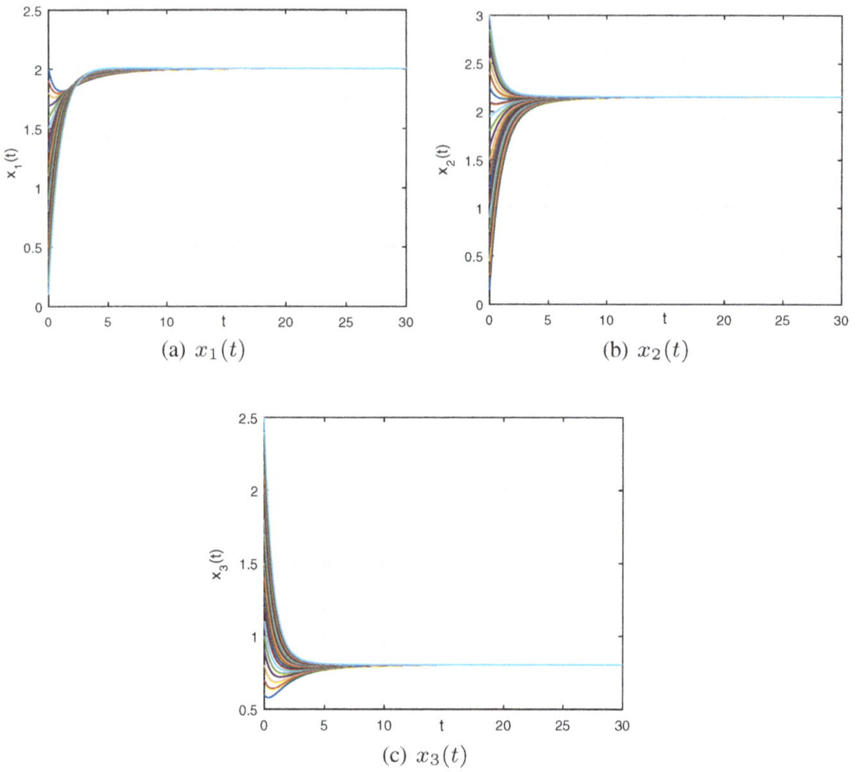

Fig. 1 State trajectories of the system with $\tau(t) = 5|\sin(0.1t)|$

Figure 1 presents 20 state trajectories of the system with a common delay $\tau_j(t) = 5|\sin(0.1t)|$. Moreover, a phase diagram of 50 state trajectories is also given in Fig. 2. It can be seen that all the conducted state trajectories converge to the positive equilibrium x_e. This validates the obtained theoretical results.

4 Exponential Stability of Positive Inertial Neural Networks with Heterogeneous Time-Varying Delays

This section considers the problem of exponential stability of positive equilibrium of positive nonlinear systems which describe inertial neural networks model with heterogeneous time-varying delays.

By utilizing comparison techniques via differential inequalities and a method of using homeomorphisms in nonlinear analysis, sufficient conditions for the existence, uniqueness and global exponential stability of a positive equilibrium are derived in terms of linear programming via M-matrices.

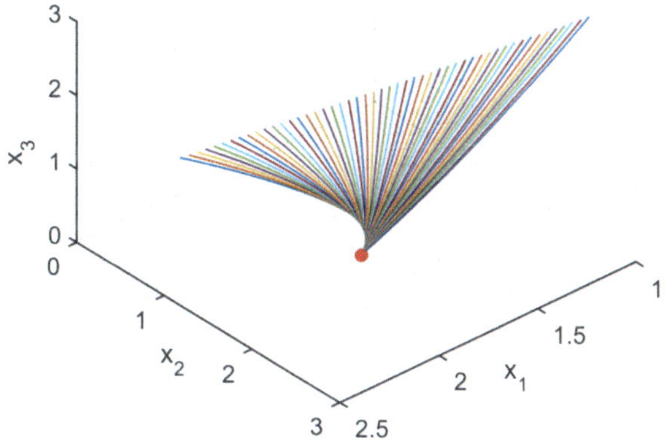

Fig. 2 Convergence of the system to the EP $x_e = (2.0059, 2.1552, 0.8057)^\top$

4.1 Model Description

Consider a model of inertial neural networks (INNs) with time-varying delays which is described by the following second-order differential system

$$\frac{d^2x_i(t)}{dt^2} = -a_i\frac{dx_i(t)}{dt} - b_ix_i(t) + \sum_{j=1}^{n} c_{ij}f_j(x_j(t)) + \sum_{j=1}^{n} d_{ij}f_j(x_j(t-\tau_{ij}(t))) + I_i,\ t \geq 0,\ i \in [n], \tag{20}$$

where n is the number of neurons, $x_i(t)$ is the state of ith neuron at time t and I_i is external input. $a_i > 0$ is the damping coefficient, $b_i > 0$ is the rate at which ith neuron resets its potential to the resting state in isolation when disconnected from the network and external inputs, c_{ij} and d_{ij} are constants representing the neuron connection weights. $f_j(.)$, $j \in [n]$, are neuro activation functions and $\tau_{ij}(t)$, $i, j \in [n]$, are heterogeneous time-varying delays satisfying $0 \leq \tau_{ij}(t) \leq \tau^+$, where $\tau^+ > 0$ is a known scalar.

We denote $x(t) = (x_i(t)) \in \mathbb{R}^n$ as the state vector, $A = \texttt{diag}\{a_1, a_2, \dots, a_n\}$, $B = \texttt{diag}\{b_1, b_2, \dots, b_n\}$, $C = (c_{ij})_{n\times n}$, $D = (d_{ij})_{n\times n}$ and $I = (I_i) \in \mathbb{R}^n$. Then, system (20) can be written in the following vector form

$$x''(t) = -Ax'(t) - Bx(t) + Cf(x(t)) + Df(x_\tau(t)) + I, \tag{21}$$

where $f(x(t)) = (f_j(x_j(t)))$ and $f(x_\tau(t)) = (f_j(x_j(t - \tau_{ij}(t))))$.

In regard to (21), each initial condition of (20) is defined by compatible vector-valued functions $\phi = (\phi_i) \in C([-\tau^+, 0], \mathbb{R}^n)$ and $\hat{\phi} = (\hat{\phi}_i) \in C([-\tau^+, 0], \mathbb{R}^n)$:

$$x_i(s) = \phi_i(s), \quad x_i'(s) = \hat{\phi}_i(s), \ s \in [-\tau^+, 0], \ i \in [n]. \tag{22}$$

By using the following state transformation

$$\eta_i y_i(t) = \frac{dx_i(t)}{dt} + \xi_i x_i(t), \ i \in [n], \tag{23}$$

where $\eta_i \neq 0$ and ξ_i, $i \in [n]$, are constants, system (20) can be represented in the following form

$$\begin{cases} x'(t) = -D_\xi x(t) + D_\eta y(t), \\ y'(t) = -D_\alpha y(t) + D_\beta x(t) + D_\eta^{-1}\left[Cf(x(t)) + Df(x_\tau(t)) + I\right], \end{cases} \tag{24}$$

where $y(t) = (y_1(t), y_2(t), \ldots, y_n(t))^\top \in \mathbb{R}^n$ and

$$\begin{aligned} D_\xi &= \mathtt{diag}\{\xi_1, \xi_2, \ldots, \xi_n\}, \quad D_\eta = \mathtt{diag}\{\eta_1, \eta_2, \ldots, \eta_n\}, \\ D_\alpha &= \mathtt{diag}\{\alpha_1, \alpha_2, \ldots, \alpha_n\}, \quad D_\beta = \mathtt{diag}\{\beta_1, \beta_2, \ldots, \beta_n\}, \\ \alpha_i &= a_i - \xi_i, \ \beta_i = \eta_i^{-1}(\alpha_i \xi_i - b_i), \ i \in [n]. \end{aligned}$$

Proposition 5 *Assume that the neuron activation functions $f_j(.)$, $j \in [n]$, of model* (20) *satisfy Assumption* (A2). *Then, for any initial condition* (22), *there exists a unique solution of system* (20) *defined on the interval* $[0, \infty)$, *which is absolutely continuous in t.*

Proof Let $x(t)$ be a solution of system (20) with initial condition (22). Then, $z(t) = (x^\top(t), y^\top(t))^\top \in \mathbb{R}^{2n}$ is a solution of system (24) with initial condition

$$x_i(s) = \phi_i(s), \ y_i(s) = \eta_i^{-1}(\xi_i \phi_i(s) + \hat{\phi}_i(s)), \ s \in [-\tau^+, 0]. \tag{25}$$

In addition, by (A2), the function $f(x) = (f_i(x_i))$, $x = (x_i)$, is Lipschitz continuous on $\mathbb{R}^n$. Thus, by utilizing fundamental results in the theory of functional differential equations [67], we can conclude that system (24) possesses a unique solution $z(t) = z(t, \phi, \hat{\phi})$ on the interval $[0, \infty)$, which is absolutely continuous in t. The proof is completed. □

4.2 Positivity of INNs with Heterogeneous Delays

A solution $x(t)$ of system (20) is said to be positive if its trajectory is confined within the first orthant, that is, $x(t) \in \mathbb{R}^n_+$ for all $t \geq 0$.

For a fixed transformation (23), where $\eta_i > 0$ and $\xi_i > 0$, $i \in [n]$, we define the following set of admissible initial functions for system (20)

$$\begin{aligned}\mathcal{A}_T = \Big\{&(\phi_i),\ (\hat{\phi}_i) \in C([-\tau^+, 0], \mathbb{R}^n) : \phi_i(s) \geq 0,\\ &\psi_i = \eta_i^{-1}(\xi_i \phi_i(s) + \hat{\phi}_i(s)) \geq 0,\ s \in [-\tau^+, 0], i \in [n]\Big\}.\end{aligned} \tag{26}$$

Note that $\mathcal{A}_T$ includes all nonnegative nondecreasing functions ϕ_i, where $\hat{\phi}_i(s) = \phi_i'(s)$.

Definition 4 System (20) is said to be *positive* if any solution $x(t)$ of (20) initiating from $\mathcal{A}_T$ with input vector $I = (I_i) \in \mathbb{R}^n_+$ is positive.

The following lemma will be use in establishing conditions for the existence of positive state transformation in (23).

Lemma 2 *For given coefficients $a_i > 0$ and $b_i > 0$, $i \in [n]$, there exists a transformation* (23) *with $\eta_i > 0$ and $\xi_i > 0$, $i \in [n]$, such that $D_\alpha \succ 0$ and $D_\beta \succ 0$ if and only if the following condition holds*

$$a_i^2 - 4b_i > 0,\ i \in [n]. \tag{27}$$

Proof (*Necessity*) If there exist $\eta_i > 0$ and $\xi_i > 0$, $i \in [n]$, satisfying $D_\alpha \succ 0$ and $D_\beta \succ 0$ then, we have

$$\xi_i(a_i - \xi_i) - b_i > 0,\ i \in [n].$$

Observe that $\xi_i(a_i - \xi_i) - b_i = \frac{1}{4}\left(a_i^2 - 4b_i\right) - \left(\xi_i - \frac{1}{2}a_i\right)^2$. Thus,

$$a_i^2 - 4b_i > (2\xi_i - a_i)^2 \geq 0,\ i \in [n].$$

(*Sufficiency*) Let condition (27) hold. Then, the constants

$$\xi_i^l = \frac{a_i - \sqrt{a_i^2 - 4b_i}}{2},\quad \xi_i^u = \frac{a_i + \sqrt{a_i^2 - 4b_i}}{2}$$

are well-defined, $\xi_i^l > 0$, and $\xi_i^u < a$. In addition, since

$$\xi_i(a_i - \xi_i) - b_i = (\xi_i^u - \xi_i)(\xi_i - \xi_i^l) > 0$$

for all $\xi_i \in (\xi_i^l, \xi_i^u)$, we have

$$\begin{cases} 0 < \xi_i < a_i, \\ -\xi_i^2 + a_i\xi_i - b_i > 0, \ i \in [n], \end{cases} \tag{28}$$

for any $\xi_i \in (\xi_i^l, \xi_i^u)$. It follows from (28) that $D_\alpha \succ 0$ and $D_\beta \succ 0$ for any $\eta_i > 0$ and $\xi_i \in (\xi_i^l, \xi_i^u)$. The proof is completed. □

In the following, we assume that the coefficients a_i and b_i, $i \in [n]$, of (20) satisfy condition (27). Then, by Lemma 2, there exists a transformation (23), where the coefficients ξ_i, η_i satisfy

$$\begin{cases} 0 < \eta_i < \xi_i, \\ \xi_i^l < \xi_i < \xi_i^u, \ i \in [n], \end{cases} \tag{29}$$

where $\xi_i^l = \frac{a_i - \sqrt{a_i^2 - 4b_i}}{2}$ and $\xi_i^u = \frac{a_i + \sqrt{a_i^2 - 4b_i}}{2}$. Hereafter, we fix a transformation satisfying the above condition. The positivity of system (20) will be characterized as in the following result.

Theorem 2 *Assume that the neuron activation functions $f_j(.)$, $j \in [n]$, of model* (20) *satisfy Assumption* (A2). *If condition* (27) *is fulfilled and the connection weight matrices $C = (c_{ij})$, $D = (d_{ij})$ are nonnegative ($C \succeq 0$, $D \succeq 0$), then system* (20) *is positive for any bounded delays.*

Proof Let $x(t)$ be a solution of system (20) initiating from $\mathcal{A}_T$ and nonnegative input vector. We will show that the corresponding solution $z(t) = (x^\top(t), y^\top(t))^\top$ of system (24) is positive. Note at first that if $y_i(t) \geq 0$, $t \in [0, t_f)$, for some $t_f > 0$, then it follows from the first equation of (24) that

$$\begin{aligned} x_i(t) &= e^{-\xi_i t}\Big(\phi_i(0) + \eta_i \int_0^t e^{\xi_i s} y_i(s) ds\Big) \\ &\geq \phi_i(0) e^{-\xi_i t} \geq 0, \quad t \in [0, t_f). \end{aligned}$$

Thus, it is only necessary to show that $y(t) = (y_i(t)) \succeq 0$ for $t \geq 0$.

Let $z_\epsilon(t) = (x_\epsilon^\top(t), y_\epsilon^\top(t))^\top$ be the solution (24) with initial functions ϕ_i and $\psi_{i\epsilon} = \psi_i + \epsilon$, where $\epsilon > 0$ is a given scalar. Since $\phi(s) = (\phi_i(s)) \succeq 0$ and $\psi_\epsilon(s) = (\psi_{i\epsilon}(s)) \succeq \epsilon \mathbf{1}_n$ for all $s \in [-\tau^+, 0]$, it follows from (24) that $y_\epsilon(t) = (y_{i\epsilon}(t)) \succ 0$, $t \in [0, t_f)$, for some small $t_f > 0$.

Suppose in contrary that there exist a $t_1 > 0$ and an index $i \in [n]$ such that

$$y_{i\epsilon}(t_1) = 0, \ \ y_{i\epsilon}(t) > 0, \ t \in [0, t_1), \tag{30}$$

and $y_\epsilon(t) \succeq 0$ for all $t \in [0, t_1]$. Then, by multiplying with $e^{\alpha_i t}$, from (24) we have

$$\frac{d}{dt}\left(y_{i\epsilon}(t)e^{\alpha_i t}\right) = \beta_i x_{i\epsilon}(t)e^{\alpha_i t} + \eta_i^{-1}\sum_{j=1}^{n} c_{ij}e^{\alpha_i t} f_j(x_{j\epsilon}(t))$$
$$+ \eta_i^{-1}\Big[\sum_{j=1}^{n} d_{ij}e^{\alpha_i t} f_j(x_{j\epsilon}(t-\tau_{ij}(t))) + I_i e^{\alpha_i t}\Big]. \quad (31)$$

Therefore, it follows from (31) and the fact $D_\beta e^{D_\alpha t}x_\epsilon(t) \succeq 0$, $t \in [0, t_1]$, that

$$D_\eta y_\epsilon(t) \succeq e^{-D_\alpha t} D_\eta \psi_\epsilon(0) + C\int_0^t e^{D_\alpha(s-t)} f(x_\epsilon(s))ds$$
$$+ D\int_0^t e^{D_\alpha(s-t)} \hat{f}(\hat{x}_\epsilon(s))ds + D_\alpha^{-1}\big(E_n - e^{-D_\alpha t}\big)I, \quad (32)$$

where $\hat{f}(\hat{x}_\epsilon(s)) = (f_j(x_{j\epsilon}(s-\tau_{ij}(s)))$ and E_n is the identity matrix in $\mathbb{R}^{n\times n}$.

Since the vector fields $F_1(x) = Cf(x)$ and $F_2(x) = Df(x)$ are order-preserving, $x_\epsilon(t) \succeq 0$ and $\hat{x}_\epsilon(t) \succeq 0$ for $t \in [0, t_1]$, from (32), we have

$$y_\epsilon(t) \succeq e^{-D_\alpha t}\psi_\epsilon(0) + D_{\alpha\eta}^{-1}\big(E_n - e^{-D_\alpha t}\big)I, \quad (33)$$

where $D_{\alpha\eta} = D_\alpha D_\eta$.

Let $t \uparrow t_1$ in (33) and note also that $E_n - e^{-D_\alpha t_1} \succ 0$, we readily obtain

$$y_\epsilon(t_1) \succeq \epsilon e^{-D_\alpha t_1}\mathbf{1}_n \succ 0,$$

which yields a contradiction with (30). Therefore, $x_\epsilon(t) \succeq 0$ and $y_\epsilon(t) \succ 0$ for all $t \geq 0$. Let $\epsilon \downarrow 0$ we then obtain $z(t) = \lim_{\epsilon\to 0} z_\epsilon(t) \succeq 0$. The proof is completed. □

Let $\tilde{\mathcal{A}}_T$ be the subset of $\mathcal{A}_T$ defined by

$$\tilde{\mathcal{A}}_T = \{\phi = (\phi_i) \in C^1([-\tau^+, 0], \mathbb{R}^n_+) : \phi_i'(s) \geq 0\}.$$

By Theorem 2, we have the following result.

Corollary 1 *Under the assumptions of Theorem 2, system* (20) *is positive with respect to initial conditions defined by* (22), *where* $(\phi_i) \in \tilde{\mathcal{A}}_T$ *and* $\hat{\phi}_i = \phi_i'$.

4.3 Positive Equilibrium

In this section, we derive conditions for the existence of positive equilibrium points for INNs in the form of (20).

Similar to Definition 1, we give the following definition of positive equilibrium of (1).

Definition 5 Given an input vector $I \in \mathbb{R}^n_+$. A vector $x_* \in \mathbb{R}^n_+$ is said to be a positive *equilibrium point* (EP) of system (20) if the following algebraic equation holds

$$- Bx_* + Cf(x_*) + Df(x_*) + I = 0. \tag{34}$$

The following auxiliary result will be used to issue the existence of positive EPs for system (20).

Lemma 3 *For given matrices $D_\eta \succ 0$ and $D_\xi \succ 0$, a vector $x_* \in \mathbb{R}^n_+$ is a positive EP of system* (20) *if and only if the augmented vector (x_*, y_*), where $y_* = D_\eta^{-1} D_\xi x_*$, is a positive EP of system* (24)*, that is,*

$$\begin{cases} -D_\xi x_* + D_\eta y_* = 0, \\ D_\eta(-D_\alpha y_* + D_\beta x_*) + Cf(x_*) + Df(x_*) + I = 0. \end{cases} \tag{35}$$

Proof Note from (34) that

$$\begin{aligned} D_\eta(-D_\alpha y_* + D_\beta x_*) &= -D_\alpha D_\xi x_* + (D_\alpha D_\xi - B)x_* \\ &= -Bx_*. \end{aligned}$$

Thus, the proof is then straightforward from (34) and (35). □

By Lemma 3, system (20) possesses an EP if and only if system (24) does.

As revealed from (35), for a given input vector $I \in \mathbb{R}^n$, an EP of (20) exists if and only if the equation $\mathcal{H}(\chi) = 0$ has a solution $\chi_* \in \mathbb{R}^{2n}$, where $\chi = \begin{bmatrix} x \\ y \end{bmatrix}$, $x, y \in \mathbb{R}^n$, and the mapping $\mathcal{H} : \mathbb{R}^{2n} \to \mathbb{R}^{2n}$ is defined by

$$\mathcal{H}(\chi) = \begin{bmatrix} -D_\xi x + D_\eta y \\ (D_{\alpha\xi} - B)x - D_{\alpha\eta} y + Cf(x) + Df(x) + I \end{bmatrix}, \tag{36}$$

where $D_{\alpha\xi} = D_\alpha D_\xi$ and $D_{\alpha\eta} = D_\alpha D_\eta$.

Clearly, the mapping $\mathcal{H}(.)$ defined in (36) is continuous on $\mathbb{R}^{2n}$. Based on Lemmas 2 and 3, we have the following result.

Theorem 3 *Assume that the assumptions of Theorem 2 are satisfied and there exists a vector $\chi_0 \in \mathbb{R}^{2n}$, $\chi_0 \succ 0$, such that*

$$\chi_0^\top \begin{pmatrix} -D_\xi & D_\eta \\ D_{\alpha\xi} - B + (C + D)L_f & -D_{\alpha\eta} \end{pmatrix} \prec 0. \tag{37}$$

Then, for each input vector $I \in \mathbb{R}^n$, there exists a unique EP $\chi_* = \begin{bmatrix} x_* \\ y_* \end{bmatrix}$ of system (24).

Proof We define the mappings $\mathcal{H}_1, \mathcal{H}_2 : \mathbb{R}^{2n} \to \mathbb{R}^{2n}$ by

$$\begin{aligned}\mathcal{H}_1(\chi) &= -D_\xi x + D_\eta y,\\ \mathcal{H}_2(\chi) &= (D_{\alpha\xi} - B)x - D_{\alpha\eta} y + (C + D)f(x) + I.\end{aligned}$$

Then, for any two vectors $\chi_1 = \begin{bmatrix} x_1 \\ y_1 \end{bmatrix}$ and $\chi_2 = \begin{bmatrix} x_2 \\ y_2 \end{bmatrix}$ in $\mathbb{R}^{2n}$, we have

$$\mathcal{H}_1(\chi_1) - \mathcal{H}_1(\chi_2) = -D_\xi(x_1 - x_2) + D_\eta(y_1 - y_2).$$

Therefore,

$$\begin{aligned}&\mathcal{S}(x_1 - x_2)[\mathcal{H}_1(\chi_1) - \mathcal{H}_1(\chi_2)]\\ &\qquad = -D_\xi|x_1 - x_2| + \mathcal{S}(x_1 - x_2)D_\eta(y_1 - y_2)\\ &\qquad \preceq -D_\xi|x_1 - x_2| + D_\eta|y_1 - y_2|,\end{aligned} \tag{38}$$

where $\mathcal{S}(x_1 - x_2) = \mathtt{diag}\{\mathtt{sgn}(x_{1i} - x_{2i})\}$.

By Assumption (A2) and $D_{\alpha\xi} - B = D_\beta D_\eta \succ 0$, similar to (38), we have

$$\begin{aligned}&\mathcal{S}(y_1 - y_2)[\mathcal{H}_2(\chi_1) - \mathcal{H}_2(\chi_2)]\\ &\quad \preceq -D_{\alpha\eta}|y_1 - y_2| + (D_{\alpha\xi} - B)|x_1 - x_2| + (C + D)|f(x_1) - f(x_2)|\\ &\quad \preceq -D_{\alpha\eta}|y_1 - y_2| + (D_{\alpha\xi} - B)|x_1 - x_2| + (C + D)L_f|x_1 - x_2|,\end{aligned} \tag{39}$$

where $L_f = \mathtt{diag}\{l_1^f, l_2^f, \ldots, l_n^f\}$.

From (38) and (39), we have

$$\mathtt{diag}\{\mathcal{S}(x_1 - x_2), \mathcal{S}(y_1 - y_2)\}[\mathcal{H}(\chi_1) - \mathcal{H}(\chi_2)] \preceq -\mathcal{M}|\chi_1 - \chi_2|, \tag{40}$$

where $\mathcal{M} = \begin{pmatrix} D_\xi & -D_\eta \\ B - D_{\alpha\xi} - (C + D)L_f & D_{\alpha\eta} \end{pmatrix}$.

On the other hand, it is clear that

$$\begin{aligned}&-\mathtt{diag}\{\mathcal{S}(x_1 - x_2), \mathcal{S}(y_1 - y_2)\}[\mathcal{H}(\chi_1) - \mathcal{H}(\chi_2)]\\ &\qquad \preceq |\mathcal{H}(\chi_1) - \mathcal{H}(\chi_2)|.\end{aligned} \tag{41}$$

Thus, combining (40) and (41) gives

$$|\mathcal{H}(\chi_1) - \mathcal{H}(\chi_2)| \succeq \mathcal{M}|\chi_1 - \chi_2|. \tag{42}$$

Let χ_0 be a vector satisfying (37). From (42), we have

$$\chi_0^\top|\mathcal{H}(\chi_1) - \mathcal{H}(\chi_2)| \succeq \chi_0^\top\mathcal{M}|\chi_1 - \chi_2|. \tag{43}$$

If $\mathcal{H}(\chi_1) - \mathcal{H}(\chi_2) = 0$ then it is clear from (37) and (43) that $\chi_1 = \chi_2$. Thus, the mapping $\mathcal{H}(.)$ is injective in $\mathbb{R}^{2n}$. In addition, inequality (43) also gives

$$\|\mathcal{H}(\chi)\|_\infty \geq \frac{1}{\|\chi_0\|_\infty}\chi_0^\top \mathcal{M}|\chi| - \|\mathcal{H}(0)\|_\infty,$$

which ensures that $\|\mathcal{H}(\chi_k)\|_\infty \to \infty$ for any sequence $\{\chi_k\}$ in $\mathbb{R}^{2n}$ satisfying $\|\chi_k\|_\infty \to \infty$. Thus, the continuous mapping $\mathcal{H}(.)$ is proper.

By Lemma 2, $\mathcal{H}(.)$ is a homeomorphism onto $\mathbb{R}^{2n}$. Consequently, the equation $\mathcal{H}(\chi) = 0$ has a unique solution $\chi_* \in \mathbb{R}^{2n}$, which is a unique EP of (24). The proof is completed. □

Remark 5 Since $\mathcal{M} = \begin{pmatrix} D_\xi & -D_\eta \\ B - D_{\alpha\xi} - (C+D)L_f & D_{\alpha\eta} \end{pmatrix}$ is an M-matrix, condition (37) holds if and only if $\mathcal{M}$ is a nonsingular M-matrix. Thus, condition (37) can be verified by various equivalent conditions for nonsingular M-matrices as given in Proposition 3.

Remark 6 Let condition (26) hold. Then, as shown in Lemma 2, there exist scalars ξ_i, η_i, $i \in [n]$, satisfying (29). Furthermore, condition (37) is feasible for a vector $\chi_0 \succ 0$ if and only if the inequality

$$v^\top\left(-B + CL_f + DL_f\right) \prec 0 \tag{44}$$

is feasible for a vector $v \in \mathbb{R}^n$, $v \succ 0$.

Indeed, if (37) holds then, by specifying $u_0 = [E_n \;\; 0_n]\chi_0$ and $v_0 = [0_n \;\; E_n]\chi_0$, we have

$$\begin{cases} u_0 - D_\alpha v_0 \prec 0, \\ D_\xi(D_\alpha v_0 - u_0) + \left[-B + L_f(C^\top + D^\top)\right] v_0 \prec 0, \end{cases}$$

which implies that (44) is feasible with $v = v_0$.

Conversely, if (44) is feasible for a vector $0 \prec v \in \mathbb{R}^n$ then, for an $\epsilon \in (0, \min_{i\in[n]} a_i)$, let $\xi_i = \xi_i^u - \epsilon$, $D_\xi^l = \mathtt{diag}\{\xi_i^l\}$, $D_\xi^u = \mathtt{diag}\{\xi_i^u\}$ and $u = D_\xi^l v \succ 0$. We have $D_\alpha = D_\xi^l + \epsilon E_n$ and $u - D_\alpha v = -\epsilon v \prec 0$.

In addition, $D_\xi(D_\alpha v - u) = \epsilon(D_\xi^u - \epsilon E_n)v \to 0$ as $\epsilon \downarrow 0$. Thus, there exists an $\epsilon > 0$ such that ξ_i, $i \in [n]$, satisfy (29) and $D_\xi(D_\alpha v - u) + [-B + L_f(C^\top + D^\top)]v \prec 0$. By selecting $0 < \eta_i < \xi_i$, $i \in [n]$, it can be verified that the vector $\chi_0 = \begin{bmatrix} u \\ v \end{bmatrix}$ satisfies (37).

4.4 Exponential Stability of Positive Equilibrium

We now prove under the Assumptions of Theorem 2 that the unique equilibrium χ_* of system (24) is positive and globally exponentially stable for any delays $\tau_j(t) \in [0, \tau^+]$.

Definition 6 An equilibrium point x_* of (20) is said to be globally exponentially stable (GES) if there exist positive constants κ, λ such that the following exponential estimate holds for any solution $x(t)$ of (20)

$$\|x(t) - x_*\|_\infty \leq \kappa \max\left\{\|\phi - x\|_C, \|\hat{\phi}\|_C\right\} e^{-\lambda t}, \quad t \geq 0. \tag{45}$$

Theorem 4 *Assume that the neuron activation functions $f_j(.)$, $j \in [n]$, satisfy Assumption* (A2). *If condition* (27) *is satisfied, $C = (c_{ij}) \succeq 0$, $D = (d_{ij}) \succeq 0$ and there exists a vector $\hat{\chi}_0 \in \mathbb{R}^{2n}$, $\hat{\chi}_0 \succ 0$, such that*

$$\begin{pmatrix} -D_\xi & D_\eta \\ D_{\alpha\xi} - B + (C + D)L_f & -D_{\alpha\eta} \end{pmatrix} \hat{\chi}_0 \prec 0. \tag{46}$$

Then, for any input vector $I \in \mathbb{R}^n_+$, there exists a unique positive EP $\chi_* = \begin{bmatrix} x_* \\ y_* \end{bmatrix} \in \mathbb{R}^{2n}_+$ of system (24) *which is GES for any delays $\tau_{ij}(t) \in [0, \tau^+]$.*

Proof The proof will be divided into four steps.

Step 1. Since $\mathcal{M}$ is an M-matrix, conditions (37) and (44) are equivalent. Thus, by Theorem 3, for any input vector $I \in \mathbb{R}^n_+$, there exists a unique EP χ_* of system (24), which satisfies

$$\begin{cases} -D_\xi x_* + D_\eta y_* = 0, \\ (D_{\alpha\xi} - B)x_* - D_{\alpha\eta} y_* + (C + D) f(x_*) + I = 0. \end{cases} \tag{47}$$

Step 2. We define the following transformations

$$\begin{aligned} u_i(t) &= |x_i(t) - x_{i*}|, \ v_i(t) = |y_i(t) - y_{i*}|, \\ \hat{u}_i(t) &= |x_i(t - \tau_i(t)) - x_{i*}|, \ i \in [n], \\ u(t) &= (u_i(t)), \ v(t) = (v_i(t)), \ \hat{u}(t) = (\hat{u}_i(t)). \end{aligned}$$

It follows from (24) and (47) that

$$\begin{aligned} D^+ u_i(t) &= \mathtt{sgn}(x_i(t) - x_{i*})(x_i(t) - x_{i*})' \\ &= -\xi_i u_i(t) + \eta_i \mathtt{sgn}(x_i(t) - x_{i*})(y_i(t) - y_{i*}) \\ &\leq -\xi_i u_i(t) + \eta_i v_i(t). \end{aligned} \tag{48}$$

Similarly, we have

$$D^+(\eta_i v_i(t)) \le -\alpha_i \eta_i v_i(t) + (\alpha_i \xi_i - b_i) u_i(t) + \sum_{j=1}^{n} c_{ij} |f_j(x_j(t)) - f_j(x_{j*})|$$
$$+ \sum_{j=1}^{n} d_{ij} |f_j(x_j(t - \tau_{ij}(t))) - f_j(x_{j*})|$$
$$\le -\alpha_i \eta_i v_i(t) + (\alpha_i \xi_i - b_i) u_i(t) + \sum_{j=1}^{n} \left(c_{ij} l_j^f u_j(t) + d_{ij} l_j^f \hat{u}_j(t) \right). \quad (49)$$

Combining (48) and (49) gives

$$D^+ \begin{bmatrix} u(t) \\ D_\eta v(t) \end{bmatrix} \preceq -\Xi_1 \begin{bmatrix} u(t) \\ v(t) \end{bmatrix} + \Xi_2 \begin{bmatrix} \hat{u}(t) \\ 0 \end{bmatrix}. \quad (50)$$

where $\Xi_1 = \begin{pmatrix} D_\xi & -D_\eta \\ B - D_{\alpha\xi} - CL_f & D_{\alpha\eta} \end{pmatrix}$ and $\Xi_2 = \begin{pmatrix} 0 & 0 \\ DL_f & 0 \end{pmatrix}$.
We will show that there exists a scalar $\lambda > 0$ such that

$$\begin{bmatrix} u(t) \\ D_\eta v(t) \end{bmatrix} \preceq \frac{1 + \|\xi\|_\infty}{\min_{1 \le i \le n} \hat{\chi}_{0i}} \|\phi_*\| e^{-\lambda t} \hat{\chi}_0, \ \forall t \ge 0, \quad (51)$$

where $\|\phi_*\| = \max \left\{ \|\phi - x_*\|_C, \|\hat{\phi}\|_C \right\}$.

Step 3. By condition (46), $(-\Xi_1 + \Xi_2)\, \hat{\chi}_0 \prec 0$. Thus, there exists a $\lambda_* > 0$ such that

$$\left(\lambda E_{2n} - \Xi_1 + e^{\lambda \tau^+} \Xi_2 \right) \hat{\chi}_0 \preceq 0 \quad (52)$$

for any $\lambda \in (0, \lambda_*]$.
As revealed in (50)–(52), we consider the scaling function $\hat{\zeta}(t) = \kappa \|\phi_*\| e^{-\lambda t} \hat{\chi}_0$, $t \ge 0$, where $\kappa = \frac{1+\|\xi\|_\infty}{\min_{1 \le i \le n} \hat{\chi}_{0i}}$.
It is clear that $\zeta(t) \preceq \kappa \|\phi_*\| \hat{\chi}_0$ for all $t \in [-\tau^+, 0]$, where

$$\zeta(t) = [u^\top(t) \quad D_\eta v^\top(t)]^\top.$$

Thus, for a fixed $\theta > 1$, $\zeta(0) \prec \theta \hat{\zeta}(0)$.
Assume in contrary that the estimation $\zeta(t) \prec \theta \hat{\zeta}(t)$ does not hold for all $t > 0$, then there exists a $t_f > 0$ and an index $i \in [2n]$ such that

$$\zeta_i(t_f) = \theta \hat{\zeta}_i(t_f), \ \zeta(t) \prec \theta \hat{\zeta}(t), \ t \in [0, t_f). \quad (53)$$

Note also that $\hat{\zeta}_j(t - \tau_{ij}(t)) = e^{\lambda \tau_{ij}(t)} \hat{\zeta}_j(t) \le e^{\lambda \tau^+} \hat{\zeta}_j(t)$ for any $j \in [n]$. Therefore, by (51), we have

$$D^{+}\hat{\zeta}(t) \succeq -\Xi_1\hat{\zeta}(t) + \Xi_2\hat{\zeta}_d(t), \tag{54}$$

where $\hat{\zeta}_d(t) = \text{col}\{\hat{\zeta}_j(t - \tau_{ij}(t)), \hat{\zeta}_{n+j}(t)\}$, $j \in [n]$.
If $i \in [n]$ then, from (48) and (54), we have

$$D^{+}\left(\zeta_i(t) - \theta\hat{\zeta}_i(t)\right) \le -\xi_i\left(\zeta_i(t) - \theta\hat{\zeta}_i(t)\right),$$

and hence, $\zeta_i(t) - \theta\hat{\zeta}_i(t) \le (\zeta_i(0) - \theta\hat{\zeta}_i(0))e^{-\xi_i t}$, $t \in [0, t_f)$.
Let $t \uparrow t_f$ we obtain

$$\zeta_i(t_f) - \theta\hat{\zeta}_i(t_f) \le (\zeta_i(0) - \theta\hat{\zeta}_i(0))e^{-\xi_i t_f} < 0,$$

which clearly contradicts with (53).
Similarly, if $i = n + k$, $k \in [n]$, then from (49) and (54), we have

$$D^{+}\left(\zeta_i(t) - \theta\hat{\zeta}_i(t)\right) \le -\alpha_k\eta_k\left(\zeta_i(t) - \theta\hat{\zeta}_i(t)\right),$$

and hence, $\zeta_i(t) - \theta\hat{\zeta}_i(t) \le (\zeta_i(0) - \theta\hat{\zeta}_i(0))e^{-\alpha_k\eta_k t}$, $t \in [0, t_f)$, which also gives a contradiction with (53) by letting $t \uparrow t_f$. This shows that $\zeta(t) \prec \theta\hat{\zeta}(t)$ for all $t \ge 0$. Let $\theta \downarrow 1$ we obtain (51), which ensures that the EP χ_* of system (24) is GES.

Step 4. Positivity of χ_*: For a given vector $u_0 \in \mathbb{R}^n_+$, let $z_0(t) = [x^\top(t) \;\; y^\top(t)]^\top$ be the solution of system (24) with constant initial condition determined as $\phi_i = u_{0i}$ and $\hat{\phi}_i = 0$. By Theorem 2, $z_0(t) \succeq 0$ for all $t \ge 0$.
On the other hand, according to the result of *Step 3*, we have

$$\texttt{diag}\{I_n, D_\eta\}|z_0(t) - \chi_*| \le \hat{\zeta}(t),$$

which converges exponentially to zero. Therefore,

$$\chi_* = \lim_{t\to\infty} z_0(t) \succeq 0.$$

This shows that the unique EP χ_* of system (24) is positive and GES. The proof is completed. □

We now summarize the results of this section in the following theorem.

Theorem 5 *Consider a model of INNs with heterogeneous delays in the form of* (20). *Assume that the neuron activation functions $f_j(.)$ of* (20) *satisfy Assumption (A2) and the following conditions hold*

(i) $a_i^2 - 4b_i > 0$, $i \in [n]$;
(ii) $C = (c_{ij}) \succeq 0$, $D = (d_{ij}) \succeq 0$;

(iii) There exists a vector $v \in \mathbb{R}^n$, $v \succ 0$, such that

$$\left(-B + CL_f + DL_f\right)v \prec 0. \tag{55}$$

Then, the following assertions are true

(a) System (20) is positive subject to initial conditions in $\mathcal{A}_T$.
(b) For a given input vector $I \in \mathbb{R}^n_+$, there exists a unique positive EP x_* of (20) which is GES for any delays $\tau_{ij}(t) \in [0, \tau^+]$. Moreover, there exists a positive constant λ_* such that any solution $x(t)$ of (20)–(21) satisfies the following exponential estimation

$$\|x(t) - x_*\|_\infty \leq \frac{(1+\sigma_0)\sigma_1}{\sigma_2} \max\left\{\|\phi - x\|_C, \|\hat{\phi}\|_C\right\} e^{-\lambda_* t}, \quad t \geq 0, \tag{56}$$

where $\sigma_0 = \max_i \frac{1}{2}\left(a_i + \sqrt{a_i^2 - 4b_i}\right)$, $\sigma_1 = \max_i \left\{v_i, \frac{1}{2}\left(a_i - \sqrt{a_i^2 - 4b_i}\right)v_i\right\}$ *and* $\sigma_2 = \min_i \left\{v_i, \frac{1}{2}\left(a_i - \sqrt{a_i^2 - 4b_i}\right)v_i\right\}$.

Remark 7 The mapping $F : \mathbb{R}^n \to \mathbb{R}^n$ defined by $F(x) = (C + D)f(x)$ is an order-preserving vector field by condition (ii) of Theorem 4. Thus, for non-zero input vector $I \in \mathbb{R}^n_+$, the positive EP x_* mentioned in statement (b) is different from the origin. Moreover, x_* is strictly positive (i.e. $x_* \succ 0$) if the input vector is positive (i.e. $I \succ 0$).

Remark 8 Conditions (55) and (i) and (ii) ensure that any solution of system (20) initiating from $\mathcal{A}_T$ is eventually positive for positive input vector $I \succ 0$. Specifically, let $x(t)$ be a solution of (20) with initial condition in $\mathcal{A}_T$. Since $x(t)$ converges exponentially to the unique EP $x_* \succ 0$, $x(t) \succ 0$ for all t in a sufficiently large interval $[t_f, \infty)$.

4.5 Numerical Examples

In this section, two numerical examples with simulations are given to illustrate the effectiveness of the results presented in the preceding section.

Example 1 Consider a class of nonlinear systems which describe cooperative-type INNs in the form of (20) with Boltzmann sigmoid activation functions [69]

$$f_j(x_j) = \frac{1 - \exp\left(-\frac{x_j}{\theta_j}\right)}{1 + \exp\left(-\frac{x_j}{\theta_j}\right)}, \quad \theta_j > 0, \ j = 1, 2, 3. \tag{57}$$

Clearly, the activation functions (57) satisfy Assumption (A2) with $l_j^f = \frac{1}{2\theta_j}$.

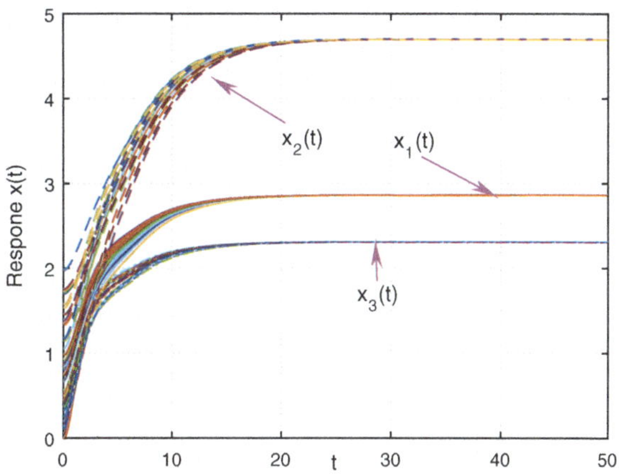

Fig. 3 Convergence of positive solutions to x_*

Let the coefficients a_i, b_i, and connection weights c_{ij}, d_{ij} be given by

$$
\begin{aligned}
a_1 &= 2.0,\ a_2 = 1.8,\ a_3 = 2.5,\\
b_1 &= 1.25,\ b_2 = 0.75,\ b_3 = 1.5,\\
C &= \begin{bmatrix} 0.3 & 0.6 & 0.7 \\ 1.0 & 0.8 & 0.2 \\ 0.1 & 0.25 & 0.4 \end{bmatrix}, \quad D = \begin{bmatrix} 0.15 & 0.5 & 0.4 \\ 0.3 & 0.5 & 0.6 \\ 0.15 & 0.35 & 0.6 \end{bmatrix}.
\end{aligned}
$$

For $\theta_1 = 2.0$, $\theta_2 = 2.0$ and $\theta_3 = 1.0$, we have $L_f = \mathtt{diag}\{0.25, 0.25, 0.5\}$. Thus,

$$
(-B + CL_f + DL_f) \begin{bmatrix} 1 \\ 2 \\ 1 \end{bmatrix} = \begin{bmatrix} -0.0375 \\ -0.125 \\ -0.6375 \end{bmatrix} \prec 0.
$$

Therefore, conditions (i) and (ii) of Theorem 4 and condition (55) are fulfilled. By Theorem 4, for an input vector $I \in \mathbb{R}^3_+$, the system has a unique positive EP $x_* \in \mathbb{R}^3_+$, which is GES for any bounded delays $\tau_j(t)$, $j = 1, 2, 3$.

To simulate the result, we take $I = (1.5, 1.0, 2.0)^\top \in \mathbb{R}^3_+$ then, by solving Eq. (27) we obtain $x_* = (2.8703, 4.7071, 2.313)^\top$. A simulation result of 20 sample state trajectories with a common delay $\tau_{ij}(t) = 1 + 4|\sin(10\pi t)|$ is represented in Fig. 3. It can be seen that all conducted state trajectories are positive and converge to the unique positive EP x_*, which support the obtained theoretical results.

Example 2 Consider the following INN model with a time-varying delay $\tau(t)$

$$
\begin{aligned}
x''(t) + 2x'(t) = &-0.9x(t) + \begin{bmatrix} 0.28 & 0.33 \\ 0.36 & 0.15 \end{bmatrix} \begin{bmatrix} \ln(1 + \exp(x_1(t))) \\ \tanh(x_2(t)) \end{bmatrix} \\
&+ \begin{bmatrix} 0.12 & 0.18 \\ 0.21 & 0.16 \end{bmatrix} \begin{bmatrix} \ln(1 + \exp(x_1(t - \tau(t)))) \\ \tanh(x_2(t - \tau(t))) \end{bmatrix} + I.
\end{aligned} \tag{58}
$$

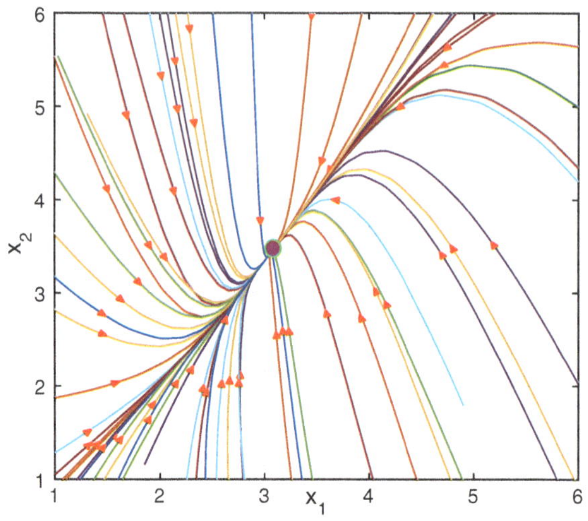

Fig. 4 Convergence of positive solutions of (58) to a unique positive EP

It is easy to verify that the activation functions of model (58) satisfy Assumption (A2) and conditions (i), (ii) of Theorem 4 are satisfied. In addition, since $B - (C + D)L_f = \begin{bmatrix} 0.5 & -0.51 \\ -0.57 & 0.59 \end{bmatrix}$ is a nonsingular M-matrix, condition (55) is also satisfied.

By Theorem 4, system (58) is positive and, for a given input vector $I \in \mathbb{R}^2_+$, there exists a unique positive EP x_* which is GES. In the simulation result given in Fig. 4, we simulate with $\tau(t) = 5|\sin(20\pi t)|$ and input vector $I = (1, 1)^\top$. It can be seen that the corresponding state trajectories converge to the unique point x_*, which strongly supports the obtained theoretical results.

5 Exponential Stability of Positive BAM Neural Networks with Multiple Delays

This section is concerned with the exponential stability of a unique equilibrium point of a class of nonlinear time-delay systems which describe positive neural networks in bidirectional associative memory (BAM) with multiple time-varying delays and nonlinear self-excitation rates.

A systematic approach involving comparison techniques via differential inequalities in combination with the use of Brouwer's fixed point theorem and M-matrix theory is presented.

5.1 Model Description and Preliminaries

Consider the following nonlinear time-delay system concerning BAM neural networks with multiple delays

$$\begin{cases} x_i'(t) = -\alpha_i \varphi_i(x_i(t)) + \sum_{j=1}^{m} a_{ij} f_j(y_j(t)) + \sum_{j=1}^{m} b_{ij} f_j(y_j(t-\sigma_j(t))) + I_i, \\ y_j'(t) = -\beta_j \psi_j(y_j(t)) + \sum_{i=1}^{n} c_{ji} g_i(x_i(t)) + \sum_{i=1}^{n} d_{ji} g_i(x_i(t-\tau_i(t))) + J_j, \end{cases} \tag{59}$$

where n and m are the number of neurons in X-layer and Y-layer, respectively; $x_i(t)$ and $y_j(t)$ are the state variables; φ_i, ψ_j are decay rate functions and $\alpha_i > 0$, $\beta_j > 0$ are self-inhibition coefficients.

For linear decay rates (that is, $\varphi_i(x_i) = x_i$ and $\psi_j(y_j) = y_j$), α_i and β_j are respectively the rates at which ith and jth neurons will reset their potential to the resting state in isolation when disconnected from the network and external inputs.

In system (59), $f_j(.)$ and $g_i(.)$ are neuron activation functions, a_{ij}, b_{ij}, c_{ji} and d_{ji} are neuron connection weights. I_i and J_j are external inputs to the ith neuron and jth neuron, respectively. $\tau_i(t)$ and $\sigma_j(t)$ denote the communication delays between neurons which satisfy

$$0 \le \tau_i(t) \le \overline{\tau},\ i \in [n],\ 0 \le \sigma_j(t) \le \overline{\sigma},\ \ j \in [m], \tag{60}$$

where $\overline{\tau}$ and $\overline{\sigma}$ are known positive constants.

Initial conditions of system (59) are specified as follows

$$x(t_0+\xi) = x^0(\xi),\ \xi \in [-\overline{\tau}, 0],\quad y(t_0+\theta) = y^0(\theta),\ \theta \in [-\overline{\sigma}, 0], \tag{61}$$

where $x(t) = (x_i(t)) \in \mathbb{R}^n$ and $y(t) = (y_j(t)) \in \mathbb{R}^m$ denote the state vectors in neuron fields F_X and F_Y, respectively, and $x^0 \in C([-\overline{\tau}, 0], \mathbb{R}^n)$, $y^0 \in C([-\overline{\sigma}, 0], \mathbb{R}^m)$ are initial functions specifying initial states of the system.

We denote the matrices $A = (a_{ij}) \in \mathbb{R}^{n\times m}$, $B = (b_{ij}) \in \mathbb{R}^{n\times m}$, $C = (c_{ji}) \in \mathbb{R}^{m\times n}$, $D = (d_{ji}) \in \mathbb{R}^{m\times n}$, $D_\alpha = \mathtt{diag}\{\alpha_1, \alpha_2, \dots, \alpha_n\}$, $D_\beta = \mathtt{diag}\{\beta_1, \beta_2, \dots, \beta_m\}$, and vectors $I = (I_i) \in \mathbb{R}^n$, $J = (J_j) \in \mathbb{R}^m$.

System (59) can be written as

$$\begin{cases} x'(t) = -D_\alpha \Phi(x(t)) + Af(y(t)) + Bf(y(t-\sigma(t))) + I, \\ y'(t) = -D_\beta \Psi(y(t)) + Cg(x(t)) + Dg(x(t-\tau(t))) + J, \end{cases} \tag{62}$$

where the associated vector-valued functions are given by

$$\begin{aligned}&\Phi(x(t))=(\varphi_i(x_i(t))),\ \Psi(y(t))=\big(\psi_j(y_j(t))\big),\\&f(y(t))=\big(f_j(y_j(t))\big),\ f(y(t-\sigma(t)))=\big(f_j(y_j(t-\sigma_j(t)))\big),\\&g(x(t))=(g_i(x_i(t))),\ g(x(t-\tau(t)))=(g_i(x_i(t-\tau_i(t)))).\end{aligned}$$

We recall here the set $\mathcal{F}$ of continuous functions $\varphi:\mathbb{R}\to\mathbb{R}$ satisfying $\varphi(0)=0$ and there exist positive scalars $l^-(\varphi)$, $\tilde{l}^+(\varphi)$ such that the following condition

$$l_\varphi^-\leq\frac{\varphi(u)-\varphi(v)}{u-v}\leq l_\varphi^+,\ u\neq v. \tag{63}$$

Assumption (A3) The decay rate functions φ_i and ψ_j, $i\in[n]$, $j\in[m]$, belong to the function class $\mathcal{F}$.

Remark 9 If the functions φ_i, ψ_j are Lipschitz continuous which have lower bounded derivatives in the sense that

$$\gamma_i=\inf_{u\in\mathbb{R}}\varphi_i'(u)>0,\ \mu_j=\inf_{u\in\mathbb{R}}\psi_j'(u)>0$$

then Assumption (A3) is satisfied, where $l_{\varphi_i}^-=\gamma_i$ and $l_{\psi_j}^-=\mu_j$. On the other hand, it follows from (63) that any function φ in $\mathcal{F}$ is also a continuous and strictly increasing function. Thus, there exists continuous inverse function φ^{-1} of φ. Moreover, φ^{-1} also belongs to $\mathcal{F}$ with $l_{\varphi^{-1}}^-=1/l_\varphi^+$ and $l_{\varphi^{-1}}^+=1/l_\varphi^-$.

Assumption (A4) The neuron activation functions $f_j(.)$, $g_i(.)$, $i\in[n]$, $j\in[m]$, are continuous, $f_j(0)=0$, $g_i(0)=0$, and there exist positive constants L_j^f, L_i^g such that

$$0\leq\frac{f_j(u)-f_j(v)}{u-v}\leq L_j^f,\quad 0\leq\frac{g_i(u)-g_i(v)}{u-v}\leq L_i^g,\quad u\neq v. \tag{64}$$

The following result will be useful for our later derivations.

Proposition 6 *Let Assumptions* (A3) *and* (A4) *hold. For any initial functions* $x^0\in C([-\overline{\tau},0],\mathbb{R}^n)$, $y^0\in C([-\overline{\sigma},0],\mathbb{R}^m)$ *specified in* (61)*, there exists a unique solution* $(x(t),y(t))$ *of system* (62) *defined on* $[t_0,\infty)$*, which is absolutely continuous in* t.

Proof We denote augmented state vector $\chi(t)=\begin{bmatrix}x(t)\\y(t)\end{bmatrix}$ and the function space

$$\mathcal{C}_d=\left\{\phi=\begin{bmatrix}x^0\\y^0\end{bmatrix}:x^0\in C([-\overline{\tau},0],\mathbb{R}^n),\ y^0\in C([-\overline{\sigma},0],\mathbb{R}^m)\right\}.$$

Define a function $F:[t_0,\infty)\times\mathcal{C}_d\to\mathbb{R}^{n+m}$ by

$$F(t,\phi)=\begin{bmatrix}-D_\alpha\Phi(x^0(0))+A\big(f_j(y_j^0(0))\big)+B\big(f_j(y_j^0(-\sigma_j(t)))\big)+I\\-D_\beta\Psi(y^0(0))+C\big(g_i(x_i^0(0))\big)+D\big(g_i(x_i^0(-\tau_i(t)))\big)+J\end{bmatrix},$$

then system (62) can be written in the following abstract functional differential equation [67]

$$\chi'(t) = F\left(t, \chi_t\right), \quad t \geq t_0, \quad \chi_{t_0} = \begin{bmatrix} x^0 \\ y^0 \end{bmatrix}, \tag{65}$$

where $\chi_t = \begin{bmatrix} x_t \\ y_t \end{bmatrix} \in \mathcal{C}_d$ and $x_t \in C([-\overline{\tau}, 0], \mathbb{R}^n)$, $y_t \in C([-\overline{\sigma}, 0], \mathbb{R}^m)$ are defined as $x_t(\xi) = x(t+\xi)$, $\xi \in [-\overline{\tau}, 0]$, $y_t(\theta) = y(t+\theta)$, $\theta \in [-\overline{\sigma}, 0]$.

By Assumptions (A3) and (A4), the function $F(t, \phi)$ is continuous and satisfies Lipschitz condition with respect to ϕ. Thus, the problem described by Eq. (65) possesses a unique solution $\chi(t)$ which is absolutely continuous in t on a maximal interval $[t_0, t_f)$.

We now show that $t_f = \infty$. In contrast, if $t_f < \infty$ then $\limsup_{t\to t_f} \|\chi(t)\| = \infty$. On the other hand, it follows from (65) that

$$\chi(t) = \chi(t_0) + \int_{t_0}^{t} F(s, \chi_s)ds, \ t \in [t_0, t_f). \tag{66}$$

In addition, we can deduce by a direct computation from the expression of function $F(t, \phi)$ that there exist positive scalars ϵ_1, ϵ_2 such that

$$\|F(t, \chi_t)\| \leq \epsilon_1 + \epsilon_2 \left(\hat{x}(t) + \hat{y}(t)\right),$$

where $\hat{x}(t) = \sup_{-\overline{\tau}\leq\xi\leq 0} \|x(t+\xi)\|$ and $\hat{y}(t) = \sup_{-\overline{\sigma}\leq\theta\leq 0} \|y(t+\theta)\|$.

Therefore, according to (66), we have

$$\|\chi(t)\| \leq \|\chi(t_0)\| + \epsilon_1(t - t_0) + \epsilon_2 \int_{t_0}^{t} \varrho(s)ds,$$

where $\varrho(t) = \hat{x}(t) + \hat{y}(t)$, which leads to

$$\varrho(t) \leq 2\|\chi(t_0)\| + 2\epsilon_1(t - t_0) + 2\epsilon_2 \int_{t_0}^{t} \varrho(s)ds. \tag{67}$$

By the Gronwall–Bellman inequality [70], from (67), we readily obtain

$$\varrho(t) \leq \frac{\epsilon_1}{\epsilon_2}\left(e^{2\epsilon_2(t-t_0)} - 1\right) + 2\|\chi(t_0)\|e^{2\epsilon_2(t-t_0)}, \ t \in [t_0, t_f),$$

which yields

$$\limsup_{t\to t_f} \|\chi(t)\| \leq \frac{\epsilon_1}{\epsilon_2}\left(e^{2\epsilon_2(t_f-t_0)} - 1\right) + 2\|\chi(t_0)\|e^{2\epsilon_2(t_f-t_0)} < \infty.$$

This contradiction shows that $t_f = \infty$. The proof is completed. □

A solution $\chi(t) = vec(x(t), y(t))$ of system (62) is said to be a positive if its trajectory is confined within the first orthant, that is, $\chi(t) \in \mathbb{R}_+^{n+m}$ for all $t \geq t_0$. We define the following admissible set of initial conditions for system (62)

$$\mathcal{A} = \left\{ \phi = \begin{bmatrix} x^0 \\ y^0 \end{bmatrix} \in C([-\overline{\tau}, 0], \mathbb{R}^n) \times C([-\overline{\sigma}, 0], \mathbb{R}^m) : \right.$$
$$\left. x^0(\xi) \succeq 0, \ \forall \xi \in [-\overline{\tau}, 0], \quad y^0(\theta) \succeq 0, \ \forall \theta \in [-\overline{\sigma}, 0] \right\}. \tag{68}$$

Definition 7 System (62) is said to be *positive* if any solution $\chi(t)$ of (62) initiating from $\mathcal{A}$ with nonnegative input vector $vec(I, J) \in \mathbb{R}_+^{n+m}$ is positive.

Definition 8 For a given vector $\mathcal{J} = vec(I, J) \in \mathbb{R}^{n+m}$, an augmented vector $\chi^* = vec(x^*, y^*)$, $x^* \in \mathbb{R}^n$, $y^* \in \mathbb{R}^m$, is said to be an *equilibrium point* (EP) of system (62) if it satisfies the following algebraic system

$$\begin{cases} -D_\alpha \Phi(x^*) + (A + B)\, f(y^*) + I = 0 \\ -D_\beta \Psi(y^*) + (C + D)\, g(x^*) + J = 0. \end{cases} \tag{69}$$

χ^* is a positive EP if it is an EP of system (62) and $\chi^* \succeq 0$.

Definition 9 A positive EP $\chi^* = vec(x^*, y^*)$ of system (62) is said to be *globally exponentially stable* (GES) if there exist positive scalars κ, γ satisfy the following exponential estimation

$$\|x(t) - x^*\| + \|y(t) - y^*\| \leq \kappa \left(\|x^0 - x^*\|_\infty + \|y^0 - y^*\|_\infty \right) e^{-\gamma(t - t_0)}, \ t \geq t_0.$$

5.2 *Positivity of BAM Neural Networks with Delays*

The positivity of BAM neural networks model (62) is presented in the following theorem.

Theorem 6 *Let Assumptions* (A3) *and* (A4). *Assume that the connection weight matrices A, B, C, and D are nonnegative (equivalently,* $\mathcal{M} = \left[\begin{array}{c|c} A & B \\ \hline C^\top & D^\top \end{array}\right] \succeq 0$*). Then, system* (62) *is positive subject to bounded delays* (60) *and admissible initial conditions in* $\mathcal{A}$.

The following auxiliary result will be employed to prove Theorem 6.

Lemma 4 *Given a function* φ *in class* $\mathcal{F}$ *and* $q(t)$ *is a nonnegative continuous function on* $[0, \infty)$. *Then, any solution of the initial value problem*

$$\begin{aligned} x'(t) &= -\varphi(x(t)) + q(t), \quad t \geq t_0, \\ x(t_0) &= x_0 > 0, \end{aligned} \tag{70}$$

satisfies $x(t) > 0$ *for* $t \geq t_0$.

Proof Suppose in contrary that there exists a $\bar{t} > t_0$ such that $x(\bar{t}) = 0$ and $x(t) > 0$ for $t \in [t_0, \bar{t})$.

By condition (63), we have

$$l_\varphi^- \leq \frac{\varphi(x(t))}{x(t)} \leq l_\varphi^+, \quad t \in [t_0, \bar{t}).$$

This, in regard to (70), leads to

$$x'(t) \geq -l_\varphi^+ x(t) + q(t), \; t \in [t_0, \bar{t}). \tag{71}$$

By taking integral both sides of (71) we then obtain

$$\begin{aligned} x(t) &\geq e^{-l_\varphi^+(t-t_0)} \left(x_0 + \int_{t_0}^{t} e^{l_\varphi^+(s-t_0)} q(s) ds \right) \\ &\geq e^{-l_\varphi^+(t-t_0)} x_0, \quad t \in [t_0, \bar{t}). \end{aligned} \tag{72}$$

Let $t \uparrow \bar{t}$, inequality (72) gives

$$0 < x_0 e^{-l_\varphi^+(\bar{t}-t_0)} \leq x(\bar{t}) = 0$$

which yields a contradiction. This shows that $x(t) > 0$ for $t \in [t_0, \infty)$. The proof of Lemma 4 is completed. □

Remark 10 It should be mentioned here that for any function $\varphi \in \mathcal{F}$ and positive scalar λ, the function $\varphi_\lambda(x) = \lambda\varphi(x)$ is also belonging to $\mathcal{F}$. Thus, the conclusion of Lemma 4 is still valid for function $\varphi_\lambda(x)$, where λ is a positive scalar.

Proof (*Proof of Theorem* 6) For an initial function $\phi \in \mathcal{A}$ and input vector $\mathcal{J} = (I, J) \in \mathbb{R}_+^{n+m}$, by Proposition 6, there exists a unique solution $\chi(t) = (x(t), y(t))$ of system (62) on interval $[t_0, \infty)$. We will show that $\chi(t)$ is a positive solution. Note at first that, by Assumption (A4), the vector fields $Af(y)$ and $Bf(y)$ are order-preserving on $\mathbb{R}_+^m$ [27]. Thus, if $y(t) \succ 0$ for $t \in [-\overline{\sigma}, t_1)$ then

$$q_i(t) \triangleq \sum_{j=1}^{m} a_{ij} f_j(y_j(t)) + \sum_{j=1}^{m} b_{ij} f_j(y_j(t - \sigma_j(t))) + I_i \geq 0, \; t \in [t_0, t_1), \; i \in [n].$$

By Lemma 4, $x_i(t) \geq 0$ for all $t \in [t_0, t_1), i \in [n]$, and therefore $\chi(t) \succeq 0, t \in [t_0, t_1)$.

Let $\chi_\epsilon(t) = vec(x_\epsilon(t), y_\epsilon(t))$ be the solution of system (62) with initial condition $\phi_\epsilon = \phi + \epsilon\mathbf{1}_{n+m}$, where $\epsilon > 0$ is fixed. Since $\chi_\epsilon(t_0) \succeq \epsilon\mathbf{1}_{n+m} \succ 0$, there exists a $t_1 > t_0$ such that $\chi_\epsilon(t) \succ 0$ for $t \in [t_0, t_1)$.

Suppose in contrary that there exist a $\tilde{t} > t_0$ and an index $j \in [m]$ such that

$$y_{j\epsilon}(\tilde{t}) = 0, \quad y_{j\epsilon}(t) > 0, \ t \in [t_0, \tilde{t}), \tag{73}$$

and $y_\epsilon(t) \succeq 0$ for $t \in [t_0, \tilde{t}]$.

By Lemma 4 and Assumption (A4), $x_\epsilon(t) \succeq 0$ for all $t \in [-\overline{\tau}, \tilde{t}]$. Therefore,

$$\widetilde{q}_j(t) \triangleq \sum_{i=1}^{n} c_{ji} g_i(x_{i\epsilon}(t)) + \sum_{i=1}^{n} d_{ji} g_i(x_{i\epsilon}(t - \tau_i(t))) + J_j \ge 0, \ t \in [t_0, \tilde{t}].$$

Similar to the proof of Lemma 4, we have

$$\begin{aligned} y_{j\epsilon}(t) &\ge e^{-\beta_j l^+_{\psi_j}(t-t_0)} \left(y_j^0(0) + \epsilon + \int_{t_0}^{t} e^{\beta_j l^+_{\psi_j}(s-t_0)} \widetilde{q}_j(s) ds \right) \\ &\ge \epsilon e^{-\beta_j l^+_{\psi_j}(t-t_0)}, \quad t \in [t_0, \tilde{t}). \end{aligned} \tag{74}$$

Let $t \uparrow \tilde{t}$ in (74), we obtain

$$y_{j\epsilon}(\tilde{t}) \ge \epsilon e^{-\beta_j l^+_{\psi_j}(\tilde{t}-t_0)} > 0$$

which contradicts with (73). Therefore, $y_\epsilon(t) \succ 0$ and $x_\epsilon(t) \succeq 0$ for all $t \ge t_0$. Let $\epsilon \downarrow 0$ we obtain $\chi(t) = \lim_{\epsilon \downarrow 0} \chi_\epsilon(t) \succeq 0$ for $t \in [t_0, \infty)$. The proof is completed. □

5.3 *Positive Equilibriums*

In this section, we utilize Brouwer's fixed point theorem, to establish the existence of a positive EP for model (62).

Noticing that a vector $\chi^* = vec(x^*, y^*) \in \mathbb{R}^{n+m}$ is an EP of model (62) if and only if it satisfies the following algebraic system

$$\begin{cases} D_\alpha^{-1}\Big((A+B)f(y^*) + I\Big) = \Phi(x^*) \\ D_\beta^{-1}\Big((C+D)g(x^*) + J\Big) = \Psi(y^*), \end{cases} \tag{75}$$

where $\mathcal{J} = (I, J) \in \mathbb{R}^{n+m}$ is a given input vector. Since the functions φ_i, ψ_j are continuous and strictly increasing, there exist inverse functions φ_i^{-1} and ψ_j^{-1}, $i \in [n]$, $j \in [m]$, which also belong to the function class $\mathcal{F}$. Let Φ^{-1} and Ψ^{-1} be the vector-valued functions defined by $\Phi^{-1}(x) = (\varphi_i^{-1}(x_i))$ and $\Psi^{-1}(x) = (\psi_j^{-1}(x_j))$. Then, system (75) can be written as

$$\begin{cases} \Phi^{-1}\left(D_\alpha^{-1}\left((A+B)f(y^*)+I\right)\right) = x^* \\ \Psi^{-1}\left(D_\beta^{-1}\left((C+D)g(x^*)+J\right)\right) = y^*. \end{cases}$$

The above system reveals that each EP $\chi_* = vec(x_*, y_*)$ of system (62) is a fixed point of the mapping $\mathcal{H} : \mathbb{R}^{n+m} \to \mathbb{R}^{n+m}$ defined by

$$\mathcal{H}(\chi) = \begin{bmatrix} \Phi^{-1}\left(D_\alpha^{-1}\left((A+B)f(y)+I\right)\right) \\ \Psi^{-1}\left(D_\beta^{-1}\left((C+D)g(x)+J\right)\right) \end{bmatrix}, \tag{76}$$

where $\chi = vec(x, y)$, $x \in \mathbb{R}^n$ and $y \in \mathbb{R}^m$.

Specifically, we rewrite the mapping $\mathcal{H}(\chi)$ defined in (76) as

$$\mathcal{H}(\chi) = \left[h_1(y) \cdots h_n(y)\ \tilde{h}_1(x) \cdots \tilde{h}_m(x)\right]^\top,$$

where

$$h_i(y) = \varphi_i^{-1}\left(\frac{1}{\alpha_i}\left(\sum_{j=1}^m \left(a_{ij}+b_{ij}\right) f_j(y_j) + I_i\right)\right), \quad i \in [n],$$

$$\tilde{h}_j(x) = \psi_j^{-1}\left(\frac{1}{\beta_j}\left(\sum_{i=1}^n \left(c_{ji}+d_{ji}\right) g_i(x_i) + J_j\right)\right), \quad j \in [m].$$

The, according to Eqs. (75) and (76), a vector $\chi^* \in \mathbb{R}^{n+m}$ is an EP of model (62) if and only if it is a fixed point of the mapping $\mathcal{H}(\chi)$, that is, $\mathcal{H}(\chi^*) = \chi^*$. Based on the Brouwer's fixed point theorem, we have the following result.

Theorem 7 *Let Assumptions* (A3) *and* (A4) *hold. Assume that*

$$\rho\left(\left[\begin{array}{c|c} 0_{n\times n} & \mathcal{K}_1 \\ \hline \mathcal{K}_2 & 0_{m\times m} \end{array}\right]\right) < 1, \tag{77}$$

where $\mathcal{K}_1 = (k_{ij}^1) \in \mathbb{R}^{n\times m}$, $\mathcal{K}_2 = (k_{ji}^2) \in \mathbb{R}^{m\times n}$ *and*

$$k_{ij}^1 = \frac{1}{\alpha_i l_{\varphi_i}^-}\left(|a_{ij}| + |b_{ij}|\right) L_j^f,$$

$$k_{ji}^2 = \frac{1}{\beta_j l_{\psi_j}^-}\left(|c_{ji}| + |d_{ji}|\right) L_i^g.$$

Then, system (62) *possesses at least an equilibrium point.*

Proof By Proposition 2 and condition (77), $I_{n+m} - \mathcal{K}$ is a nonsingular M-matrix, where $\mathcal{K} = \begin{bmatrix} 0_{n\times n} & \mathcal{K}_1 \\ \mathcal{K}_2 & 0_{m\times m} \end{bmatrix}$.

Thus, there exists nonnegative inverse matrix $(I_{n+m} - \mathcal{K})^{-1} \succeq 0$. In addition, there exists a vector $\eta \in \mathbb{R}^{n+m}$, $\eta \succ 0$, such that $(I_{n+m} - \mathcal{K})\eta \succ 0$.

We now define the following vector

$$\begin{bmatrix} \delta \\ \varrho \end{bmatrix} = (I_{n+m} - \mathcal{K})^{-1} \begin{bmatrix} \theta_1 \\ \theta_2 \end{bmatrix} + \underbrace{\begin{bmatrix} \eta_1 \\ \eta_2 \end{bmatrix}}_{\eta}, \tag{78}$$

where $\theta_1 = (\theta_i^1) \in \mathbb{R}^n$, $\theta_2 = (\theta_j^2) \in \mathbb{R}^m$ and

$$\theta_i^1 = \frac{1}{\alpha_i l_{\varphi_i}^-}|I_i|, \; i \in [n], \quad \theta_j^2 = \frac{1}{\beta_j l_{\psi_j}^-}|J_j|, \; j \in [m].$$

It is obvious that $\delta \succ 0$ and $\varrho \succ 0$.

Furthermore, by (78), we have

$$\begin{bmatrix} \delta \\ \varrho \end{bmatrix} = \mathcal{K} \begin{bmatrix} \delta \\ \varrho \end{bmatrix} + \begin{bmatrix} \theta_1 \\ \theta_2 \end{bmatrix} + (I_{n+m} - \mathcal{K})\,\eta. \tag{79}$$

Based on Eq. (79), we define the following convex compact subset

$$\mathcal{B} = \left\{ \chi \in \mathbb{R}^{n+m} \;\middle|\; |\chi| \preceq \begin{bmatrix} \delta \\ \varrho \end{bmatrix} \right\}.$$

For any $\chi = vec(x, y) \in \mathcal{B}$, $x \in \mathbb{R}^n$, $y \in \mathbb{R}^m$, we have $|x| \preceq \delta$ and $|y| \preceq \varrho$. Thus,

$$|x_i| \le \delta_i, \; i \in [n], \quad |y_j| \le \varrho_j, \; j \in [m].$$

On the other hand, since $\varphi_i \in \mathcal{F}$, we have

$$\left|\varphi_i^{-1}(u)\right| \le (1/l_{\varphi_i}^-)|u|.$$

This, in combination with Assumption (A4), gives

$$\begin{aligned}
|h_i(y)| &= \left|\varphi_i^{-1}\left(\frac{1}{\alpha_i}\left(\sum_{j=1}^{m}\left(a_{ij}+b_{ij}\right)f_j(y_j)+I_i\right)\right)\right| \\
&\le \frac{1}{\alpha_i l_{\varphi_i}^-}\left|\sum_{j=1}^{m}\left(a_{ij}+b_{ij}\right)f_j(y_j)+I_i\right| \\
&\le \frac{1}{\alpha_i l_{\varphi_i}^-}\left(\sum_{j=1}^{m}\left(|a_{ij}|+|b_{ij}|\right)|f_j(y_j)|+|I_i|\right) \\
&\le \frac{1}{\alpha_i l_{\varphi_i}^-}\left(\sum_{j=1}^{m}\left(|a_{ij}|+|b_{ij}|\right)L_j^f|y_j|+|I_i|\right) \\
&\le \sum_{j=1}^{m}k_{ij}^1\varrho_j+\theta_i^1, \quad i\in[n]. \qquad (80)
\end{aligned}$$

Similarly, we also have

$$\begin{aligned}
|\tilde{h}_j(x)| &= \left|\psi_j^{-1}\left(\frac{1}{\beta_j}\left(\sum_{i=1}^{n}\left(c_{ji}+d_{ji}\right)g_i(x_i)+J_j\right)\right)\right| \\
&\le \frac{1}{\beta_j l_{\psi_j}^-}\left(\sum_{i=1}^{n}\left(|c_{ji}|+|d_{ji}|\right)L_i^g|x_i|+|J_j|\right) \\
&\le \sum_{i=1}^{n}k_{ji}^2\delta_i+\theta_j^2, \quad j\in[m]. \qquad (81)
\end{aligned}$$

It follows from (76) and (80)–(81) that

$$\begin{aligned}
|\mathcal{H}(\chi)| &\preceq \begin{bmatrix}\mathcal{K}_1\varrho \\ \hdashline \mathcal{K}_2\delta\end{bmatrix}+\begin{bmatrix}\theta_1 \\ \hdashline \theta_2\end{bmatrix} \\
&= \mathcal{K}\begin{bmatrix}\delta \\ \hdashline \varrho\end{bmatrix}+\begin{bmatrix}\theta_1 \\ \hdashline \theta_2\end{bmatrix}. \qquad (82)
\end{aligned}$$

From Eqs. (79) and (82) we readily obtain

$$|\mathcal{H}(\chi)| \prec \begin{bmatrix}\delta \\ \hdashline \varrho\end{bmatrix}$$

for any $\chi\in\mathcal{B}$ meaning that $\mathcal{H}(\mathcal{B})\subset\mathcal{B}$. Therefore, the continuous mapping $\mathcal{H}(\chi)$ defined by (76) maps the convex compact subset $\mathcal{B}\subset\mathbb{R}^{n+m}$ into its self. By Brouwer's fixed point theorem [65], the mapping $\mathcal{H}$ has at least one fixed point $\chi^*\in\mathcal{B}$ (i.e. $\mathcal{H}(\chi^*)=\chi^*$), which is also an EP of system (62). The proof is completed. □

Remark 11 In the proof of Theorem 7, if the connection weight matrices A, B, C, D, and input vectors I, J are nonnegative, then

$$u_i \triangleq \frac{1}{\alpha_i}\left(\sum_{j=1}^{m}\left(a_{ij}+b_{ij}\right) f_j(y_j)+I_i\right) \geq 0,\ i \in [n],$$

and

$$v_j \triangleq \frac{1}{\beta_j}\left(\sum_{i=1}^{n}\left(c_{ji}+d_{ji}\right) g_i(x_i)+J_j\right) \geq 0,\ j \in [m],$$

for any $x \succeq 0$ and $y \succeq 0$. Thus, $h_i(y) = \varphi^{-1}(u_i) \geq 0$ and $\tilde{h}_j(x) = \psi_j^{-1}(v_j) \geq 0$ for $\chi = vec(x, y) \succeq 0$. Therefore, $\mathcal{H}\left(\mathbb{R}_+^{n+m}\right) \subset \mathbb{R}_+^{n+m}$. This shows that $\mathcal{H} : \mathcal{B}_+ \mapsto \mathcal{B}_+$, where $\mathcal{B}_+ = \mathcal{B} \cap \mathbb{R}_+^{n+m}$. Since

$$\mathcal{B}_+ = \left\{\chi \in \mathbb{R}^{n+m} \;\middle|\; 0 \preceq \chi \preceq \begin{bmatrix} \delta \\ \hdashline \varrho \end{bmatrix}\right\}$$

is also a convex compact subset of $\mathbb{R}^{n+m}$, by Brouwer's fixed point theorem, the continuous mapping $\mathcal{H}$ has at least one fixed point $\chi_+^* \in \mathcal{B}_+$, which is a positive EP of system (62). We summarize this result in the following corollary.

Corollary 2 *Under the assumptions of Theorem 7, if the connection weight matrices A, B, C and D are nonnegative, then for a given input vector $\mathcal{J} = vec(I, J) \in \mathbb{R}_+^{n+m}$, system* (62) *has at least one positive EP $\chi^* \in \mathbb{R}_+^{n+m}$.*

Remark 12 Let Ω be an $(n+m) \times (n+m)$ block matrix of the form

$$\Omega = \begin{bmatrix} U & V \\ W & Z \end{bmatrix}$$

where $U \in \mathbb{R}^{n\times n}$, $V \in \mathbb{R}^{n\times m}$, $W \in \mathbb{R}^{m\times n}$ and $Z \in \mathbb{R}^{m\times m}$.

If Z is nonsingular, then we have

$$\left[\begin{array}{c:c} U - VZ^{-1}W & V \\ \hdashline 0_{m\times n} & Z \end{array}\right] = \begin{bmatrix} U & V \\ W & Z \end{bmatrix} \left[\begin{array}{c:c} I_n & 0_{n\times m} \\ \hdashline -Z^{-1}W & I_m \end{array}\right].$$

Therefore,

$$\det(\Omega) = \det(U - VZ^{-1}W)\det(Z). \tag{83}$$

By utilizing equality (83), for any $\lambda \neq 0$, we have

$$\begin{aligned}\det(\lambda I_{n+m}-\mathcal{K}) &= \det\begin{bmatrix}\lambda I_n & -\mathcal{K}_1\\ -\mathcal{K}_2 & \lambda I_m\end{bmatrix}\\ &= \lambda^m \det\left(\lambda I_n - \frac{1}{\lambda}\mathcal{K}_1\mathcal{K}_2\right)\\ &= \lambda^{m-n}\det(\lambda^2 I_n - \mathcal{K}_1\mathcal{K}_2).\end{aligned}$$

The above identity shows that $\lambda \in \sigma(\mathcal{K}) \setminus \{0\}$ if and only if $\mu = \lambda^2 \in \sigma(\mathcal{K}_1\mathcal{K}_2) \setminus \{0\}$. Consequently,

$$\rho(\mathcal{K}) < 1 \Longleftrightarrow \rho(\mathcal{K}_1\mathcal{K}_2) < 1.$$

Based on the above observation, we have the following results.

Proposition 7 *Condition* (77) *holds if and only if one of the two following conditions does*

$$\rho\left(\left(\frac{1}{\alpha_i l^-_{\varphi_i}}\sum_{j=1}^{m}\frac{1}{\beta_j} l^{-1}(\psi_j)\left(|a_{ij}|+|b_{ij}|\right)\left(|c_{jk}|+|d_{jk}|\right)L^f_j L^g_k\right)_{n\times n}\right) < 1, \quad (84)$$

$$\rho\left(\left(\frac{1}{\beta_j l^-_{\psi_j}}\sum_{i=1}^{n}\frac{1}{\alpha_i} l^{-1}(\varphi_i)\left(|c_{ji}|+|d_{ji}|\right)\left(|a_{ik}|+|b_{ik}|\right)L^f_k L^g_i\right)_{m\times m}\right) < 1. \quad (85)$$

Corollary 3 *Let Assumptions* (A3) *and* (A4) *hold. Assume that* $\left[\begin{array}{c|c} A & B\\ \hline C^\top & D^\top\end{array}\right] \succeq 0$ *and either condition* (84) *or condition* (85) *is satisfied. Then, for a given input vector* $\mathcal{J} = vec(I, J) \in \mathbb{R}^{n+m}_+$, *system* (62) *has at least a positive equilibrium.*

5.4 Exponential Stability of Positive Equilibrium

In this section, we will prove that under the assumptions of Theorems 6 and 7 system (62) has a unique positive EP χ^*, which is globally exponentially stable.

Theorem 8 *Let Assumptions* (A3) *and* (A4) *hold. Assume that the connection weight matrices are nonnegative and one of the three conditions* (77), (84), *or* (85) *is satisfied. Then, for any input vector* $\mathcal{J} = vec(I, J) \in \mathbb{R}^{n+m}_+$, *system* (62) *has a unique positive EP* $\chi^* \in \mathbb{R}^{n+m}_+$, *which is GES for any delays* $\tau_i(t) \in [0, \overline{\tau}]$ *and* $\sigma_j(t) \in [0, \overline{\sigma}]$.

Proof By Theorem 7, for a given input vector $\mathcal{J} = vec(I, J) \in \mathbb{R}_+^{n+m}$, there exists a positive EP $\chi^* = vec(x^*, y^*)$ of system (62) which satisfies the following algebraic system

$$\begin{cases} -D_\alpha \Phi(x^*) + (A + B) f(y^*) + I = 0 \\ -D_\beta \Psi(y^*) + (C + D) g(x^*) + J = 0. \end{cases} \tag{86}$$

We now prove exponential attractivity of the EP χ^*. Since conditions (77), (84) and (85) are equivalent to the one that $I_{n+m} - \mathcal{K}$ is a nonsingular M-matrix, by Proposition 2, there exists a positive vector $\xi = vec(\xi^1, \xi^2) \succ 0$ such that $(I_{n+m} - \mathcal{K})\xi \succ 0$, where $\xi^1 = (\xi_i^1) \in \mathbb{R}^n$ and $\xi^2 = (\xi_j^2) \in \mathbb{R}^m$. Thus, for sufficiently small $\epsilon > 0$, we have

$$-\alpha_i l_{\varphi_i}^- \xi_i^1 + \sum_{j=1}^m (a_{ij} + b_{ij}) L_j^f \xi_j^2 < -\epsilon, \quad i \in [n], \tag{87}$$

$$-\beta_j l_{\psi_j}^- \xi_j^2 + \sum_{i=1}^n (c_{ji} + d_{ji}) L_i^g \xi_i^1 < -\epsilon, \quad j \in [m]. \tag{88}$$

Let $\chi(t) = vec(x(t), y(t))$ be a solution of system (62). It follows from Eqs. (62) and (86) that

$$\begin{cases} \left(x(t) - x^*\right)' = & - D_\alpha \left(\Phi(x(t)) - \Phi(x^*)\right) + A(f(y(t)) - f(y^*)) \\ & + B(f(y(t - \sigma(t))) - f(y^*)) \\ \left(y(t) - y^*\right)' = & - D_\beta \left(\Psi(y(t)) - \Psi(y^*)\right) + C(g(x(t)) - g(x^*)) \\ & + D(g(x(t - \tau(t))) - g(x^*)). \end{cases} \tag{89}$$

System (89) can be written in detail as

$$\begin{aligned} (x_i(t) - x_i^*)' = & -\alpha_i \left(\varphi_i(x_i(t)) - \varphi_i(x_i^*)\right) + \sum_{j=1}^m a_{ij} \left(f_j(y_j(t)) - f_j(y_j^*)\right) \\ & + \sum_{j=1}^m b_{ij} \left(f_j(y_j(t - \sigma_j(t))) - f_j(y_j^*)\right), \ i \in [n], \end{aligned} \tag{90}$$

and

$$\begin{aligned} (y_j(t) - y_j^*)' = & -\beta_j \left(\psi_j(y_j(t)) - \psi_j(y_j^*)\right) + \sum_{i=1}^n c_{ji} \left(g_i(x_i(t)) - g_i(x_i^*)\right) \\ & + \sum_{i=1}^n d_{ji} \left(g_i(x_i(t - \tau_i(t))) - g_i(x_i^*)\right), \ j \in [m]. \end{aligned} \tag{91}$$

Next, we define the following functions

$$\zeta_i(t) = |x_i(t) - x_i^*| \text{ and } \eta_j(t) = |y_j(t) - y_j^*| \text{ for } t \geq t_0,\ i \in [n],\ j \in [m].$$

By Assumption (A1),

$$\begin{aligned} \mathtt{sgn}(x_i(t) - x_i^*)\left(\varphi_i(x_i(t)) - \varphi_i(x_i^*)\right) &= \left|\varphi_i(x_i(t)) - \varphi_i(x_i^*)\right| \\ &\geq l_{\varphi_i}^- |x_i(t) - x_i^*|. \end{aligned}$$

Thus, from Eq. (90), we have

$$\begin{aligned} D^+\zeta_i(t) &= \mathtt{sgn}(x_i(t) - x_i^*)(x_i(t) - x_i^*)' \\ &= -\alpha_i \mathtt{sgn}(x_i(t) - x_i^*)\left(\varphi_i(x_i(t)) - \varphi_i(x_i^*)\right) \\ &\quad + \sum_{j=1}^m a_{ij} \mathtt{sgn}(x_i(t) - x_i^*)\left(f_j(y_j(t)) - f_j(y_j^*)\right) \\ &\quad + \sum_{j=1}^m b_{ij} \mathtt{sgn}(x_i(t) - x_i^*)\left(f_j(y_j(t - \sigma_j(t))) - f_j(y_j^*)\right) \\ &\leq -\alpha_i l_{\varphi_i}^- \zeta_i(t) + \sum_{j=1}^m L_j^f \left(a_{ij}\eta_j(t) + b_{ij}\eta_j(t - \sigma_j(t))\right),\ i \in [n], \end{aligned} \tag{92}$$

where D^+ denotes the upper-right Dini derivative.

Similarly, we have

$$D^+\eta_j(t) \leq -\beta_j l_{\psi_j}^- \eta_j(t) + \sum_{i=1}^n L_i^g \left(c_{ji}\zeta_i(t) + d_{ji}\zeta_i(t - \tau_i(t))\right),\ j \in [m]. \tag{93}$$

Based on the above differential inequalities, we now derive an exponential domination of positive solutions of system (62). Let us consider the following function for each $i \in [n]$

$$\hbar_i(\lambda) = (\lambda - \alpha_i l_{\varphi_i}^-)\xi_i^1 + \sum_{j=1}^m L_j^f \left(a_{ij} + b_{ij}e^{\lambda\bar{\sigma}}\right)\xi_j^2,\ \lambda \in [0, \infty).$$

The function $\hbar_i(\lambda)$ is continuous and strictly increasing on $[0, \infty)$, $\hbar_i(0) < -\epsilon$ and $\hbar_i(\lambda) \to \infty$ as $\lambda \to \infty$. Thus, there exists a unique positive scalar λ_i such that $\hbar_i(\lambda_i) = 0$ and $\hbar(\lambda) < 0$ for all $\lambda \in [0, \lambda_i)$.

By the same arguments we can conclude that, for each $j \in [m]$, there exists a unique positive scalar $\hat{\lambda}_j$ such that $\hat{\hbar}_j(\hat{\lambda}_j) = 0$, where

$$\hat{\hbar}_j(\lambda) = (\lambda - \beta_j l^-_{\psi_j})\xi^2_j + \sum_{i=1}^{n} L^g_i \left(c_{ji} + d_{ji}e^{\lambda\bar{\tau}}\right)\xi^1_i, \ \lambda \in [0, \infty).$$

Let $\lambda_0 = \min_{i,j}\left\{\lambda_i, \hat{\lambda}_j\right\}$, then $\lambda_0 > 0$ and $\hbar_i(\lambda_0) \leq 0$, $\hat{\hbar}_j(\lambda_0) \leq 0$ for all $i \in [n]$, $j \in [m]$. Therefore,

$$-\alpha_i l^-_{\varphi_i}\xi^1_i + \sum_{j=1}^{m} L^f_j \left(a_{ij} + b_{ij}e^{\lambda\bar{\sigma}}\right)\xi^2_j \leq -\lambda\xi^1_i, \ i \in [n], \tag{94}$$

$$-\beta_j l^-_{\psi_j}\xi^2_j + \sum_{i=1}^{n} L^g_i \left(c_{ji} + d_{ji}e^{\lambda\bar{\tau}}\right)\xi^1_i \leq -\lambda\xi^2_j, \ j \in [m], \tag{95}$$

for any $\lambda \in (0, \lambda_0]$.

Suggested by inequalities (94) and (95), we denote $C(\xi) = 1/\min_{i,j}\left\{\xi^1_i, \xi^2_j\right\}$ and consider the following domination functions

$$\hat{\zeta}(t) = C(\xi)\|\phi^*\|_\infty e^{-\lambda(t-t_0)}\xi^1, \quad \hat{\eta}(t) = C(\xi)\|\phi^*\|_\infty e^{-\lambda(t-t_0)}\xi^2, \ t \geq t_0, \tag{96}$$

where $\|\phi^*\|_\infty = \|x^0 - x^*\|_C + \|y^0 - y^*\|_C$ and a constant $\lambda \in (0, \lambda_0]$. For convenience, we also denote

$$\upsilon(t) = \begin{bmatrix}\zeta(t)\\ \eta(t)\end{bmatrix}, \quad \hat{\upsilon}(t) = \begin{bmatrix}\hat{\zeta}(t)\\ \hat{\eta}(t)\end{bmatrix} = C(\xi)\|\phi^*\|_\infty e^{-\lambda(t-t_0)}\xi.$$

Then, for any $t \geq t_0$, we have $D^+\hat{\upsilon}(t) = -\lambda\hat{\upsilon}(t)$ and

$$\hat{\upsilon}_d(t) \triangleq \begin{bmatrix}\hat{\zeta}_i(t - \tau_i(t))\\ \hat{\eta}_j(t - \sigma_j(t))\end{bmatrix} \preceq \begin{bmatrix}e^{\lambda\bar{\tau}}\hat{\zeta}_i(t)\\ e^{\lambda\bar{\sigma}}\hat{\eta}_j(t)\end{bmatrix}. \tag{97}$$

We now show that $\upsilon(t) \preceq \hat{\upsilon}(t)$ for all $t \geq t_0$. Note that, for $t \leq t_0$, $\upsilon(t) \preceq C(\xi)\|\phi^*\|_\infty\xi$ as $C(\xi)\xi \succeq \mathbf{1}_{n+m}$. Thus, for a fixed $\theta > 1$, $\upsilon(t_0) \prec \theta\hat{\upsilon}(t_0)$.

Suppose that $\upsilon(t) \prec \theta\hat{\upsilon}(t)$ does not hold for all $t > t_0$. Then, there exist a $\bar{t} > t_0$ and an index $k \in [n+m]$ such that

$$\upsilon_k(\bar{t}) = \theta\hat{\upsilon}_k(\bar{t}), \quad \upsilon_k(t) < \theta\hat{\upsilon}_k(t), \ t \in [t_0, \bar{t}), \quad \upsilon(t) \preceq \theta\hat{\upsilon}(t), \ t \in [t_0, \bar{t}]. \tag{98}$$

Without loss of generality, we can assume $k \in [n]$. Then, from equations (92), (94) and (97), we have

$$
\begin{aligned}
D^+\upsilon_k(t) &\le -\alpha_k l^-_{\varphi_k}\upsilon_k(t) + \sum_{j=1}^{m} L^f_j\left(a_{kj}\eta_j(t) + b_{kj}\eta_j(t-\sigma_j(t))\right)\\
&\le -\alpha_k l^-_{\varphi_k}\upsilon_k(t) + \sum_{j=1}^{m} \theta L^f_j\left(a_{kj}\hat{\eta}_j(t) + b_{kj}\hat{\eta}_j(t-\sigma_j(t))\right)\\
&\le -\alpha_k l^-_{\varphi_k}\upsilon_k(t) + \theta C(\xi)\|\phi^*\|_\infty\left(\sum_{j=1}^{m} L^f_j\left(a_{kj} + b_{kj}e^{\lambda\bar{\sigma}}\right)\xi^2_j\right)e^{-\lambda(t-t_0)}\\
&\le -\alpha_k l^-_{\varphi_k}\upsilon_k(t) + \theta C(\xi)\|\phi^*\|_\infty\left(-\lambda + \alpha_k l^-_{\varphi_k}\right)\xi^1_k e^{-\lambda(t-t_0)}\\
&= -\alpha_k l^-_{\varphi_k}\left(\upsilon_k(t) - \theta\hat{\upsilon}_k(t)\right) - \lambda\theta\hat{\upsilon}_k(t), \quad t \in [t_0, \bar{t}).
\end{aligned}
\tag{99}
$$

It can be deduced from inequality (99) that

$$
D^+\left(\upsilon_k(t) - \theta\hat{\upsilon}_k(t)\right) \le -\alpha_k l^-_{\varphi_k}\left(\upsilon_k(t) - \theta\hat{\upsilon}_k(t)\right), \quad t \in [t_0, \bar{t}),
$$

and hence

$$
\upsilon_k(t) - \theta\hat{\upsilon}_k(t) \le \left(\upsilon_k(t_0) - \theta\hat{\upsilon}_k(t_0)\right)e^{-\alpha_k l^-_{\varphi_k}(t-t_0)}, \ t \in [t_0, \bar{t}). \tag{100}
$$

Let $t \uparrow \bar{t}$ in inequality (100) we obtain

$$
0 = \upsilon_k(\bar{t}) - \theta\hat{\upsilon}_k(\bar{t}) \le \left(\upsilon_k(t_0) - \theta\hat{\upsilon}_k(t_0)\right)e^{-\alpha_k l(\varphi_k)(\bar{t}-t_0)} < 0.
$$

This contradiction shows that $\upsilon(t) \prec \theta\hat{\upsilon}(t)$ for all $t \ge t_0$. Let $\theta \downarrow 1$ we get $\upsilon(t) \preceq \hat{\upsilon}(t)$. Since

$$
C(\xi)\xi \preceq C(\xi)\max_{i,j}\left\{\xi^1_i, \xi^2_j\right\}\mathbf{1}_{n+m},
$$

we have

$$
\|\upsilon(t)\| \le C(\xi)\max_{i,j}\left\{\xi^1_i, \xi^2_j\right\}\|\phi^*\|_\infty e^{-\lambda(t-t_0)}
$$

and, therefore,

$$
\|x(t) - x^*\| + \|y(t) - y^*\| \le 2\|\upsilon(t)\| \le \kappa\|\phi^*\|_\infty e^{-\lambda(t-t_0)}, \ t \ge t_0,
$$

where $\kappa = 2C(\xi)\max_{i,j}\left\{\xi^1_i, \xi^2_j\right\}$. This shows that the EP χ^* of (62) is exponentially attractive.

Finally, we will prove the uniqueness of positive EP. Assume that $\bar{\chi}^* = vec(\bar{x}^*, \bar{y}^*)$ is also a positive EP of system (62). Since $\bar{\chi}^*$ can be regarded as a stationary solution of (62) with constant initial condition $\phi = \bar{\chi}^*$, by the aforementioned exponential attractivity, we have

$$
\left(\|\bar{x}^* - x^*\| + \|\bar{y}^* - y^*\|\right)\left(1 - \kappa e^{-\lambda(t-t_0)}\right) \le 0.
$$

Let $t \to \infty$ we obtain $\bar{\chi}^* = \chi^*$, which shows the uniqueness of χ^*. The proof is completed. □

Remark 13 As the mappings $F(y) = (A + B)f(y)$ and $G(x) = (C + D)g(x)$ are order-preserving vector fields, it can be seen from the proof of Theorem 8 that, for any input vectors $I \in \mathbb{R}^n_+$, $J \in \mathbb{R}^m_+$ such that I and J are not simultaneously zero, the unique positive EP χ^* of system (62) satisfies

$$\chi^* \succeq \begin{bmatrix} \Phi^{-1}\left(D_\alpha^{-1} I\right) \\ \Psi^{-1}\left(D_\beta^{-1} J\right) \end{bmatrix}$$

and, therefore, $\chi^* \in \mathbb{R}^{n+m}_+ \setminus \{0\}$. Moreover, if $I \succ 0$ and $J \succ 0$ then χ^* is strictly positive (i.e. $\chi^* \succ 0$).

Remark 14 The result of Theorem 8 also guarantees that all state trajectories of system (62) are eventually positive for positive input vectors $I \succ 0$, $J \succ 0$. More precisely, let $\chi(t) = vec(x(t), y(t))$ be a general solution of system (62) (the initial condition ϕ is not necessary to be positive). Since $\chi(t)$ converges exponentially to the unique EP χ^* which is strictly positive, there exists a $t_f > t_0$ such that $\chi(t) \succ 0$ for all $t \in [t_f, \infty)$.

6 Simulations

Consider a class of nonlinear time-delay systems which describe cooperative-type BAM neural networks with the following Boltzmann sigmoid activation functions

$$f_j(y_j) = \frac{1 - \exp\left(-\frac{y_j}{\theta_j^f}\right)}{1 + \exp\left(-\frac{y_j}{\theta_j^f}\right)}, \quad g_i(x_i) = \frac{1 - \exp\left(-\frac{x_i}{\theta_i^g}\right)}{1 + \exp\left(-\frac{x_i}{\theta_i^g}\right)}, \quad i, j = 1, 2,$$

where $\theta_j^f > 0$ and $\theta_i^g > 0$ are weighted coefficients, and decay rate functions

$$\varphi_i(x) = 2x + \sin^2(0.25x), \quad \psi_j(y) = 2y - \sin^2(0.25y), \; i, j = 1, 2. \qquad (101)$$

Clearly, the decay rate functions (101) satisfy Assumption (A3) with $l^-_{\varphi_i} = l^-_{\psi_j} = 1.75$ and $l^+_{\varphi_i} = l^+_{\psi_j} = 2.25$ $(i, j = 1, 2)$. In addition, it is easy to verify that Assumption (A4) is satisfied with $L_j^f = \frac{1}{2\theta_j^f}$ and $L_i^g = \frac{1}{2\theta_i^g}$.

For illustrative purpose, the parameters θ_j^f, θ_i^g and connection weight matrices are given as

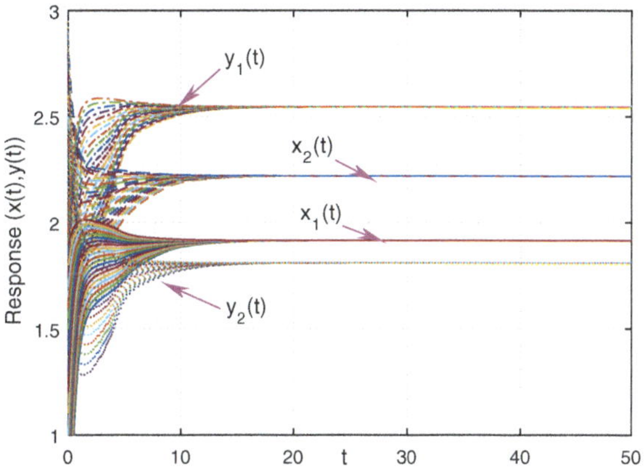

Fig. 5 Convergence of state trajectories to the unique positive EP χ^* for the input vector $\mathcal{J} = (1.0, 2.0, 2.0, 1.0)^\top$ and delays $\tau_i(t) = 1 + 3|\sin(10\pi t)|$, $\sigma_j(t) = 4|\cos(5\pi t)|$

$$\theta_j^f = 1.6, \quad \theta_i^g = 1.2, \ i, j = 1, 2, \quad D_\alpha = D_\beta = \texttt{diag}\{1, 1\},$$
$$A = \begin{bmatrix} 2.25 & 1.18 \\ 0.95 & 1.26 \end{bmatrix}, \quad B = \begin{bmatrix} 0.54 & 1.16 \\ 1.46 & 0.94 \end{bmatrix},$$
$$C = \begin{bmatrix} 0.64 & 1.48 \\ 1.16 & 0.37 \end{bmatrix}, \quad D = \begin{bmatrix} 1.25 & 0.56 \\ 0.54 & 1.42 \end{bmatrix}.$$

Then, we have $\mathcal{K}_1 = \begin{bmatrix} 0.4982 & 0.4179 \\ 0.4304 & 0.3929 \end{bmatrix}$, $\mathcal{K}_2 = \begin{bmatrix} 0.45 & 0.4857 \\ 0.4048 & 0.4262 \end{bmatrix}$, and hence

$$I_4 - \mathcal{K} = \left[\begin{array}{cc|cc} 1.0 & 0 & -0.4982 & -0.4179 \\ 0 & 1.0 & -0.4304 & -0.3929 \\ \hline -0.45 & -0.4857 & 1.0 & 0 \\ -0.4048 & -0.4262 & 0 & 1.0 \end{array}\right]$$

is a nonsingular M-matrix.

By Theorem 8, for a given input vector $\mathcal{J} = vec(I, J) \in \mathbb{R}_+^4$, the system has a unique positive EP χ^* which is GES for any bounded time-varying delays.

For $\mathcal{J} = (1.0, 2.0, 2.0, 1.0)^\top$, by solving system (69) with Matlab Toolbox, we obtain $\chi^* = (1.9164, 2.222, 2.5467, 1.8116)^\top$.

A simulation result of 20 sample state trajectories for the delays $\tau_i(t) = 1 + 3|\sin(10\pi t)|$ and $\sigma_j(t) = 4|\cos(5\pi t)|$ is portrayed in Fig. 5. The corresponding phase diagram is also presented in Fig. 6. Clearly, the conducted state trajectories of the system are positive and converge to the unique positive EP χ^*, which validates the obtained theoretical results.

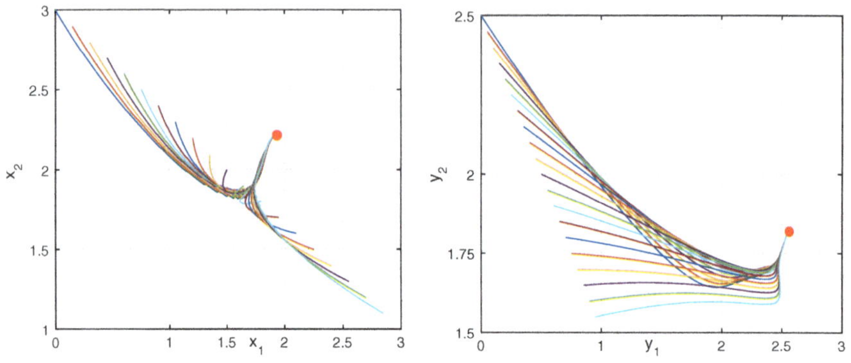

(a) x_1 and x_2 (b) y_1 and y_2

Fig. 6 Convergence of solutions to the positive EP χ^* in phase space

7 Conclusion

This chapter has dealt with the positivity and exponential stability of positive non-linear systems which describe dynamics of various models of neural networks with multiple time-varying delays. A systematic approach involving comparison techniques via differential and integral inequalities in combination with nonlinear analysis has been presented. By utilizing the proposed method, LP-based conditions for the existence and exponential stability of positive equilibriums have been derived in LP setting for M-matrix, which can be effectively solved by various convex algorithms. Numerical simulations have been presented to demonstrate the applicability and effectiveness of the derived theoretical results.

References

1. Soulié, F.F., Gallinari, P.: Industrial Applications of Neural Networks. World Scientific Publishing, Singapore (1998)
2. Venketesh, P., Venkatesan, R.: A survey on applications of neural networks and evolutionary techniques in web caching. IETE Tech. Rev. **26**, 171–180 (2009)
3. Raschman, E., Záluský, R., Ďuračková, D.: New digital architecture of CNN for pattern recognition. J. Electrical Eng. **61**, 222–228 (2010)
4. Huang, G., Huang, G.B., Song, S., You, K.: Trends in extreme learning machines: a review. Neural Netw. **61**, 32–48 (2015)
5. Mrugalski, M., Luzar, M., Pazera, M., Witczak, M., Aubrun, C.: Neural network-based robust actuator fault diagnosis for a non-linear multi-tank system. ISA Trans. **61**, 318–328 (2016)
6. Witczak, P., Patan, K., Witczak, M., Mrugalski, M.: A neural network approach to simultaneous state and actuator fault estimation under unknown input decoupling. Neurocomputing **250**, 65–75 (2017)
7. Kiakojoori, S., Khorasani, K.: Dynamic neural networks for gas turbine engine degradation prediction, health monitoring and prognosis. Neural Comput. Appl. **27**, 2157–2192 (2016)

8. Gong, M., Zhao, J., Liu, J., Miao, Q., Jiao, J.: Change detection in synthesis aperture radar images based on deep neural networks. IEEE Trans. Neural Netw. Learn. Syst. **27**, 125–138 (2016)
9. Zhang, H., Wang, Z., Liu, D.: A comprehensive review of stability analysis of continuous-time recurrent neural networks. IEEE Trans. Neural Netw. Learn. Syst. **25**, 1229–1262 (2014)
10. Baldi, P., Atiya, A.F.: How delays affect neural dynamics and learning. IEEE Trans. Neural Netw. **5**, 612–621 (1995)
11. Lu, H.: Chaotic attractors in delayed neural networks. Phys. Lett. A **298**, 109–116 (2012)
12. Park, J.H., Lee, T.H., Liu, Y., Chen, J.: Dynamic Systems with Time Delays: Stability and Control. Springer-Nature, Singapore (2019). https://doi.org/10.1007/978-981-13-9254-2
13. Arik, S.: An improved robust stability result for uncertain neural networks with multiple time delays. Neural Netw. **54**, 1–10 (2014)
14. Park, J.H., Kwon, O.M., Lee, S.M.: LMI optimization approach on stability for delayed neural networks of neutral-type. Appl. Math. Comput. **196**(1), 236–244 (2008)
15. Hien, L.V., Loan, T.T., Huyen Trang, B.T., Trinh, H.: Existence and global asymptotic stability of positive periodic solution of delayed Cohen-Grossberg neural networks. Appl. Math. Comput. **240**, 200–212 (2014)
16. Huang, H., Huang, T., Chen, X.: Further result on guaranteed $\mathcal{H}_\infty$ performance state estimation of delayed static neural networks. IEEE Trans. Neural Netw. Learn. Syst. **26**, 1335–1341 (2015)
17. Hien, L.V., Son, D.T.: Finite-time stability of a class of non-autonomous neural networks with heterogeneous proportional delays. Appl. Math. Comput. **251**, 14–23 (2015)
18. He, Y., Ji, M.D., Zhang, C.K., Wu, M.: Global exponential stability of neural networks with time-varying delay based on free-matrix-based integral inequality. Neural Netw. **77**, 80–86 (2016)
19. Arik, S.: Dynamical analysis of uncertain neural networks with multiple time delays. Int. J. Syst. Sci. **47**, 730–739 (2016)
20. Liu, B.: Global exponential convergence of non-autonomous SICNNs with multi-proportional delays. Neural Comput. Appl. **28**, 1927–1931 (2017)
21. Manivannan, R., Samidurai, R., Sriraman, R.: An improved delay-partitioning approach to stability criteria for generalized neural networks with interval time-varying delays. Neural Comput. Appl. **28**, 3353–3369 (2017)
22. Hai-An, L.D., Hien, L.V., Loan, T.T.: Exponential stability of non-autonomous neural networks with heterogeneous time-varying delays and destabilizing impulses. Vietnam J. Math. **45**, 425–440 (2017)
23. Lee, T.H., Trinh, H., Park, J.H.: Stability analysis of neural networks with time-varying delay by constructing novel Lyapunov functionals. IEEE Trans. Neural Netw. Learn. Syst. **29**, 4238–4247 (2018)
24. Ge, C., Park, J.H., Hua, C., Shi, C.: Robust passivity analysis for uncertain neural networks with discrete and distributed time-varying delays. Neurocomputing **364**, 330–337 (2019)
25. He, J., Liang, Y., Yang, F., Yang, F.: New $\mathcal{H}_\infty$ state estimation criteria of delayed static neural networks via the Lyapunov-Krasovskii functional with negative definite terms. Neural Netw. **123**, 236–247 (2020)
26. Farina, L., Rinaldi, S.: Positive Linear Systems: Theory and Applications. Wiley, New York (2000)
27. Smith, H.: Monotone Dynamical Systems: An Introduction to the Theory of Competitive and Cooperative Systems. American Mathematical Society, Providence (2008)
28. Liu, X., Yu, W., Wang, L.: Stability analysis for continuous-time positive systems with time-varying delays. IEEE Trans. Autom. Control **55**, 1024–1028 (2010)
29. Briat, C.: Robust stability and stabilization of uncertain linear positive systems via integral linear constraints: $\mathcal{L}_1$-gain and $\mathcal{L}_\infty$-gain characterization. Int. J. Robust Nonlinear Control **23**, 1932–1954 (2013)
30. Feyzmahdavian, H.R., Charalambous, T., Johanson, M.: Exponential stability of homogeneous positive systems of degree one with time-varying delays. IEEE Trans. Autom. Control **59**, 1594–1599 (2014)

31. Zaidi, I., Chaabane, M., Tadeo, F., Benzaouia, A.: Static state-feedback controller and observer design for interval positive systems with time delay. IEEE Trans. Circ. Syst-II **62**, 506–510 (2015)
32. Hien, L.V.: An LP approach to full-order and reduced-order state estimations of positive Markov jump systems with delay. Int. J. Syst. Sci. **48**, 2534–2543 (2017)
33. Hien, L.V., Kinh, C.T.: Decentralised stabilization of positive fractional-order interconnected systems. IET Control Theory Appl. **11**, 2391–2395 (2017)
34. Hien, L.V., Kinh, C.T.: Robust control of positive fractional-order interconnected systems with heterogeneous delays. Asian J. Control **21**, 596–608 (2019)
35. Hien, L.V., Hong, M.T.: An optimization approach to static output-feedback control of LTI positive systems with delayed measurements. J. Frank. Inst. **356**, 5087–5103 (2019)
36. Mózaryn, J., Kurek, J.E.: Design of a neural network for an identification of a robot model with a positive definite inertia matrix. In: Artificial Intelligence and Soft Computing. Springer, Berlin (2010)
37. Ma, G.J., Wu, S., Cai, G.Q.: Neural networks control of the Ni-MH power battery positive mill thickness. Appl. Mech. Mater. **411–414**, 1855–1858 (2013)
38. Ameloot, T.J., den Bussche, J.V.: Positive neural networks in discrete time implement monotone-regular behaviors. Neural Comput. **27**, 2623–2660 (2015)
39. Abiodun, O.I., Jantan, A., Omolara, A.E., Dada, K.V., Mohamed, N.A., Arshad, H.: State-of-the-art in artificial neural network applications: a survey. Heliyon **4**, e00938 (2018)
40. Wheeler, D.W., Schieve, W.C.: Stability and chaos in an inertial two neuron system. Phys. D: Nonlinear Phenom **105**, 267–284 (1997)
41. Koch, C.: Cable theory in neurons with active linearized membrane. Biol. Cybern. **50**, 15–33 (1984)
42. Babcock, K.L., Westervelt, R.M.: Stability and dynamics of simple electronic neural networks with added inertia. Phys. D: Nonlinear Phenom **23**, 464–469 (1986)
43. Tu, Z., Cao, J., Hayat, T.: Matrix measure based dissipativity analysis for inertial delayed uncertain neural networks. Neural Netw. **75**, 47–55 (2016)
44. Wan, P., Jian, J.: Global convergence analysis of impulsive inertial neural networks with time-varying delays. Neurocomputing **245**, 68–76 (2017)
45. Tu, Z., Cao, J., Alsaedi, A., Alsaadi, F.: Global dissipativity of memristor-based neutral type inertial neural networks. Neural Netw. **88**, 125–133 (2017)
46. Zhang, G., Zeng, Z., Hu, J.: New results on global exponential dissipativity analysis of memristive inertial neural networks with distributed time-varying delays. Neural Netw. **97**, 183–191 (2018)
47. Tu, Z., Cao, J., Hayat, T.: Global exponential stability in Lagrange sense for inertial neural networks with time-varying delays. Neurocomputing **171**, 524–531 (2016)
48. Wang, J., Tian, L.: Global Lagrange stability for inertial neural networks with mixed time-varying delays. Neurocomputing **235**, 140–146 (2017)
49. Wang, L., Huang, T.W., Xiao, Q.: Lagrange stability of delayed switched inertial neural networks. Neurocomputing **381**, 52–60 (2020)
50. Wang, J., Tian, L.: Stability of inertial neural network with time-varying delays via sampled-data control. Neural Process Lett. **50**, 1123–1138 (2019)
51. Kosko, B.: Bi-directional associative memory. IEEE Trans. Syst. Man Cybern. **18**, 49–60 (1998)
52. Chen, A., Cao, J., Huang, L.: Exponential stabilityof BAM neural networks with transmission delays. Neurocomputing **57**, 435–454 (2004)
53. Liao, X., Yu, J., Chen, G.: Novel stability criteria for bidirectional associative memory neural networks with time delays. Int. J. Circuit Theor. Appl. **30**, 519–546 (2002)
54. Park, J.H., Kwon, O.M.: Delay-dependent stabilility criterion for bidirectional associative memory neural networks with interval time-varying delays. Modern Phys. Lett. B **23**, 35–46 (2009)
55. Lakshmanan, S., Park, J.H., Lee, T.H., Jung, H.Y., Rakkiyappan, R.: Stability criteria for BAM neural networks with leakage delays and probabilistic time-varying delays. Appl. Math. Comput. **219**, 9408–9423 (2013)

56. Thipcha, J., Niamsup, P.: New exponential passivity of BAM neural networks with time-varying delays. Neural Comput. Appl. **29**, 1593–1600 (2018)
57. Park, J.H., Park, C.H., Kwon, O.M., Lee, S.M.: A new stability criterion for bidirectional associative memory neural networks of neutral-type. Appl. Math. Comput. **199**(2), 716–722 (2008)
58. Mathiyalagan, K., Park, J.H., Sakthivel, R.: Synchronization for delayed memristive BAM neural networks using impulsive control with random nonlinearities. Appl. Math. Comput. **259**, 967–979 (2015)
59. Sader, M., Abdurahman, A., Jiang, H.: General decay synchronization of delayed BAM neural networks via nonlinear feedback control. Appl. Math. Comput. **337**, 302–314 (2018)
60. Li, L., Yang, Y.Q., Lin, G.: The stabilization of BAM neural networks with time-varying delays in the leakage terms via sampled-data control. Neural Comput. Appl. **27**, 447–457 (2016)
61. Chinnathambi, R., Rihan, F.A., Shanmugam, L.: Stabilization of delayed Cohen-Grossberg BAM neural networks. Math. Meth. Appl. Sci. **41**, 593–605 (2018)
62. Zhu, Q., Cao, J.: Stability analysis of Markovian jump stochastic BAM neural networks with impulse control and mixed time delays. IEEE Trans. Neural Netw. Learn. Syst. **23**, 467–479 (2012)
63. Aouiti, C., Assali, E.A.: Stability analysis for a class of impulsive bidirectional associative memory (BAM) neural networks with distributed delays and leakage time-varying delays. Neural Process Lett. **50**, 851–885 (2019)
64. Forti, M., Tesi, A.: New conditions for global stability of neural networks with application to linear and quadratic programming problems. IEEE Trans. Circuits Syst-I: Fund. **42**, 354–366 (1995)
65. Zeidler, E.: Nonlinear Functional Analysis and its Applications-I: Fixed-Point Theorems. Springer, New York (1986)
66. Plemmons, R.J.: M-matrix characterizations I-Nonsingular M-matrices. Linear Alg. Appl. **18**, 175–188 (1977)
67. Arino, O., Hbid, M.L., Ait Dads, E.: Delay Differential Equations and Applications. Springer, Dordrecht (2002)
68. Berman, A., Plemmons, R.J.: Nonnegative Matrices in the Mathematical Sciences. SIAM, Philadelphia (1984)
69. Haykin, S.: Neural Networks: A Comprehensive Foundation. Prentice Hall, Upper Saddle River (1999)
70. Pachpatte, B.G.: Inequalities for Differential and Integral Equations. Academic, San Diego (1998)

Fault Detection and Isolation Methodology in Finite Frequency Domain for Constrained Networked Systems

Yue Long and Ju H. Park

Abstract This chapter is concerned with transmission-dependent fault detection and isolation strategy in a finite frequency domain for networked systems with constrained communication capacities. By considering the possible network-induced phenomena, a switched system with multiple stochastic parameters is firstly derived to model the considered features. Then, drawing support from the geometric method and the finite frequency stochastic performance indices, a transmission-dependent fault detection and isolation scheme is developed in a finite frequency domain. Sufficient conditions are established to guarantee the desired performances and characterize the filter gains. The proposed filters can complete the mission of detection and isolation well only by partially available measurements, which is shown by an application to the VTOL aircraft.

Keywords Fault detection and isolation · Finite frequency · Transmission dependent · Constrained capacity

1 Introduction

The increasing scale together with the corresponding requirement on high security of the modern control systems has promoted the researches on networked systems [1–6] and fault detection (FD) technologies. One the one hand, the development of the networked systems introduced some advantages such as decreased system wiring, easy reconfigurability and so on [8], it also brings some challenges [3, 9–11]. An

Y. Long (✉)
School of Automation Engineering, University of Electronic Science and Technology of China, Chengdu 610054, People's Republic of China
e-mail: longyueyl@163.com

J. H. Park
Department of Electrical Engineering, Yeungnam University, 38541 Gyeongsan, Republic of Korea
e-mail: jessie@ynu.ac.kr

J. H. Park (ed.), *Recent Advances in Control Problems of Dynamical Systems and Networks*, Studies in Systems, Decision and Control 301,
https://doi.org/10.1007/978-3-030-49123-9_16

inevitably one is communication constraint [1], which is a restriction on the quantity of the transmitted data [7]. Consequently, there have been many researches concerned about such kind of problems, one can see [8, 9, 12, 13] for more information.

On the other hand, great efforts have been made to the FD technologies and the related fault-tolerant control techniques, for such issue, one can see [14–18] and the references therein for more information. We can see from the above statements, it is natural to proceed with the researches on FD methods for a networked system. As a result, there are fruitful results concerned with such issuees. For example, in [19, 20], the FD schemes against to network-induced time-delays have been proposed, and with consideration of the packet dropout, FD methods have been developed in [21–24].

Nevertheless, the referred FD methods above have been proposed in the time-domain. In practical systems, the faults always occurred in low-frequency range [25]. With aid of the GKYP (generalized Kalman-Yakubovich-Popov) lemma, many FD methods have been proposed in a finite frequency domain [25–27]. Even though there have been some results concerned on FD techniques for networked systems, such as [28]. However, the fault dectetion isolation (FDI) method for networked systems with respect to frequency features has been in its infancy.

Motivated by the above discussions, in this chapter, an FDI strategy is developed for networked systems under constraint channels. A switched system with multiple parameters is firstly derived to model the whole networked system. Then, with the help of the geometric method, a novel lemma is developed to ensure the decoupling of the faults to fault detection and isolation filters and ensure the required finite frequency performances of the derived systems. Subsequently, with aid of the derived lemma, a set of filters can be constructed corresponding to the faulty modes and transmission intervals.

Notations: The following notations are used throughout this chapter. l_2 denotes the Hilbert space of square-integrable functions; $\mathbf{E}(\cdot)$ denotes the expectation operator; I represents the identity matrix with compatible dimensions. $\star$ is utilized to represent a term that is induced by symmetry; $A^{'}$ represents the generalized inverse of A; A^* means the transpose of A. Denoting $\mathrm{He}(A) := A + A^T$ be the sum of a square matrix A and its transpose A^T.

2 System Modeling and Problem Preliminaries

The considered networked system is with the structure as Fig. 1, wherein the plant is as the form of

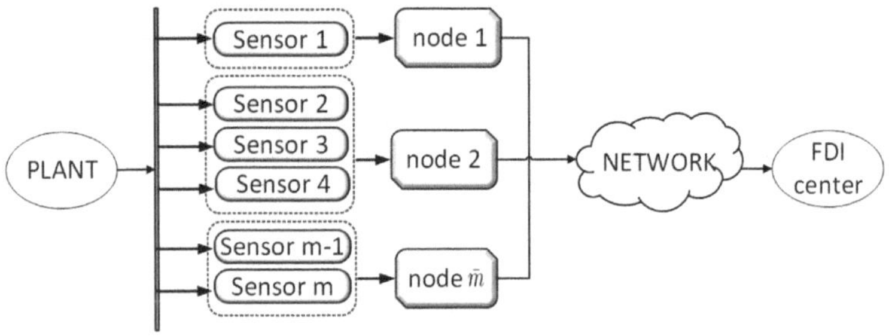

Fig. 1 Structure of the Networked Systems

$$
\begin{aligned}
\dot{x}_t &= A_o x_t + B_c w_t + \sum_{\iota=1}^{L} F_o^{\iota} f_{\iota t} \\
y_t^1 &= c_1 x_t \\
&\vdots \\
y_t^m &= c_m x_t
\end{aligned}
\tag{1}
$$

where $x_t \in \mathcal{X}$ with dimension n and $y_t^i \in \mathcal{Y}^i$, $i = 1, 2, \ldots, m$, with dimension 1 are systems states and the measured outputs, $w_t \in \mathbf{R}^w$ is the external disturbances, which belongs to $l_2[0, \infty)$, $f_{\iota t} \in \mathcal{F}_\iota$ with dimension p represents the possible failures, and F_o^ι denotes the fault signature. A_o, B_o, F_o^ι and c_i are known matrices.

Here, a general assumption, that is A_c *is stable*, is made to ensure the stability of the filtering error dynamics. Further, during the transmission, it is assumed that m measurements y^i, $i = 1, 2, \ldots, m$ are packaged into $\bar{m}$, $(\bar{m} \leq m)$, nodes for transmission in a finite capacity networks, wherein only one node can get the accession to the channel at each transmission instant t_k. And then this transmitted node can be received by the FDI center at the arrival instant $r_k = t_k + \tau_k$. As in [29], in this chapter, it is also assumed a new transmission will begin after the last one is completed, i.e. $t_k \leq r(k) \leq t_{k+1}$, in which $t_0 = 0$. In addition, as shown in Fig. 2, the transmission-induced delay $\tau_k = r_k - t_k$ and transmission interval $h_k = t_{k+1} - t_k$ are time-varying. Without loss of generality, it is assumed that τ_k and h_k are i.i.d. and conformed to $h_k \subseteq \{h_{\sigma_k}\}$, $0 \leq \tau_k < h_k$, $\sigma_k \in \mathbb{H} = \{1, 2, \ldots, N\}$ for $k \in \mathbf{N}^+$.

Remark 1 By adding a buffer at the front end of the channel, the assumption $h_k \subseteq \{h_{\sigma_k}\}$ can be realized. Under such assumption, if $t - t_k > \bar{h}$, the packet is set to be lost, where $\bar{h} = \max\{h_{\sigma_k}\}$.

It is supposed the transmission of every data node is scheduled by a stochastic protocol, which is directed by $\bar{m} + 1$ (i.e. $0, 1, \ldots, \bar{m}$) Bernoulli distributed white sequences with $\mathbf{P}(\varrho_{ki} = 1) = \mathbf{E}(\varrho_{ki}) = \phi_i$, and $\mathbf{P}(\varrho_{ki} = 0) = 1 - \mathbf{E}(\varrho_{ki}) = 1 - \phi_i$, where $\varrho_{ki} = 1$ implies the ith node is accessed to the networked channel at the kth transmission instant and $\varrho_{k0} = 1$ means packet loss. It is not difficult to derive

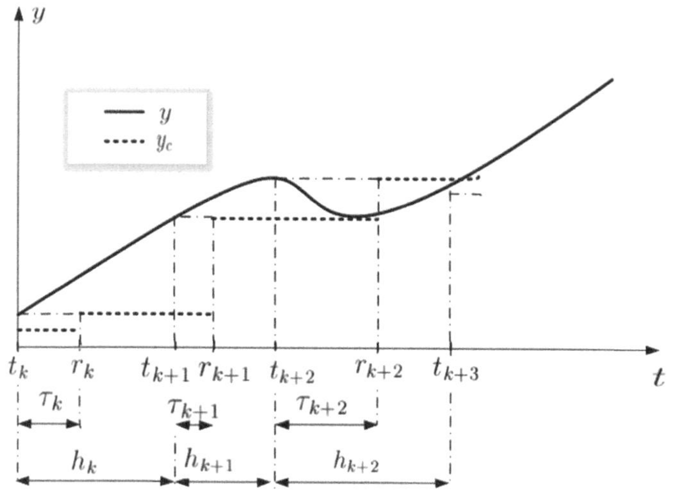

Fig. 2 Illustration of the transmission

$\sum_{i=0}^{\bar{m}} \phi_i = 1$ from the practical point of view. Thus, the received data by the FDI center can be renewed by

$$y_{ct} = \sum_{i=0}^{\bar{m}} \varrho_{ki} \mathfrak{T}_i y_{t_k} + \Big(I - \sum_{i=0}^{\bar{m}} \varrho_{ki} \mathfrak{T}_i\Big) y_{ct_k} + v_{t_k} \tag{2}$$

$\forall t \in [r_k, r_{k+1})$, where v_{tk} denotes the channel noise, and $\mathfrak{T}_i$ is defined as $\mathfrak{T}_i = \mathrm{diag}\{\mathfrak{D}_{ki}(1), \mathfrak{D}_{ki}(2), \ldots, \mathfrak{D}_{ki}(\iota), \ldots \mathfrak{D}_{ki}(m)\}$ where $\mathfrak{D}_{ki}(\iota) = 1$ if the corresponded output y_k^ι is transmitted in node i, otherwise $\mathfrak{D}_{ki}(\cdot) = 0$.

Remark 2 During recent decades, there appeared many researches, for example [30, 31], concerning the data processing protocols. In which how to choose the accessing probability has been deeply discussed. Among them, the stochastic one modelled by Bernoulli distribution is an easier protocol to be realized. Thus it is natural to assume that the probability ϕ_i is designed in advance, i.e. they are known beforehand.

Then, defining $y = [y_1^T, \ldots, y_m^T]^T$ and $C = [c_1^T, \ldots, c_m^T]^T$, the discretely dynamics between t_k and $t_{k+1} = t_k + h_k$ of (1) can be derived as

$$\begin{aligned} x_{k+1} &= A_{\sigma_k} x_k + B_{\sigma_k} w_k + \sum_{\iota=1}^{L} F_{\sigma_k}^{\iota} f_{\iota k} \\ y_k &= C x_k \end{aligned} \tag{3}$$

for $h_k = h_{\sigma_k}$, where $x_k = x_{t_k}$, $A_{\sigma_k} = e^{A_c h_k}$, $B_{\sigma_k} = \int_0^{h_k} e^{A_c s} B_c ds$ and $F^{\iota}_{\sigma_k} = \int_0^{h_k} e^{A_c s} F^{\iota}_c ds$. It is clearly that (3) is switching in $\mathbb{H}$ due to its reliance on h_k.

For convenience of understanding, some definitions are firstly given for the following switched system.

$$\begin{aligned} x_{k+1} &= A_\varsigma x_k, \\ y_k &= C_\varsigma x_k. \end{aligned} \tag{4}$$

Definition 1 ([32]) For system (4) with $\varsigma \in \mathbb{N}$ and the subspace $\mathcal{Q} \subseteq \mathcal{X}$, if $A_\varsigma \mathcal{Q} \subseteq \mathcal{Q}$, the subspace $\mathcal{Q}$ is defined to be A_ς-invariant.

Denoting $\ll \mathcal{Q}|A_\varsigma \gg_{\varsigma\in\mathbb{N}}$ as the largest A_ς−invariant subspace involved in $\mathcal{Q}$, the following definition is given.

Definition 2 ([33]) A subspace $\mathfrak{S} \subseteq \mathcal{X}$ is called to be finite unobservable subspace (f.u.o.s.) for system (4) if $\mathfrak{S} = \ll \cap_{\varsigma=1}^{N} \ker \mathcal{H}_\varsigma C_{\sigma_k} | A_\varsigma + \mathcal{G}_\varsigma C_\varsigma \gg$, $\varsigma \in \mathbb{H}$ holds for maps $\mathcal{G}_\varsigma : \mathcal{Y} \to \mathcal{X}$ and $\mathcal{H}_\varsigma : \mathcal{Y} \to \mathcal{Y}$.

Defining a class of f.u.o.s. of $\mathcal{X}$ for system (3) as $\underline{\mathfrak{S}}(\mathcal{X})$. It is easy to see $\underline{\mathfrak{S}}(\mathcal{X})$ is a nonempty close set. Thus the set of f.u.o.s. containing $\mathcal{F}$ can be signified by $\underline{\mathfrak{S}}(\mathcal{F})$ for arbitrary $\mathfrak{F} \subseteq \mathcal{X}$, which has the smallest unit denoting by $\mathfrak{S}^\star = \inf \underline{\mathfrak{S}}(\mathcal{F})$ and could be computed by following the procedure of the Algorithm in [34, 35]. For calculated f.u.o.s. $\mathfrak{S}^\star \subset \mathcal{X}$, we subsequently introduce the factor system of (4), which could be expressed by

$$\begin{aligned} \breve{x}_{k+1} &= \bar{A}_\varsigma \breve{x}_k, \\ \breve{y}(k) &= M_\varsigma \breve{x}_k, \end{aligned} \tag{5}$$

where $\breve{x} \in \mathfrak{S}^\star \subset \mathcal{X}$ is the state with dimension $n - \dim(\mathfrak{S}^\star)$, $\bar{A}_\varsigma = (A_\varsigma + \mathcal{G}_\varsigma C_\varsigma : \mathcal{X}/\mathfrak{S}^\star)$ and M_ς is the solution of $M_\varsigma \mathcal{T} = \mathcal{H}_\varsigma C_\varsigma$, in which $\mathcal{T} : \mathcal{X} \to \mathcal{X}/\mathfrak{S}^\star$ is the canonical projection satisfying $\mathcal{T}(A_\varsigma + \mathcal{G}_\varsigma C_\varsigma) = \bar{A}_\varsigma \mathcal{T}$.

In this chapter, the set of the residuals $r_{\iota k}$ with dimension 1 for fault detection and isolation are generated by

$$\begin{aligned} \tilde{x}_{\iota k+1} &= A^{\iota}_{f\sigma_k} \tilde{x}_{\iota k} + B^{\iota}_{f\sigma_k} y_{ck}, \\ r_{\iota k} &= C^{\iota}_{f\sigma_k} \tilde{x}_{\iota k}, \end{aligned} \tag{6}$$

where $\tilde{x}_\iota \in \mathcal{X}$ is the state of the filter, its dimension is same with the factor system of (3) calculated below, $A^{\iota}_{f\sigma_k}$, $B^{\iota}_{f\sigma_k}$ and $C^{\iota}_{f\sigma_k}$ are filter gains to be determined, and the value of y_{ck} is $y_{ck} = \lim_{t\to r_k^+} y_{ct}$.

For a certain f.u.o.s. $\mathfrak{S}^{\iota\star}_{\sigma_k}$, we can find two maps $\mathcal{G}^{\iota}_{\sigma_k}$ and $\mathcal{H}^{\iota}_{\sigma_k}$ such that $\mathfrak{S}^{\iota\star}_{\sigma_k} = \ll \cap_{\sigma_k=1}^{N} \ker \mathcal{H}^{\iota}_{\sigma_k} C | A_{\sigma_k} + \mathcal{G}^{\iota}_{\sigma_k} C \gg$, where $\mathcal{H}^{\iota}_{\sigma_k}$ can be derived by $\ker \mathcal{H}^{\iota}_{\sigma_k} C = \mathfrak{S}^{\iota\star}_{\sigma_k} + \ker C$. Further, denoting $\mathcal{T}^{\iota}_{\sigma_k} : \mathcal{X} \to \mathcal{X}/\mathfrak{S}^{\iota\star}_{\sigma_k}$, the factor system of (3) corresponded to $\mathfrak{S}^{\iota\star}_{\sigma_k}$ can be obtained by $\mathcal{M}^{\iota}_{\sigma_k} \mathcal{T}^{\iota}_{\sigma_k} = \mathcal{H}^{\iota}_{\sigma_k} C$ and $\mathcal{T}^{\iota}_{\sigma_k}(A_{\sigma_k} + \mathcal{G}^{\iota}_{\sigma_k} C) = A^{\iota}_{\sigma_k} \mathcal{T}^{\iota}_{\sigma_k}$ with $A^{\iota}_{\sigma_k} = (A_{\sigma_k} + \mathcal{G}^{\iota}_{\sigma_k} C : \mathcal{X}/\mathfrak{S}^{\iota\star}_{\sigma_k})$ and $B^{\iota}_{\sigma_k} = \mathcal{T}^{\iota}_{\sigma_k} B_{\sigma_k}$. Subsequently, defining the transmission error as $e_t = y_{ct} - y_t$, the dynamics of the whole FDI system can be obtained by

$$
\begin{aligned}
\xi_{\iota k+1} &= \mathbf{A}^{\iota}_{\sigma_k}\xi_{\iota k} + \mathbf{B}^{\iota}_{\sigma_k} d_k + \mathcal{Z}_3 \mathcal{T}^{\iota}_{\sigma_k} \sum_{\iota=1}^{L} F^{\iota}_{\sigma_k} f_{\iota k}, \\
r_{\iota k} &= \mathbf{C}^{\iota}_{\sigma_k}\xi_{\iota k},
\end{aligned} \tag{7}
$$

where $\xi_{\iota k} = \left[\left(\mathcal{T}^{\iota}_{\sigma_k} x_k\right)^T \; \tilde{x}^T_{\iota k} \; e^T_k \right]^T$, $d_k = \left[w^T_k \; v^T_k \right]^T$, $\mathbf{C}^{\iota}_{\sigma_k} = \left[\mathcal{Z}^T_2 \mathcal{C}^{\iota}_{\sigma_k} \; 0 \right]$, and

$$
\mathbf{A}^{\iota}_{\sigma_k} = \begin{bmatrix} \mathcal{A}^{\iota}_{\sigma_k} + \mathcal{B}^{\iota}_{\sigma_k}\mathcal{C}^{\iota}_{\sigma_k} & \mathbf{A}^{\iota}_{\sigma_k 12} \\ \mathcal{Z}^T_1 \mathcal{C}^{\iota}_{\sigma_k}\left(I - \mathcal{A}^{\iota}_{\sigma_k} - \mathcal{B}^{\iota}_{\sigma_k}\mathcal{C}^{\iota}_{\sigma_k}\right) & \mathbf{A}^{\iota}_{\sigma_k 22} \end{bmatrix}, \; \mathbf{B}^{\iota}_{\sigma_k} = \begin{bmatrix} \mathcal{E}^{\iota}_{\sigma_k}\mathcal{Z}_1 & \mathcal{B}^{\iota}_{\sigma_k}\mathcal{Z}_1 \\ -\mathcal{Z}^T_1 \mathcal{C}^{\iota}_{\sigma_k}\mathcal{E}^{\iota}_{\sigma_k}\mathcal{Z}_1 & I \end{bmatrix},
$$

with

$$
\begin{aligned}
\mathcal{Z}_3 &= \left[I \; 0 \; -(\mathcal{M}^{\iota}_{\sigma_k})^T \right]^T, \\
\mathbf{A}^{\iota}_{\sigma_k 12} &= \mathcal{B}^{\iota}_{\sigma_k}\mathcal{Z}_1 \mathcal{H}^{\iota}_{\sigma_k}\mathcal{H}^{\iota\prime}_{\sigma_k} - \mathcal{B}^{\iota}_{\sigma_k}\mathcal{Z}_1\mathcal{H}^{\iota}_{\sigma_k}\left(\sum_{i=0}^{\bar{m}} \varrho_{ki}\mathfrak{T}_i\right)\mathcal{H}^{\iota\prime}_{\sigma_k}, \\
\mathbf{A}^{\iota}_{\sigma_k 22} &= \mathcal{H}^{\iota}_{\sigma_k}\mathcal{H}^{\iota\prime}_{\sigma_k} - \mathcal{H}^{\iota}_{\sigma_k}\left(\sum_{i=0}^{\bar{m}} \varrho_{ki}\mathfrak{T}_i\right)\mathcal{H}^{\iota\prime}_{\sigma_k},
\end{aligned}
$$

and

$$
\left[\begin{array}{c|c|c} \mathcal{A}^{\iota}_{\sigma_k} & \mathcal{B}^{\iota}_{\sigma_k} & \mathcal{Z}_1 \\ \hline \mathcal{C}^{\iota}_{\sigma_k} & \mathcal{E}^{\iota}_{\sigma_k} & \mathcal{Z}_2 \end{array}\right] = \left[\begin{array}{cc|cc|c} A^{\iota}_{\sigma_k} & 0 & 0 & 0 & I \\ 0 & A^{\iota}_{f\sigma_k} & B^{\iota}_{f\sigma_k} & 0 & 0 \\ \hline \mathcal{M}^{\iota}_{\sigma_k} & 0 & B^{\iota}_{\sigma_k} & 0 & 0 \\ 0 & C^{\iota}_{f\sigma_k} & 0 & 0 & I \end{array}\right].
$$

After the above manipulations, the phenomena induced by networked transmission were transformed into the switching of the system (7) and its stochastic parameter ϱ_{ki}. Next, we are going to make the following general assumption.

Assumption The sequences $\{\varrho_{ki}, i \in \{0, 1, \ldots, \bar{m}\}\}$ are independent from the initial state x_0 and the noises w_k and v_k. □

Now, with aid of the Definition 3 in [36], the FDI methodology addressed in this chapter can be concluded as: designing a set of FDI filters corresponding to h_k and the faults as the form of (6), such that the system (7) is stochastic stable. And the residual $r_{\iota k}$, $\iota \in L$ is sensitive to the fault $f_{\iota k}$ in certain frequency range ensuring by $\|G_{rf}\|^{E}_{\iota k-} > \beta$, and decoupled from the others $f_{\kappa k}$, $\kappa \in \{1, 2, \ldots, L\}$ and $\kappa \neq \iota$. In addition, $r_{\iota k}$ is also designed to the robust to d_k by achieving

$$
\|r_{\iota k}\|^{\mathbb{E}}_2 \leq \gamma^2 \|d_k\|^{\mathbb{E}}_2. \tag{8}
$$

3 Fault Detection and Isolation Filter Synthesis

To make convenience of the analysis, the system (7) is transformed into the following form firstly

$$\begin{aligned}\eta_{\iota k+1} &= \left(\mathscr{A}^{\iota}_{\sigma_k} + \sum_{i=0}^{\bar{m}}(\phi_i - \varrho_{ki})\tilde{\mathscr{A}}^{\iota}_{\sigma_k i}\right)\eta_{\iota k} + \mathscr{B}^{\iota}_{\sigma_k} d_k + \mathcal{Z}_3 \mathcal{T}^{\iota}_{\sigma_k}\sum_{\iota=1}^{L} F^{\iota}_{\sigma_k} f_{\iota k}, \\ r_{\iota k} &= \mathscr{C}^{\iota}_{\sigma_k}\eta_{\iota k},\end{aligned} \tag{9}$$

where $\tilde{\mathscr{A}}_{\sigma_k i} = \begin{bmatrix} 0 & \mathcal{B}^{\iota}_{\sigma_k}\mathcal{Z}_1\mathcal{H}^{\iota}_{\sigma_k}\mathfrak{T}_i\mathcal{H}^{\iota\prime}_{\sigma_k} \\ 0 & \mathcal{H}^{\iota}_{\sigma_k}\mathfrak{T}_i\mathcal{H}^{\iota\prime}_{\sigma_k}\end{bmatrix}$, $\mathscr{B}^{\iota}_{\sigma_k} = \mathbf{B}^{\iota}_{\sigma_k}$, $\mathscr{C}^{\iota}_{\sigma_k} = \mathbf{C}^{\iota}_{\sigma_k}$, and

$$\left[\mathscr{A}^{\iota}_{\sigma_k}\right] = \begin{bmatrix} \mathcal{A}^{\iota}_{\sigma_k} + \mathcal{B}^{\iota}_{\sigma_k}\mathcal{C}^{\iota}_{\sigma_k} & \mathscr{A}^{\iota}_{\sigma_k 12} \\ \mathcal{Z}_1^T\mathcal{C}^{\iota}_{\sigma_k}\left(I - \mathcal{A}^{\iota}_{\sigma_k} - \mathcal{B}^{\iota}_{\sigma_k}\mathcal{C}^{\iota}_{\sigma_k}\right) & \mathscr{A}^{\iota}_{\sigma_k 22}\end{bmatrix}$$

with $\mathscr{A}^{\iota}_{\sigma_k 12} = \mathcal{B}^{\iota}_{\sigma_k}\mathcal{Z}_1\mathcal{H}^{\iota}_{\sigma_k}\mathcal{H}^{\iota\prime}_{\sigma_k} - \mathcal{B}^{\iota}_{\sigma_k}\mathcal{Z}_1\mathcal{H}^{\iota}_{\sigma_k}\left(\sum_{i=0}^{\bar{m}}\phi_i\mathfrak{T}_i\right)\mathcal{H}^{\iota\prime}_{\sigma_k}$ and $\mathscr{A}^{\iota}_{\sigma_k 22} = \mathcal{H}^{\iota}_{\sigma_k}\mathcal{H}^{\iota\prime}_{\sigma_k} - \mathcal{H}^{\iota}_{\sigma_k}\left(\sum_{i=0}^{\bar{m}}\phi_i\mathfrak{T}_i\right)\mathcal{H}^{\iota\prime}_{\sigma_k}$.

In the following, the subscripts corresponded to σ_k and σ_{k+1} are respectively replaced by ι and $\jmath$.

Lemma 1 *For stochastic stabled system (9), if there exist matrices* $\mathcal{P}^{\iota}_{\iota} = (\mathcal{P}^{\iota}_{\iota})^T > 0$, $\mathcal{P}^{\iota}_{\jmath} = (\mathcal{P}^{\iota}_{\jmath})^T > 0$, $\mathcal{Q}^{\iota}_{\iota} = (\mathcal{Q}^{\iota}_{\iota})^T > 0$ *and f.u.o.s.* $\mathfrak{S}^{\iota\star}_{\iota} = \inf \underline{\mathcal{S}}\left(\sum_{\kappa=1,\kappa\neq\iota}^{L}\mathfrak{F}^{\kappa}_{\iota}\right)$, $\iota = 1, 2, \ldots, L$, *such that*

$$\mathfrak{S}^{\iota\star}_{\iota} \cap \mathfrak{F}^{\iota}_{\iota} = 0, \qquad \iota = 1, 2, \ldots, L, \tag{10}$$

and

$$\begin{bmatrix} \mathscr{A}^{\iota}_{\iota} & \mathscr{F}^{\iota}_{\iota} \\ \Psi_{\tilde{\mathscr{A}}^{\iota}_{\iota}} & 0 \\ I & 0 \end{bmatrix}^T \Xi^{\iota}_{\iota} \begin{bmatrix} \mathscr{A}^{\iota}_{\iota} & \mathscr{F}^{\iota}_{\iota} \\ \Psi_{\tilde{\mathscr{A}}^{\iota}_{\iota}} & 0 \\ I & 0 \end{bmatrix} + \begin{bmatrix} \mathscr{C}^{\iota}_{\iota} & 0 \\ 0 & 0 \\ 0 & I \end{bmatrix}^T \Pi \begin{bmatrix} \mathscr{C}^{\iota}_{\iota} & 0 \\ 0 & 0 \\ 0 & I \end{bmatrix} < 0 \tag{11}$$

holds $\forall \iota, \jmath \in \mathbb{H}$, *the residual* r_{ι} *will be with a finite frequency stochastic* $\mathcal{H}_-$ *index* β ($\beta > 0$) *to* $f_{\iota k}$ *when* $d_k = 0$, *and decoupled from* f_{κ}, $\kappa \in \{1, \ldots, L\}$, $\kappa \neq \iota$, *where* $\mathfrak{F}^{\kappa}_{\iota}$ *is the subspace containing* F^{κ}_{ι}, $\kappa \in \{1, 2, \ldots, L\}$, $\mathscr{F}^{\iota}_{\iota} = \mathcal{Z}_3\mathcal{T}^{\iota}_{\iota}F^{\iota}_{\iota}$, $\Psi_{\tilde{\mathscr{A}}^{\iota}_{\iota}} = \left[\alpha_0^2\tilde{\mathscr{A}}^{\iota T}_{0\iota}\ \alpha_1^2\tilde{\mathscr{A}}^{\iota T}_{1\iota}\ \cdots\ \alpha_{\bar{m}}^2\tilde{\mathscr{A}}^{\iota T}_{\bar{m}\iota}\right]^T$,

$$\Pi = \begin{bmatrix} -I & 0 & 0 \\ 0 & -\Psi'' & 0 \\ 0 & 0 & \beta^2 I \end{bmatrix}$$

and

(i) In LF range

$$\Xi_{\iota}^{\iota} = \begin{bmatrix} -\mathcal{P}_{J}^{\iota} & 0 & \mathcal{Q}_{\iota}^{\iota} \\ 0 & -\Psi_{\iota}^{\iota} & 0 \\ \mathcal{Q}_{\iota}^{\iota} & 0 & \mathcal{P}_{\iota}^{\iota} - 2\cos\vartheta_{l}\mathcal{Q}_{\iota}^{\iota} \end{bmatrix}. \tag{12}$$

(ii) In MF range

$$\Xi_{\iota}^{\iota} = \begin{bmatrix} -\mathcal{P}_{J}^{\iota} & 0 & e^{j\vartheta_c}\mathcal{Q}_{\iota}^{\iota} \\ 0 & -\Psi_{\iota}^{\iota} & 0 \\ e^{-j\vartheta_c}\mathcal{Q}_{\iota}^{\iota} & 0 & \mathcal{P}_{\iota}^{\iota} - 2\cos\vartheta_{w}\mathcal{Q}_{\iota}^{\iota} \end{bmatrix} \tag{13}$$

where $\vartheta_c = (\vartheta_2 + \vartheta_1)/2$, $\vartheta_w = (\vartheta_2 - \vartheta_1)/2$.

(iii) In HF range

$$\Xi_{\iota}^{\iota} = \begin{bmatrix} -\mathcal{P}_{J}^{\iota} & 0 & -\mathcal{Q}_{\iota}^{\iota} \\ 0 & -\Psi_{\iota}^{\iota} & 0 \\ -\mathcal{Q}_{\iota}^{\iota} & 0 & \mathcal{P}_{\iota}^{\iota} + 2\cos\vartheta_{h}\mathcal{Q}_{\iota}^{\iota} \end{bmatrix} \tag{14}$$

where $\Psi'' = \mathrm{diag}\left\{\alpha_0^2 I\ \alpha_1^2 I\ \cdots\ \alpha_{\bar{m}}^2 I\right\}$, $\Psi_{\iota}^{\iota} = \mathrm{diag}\left\{\alpha_0^2\mathcal{P}_{J}^{\iota}\ \alpha_1^2\mathcal{P}_{J}^{\iota}\ \cdots\ \alpha_{\bar{m}}^2\mathcal{P}_{J}^{\iota}\right\}$ *with* $\alpha_i^2 = \mathbf{E}\{(\varrho_{ki} - \phi_i)^2\} = \phi_i(1 - \phi_i)$.

Proof Since $\mathfrak{F}_{\iota}^{\kappa} \in \mathfrak{S}_{\iota}^{\iota\star}$ for $\kappa \neq \iota$, we can see $\mathcal{T}_{\iota}^{\iota}F_{\iota}^{\kappa} = 0$, $\kappa \neq \iota$, from (10). Then, we can rewrite the system (9) into

$$\begin{aligned} \eta_{\iota k+1} &= \left(\mathscr{A}_{\iota}^{\iota} + \sum_{i=0}^{\bar{m}}(\phi_i - \varrho_{ki})\tilde{\mathscr{A}}_{\iota i}^{\iota}\right)\eta_{\iota k} + \mathscr{B}_{\iota}^{\iota}d_k + \mathcal{Z}_3\mathcal{T}_{\iota}^{\iota}F_{\iota}^{\iota}f_{\iota k} \\ r_{\iota k} &= \mathscr{C}_{\iota}^{\iota}\eta_{\iota k}. \end{aligned} \tag{15}$$

From the above equations, it is easy to see the f_{κ}, $\kappa \in \{1, 2, \ldots, L\}$, $\kappa \neq \iota$ are decoupled from the ιth FDIF.

Next, we are going to prove the finite frequency stochastic $\mathcal{H}_-$ index to $f_{\iota k}$. We consider the LF case firstly for system (15), deriving from Euler's formula and by some calculations, the restriction of the system states in LF case can be written into

$$\sum_{k=0}^{\infty}\left(\eta_{\iota k}\eta_{\iota k+1}^T + \eta_{\iota k+1}\eta_{\iota k}^T - 2cos\vartheta_{w}^{\iota}\eta_{\iota k}\eta_{\iota k}^T\right) \geq 0. \tag{16}$$

Recalling the property $\mu^T\mathcal{Q}_{\iota}^{\iota}\nu = \mathbf{tr}(\nu\mu^T\mathcal{Q}_{\iota}^{\iota})$ for arbitrary vectors μ and ν, where $\mathbf{tr}(\cdot)$ denotes the trace of the matrix. Then we have

$$\begin{aligned} \mathfrak{J}_c(k) &= \mathbf{tr}\left\{\left(\eta_{\iota k}\eta_{\iota k+1}^T + \eta_{\iota k+1}\eta_{\iota k}^T - 2cos\vartheta_{w}^{\iota}\eta_{\iota k}\eta_{\iota k}^T\right)\mathcal{Q}_{\iota}^{\iota}\right\} \\ &= \eta_{\iota k+1}^T\mathcal{Q}_{\iota}^{\iota}\eta_{\iota k} + \eta_{\iota k}^T\mathcal{Q}_{\iota}^{\iota}\eta_{\iota k+1} - 2\cos\vartheta_{l}\eta_{\iota k}^T\mathcal{Q}_{\iota}^{\iota}\eta_{\iota k} \\ &= \zeta_{\iota k}^T\beth_1\zeta_{\iota k} \geq 0 \end{aligned} \tag{17}$$

in which $\zeta_{\iota k} = \left[\, \eta_{\iota k}^T \;\, f_{\iota k}^T \,\right]^T$ and

$$\beth_1 = \begin{bmatrix} \beth_1^{11} & \mathcal{Q}_\iota^\iota \mathcal{Z}_3 \mathcal{T}_\iota^\iota F_\iota^\iota \\ (\mathcal{Z}_3 \mathcal{T}_\iota^\iota F_\iota^\iota)^T \mathcal{Q}_\iota^\iota & 0 \end{bmatrix}$$

where $\beth_1^{11} = \left(\mathscr{A}_\iota^\iota + \sum_{i=0}^{\bar{m}}(\phi_i - \varrho_{ki})\tilde{\mathscr{A}}_{\iota i}^\iota\right)^T \mathcal{Q}_\iota^\iota + \mathcal{Q}_\iota^\iota\left(\mathscr{A}_\iota^\iota + \sum_{i=0}^{\bar{m}}(\phi_i - \varrho_{ki})\tilde{\mathscr{A}}_{\iota i}^\iota\right) - 2cos\vartheta_l \mathcal{Q}_\iota^\iota$.

To achieve our purpose, considering the index as follows,

$$\mathfrak{J}_{r,f} = \mathbf{E}\left\{\sum_{k=0}^{\infty}\left(-r_{\iota k}^T r_{\iota k} + \beta^2 f_{\iota k}^T f_{\iota k}\right)\right\}. \tag{18}$$

We define the functional $V_s = \eta_{\iota k}^T(k)\mathcal{P}_\iota^\iota \eta_{\iota k}$ and $\triangle V_s = \mathbf{E}(V_s) - V_s$. From the process to prove the stability of the system, i.e. Theorem 2, one can obtain $V_s > 0$ and $\triangle V_s < 0$ for $\mathcal{P}_\iota^\iota > 0$. Then, we have

$$\begin{aligned} \mathfrak{J}_{r,f} &\leq \mathbf{E}\left\{\sum_{k=0}^{\infty}\left(-r_{\iota k}^T r_{\iota k} + \beta^2 f_{\iota k}^T f_{\iota k} - \triangle V_s\right)\right\} \\ &\leq \mathbf{E}\left\{\sum_{k=0}^{\infty}\left(-r_{\iota k}^T r_{\iota k} + \beta^2 f_{\iota k}^T f_{\iota k} - \triangle V_s + V_s\right)\right\} \\ &= \mathbf{E}\left\{\sum_{k=0}^{\infty}\left(\zeta^T(k)(\beth_1 + \beth_2)\zeta(k)\right)\right\} \end{aligned} \tag{19}$$

in which

$$\beth_2 = \begin{bmatrix} \beth_2^{11} + \mathcal{P}_\iota^\iota - \mathscr{C}_\iota^{\iota *}\mathscr{C}_\iota^\iota & \beth_2^{12} \\ \beth_2^{21} & \beth_2^{22} \end{bmatrix}$$

where $\beth_2^{11} = -\left(\mathscr{A}_\iota^\iota + \sum_{i=0}^{\bar{m}}(\phi_i - \varrho_{ki})\tilde{\mathscr{A}}_{\iota i}^\iota\right)^T \mathcal{P}_J^\iota\left(\mathscr{A}_\iota^\iota + \sum_{i=0}^{\bar{m}}(\phi_i - \varrho_{ki})\tilde{\mathscr{A}}_{\iota i}^\iota\right)$, $\beth_2^{12} = -\left(\mathscr{A}_\iota^\iota + \sum_{i=0}^{\bar{m}}(\phi_i - \varrho_{ki})\tilde{\mathscr{A}}_{\iota i}^\iota\right)^T \mathcal{P}_J^\iota \mathcal{Z}_3 \mathcal{T}_\iota^\iota F_\iota^\iota$, $\beth_2^{21} = -(\mathcal{Z}_3 \mathcal{T}_\iota^\iota F_\iota^\iota)^T \mathcal{P}_J^\iota\left(\mathscr{A}_\iota^\iota + \sum_{i=0}^{\bar{m}}(\phi_i - \varrho_{ki})\tilde{\mathscr{A}}_{\iota i}^\iota\right)$ and $\beth_2^{22} = -(\mathcal{Z}_3 \mathcal{T}_\iota^\iota F_\iota^\iota)^T \mathcal{P}_J^\iota (\mathcal{Z}_3 \mathcal{T}_\iota^\iota F_\iota^\iota) + \beta^2 I$.

In addition, recalling the properties of $\mathbf{E}\left\{\sum_{i=0}^{\bar{m}}(\phi_i - \varrho_{ki})\right\} = 0$ and $\mathbf{E}\left\{(\phi_i - \varrho_{ki})^2\right\} = \phi_i(1 - \phi_i)$. It is easy to get (11) is equivalent to $\beth_2 + \beth_1 < 0$ with $\mathbf{E}\left\{(\phi_i - \varrho_{ki})^2\right\} = \alpha_i^2$, which yields to

$$\mathfrak{J}_{r,f} = \mathbf{E}\left\{\sum_{k=0}^{\infty}\left(-r_{\iota k}^T r_{\iota k} + \beta^2 f_{\iota k}^T f_{\iota k}\right)\right\} \leq 0 \tag{20}$$

which means $\|G_{rf}\|_{\iota k-}^E > \beta$.

Furthermore, the conclusions for MF and HF cases can be developed similarly. This proof is completed. □

Even though sufficient conditions make sure the desired performance and decoupling of the faults was developed in the above lemma. However, it is obvious that the conditions contain nonlinear items and cannot be solved directly. In the following, we will deal with this problem.

Theorem 1 *Assume the system (9) is stochastic stable, the residual $r_{\iota k}$ is decoupled from $f_{\kappa k}$, $\kappa \neq \iota$, and is with a finite frequency stochastic $\mathcal{H}_-$ index β to $f_{\iota k}$ when $d_k = 0$ if there exist f.u.o.s. $\mathfrak{S}_\iota^{\iota\star} = \inf \underline{\mathcal{Q}}(\sum_{\kappa=1,\kappa\neq\iota}^{L} \mathfrak{F}_\iota^\kappa)$, $\iota = 1, 2, \ldots, L$, matrices $\mathscr{X}_\iota^\iota$, $\mathscr{G}_\iota^\iota$, $\mathscr{Y}_\iota^\iota$, $\mathscr{N}_\iota^\iota$, $\mathscr{R}_\iota^\iota$, $\mathscr{A}_{f\iota}^\iota$, $\mathscr{B}_{f\iota}^\iota$, $\mathscr{C}_{f\iota}^\iota$ and $\mathcal{P}_\iota^{\iota T} = \mathcal{P}_\iota^\iota = \begin{bmatrix} \mathscr{P}_{\iota 11}^\iota & * & * \\ \mathscr{P}_{\iota 21}^\iota & \mathscr{P}_{\iota 22}^\iota & * \\ \mathscr{P}_{\iota 31}^\iota & \mathscr{P}_{\iota 32}^\iota & \mathscr{P}_{\iota 33}^\iota \end{bmatrix} > 0$, $\mathcal{P}_\jmath^{\iota T} =$ $\mathcal{P}_\jmath^\iota = \begin{bmatrix} \mathscr{P}_{\jmath 11}^\iota & * & * \\ \mathscr{P}_{\jmath 21}^\iota & \mathscr{P}_{\jmath 22}^\iota & * \\ \mathscr{P}_{\jmath 31}^\iota & \mathscr{P}_{\jmath 32}^\iota & \mathscr{P}_{\jmath 33}^\iota \end{bmatrix} > 0$, $\mathcal{Q}_\iota^{\iota T} = \mathcal{Q}_\iota^\iota = \begin{bmatrix} \mathscr{Q}_{\iota 11}^\iota & * & * \\ \mathscr{Q}_{\iota 21}^\iota & \mathscr{Q}_{\iota 22}^\iota & * \\ \mathscr{Q}_{\iota 31}^\iota & \mathscr{Q}_{\iota 32}^\iota & \mathscr{Q}_{\iota 33}^\iota \end{bmatrix} > 0$, such that (10) and*

$$\begin{bmatrix} -\mathcal{P}_\jmath^\iota & * & * & * & * \\ 0 & -\Psi_\iota^\iota & * & * & * \\ \Upsilon_{\iota 1}^\iota & 0 & \Upsilon_{\iota 4}^\iota & * & * \\ \Upsilon_{\iota 2}^\iota & \Upsilon_{\iota 3}^\iota & \Upsilon_{\iota 5}^\iota & \beta^2 I + \mathbf{He}(I_\upsilon^T \mathscr{X}_\iota^\iota T_\iota^\iota \mathcal{F}_\iota^\iota) & * \\ 0 & 0 & \Upsilon_{\iota 6}^\iota & 0 & -I \end{bmatrix} < 0 \tag{21}$$

hold $\forall \iota, \jmath \in \mathbb{H}$ for given scalars $\beta > 0$ and ϵ_i, $i = 1, \ldots, \bar{m}$, where

$$\Upsilon_{\iota 1}^\iota = \begin{bmatrix} \mathscr{Q}_{\iota 11}^\iota - \mathscr{X}_\iota^\iota & \mathscr{Q}_{\iota 21}^{\iota T} - \mathscr{Y}_\iota^\iota & \mathscr{Q}_{\iota 31}^{\iota T} \\ \mathscr{Q}_{\iota 21}^\iota - \mathscr{G}_\iota^\iota & \mathscr{Q}_{\iota 22}^\iota - \mathscr{Y}_\iota^\iota & \mathscr{Q}_{\iota 32}^{\iota T} \\ \mathscr{Q}_{\iota 31}^\iota & \mathscr{Q}_{\iota 32}^\iota & \mathscr{Q}_{\iota 33}^\iota - \mathscr{N}_\iota^\iota \end{bmatrix},$$

$$\Upsilon_{\iota 4}^\iota = \begin{bmatrix} \Upsilon_{\iota 4}^{\iota 11} & * & * \\ \Upsilon_{\iota 4}^{\iota 21} & \Upsilon_{\iota 4}^{\iota 22} & * \\ \Upsilon_{\iota 4}^{\iota 31} & \Upsilon_{\iota 4}^{\iota 32} & \Upsilon_{\iota 4}^{\iota 33} \end{bmatrix},$$

$$\Upsilon_{\iota 2}^\iota = \begin{bmatrix} -I_\upsilon^T \mathscr{X}_\iota^\iota & -I_\upsilon^T \mathscr{Y}_\iota^\iota & 0 \end{bmatrix},$$

$$\Upsilon_{\iota 3}^\iota = \begin{bmatrix} \epsilon_0^2 \Upsilon_{\iota 2}^\iota & \epsilon_1^2 \Upsilon_{\iota 2}^\iota & \cdots & \epsilon_{\bar{m}}^2 \Upsilon_{\iota 2}^\iota \end{bmatrix},$$

$$\Upsilon_{\iota 6}^\iota = \begin{bmatrix} 0 & \mathscr{R}_\iota^{\iota T} I_\upsilon^T & 0 \end{bmatrix},$$

$$\Upsilon_{\iota 5}^\iota = \begin{bmatrix} \Upsilon_{\iota 5}^{\iota'} & F_\iota^{\iota T} T_\iota^{\iota T} \mathscr{G}_\iota^{\iota T} + I_\upsilon^T \mathscr{A}_{f\iota}^\iota & \Upsilon_{\iota 5}^{\iota''} \end{bmatrix}$$

with

$$\begin{aligned}
\Upsilon_{\iota 4}^{\iota 11} &= \mathscr{P}_{\iota 11}^{\iota} - 2\cos\vartheta_l \mathscr{Q}_{\iota 11}^{\iota} + \mathbf{He}(\mathscr{X}_{\iota}^{\iota} A_{\iota}^{\iota} + \mathscr{B}_{f\iota}^{\iota} \mathcal{M}_{\iota}^{\iota}),\\
\Upsilon_{\iota 4}^{\iota 21} &= \mathscr{P}_{\iota 21}^{\iota} - 2\cos\vartheta_l \mathscr{Q}_{\iota 21}^{\iota} + \mathscr{G}_{\iota}^{\iota} A_{\iota}^{\iota} + \mathscr{B}_{f\iota}^{\iota} \mathcal{M}_{\iota}^{\iota} + \mathscr{A}_{f\iota}^{\iota T} + \mathscr{C}_{f\iota}^{\iota T} I_{\upsilon}^{T},\\
\Upsilon_{\iota 4}^{\iota 22} &= \mathscr{P}_{\iota 22}^{\iota} - 2\cos\vartheta_l \mathscr{Q}_{\iota 22}^{\iota} + \mathbf{He}(\mathscr{A}_{f\iota}^{\iota})\\
\Upsilon_{\iota 4}^{\iota 31} &= \mathscr{P}_{\iota 31}^{\iota} - 2\cos\vartheta_l \mathscr{Q}_{\iota 31}^{\iota} + \mathscr{N}_{\iota}^{\iota} \mathcal{M}_{\iota}^{\iota} - \mathscr{N}_{\iota}^{\iota} \mathcal{M}_{\iota}^{\iota} A_{\iota}^{\iota} + \mathcal{H}_{\sigma_k}^{\iota\prime T} \mathcal{H}_{\sigma_k}^{\iota T} \mathscr{B}_{f\iota}^{\iota T}\\
&\quad - \mathcal{H}_{\sigma_k}^{\iota\prime T} \sum_{i=0}^{\bar{m}} \phi_i \mathfrak{T}_i \mathcal{H}_{\sigma_k}^{\iota T} \mathscr{B}_{f\iota}^{\iota T},\\
\Upsilon_{\iota 4}^{\iota 32} &= \mathscr{P}_{\iota 32}^{\iota} - 2\cos\vartheta_l \mathscr{Q}_{\iota 32}^{\iota} + \mathcal{H}_{\sigma_k}^{\iota\prime T} \mathcal{H}_{\sigma_k}^{\iota T} \mathscr{B}_{f\iota}^{\iota T} - \mathcal{H}_{\sigma_k}^{\iota\prime T} \sum_{i=0}^{\bar{m}} \phi_i \mathfrak{T}_i \mathcal{H}_{\sigma_k}^{\iota T} \mathscr{B}_{f\iota}^{\iota T},\\
\Upsilon_{\iota 4}^{\iota 33} &= \mathscr{P}_{\iota 33}^{\iota} - 2\cos\vartheta_l \mathscr{Q}_{\iota 33}^{\iota} + \mathbf{He}(\mathscr{N}_{\iota}^{\iota} \mathcal{H}_{\sigma_k}^{\iota} \mathcal{H}_{\sigma_k}^{\iota\prime} - \mathscr{N}_{\iota}^{\iota} \mathcal{H}_{\sigma_k}^{\iota} \sum_{i=0}^{\bar{m}} \phi_i \mathfrak{T}_i \mathcal{H}_{\sigma_k}^{\iota\prime})\\
\Upsilon_{\iota 5}^{\iota\prime} &= F_{\iota}^{\iota T} \mathcal{T}_{\iota}^{\iota T} \mathscr{X}_{\iota}^{\iota T} + I_{\upsilon}^{T} \mathscr{X}_{\iota}^{\iota} A_{\iota}^{\iota} + I_{\upsilon}^{T} \mathscr{B}_{f\iota}^{\iota} \mathcal{M}_{\iota}^{\iota},\\
\Upsilon_{\iota 5}^{\iota\prime\prime} &= -F_{\iota}^{\iota T} \mathcal{T}_{\iota}^{\iota T} \mathcal{M}_{\iota}^{\iota T} \mathscr{N}_{\iota}^{\iota T} + I_{\upsilon}^{T} \mathscr{B}_{f\iota}^{\iota} \mathcal{H}_{\sigma_k}^{\iota} \mathcal{H}_{\sigma_k}^{\iota\prime} - I_{\upsilon}^{T} \mathscr{B}_{f\iota}^{\iota} \sum_{i=0}^{\bar{m}} \mathcal{H}_{\sigma_k}^{\iota} (\phi_i \mathfrak{T}_i\\
&\quad - \epsilon_i^2 \alpha_i^2 \mathfrak{T}_i) \mathcal{H}_{\sigma_k}^{\iota\prime}.
\end{aligned}$$

Proof It is seen that the decoupling of the fault $f_{\kappa k}$ form the ιth FDI filter, $\kappa \neq \iota$, has been proved in Lemma 1. Subsequently, we mainly focus on deriving the linear conditions to ensure the finite frequency sensitivity performance. As shown in Lemma 1, this performance in LF range can be captured by (11), which can be transformed into

$$\begin{bmatrix} \mathscr{A}_{\iota}^{\iota T} & \Psi_{\tilde{\mathscr{A}}_{\iota}^{\iota}}^{T} & I & 0 \\ \mathscr{F}_{\iota}^{\iota T} & 0 & 0 & I \end{bmatrix} \Xi_{\iota}^{\iota} \begin{bmatrix} \mathscr{A}_{\iota}^{\iota T} & \Psi_{\tilde{\mathscr{A}}_{\iota}^{\iota}}^{T} & I & 0 \\ \mathscr{F}_{\iota}^{\iota T} & 0 & 0 & I \end{bmatrix}^{T} < 0 \tag{22}$$

wherein

$$\Xi_{\iota}^{\iota} = \aleph_{\iota}^{\iota T} \begin{bmatrix} -\mathcal{P}_{\jmath}^{\iota} & 0 & \mathcal{Q}_{\iota}^{\iota} & 0 & 0 \\ 0 & -\Psi_{\iota}^{\iota} & 0 & 0 & 0 \\ \mathcal{Q}_{\iota}^{\iota} & 0 & \mathcal{P}_{\iota}^{\iota} - 2\cos\vartheta_l \mathcal{Q}_{\iota}^{\iota} & 0 & 0 \\ 0 & 0 & 0 & -I & 0 \\ 0 & 0 & 0 & 0 & \beta^2 I \end{bmatrix} \aleph_{\iota}^{\iota}$$

with $\aleph_{\iota}^{\iota} = \begin{bmatrix} I & 0 & 0 & 0 \\ 0 & I & 0 & 0 \\ 0 & 0 & I & 0 \\ 0 & 0 & \mathscr{C}_{\iota}^{\iota} & 0 \\ 0 & 0 & 0 & I \end{bmatrix}$.

By exploiting Finsler's Lemma [37] on it, one has (11) holds if and only if

$$\Xi_{\iota}^{\iota} + \mathbf{He}\left(\begin{bmatrix} -I & 0 \\ 0 & -I \\ \mathscr{A}_{\iota}^{\iota T} & \Psi_{\tilde{\mathscr{A}}_{\iota}^{\iota T}} \\ \mathscr{F}_{\iota}^{\iota T} & 0 \end{bmatrix}^{T} \begin{bmatrix} 0 & 0 \\ 0 & 0 \\ \mathcal{W}_{\iota}^{\iota} & 0 \\ I_{v}^{T}\mathcal{W}_{\iota}^{\iota} & \epsilon^{'} I_{v}^{T}\mathcal{W}_{\iota}^{\iota} \end{bmatrix}\right) < 0 \tag{23}$$

where $\mathcal{W}_{\iota}^{\iota}$ is the introduced slack matrix variable, $\epsilon^{'} = \begin{bmatrix} \epsilon_0^2 I & \epsilon_1^2 I & \cdots & \epsilon_{\bar{m}}^2 I \end{bmatrix}^T$. Sorting out the above inequality, we can see

$$\begin{bmatrix} -\mathcal{P}_{J}^{\iota} & * & * & * \\ 0 & -\Psi_{\iota}^{\iota} & * & * \\ \mathcal{Q}_{\iota}^{\iota} - \mathcal{W}_{\iota}^{\iota} & 0 & \mathfrak{El}_{33} & * \\ -I_{v}^{T}\mathcal{W}_{\iota}^{\iota} & -I_{v}^{T}\mathcal{W}_{\iota}^{\iota}\epsilon^{'T} & \mathfrak{El}_{43} & \mathfrak{El}_{44} \end{bmatrix} < 0 \tag{24}$$

which performs as a sufficient condition of (22), where

$$\begin{aligned} \mathfrak{El}_{33} &= \mathfrak{El}_{33}^{'} - \mathscr{C}_{\iota}^{\iota T}\mathscr{C}_{\iota}^{\iota}; \\ \mathfrak{El}_{43} &= \mathscr{F}_{\iota}^{\iota T}\mathcal{W}_{\iota}^{\iota T} + I_{v}^{T}\mathcal{W}_{\iota}^{\iota}\mathscr{A}_{\iota}^{\iota} + I_{v}^{T}\mathcal{W}_{\iota}^{\iota}\epsilon^{'T}\Psi_{\tilde{\mathscr{A}}_{\iota}^{\iota}}; \\ \mathfrak{El}_{44} &= \beta^2 I + \mathbf{He}(I_{v}^{T}\mathcal{W}_{\iota}^{\iota}\mathscr{F}_{\iota}^{\iota}), \end{aligned}$$

in which $\mathfrak{El}_{33}^{'} = \mathcal{P}_{\iota}^{\iota} - 2\cos\vartheta_{l}\mathcal{Q}_{\iota}^{\iota} + \mathbf{He}(\mathcal{W}_{\iota}^{\iota}\mathscr{A}_{\iota}^{\iota})$.

After the finished treatments, there have still been some coupled items $-\mathscr{C}_{\iota}^{\iota T}\mathscr{C}_{\iota}^{\iota}$. To tackle with this problem, we rewrite (24) into

$$\Delta_{\iota}^{\iota T}\Lambda_{\iota}^{\iota}\Delta_{\iota}^{\iota} < 0 \tag{25}$$

where $\Delta_{\iota}^{\iota T} = \begin{bmatrix} I & 0 & 0 & 0 & 0 \\ 0 & I & 0 & 0 & 0 \\ 0 & 0 & I & 0 & \mathscr{C}_{\iota}^{\iota T} \\ 0 & 0 & 0 & I & 0 \end{bmatrix}$ and

$$\Lambda_{\iota}^{\iota} = \begin{bmatrix} -\mathcal{P}_{J}^{\iota} & * & * & * & * \\ 0 & -\Psi_{\iota}^{\iota} & * & * & * \\ \mathcal{Q}_{\iota}^{\iota} - \mathcal{W}_{\iota}^{\iota} & 0 & \mathfrak{El}_{33}^{'} & * & * \\ -I_{v}^{T}\mathcal{W}_{\iota}^{\iota} & -I_{v}^{T}\mathcal{W}_{\iota}^{\iota}\epsilon^{'T} & \mathscr{F}_{\iota}^{\iota T}\mathcal{W}_{\iota}^{\iota T} + I_{v}^{T}\mathcal{W}_{\iota}^{\iota}\epsilon^{'T}\Psi_{\tilde{\mathscr{A}}_{\iota}^{\iota}} & \beta^2 I + \mathbf{He}(I_{v}^{T}\mathcal{W}_{\iota}^{\iota}\mathscr{F}_{\iota}^{\iota}) & * \\ 0 & 0 & 0 & 0 & -I \end{bmatrix}.$$

Performing Finsler's Lemma again, one has

$$\Lambda_{\iota}^{\iota} + \mathbf{He}\left(\begin{bmatrix} 0 & 0 & \mathscr{C}_{\iota}^{\iota} & 0 & -I \end{bmatrix}^{T} \begin{bmatrix} 0 & 0 & \mathscr{R}_{\iota}^{\iota T} & I_{v}^{T} & 0 & 0 \end{bmatrix}\right) < 0. \tag{26}$$

Restricting the form of $\mathcal{W}_\iota^\iota$ to be $\mathcal{W}_\iota^\iota = \begin{bmatrix} \mathscr{X}_\iota^\iota & \mathscr{Y}_\iota^\iota & 0 \\ \mathscr{G}_\iota^\iota & \mathscr{Y}_\iota^\iota & 0 \\ 0 & 0 & \mathscr{N}_\iota^\iota \end{bmatrix}$ and defining

$$\mathscr{A}_{f\iota}^\iota = \mathscr{Y}_\iota^\iota A_{f\iota}^\iota,\ \mathscr{B}_{f\iota}^\iota = \mathscr{Y}_\iota^\iota B_{f\iota}^\iota,\ \mathscr{C}_{f\iota} = \mathscr{R}_\iota^\iota C_{f\iota}^\iota, \tag{27}$$

(21) is achieved. By performing the same procedure, the conditions with respect to MF and HF domain could be developed, we do not explained here. The proof is completed. □

Theorem 2 *The system (9) is stochastic stable, and robust to d_k with level γ ($\gamma > 0$), i.e., (8) is ensured, if there exist matrices $\mathscr{X}_\iota^\iota$, $\mathscr{G}_\iota^\iota$, $\mathscr{Y}_\iota^\iota$, $\mathscr{N}_\iota^\iota$, $\mathscr{R}_\iota^\iota$, $\mathscr{A}_{f\iota}^\iota$, $\mathscr{B}_{f\iota}^\iota$, $\mathscr{C}_{f\iota}^\iota$ and*

$$\mathcal{P}_{d\iota}^{\iota T} = \mathcal{P}_{d\iota}^\iota = \begin{bmatrix} \mathscr{P}_{d\iota 11}^\iota & * & * \\ \mathscr{P}_{d\iota 21}^\iota & \mathscr{P}_{d\iota 22}^\iota & * \\ \mathscr{P}_{d\iota 31}^\iota & \mathscr{P}_{d\iota 32}^\iota & \mathscr{P}_{d\iota 33}^\iota \end{bmatrix} > 0,$$

$$\mathcal{P}_{dj}^{\iota T} = \mathcal{P}_{dj}^\iota = \begin{bmatrix} \mathscr{P}_{dj11}^\iota & * & * \\ \mathscr{P}_{dj21}^\iota & \mathscr{P}_{dj22}^\iota & * \\ \mathscr{P}_{dj31}^\iota & \mathscr{P}_{dj32}^\iota & \mathscr{P}_{dj33}^\iota \end{bmatrix} > 0,$$

such that

$$\begin{bmatrix} \Omega_\iota^{\iota 11} & * & * & * & * \\ 0 & \Omega_\iota^{\iota 22} & * & * & * \\ \Omega_\iota^{\iota 31} & \Omega_\iota^{\iota 32} & -\mathcal{P}_{d\iota}^\iota & * & * \\ \Omega_\iota^{\iota 41} & 0 & 0 & -\gamma^2 I & * \\ 0 & 0 & \Omega_\iota^{\iota 53} & 0 & I - \boldsymbol{He}(\mathscr{R}_\iota^\iota) \end{bmatrix} < 0 \tag{28}$$

hold $\forall \iota, j \in \mathbb{H}$ with zero initial conditions for given ν_i, $i = 1, \ldots, \bar{m}$, when $f_{\iota k} = 0$, where

$$\Omega_\iota^{\iota 11} = \begin{bmatrix} \mathscr{P}_{dj11}^\iota - \boldsymbol{He}(\mathscr{X}_\iota^\iota) & * & * \\ \mathscr{P}_{dj21}^\iota - \mathscr{Y}_\iota^{\iota T} - \mathscr{G}_\iota^\iota & \mathscr{P}_{dj22}^\iota - \boldsymbol{He}(\mathscr{Y}_\iota^\iota) & * \\ \mathscr{P}_{dj31}^\iota & \mathscr{P}_{dj32}^\iota & \mathscr{P}_{dj33}^\iota - \boldsymbol{He}(\mathscr{N}_\iota^\iota) \end{bmatrix},$$

$$\Omega_\iota^{\iota 22} = -\operatorname{diag}\left\{ \Omega_{\iota 0}^{\iota 22}\ \Omega_{\iota 1}^{\iota 22} \cdots \Omega_{\iota \bar{m}}^{\iota 22} \right\},$$

$$\Omega_\iota^{\iota 32} = \left[\alpha_0^2 \nu_0^2 \Omega_{\iota 0}^{\iota 32}\ \alpha_1^2 \nu_1^2 \Omega_{\iota 1}^{\iota 32} \cdots \alpha_{\bar{m}}^2 \nu_{\bar{m}}^2 \Omega_{\iota \bar{m}}^{\iota 32} \right],$$

$$\Omega_\iota^{\iota 31} = \begin{bmatrix} A_\iota^{\iota T} \mathscr{X}_\iota^{\iota T} + \mathcal{M}_\iota^{\iota T} \mathscr{B}_{f\iota}^T & A_\iota^{\iota T} \mathscr{G}_\iota^{\iota T} + \mathcal{M}_\iota^{\iota T} \mathscr{B}_{f\iota}^T & \mathcal{M}_\iota^{\iota T} \mathscr{N}_\iota^{\iota T} - A_\iota^{\iota T} \mathcal{M}_\iota^{\iota T} \mathscr{N}_\iota^{\iota T} \\ \mathscr{A}_{f\iota}^{\iota T} & \mathscr{A}_{f\iota}^{\iota T} & 0 \\ \mathscr{B}_{f\iota}^{\iota T} - \sum_{i=0}^{\bar{m}} \phi_i \Sigma_i \mathscr{B}_{f\iota}^{\iota T} & \mathscr{B}_{f\iota}^{\iota T} - \sum_{i=0}^{\bar{m}} \phi_i \Sigma_i \mathscr{B}_{f\iota}^{\iota T} & \mathscr{N}_\iota^{\iota T} - \sum_{i=0}^{\bar{m}} \phi_i \Sigma_i \mathscr{N}_\iota^{\iota T} \end{bmatrix},$$

$$\Omega_\iota^{\iota 41} = \begin{bmatrix} B_\iota^{\iota T} \mathscr{X}_\iota^{\iota T} & B_\iota^{\iota T} \mathscr{G}_\iota^{\iota T} & -B_\iota^{\iota T} \mathcal{M}_\iota^{\iota T} \mathscr{N}_\iota^{\iota T} \\ \mathscr{B}_{f\iota}^{\iota T} & \mathscr{B}_{f\iota}^{\iota T} & \mathscr{N}_\iota^{\iota T} \end{bmatrix},\quad \Omega_\iota^{\iota 53} = \left[0\ \mathscr{C}_{f\iota}^\iota\ 0 \right],$$

with

$$\Omega_{\iota i}^{\iota 22} = \begin{bmatrix} \alpha_i^2 \mathscr{P}_{dj11}^{\iota} - \nu_i^2 \boldsymbol{He}(\mathscr{X}_{\iota}^{\iota}) & * & * \\ \alpha_i^2 \mathscr{P}_{dj21}^{\iota} - \nu_i^2 (\mathscr{Y}_{\iota}^{\iota T} + \mathscr{G}_{\iota}^{\iota}) & \alpha_i^2 \mathscr{P}_{dj22}^{\iota} - \nu_i^2 \boldsymbol{He}(\mathscr{Y}_{\iota}^{\iota}) & * \\ \alpha_i^2 \mathscr{P}_{dj31}^{\iota} & \alpha_i^2 \mathscr{P}_{dj32}^{\iota} & \alpha_i^2 \mathscr{P}_{dj33}^{\iota} - \nu_i^2 \boldsymbol{He}(\mathscr{N}_{\iota}^{\iota}) \end{bmatrix}$$

$$\Omega_{\iota i}^{\iota 32} = \begin{bmatrix} 0 & 0 & 0 \\ 0 & 0 & 0 \\ \mathfrak{T}_i^T \mathscr{B}_{f\iota}^{\iota T} & \mathfrak{T}_i^T \mathscr{B}_{f\iota}^{\iota T} & \mathfrak{T}_i^T \mathscr{N}_{\iota}^{\iota T} \end{bmatrix}.$$

Proof We first prove the stochastic stability for (9).

Choosing the Lyapunov functional as

$$V_{\iota}(\eta_{\iota k}, k) = \eta_{\iota k}^T \mathcal{P}_{d\iota}^{\iota} \eta_{\iota k} \tag{29}$$

where $\mathcal{P}_{d\iota}^{\iota} > 0$.

Then, one has

$$\begin{aligned} \Delta V_{\iota}(\eta_{\iota k}, k) &= \mathbf{E}\big\{V_{\iota}(\eta_{\iota k+1}, k+1)|\eta_{\iota k}, k\big\} - V_{\iota}(\eta_{\iota k}, k) \\ &= \mathbf{E}\{\eta_{\iota k+1}^T \mathcal{P}_{dj}^{\iota} \eta_{\iota k+1}\} - \eta_{\iota k}^T \mathcal{P}_{d\iota}^{\iota} \eta_{\iota k}. \end{aligned} \tag{30}$$

Applying the mathematical expectation on it, one can derive

$$\begin{aligned} &\Delta V_{\iota}(\eta_{\iota k}, k) \\ &= \eta_{\iota k}^T \left(\mathscr{A}_{\iota}^{\iota T} \mathcal{P}_{dj}^{\iota} \mathscr{A}_{\iota}^{\iota} + 2\mathbf{E}\left\{ \sum_{i=0}^{\bar{m}} (\phi_i - \varrho_{ki}) \tilde{\mathscr{A}}_{\iota}^{\iota T} \mathcal{P}_{dj}^{\iota} \tilde{\mathscr{A}}_{\iota}^{\iota} \right\} \right. \\ &\quad \left. + \mathbf{E}\left\{ \sum_{i=0}^{\bar{m}} (\phi_i - \varrho_{ki})^2 \tilde{\mathscr{A}}_{\iota}^{\iota T} \mathcal{P}_{dj}^{\iota} \tilde{\mathscr{A}}_{\iota}^{\iota} \right\} - \mathcal{P}_{d\iota}^{\iota} \right) \eta_{\iota k}. \end{aligned} \tag{31}$$

On the other hand, it is not difficult to get $\mathbf{E}\big\{\sum_{i=0}^{\bar{m}}(\phi_i - \varrho_{ki})\big\} = 0$ and $\mathbf{E}\big\{(\phi_i - \varrho_{ki})^2\big\} = \phi_i(1-\phi_i)$. Together with $\phi_i(1-\phi_i) = \alpha_i^2$, (31) is equivalent to

$$\begin{aligned} &\Delta V_{\iota}(\eta_{\iota k}, k) = \eta_{\iota k}^T \mathscr{S} \eta_{\iota k} \\ &= \eta_{\iota k}^T \left(\mathscr{A}_{\iota}^{\iota T} \mathcal{P}_{dj}^{\iota} \mathscr{A}_{\iota}^{\iota} + \sum_{i=0}^{\bar{m}} (\alpha_i^2 \tilde{\mathscr{A}}_{\iota}^{\iota T} \mathcal{P}_{dj}^{\iota} \tilde{\mathscr{A}}_{\iota}^{\iota}) - \mathcal{P}_{d\iota}^{\iota} \right) \eta_{\iota k} \end{aligned} \tag{32}$$

where $\mathscr{S} = \mathscr{A}_{\iota}^{\iota T} \mathcal{P}_{dj}^{\iota} \mathscr{A}_{\iota}^{\iota} + \sum_{i=0}^{\bar{m}} (\alpha_i^2 \tilde{\mathscr{A}}_{\iota}^{\iota T} \mathcal{P}_{dj}^{\iota} \tilde{\mathscr{A}}_{\iota}^{\iota}) - \mathcal{P}_{d\iota}^{\iota}$.

From (28), one can see $\mathscr{S} < 0$ by utilizing Schur's complement lemma. Therefore, the system (9) is stochastic stable when $d_k = 0$ and $f_{\iota k} = 0$.

Next, we will consider the index as follows for $d_k \neq 0$ when $\eta(0) = 0$

$$\begin{aligned}\mathbf{J}(r_{\iota k}, d_k) &= \sum_{k=0}^{\infty}\Big\{\mathbf{E}\big(r_{\iota k}^T r_{\iota k} - \gamma^2 d_k^T d_k\big)\Big\} \\ &= \sum_{k=0}^{\infty}\Big\{\mathbf{E}\big(r_{\iota k}^T r_{\iota k} - \gamma^2 d_k^T d_k\big) + \delta V(\eta_{\iota k}, k)\Big\} \\ &\quad + \mathbf{E}\big\{V(\eta_{\iota 0}, 0)\big\} - \mathbf{E}\big\{V(\eta_{\iota\infty}, \infty)\big\} \\ &\leq \sum_{k=0}^{\infty}\Big\{\mathbf{E}\big(r_{\iota k}^T r_{\iota k} - \gamma^2 d_k^T d_k\big) + \delta V(\eta_{\iota k}, k)\Big\},\end{aligned} \tag{33}$$

one has

$$\begin{aligned}&\mathbf{E}\big(r_{\iota k}^T r_{\iota k} - \gamma^2 d_k^T d_k\big) + \delta V(\eta_{\iota k}, k) \\ &= \eta_{\iota k}^T\Big(\mathscr{A}_\iota^{\iota T}\mathcal{P}_{dj}^{\iota}\mathscr{A}_\iota^{\iota} + \sum_{i=0}^{\bar{m}}\big(\alpha_i^2\tilde{\mathscr{A}}_\iota^{\iota T}\mathcal{P}_{dj}^{\iota}\tilde{\mathscr{A}}_\iota^{\iota}\big) + \mathscr{C}_\iota^{\iota T}\mathscr{C}_\iota^{\iota} \\ &\quad - \mathcal{P}_{d\iota}^{\iota}\Big)\eta_{\iota k} + d_k^T\Big(\mathscr{B}_\iota^{\iota T}\mathcal{P}_{dj}^{\iota}\mathscr{B}_\iota^{\iota} - \gamma^2 I\Big)d_k \\ &= \xi_{\iota k}^T\mathfrak{H}_\iota^{\iota}\xi_{\iota k}\end{aligned} \tag{34}$$

where $\xi_{\iota k} = \begin{bmatrix}\eta_{\iota k}^T & d_k^T\end{bmatrix}^T$ and $\mathfrak{H}_\iota^{\iota} = \begin{bmatrix}\mathfrak{H}_{\iota 11}^{\iota} & * \\ \mathscr{B}_\iota^{\iota T}\mathcal{P}_{dj}^{\iota}\mathscr{A}_\iota^{\iota} & \mathscr{B}_\iota^{\iota T}\mathcal{P}_{dj}^{\iota}\mathscr{B}_\iota^{\iota} - \gamma^2 I\end{bmatrix}$ with $\mathfrak{H}_{\iota 11}^{\iota} = \mathscr{A}_\iota^{\iota T}\mathcal{P}_{dj}^{\iota}\mathscr{A}_\iota^{\iota} + \sum_{i=0}^{\bar{m}}\Big(\alpha_i^2\tilde{\mathscr{A}}_\iota^{\iota T}\mathcal{P}_{dj}^{\iota}\tilde{\mathscr{A}}_\iota^{\iota}\Big) - \mathcal{P}_{d\iota}^{\iota} + \mathscr{C}_\iota^{\iota T}\mathscr{C}_\iota^{\iota}$. Then by performing the similar procedure of the proof for Theorem 1, we have $\mathbf{E}\big(r_{\iota k}^T r_{\iota k} - \gamma^2 d_k^T d_k\big) + \delta V(\eta_{\iota k}, k) < 0$ from (28). Summing the above equality with respect to k from $k = 0$ to ∞, one has $\mathbf{J}(r_{\iota k}, d_k) < 0$, which is equal to (8). This proof is completed. □

Remark 3 The conditions in the above theorem are investigated without consideration of the frequency due to the fact that the disturbances may be appeared with uncertain frequency. If its range can be known in advance, it is not complex to develop the conditions in a finite frequency domain by following the line of the Lemma 1 and Theorem 1.

Till now, according to Theorems 1–2, the gains of the FDI filter can be calculated out through (26) with the feasible solutions of the following optimization problem

$$\begin{aligned}&\max \ \beta \\ &s.t. \ (16.10), (16.21), (16.28), \quad \iota, j \in \mathbb{N}\end{aligned} \tag{35}$$

for acceptable robustness level γ to the disturbance d_k.

Remark 4 For the sake of getting linear conditions, some scalars ϵ_i and ν_i need to be given beforehand in Theorems 1–2, which are utilized to offset the effectiveness of the parameters ϱ_{ki} on the system. They could be decided by the heuristic method. Without loss of generality, they can be chosen as $\epsilon_i^2 = \nu_i^2 = \phi_i(1 - \phi_i)$.

After getting $r_{\iota k}$, $\iota = 1, \ldots, L$, successfully, the next step is determining the evaluation function and making the judgement. In this chapter, the detection logic is chosen as

$$J_r^{\iota}(\tau) > J_{th}^{\iota} \Rightarrow f_{\iota k} \neq 0, \qquad \iota = 1, \ldots, L.$$

where the evaluation functions J_r^{ι} and the thresholds J_{th}^{ι} are determined by

$$J_r^{\iota}(\tau) = \sqrt{\frac{1}{\tau}\sum_{k=k_0}^{k_\tau} r_{\iota k}^T r_{\iota k}} \tag{36}$$

and

$$J_{th}^{\iota} = \sup_{d_k \in l_2, f_{\iota k}=0} J_{r_\iota}(\tau), \tag{37}$$

where $[k_0, k_\tau]$ implies the evaluation time window.

4 Simulation Example

In this section, a linearized longitudinal model of a VTOL aircraft will be applied to show the effectiveness of the obtained results.

The applied system borrowed from [38] is with the following form:

$$\begin{aligned}
\dot{x}_t &= \begin{bmatrix} -9.9477 & -0.7476 & 0.2632 & 5.0337 \\ 52.1659 & 2.7452 & 5.5532 & -24.4221 \\ 26.0922 & 2.6361 & -4.1975 & -19.2774 \\ 0 & 0 & 1 & 0 \end{bmatrix} x_t + \begin{bmatrix} 0 & 0 \\ 0 & 1 \\ 1 & 0 \\ 0 & 0 \end{bmatrix} w_t \\
&\quad + \begin{bmatrix} -2.035 \\ 0.4847 \\ 3.3051 \\ 0.8451 \end{bmatrix} f_{1t} + \begin{bmatrix} 0.1761 \\ 7.5922 \\ 4.4900 \\ 0 \end{bmatrix} f_{2t}, \\
y_t^1 &= \begin{bmatrix} 0 & 1 & 0 & 1 \end{bmatrix} x_t, \\
y_t^2 &= \begin{bmatrix} 1 & 0 & 1 & 1 \end{bmatrix} x_t,
\end{aligned}$$

wherein each element of the states denotes the horizontal velocity, vertical velocity, pitch rate and pitch angle, respectively. Also, $F_c^1 = \begin{bmatrix} -2.035 & 0.4847 & 3.3051 & 0.8451 \end{bmatrix}^T$ and $F_c^2 = \begin{bmatrix} 0.1761 & 7.5922 & 4.4900 & 0 \end{bmatrix}^T$ selecting as the first and third column of the input matrix denotes the fault signatures, therefore they can expresses the faults occurring on the first and third actuator, respectively.

Further, the two outputs are assumed to be transmitted into the networked channel by two nodes. It means, in company with the situation of packet dropout, the matrices $\mathfrak{T}_i$, $i = 0, 1, 2$ can be obtained as

Table 1 Comparison of β for different $\mathbb{H}$ and ϕ_i

	Rate-1	**Rate-2**	**Rate-3**	**Rate-4**
TR-1	$\beta = 2.3885$	$\beta = 2.3657$	$\beta = 2.3029$	Infeasible
TR-2	$\beta = 0.4789$	$\beta = 0.4682$	$\beta = 0.4091$	Infeasible

Table 2 β corresponded to different γ

C-1	**C-2**	**C-3**	**C-4**
$\gamma = 0.25$	$\gamma = 0.68$	$\gamma = 1$	$\gamma = 3$
$\beta = 0.6729$	$\beta = 1.1508$	$\beta = 2.3885$	$\beta = 8.0936$

$$\mathfrak{T}_0 = \begin{bmatrix} 0 & 0 \\ 0 & 0 \end{bmatrix}, \quad \mathfrak{T}_1 = \begin{bmatrix} 1 & 0 \\ 0 & 0 \end{bmatrix} \quad \mathfrak{T}_2 = \begin{bmatrix} 0 & 0 \\ 0 & 1 \end{bmatrix}.$$

Also, σ_k is assumed to be varying in the set $\mathbb{H} = \{1, 2, 3, 4\}$. In addition, the f.u.o.s. in Theorems 1–2 can be obtained as $\mathfrak{S}_l^{1\star} = \mathfrak{F}_2$ and $\mathfrak{S}_l^{2\star} = \mathfrak{F}_1$. Subsequently, the corresponded factor systems as form of (5) along with the derived subspaces can be calculated. Then, aiming at indicating the relationships between networked phenomena and the system performances, the optimization problem (35) is firstly solved for $\vartheta_l = 0$ and an admissible robust level $\gamma = 1$, and different $\mathbb{H}$ and ϕ_i, the optimal solutions of β are given in the following Table 1.

Where 'TR' represents 'transmission intervals' h_{σ_k} varying in $\mathbb{H}$, and

$$\left[\begin{array}{l} \textbf{TR-1}: \;\; h_1 = 1s;\, h_2 = 1.2s;\, h_3 = 2.5s;\, h_4 = 3s. \\ \hline \textbf{TR-2}: h_1 = 0.8s;\, h_2 = 2s;\, h_3 = 2.5s;\, h_4 = 3.5s. \end{array}\right]$$

Also, 'Rate' means 'accessing rate' ϕ_i, and

$$\left[\begin{array}{ll} \textbf{Rate-1}: & \phi_0 = 0.02;\; \phi_1 = 0.49;\; \phi_2 = 0.49; \\ \hline \textbf{Rate-2}: & \phi_0 = 0.29;\; \phi_1 = 0.36;\; \phi_2 = 0.35; \\ \hline \textbf{Rate-3}: & \phi_0 = 0.8;\; \phi_1 = 0.1;\; \phi_2 = 0.1; \\ \hline \textbf{Rate-4}: & \phi_0 = 1;\; \phi_1 = 0;\; \phi_2 = 0. \end{array}\right].$$

From the calculated results, it is easy to see that the performance index β is related to h_k and ϕ_i closely, i.e., for fixed h_k, lower ϕ_0 results in larger β, and for certain ϕ_0, longer h_k results in smaller β. It is to say, higher rate of packet dropout or longer transmission delays will lead to poor performance of fault detection.

Next, it is the step to validate the relationship between the robustness level γ and sensitiveness level β. Specifically, choosing **Rate-1** and **TR-1** to be a general case, various β can be gotten for different γ, which are presented in Table 2.

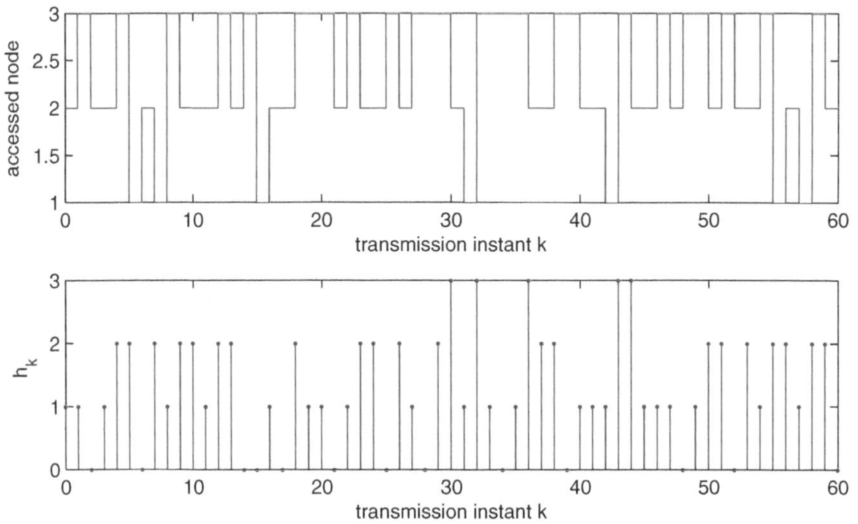

Fig. 3 The accessed node and h_k at each transmission instant in CASE-1

It is obviously from Table 2, that β is strongly influenced by γ, i.e., larger γ results in larger β, that implies better detection ability can be reached by some expense of robustness.

Without loss of generality, the simulation results will be performed in the following for the situation of **Rate-1**, **TR-1** and **C-3**. Additionally, the comparison with the existed FD method proposed in [39] will be done to illustrate the advantages of the developed FDI methodology. In [39], the authors utilized a weighting matrix to restrict the signal frequency, an $\mathcal{H}_\infty$ based FD scheme without consideration of fault isolation has been further addressed. For the purpose of fairy, the simulation results of [39] would be issued in the fault isolation framework of this chapter. Then, the simulation will be done under the two **Case**s as follows with the zero-initial condition. In addition, the disturbance w_{1k} and the channel noise v_k are chosen as bound-limited white noise with a power of 0.8 and 0.2, respectively. The disturbance w_{2k} is selected as $w_{2k} = 0.1\sin(0.2k)$ when $0s < t < 100$ s.

CASE-1: The transmitted node i and h_k are varying as shown in Fig. 3. In this CASE, the following two faulty modes will be simulated.

CASE-1-Fmode1: Stuck failure with value 1.5 acting upon the 1st actuator when $30s < t < 70s$, and the 3rd actuator works normally.

CASE-1-Fmode2: The 1st actuator is in failure-free mode, and the stuck failure with value 1.5 occurred on the 3rd actuator when $30s < t < 70s$.

Figures 4 and 5 has shown the generated residuals r_{i_k} and the related evaluation functions $J_r^i(\tau)$ for each fault case in CASE-1, respectively. In Fig. 4, we can see

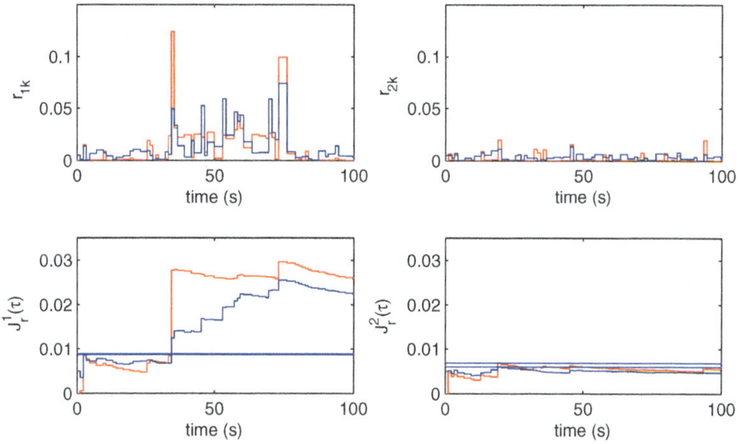

Fig. 4 The residuals and the corresponding evaluation function for *CASE-1-Fmode1 (red line—by our method, blue line—by existing method)*

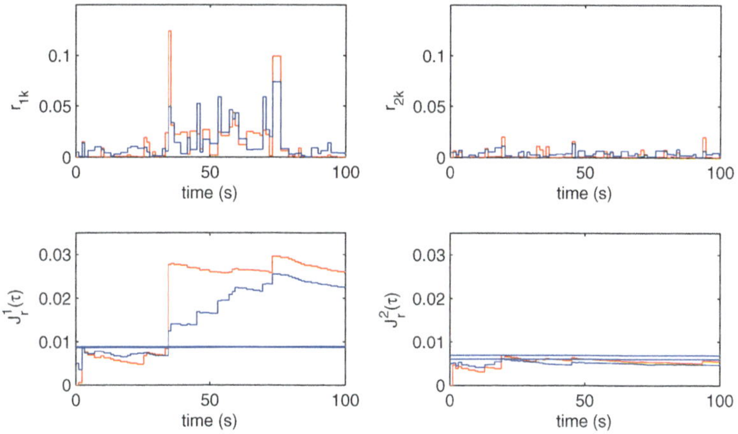

Fig. 5 The residuals and the corresponding evaluation function for *CASE-1-Fmode2 (red line—by our method, blue line—by existing method)*

$J_r^1(\tau) > J_{th}^1$ when $t = 34s$ by our scheme and when $t = 35s$ by existing one, and $J_r^2(\tau) < J_{th}^2$ over the running time. In addition, in Fig. 5, it is seen $J_r^1(\tau) < J_{th}^1$ over all the running time, and $J_r^2(\tau) > J_{th}^2$ when $t = 31s$ by our methodology and when $t = 34s$ by existing one. That is to say, by the developed geometric FDI scheme, the occurred failures could be isolated effectively and detected earlier.

CASE-2: The transmitted node i and h_k are varying as shown in Fig. 6. In this CASE, the following two faulty modes will be considered.

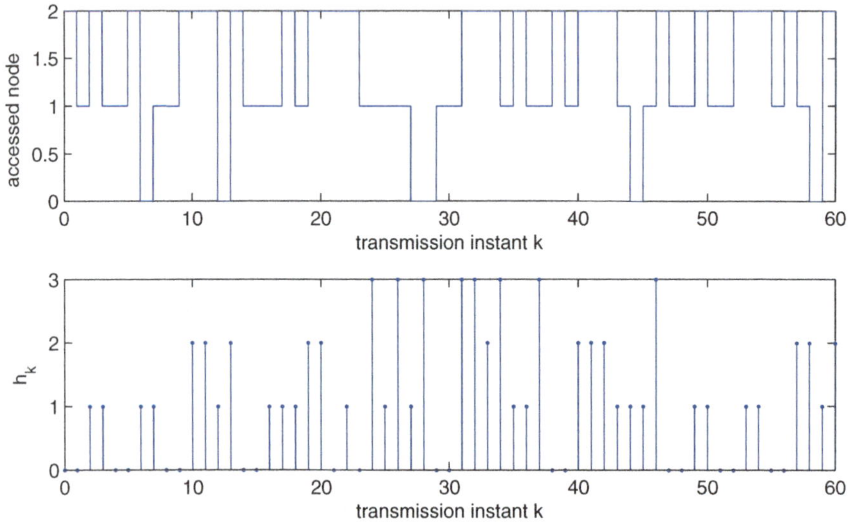

Fig. 6 The accessed node and h_k at each transmission instant in CASE-2

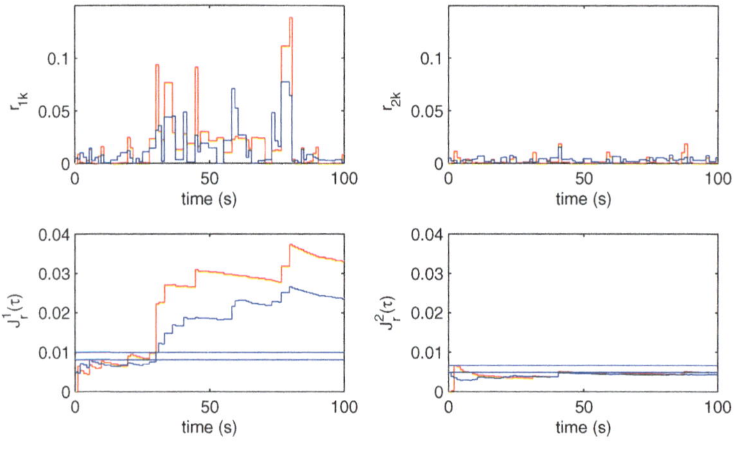

Fig. 7 The residuals and the corresponding evaluation function for *CASE-2-Fmode1 (red line—by our method, blue line—by existing method)*

CASE-2-Fmode1: Stuck failure with value 1.5 acting upon the 1st actuator when $30s < t < 70s$, and the 3rd actuator works normally.

CASE-2-Fmode2: The 1st actuator is in failure-free mode, and the stuck failure with value 1.5 occurred on the 3rd actuator when $30s < t < 70s$.

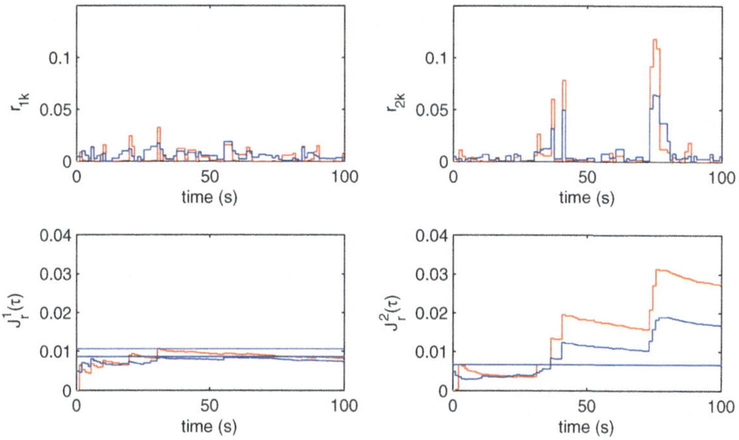

Fig. 8 The residuals and the corresponding evaluation function for *CASE-2-Fmode2 (red line—by our method, blue line—by existing method)*

Similarly, Figs. 7 and 8 showed the generated residual and its evaluation functions for the two fault cases in CASE-2. In Fig. 7, it is obvious $J_r^1(\tau) > J_{th}^1$ when $t = 31$ s by the proposed method and when $t = 31$ s by existing one, also $J_r^2(\tau) < J_{th}^2$ over all the running time. Further, from Fig. 8, we can seen $J_r^1(\tau) < J_{th}^1$ all the time and $J_r^2(\tau) > J_{th}^2$ when $t = 31$ s by our scheme and $t = 32$ s by existing one. The results in this CASE also showed the effectiveness of the FDI methodology investigated in this chapter.

5 Conclusions

This chapter has investigated a transmission-dependent fault detection and isolation methodology for constrained networked systems in a finite frequency domain. A switched system with multiple stochastic parameters was firstly derived to model the whole system. Then, with aid of the geometric approaches and the definitions of finite frequency performances, a novel FDI scheme has been proposed. Some sufficient conditions have been derived further to determine the FDI gains. Finally, the effectiveness and the advantages of the proposed methodology has been illustrated by an application to a VTOL aircraft.

Acknowledgements This work of J.H. Park was supported by the National Research Foundation of Korea (NRF) grant funded by the Korea government (Ministry of Science and ICT) (No. 2019R1A5A8080290 11).

References

1. Shi, D.W., Chen, T.W., Darouach, M.: Event-based state estimation of linear dynamics systems with unknown exogenous inputs. Automatica **69**, 275–288 (2016)
2. Qiu, J.B., Gao, H.J., Ding, S.X.: Recent advances on fuzzy-model-based nonlinear networked control systems: a survey. IEEE Trans. Ind. Electron. **63**(2), 1207–1217 (2016)
3. Li, H.Y., Wu, C.W., Shi, P., Gao, Y.B.: Control of nonlinear networked systems with packet dropouts: Interval Type-2 fuzzy model-based approach. IEEE Trans. Cybern. **45**(11), 2378–2389 (2015)
4. Lee, T.H., Xia, J., Park, J.H.: Networked control system with asynchronous samplings and quantizations in both transmission and receiving channels. Neurocomputing **237**, 25–38 (2017)
5. Mathiyalagan, K., Park, J.H., Sakthivel, R.: Finite-time boundedness and dissipativity analysis of networked cascade control systems. Nonlinear Dyn. **84**(4), 2149–2160 (2016)
6. Tang, Z., Park, J.H., Lee, T.H.: Dynamic output-feedback-based $\mathcal{H}_\infty$ design for networked control systems with multipath packet dropouts. Appl. Math. Comput. **275**, 121–133 (2016)
7. Park, J.H., Shen, H., Chang, X.H., Lee, T.H.: Recent Advances in Control and Filtering of Dynamic Systems with Constrained Signals. Springer, Cham (2018). https://doi.org/10.1007/978-3-319-96202-3
8. Zhang, W.A., Yu, L., Feng, G.: Stabilization of linear discrete-time networked control systems via protocol and controller co-design. Int. J. Robust Nonlinear Control **25**, 3072–3085 (2015)
9. Sun, X.M., Liu, K.Z., Wen, C.Y., Wang, W.: Predictive control of nonlinear continuous networked control systems with large time-varying transmission delays and transmission protocols. Automatica **64**, 76–85 (2016)
10. Zhang, D., Wang, Q.G., Yu, L., Shao, Q.K.: $\mathcal{H}_\infty$ filtering for networked systems with multiple time-varying transmissions and random packet dropouts. IEEE Trans. Ind. Inf. **9**(3), 1705–1716 (2013)
11. Che, W.W., Wang, J.L., Yang, G.H.: Quantised $\mathcal{H}_\infty$ filtering for networked systems with random sensor packet losses. IET Control Theory Appl. **4**(8), 1339–1352 (2010)
12. Antunes, D., Hespanha, J.P., Silvestre, C.: Stochastic networked control systems with dynamic protocols. Asian J. Control **17**(1), 99–110 (2015)
13. Qiu, L., Li, S.B., Xu, B.G., Xu, G.: $\mathcal{H}_\infty$ control of networked control systems based on Markov jump unified model. Int. J. Robust Nonlinear Control **25**, 2770–2786 (2015)
14. Pourbabaee, B., Meskin, N., Khorasani, K.: Sensor fault detection, isolation and identification using multiple-model-based hybrid kalman filter for gas turbine engines. IEEE Trans. Control Syst. Technol. **20**(4), 1184–1200 (2016)
15. Hwang, I., Kim, S., Kim, Y., Seah, C.E.: A survey of fault detection, isolation, and reconfiguration methods. IEEE Trans. Control Syst. Technol. **18**(3), 636–653 (2010)
16. Samy, I., Postlethwaite, I., Gu, D.W.: Survey and application of sensor fault detection and isolation schemes. Control Eng. Pract. **19**(7), 658–674 (2011)
17. Hassani, H., Zarei, J., Chadli, M., Qiu, J.B.: Unknown input observer design for interval Type-2 T-S fuzzy systems with immeasurable premise variables. IEEE Trans. Cybern. **47**(9), 2639–2650 (2017)
18. Su, X.J., Shi, P., Wu, L.G., Song, Y.D.: Fault detection filtering for nonlinear switched stochastic systems. IEEE Trans. Autom. Control **61**(5), 1310–1315 (2016)
19. Wang, Y.Q., Ding, S.X., Ye, H., Wang, G.Z.: A new fault detection scheme for networked control systems subject to uncertain time-varying delay. IEEE Trans. Signal Process. **56**(10), 5258–5268 (2008)
20. He, X., Wang, Z., Zhou, D.H.: Robust fault detection for networked systems with communication delay and data missing. Automatica **45**(11), 2634–2639 (2009)
21. Yan, H.C., Qian, F.F., Zhang, H., Yang, F.W., Guo, G.: $\mathcal{H}_\infty$ fault detection filtering for mechanical spring-mass systems over networked systems with incomplete information. IEEE Trans. Ind. Electron. **63**(9), 5622–5631 (2016)

22. Mao, Z.H., Jiang, B., Shi, P.: $\mathcal{H}_\infty$ fault detection filter design for networked control systems modelled by discrete Markovian jump systems. IET Control Theory Appl. **1**(5), 1336–1343 (2007)
23. Gao, H.J., Chen, T.W., Wang, L.: Robust fault detection with missing measurements. Int. J. Control **81**(5), 804–819 (2008)
24. Zhang, Y., Wang, Z., Zou, L., Liu, Z.X.: Fault detection filter design for networked multi-rate systems with fading measurements and randomly occurring faults. IET Control Theory Appl. **10**(5), 573–581 (2016)
25. Li, X.J., Yang, G.H.: Fault detection in finite frequency domain for Takagi-Sugeno fuzzy systems with sensor faults. Trans. Cybern. **44**(8), 1446–1458 (2014)
26. Wang, H., Yang, G.H.: Integrated fault detection and control for LPV systems. Int. J. Robust Nonlinear Control **19**(3), 341–363 (2009)
27. Yang, H.J., Xia, Y.Q., Liu, B.: Fault detection for T-S fuzzy discrete systems in finite-frequency domain. IEEE Trans. Syst. Man Cybern.-Part B: Cybern. **41**(4), 911–920 (2011)
28. Long, Y., Yang, G.H.: Fault detection for a class of networked control systems with finite-frequency servo inputs and random packet dropouts. IET Control Theory Appl. **6**(15), 2397–2408 (2012)
29. Heemels, W.P.M.H., Teel, A.R., van de Wouw, N., Něsić, D.: Networked control systems with communication constraints: tradeoffs between transmission intervals, delays and performance. IEEE Trans. Autom. Control **55**(8), 1781–1796 (2010)
30. Dačić, D., Nešić, D.: Quadratic stabilization of linear networked control systems via simulateneous protocol and controller design. Automatica **43**(7), 1145–1155 (2007)
31. Tabbara, M., Nešić, D.: Input-output stability of networked control systems with stochstic protocols and channels. IEEE Trans. Autom. Control **53**(5), 1160–1175 (2008)
32. Basile, G., Marro, G.: Controlled and Conditioned Invariant Subspaces in Linear System Theory. PrenticeHall, Upper Saddle River (1992)
33. Meskin, N., Khorasani, K.: Fault detection and isolation of distributed time-delay systems. IEEE Trans. Autom. Control **54**(11), 2680–2685 (2009)
34. Long, Y., Park, J.H., Ye, D.: Asynchronous fault detection and isolation for Markov jump systems with actuator failures under networked environment. IEEE Trans. Syst. Man Cybern.:Syst. (2019). https://doi.org/10.1109/TSMC.2019.2930995
35. Meskin N, Khorasani K (2009) Unobservability subspaces for continuous-time markovian jump systems with application to faule diagnosis. In: Proceeding of the 7th IFAC Symposium on Fault Detection, Supervision and Safety of Technical Process, pp. 953-958
36. Long, Y., Park, J.H., Ye, D.: A co-design methodology for constrained networked systems with frequency specifications. IEEE Trans. Control Syst. Technol. (2019). https://doi.org/10.1109/TCST.2019.2921942
37. Ding, D.W., Yang, G.H.: Fuzzy filter design for nonlinear systems in finite-frequency domain. IEEE Trans. Fuzzy Syst. **18**(5), 935–945 (2010)
38. Wang HB, Wang JL, Lam J (2004) An optimization approach for worst-case fault detection observer design. In: Proceedings of American Control Conference, Boston, MA, pp. 2475–2480
39. Wang, D., Wang, W., Shi, P.: Robust fault detection for switched linear systems with state delays. IEEE Trans. Syst. Man Cybern.-Part B: Cybern. **39**(3), 800–805 (2009)

Mean-Square Stochastic Stability of Delayed Hybrid Stochastic Inertial Neural Networks

R. Krishnasamy, A. Manivannan, and Raju K. George

Abstract This work considers the problem of mean-square stochastic stability analysis of hybrid stochastic inertial neural networks (INNs) with the effect of time-delays. Here, the hybrid stochastic INNs are represented as the combination of a two-level system in which the first level is directed by the system of second-order differential equations and the second level is directed by the discrete set representing the switch (jump) nodes. Initially, sufficient delay-dependent mean-square stability conditions for the hybrid INNs with time-delays are established using variable transformation techniques and some integral inequalities. This process is carried out mainly based on Lyapunov theory and linear matrix inequality technique. Some general cases on time-delay such as time-varying delay, interval time-varying delay, and mode-dependent delay are considered and corresponding results are established. Finally, numerical examples pertaining to the validation and effectiveness of the derived theoretical results are demonstrated.

Keywords Inertial neural networks · Lyapunov stability · Markovian jump parameters · Brownian motion · Linear matrix inequality

---

R. Krishnasamy (✉)
Department of Applied Mathematics and Computational Sciences, PSG College of Technology, Coimbatore, Tamil Nadu, India
e-mail: rksamy123@gmail.com

A. Manivannan
School of Advanced Sciences - Division of Mathematics, Vellore Institute of Technology-Chennai Campus, Chennai 600127, India
e-mail: manivannan.a@vit.ac.in

R. K. George
Department of Mathematics, Indian Institute of Space Science and Technology, Thiruvananthapuram, Kerala, India
e-mail: george@iist.ac.in

J. H. Park (ed.), *Recent Advances in Control Problems of Dynamical Systems and Networks*, Studies in Systems, Decision and Control 301,
https://doi.org/10.1007/978-3-030-49123-9_17

1 Introduction

One of the most important characteristics of any dynamical system is stability which enables us to analyze or predict how the system behaves as time increases. More insight can be gathered if the system is stable since it gives us the information about how the dynamics of the system depends on the initial data, that is continuous dependence on initial conditions. Also, in some cases, some other characteristics such as stabilizability, synchronization, state estimation, etc. can be applied to understand their behavior. In dynamical systems, there may exist delays or time-delays between the applied input and its response. These types of systems can be modeled with an idea of time-delay and they are called dynamical systems with time-delay or simply time-delay systems [1–5].

There are many factors that affects the stability of time-delay systems. Some of them are switching effects, impulses, parameter/system uncertainties, disturbances, etc. Above factors inspired many researchers to focus their investigation on the qualitative aspects of the dynamical systems, for example see [6–13] and therein.

The human brain consists of one billion neurons that are interconnected to themselves in a complex manner. They constitute a network called neural network (NN). In the same manner analogy to NN, one can design a network called artificial NN (ANN) in order to solve many complex problems occurring in real-life situations. ANNs can be viewed as a dynamical system and also on the application point of view, the dynamical behavior of ANNs is carried out. There are different types of ANNs which include Cohen-Grossberg NNs, bidirectional associative NNs, cellular NNs, etc.

Dynamical analysis for the above types of NNs are carried out extensively by numerous researchers due to its relevance in the real-life situations, for details see, [3, 8, 14–17] and references therein.

Switched or jump dynamical system is a class of dynamical system which consists of a finite number of subsystems and a finite number of switches or modes. Here, the switch from one mode to other is coordinated by a switching law and there are two types of switching namely arbitrary switching and constrained switching. Markovian switching is a type of arbitrary switching where the switching is taken to be random among the subsystems.

During switching, there will be associated a time-delay when the dynamics of the system changes from one subsystem to another subsystem and those type of time-delay is called mode-dependent time-delay, for details, see [5, 15, 18–23] and references therein.

During recent years, a special class of ANNs namely INNs got much attention due to its physical and biological relevance. The dynamics of INNs model is somewhat complicated related to the ordinary ANNs model due to the presence of second-order derivative term. Hence, there is an increasing interest among the research community to analyze the behavior of INNs and also the factors that affect the various dynamics. As mentioned before, the stability of INNs is also affected by numerous factors namely time-delays, impulses, uncertainties (which occur in the system parameters), external noise, internal disturbances (which occurs in system parameters), switching effects and so on, see [24–26]. Also, there are different kinds of time-delay namely discrete time-delay, continuous time-delay, neutral time-delay, mode-dependent time-delay, and these time-delays can be constant, time-varying (which may be continuous and either differentiable or non-differentiable), interval time-varying, proportional and some times random also, for more results, see [14, 24, 27–30].

Stability of INNs with these effects can be analyzed by transforming the INNs into a system of differential equations using a variable transformation technique. Here, INNs will be converted into a set of differential equations of first-order type. On the other hand, it can be handled without transforming the original system also, for details see Refs. [31, 32]. In both of these techniques, stability INNs can be carried out with the concepts of Lyapunov stability theory and linear matrix inequality (LMI) [6]. Here, the idea is to investigate the stability of the concerned without actually solving it. The problem of stability is converted into a convex optimization problem which can be solved by using the LMI technique, refer [6]. Besides Lyapunov technique, there are also other techniques to investigate stability namely Lagrangian method, M-matrix approach, etc.

Authors in [33] analyzed the stability for the problem of inertial Cohen-Grossberg NNs subject to the effects of Markovian jump parameters (MJPs) that are described by a continuous-time finite-state Markov chain. While the authors in [34] studied the synchronization of INNs subject to MJPs and implemented their results in image processing as an application. Authors in [35] obtained the reachable set estimation for inertial Markov jump BAM NN with bounded input disturbances in which transition rates are assumed to be partially unknown. Authors in [36] established stability results for INNs in which time-delay is assumed to be mode-dependent. Authors in [37] discussed the exponential stability of time-delay NNs in which both MJPs and Brownian motion are taken into account.

Also, construction of LKF and estimation of integral terms play a vital role in obtaining stability conditions of the system under consideration, for details see [38] and [39]. Hence, from the above discussions, it can be seen that the results are available for the stability of INNs pertaining to MJPs but the literature related to the problem of stability of INNs with both MJPs and Brownian motion is to be investigated which motivates this problem.

Through the above discussions, we can conclude that the problem of hybrid stochastic INNs with time-delay is not fully investigated in the literature. Thus, it is essential to analyze the stability behavior of INNs with the effects of disturbances and time-delay. Here, the special case of time-delay, that is mode-dependent time-delay is also considered to investigate the stability of INNs with time-delay and MJPs in the absence of disturbances.

Following are key factors taken into consideration while obtaining the stability results:

- Activation function is taken to be bounded and monotonic;

- Effect of intrinsic noise is considered with the help of Markovian jump parameters, extrinsic noise is analyzed with the help of stochastic calculus;
- Time delay and its derivatives are assumed to be bounded;
- Integral inequalities such as Jensen's and Wirtinger's inequality are applied to evaluate some integral terms.

Moreover, sufficient stability results are derived based on constructing a piecewise positive functional which incorporates the information about the bounds of time-delay. Manipulated results are stated in terms of LMIs that depend both on the time-delay, derivatives and also jump modes. Further, formulated conditions are examined through sufficient numerical examples.

The contents of this chapter are organized in the following manner. Section 2 gives the model description, and the lemmas, definitions, and assumptions which are used in the sequel are presented. Section 3 presents the corresponding stability conditions for the problem under study. Section 4 demonstrates the effectiveness of the derived results through numerical examples. Section 5 draws the conclusion.

Notations $\mathbb{R}^n$ and $\mathbb{R}^{n\times m}$ denote n-dimensional Euclidean space and the set of all $n \times m$ matrices with real entries respectively; $M \geq N$ $(M > N)$ represent that $M - N$ is positive-semidefinite (positive-definite) matrix, where M and N are symmetric matrices. A^T indicates the transpose of the matrix A and the symbol $*$ denotes the transposed elements in the symmetric matrix.; I is the identity matrix; $\|\cdot\|$ denotes Euclidean norm in $\mathbb{R}^n$. Matrices that are not defined explicitly are considered to have appropriate dimensions. $(\Omega, \mathcal{F}, \{\mathcal{F}_t\}_{t\geq 0}, \mathcal{P})$ denotes the probability space. $\{\mathcal{F}_t\}_{t\geq 0}$ fulfilling the usual conditions. $\mathcal{L}^2_{\mathcal{F}_0}((-\infty, 0]; \mathbb{R}^n)$ represents the collection of all $\mathcal{F}_0$-measurable. $\mathbb{E}(\cdot)$ denotes the mathematical expectation with respect to the given probability measure $\mathcal{P}$.

2 Preliminaries

INNs model incorporated with the effects of time-delay is described by a system of differential equations as

$$\begin{aligned}\ddot{m}_i(t) = &- a_i\dot{m}_i(t) - b_i m_i(t) + \sum_{j=1}^{n} c_{ij} f_j(m_j(t)) \\ &+ \sum_{j=1}^{n} d_{ij} f_j(m_j(t-\tau(t)) + I_i,\ i = 1, 2, \ldots, n,\end{aligned} \tag{1}$$

where $\ddot{m}_i(t)$ denotes the inertial term of the ith neuron at time t; $m_i(t)$ is the ith neuron's state at time t. $f_j(\cdot)$ denotes the activation function of jth neuron at time t and $f_j(0) = 0$, $j \in \{1, 2, \ldots, n\}$; c_{ij} and d_{ij} are the connection weights associated to the neurons with out and with time-delays respectively. a_i and b_i are positive constants. b_i is the rate of passive decay of the ith neuron. I_i represents the extraneous input on the ith neuron at time t.

Time-varying delay $\tau(t)$ satisfies

$$0 \leq \tau_1 \leq \tau(t) \leq \tau_2, \quad \dot{\tau}(t) \leq \mu < 1, \tag{2}$$

where τ_1, τ_2 and μ are positive constants. Initial condition of (1) is given by $m_i(s) = \phi_i(s)$ and $\frac{d}{dt}(m_i(s)) = \psi_i(s)$, for $-\tau_2 \leq s \leq 0$ where $\phi_i(s)$ and $\psi_i(s)$ are continuous and bounded.

Now, let

$$p_i(t) = \dot{m}_i(t) + \gamma_i m_i(t), \; i = 1, 2, \ldots, n. \tag{3}$$

Next, using the transformation (3), we transform system (1) into a set of first order differential equations and the transformed system is given as

$$\begin{cases} \dot{m}_i(t) = -\gamma_i m_i(t) + p_i(t), \\ \dot{p}_i(t) = -(a_i - \gamma_i)p_i(t) - (b_i - (a_i - \gamma_i)\gamma_i)m_i(t) + \sum\limits_{j=1}^{n} c_{ij} f_j(m_j(t)) \\ \qquad + \sum\limits_{j=1}^{n} d_{ij} f_j(m_j(t - \tau(t))) + I_i. \end{cases} \tag{4}$$

Initial conditions of (4) is given by $m_i(s) = \phi_i(s)$ and $p_i(s) = \psi_i(s) + \phi_i(s)$, for $-\tau_2 \leq s \leq 0$.

Let us shift the equilibrium point of (4) to the origin using the transformation $x(t) = m(t) - \bar{m}$ and $y(t) = p(t) - \bar{p}$, where $(\bar{m}, \bar{p})$ is the equilibrium point of (4). Hence, the resulting system in vector form is given as

$$\begin{cases} \dot{x}(t) = -\Gamma x(t) + y(t) \\ \dot{y}(t) = -Ay(t) - Bx(t) + Cg(x(t)) + Dg(x(t - \tau(t))), \end{cases} \tag{5}$$

where $x(s) = \phi(s) - m^*$ and $y(s) = \psi(s) + \phi(s) - p^*$, $s \in [-\tau_2, 0]$ denote the initial condition. Here, $x(t) \in \mathbb{R}^{n\times 1}$ and $y(t) \in \mathbb{R}^{n\times 1}$ are the state vectors of (5), $g(x(t)) = [g_1(x_1(t)), \ldots, g_n(x_n(t))]^T$ with $g(x(t)) = f(x(t) + m^*) - f(m^*)$, $A = diag\{(a_1 - \gamma_1), \ldots, (a_n - \gamma_n)\}$, $B = diag\{b_1 - (a_1 - \gamma_1)\gamma_1, \ldots, b_n - (a_n - \gamma_n)\gamma_n\}$, $C = (c_{ij})_{n\times n}$ and $D = (d_{ij})_{n\times n}$.

Dynamics of INNs (5) can be affected by the effects of noises which are classified into two types namely intrinsic and extrinsic noises. Extrinsic noise which occurs as an external disturbance is taken as the white noise since it gives constant power spectral density at different frequencies. Here, the white noise is specifically assumed to be Gaussian white noise.

Basically, the INN model represents the electronic circuit model with elements inductor, capacitor, and resistor. Further any electric circuit system may subject to disturbances which will spoil the nature of the system, refer [17]. Hence, the existence of the model considered in this work is guaranteed. Thus, taking these effects into account, system (5) can be written as

$$\begin{cases} dx(t) = [-\Gamma x(t) + y(t)]dt, \\ dy(t) = [-A_{r(t)}y(t) - B_{r(t)}x(t) + C_{r(t)}g(x(t)) + D_{r(t)}g(x(t-\tau(t)))]dt \\ \qquad +[E_{r(t)}x(t) + F_{r(t)}y(t)]\eta(t)dt. \end{cases} \tag{6}$$

Here, $\{r(t), t \geq 0\}$ is a Markov chain which is right-continuous on the probability space and takes value in a finite space $S = \{1, 2, \ldots, N\}$ with generator $\Pi = (\pi_{\bar{i}\bar{j}})_{N\times N}$ given by

$$Pr\{r(t+\delta t) = \bar{j}|r(t) = \bar{i}\} = \begin{cases} \pi_{\bar{i}\bar{j}}\delta t + o(\delta t), & if\ \bar{i} \neq \bar{j}, \\ 1 + \pi_{\bar{i}\bar{i}}\delta t + o(\delta t), & if\ \bar{i} = \bar{j}, \end{cases}$$

where $\delta t > 0$ and $\lim\limits_{\delta t\to 0} \frac{o(\delta t)}{\delta t} = 0$, $\pi_{\bar{i}\bar{j}} \geq 0$ represents the transition rate from mode $\bar{i}$ at time t to mode $\bar{j}$ at time $t + \delta t$ if $\bar{i} \neq \bar{j}$ and $\pi_{\bar{i}\bar{i}} = -\sum\limits_{\bar{i}\neq\bar{j}} \pi_{\bar{i}\bar{j}}$. $\eta(t)$ denotes the 1-dimensional white noise process.

Using the fact that the white noise process is the time derivative of a Brownian motion, we have $\eta(t)dt = d\omega(t)$, where $\omega(t)$ is an one-dimensional Brownian motion defined on $(\Omega, \mathcal{F}, \{\mathcal{F}_t\}_{t\geq 0}, \mathcal{P})$ which satisfies $\mathbb{E}[d\omega(t)] = 0$ and $\mathbb{E}[d\omega(t)]^2 = dt$.

Here, it is assumed that Markov chain $r(t)$ and Brownian motion $\omega(t)$ are independent of each other. For simplicity, denote $r(t) = k$, then system (6) is written as

$$\begin{cases} dx(t) = [-\Gamma x(t) + y(t)]dt, \\ dy(t) = g_{1k}(t)dt + g_{2k}d\omega(t), \end{cases} \tag{7}$$

where $g_{1k}(t) = -A_k y(t) - B_k x(t) + C_k g(x(t)) + D_k g(x(t-\tau(t)))$ and $g_{2k} = E_k x(t) + F_k y(t)$.

In order to prove our main results, we state an assumption, some lemmas and a definition.

Assumption 1 Activation function of the neuron $g_j(\cdot)$ is assumed to satisfy the following condition with $g_j(0) = 0$

$$l_j^- \leq \frac{g_j(u) - g_j(v)}{u - v} \leq l_j^+, \ \forall\, u, v \in \mathbb{R} \text{ and } u \neq v,$$

where $j = 1, 2, \ldots, n$, l_j^- and l_j^+ are some real scalars. □

Lemma 1 ([3]) *For any $Z_1, Z_2 \in \mathbb{R}^n$ and a diagonal matrix $Q > 0$ with compatible dimension, the following condition satisfies*

$$\pm 2Z_1^T Z_2 \leq Z_1^T Q^{-1} Z_1 + Z_2^T Q Z_2.$$

Lemma 2 ([3]) *For any positive-definite matrix M, any scalars a and b with $b > a$, and a vector-valued function $\dot{y}(t) : [a, b] \rightarrow \mathbb{R}^n$ such that the integrals in the following inequality are well defined, then*

$$\int_a^b \dot{y}^T(s) M \dot{y}(s) ds \geq (b-a) \Big[\int_a^b \dot{y}(s) ds \Big]^T M \Big[\int_a^b \dot{y}(s) ds \Big].$$

Lemma 3 ([38]) *For any positive matrix M, and vector function $\dot{x}(t) : [a, b] \rightarrow \mathbb{R}^n$, the following inequality holds*

$$\int_a^b \dot{x}^T(s) M \dot{x}(s) ds \geq \frac{1}{(b-a)} \Upsilon(\cdot) \begin{bmatrix} 4M & 2M & -6M \\ * & 4M & -6M \\ * & * & 12M \end{bmatrix} \Upsilon^T(\cdot),$$

where

$$\Upsilon(\cdot) = \begin{bmatrix} x^T(b) \; x^T(a) \; \frac{1}{(b-a)} \left(\int_a^b x(s) ds \right)^T \end{bmatrix}.$$

Definition 1 The zero solution of (7) is said to be mean-square asymptotically stable if for every system mode and all finite states $\phi(t)$ and $\psi(t)$ then it satisfy

$$\lim_{t \to \infty} \mathbb{E}(\|x(t)\|^2 + \|y(t)\|^2) = 0.$$

3 Stability Results

In this section, mean-square asymptotic stability of INNs with time-delay, disturbances and MJPs is analyzed.

If, in Assumption 1, $l_j^- = 0$ and $l_j^+ > 0$, then $g_j(\cdot)$ is said to be of Lurie type activation function.

Further, if $l_j^+ + l_j^- = 0$ for all $j = 1, 2, \ldots, n$, then Assumption 1 becomes

$$|g_j(u) - g_j(v)| \leq l_j |u - v|, \forall u, v \in \mathbb{R} \text{ and } u \neq v, \tag{8}$$

which is a Lipschitz type activation function.

This section presents stability results of system (7) corresponding to the above two cases.

Now, we present the first main result.

Theorem 1 *If the the time-delay $\tau(t)$ and activation function $g(\cdot)$ satisfy conditions (2) and (8) respectively, then system (7) is said to be mean-square asymptotically stable if there exist symmetric positive-definite matrices P_{1k}, P_{2k}, Q_1, Q_2, Q_3, R_1, positive diagonal matrices S_1, S_2 such that the following symmetric LMI holds for all $k \in S$:*

$$\begin{bmatrix} \Phi^k & M_{1k}^T & M_{2k}^T & M_{3k}^T \\ * & -S_1 & 0 & 0 \\ * & * & -S_2 & 0 \\ * & * & * & -P_{2k} \end{bmatrix} < 0, \tag{9}$$

where $\Phi^k = (\phi^k_{p,q})_{5n\times 5n}$, with

$$\begin{aligned}
\phi^k_{1,1} &= -P_{2k}A_k - A_k^T P_{2k} + \tau_{21}R_1 + \sum_{\bar{i}=1}^{N} \pi_{k\bar{i}} P_{2k},\\
\phi^k_{1,2} &= -P_{2k}B_k + P_{1k}\Gamma + \tau_{21}R_1,\\
\phi^k_{2,2} &= -2P_{1k}\Gamma + Q_1 + Q_2 + Q_3 + \tau_{21}R_1 + L^T S_1 L + \sum_{\bar{i}=1}^{N} \pi_{k\bar{i}} P_{1k},\\
\phi^k_{3,3} &= -Q_1 - \frac{1}{\tau_{21}}R_1, \quad \phi^k_{3,4} = \frac{1}{\tau_{21}}R_1, \quad \phi^k_{4,4} = -Q_2 - \frac{1}{\tau_{21}}R_1,\\
\phi^k_{5,5} &= -(1-\mu)Q_3 + L^T S_2 L, \quad M_{1k} = \left[C_k^T P_{2k}\, 0\, 0\, 0\, 0\right],\\
M_{2k} &= \left[D_k^T P_{2k}\, 0\, 0\, 0\, 0\right], \quad M_{3k} = \left[F_k\; E_k\; 0\, 0\, 0\right],\\
\tau_{21} &= \tau_2 - \tau_1,
\end{aligned}$$

and remaining entries in the symmetric block are zero.

Proof Consider the following LKF

$$\begin{aligned}
V(k, x(t), y(t)) =& x^T(t)P_{1k}x(t) + y^T(t)P_{2k}y(t) + \int_{t-\tau_1}^{t} x^T(s)Q_1x(s)ds\\
&+ \int_{t-\tau_2}^{t} x^T(s)Q_2x(s)ds + \int_{t-\tau(t)}^{t} x^T(s)Q_3x(s)ds\\
&+ \int_{-\tau_2}^{-\tau_1}\int_{t+\theta}^{t} \dot{x}^T(s)R_1\dot{x}(s)dsd\theta,
\end{aligned} \tag{10}$$

where P_{1k}, P_{2k}, Q_1, Q_2, Q_3 and R_1 are unknown symmetric positive-definite matrices.

Now, stochastic derivative of (10) along the state trajectory of (7) using Itô formula can be obtained as follows:

$$dV(k, x(t), y(t)) = \mathcal{L}V(k, x(t), y(t)) + 2y^T(t)P_{2k}g_{2k}(t)d\omega(t), \tag{11}$$

where $\mathcal{L}$ represents the diffusion operator and

$$\begin{aligned}\mathcal{L}V(k, x(t), y(t)) \leq & 2x^T(t)P_{1k}\dot{x}(t) + 2y^T(t)P_{2k}g_{1k}(t)\\ & + x^T(t)(Q_1 + Q_2 + Q_3)x(t)\\ & - x^T(t-\tau_1)Q_1x(t-\tau_1) - x^T(t-\tau_2)Q_2x(t-\tau_2)\\ & - (1-\mu)x^T(t-\tau(t))Q_3x(t-\tau(t)) + \tau_{21}\dot{x}^T(t)R_1\dot{x}(t)\\ & - \int_{t-\tau_2}^{t-\tau_1} \dot{x}^T(s)R_1\dot{x}(s)ds + x^T(t)\sum_{\bar{i}=1}^{N}\pi_{k\bar{i}}P_{1k}x(t)\\ & + y^T(t)\sum_{\bar{i}=1}^{N}\pi_{k\bar{i}}P_{2k}y(t) + trace(g_{2k}^T(t)P_{2k}g_{2k}(t)).\end{aligned} \tag{12}$$

From Lemma 2 and condition (8), we have

$$\begin{aligned}2y^T(t)P_{2k}C_kg(x(t)) &\leq y^T(t)P_{2k}C_kS_1^{-1}C_k^TP_{2k}y(t)\\ &\quad + x^T(t)L^TS_1Lx(t), \qquad (13)\\ 2y^T(t)P_{2k}D_kg(x(t-\tau(t))) &\leq y^T(t)P_{2k}CD_kS_2^{-1}D_k^TP_{2k}y(t)\\ &\quad + x^T(t-\tau(t))L^TS_2Lx(t-\tau(t)), \qquad (14)\end{aligned}$$

where S_1 and S_2 are positive-definite diagonal matrices and $L = diag\{l_1, l_2, \ldots, l_n\}$.

Then, Lemma 3 yields,

$$\begin{aligned}&\int_{t-\tau_2}^{t-\tau_1} \dot{x}^T(s)R_1\dot{x}(s)ds \geq\\ &\quad \frac{1}{\tau_{21}}\left[x^T(t-\tau_1)\; x^T(t-\tau_2)\right]\begin{bmatrix} R_1 & -R_1\\ * & R_1\end{bmatrix}\begin{bmatrix} x(t-\tau_1)\\ x(t-\tau_2)\end{bmatrix}.\end{aligned} \tag{15}$$

Last term of inequality (12) can be written as

$$g_{2k}^T(t)P_{2k}g_{2k}(t) = \left[y^T(t)\; x^T(t)\right]\begin{bmatrix} F_k^T\\ E_k^T\end{bmatrix}P_{2k}\left[F_k\; E_k\right]\begin{bmatrix} y(t)\\ x(t)\end{bmatrix}. \tag{16}$$

Combining (12)-(16), we have

$$\mathcal{L}V(k, x(t), y(t)) \leq \\ \xi^T(t)\left[\Phi^k + M_{1k}^T S_1^{-1} M_{1k} + M_{2k}^T S_2^{-1} M_{2k} + M_{3k}^T P_{2k}^{-1} M_{3k}\right]\xi(t), \tag{17}$$

where $\xi^T(t) = \left[y^T(t) \;\; x^T(t) \;\; x^T(t-\tau_1) \;\; x^T(t-\tau_2) \;\; x^T(t-\tau(t))\right]$.

Next, taking mathematical expectation on both sides of inequality (11) gives

$$\mathbb{E}dV(k, x(t), y(t)) \leq \mathcal{L}V(k, x(t), y(t)). \tag{18}$$

Then, according to Lyapunov stability theorem, for the system (7) to be asymptotically stable in the mean-square, we must have $\mathbb{E}dV(k, x(t), y(t)) < 0$ for all non-zero $x(t)$ and $y(t)$.

Thus, $\mathbb{E}dV(k, x(t), y(t)) < 0$ implies $\mathcal{L}V(k, x(t), y(t)) < 0$ which is true only if

$$\Phi^k + M_{1k}^T S_1^{-1} M_{1k} + M_{2k}^T S_2^{-1} M_{2k} + M_{3k}^T P_{2k}^{-1} M_{3k} < 0. \tag{19}$$

Since inequality (19) is non-linear, apply Schur complement lemma [3] which gives inequality (9).

Following the similar proof of Theorem 1 in [37], we can obtain $\lim_{t\to\infty} \mathbb{E}(\|x(t)\|^2 + \|y(t)\|^2) = 0$. Thus, according to Definition 1, system (7) is mean-square asymptotically stable which completes the proof of this theorem. □

Remark 1 Construction of LKF and the inequalities which we use to estimate the bound of integral terms play a significant role in obtaining conservative results on time-delay. In general, Wirtinger's inequality is preferable over Jensen's inequality to estimate the bound of an integral term in the problem of stability analysis of dynamical time-delay systems so as to obtain less conservative results, for reference see [39]. This is obvious from the existing literature related to the qualitative behavior of dynamical systems. Taking this factor into account, Wirtinger's inequality is used to estimate the bound of integral term $\int_{t-\tau_2}^{t-\tau_1} \dot{x}^T(s) R_1 \dot{x}(s) ds$. Thus, according to Lemma 3, we have

$$-\int_{t-\tau_2}^{t-\tau_1} \dot{x}^T(s) R_1 \dot{x}(s) ds \leq -\frac{1}{\tau_{21}}\left[x^T(t-\tau_2)\; x^T(t-\tau_1)\; \frac{1}{\tau_{21}}\left(\int_{t-\tau_2}^{t-\tau_1} x(s)ds\right)^T\right] \\ \times \begin{bmatrix} 4M & 2M & -6M \\ * & 4M & -6M \\ * & * & 12M \end{bmatrix} \begin{bmatrix} x(t-\tau_2) \\ x(t-\tau_1) \\ \frac{1}{\tau_{21}}\int_{t-\tau_2}^{t-\tau_1} x(s)ds \end{bmatrix}. \tag{20}$$

This effects to the following changes:

$$\bar{\xi}^T(t) = \Big[y^T(t) \;\; x^T(t) \;\; x^T(t-\tau_1) \;\; x^T(t-\tau_2) \;\; x^T(t-\tau(t)) \;\; \Big(\int_{t-\tau_2}^{t-\tau_1} x(s)ds \Big)^T \Big],$$

and

$$\bar{M}_{1k} = \big[C_k^T P_{2k}\; 0\,0\,0\,0\,0\big], \quad \bar{M}_{2k} = \big[D_k^T P_{2k}\; 0\,0\,0\,0\,0\big], \quad \bar{M}_{3k} = [F_k \; E_k \; 0\,0\,0\,0].$$

It is noted that the corresponding results are summarized in the following corollary. These results can be directly obtained from Theorem 1 and hence it is omitted.

Corollary 1 *Suppose conditions (2) and (8) hold, system (7) is mean-square asymptotically stable if there exist symmetric positive-definite matrices* P_{1k}*,* P_{2k}*,* Q_1*,* Q_2*,* Q_3*,* R_1*, diagonal matrices* S_1*,* S_2 *such that symmetric LMI holds for all* $k \in S$*:*

$$\begin{bmatrix} \bar{\Phi}^k & \bar{M}_{1k}^T & \bar{M}_{2k}^T & \bar{M}_{3k}^T \\ * & -S_1 & 0 & 0 \\ * & * & -S_2 & 0 \\ * & * & * & -P_{2k} \end{bmatrix} < 0,$$

where $\bar{\Phi}^k = (\bar{\phi}^k_{p,q})_{6n\times 6n}$*,*

$$\bar{\phi}^k_{1,1} = \phi^k_{1,1}, \quad \bar{\phi}^k_{1,2} = \phi^k_{1,2}, \quad \bar{\phi}^k_{2,2} = \phi^k_{2,2}, \quad \bar{\phi}^k_{3,3} = -Q_1 - \frac{4}{\tau_{21}}R_1,$$
$$\bar{\phi}^k_{3,4} = -\frac{2}{\tau_{21}}R_1, \quad \bar{\phi}^k_{3,6} = \bar{\phi}^k_{4,6} = \frac{6}{\tau_{21}^2}R_1, \quad \bar{\phi}^k_{4,4} = -Q_2 - \frac{4}{\tau_{21}}R_1,$$
$$\bar{\phi}^k_{5,5} = \phi^k_{5,5}, \quad \bar{\phi}^k_{6,6} = -\frac{12}{\tau_{21}^3}R_1,$$

and all other entries are zero.

Remark 2 If the activation function is assumed to satisfy Assumption 1, then the corresponding asymptotic stability condition of system (7) can be obtained with construction of the following LKF

$$\hat{V}(k, x(t), y(t)) = V(k, x(t), y(t)) + 2\sum_{i=1}^{n}\left[\lambda_{1i}\int_0^{x_i(s)} \big(g_i(s) - l_i^- s\big)\, ds\right] + 2\sum_{i=1}^{n}\left[\lambda_{2i}\int_0^{x_i(s)} \big(l_i^+ s - g_i(s)\big)\, ds\right]. \tag{21}$$

Now, according to Assumption, there exist positive diagonal matrices T_1 and T_2 such that

$$-2\{g^T(x(t))T_1g(x(t)) - x^T(t)(L_1 + L_2)g(x(t)) + x^T(t)L_1T_1L_2x(t)\} \geq 0, \tag{22}$$

and

$$\begin{aligned} &-2\{g^T(x(t-\tau(t)))T_2g(x(t-\tau(t))) - x^T(t-\tau(t))(L_1 + L_2)g(x(t-\tau(t))\\ &\quad + x^T(t-\tau(t))L_1T_2L_2x(t-\tau(t))\} \geq 0. \end{aligned} \tag{23}$$

Following theorem provides asymptotic mean-square stability conditions under above situations.

Theorem 2 *Suppose condition (2) and Assumption 1 hold, system (7) is said to be mean-square asymptotically stable if there exist symmetric positive-definite matrices* P_{1k}, P_{2k}, Q_1 Q_2, Q_3, R_1, *diagonal matrices* Λ_1, Λ_2 *then the following symmetric LMI satisfies for all* $k \in S$:

$$\hat{\Phi}^k < 0,$$

where $\hat{\Phi}^k = (\hat{\phi}^k_{p,q})_{7n\times 7n}$,

$$\begin{aligned} &\hat{\phi}^k_{1,1} = \phi^k_{1,1},\quad \hat{\phi}^k_{1,2} = \phi^k_{1,2} + \Lambda_2L_2,\quad \hat{\phi}^k_{1,5} = P_{2k}C_k,\quad \hat{\phi}^k_{1,6} = P_{2k}D_k + \Lambda_1 - \Lambda_2,\\ &\hat{\phi}^k_{2,2} = -2P_{1k}\Gamma + Q_1 + Q_2 + Q_3 + \tau_{21}R_1 + 2L_1^T\Lambda_1 - 2L_2^T\Lambda_2 - 2L_1T_1L_2,\\ &\hat{\phi}^k_{2,6} = -\Lambda_1 + \Lambda_2 + L_1 + L_2,\quad \hat{\phi}^k_{3,3} = \phi^k_{3,3},\quad \hat{\phi}^k_{3,4} = \phi^k_{3,4},\quad \hat{\phi}^k_{4,4} = \phi^k_{4,4},\\ &\hat{\phi}^k_{5,5} = -(1-\mu)Q_3 + L_1 + L_2 - L_1T_2L_2,\quad \hat{\phi}^k_{6,6} = -2T_1,\quad \hat{\phi}^k_{7,7} = -2T_2 \end{aligned}$$

and all other entries are zero.

Proof This theorem follows directly from Theorem 1 with the help of (21)-(23). Hence, the detail is omitted. □

Here, we have another result that follows directly from Theorem 2.

Corollary 2 *Suppose conditions (2) and (20), and Assumption 1 hold, system (7) is said to be mean-square asymptotically stable if there exist positive-definite symmetric matrices* P_{1k}, P_{2k}, Q_1 Q_2, Q_3, R_1, *diagonal matrices* Λ_1, Λ_2 *such that the following symmetric LMI holds for all* $k \in S$:

$$\tilde{\Phi}^k < 0,$$

where $\tilde{\Phi}^k = (\tilde{\phi}^k_{p,q})_{8n\times 8n}$,

$$
\begin{aligned}
\tilde{\phi}^k_{1,1} =& \hat{\phi}^k_{1,1}, \quad \tilde{\phi}^k_{1,2} = \hat{\phi}^k_{1,2}, \quad \tilde{\phi}^k_{1,5} = \hat{\phi}^k_{1,5}, \quad \tilde{\phi}^k_{1,6} = \hat{\phi}^k_{1,6}, \\
\tilde{\phi}^k_{2,2} =& \hat{\phi}^k_{2,2}, \quad \tilde{\phi}^k_{2,6} = \hat{\phi}^k_{2,6}, \quad \tilde{\phi}^k_{3,3} = \bar{\phi}^k_{3,3}, \quad \tilde{\phi}^k_{3,4} = \bar{\phi}^k_{3,4}, \\
\tilde{\phi}^k_{3,8} =& \frac{6}{\tau^2_{21}} R_1, \quad \tilde{\phi}^k_{4,4} = \bar{\phi}^k_{4,4}, \quad \tilde{\phi}^k_{4,8} = \frac{6}{\tau^2_{21}} R_1, \tilde{\phi}^k_{5,5} = \hat{\phi}^k_{5,5}, \\
\tilde{\phi}^k_{6,6} =& \hat{\phi}^k_{6,6}, \quad \tilde{\phi}^k_{7,7} = \hat{\phi}^k_{7,7}, \quad \tilde{\phi}^k_{8,8} = -\frac{12}{\tau^3_{21}} R_1
\end{aligned}
$$

and all other entries are zero.

Remark 3 In this work, the activation function is assumed to be bounded and monotone, and a special case namely Lipschitz continuity is also assumed while deriving the sufficient conditions of the concerned system. Required delay-dependent mean-square asymptotic conditions are given for each of these cases separately.

Remark 4 The problem considered in this work is distinct from those in the existing literature in the following aspects:

(i) here both types of disturbances namely internal and external types are taken into account whereas in the existing literature either external or internal type, i.e., INNs with Brownian motion or MJPs are considered;

(ii) two types of assumptions are made on the activation functions and the conditions for each case are summarized separately.

Remark 5 Letting $g_{2k} = 0$, $\Gamma = I$ and $\tau(t) = \tau_k(t)$, the system (7) changes to the following form

$$
\begin{cases}
dx(t) = [-x(t) + y(t)]dt, \\
dy(t) = -A_k y(t) - B_k x(t) + C_k g(x(t)) + D_k g(x(t - \tau_k(t)))dt.
\end{cases} \tag{24}
$$

$\tau_k(t)$ is the mode-dependent time-varying delay which satisfies $0 \leq \tau_k(t) \leq \tau_{2k}$, $\dot{\tau}_k(t) \leq \mu_k < 1$, $\forall\, k \in S$, where τ_{2k} and μ_k are constants for any $k \in S$.

Corresponding stability results for the above system is summarized in [36].

Remark 6 In general, the presence of more cross-terms and appropriate manipulation of integral inequalities will reduce the conservatism on time-delay. Cross terms can be increased through the usage of free-weighting matrix technique. This involves zero equations with some free weighting matrices. Delay-dependent stability conditions summarized in Sect. 3 can be still made less conservative with the above approach. Even though this method will give less conservative results, on the other hand, it will automatically increase the computation complexity also. Hence, in this work, we focus our attention towards Wiritnger's inequality technique to obtain less conservative results.

4 Numerical Example

In this section, numerical examples of stochastic INNs with MJPs is demonstrated to show the effectiveness and reliability of the obtained theoretical results.

Sufficient LMI conditions obtained in Sect. 3 are verified through MATLAB LMI solvers.

Example 1 Consider system (7) with two jump modes whose parameters are described as follows

Mode 1

$$A_1 = \begin{bmatrix} 5 & 0 & 0 \\ 0 & 5.5 & 0 \\ 0 & 0 & 7 \end{bmatrix}, \quad B_1 = \begin{bmatrix} 0.2 & 0 & 0 \\ 0 & 0.9 & 0 \\ 0 & 0 & 0.4 \end{bmatrix},$$

$$C_1 = \begin{bmatrix} -2.2 & 0.4 & 0 \\ 1 & -1.6 & 0 \\ -0.5 & 0 & -0.8 \end{bmatrix}, \quad D_1 = \begin{bmatrix} 1 & 1.2 & 0 \\ 0.9 & -1.5 & 0 \\ 0.3 & 0.5 & -1.0 \end{bmatrix},$$

Mode 2

$$A_2 = \begin{bmatrix} 2.5 & 0 & 0 \\ 0 & 4 & 0 \\ 0 & 0 & 4 \end{bmatrix}, \quad B_2 = \begin{bmatrix} 0.2 & 0 & 0 \\ 0 & 0.9 & 0 \\ 0 & 0 & 0.1 \end{bmatrix},$$

$$C_2 = \begin{bmatrix} -1.5 & 0.4 & 0 \\ 1 & -1.1 & 0 \\ -0.7 & 0 & -2.2 \end{bmatrix}, \quad D_2 = \begin{bmatrix} 2.1 & 1.6 & 0 \\ 1.6 & -1.1 & 0 \\ 0.4 & 0.8 & -1.2 \end{bmatrix},$$

with

$$\Pi = \begin{bmatrix} -5 & 5 \\ 4 & -4 \end{bmatrix}, \quad E_1 = 0.1I, \quad F_1 = 0.2I,$$
$$E_2 = diag\{0.1, 0.3, 0.2\}, \quad F_2 = diag\{0.2, 0.5, 0.2\}.$$

If we choose $\tau_2 = 0.3$, $\tau_1 = 0$, $\mu = 0.5$ and $L = diag\{0.1, 0.1, 0.1\}$, then the mean-square asymptotic stability conditions for the system (7) with the above parameters can be verified based on the LMI condition specified in Theorem 1.

Solving the LMI condition (9) results in the following feasible matrices

$$P_{11} = \begin{bmatrix} 160.8966 & 9.9924 & -30.6903 \\ 9.9924 & 132.7698 & 1.9028 \\ -30.6903 & 1.9028 & 148.6468 \end{bmatrix},$$

$$P_{21} = \begin{bmatrix} 163.7432 & 10.4369 & -29.8908 \\ 10.4369 & 134.9727 & 1.8454 \\ -29.8908 & 1.8454 & 150.7557 \end{bmatrix},$$

$$P_{12} = \begin{bmatrix} 44.3752 & 0.9817 & -6.2472 \\ 0.9817 & 46.9122 & 0.4032 \\ -6.2472 & 0.4032 & 38.6418 \end{bmatrix},$$

$$P_{22} = \begin{bmatrix} 65.7336 & 3.6342 & -8.8578 \\ 3.6342 & 66.2666 & 0.3200 \\ -8.8578 & 0.3200 & 57.4530 \end{bmatrix},$$

and

$$Q_1 = \begin{bmatrix} 34.5425 & 3.0359 & -9.1952 \\ 3.0359 & 43.7191 & 0.7996 \\ -9.1952 & 0.7996 & 45.1519 \end{bmatrix},$$

$$Q_2 = \begin{bmatrix} 34.5425 & 3.0359 & -9.1952 \\ 3.0359 & 43.7191 & 0.7996 \\ -9.1952 & 0.7996 & 45.1519 \end{bmatrix},$$

$$Q_3 = \begin{bmatrix} 57.0554 & 5.0476 & -15.4378 \\ 5.0476 & 73.3584 & 1.2818 \\ -15.4378 & 1.2818 & 75.9318 \end{bmatrix},$$

$$R_1 = \begin{bmatrix} 14.2899 & 0.3920 & -1.1768 \\ 0.3920 & 19.4208 & 0.1662 \\ -1.1768 & 0.1662 & 18.7159 \end{bmatrix}.$$

These feasible matrices obtained using MATLAB LMI solvers ensures that the LMI condition (9) given in Theorem 1 is negative-definite which corresponds to the LKF (10). Hence, according to Definition 1 and LMI condition (9), the considered system (7) corresponding to the considered parameters mean-square asymptotically stable.

In order to compare the results obtained in Theorem 1, Corollary 1 and Theorem 2 of Sect. 3, we found the maximum allowable upper bound of time-delay for the above-mentioned parameters. Maximum bound of time-delay is found to be 2.73, 2.98 and 2.87 for Theorem 1, Corollary 1 and Theorem 2, respectively.

Also, from the LMI conditions, one can see that the size of the LMI obtained in Theorem 2 is less compared to Theorem 1.

Further, numerical simulation is carried out for the considered system with the same set of parameters expect the time-delay which is taken to be $0.9 + \sin^2(t)$. Figures 1 and 2 shows the state trajectories of subsystem 1 and 2 respectively whereas Fig. 3 shows the behavior of the Markov chain $r(t)$ for the state transition probability matrix Π.

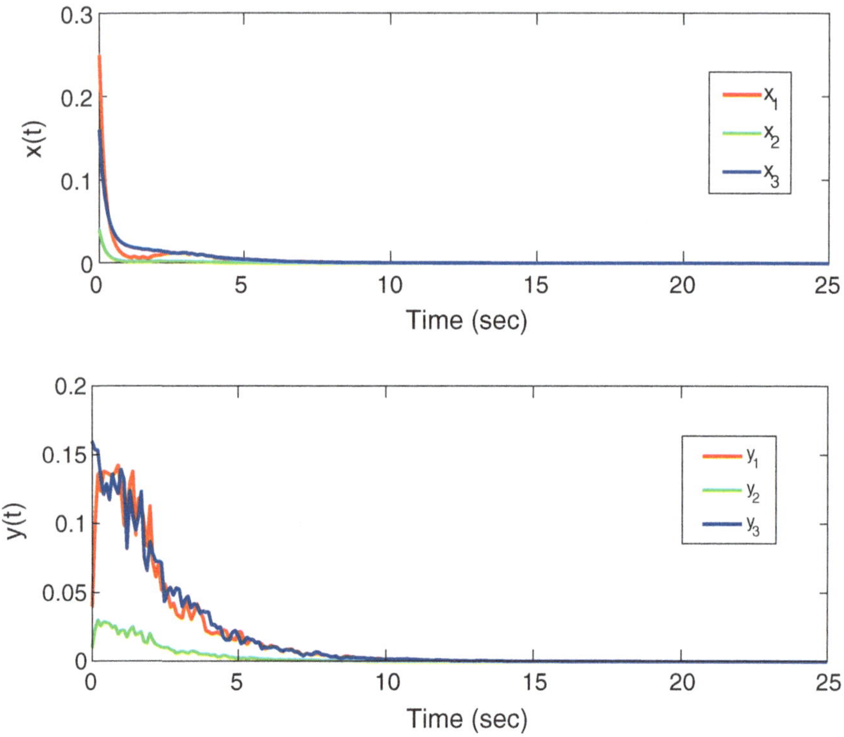

Fig. 1 State trajectories of subsystem 1

The numerical simulation clearly depicts that both the subsystems are asymptotically stable in the mean-square.

Example 2 Consider the INNs (24) with two modes described by the following parameters

Mode 1:

$$A_1 = \begin{bmatrix} 0.7 & 0 \\ 0 & 2.1 \end{bmatrix},\ B_1 = \begin{bmatrix} -0.2 & 0 \\ 0 & -0.4 \end{bmatrix},$$
$$C_1 = \begin{bmatrix} -1.43 & 0.1 \\ 0.4 & -1.1 \end{bmatrix},\ D_1 = \begin{bmatrix} 0 & 0.2 \\ -0.02 & -0.1 \end{bmatrix},$$

Mode 2:

$$A_2 = \begin{bmatrix} 0.9 & 0 \\ 0 & 0.8 \end{bmatrix},\ B_2 = \begin{bmatrix} -0.4 & 0 \\ 0 & -0.4 \end{bmatrix},$$
$$C_2 = \begin{bmatrix} -2.1 & 0.5 \\ 0.5 & -1 \end{bmatrix},\ D_2 = \begin{bmatrix} 0.4 & 0.6 \\ -0.1 & -0.6 \end{bmatrix};$$

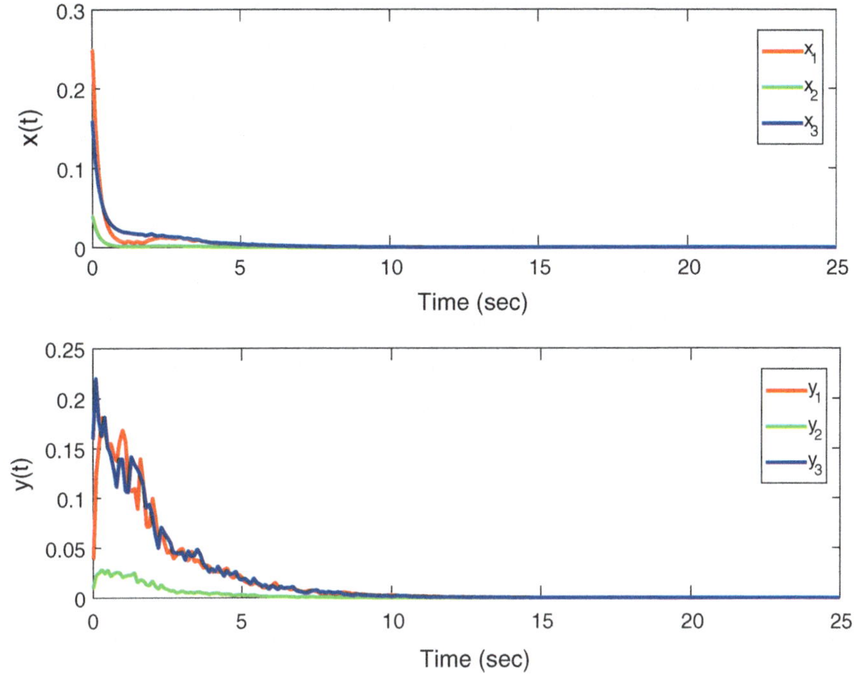

Fig. 2 State trajectories of subsystem 2

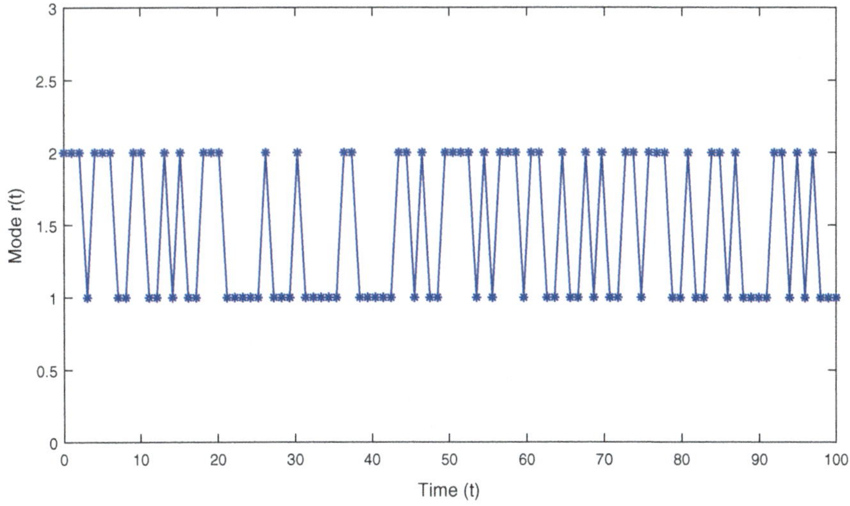

Fig. 3 Markovian jump modes

with $h_2 = 11$, $\mu_1 = 0.4$, $\mu_2 = 0.3$, $L_1 = 0$, $L_2 = I$.

Also, the transition probability matrix is assumed as $\Pi = \begin{bmatrix} -0.3 & 0.3 \\ 0.7 & -0.7 \end{bmatrix}$.

Stochastic stability condition given in for the system with the above parameters is validated through MATLAB LMI solvers and the corresponding feasible matrices are listed as follows

$$P_{11} = \begin{bmatrix} 0.1032 & 0.0015 \\ 0.0015 & 0.1181 \end{bmatrix}, \; P_{12} = \begin{bmatrix} 0.1644 & -0.0031 \\ -0.0031 & 0.1605 \end{bmatrix},$$
$$P_{21} = \begin{bmatrix} 0.0231 & -0.0006 \\ -0.0006 & 0.0087 \end{bmatrix}, \; P_{22} = \begin{bmatrix} 0.0242 & 0.0141 \\ 0.0141 & 0.0342 \end{bmatrix},$$
$$Q_1 = \begin{bmatrix} 0.0524 & -0.0002 \\ -0.0002 & 0.0510 \end{bmatrix}, \; Q_2 = \begin{bmatrix} 0.0237 & -0.0000 \\ -0.0000 & 0.0229 \end{bmatrix},$$
$$Q_3 = \begin{bmatrix} 0.0084 & -0.0001 \\ -0.0001 & 0.0079 \end{bmatrix}, \; Q_4 = \begin{bmatrix} 0.0085 & -0.0000 \\ -0.0000 & 0.0083 \end{bmatrix},$$
$$Q_5 = \begin{bmatrix} 0.0107 & -0.0001 \\ -0.0001 & 0.0098 \end{bmatrix}, \; R_1 = \begin{bmatrix} 0.0129 & 0.0001 \\ 0.0001 & 0.0159 \end{bmatrix},$$
$$R_2 = \begin{bmatrix} 1.2212 & -0.0017 \\ -0.0017 & 1.2866 \end{bmatrix}.$$

For the purpose of simulation, the mode-dependent time-varying delay is assumed to be $\tau_1(t) = 0.5 + 0.\sin^2(t)$ when the system is in Mode 1 and $\tau_2(t) = 0.75 + 0.2\sin^2(t)$ when the system is in Mode 2 and the activation function is assumed to be $\tanh(x(t))$ for both the modes.

Other parameters are taken as given above in the problem. Corresponding state trajectories for two different time-varying delay is simulated and are given below remaining parameters are chosen as above.

Figures 4 and 5 confirms that both subsystems converge to the equilibrium point. Hence, the considered system with the given parameters is stable in the sense of Lyapunov. Jump modes of (24) for the above parameters is simulated and it is given in Fig. 6.

Now, when there is no jump in the system dynamics, that is, there are no subsystems, stability of the system (24) is investigated by taking $A_1 = A_2$, $B_1 = B_2$, $C_1 = C_2$, $D_1 = D_2$, $\mu_1 = \mu_2$ and $\Pi = 0$.

Corresponding feasible matrices obtained are given below

$$P_{11} = \begin{bmatrix} 0.1291 & 0.0000 \\ 0.0000 & 0.1290 \end{bmatrix}, \; P_{21} = \begin{bmatrix} 0.0362 & -0.0005 \\ -0.0005 & 0.0091 \end{bmatrix},$$
$$Q_1 = \begin{bmatrix} 0.0727 & -0.0006 \\ -0.0006 & 0.0660 \end{bmatrix}, \; Q_2 = \begin{bmatrix} 0.0300 & -0.0001 \\ -0.0001 & 0.0281 \end{bmatrix},$$
$$Q_3 = \begin{bmatrix} 0.0850 & -0.0009 \\ -0.0009 & 0.0762 \end{bmatrix}, \; Q_4 = \begin{bmatrix} 0.0725 & -0.0001 \\ -0.0001 & 0.0718 \end{bmatrix},$$

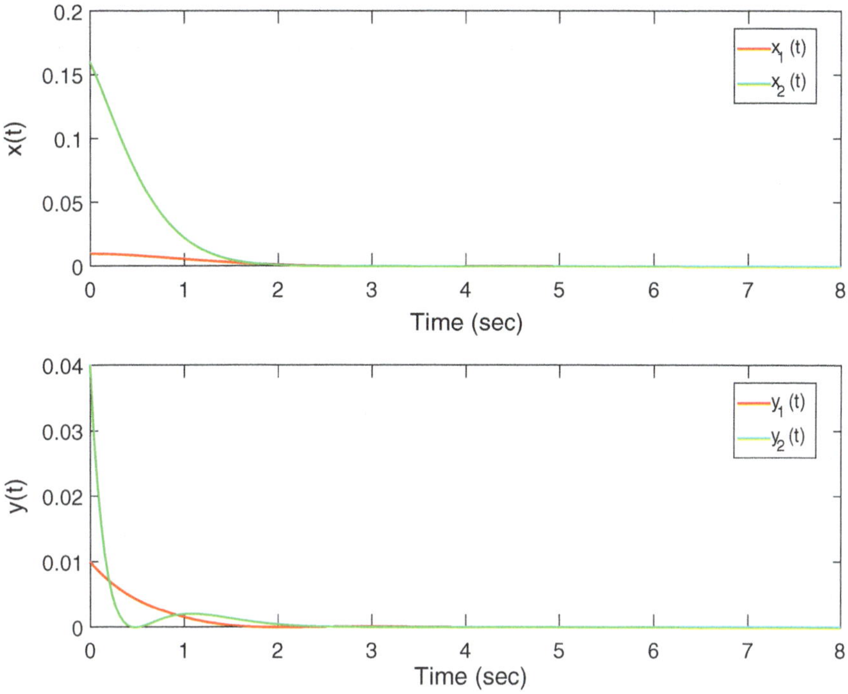

Fig. 4 State trajectories of subsystem 1

$$Q_5 = \begin{bmatrix} 0.1032 & -0.0010 \\ -0.0010 & 0.0914 \end{bmatrix}, \quad R_1 = \begin{bmatrix} 0.0157 & 0.0002 \\ 0.0002 & 0.0173 \end{bmatrix},$$
$$R_2 = \begin{bmatrix} 1.4532 & 0.0020 \\ 0.0020 & 1.4487 \end{bmatrix}.$$

Remark 7 The work [40] analyzed the synchronization behavior of the switched NNs with the effects of mode-dependent time-delay and impulses. Also, [36] investigated the problem of stability analysis for INNs with mode-dependent time-delays in which the disturbance is taken to be intrinsic noise.

Literature review reveals that there exists no work in the area of stability of INNs with both intrinsic and extrinsic noises. Hence, the effect of these noises for the stability problem of INNs is taken into an investigation in this work which is the generalization of [36].

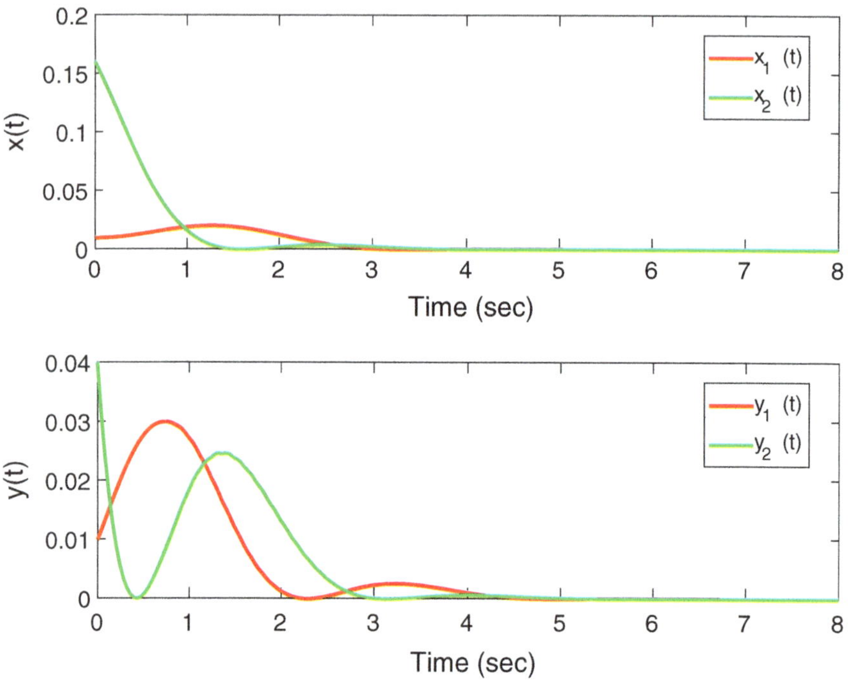

Fig. 5 State trajectories of subsystem 2

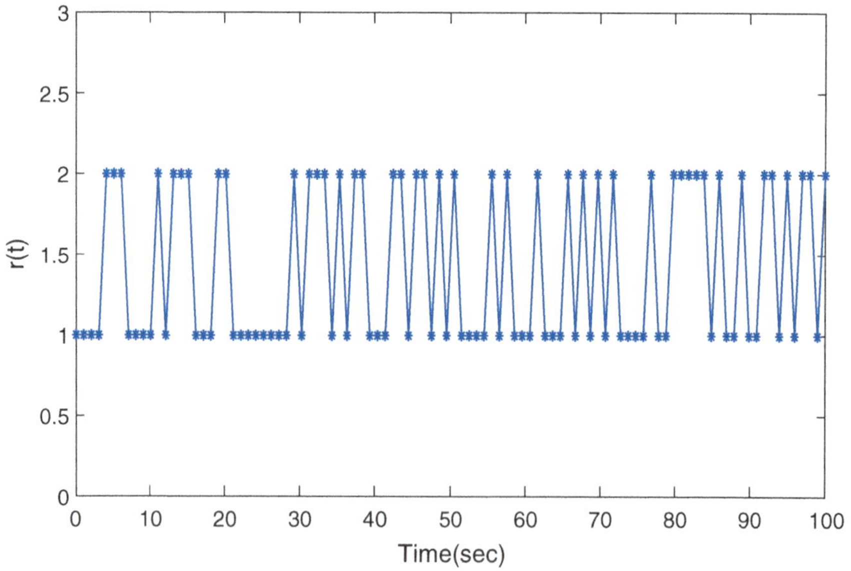

Fig. 6 Markovian jump modes

5 Conclusion

The problem of stochastic stability of INNs along with time-delay, disturbances, and MJPs is investigated in this chapter. Sufficient delay-dependent stochastic stability conditions are obtained in terms of LMIs which can be solved using LMI solvers through MATLAB software. Here, both types of noises namely intrinsic and extrinsic noises are taken in to account which is different from the existing literature concerning INNs. Also, stability results for mode-dependent Markovian jump INNs are investigated in the absence of external disturbances. INNs Numerical example illustrating the merits of the proposed method are given. Problem considered in this work generalizes some of the results of existing literature in the domain of INNs. Also, the problem discussed in this work can be extended with the assumption that some of the transition rates in the transition matrix are unknown. Further, it can be generalized with the incorporation of uncertainties in both system matrices and transition rates.

Acknowledgements Part of this work was done when the first author was a Post-Doctoral fellow in the Department of Mathematics, Indian Institute of Space Science and Technology, Thiruvananthapuram, Kerala and it was supported by DST-SERB through NPDF scheme (file No. PDF/2016/002992 dated 01/04/2017).

References

1. Kuang, Y.: Delay Differential Equations with Applications in Population Dynamics. Academic, New York (1993)
2. Niculescu SI (2001) Delay Effects on Stability: A Robust Control Approach. Springer, Berlin
3. Gu, K., Kharitonov, L., Chen, J.: stability of Time Delay Systems. Birkhauser (2003)
4. Hale, J.K., Lunel, S.M.V.: Introduction to Functional Differential Equations. Springer, Berlin (1993)
5. Park, J.H., Lee, T.H., Liu, Y., Chen, J.: Dynamic Systems with Time Delays: Stability and Control. Springer-Nature, Singapore (2019)
6. Boyd S, Ghaoui L, Feron E, Balakrishnan V (1994) Linear Matrix Inequalities in System and Control Theory. SIAM
7. Khalil, H.K.: Nonlinear Systems. Prentice Hall, Upper Saddle River (2002)
8. Lakshmanan, M., Senthilkumar, D.V.: Dynamics of Nonlinear Time-Delay Systems. Springer, Berlin (2010)
9. Lakshmikanthan, V., Bainov, D.D., Simeonov, P.S.: Theory of Impulsive Differential Equations. World Scientific, Singapore (1989)
10. Scheinerman, E.R.: Invitation to Dynamical Systems. Dover, Mineola (1996)
11. Xu, S., Lam, J.: Robust Control and Filtering of Singular Systems. Springer , Berlin (2006)
12. Zhong, Q.C.: Robust Control of Time-Delay Systems. Springer, Berlin (2006)
13. Stamova, I.: Stability Analysis of Impulsive Functional Differential Equations. Walter de Gruyter, Berlin (2009)
14. Rakkiyappan, R., Udhaya Kumari, E., Chandrasekar, A., Krishnasamy, R.: Synchronization and periodicity of coupled inertial memristive neural networks with supremums. Neurocomputing **214**, 739–749 (2016)

15. Rakkiyappan, R., Gayathri, D., Velmurugan, G., Cao, J.: Exponential synchronization of inertial memristor-based neural networks with time delay using average impulsive interval approach. Neural Process. Lett. **50**, 2053–2071 (2019)
16. Lakshmanan, S., Prakash, M., Rakkiyappan, R., Joo, Y.H.: Adaptive synchronization of reaction-diffusion neural networks and its application to secure communication. IEEE Trans. Cybern. **50**, 911–922 (2020)
17. Mao X (2011) Stochastic Differential Equations and Applications. Woodhead Publishing
18. Liberzon D (1973) Switching in Systems and Control. Birkh$\ddot{a}$user
19. Mahmoud, M.S., Shi, P.: Methodologies for Control of Jump Time-Delay Systems. Kluwer Academic Publishers, Dordrecht (2003)
20. Shorten, S., Wirth, F., Mason, O., Wulff, K., King, C.: Stability criteria for switched and hybrid systems. SIAM Rev. **49**, 545–592 (2007)
21. Sun, Z., Ge, S.S.: Stability Theory of Switched Dynamical Systems. Springer, Berlin (2011)
22. Park, J.H., Shen, H., Chang, Xh, Lee, T.H.: Recent Advances in Control and Filtering of Dynamic Systems with Constrained Signals. Springer, Cham, Switzerland (2018). https://doi.org/10.1007/978-3-319-96202-3
23. Shen, H., Park, J.H., Zhang, L., Wu, Z.G.: Robust extended dissipative control for sampled-data Markov jump systems. Int. J. Control **87**(8), 1549–1564 (2014)
24. Yu, T., Wang, H., Su, M., Cao, D.: Distributed-delay-dependent exponential stability of impulsive neural networks with inertial term. Neurocomputing **313**, 220–228 (2018)
25. Zhang, W., Huang, T., He, X., Li, C.: Global exponential stability of inertial memristor-based neural networks with time-varying delayed and impulses. Neural Netw. **95**, 102–109 (2017)
26. Zhang, H., Qiu, Z., Xiong, L.: Stochastic stability criterion of neutral-type neural networks with additive time-varying delay and uncertain semi-Markov jump. Neurocomputing **333**, 395–406 (2019)
27. Wang, L., Ge, M.F., Hu, J., Zhang, G.: Global stability and stabilization for inertial memristive neural networks with unbounded distributed delays. Nonlinear Dyn. **95**, 943–955 (2019)
28. Maharajan, C., Raja, R., Cao, J., Rajchakit, G.: Novel global robust exponential stability criterion for uncertain inertial-type BAM neural networks with discrete and distributed time-varying delays via Lagrange sense. J. Frankl. Inst. **355**, 4727–4754 (2018)
29. Li, H., Li, C., Zhang, W., Xu, J.: Global dissipativity of inertial neural networks with proportional delay via new generalized Halanay inequalities. Neural Process. Lett. **48**, 1543–1561 (2018)
30. Zhou, L.: Global asymptotic stability of cellular neural networks with proportional delays. Nonlinear Dyn. **77**, 41–47 (2014)
31. Li, X., Li, X., Hu, C.: Some new results on stability and synchronization for delayed inertial neural networks based on non-reduced order method. Neural Netw. **96**, 91–100 (2017)
32. Huang, C., Liu, B.: New studies on dynamic analysis of inertial neural networks involving non-reduced order method. Neurocomputing **325**, 283–287 (2019)
33. Huang, Q., Cao, J.: Stability analysis of inertial Cohen-Grossberg neural networks with Markovian jumping parameters. Neurocomputing **282**, 89–97 (2018)
34. Prakash, M., Balasubramaniam, P., Lakshmanan, S.: Synchronization of Markovian jumping inertial neural networks and its applications in image encryption. Neural Netw. **83**, 86–93 (2016)
35. Ji, H., Zhang, H., Senping, T.: Reachable set estimation for inertial Markov jump BAM neural network with partially unknown transition rates and bounded disturbances. J. Frankl. Inst. **354**, 7158–7182 (2017)
36. Krishnasamy, R., George Raju, K.: Stochastic stability of mode-dependent Markovian jump inertial neural networks. J. Anal. **27**, 179–196 (2019)
37. Zhu, Q., Cao, J.: Exponential stability of stochastic neural networks with both Markovian jump parameters and mixed time delays. IEEE Trans. Syst. Man Cybern.-Part B: Cybern. **41**, 341–353 (2011)
38. Lee, T.H., Park, J.H.: A novel Lyapunov functional for stability of time-varying delay systems via matrix-refined-function. Automatica **80**, 239–242 (2017)

39. Liu, K., Fridman, E.: Wirtinger's inequality and Lyapunov-based sampled-data stabilization. Automatica **48**, 102–108 (2012)
40. Xie, X., Liu, X., Xu, H.: Synchronization of delayed coupled switched neural networks: Mode-dependent average impulsive interval. Neurocomputing **365**, 261–272 (2019)

Exponential Stability and Stabilization of Stochastic Neural Network Systems via Switching and Impulsive Control

Ticao Jiao and Ju H. Park

Abstract The exponential stability and stabilization problem of stochastic neural network systems via a switching and impulsive control are dealt with in the chapter. To achieve the desired performance, a new type of switching and impulsive controllers are designed. With comparison to the previous results, the main characteristics of our proposed controller lies in the asynchronous behavior between switchings and impulses. That is, the occurrence instants of switchings and impulses are not the same. To describe the generation mechanisms of switchings and impulses, the concepts of dwell time and impulsive interval are employed. Then, based on multiple Lyapunov functions approach, sufficient conditions for mean square exponential stability are first established in terms of linear matrix inequalities (LMIs), based on which gain matrices of switching and impulsive controllers are presented. We, at last, provide two numerical examples to verify the applicability of the proposed theoretical results.

Keywords Neural network systems · Switching · Impulsive · Dwell time · Impulsive interval

1 Introduction

During past decades, neural network systems have been widely applied in practical fields such as signal processing, image reconstruction, artificial intelligence and associative memory, etc. These applications strongly rely on the stability, asymptotic stability, and synchronization of neural network systems. Therefore, the great

---

T. Jiao (✉)
School of Electrical and Electronic Engineering, Shandong University of Technology, Shandong 255000, People's Republic of China
e-mail: ticaojiao@gmail.com

J. H. Park
Department of Electrical Engineering, Yeungnam University, 280 Daehak-Ro, Kyongsan 38541, Republic of Korea
e-mail: jessie@ynu.ac.kr

J. H. Park (ed.), *Recent Advances in Control Problems of Dynamical Systems and Networks*, Studies in Systems, Decision and Control 301,
https://doi.org/10.1007/978-3-030-49123-9_18

interest of researchers have been attracted to the dynamical behaviors of neural network systems and a large number of articles and monographs concerning Hopfield, stochastic, switching, uncertain, and reaction-diffusion network models have been published (see, e.g. [1–7] and references therein).

As is well known, many physical and industrial models can be well described by switched systems, for instance, the continuous stirred tank reactor with two modes feed stream and mass-spring-damper systems in [8, 9]. Switched (neural network) systems have gone through extensive studies in the literature. For instance, in [10], standard $\mathcal{H}_\infty$ control of switched systems was investigates, in which any internal stability on switched subsystems is not required. By constructing an impulse-time-dependent discretized copositive Lyapunov function, [11] addressed the stability analysis and $\mathcal{L}_\infty$-gain characterization for positive Takagi-Sugeno fuzzy systems subject to impulse effects. Via the method of mode-dependent average dwell time, exponential stability criteria of delayed switched neural networks were obtained in [12]. In [13], a novel method was introduced to describe the switching law, based on which the state estimation problem for a class of discrete-time switched neural networks was considered.

On the other hand, some real-world systems usually undergo instantaneous state changes at discrete times. The natural and reasonable way to model such a class of dynamical systems is impulsive differential equations, which have shown their enormous power in the fields of theory and applications during the past decades. Especially, the increasing popularity has been achieved in neural networks [2, 14, 15]. In [15], from the viewpoint of destabilizing impulses, new criteria on stochastic stability with the impulse effects were derived if the feasible bound of impulses is imposed. Then, a general class of neural networks with heterogeneous impulsive effects was introduced in [16] and the synchronization problem was dealt with.

Combining neural network systems, impulsive systems with switched systems, it is natural to develop a more all-around system, i.e., switching and impulsive neural networks. Despite the apparent abundance of applications, switching and impulsive neural networks have not received the attention they deserve in the past decade [2]. In [17], the asymptotic synchronization problem for a class of uncertain hybrid switching and impulsive complex networks was investigated by partitioning the dwell time. When time delay is considered, by means of Halanay inequality and Lyapunov-Krasovskii functionals, stability issue of impulsive and switching systems was studied in [18, 19].

It is worth noting that in the works [2, 17–21], the switching signal and the impulsive law have been integrated as an impulsive and switching law, which is described by using the concept of (average) dwell time. In other words, these results assume that the switchings and impulses occur at the same time instants. Note that such an assumption is much too conservative and out of step with reality. For instance, when switchings and state-dependent impulses are simultaneously considered, the resulting equivalent system by using B-equivalence method, in fact, owns different impulsive and switching instants [22, 23]. Another example is the switched networks with mode-dependent impulsive effects. In each switching interval, impulses have happened multiple times. In this case, it means that switchings and impulses are asynchronous [16]. Therefore, compared with [2, 17–19], it is more reasonable and practical to make an investigation of dynamic systems with asynchronous switchings and impulses [25]. To the authors' knowledge, the research of such systems has not received sufficient attention and many open problems have not been successfully solved.

Motivated by the above analysis, we in this chapter address the issue of exponential stability and stabilization of stochastic neural network systems via switching and impulsive control. Here, the switching instants differ from the impulsive ones. Our main contributions in the chapter lie in the following aspects:

1. In contrast with the neural networks investigated in the previous works (e.g. [2, 17–19]), the neural networks studied in the chapter take a general form, which allows the asynchronism of switchings and impulses.
2. The approach of average dwell time and (inverse) average impulsive interval is employed to control the occurrence frequency of switchings and impulses. Then, combining it with multiple Lyapunov functions approach, sufficient criteria ensuring mean square exponential stability are developed in terms of LMIs, based on which gain matrices of switching and impulsive controllers are proposed.

The rest of this chapter is organized as follows. Section 2 presents the preliminary results. The main results are given in Sect. 3. Through two examples the obtained results are studied in Sect. 4. Finally, the concluding remarks are provided in Sect. 5.

2 Preliminary Results

Consider the following class of stochastic neural networks with asynchronous impulses and switchings (SNNAIWs) described by

$$\begin{aligned} \mathrm{d}x(t) &= \big(-C_{r(t)}x + D_{r(t)}F_{r(t)}(x) + u_1(t)\big)\,\mathrm{d}t + G_{r(t)}(x,t)\,\mathrm{d}\omega(t),\ t \neq t_k,\\ x(t_k) &= Ax(t_k^-) + u_2(t_k^-),\ t = t_k,\\ x(t_0) &= x_0, r(t_0) = r_0, \end{aligned} \tag{1}$$

where $x(t) \in \mathbb{R}^n$ denotes the neuron systems state vector; $r(t)$ is a right-continuous function, which controls the switching sequences among m networks; the discontinuous point set $\{t_k\}_{k\in\mathbb{N}^+}$ is the impulsive instants; $x(t_k^-) = \lim_{t\to t_k^-} x(t)$ and $u_1(t)$, $u_2(t)$ are control input signals.

For $i \in \mathcal{M} = \{1, \ldots, m\}$, $C_i = \mathrm{diag}\{c_{i1}, \ldots, c_{in}\}$ with $c_{ij} > 0$ $(j = 1, \ldots, n)$ and $G_i \in \mathbb{R}^{n\times n}$, $A \in \mathbb{R}^{n\times n}$ are constant matrices; $F_i(\cdot)$, $G_i(\cdot)$ are continuous functions with proper dimensions. $\omega(t)$ is a standard Wiener process vector defined on the complete probability space $(\Omega, \mathcal{F}, \mathcal{F}_t, \mathcal{P})$ with a filtration $\mathcal{F}_t$ satisfying the usual conditions.

The following assuming conditions are imposed on the architecture defining system (1).

Assumption 1 ([18]) For $i \in \mathcal{M}$, the neuron activation function $F_i(x) = [F_{i1}(x_1), \ldots, F_{in}(x_n)]^T$ satisfying:

$$F_{ij}(0) = 0, \ \underline{b}_{ij} \le \frac{F_{ij}(y)}{y} \le \bar{b}_{ij}, \ j = 1, \ldots, n, \tag{2}$$

where $y \in \mathbb{R}\backslash\{0\}$, $\underline{b}_{ij}$ and $\bar{b}_{ij}$ are known constants.

Assumption 2 For the noise density function $G_i(x, t)$, there exist positive definite matrices Q_i such that

$$\mathtt{trace}\{G_i^T(x, t)G_i(x, t)\} \le x^T Q_i x. \tag{3}$$

Remark 1 It is worth stressing that the introduced SNNAIW model (1) is more general than those in some existing works. For example, a special case of model (1) without stochastic noises and asynchronous impulses and switchings is investigated in [2, 17–19].

For the investigation of exponential stabilization of system (1), the following stochastic stability and stabilization definitions are recalled.

Definition 1 ([26]) The trivial solution $x = 0$ of system (1) with $u_1(t) = 0, u_2(t) = 0$ is said to be exponentially stable in mean (ES-M) if there exist a constant $\gamma > 0$ and a class $\mathcal{K}_\infty$ function α, such that for all $x_0 \in \mathbb{R}^n\backslash\{0\}$,

$$\mathcal{E}|x(t)| \le \alpha(|x(t_0)|)e^{-\gamma(t-t_0)}. \tag{4}$$

Definition 2 The trivial solution $x = 0$ of system (1) is said to be exponentially stabilized in mean (E-S-M) if there exist control inputs $u_1(t)$ and $u_2(t)$ such that (4) holds.

3 Main Results

In this section, our attention is first focused on the stability analysis of system (1) with $u_1(t) = 0$ and $u_2(t) = 0$, then, on the basis of the developed stability criteria, the asynchronous switching and impulsive controller is to be designed.

3.1 Stability Analysis

3.1.1 Every Subsystem Is ES-M

In this subsection, we consider the case where every subsystem is ES-M but the impulse is destabilizing. To establish the main results, let us recall the definitions of average dwell time and average impulsive interval, respectively.

Definition 3 (*Average Dwell Time* [27])Let $N_r(t', t'')$ denote the number of switchings on an interval $[t', t'')$. If there exists two positive numbers N_{r0} and η_s, such that

$$N_r(t', t'') \leq N_{r0} + \frac{t'' - t'}{\eta_s}, \tag{5}$$

then we say that the switching signal $r(t)$ has an average dwell time η_s.

Definition 4 (*Average Impulsive Interval* [28]) For the impulsive sequences $\{t_k\}_{k\in\mathbb{N}^+}$, let $N_i(t', t'')$ be the number of impulses on an interval $[t', t'')$. If there exist two numbers $N_{i0} > 0$, $\eta_i > 0$ such that

$$N_i(t', t'') \leq N_{i0} + \frac{t'' - t'}{\eta_i}, \tag{6}$$

then the impulsive sequence $\{t_k\}_{k\in\mathbb{N}^+}$ is said to have an average impulsive interval η_i.

Now, we have the following main result.

Theorem 1 *Under Assumptions 1 and 2, if there exist positive definite matrices P_i, real parameters $\mu \geq 1$, $\delta > 1$, $\bar{\gamma} > 0$, $c_i > 0$, $\rho_{ij} > 0\,(i, j \in \mathcal{M})$ satisfying*

$$\begin{bmatrix} -P_iC_i - C_iP_i - \underline{B}_{i1}\Lambda_i + c_iQ_i + \bar{\gamma}P_i & P_iD_i + \bar{B}_{i2}\Lambda_i \\ * & -\Lambda_i \end{bmatrix} < 0, \tag{7}$$

$$P_i \leq c_iI_n, \tag{8}$$

$$P_i \leq \mu P_j,\ i \neq j, \tag{9}$$

$$A^TP_iA \leq \delta P_i, \tag{10}$$

$$\bar{\gamma} - \frac{\ln\mu}{\eta_s} - \frac{\ln\delta}{\eta_i} > 0, \tag{11}$$

where

$$\Lambda_i = diag\{\rho_{i1}, \ldots, \rho_{in}\},$$
$$\underline{B}_{i1} = diag\{\underline{b}_{i1}\bar{b}_{i1}, \ldots, \underline{b}_{in}\bar{b}_{in}\},$$
$$\underline{B}_{i2} = diag\{\frac{\underline{b}_{i1} + \bar{b}_{i1}}{2}, \ldots, \frac{\underline{b}_{in} + \bar{b}_{in}}{2}\},$$

then system (1) is ES-M when the switching signal $r(t)$ *has an average dwell time* η_s *and the impulsive sequence* $\{t_k\}_{k\in\mathbb{N}^+}$ *an average impulsive interval* η_i.

Proof Fix any $t > t_0$. On the interval $[t_0, t)$, assume that the switching sequence is $\{t_1^s, t_2^s, \ldots, t_{l_1}^s\}$ and the impulsive sequence $\{t_1, t_2, \ldots, t_{l_2}\}$, which satisfy $t_0 < t_1^s < t_2^s < \cdots < t_{j_1}^s < t_1 < t_{j_1+1}^s < \cdots < t_{j_2}^s < t_2 < t_{j_2+1}^s < \cdots < t_{j_k}^s < t_{l_2} < t_{j_k+1}^s < \cdots < t_{l_1}^s < t$. Therefore, $N_i(t_0, t) = l_2$, $N_s(t_0, t) = l_1$.

Moreover, the candidate Lyapunov function is defined as

$$V_{r(t)}(x(t), t) = x^T(t)P_{r(t)}x(t).$$

From system (1), we have

$$\begin{aligned}
&\mathcal{L}V_{r(t)}(x(t), t) + \bar{\gamma}V_{r(t)}(x(t), t)\\
&\quad = 2x^T(t)P_{r(t)}\big(-C_{r(t)}x + D_{r(t)}F_{r(t)}(x)\big)\\
&\quad + \mathtt{trace}\{G^T_{r(t)}(x, t)P_{r(t)}G_{r(t)}(x, t)\} + \bar{\gamma}x^T(t)P_{r(t)}x(t).
\end{aligned} \tag{12}$$

It follows from Assumption 2 and condition (8) that

$$\begin{aligned}
\mathtt{trace}\{G^T_{r(t)}(x, t)P_{r(t)}G_{r(t)}(x, t)\} &\le c_i\mathtt{trace}\{G^T_{r(t)}(x, t)G_{r(t)}(x, t)\}\\
&\le c_i x^T(t)Q_i x(t).
\end{aligned} \tag{13}$$

On the other hand, Assumption 1 implies that

$$0 \le (F_{ij}(x_j) - \underline{b}_{ij}x_j)(\bar{b}_{ij}x_j - F_{ij}(x_j)).$$

Thus, for any $\rho_{ij} > 0$, it holds

$$0 \le \sum_{j=1}^{n} \rho_{ij}(F_{ij}(x_j) - \underline{b}_{ij}x_j)(\bar{b}_{ij}x_j - F_{ij}(x_j)), \tag{14}$$

which can be rewritten as the following compact form

$$0 \le \begin{bmatrix} x^T & F_i^T(x) \end{bmatrix} \begin{bmatrix} -\underline{B}_{i1}\Lambda_i & \bar{B}_{i2}\Lambda_i \\ * & -\Lambda_i \end{bmatrix} \begin{bmatrix} x \\ F_i(x) \end{bmatrix}. \tag{15}$$

Substituting (13) and (15) into (12) leads to

$$\begin{aligned}
&\mathcal{L}V_{r(t)}(x(t),t)+\bar{\gamma}V_{r(t)}(x(t),t)\\
&\quad=2x^T(t)P_{r(t)}\big(-C_{r(t)}x+D_{r(t)}F_{r(t)}(x)\big)\\
&\quad+c_ix^T(t)Q_ix(t)+\bar{\gamma}x^T(t)P_{r(t)}x(t)\\
&\quad+\big[x^T\ F_i^T(x)\big]\begin{bmatrix}-\underline{B}_{i1}\Lambda_i & \bar{B}_{i2}\Lambda_i\\ * & -\Lambda_i\end{bmatrix}\begin{bmatrix}x\\ F_i(x)\end{bmatrix},
\end{aligned}\tag{16}$$

by which as well as condition (7), we have

$$\mathcal{L}V_{r(t)}(x(t),t)+\bar{\gamma}V_{r(t)}(x(t),t)\leq 0,$$

that is,

$$\mathcal{L}V_{r(t)}(x(t),t)\leq-\bar{\gamma}V_{r(t)}(x(t),t).\tag{17}$$

We next introduce a new function

$$\bar{W}(x(t))=e^{\bar{\gamma}t}V_{r(t)}(x(t),t).$$

By Itô formula and (17), one has

$$\begin{aligned}
\mathcal{L}\bar{W}(x(t))&=\bar{\gamma}e^{\bar{\gamma}t}V_{r(t)}(x(t),t)+e^{\bar{\gamma}t}\mathcal{L}V_{r(t)}(x(t),t)\\
&\leq\bar{\gamma}e^{\bar{\gamma}t}V_{r(t)}(x(t),t)-\bar{\gamma}e^{\bar{\gamma}t}V_{r(t)}(x(t),t)\\
&=0,
\end{aligned}\tag{18}$$

from which it follows that $\bar{W}(x(t))$ is a continuous non-negative supermartingale.

Thus, for $\forall t_0<s_1<s_2$,

$$\mathbb{E}[\bar{W}(x(s_2))|\mathcal{F}_{s_1}]\leq\bar{W}(x(s_1)).\tag{19}$$

Then, by taking expectation on both sides of (19), we get

$$\mathbb{E}\bar{W}(x(s_2))\leq\mathbb{E}\bar{W}(x(s_1)).\tag{20}$$

Therefore, we have

$$\mathbb{E}V_{r(s_2)}(x(s_2),s_2)\leq e^{-\bar{\gamma}(s_2-s_1)}\mathbb{E}V_{r(s_1)}(x(s_1),s_1).\tag{21}$$

With this in mind, we first focus on the interval $[t_0,t_1^s)$. It is obvious that

$$\mathbb{E}V_{r(t_0)}(x(t_1),t_1^s)\leq e^{-\bar{\gamma}(t_1^s-t_0)}\mathbb{E}V_{r(t_0)}(x(t_0),t_0).\tag{22}$$

At the switching instant t_1^s from (8), it yields

$$\mathbb{E}V_{r(t_1^s)}(x(t_1^s), t_1^s) \leq \mu \mathbb{E}V_{r(t_0)}(x(t_1^s), t_1^s). \tag{23}$$

Combining (22) with (23), one obtains

$$\mathbb{E}V_{r(t_1)}(x(t_1^s), t_1^s) \leq \mu e^{-\bar{\gamma}(t_1^s - t_0)} \mathbb{E}V_{r(t_0)}(x(t_0), t_0). \tag{24}$$

By repeating the above derivation procedure until the switching instant $t_{j_1}^s$, we arrive at

$$\mathbb{E}V_{r(t_{j_1}^s)}(x(t_{j_1}^s), t_{j_1}^s) \leq \mu^{j_1} e^{-\bar{\gamma}(t_{j_1}^s - t_0)} \mathbb{E}V_{r(t_0)}(x(t_0), t_0). \tag{25}$$

Note that at the impulsive point t_1, by (10), it can be deduced that

$$\begin{aligned} \mathbb{E}V_{r(t_{j_1}^s)}(x(t_1), t_1) &= \mathbb{E}x^T(t_1) A^T P_{r(t_{j_1}^s)} A x(t_1) \\ &\leq \delta \mathbb{E}x^T(t_1^-) P_{r(t_{j_1}^s)} x(t_1^-). \end{aligned} \tag{26}$$

Thus, we further get

$$\mathbb{E}V_{r(t_{j_1}^s)}(x(t_{j_1+1}^s), t_{j_1+1}^s) \leq \delta \mu^{j_1} e^{-\bar{\gamma}(t_{j_1+1}^s - t_0)} \mathbb{E}V_{r(t_0)}(x(t_0), t_0). \tag{27}$$

By recurrence method, we finally get

$$\begin{aligned} \mathbb{E}V_{r(t)}(x(t), t) &\leq \delta^{N_i(t_0,t)} \mu^{N_s(t_0,t)} e^{-\bar{\gamma}(t-t_0)} \mathbb{E}V_{r(t_0)}(x(t_0), t_0) \\ &= e^{\ln \delta N_i(t_0,t) + \ln \mu N_s(t_0,t) - \bar{\gamma}(t-t_0)} \mathbb{E}V_{r(t_0)}(x(t_0), t_0) \end{aligned} \tag{28}$$

Using Definitions 3 and 4, inequality (28) can be computed as

$$\begin{aligned} \mathbb{E}V_{r(t)}(x(t), t) \leq\, & e^{\ln \delta N_{i0} + \frac{\ln \delta}{\eta_i}(t-t_0) + \ln \mu N_{s0} + \frac{\ln \mu}{\eta_i}(t-t_0) - \bar{\gamma}(t-t_0)} \\ & \times \mathbb{E}V_{r(t_0)}(x(t_0), t_0) \\ =\, & e^{\ln \delta N_{i0} + \ln \mu N_{s0}} \mathbb{E}V_{r(t_0)}(x(t_0), t_0) e^{-(\bar{\gamma} - \frac{\ln \delta}{\eta_i} - \frac{\ln \mu}{\eta_i})(t-t_0)}. \end{aligned} \tag{29}$$

From the definition $V_{r(t)}(x(t), t)$, it derives from (29) that

$$\mathbb{E}|x(t)|^2 \leq \frac{\lambda_{\max}(P_i)}{\lambda_{\min}(P_j)} e^{\ln \delta N_{i0} + \ln \mu N_{s0}} \mathbb{E}|x(t_0)|^2 e^{-(\bar{\gamma} - \frac{\ln \delta}{\eta_i} - \frac{\ln \mu}{\eta_i})(t-t_0)}. \tag{30}$$

Via moment inequality (30) can be further calculated that

$$\mathbb{E}|x(t)| \leq \sqrt{\frac{\lambda_{\max}(P_i)}{\lambda_{\min}(P_j)} e^{\ln \delta N_{i0} + \ln \mu N_{s0}} \mathbb{E}|x(t_0)|^2}\; e^{-\frac{-(\bar{\gamma} - \frac{\ln \delta}{\eta_i} - \frac{\ln \mu}{\eta_i})(t-t_0)}{2}}. \tag{31}$$

Therefore, system (1) is ES-M based on Definition 1. The proof is complete. □

Remark 2 From (7) and (17), we can derive that continuous subsystems are stable. However, the impulse is destabilizing according to (10). To ensure ES-M of the system (1), both a switching signal and an impulsive signal need to be skillfully designed satisfying (11).

3.1.2 No Subsystem Is ES-M

We now turn our attention to the case where any subsystem is not ES-M but the impulse is stabilizing.

To achieve this goal, the approach of an inverse average impulsive interval is applied.

Definition 5 (*Inverse Average Impulsive Interval* [28]) For the impulsive sequences $\{t_k\}_{k\in\mathbb{N}^+}$, let $N_i(t', t'')$ be the number of impulses on an interval $[t', t'')$. If there exist two numbers $N_{i0} > 0$, $\eta_i > 0$ such that

$$N_i(t', t'') \geq -N_{i0} + \frac{t'' - t'}{\eta_i}, \tag{32}$$

then the impulsive sequence $\{t_k\}_{k\in\mathbb{N}^+}$ is said to have an inverse average impulsive interval η_i.

Theorem 2 *Under Assumptions 1 and 2, if there exist positive definite matrices P_i, real parameters $\mu \geq 1$, $0 < \delta < 1$, $c_i > 0$, $\rho_{ij} > 0\,(i, j \in \mathcal{M})$ satisfying*

$$\begin{bmatrix} -P_iC_i - C_iP_i - \underline{B}_{i1}\Lambda_i + c_iQ_i - \bar{\gamma}P_i & P_iD_i + \bar{B}_{i2}\Lambda_i \\ * & -\Lambda_i \end{bmatrix} < 0, \tag{33}$$

$$P_i \leq c_iI_n, \tag{34}$$

$$P_i \leq \mu P_j,\ i \neq j, \tag{35}$$

$$A^TP_iA \leq \delta P_i, \tag{36}$$

$$\bar{\gamma} + \frac{\ln\mu}{\eta_s} + \frac{\ln\delta}{\eta_i} < 0, \tag{37}$$

stop where Λ_i, $\underline{B}_{i1}$, $\underline{B}_{i2}$ have been defined in Theorem 1, then system (1) is ES-M when the switching signal $r(t)$ has an average dwell time η_s and the impulsive sequence $\{t_k\}_{k\in\mathbb{N}^+}$ an inverse average impulsive interval η_i.

Proof As in Theorem 1, for any $t > t_0$, consider the interval $[t_0, t)$ and the switching sequence and the impulsive sequence are assumed to be $\{t_1^s, t_2^s, \ldots, t_{l_1}^s\}$ and $\{t_1, t_2, \ldots, t_{l_2}\}$, which satisfy $t_0 < t_1^s < t_2^s < \cdots < t_{j_1}^s < t_1 < t_{j_1+1}^s < \cdots < t_{j_2}^s < t_2 < t_{j_2+1}^s < \cdots < t_{j_k}^s < t_{l_2} < t_{j_k+1}^s < \cdots < t_{l_1}^s < t$.

Then, the candidate Lyapunov function is defined as

$$V_{r(t)}(x(t), t) = x^T(t)P_{r(t)}x(t).$$

From system (1), we have

$$\begin{aligned}
&\mathcal{L}V_{r(t)}(x(t),t)-\bar{\gamma}V_{r(t)}(x(t),t)\\
&\quad=2x^T(t)P_{r(t)}\big(-C_{r(t)}x+D_{r(t)}F_{r(t)}(x)\big)\\
&\quad+\mathtt{trace}\{G^T_{r(t)}(x,t)P_{r(t)}G_{r(t)}(x,t)\}-\bar{\gamma}x^T(t)P_{r(t)}x(t).
\end{aligned}\tag{38}$$

Substituting (13) and (15) into (38) leads to

$$\begin{aligned}
&\mathcal{L}V_{r(t)}(x(t),t)-\bar{\gamma}V_{r(t)}(x(t),t)\\
&\quad=2x^T(t)P_{r(t)}\big(-C_{r(t)}x+D_{r(t)}F_{r(t)}(x)\big)\\
&\quad+c_ix^T(t)Q_ix(t)-\bar{\gamma}x^T(t)P_{r(t)}x(t)\\
&\quad+\big[x^T\ \ F_i^T(x)\big]\begin{bmatrix}-\underline{B}_{i1}\Lambda_i & \bar{B}_{i2}\Lambda_i\\ * & -\Lambda_i\end{bmatrix}\begin{bmatrix}x\\ F_i(x)\end{bmatrix},
\end{aligned}\tag{39}$$

by which as well as condition (33), we have

$$\mathcal{L}V_{r(t)}(x(t),t)-\bar{\gamma}V_{r(t)}(x(t),t)\le 0,$$

that is,

$$\mathcal{L}V_{r(t)}(x(t),t)\le\bar{\gamma}V_{r(t)}(x(t),t).\tag{40}$$

Let $\bar{W}(x(t))=e^{-\bar{\gamma}t}V_{r(t)}(x(t),t)$. With the means of Itô formula and (40), one has

$$\begin{aligned}
\mathcal{L}\bar{W}(x(t))&=-\bar{\gamma}e^{-\bar{\gamma}t}V_{r(t)}(x(t),t)+e^{-\bar{\gamma}t}\mathcal{L}V_{r(t)}(x(t),t)\\
&\le-\bar{\gamma}e^{\bar{\gamma}t}V_{r(t)}(x(t),t)+\bar{\gamma}e^{\bar{\gamma}t}V_{r(t)}(x(t),t)\\
&=0,
\end{aligned}\tag{41}$$

from which it follows that $\bar{W}(x(t))$ is a continuous non-negative supermartingale.

Thus, for $\forall t_0<s_1<s_2$,

$$\mathbb{E}[\bar{W}(x(s_2))|\mathcal{F}_{s_1}]\le\bar{W}(x(s_1)).\tag{42}$$

Then, taking expectation on both sides of (19) gives

$$\mathbb{E}\bar{W}(x(s_2))\le\mathbb{E}\bar{W}(x(s_1)).\tag{43}$$

Therefore,

$$\mathbb{E}V_{r(s_2)}(x(s_2),s_2)\le e^{\bar{\gamma}(s_2-s_1)}\mathbb{E}V_{r(s_1)}(x(s_1),s_1).\tag{44}$$

With this in mind, we first focus on the interval $[t_0, t_1^s)$. It is obvious that

$$\mathbb{E}V_{r(t_0)}(x(t_1), t_1^s) \leq e^{\bar{\gamma}(t_1^s - t_0)}\mathbb{E}V_{r(t_0)}(x(t_0), t_0). \tag{45}$$

At the switching instant t_1^s from (8), it yields

$$\mathbb{E}V_{r(t_1^s)}(x(t_1^s), t_1^s) \leq \mu\mathbb{E}V_{r(t_0)}(x(t_1^s), t_1^s). \tag{46}$$

Combining (45) with (46), one obtains

$$\mathbb{E}V_{r(t_1)}(x(t_1^s), t_1^s) \leq \mu e^{\bar{\gamma}(t_1^s - t_0)}\mathbb{E}V_{r(t_0)}(x(t_0), t_0). \tag{47}$$

By repeating the above derivation procedure until the switching instant $t_{j_1}^s$, we arrive at

$$\mathbb{E}V_{r(t_{j_1}^s)}(x(t_{j_1}^s), t_{j_1}^s) \leq \mu^{j_1} e^{-\bar{\gamma}(t_{j_1}^s - t_0)}\mathbb{E}V_{r(t_0)}(x(t_0), t_0). \tag{48}$$

Considering the relation proposed in (26), then we get

$$\mathbb{E}V_{r(t_{j_1}^s)}(x(t_{j_1+1}^s), t_{j_1+1}^s) \leq \delta\mu^{j_1} e^{-\bar{\gamma}(t_{j_1+1}^s - t_0)}\mathbb{E}V_{r(t_0)}(x(t_0), t_0). \tag{49}$$

After a simple recurrence, we finally have

$$\begin{aligned}\mathbb{E}V_{r(t)}(x(t), t) &\leq \delta^{N_i(t_0,t)}\mu^{N_s(t_0,t)}e^{\bar{\gamma}(t-t_0)}\mathbb{E}V_{r(t_0)}(x(t_0), t_0)\\ &= e^{\ln\delta N_i(t_0,t)+\ln\mu N_s(t_0,t)+\bar{\gamma}(t-t_0)}\mathbb{E}V_{r(t_0)}(x(t_0), t_0).\end{aligned} \tag{50}$$

Using Definitions 3 and 5, the inequality (28) can be computed as

$$\begin{aligned}\mathbb{E}V_{r(t)}(x(t), t) \leq &\, e^{-\ln\delta N_{i0}+\frac{\ln\delta}{\eta_i}(t-t_0)+\ln\mu N_{s0}+\frac{\ln\mu}{\eta_i}(t-t_0)+\bar{\gamma}(t-t_0)}\\ &\times \mathbb{E}V_{r(t_0)}(x(t_0), t_0)\\ =&\, e^{-\ln\delta N_{i0}+\ln\mu N_{s0}}\mathbb{E}V_{r(t_0)}(x(t_0), t_0)e^{(\bar{\gamma}+\frac{\ln\mu}{\eta_S}+\frac{\ln\delta}{\eta_i})(t-t_0)}.\end{aligned} \tag{51}$$

Based on the definition $V_{r(t)}(x(t), t)$, it derives from (29) that

$$\mathbb{E}|x(t)|^2 \leq \frac{\lambda_{\max}(P_i)}{\lambda_{\min}(P_j)}e^{-\ln\delta N_{i0}+\ln\mu N_{s0}}\mathbb{E}|x(t_0)|^2 e^{(\bar{\gamma}+\frac{\ln\mu}{\eta_S}+\frac{\ln\delta}{\eta_i})(t-t_0)}, \tag{52}$$

which together with moment inequality results in

$$\mathbb{E}|x(t)| \leq \sqrt{\frac{\lambda_{\max}(P_i)}{\lambda_{\min}(P_j)}e^{-\ln\delta N_{i0}+\ln\mu N_{s0}}\mathbb{E}|x(t_0)|^2}\; e^{\frac{(\bar{\gamma}+\frac{\ln\mu}{\eta_S}+\frac{\ln\delta}{\eta_i})(t-t_0)}{2}}. \tag{53}$$

Therefore, it is concluded that system (1) is ES-M. The proof is complete.

Remark 3 It is obvious that continuous subsystems are unstable based on (33) and (40). However, condition (36) implies the stabilizing effect of impulses. By designing both a switching signal and an impulsive signal satisfying (37), ES-M of the system (1) can still be guaranteed.

Inspired by the above two proposed cases, we next consider a special case, i.e., stable subsystems with stabilizing impulses.

Theorem 3 *Under Assumptions 1 and 2, if there exist positive definite matrices P_i, real parameters $\mu \geq 1, 0 < \delta < 1, c_i > 0, \rho_{ij} > 0\,(i, j \in \mathcal{M})$ satisfying*

$$\begin{bmatrix} -P_iC_i - C_iP_i - \underline{B}_{i1}\Lambda_i + c_iQ_i + \bar{\gamma}P_i & P_iD_i + \bar{B}_{i2}\Lambda_i \\ * & -\Lambda_i \end{bmatrix} < 0, \tag{54}$$

$$P_i \leq c_i I_n, \tag{55}$$

$$P_i \leq \mu P_j,\ i \neq j, \tag{56}$$

$$A^T P_i A \leq \delta P_i, \tag{57}$$

$$\bar{\gamma} - \frac{\ln \mu}{\eta_s} - \frac{\ln \delta}{\eta_i} > 0, \tag{58}$$

where Λ_i, $\underline{B}_{i1}$, $\underline{B}_{i2}$ take the same form defined in Theorem 1, then system (1) is ES-M when the switching signal $r(t)$ has an average dwell time η_s and the impulsive sequence $\{t_k\}_{k\in\mathbb{N}^+}$ an inverse average impulsive interval η_i.

Proof It can be proven by referring to the proof processes of Theorems 1 and 2 and therefore the details are omitted here. □

Remark 4 Although conditions (11) and (58) are expressed exactly the same, they are fundamentally different caused by the parameter δ. To see it clearly, from (11) or (58) we have $\bar{\gamma} - \frac{\ln \delta}{\eta_i} > \frac{\ln \mu}{\eta_s}$. Note that $\bar{\gamma} - \frac{\ln \delta}{\eta_i}$ with $0 < \delta < 1$ is bigger than $\bar{\gamma} - \frac{\ln \delta}{\eta_i}$ with $\delta > 1$, which illustrates that condition (58) provides more freedom with the implementation of the switching signal $r(t)$.

3.2 Controller Design

The issue of controller design for the system (1) is to be dealt with by feat of the results developed in Sect. 3.1. The asynchronous switching and impulsive controllers $u_1(t)$ and $u_2(t)$ are designed as

$$\begin{aligned} u_1(t) =& K_{r(t)}x(t), \\ u_2(t_k^-) =& K_2 x(t_k^-). \end{aligned} \tag{59}$$

By substituting (59) into (1), we immediately have the following closed-loop system

$$\begin{aligned} dx(t) &= \big(-(C_{r(t)} - K_{r(t)})x + D_{r(t)}F_{r(t)}(x) \big)\, dt \\ &\quad + G_{r(t)}(x,t)\, d\omega(t),\ t \neq t_k, \\ x(t_k) &= (A + K_2)x(t_k^-),\ t = t_k, \\ x(t_0) &= x_0, r(t_0) = r_0. \end{aligned} \tag{60}$$

The controller gain matrices can be obtained by the following result.

Theorem 4 *Under Assumptions 1 and 2, if there exist positive definite matrices P_i, matrices H_i, H_2, real parameters $\mu \geq 1, 0 < \delta < 1, c_i > 0, \rho_{ij} > 0\, (i, j \in \mathcal{M})$ satisfying*

$$\begin{bmatrix} \begin{pmatrix} -P_i(C_i - H_i) - (C_i - H_i)P_i \\ -\underline{B}_{i1}\Lambda_i + c_iQ_i + \bar{\gamma}P_i \end{pmatrix} & P_iD_i + \bar{B}_{i2}\Lambda_i \\ * & -\Lambda_i \end{bmatrix} < 0, \tag{61}$$

$$P_i \leq c_i I_n, \tag{62}$$

$$P_i \leq \mu P_j,\ i \neq j, \tag{63}$$

$$\begin{bmatrix} -\delta P_i & A^T P_i + H_2 \\ * & -P_i \end{bmatrix} < 0, \tag{64}$$

$$\bar{\gamma} - \frac{\ln \mu}{\eta_s} - \frac{\ln \delta}{\eta_i} > 0, \tag{65}$$

where $\Lambda_i, \underline{B}_{i1}, \underline{B}_{i2}$ take the same form defined in Theorem 1, then system (1) is E-S-M when the switching signal $r(t)$ has an average dwell time η_s and the impulsive sequence $\{t_k\}_{k\in\mathbb{N}^+}$ an inverse average impulsive interval η_i. Moreover, the controller gain matrices are $K_i = H_i$, $K_2 = P_i^{-1}H_2^T$.

Proof We can prove it by simply using Theorem 3 and hence omit the proof details.

4 Numerical Examples

In this section, two simulation examples are provided to illustrate the effectiveness of our results.

Example 1 Consider system (1) with the following parameters

$$C_1 = \begin{bmatrix} 6 & 0 & 0 \\ 0 & 6 & 0 \\ 0 & 0 & 6 \end{bmatrix}, D_1 = \begin{bmatrix} -1 & 0 & 0.3 \\ 0 & -1 & 0.6 \\ 1.4 & 0 & 2 \end{bmatrix}, G_1 = \begin{bmatrix} -1 & 0.8 & 0 \\ -3 & -0.5 & 0 \\ 0.1 & 0.2 & -0.1 \end{bmatrix},$$

$$C_2 = \begin{bmatrix} 3 & 0 & 0 \\ 0 & 3 & 0 \\ 0 & 0 & 3 \end{bmatrix}, D_2 = \begin{bmatrix} 0.75 & -0.5 & 0 \\ 1.1 & -0.1 & 0 \\ 1.4 & 0 & 2 \end{bmatrix}, G_2 = \begin{bmatrix} 1 & -0.7 & 0.1 \\ 0.8 & 1 & -0.5 \\ 0.1 & 0.2 & -0.1 \end{bmatrix},$$

$$A = 1.4 * \begin{bmatrix} 1 & 0 & 0 \\ 0 & 1 & 0 \\ 0 & 0 & 1 \end{bmatrix},$$

and the neuron activation functions $F_{ij}(x_j)$ are defined as

$$F_{ij}(x_j) = f_{ij}(|x_j + 1| - |x_j - 1|)$$

where $f_{ij} \in \mathbb{R}$.

By choosing $f_{11} = 1.1$, $f_{12} = f_{13} = 1.2$, $f_{21} = f_{22} = f_{23} = 0.7$, it is easy to verify the stability of the continuous subsystems. Moreover, from the definition of matrix A, we have the destabilizing impulses. In order to examine the stability of system (1) defined by the above parameters, we next turn our attention to Theorem 1.

To obtain a set of feasible solutions satisfying conditions (7)–(10), let $\mu = 2$, $\delta = 2$, $c_1 = 1.3$, $c_2 = 2$, $\bar{\gamma} = 1.4$, and then we have the following results:

$$P_1 = \begin{bmatrix} 1.2858 & -0.0451 & -0.0057 \\ -0.0451 & 1.1125 & -0.0409 \\ -0.0057 & -0.0409 & 1.1053 \end{bmatrix},$$

$$P_2 = \begin{bmatrix} 1.8788 & 0.0430 & 0.0894 \\ 0.0430 & 1.5101 & -0.3957 \\ 0.0894 & -0.3957 & 1.2994 \end{bmatrix},$$

$$\Lambda_1 = \begin{bmatrix} 2.5694 & 0 & 0 \\ 0 & 5.4837 & 0 \\ 0 & 0 & 22.2668 \end{bmatrix},$$

$$\Lambda_2 = \begin{bmatrix} 11.7520 & 0 & 0 \\ 0 & 10.1126 & 0 \\ 0 & 0 & 5.4046 \end{bmatrix},$$

which implies the feasibility of Theorem 1.

To perform the simulation results, choose the initial conditions $x(0) = [3; -3.8; -3]$, $r(0) = 1$, and $\eta_s = 2$, $\eta_i = 1.3$, which satisfy condition (11). The system states are plotted in Fig. 1, from which we have that the stability of system (1) can still be guaranteed if the destabilizing impulses do not happen frequently. The corresponding switching signal $r(t)$ and impulsive instants are shown in Figs. 2 and 3, respectively.

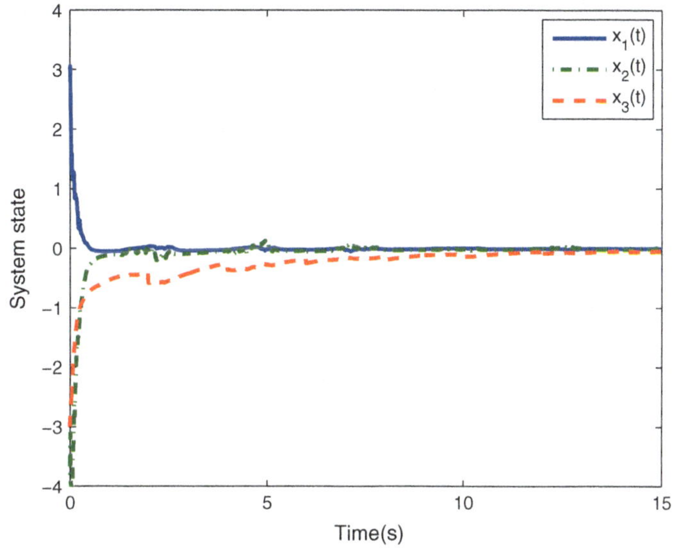

Fig. 1 Time variation of system states

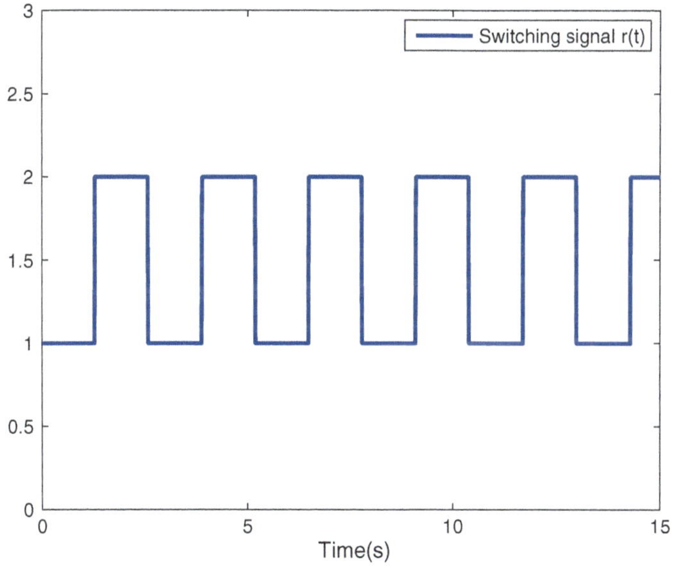

Fig. 2 The evolutionary process of switching signal $r(t)$

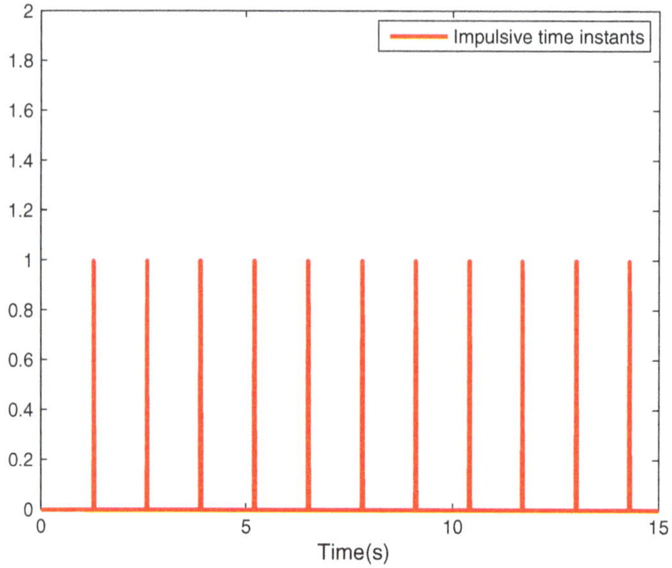

Fig. 3 Times-series of impulsive instants

Example 2 We next consider system (1) defined by the following parameters

$$C_1 = \begin{bmatrix} 1.2 & 0 & 0 \\ 0 & 2 & 0 \\ 0 & 0 & 2 \end{bmatrix}, D_1 = \begin{bmatrix} 0.3 & 0.1 & 0 \\ 0 & 0.4 & 0 \\ 1 & 2 & 1 \end{bmatrix}, G_1 = \begin{bmatrix} 1 & 0 & 0 \\ 1 & 1 & -0.1 \\ -0.7 & -0.4 & 1 \end{bmatrix},$$

$$C_2 = \begin{bmatrix} 2.3 & 0 & 0 \\ 0 & 2.3 & 0 \\ 0 & 0 & 2.3 \end{bmatrix}, D_2 = \begin{bmatrix} 0.3 & 0.1 & 0 \\ 0 & 0.4 & 0 \\ 1 & 2 & 1 \end{bmatrix},$$

$$G_2 = \begin{bmatrix} 1 & 0 & 0 \\ 0.1 & 1 & -0.1 \\ -0.7 & -0.4 & 1 \end{bmatrix}, A = 0.1 * \begin{bmatrix} 1 & 0 & 0 \\ 0 & 1 & 0 \\ 0 & 0 & 1 \end{bmatrix},$$

and the neuron activation functions $F_{ij}(x_j)$ are the same with that defined Example 1.

By choosing $f_{11} = f_{12} = f_{13} = 1.2$, $f_{21} = f_{22} = f_{23} = 2$, from the simulation results plotted in Figs. 4 and 5, we can see that the continuous subsystems are unstable. Note that the definition of matrix A implies the stabilizing effectiveness of impulses. Therefore, we next apply Theorem 2 to examine the stability of system (1) defined by the above parameters.

Letting $\mu = 3$, $\delta = 0.2$, $c_1 = 0.3$, $c_2 = 0.2$, $\bar{\gamma} = 0.4$, we have the following results satisfying conditions (33)–(36),

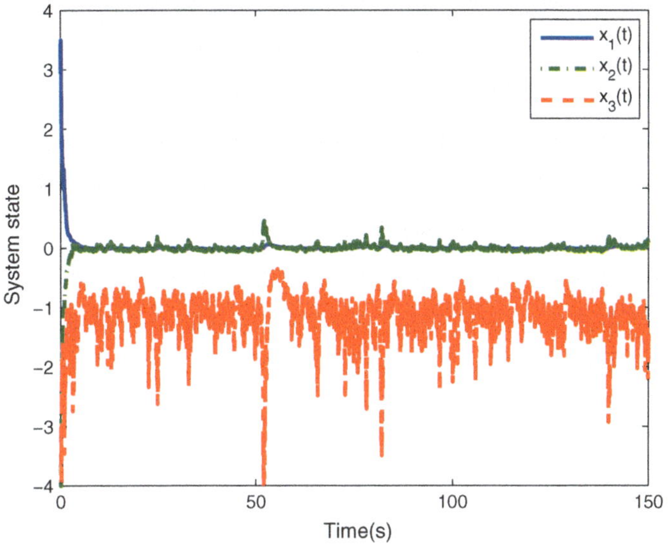

Fig. 4 Time variation of the first subsystem states

$$P_1 = \begin{bmatrix} 0.2631 & 0.0142 & -0.0311 \\ 0.0142 & 0.2595 & 0.0172 \\ -0.0311 & 0.0172 & 0.2151 \end{bmatrix},$$

$$P_2 = \begin{bmatrix} 0.1716 & 0.0060 & -0.0122 \\ 0.0060 & 0.1806 & 0.0092 \\ -0.0122 & 0.0092 & 0.1531 \end{bmatrix},$$

$$\Lambda_1 = \begin{bmatrix} 0.6676 & 0 & 0 \\ 0 & 1.7864 & 0 \\ 0 & 0 & 0.5230 \end{bmatrix},$$

$$\Lambda_2 = \begin{bmatrix} 0.9148 & 0 & 0 \\ 0 & 1.2015 & 0 \\ 0 & 00.3511 & \end{bmatrix},$$

which implies the ES of system (1).

To perform the simulation results, choose $\eta_s = 2$, $\eta_i = 0.3$, and obviously condition (37) holds.

Choosing the initial conditions $x(0) = [3; -3.8; -3]$ and $r(0) = 1$, we plot the system states in Fig. 6, the corresponding switching signal $r(t)$ in Fig. 7 and impulsive instants in Fig. 8, from which it is shown that the stability of system (1) with the unstable continuous subsystems can be achieved if the stabilizing impulses do happen frequently.

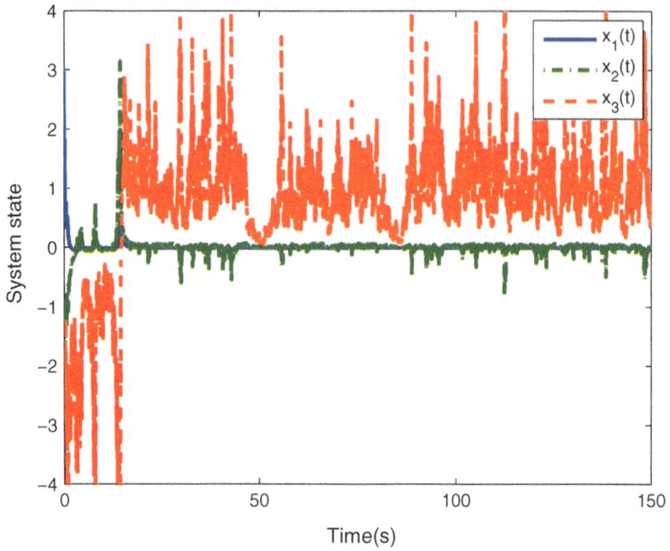

Fig. 5 Time variation of the second subsystem states

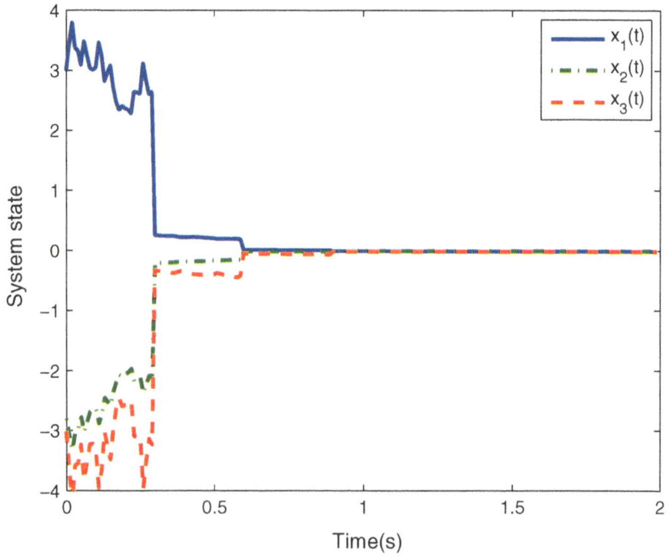

Fig. 6 Times-trajectories of system states

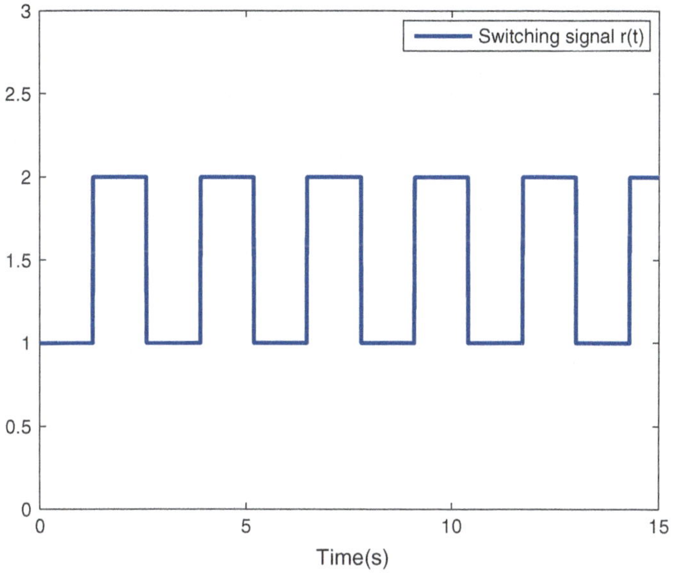

Fig. 7 The evolutionary process of switching signal $r(t)$

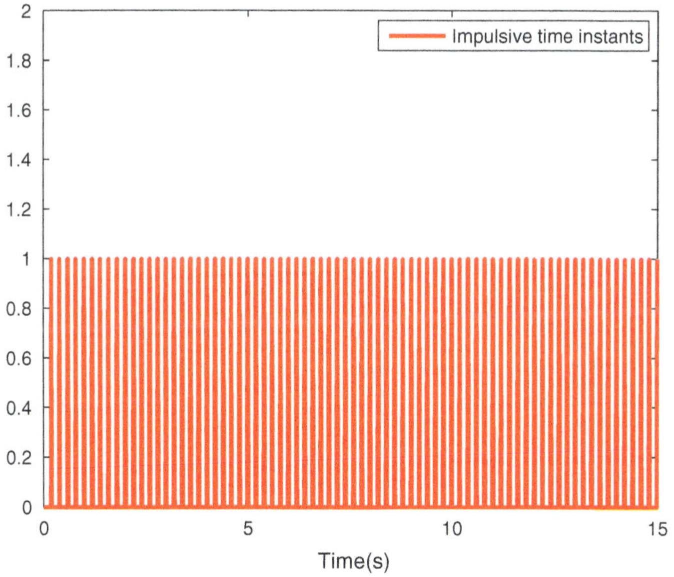

Fig. 8 Times-series of impulsive instants

5 Conclusion

In this chapter, the exponential stability and stabilization problem for a class of stochastic neural network systems has been addressed. Comparing with the previous results, the main distinction here is the asynchronization of switchings and impulses. By employing the concepts of dwell time and impulsive interval as well as multiple Lyapunov functions approach, sufficient conditions for mean square exponential stability and stabilization have been proposed in terms of LMIs.

Acknowledgements This work of T. Jiao was supported by National Natural Science Foundation of China (61703249) and China Postdoctoral Science Foundation (2019M652351). Also, the work of J.H. Park was supported by the National Research Foundation of Korea (NRF) grant funded by the Korea government (Ministry of Science and ICT) (No. 2020R1A2B5B02002002).

References

1. Wang, J., Huang, T., Wu, H., Ren, S.: Analysis and Control of Coupled Neural Networks with Reaction-Diffusion Terms. Springer, Berlin (2018)
2. Li C, Feng G, Huang T (2008) On hybrid impulsive and switching neural networks. IEEE Trans. Syst. Man Cybern. Part B (Cybern.) **38**(6), 1549–1560
3. Zhang, B., Xu, S., Zong, G., Zou, Y.: Delay-dependent exponential stability for uncertain stochastic Hopfield neural networks with time-varying delays. IEEE Trans. Circuits Syst. I: Regul. Pap. **56**(6), 1241–1247 (2008)
4. Park, J.H., Shen, H., Chang, X.H., Lee, T.H.: Recent Advances in Control and Filtering of Dynamic Systems with Constrained Signals. Springer, Cham, Switzerland (2018). https://doi.org/10.1007/978-3-319-96202-3
5. Shen, H., Zhu, Y., Zhang, L., Park, J.H.: Extended dissipative state estimation for Markov jump neural networks with unreliable links. IEEE Trans. Neural Netw. Learn. Syst. **28**(2), 346–358 (2017)
6. Park, J.H., Kwon, O.M., Lee, S.M.: LMI optimization approach on stability for delayed neural networks of neutral-type. Appl. Math. Comput. **196**(1), 236–244 (2008)
7. Sun, H., Zong, G., Ahn, C.K.: Quantized decentralized adaptive neural network PI tracking control for uncertain interconnected nonlinear systems with dynamic uncertainties. IEEE Trans. Syst. Man. Cybern.: Syst. (2019). https://doi.org/10.1109/TSMC.2019.2918142
8. Niu, B., Zhao, X., Fan, X., Cheng, Y.: A new control method for state-constrained nonlinear switched systems with application to chemical process. Int. J. Control **88**(9), 1693–1701 (2015)
9. Gordillo G.L.O. (2015) Observers design for uncertain descriptor systems: application to control and diagnosis. Ph.D. thesis
10. Fu, J., Ma, R., Chai, T., Hu, Z.: Dwell-time-based standard $\mathcal{H}_\infty$ control of switched systems without requiring internal stability of subsystems. IEEE Trans. Autom. Control **64**(7), 3019–3025 (2019)
11. Zhu, B., Suo, M., Chen, L., Li, S.: Stability and L_1-gain analysis for positive Takagi-Sugeno fuzzy systems with impulse. IEEE Trans. Fuzzy Syst. **26**(6), 3893–3901 (2018)
12. Liu, C., Yang, Z., Sun, D., Liu, X., Liu, W.: Stability of switched neural networks with time-varying delays. Neural Comput. Appl. **30**(7), 2229–2244 (2018)
13. Zhang, L., Zhu, Y., Zheng, W.X.: State estimation of discrete-time switched neural networks with multiple communication channels. IEEE Trans. Cybern. **47**(4), 1028–1040 (2016)
14. Xia, J., Chen, G., Sun, W.: Extended dissipative analysis of generalized Markovian switching neural networks with two delay components. Neurocomputing **260**, 275–283 (2017)

15. Huang, T., Li, C., Duan, S., Starzyk, J.A.: Robust exponential stability of uncertain delayed neural networks with stochastic perturbation and impulse effects. IEEE Trans. Neural Netw. Learn. Syst. **23**(6), 866–875 (2012)
16. Zhang, W., Tang, Y., Wu, X., Fang, J.: Synchronization of nonlinear dynamical networks with heterogeneous impulses. IEEE Trans. Circuits Syst. I: Regul. Pap. **61**(4), 1220–1228 (2013)
17. Yang, X., Lu, J., Ho, D.W., Song, Q.: Synchronization of uncertain hybrid switching and impulsive complex networks. Appl. Math. Modell. **59**, 379–392 (2018)
18. Li, C., Ma, F., Feng, G.: Hybrid impulsive and switching time-delay systems. IET Control Theory Appl. **3**(11), 1487–1498 (2009)
19. Liu, J., Liu, X., Xie, W.C.: Input-to-state stability of impulsive and switching hybrid systems with time-delay. Automatica **47**(5), 899–908 (2011)
20. Zong, G., Yang, D.: $\mathcal{H}_\infty$ synchronization of switched complex networks: a switching impulsive control method. Commun. Nonlinear Sci. Numer. Simul. **77**, 338–348 (2019)
21. Jiao, T., Park, J.H., Zong, G., Zhao, Y., Du, Q.: On stability analysis of random impulsive and switching neural networks. Neurocomputing **350**, 146–154 (2019)
22. Zhang, X., Li, C., Huang, T.: Hybrid impulsive and switching Hopfield neural networks with state-dependent impulses. Neural Netw. **93**, 176–184 (2017)
23. Ren, W., Xiong, J.: Stability analysis of impulsive switched time-delay systems with state-dependent impulses. IEEE Trans. Autom. Control **64**(9), 3928–3935 (2019)
24. Zhang, W., Tang, Y., Miao, Q., Du, W.: Exponential synchronization of coupled switched neural networks with mode-dependent impulsive effects. IEEE Trans. Neural Netw. Learn. Syst. **24**(8), 1316–1326 (2013)
25. Jiao, T., Park, J.H., Zong, G.: Stability criteria of stochastic nonlinear systems with asynchronous impulses and switchings. Nonlinear Dyn. **97**(1), 135–149 (2019)
26. Mao X, Yuan C (2006) Stochastic Differential Equations with Markovian Switching. Imperial college press
27. Zhao, X., Zhang, L., Shi, P., Liu, M.: Stability and stabilization of switched linear systems with mode-dependent average dwell time. IEEE Trans. Autom. Control **57**(7), 1809–1815 (2011)
28. Lu, J., Ho, D.W., Cao, J.: A unified synchronization criterion for impulsive dynamical networks. Automatica **46**(7), 1215–1221 (2010)

Hybrid-Triggered Synchronization of Delayed Complex Dynamical Networks Subject to Stochastic Cyber-Attacks

Xiaojian Yi and Yajuan Liu

Abstract This chapter is concerned with the synchronization problem for a class of delayed complex networks with random occurring cyber-attacks. First, a hybrid event-triggered communication scheme is proposed to reduce the transmission load of networks, which includes the event-triggered scheme and the time-triggered scheme in one framework. Then, a synchronization error dynamic system is obtained for the considered networks, and a new sufficient condition is derived such that the synchronization error system is asymptotically stable in the mean-square sense. Furthermore, it is also shown that the controller gain matrices can be determined by solving some matrix inequalities. Finally, two simulation examples are introduced to validate the main results.

Keywords Synchronization · Complex dynamical networks · Hybrid-triggered scheme · Stochastic cyber-attacks

1 Introduction

Many natural and man-made systems can be expressed as networks, where many nodes interact through the network link [1–4]. Complex dynamical networks have attracted increasing attention in various fields such as social networks, the World Wide Web, food webs, computer networks, and so on [5]. A complex network refers to a set of coupled interconnected nodes, where each node can be considered as a dynamical system. One of the important research topics of complex dynamical

X. Yi
School of Mechanical Engineering, Beijing Institute of Technology, Beijing, People's Republic of China
e-mail: yixiaojianbit@sina.cn

Y. Liu (✉)
School of Control and Computer Engineering, North China Electric Power University, Beijing, People's Republic of China
e-mail: yajuan.liu.12@gmail.com

J. H. Park (ed.), *Recent Advances in Control Problems of Dynamical Systems and Networks*, Studies in Systems, Decision and Control 301,
https://doi.org/10.1007/978-3-030-49123-9_19

networks is synchronization. Synchronization is regarded as a kind of collective behavior which is exhibited in many practical systems, see, e.g., chemical reactions, biological systems, and secure communication [6–8]. Thus, the problem of synchronization of complex dynamical networks has been studied by many researchers and many important works have been reported in recent years [9–11].

In general, not all nodes in complex dynamical networks can synchronize up to each other because of the complexities of node dynamics and also the topological structure of the network. Therefore, the synchronization control problem of complex networks deserves more attention. In the last a few years, various control schemes have been successfully applied to achieve synchronization of complex dynamical networks, see, e.g., adaptive control [10], impulsive control [9] and non-fragile control [11]. However, it should be noted that all results mentioned above are only focused on continuous-time controllers, which may not be easily implemented in real-world sometimes [12]. In order to solve this situation, sampled-data [13, 14] and event-triggered control [15] schemes are proposed to reduce the transmission load of network more effectively. Furthermore, the hybrid event-trigger scheme was established for stabilization of networked control systems [16]. Recently, the synchronization for complex dynamical networks with sampled-data was investigated in [17, 18]. In addition, the event-triggered synchronization for complex dynamical networks has also been studied in [19, 20]. However, the hybrid-driven communication scheme for synchronization for the complex dynamical network has not been investigated yet, which is one aim of this study.

On the other hand, nodes in the network usually communicate with each via communication networks. For example, the vehicle in transportation system communicates with other vehicles via wireless networks. However, the network not only brings convenience and but also introduces security threats [21–23]]. Recently, the security issue has attracted increasing attention. For example, in [24], the deception attacks were taken into account in dealing with the problem of consensus for multi-agent systems. The consensus of heterogeneous multi-agent systems with DoS attach was addressed in [25]. The problem of $\mathcal{H}_\infty$ control of networked cascade control systems with stochastic cyber attacks and hybrid-driven-based transmission scheme was investigated in [26]. The problem of quantized stabilization for Takagi–Sugeno fuzzy systems with stochastic cyber-attack were studied in [27]. However, for the synchronization of complex dynamical networks with cyber-attacks, few results have been published, especially when the hybrid-triggered scheme is used, which is another motivation of this study.

Motivated by the above discussions, this paper is concerned with the study of hybrid-triggered synchronization for delayed complex dynamical networks stochastic cyber-attacks. A hybrid-triggered scheme can be represented by Bernoulli variable and a type of nonlinear function is used to describe the features of cyber-attacks.

The main contributions of this chapter are listed:

(1) The hybrid triggered scheme is first considered in the study of synchronization control of complex dynamical networks while taking the effect of stochastic cyber attacks into account;

(2) A closed-loop synchronization error system model is first established by considering the limited network resources and stochastic cyber-attacks;
(3) According to the constructed model, the synchronization criterion for the considered complex dynamical networks is obtained by using the Lyapunov stability theory. Moreover, the controller gain matrices are calculated by solving a set of liner matrix inequality (LMIs).

Notations. Throughout this paper, I denotes the identity matrix with appropriate dimensions, ^{T} is the matrix transposition. $\mathbb{R}^n$ denotes the n-dimensional Euclidean space, and $\mathbb{R}^{m\times n}$ is the set of all $m \times n$ real matrices, $*$ denotes the elements below the main diagonal of a symmetric block matrix. For A, $A > 0$ means that the matrix A is a real symmetric positive definite.

2 Problem Statement

Suppose we have a network with N nodes, and the it is described by

$$\dot{x}_i(t) = \phi(x_i(t)) + c\sum_{j=1}^{N} g_{ij}Ax_j(t-\tau(t)) + u_i(t), \tag{1}$$

where $x_i(t)$ denotes the state vector and $u_i(t)$ denotes the control input; $\phi : \mathbb{R}^n \to \mathbb{R}^n$ represents a continuous vector-valued function; the positive scalar constant c is the coupling strength; $\tau(t)$ describes a time-varying delay, which satisfies $0 \le \tau(t) \le \tau_M$, $\dot{\tau}(t) \le \mu$, where τ_M and μ are known real constants; $A = (a_{ij})_{n\times n}$ is the inner-coupling matrix; $\mathcal{G} = (g_{ij})_{N\times N}$ represents an outer-coupling configuration matrix where G_{ij} is defined as: if there exists a connection between node i and j, then $g_{ij} > 0$; $g_{ij} = 0$, otherwise, the diagonal elements of matrix G are given as

$$g_{ii} = -\sum_{j=1, j\neq i}^{N} g_{ij} \quad (i = 1, 2, \ldots, N). \tag{2}$$

For constant matrices U_1 and U_2 with appropriate dimensions. The nonlinear function $\phi : \mathbb{R}^n \to \mathbb{R}^n$ meet the following sector-bound condition [12]:

$$\begin{aligned}&[\phi(x_1) - \phi(x_2) - U_1(x_1 - x_2)]^T \times \\ &\quad [\phi(x_1) - \phi(x_2) - U_2(x_1 - x_2)] \le 0, \forall x_1, x_2 \in \mathbb{R}^n,\end{aligned} \tag{3}$$

We now define the synchronization error signal as $\rho e(t) = x_i(t) - s(t)$, where $s(t) \in \mathbb{R}^n$ the state of the target node with $\dot{s}(t) = \phi(s(t))$. Thus, the error dynamics can be described by as

$$\dot{\rho}_i(t) = f(\rho_i(t)) + c\sum_{j=1}^{N} g_{ij}A\rho_j(t-\tau(t)) + u_i(t), \tag{4}$$

where $i = 1, 2, \ldots, N$ and $f(\rho_i(t)) = \phi(x_i(t)) - \phi(s)$.

In this chapter, the hybrid-driven-based scheme including event-triggered scheme and time-triggered scheme is established, and it is employed to reduce the communication load. In the following part, the two cases are discussed in detail.

2.1 *Time-Triggered Scheme*

When the network resource is not limited, we choose to use the 'time-triggered' mechanism to transmit the data, then the control input is calculated based on the following data:

$$\rho_{i1}(t) = \rho_i(t_k h), t \in [t_k h + d_{t_k}, t_{k+1}h + d_{t_{k+1}}) \tag{5}$$

where h denotes the sampling period, t_k denotes some integers satisfying $\{t_1, t_2, \ldots\} \in \{0, 1, 2, \ldots\}$, d_{t_k} is the communication delay [28].

Similar to [18], Eq. (5) can be rewritten as

$$\rho_{i1}(t) = \rho_i(t - d(t)) \tag{6}$$

where $d(t) \in [0, d_M]$, d_M is the upper bound of $d(t)$.

2.2 *Event-Triggered Scheme*

When the network resource is not limited, we choose to use the 'event-triggered' mechanism.

The release time under the event-triggered scheme can be expressed as follows [15]

$$t_{k+1} = t_k h + inf_{j\geq 1}\{jh | r_{ik}^T(t)\Omega r_{ik}(t) \leq \sigma\rho_i^T(t_k h)\Omega\rho_i(t_k h)\}, \tag{7}$$

where $r_{ik}(t) = \rho_i(t_k h) - \rho_i(t_k h + jh)$ denotes the threshold error, Ω is positive, $\sigma \in [0, 1)$, and $j = 1, 2, \ldots$.

In order to analyze more easily, similar to the method in [27], dividing the interval $[t_k h + d_{t_k}, t_{k+1}h + d_{t_{k+1}})$ into several intervals, and assuming that there exists a constant N satisfying $[t_k h + d_{t_k}, t_{k+1}h + d_{t_{k+1}}) = \bigcup_{j=1}^{N} \prod_j$, where $\prod_j = [t_k h + jh + d_{t_k+j}, t_k h + jh + h + d_{t_k+j+1}]$, $j = \{1, 2, \ldots, N\}$, $N = t_{k+1} - t_k - 1$, and define

$\eta(t) = t - t_k h - jh, 0 \leq d_{t_k} \leq \eta(t) \leq h + d_{t_k+j+1} = \eta_M$, then the error data can be rewritten as

$$\rho_{i2}(t) = \rho_i(t - \eta(t)) + r_{ik}(t). \tag{8}$$

Remark 1 In the event-triggered scheme (7), the parameter σ is a positive scalar, which is related to the transmission frequency. Furthermore, when $\sigma = 0$, the event-triggered scheme reduces to the time-triggered scheme.

Inspired by the work [26], the time-triggered scheme combined with the event-triggered scheme are adopted to reduce the communication burden, and a stochastic variable $\alpha(t)$ is introduced to trigger one of the above transmission scheme:

$$\begin{aligned}\bar{\rho}_i(t) &= \alpha(t)\rho_{i1}(t) + (1 - \alpha(t))\rho_{i2}(t)\\ &= \alpha(t)\rho_i(t - d(t)) + (1 - \alpha(t))(\rho_i(t - \eta(t)) + r_{ik}(t)),\end{aligned} \tag{9}$$

where $\alpha(t)$ take value on $[0, 1]$ with $\mathbb{E}\{\alpha(t)\} = \bar{\alpha}$.

In practical systems, the transmitted date over communication networked is susceptible to be attacked, which may cause degradation of the system performance.

As in [27], the cyber-attack is modeled as nonlinear function $g(\rho_i(t))$ and the corresponding time delay is supposed as $\theta(t) \in [0, \theta_M]$.

Inspired by the work in [26], a Bernoulli random variable $\beta(t)$, which is independent to $\alpha(t)$, is used to express the stochastic cyber attacks. Therefore, the signal transmitted to the controller, can be expressed as follows

$$\begin{aligned}\hat{\rho}_i(t) = \beta(t)g(\rho_i(t - \theta(t))) + (1 - \beta(t))\{\alpha(t)\rho_i(t - d(t))\\ +(1 - \alpha(t))[\rho_i(t - \eta(t)) + r_{ik}(t)]\}.\end{aligned} \tag{10}$$

where the cyber-attacks $g(u)$ are supposed to satisfy

$$||g(x)||_2 \leq ||Vx||_2 \tag{11}$$

with V is a constant matrix describing the upper bound of nonlinearity.

Remark 2 Inspired by the work [27], a Bernoulli variable is used to represent the probability of stochastic cyber-attack. When $\beta(t) = 1$, which means that the cyber-attacks are implemented successfully. Otherwise, when $\beta(t) = 0$, it means that the data is transmitted normally.

Then, the controller can be written as

$$\begin{aligned}u_i(t) = \beta(t)K_i g(\rho_i(t - \theta(t))) + (1 - \beta(t))K_i\{\alpha(t)\rho_i(t - d(t))\\ +(1 - \alpha(t))[\rho_i(t - \eta(t)) + r_{ik}(t)]\}.\end{aligned} \tag{12}$$

Subsisting (12) into (4), the closed-loop system (4) can be described as

$$\begin{aligned}\dot{\rho}_i(t) = f(\rho_i(t)) + c\sum_{j=1}^{N} g_{ij} A\rho_j(t-\tau(t)) + \beta(t)K_i g(\rho_i(t-\theta(t)))\\ +(1-\beta(t))K_i\{\alpha(t)\rho_i(t-d(t))\\ +(1-\alpha(t))[\rho_i(t-\eta(t)) + r_{ik}(t)]\}\end{aligned} \tag{13}$$

where $i = 1, 2, \ldots, N$.

Based on Kronecker product, the system (14) can be rewritten as

$$\begin{aligned}\dot{\rho}(t) = \bar{f}(\rho(t)) + c(G\otimes A)\rho(t-\tau(t)) + \beta(t)Kg(\rho(t-\theta(t)))\\ +(1-\beta(t))K\{\alpha(t)\rho(t-d(t))\\ +(1-\alpha(t))[\rho(t-\eta(t)) + r_k(t)]\}\end{aligned} \tag{14}$$

where

$$\begin{aligned}&\rho(t) = [\rho_1^T(t), \rho_2^T(t), \ldots, \rho_N^T(t)]^T,\\ &\bar{f}(\rho(t)) = [f^T(\rho_1(t)), f^T(\rho_2(t)), \ldots, f^T(\rho_N(t))]^T,\\ &\bar{g}(\rho(t)) = [g^T(\rho_1(t)), g^T(\rho_2(t)), \ldots, g^T(\rho_N(t))]^T,\\ &K = diag[K_1, K_2, \ldots, K_N],\\ &r_k(t) = [r_{1k}^T(t), r_{2k}^T(t), \ldots, r_{Nk}^T(t)]^T.\end{aligned} \tag{15}$$

The following lemmas will be given to obtain the main results of this work.

Lemma 1 ([29]) *For $M > 0 \in \mathbb{R}^{n\times n}$ and $\mathfrak{a} \leq s \leq \mathfrak{b}$, the following inequality holds:*

$$-(\mathfrak{b}-\mathfrak{a})\int_{\mathfrak{a}}^{\mathfrak{b}} \dot{\delta}^T(s)M\dot{\delta}(s)ds \leq -[\delta(\mathfrak{b}) - \delta(\mathfrak{a})]^T M [\delta(\mathfrak{b}) - \delta(\mathfrak{a})].$$

Lemma 2 ([30]) *Let $D \in \mathbb{R}^m$ is an open subset, and $\mathfrak{f}_1, \mathfrak{f}_2, \cdots, \mathfrak{f}_N : \mathbb{R}^m \to \mathbb{R}$ have positive values D. Then, one has*

$$\min_{\{\alpha_i|\alpha_i>0, \sum_i \alpha_i = 1\}} \sum_i \frac{1}{\alpha_i}\mathfrak{f}_i(t) = \sum_i \mathfrak{f}_i(t) + \max_{\mathfrak{m}_{ij}(t)} \sum_{i\neq j} \mathfrak{m}_{ij}(t)$$

subject to

$$\left\{\mathfrak{m}_{ij} : \mathbb{R}^m \to \mathbb{R}, \mathfrak{m}_{j,i}(t) \triangleq \mathfrak{m}_{i,j}(t), \begin{pmatrix} \mathfrak{f}_i(t) & \mathfrak{m}_{i,j}(t)\\ \mathfrak{m}_{i,j}(t) & \mathfrak{f}_j(t)\end{pmatrix} \geq 0\right\}.$$

Lemma 3 ([31]) *For $R > 0$, there exists a differentiable function $\{y(v)|v \in [\mathfrak{a}, \mathfrak{b}]\}$,*

$$-(\mathfrak{b}-\mathfrak{a})\int_{\mathfrak{a}}^{\mathfrak{b}} \dot{y}^T(s)R\dot{y}(s)ds \leq -\Omega_1^T R\Omega_1 - 3\Omega_2^T R\Omega_2,$$

where

$$\Omega_1 = y(\mathfrak{b}) - y(\mathfrak{a}),$$
$$\Omega_2 = y(\mathfrak{b}) + y(\mathfrak{a}) - \frac{2}{\mathfrak{b}-\mathfrak{a}}\int_{\mathfrak{a}}^{\mathfrak{b}} y(s)ds.$$

3 Main Results

In this section, a synchronization criterion for complex dynamical networks with hybrid event-triggered scheme and stochastic cyber-attack will be discussed in Theorem 1.

For simplify, $\hat{e}_i \in \mathbb{R}^{(19n)\times n}$ denotes block entry matrix. For example, $e_2 = [0 \ I \ \underbrace{0 \ \ldots \ 0}_{16}, 0]^T$. And define some notations as follows:

$$\Psi = \Big[0 \ cS(G\otimes A) \ 0 \ -S \ (1-\bar{\beta})\bar{\alpha}H \ 0 \ 0 \ 0 \ (1-\bar{\beta})(1-\bar{\alpha})H$$
$$\underbrace{0 \ \ldots \ 0}_{7} \ S \ \bar{\beta}H \ (1-\bar{\beta})(1-\bar{\alpha})H\Big],$$
$$\bar{U}_1 = \frac{(I\otimes U_1)^T(I\otimes U_2)}{2} + \frac{(I\otimes U_2)^T(I\otimes U_1)}{2},$$
$$\bar{U}_2 = -\frac{(I\otimes U_1)^T + (I\otimes U_2)^T}{2},$$
$$\xi^T(t) = \Big[\rho^T(t) \ \rho^T(t-\tau(t)) \ \rho^T(t-\tau_M) \ \dot{\rho}(t) \ \rho^T(t-d(t))$$
$$\rho^T(t-d_M) \ \ \frac{1}{d(t)}\int_{t-d(t)}^{t}\rho^T(s)ds \ \ \frac{1}{d_M-d(t)}\int_{t-d_M}^{t-d(t)}\rho^T(s)ds$$
$$\rho^T(t-\eta(t)) \ \rho^T(t-\eta_M) \ \ \frac{1}{\eta(t)}\int_{t-\eta(t)}^{t}\rho^T(s)ds$$
$$\frac{1}{\eta_M-\eta(t)}\int_{t-\eta_M}^{t-\eta(t)}\rho^T(s)ds \ \ \rho^T(t-\theta(t)) \ \rho^T(t-\theta_M)$$
$$\frac{1}{\theta(t)}\int_{t-\theta(t)}^{t}\rho^T(s)ds \ \ \frac{1}{\theta_M-\theta(t)}\int_{t-\theta_M}^{t-\theta(t)}\rho^T(s)ds$$

$$\bar{f}(\rho(t))\ \bar{g}(\rho(t-\theta(t)))\ r_k(t)\Big],$$

$$\begin{aligned}
\Pi_1 &= [e_1 - e_5\ \ e_1 + e_5 - 2e_7],\\
\Pi_2 &= [e_5 - e_6\ \ e_5 + e_6 - 2e_8],\\
\Pi_3 &= [e_1 - e_9\ \ e_1 + e_9 - 2e_{11}],\\
\Pi_4 &= [e_9 - e_{10}\ \ e_9 + e_{10} - 2e_{12}],\\
\Pi_5 &= [e_1 - e_{13}\ \ e_1 + e_{13} - 2e_{15}],\\
\Pi_6 &= [e_{13} - e_{14}\ \ e_{13} + e_{14} - 2e_{16}],\\
\bar{R}_1 &= \begin{bmatrix} -R_1 & R_1 - T_1 & T_1 \\ * & -2R_1 + T_1 + T_1^T & R_1 - T_1 \\ * & * & -R_1 \end{bmatrix},
\end{aligned}$$

$$\begin{aligned}
\Sigma = {} & e_1 P e_4^T + e_4 P e_1^T + e_1 Q_1 e_1^T - (1-\mu) e_2 Q_1 e_2^T + e_1(Q_2 + Q_3)e_1^T \\
& - e_3 Q_2 e_3^T - e_6 Q_3 e_6^T + e_1(Q_4 + Q_5)^T - e_{10} Q_4 e_{10}^T \\
& - e_{14} Q_5 e_{14}^T + \tau^2 e_4 R_1 e_4^T - [e_1\ \ e_2\ \ e_3]\bar{R}_1[e_1\ \ e_2\ \ e_3]^T \\
& + d_M^2 e_4 R_2 e_4^T - [\Pi_1\ \ \Pi_2]\begin{bmatrix} \bar{R}_2 & T_2 \\ * & \bar{R}_2 \end{bmatrix}[\Pi_1\ \ \Pi_2]^T \\
& + \eta_M^2 e_4 R_3 e_4^T - [\Pi_3\ \ \Pi_4]\begin{bmatrix} \bar{R}_3 & T_3 \\ * & \bar{R}_3 \end{bmatrix}[\Pi_3\ \ \Pi_4]^T \\
& + \theta_M^2 e_4 R_4 e_4^T - [\Pi_5\ \ \Pi_6]\begin{bmatrix} \bar{R}_4 & T_4 \\ * & \bar{R}_4 \end{bmatrix}[\Pi_5\ \ \Pi_6]^T \\
& + [e_1 + \epsilon e_4]\Psi + \Psi^T[e_1 + \epsilon e_4]^T - \epsilon_1[e_1\ \ e_{17}]\begin{bmatrix} \bar{U}_1 & \bar{U}_2 \\ * & I \end{bmatrix}[e_1\ \ e_{17}]^T \\
& + \epsilon_2 e_{13}(I \otimes V)^T(I \otimes V)e_{13}^T - \epsilon_2 e_{18} e_{18}^T \\
& + \sigma e_9(I \otimes \Omega)e_9^T - e_{19}(I \otimes \Omega)e_{19}^T.
\end{aligned}$$

Theorem 1 *For given scalars $\tau > 0$, μ, $d_M > 0$, $\eta_M > 0$, $\theta_M > 0$, $\bar{\alpha}$, $\bar{\beta}$, $\epsilon > 0$, $\epsilon_1 > 0$, $\epsilon_2 > 0$, $\sigma > 0$, the closed loop system (14) is asymptotically stable, if there exist symmetric positive definite matrices $P > 0, Q_1, Q_2, Q_3, Q_4, Q_5,, R_1, R_2, R_3, R_4$, any matrices T_1, T_2, T_3, T_4, $S = diag\{S_1, S_2, \ldots, S_N\}$, $H = diag\{H_1, H_2, \ldots, H_N\}$, with appropriate dimensions such that the following LMIs hold*

$$\Sigma < 0, \tag{16}$$

$$\begin{bmatrix} \bar{R}_i & T_i \\ * & \bar{R}_i \end{bmatrix} \geq 0,\ i = 1, 2, 3, 4. \tag{17}$$

Moreover, the designed controller gain could be calculated as

$$K_j = S_j^{-1} H_j, \; j = 1, 2, \ldots, N. \tag{18}$$

Proof Construct the following Lyapunov function candidate

$$V(t) = \sum_{i=1}^{8} V_i(t), \tag{19}$$

where

$$\begin{aligned}
V_1(t) &= \rho^T(t) P \rho(t), \\
V_2(t) &= \int_{t-\tau(t)}^{t} \rho^T(s) Q_1 \rho(s) ds, \\
V_3(t) &= \int_{t-\tau_M}^{t} \rho^T(s) Q_2 \rho(s) ds + \int_{t-d_M}^{t} \rho^T(s) Q_3 \rho(s) ds, \\
V_4(t) &= \int_{t-\eta_M}^{t} \rho^T(s) Q_4 \rho(s) ds + \int_{t-\theta_M}^{t} \rho^T(s) Q_5 \rho(s) ds, \\
V_5(t) &= \tau_M \int_{t-\tau_M}^{t} \int_{s}^{t} \dot{\rho}^T(u) R_1 \dot{\rho}(u) du ds, \\
V_6(t) &= d_M \int_{t-d_M}^{t} \int_{s}^{t} \dot{\rho}^T(u) R_2 \dot{\rho}(u) du ds, \\
V_7(t) &= \eta_M \int_{t-\eta_M}^{t} \int_{s}^{t} \dot{\rho}^T(u) R_3 \dot{\rho}(u) du ds, \\
V_8(t) &= \theta_M \int_{t-\theta_M}^{t} \int_{s}^{t} \dot{\rho}^T(u) R_4 \dot{\rho}(u) du ds.
\end{aligned}$$

By calculating the time derivative $V(t)$ along the trajectory of (14), one has

$$\mathbb{E}\{V_1(t)\} = 2\rho^T(t) P \dot{\rho}(t), \tag{20}$$

$$\mathbb{E}\{V_2(t)\} \leq \rho^T(t-\tau(t)) Q_1 \rho(t-\tau(t)) - (1-\mu)\rho^T(t) Q_1 \rho(t), \tag{21}$$

$$\begin{aligned}
\mathbb{E}\{V_3(t)\} &= \rho^T(t)(Q_2 + Q_3)\rho(t) \\
&\quad -\rho^T(t-\tau_M) Q_2 \rho(t-\tau_M) - \rho^T(t-d_M) Q_3 \rho(t-d_M),
\end{aligned} \tag{22}$$

$$\mathbb{E}\{V_4(t)\} = \rho^T(t)(Q_4 + Q_5)\rho(t) \tag{23}$$

$$-\rho^T(t-\eta_M) Q_4 \rho(t-\eta_M) - \rho^T(t-\theta_M) Q_5 \rho(t-\theta_M), \tag{24}$$

$$\mathbb{E}\{V_5(t)\} = \tau_M^2 \dot{\rho}^T(t) R_1 \dot{\rho}(t) - \tau_M \int_{t-\tau_M}^{t} \dot{\rho}^T(s) R_1 \dot{\rho}(s) ds, \tag{25}$$

$$\mathbb{E}\{V_6(t)\} = d_M^2 \dot{\rho}^T(t) R_2 \dot{\rho}(t) - d_M \int_{t-d_M}^{t} \dot{\rho}^T(s) R_2 \dot{\rho}(s) ds, \tag{26}$$

$$\mathbb{E}\{V_7(t)\} = \eta_M^2 \dot{\rho}^T(t) R_3 \dot{\rho}(t) - \eta_M \int_{t-\eta_M}^{t} \dot{\rho}^T(s) R_3 \dot{\rho}(s) ds, \tag{27}$$

$$\mathbb{E}\{V_8(t)\} = \theta_M^2 \dot{\rho}^T(t) R_4 \dot{\rho}(t) - \theta_M \int_{t-\theta_M}^{t} \dot{\rho}^T(s) R_4 \dot{\rho}(s) ds. \tag{28}$$

In view of Lemmas 1 and 2, for any matrix T_1, one has

$$\begin{aligned} &-\tau_M \int_{t-\tau_M}^{t} \dot{\rho}^T(s) R_1 \dot{\rho}(s) ds \\ &= -\tau_M \int_{t-\tau(t)}^{t} \dot{\rho}^T(s) R_1 \dot{\rho}(s) ds - \tau_M \int_{t-\tau_M}^{t-\tau(t)} \dot{\rho}^T(s) R_1 \dot{\rho}(s) ds \\ &\leq \mathcal{X}^T(t) \bar{R}_1 \mathcal{X}(t) \end{aligned} \tag{29}$$

where $\mathcal{X}(t) = [\rho^T(t) \;\; \rho^T(t-\tau(t)) \;\; \rho^T(t-\tau_M)]^T$ and

$$\bar{R}_1 = \begin{bmatrix} -R_1 & R_1 - T_1 & T_1 \\ * & -2R_1 + T_1 + T_1^T & R_1 - T_1 \\ * & * & -R_1 \end{bmatrix}.$$

Applying Lemmas 1 and 2 gives

$$\begin{aligned} &-d_M \int_{t-d_M}^{t} \dot{\rho}^T(s) R_2 \dot{\rho}(s) ds \\ &= -d_M \int_{t-d(t)}^{t} \dot{\rho}^T(s) R_2 \dot{\rho}(s) ds - d_M \int_{t-d_M}^{t-d(t)} \dot{\rho}^T(s) R_2 \dot{\rho}(s) ds \\ &\leq -\frac{d_M}{d(t)} \bar{\gamma}_1^T \hat{R}_2 \bar{\gamma}_1 - \frac{d_M}{d_M - d(t)} \bar{\gamma}_2^T \bar{R}_2 \bar{\gamma}_2 \\ &\leq -\begin{bmatrix} \bar{\gamma}_1 \\ \bar{\gamma}_2 \end{bmatrix}^T \begin{bmatrix} \bar{R}_2 & T_2 \\ * & \bar{R}_2 \end{bmatrix} \begin{bmatrix} \bar{\gamma}_1 \\ \bar{\gamma}_2 \end{bmatrix}, \end{aligned} \tag{30}$$

where

$$\begin{aligned} &\bar{\gamma}_1 = [\gamma_1^T \;\; \gamma_1^T]^T, \;\; \bar{\gamma}_2 = [\gamma_3^T \;\; \gamma_4^T]^T, \\ &\gamma_1 = \rho(t) - \rho(t - d(t)), \\ &\gamma_2 = \rho(t) + \rho(t - d(t)) - \frac{2}{d(t)} \int_{t-d(t)}^{t} \rho(s) ds, \\ &\gamma_3 = \rho(t - d(t)) - \rho(t - d_M), \\ &\gamma_4 = \rho(t - d(t)) + \rho(t - d_M) - \frac{2}{d_M - d(t)} \int_{t-d_M}^{t-d(t)} \rho(s) ds. \end{aligned}$$

Similar to (30), we can obtain

$$-\eta_M \int_{t-\eta_M}^{t} \dot{e}^T(s)R_3\dot{e}(s)ds \leq -\begin{bmatrix} \bar{\gamma_3} \\ \bar{\gamma_4} \end{bmatrix}^T \begin{bmatrix} \bar{R}_3 & T_3 \\ * & \bar{R}_3 \end{bmatrix} \begin{bmatrix} \bar{\gamma_3} \\ \bar{\gamma_4} \end{bmatrix}, \tag{31}$$

where

$$\begin{aligned}
&\bar{\gamma_3} = [\gamma_5^T \ \ \gamma_6^T \ ]^T, \ \bar{\gamma_4} = [\gamma_7^T \ \ \gamma_8^T \ ]^T, \\
&\gamma_5 = \rho(t) - \rho(t-\eta(t)), \\
&\gamma_6 = \rho(t) + \rho(t-\eta(t)) - \frac{2}{\eta(t)} \int_{t-\eta(t)}^{t} \rho(s)ds, \\
&\gamma_7 = \rho(t-\eta(t)) - \rho(t-\eta_M), \\
&\gamma_8 = \rho(t-\eta(t)) + \rho(t-\eta_M) - \frac{2}{\eta_M - \eta(t)} \int_{t-\eta_M}^{t-\eta(t)} \rho(s)ds.
\end{aligned}$$

And,

$$-\theta_M \int_{t-\theta_M}^{t} \dot{e}^T(s)R_4\dot{e}(s)ds \leq -\begin{bmatrix} \bar{\gamma_5} \\ \bar{\gamma_6} \end{bmatrix}^T \begin{bmatrix} \bar{R}_4 & T_4 \\ * & \bar{R}_4 \end{bmatrix} \begin{bmatrix} \bar{\gamma_5} \\ \bar{\gamma_6} \end{bmatrix}, \tag{32}$$

where

$$\begin{aligned}
&\bar{\gamma_5} = [\gamma_9^T \ \ \gamma_{10}^T \ ]^T, \ \bar{\gamma_6} = [\gamma_{11}^T \ \ \gamma_{12}^T]^T, \\
&\gamma_9 = \rho(t) - \rho(t-\theta(t)), \\
&\gamma_{10} = \rho(t) + \rho(t-\theta(t)) - \frac{2}{\theta(t)} \int_{t-\theta(t)}^{t} \rho(s)ds, \\
&\gamma_{11} = \rho(t-\theta(t)) - \rho(t-\theta_M), \\
&\gamma_{12} = \rho(t-\theta(t)) + \rho(t-\theta_M) - \frac{2}{\theta_M - \theta(t)} \int_{t-\theta_M}^{t-\eta(t)} \rho(s)ds.
\end{aligned}$$

Furthermore, based on system (14), for a matrix S with any appropriate dimension and a scalar ϵ, it can be concluded that:

$$\begin{aligned}
&2[\rho^T(t)S + \epsilon\dot{\rho}^T(t)S]\Big\{-\dot{\rho}(t) + \bar{f}(\rho(t)) + c(G \otimes A)\rho(t-\tau(t)) \\
&+\beta(t)Kg(\rho(t-\theta(t))) + (1-\beta(t))K\{\alpha(t)\rho(t-d(t)) \\
&+(1-\alpha(t))[\rho(t-\eta(t)) + r_k(t)]\}\Big\} = 0.
\end{aligned} \tag{33}$$

On the other hand, one can obtain

$$\varepsilon_1[f(\rho_i(t)) - U_1\rho_i(t)]^T[f(\rho_i(t)) - U_2\rho_i(t)] \leq 0, \tag{34}$$

which is equivalent to

$$-\varepsilon_1 \begin{bmatrix} \rho(t) \\ \bar{f}(\rho(t)) \end{bmatrix}^T \begin{bmatrix} \bar{U}_1 & \bar{U}_2 \\ * & I \end{bmatrix} \begin{bmatrix} \rho(t) \\ \bar{f}(\rho(t)) \end{bmatrix} \geq 0. \tag{35}$$

From (11), one has

$$\begin{aligned} &\varepsilon_2 \rho_i(t-\theta(t)) V_i^T V_i \rho_i(t-\theta(t)) \\ &\quad -\varepsilon_2 g(\rho_i(t-\theta(t))) g(\rho_i(t-\theta(t))) \geq 0. \end{aligned} \tag{36}$$

Combining (20) to (36) and taking the event-triggered condition (7) into account yield

$$\mathbb{E}\{\dot{V}(t)\} \leq \xi^T(t) \Sigma \xi(t). \tag{37}$$

If LMIs (16)–(17) holds. the closed loop system (14) is asymptotically stable.

4 Numerical Example

In this section, two numerical examples are provided to illustrate the effectiveness and feasibility of the designed approach.

Example 1 A CDN with three nodes (1) is considered in this example. The outer-coupling matrix is assumed to be $G = (G_{ij})_{N \times N}$ with

$$G = \begin{bmatrix} -1 & 0 & 1 \\ 0 & -1 & 1 \\ 1 & 1 & -2 \end{bmatrix}.$$

The inner-coupling matrix A is given as

$$A = \begin{bmatrix} 1 & 0 \\ 0 & 1 \end{bmatrix}.$$

The nonlinear function $f(\cdot)$ is taken as

$$f(x_i(t)) = \begin{bmatrix} -0.5x_{i1} + \tanh(0.2x_{i1}) + 0.2x_{i2} \\ 0.95x_{i2} - \tanh(0.75x_{i2}) \end{bmatrix},$$

which implies that $f(\cdot)$ satisfies (3) with

$$U_1 = \begin{bmatrix} -0.5 & 0.2 \\ 0 & 0.95 \end{bmatrix}, \quad U_2 = \begin{bmatrix} -0.3 & 0.2 \\ 0 & 0.2 \end{bmatrix}.$$

Letting the nonlinear signals of cyber attacks is $g(x_i(t)) = [-\tanh^T(x_{i1}(t)) \;\; -\tanh^T(x_{i2}(t))]^T$, the parameter of event-triggered scheme is $\sigma = 0.02$, the parameters of delay is given as $\tau = 0.25$, $\mu = 0.5$, $d_M = 0.1$, $\eta_M = 0.2$, $\theta_M = 0.1$.

Our aim is to demonstrate the effectiveness of the derived controller design approach. Let $\bar{\alpha} = 0.2$, $\bar{\beta} = 0.4$, which means that the control system is subject to hybrid-triggered scheme with probability of 20% and the switch probability of cyber-attack is 40%.

By solving the LMIs (16) and (17), the controllers gains and the event-triggered gain matrices are obtained as

$$K_1 = \begin{bmatrix} -0.3049 & -0.4032 \\ -0.3799 & -2.5177 \end{bmatrix}, \quad K_2 = \begin{bmatrix} -0.3049 & -0.4032 \\ -0.3799 & -2.5177 \end{bmatrix},$$
$$K_3 = \begin{bmatrix} -0.2912 & -0.3574 \\ -0.3347 & -2.2603 \end{bmatrix}, \quad \Omega_1 = \begin{bmatrix} 0.1713 & -0.0110 \\ -0.0110 & 0.0681 \end{bmatrix},$$
$$\Omega_2 = \begin{bmatrix} 0.1718 & -0.0115 \\ -0.0115 & 0.0685 \end{bmatrix}, \quad \Omega_3 = \begin{bmatrix} 0.1789 & -0.0118 \\ -0.0118 & 0.0691 \end{bmatrix}.$$

For the initial condition $x_1(0) = [8 \;\; -3]^T$, $x_2(0) = [4 \;\; -8]^T$, $x_3(0) = [-5 \;\; 6]^T$ and $s(0) = [1 \;\; -0]^T$, the state response of error systems with event-triggered and cyber-attacks is given in Fig. 1. Figure 2 presents the switching probability between the time-triggered and event-triggered scheme under $\bar{\alpha} = 0.2$. Figure 3 shows the switching probabilities between the cyber-attacks under $\bar{\beta} = 0.4$. Figure 4 describes the graph of event-triggered instants and released intervals.

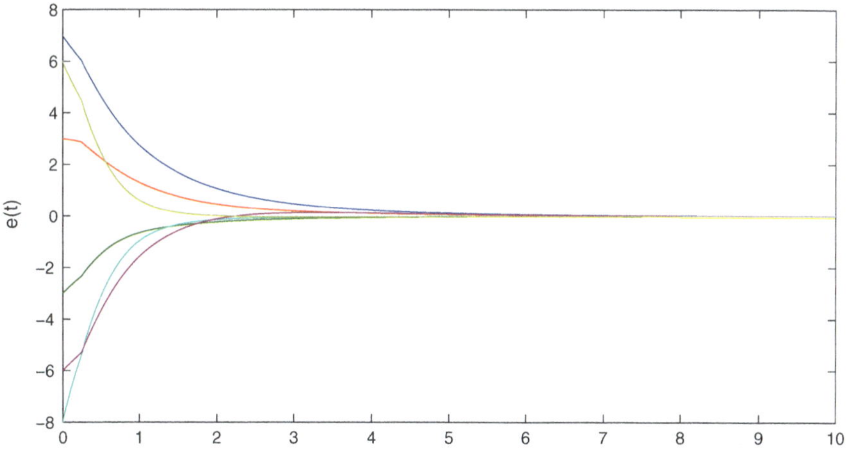

Fig. 1 State response of $e(t)$

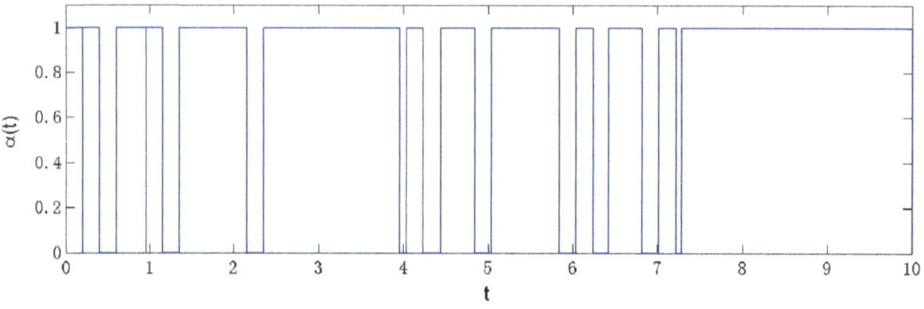

Fig. 2 The signal $\alpha(t)$ in Example 1

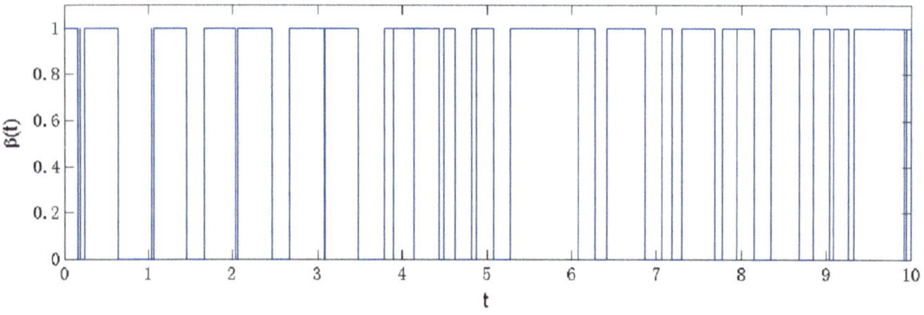

Fig. 3 The signal $\beta(t)$ in Example 1

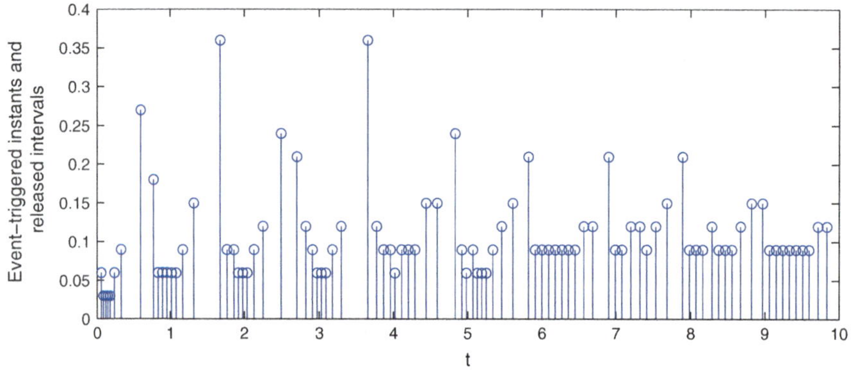

Fig. 4 Event-triggered instants and released intervals

Obviously, the above figures effectively demonstrate the usefulness of the designed synchronization method for complex dynamical networks with stochastic cyber-attacks and hybrid-triggered scheme.

Example 2 Consider the Chua's circuit as the unforced isolated node of CDN (1), which is described by the following model

$$\dot{s}_1 = \sigma_1(-s_1 + s_2 - v(s_1))$$
$$\dot{s}_2 = s_1 - s_2 + s_3$$
$$\dot{s}_3 = -\sigma_2 s_2$$

where $\sigma_1 = 10, \sigma_2 = 14.87$, and $v(s_1) = bs_1 + 0.5(a - b)r(s_1)$, in which $a = -1.27, b = -0.68$, and $r(s_1) = (|s_1 + 1| - |s_1 - 1|)$.

Denote $s = [s_1 \ \ s_2 \ \ s_3]^T$ and

$$f(s) = \begin{bmatrix} -\sigma_1 - \sigma_1 b & \sigma_1 & 0 \\ 1 & -1 & 1 \\ 0 & -\sigma_2 & 0 \end{bmatrix} + \begin{bmatrix} -0.5\sigma_1(a-b)r(s_1) & 0 & 0 \\ 0 & 0 & 0 \\ 0 & 0 & 0 \end{bmatrix}.$$

It is easy to know that $f(s)$ satisfies with

$$U = \begin{bmatrix} 2.7 & 10 & 0 \\ 1 & -1 & 1 \\ 0 & -14.87 & 0 \end{bmatrix}, \quad V = \begin{bmatrix} -3.2 & 10 & 0 \\ 1 & -1 & 1 \\ 0 & -14.87 & 0 \end{bmatrix}.$$

And, the inner-coupling matrix A and out-coupling matrix G are given as

$$A = \begin{bmatrix} 1 & 0 & 0 \\ 0 & 1 & 0 \\ 0 & 0 & 1 \end{bmatrix}$$

and

$$G = \begin{bmatrix} -3 & 1 & 1 & 1 \\ 1 & -2 & 1 & 0 \\ 1 & 1 & -2 & 0 \\ 1 & 0 & 0 & -1 \end{bmatrix}.$$

Let the nonlinear signals of cyber attacks be

$$g(x_i(t)) = [-\tanh^T(x_{i1}(t)) \quad -\tanh^T(x_{i2}(t)) \quad -\tanh^T(x_{i3}(t))]^T.$$

Also, the parameter of event-triggered scheme is $\sigma = 0.03$. The parameters of delay are given as $\tau = 0.1$, $\mu = 0.5$, $d_M = 0.1$, $\eta_M = 0.3$, and $\theta_M = 0.2$.

Here, the objective of this example is to demonstrate the effectiveness of the derived controller design approach. Let $\bar{\alpha} = 0.3$, $\bar{\beta} = 0.6$, which means that the control system is subject to hybrid-triggered scheme with probability of 30% and the switch probability of cyber-attack is 60%.

By solving the LMIs (16) and (17), the controllers gains and the event-triggered gain matrices are obtained as

$$K_1 = \begin{bmatrix} -8.2740 & -8.3071 & 0.9043 \\ -1.3559 & -6.8659 & -0.5217 \\ 1.0260 & 12.3906 & -8.5663 \end{bmatrix},$$

$$K_2 = \begin{bmatrix} -8.1898 & -8.4044 & 0.9417 \\ -1.4055 & -6.7465 & -0.4822 \\ 1.0474 & 12.5701 & -8.5101 \end{bmatrix},$$

$$K_3 = \begin{bmatrix} -8.1919 & -8.4040 & 0.9413 \\ -1.4059 & -6.7445 & -0.4812 \\ 1.0472 & 12.5700 & -8.5112 \end{bmatrix},$$

$$K_4 = \begin{bmatrix} -8.2183 & -8.4863 & 0.9651 \\ -1.4434 & -6.7287 & -0.4545 \\ 1.0557 & 12.7148 & -8.5572 \end{bmatrix},$$

$$\Omega_1 = \begin{bmatrix} 1.5623 & 0.1428 & 0.5632 \\ 0.1507 & 2.3389 & -0.1202 \\ 0.5652 & -0.1202 & 2.2105 \end{bmatrix},$$

$$\Omega_2 = \begin{bmatrix} 1.1679 & 0.1604 & 0.5728 \\ 0.1608 & 2.3427 & -0.1208 \\ 0.5629 & -0.1208 & 2.2402 \end{bmatrix},$$

$$\Omega_3 = \begin{bmatrix} 1.1625 & 0.1543 & 0.5728 \\ 0.1617 & 2.3478 & -0.1304 \\ 0.5783 & -0.1304 & 2.3482 \end{bmatrix}.$$

$$\Omega_4 = \begin{bmatrix} 1.1723 & 0.1658 & 0.5826 \\ 0.1658 & 2.3562 & -0.1317 \\ 0.5713 & -0.1317 & 2.3614 \end{bmatrix}.$$

For the initial condition $x_1(0) = [0 \quad -3 \quad 0]^T, x_2(0) = [5 \quad -6 \quad 1]^T, x_3(0) = [1 \quad -2 \quad -4]^T, x_4(0) = [-1 \quad -1 \quad 6]^T$ and $s(0) = [1 \quad -1 \quad 0]^T$ the state response of error systems with event-triggered and cyber-attacks is shown in Fig. 5.

Figure 6 describes the switching probability between the time-triggered and the event-triggered scheme under $\bar{\alpha} = 0.3$. Figure 7 represents the switching probabilities between the cyber-attacks under $\bar{\beta} = 0.6$. Figure 8 gives the graph of event-triggered instants and released intervals.

It can be seen that the above figures effectively demonstrate the usefulness of the designed synchronization method for complex dynamical networks with stochastic cyber-attacks and hybrid-triggered scheme.

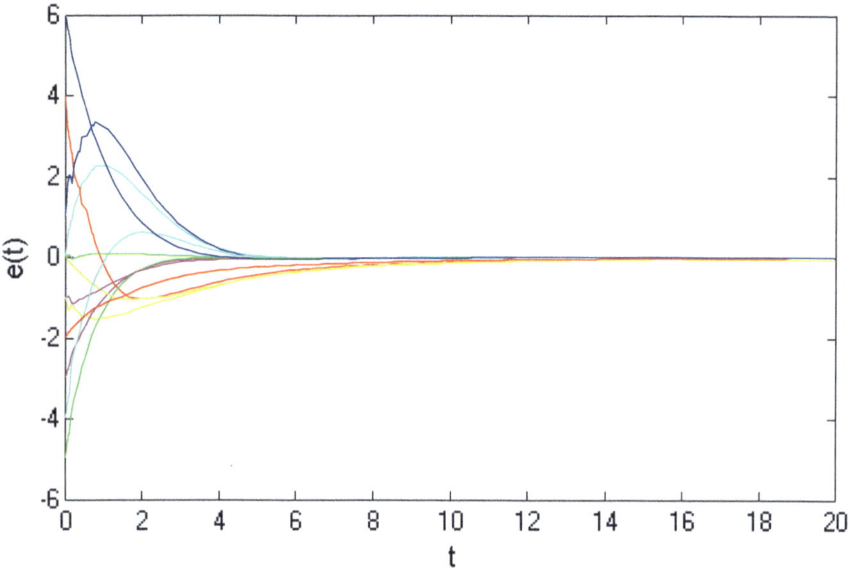

Fig. 5 State response of $e(t)$

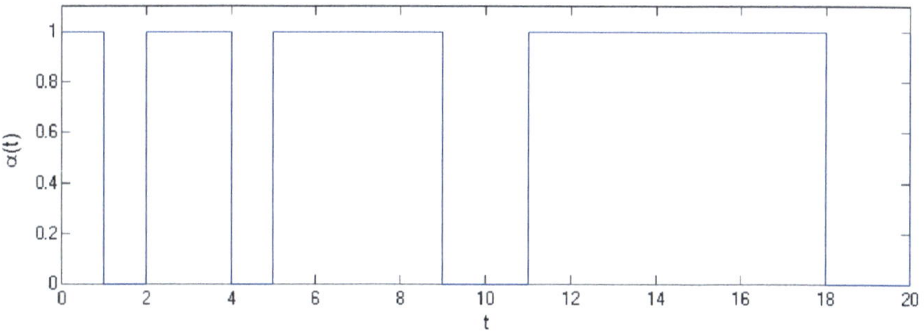

Fig. 6 The signal $\alpha(t)$ in Example 2

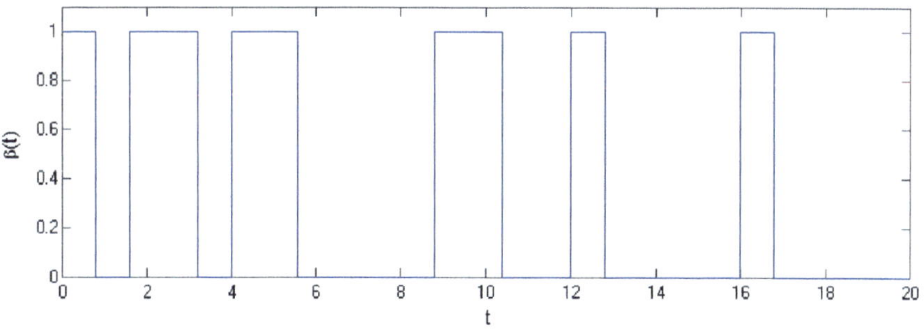

Fig. 7 The signal $\beta(t)$ in Example 2

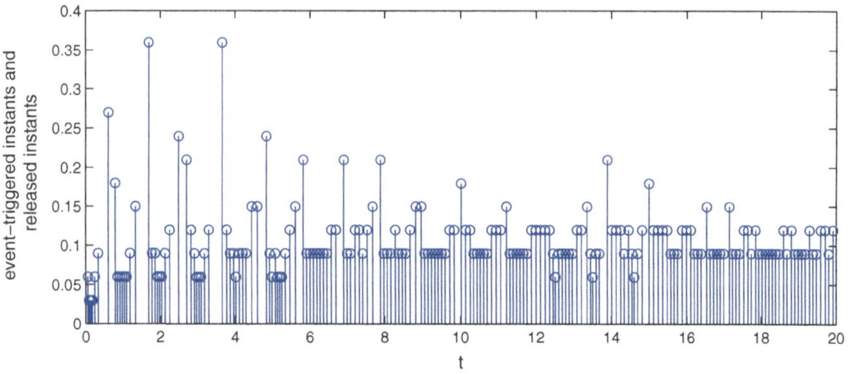

Fig. 8 Event-triggered instants and released intervals

5 Conclusions

We have studied the synchronization control of delayed complex networks with stochastic cyber-attacks. In order to alleviate the communication burden, we have adopted the hybrid-triggered transmission scheme. Based on the Lyapunov stability theory and matrix inequality method, sufficient conditions were obtained such that the synchronization error system is asymptotically stable in the mean-square sense and the controller gain matrices were calculated by solving some matrix inequalities. Finally, two simulation studies have been presented to validate the proposed method.

Acknowledgements This work was supported by the National Natural Science Foundation of China under the grant (71801196, 61803153).

References

1. Park, J.H., Shen, H., Chang, X.H., Lee, T.H.: Recent Advances in Control and Filtering of Dynamic Systems with Constrained Signals. Springer, Cham (2018). https://doi.org/10.1007/978-3-319-96202-3. ISBN 978-3-319-96202-3
2. Boccaletti, S., Latora, V., Moreno, Y., Chavez, M., Hwang, D.: Complex networks: Structure and dynamics. Phys. Rep. **424**(4), 175–308 (2006)
3. Chang, Q., Yang, Y., Sui, X., et al.: The optimal control synchronization of complex dynamical networks with time-varying delay using PSO. Neurocomputing **333**, 1–10 (2019)
4. Xu, M., Wang, J.-L., Huang, Y.-L., Wei, P., Wang, S.: Pinning synchronization of complex dynamical networks with and without time-varying delay. Neurocomputing **266**, 263–273 (2017)
5. Tang, Z., Park, J.H., Lee, T.H.: Distributed adaptive pining control for cluster synchronization of nonlinear coupled Lur'e networks. Commun. Nonlinear Sci. Numer. Simul. **39**, 7–20 (2016)
6. Strogatz, S.: Exploring complex networks. Nature **410**(6225), 268–276 (2001)
7. Wu, W., Zhou, W., Chen, T.: Cluster synchronization of linearly coupled complex networks under pinning control. IEEE Trans. Circuits Syst. I: Regul. Pap. **56**(4), 829–839 (2009)

8. Pecora, L., Carroll, T.: Synchronization in chaotic systems. Phys. Rev. Lett. **64**(8), 821–824 (1990)
9. Wang, J., Wu, H.: Synchronization criteria for impulsive complex dynamical networks with time-varying delay. Nonlinear Dyn. **70**, 13–24 (2012)
10. Zhao, J.: Adaptive Q-S synchronization between coupled chaotic systems with stochastic perturbation and delay. Appl. Math. Model. **36**, 3312–3319 (2012)
11. Chen, W., Jiang, Z., Lu, X., et al.: $\mathcal{H}_\infty$ synchronization for complex dynamical networks with coupling delays using distributed impulsive control. Nonlinear Anal.: Hybrid Syst. **17**, 111–127 (2015)
12. Wu, Z., Shi, P., Su, H., et al.: Sampled-data exponential synchronization of complex dynamical networks with time-varying coupling delay. IEEE Trans. Neural Netw. Learn. Syst. **219**(15), 1177–1187 (2013)
13. Fridman, E.: A refined input delay approach to sampled-data control. Automatica **46**(2), 421–427 (2010)
14. Zhang, D., Shi, P., Yu, L.: Containment control of linear multiagent systems with aperiodic sampling and measurement size reduction. IEEE Trans. Neural Netw. Learn. Syst. **29**(10), 5020–5029 (2018)
15. Yue, D., Tian, E., Han, Q.: A delay system method for designing event-triggered controllers of networked control systems. IEEE Trans. Autom. Control **58**(2), 475–481 (2013)
16. Liu, J., Zha, L., Cao, J., et al.: Hybrid-driven-based stabilisation for networked control systems. IET Control Theory Appl. **10**(17), 149–153 (2016)
17. Liu, Y., Guo, G., Park, J.H., et al.: Nonfragile exponential synchronization of delayed complex dynamical networks with memory sampled-data control. IEEE Trans. Neural Netw. Learn. Syst. **29**(1), 118–128 (2018)
18. Zhang, R., Zeng, D., Park, J.H., et al.: Non-fragile sampled-data synchronization for delayed complex dynamical networks with randomly occurring controller gain fluctuations. IEEE Trans. Syst. Man Cybern.: Syst. **48**(2), 2271–2281 (2018)
19. Wang, A., Dong, T., Liao, X.: Event-triggered synchronization strategy for complex dynamical networks with the Markovian switching topologies. Neural Netw. **74**, 52–57 (2016)
20. Liu, D., Yang, G.: Event-triggered synchronization control for complex networks with actuator saturation. Neurocomputing **275**, 2209–2219 (2018)
21. Farraj, A., Hammad, E., Kundur, D.: On the impact of cyber attacks on data integrity in storage-based transient stability control. IEEE Trans. Ind. Inform. **13**(6), 3322–3333 (2017)
22. Kwon, C., Hwang, I.: Reachability analysis for safety assurance of cyber-physical systems against cyber attacks. IEEE Trans. Autom. Control **63**(7), 2272–2279 (2018)
23. Zha, L., Tian, E., Xie, X., et al.: Decentralized event-triggered $\mathcal{H}_\infty$ control for neural networks subject to cyber-attacks. Inf. Sci. **457–458**, 141–155 (2018)
24. Ding, D., Wang, Z., Ho, D., et al.: Observer-based event-triggering consensus control for multiagent systems with lossy sensors and cyber-attacks. IEEE Trans. Cybern. **47**(8), 1936–1947 (2017)
25. Zhang, D., Liu, L., Feng, G.: Consensus of heterogeneous linear multiagent systems subject to aperiodic sampled-data and DoS attack. IEEE Trans. Cybern. **49**(4), 1501–1511 (2018)
26. Liu, J., Gu, Y., Xie, X., et al.: Hybrid-driven-based $\mathcal{H}_\infty$ control for networked cascade control systems with actuator saturations and stochastic cyber attacks. IEEE Trans. Syst. Man Cybern.: Syst. (2018). https://doi.org/10.1109/TSMC.2018.2875484
27. Liu, J., Wei, J., Xie, X., et al.: Quantized stabilization for T-S fuzzy systems with hybrid-triggered mechanism and stochastic cyber-attacks. IEEE Trans. Fuzzy Syst. **26**(2), 3820–3834 (2018)
28. Park, J.H., Lee, T.H., Liu, Y., Chen, J.: Dynamic Systems with Time Delays: Stability and Control. Springer-Nature, Singapore (2019). https://doi.org/10.1007/978-981-13-9254-2

29. Li, N., Zhang, Y., Hu, J., et al.: Synchronization for general complex dynamical networks with sampled-data. Neurocomputing **74**(5), 805–811 (2011)
30. Park, P., Ko, J., Jeong, C.: Reciprocally convex approach to stability of systems with time-varying delays. Automatica **47**(1), 235–238 (2011)
31. Seuret, A., Gouaisbaut, F.: Wirtinger-based integral inequality: application to time-delay systems. Automatica **49**(8), 2860–2866 (2013)

Cluster Synchronization on Derivative Coupled Lur'e Networks: Impulsive Pinning Strategy

Ze Tang, Dong Ding, and Ju H. Park

Abstract This chapter is concerned with the global and exponential synchronization issue for a class of nonidentically delay coupled Lur'e networks with stochastic disturbance and derivative couplings. Considering that the topological structure of the coupled Lur'e networks consists of several groups, cluster synchronization instead of complete synchronization is discussed. In order to present the synchronization criteria, a novel controller combining with the impulsive control and pinning control strategies is designed, which will be imposed on those Lur'e systems in the cluster but have directed connections with the Lur'e systems in the other clusters. By applying the extended comparison principle of the impulsive differential equations, the definition of the average impulsive interval, the mathematical taxonomy on impulsive parameters simultaneously, the achievement of the cluster synchronization on nonidentically derivative coupled Lur'e networks are guaranteed. Furthermore, the exponential convergence velocity of the derivative coupled Lur'e networks with stochastic disturbance is precisely estimated. Eventually, several numerical simulations demonstrate the effectiveness and applicability of the established synchronization technique.

Keywords Cluster synchronization · Nonidentical Lur'e systems · Stochastic disturbance · Derivative couplings · Impulsive control · Complex networks.

Z. Tang (✉) · D. Ding
Engineering Research Center of Internet of Things Technology Applications (Ministry of Education), Jiangnan University, Wuxi 214122, People's Republic of China
e-mail: tangze0124@gmail.com

J. H. Park
Department of Electrical Engineering, Yeungnam University, 280 Daehak-Ro, Gyeongsan 38541, Republic of Korea
e-mail: jessie@ynu.ac.kr

J. H. Park (ed.), *Recent Advances in Control Problems of Dynamical Systems and Networks*, Studies in Systems, Decision and Control 301,
https://doi.org/10.1007/978-3-030-49123-9_20

1 Introduction

Since the last two decades of the last century, with the quick development of information technology, the Internet and human society have stride forward into the era of networks. In daily lives, typical examples of complex networks could be found in many situations, such as the Internet, WWW, giant electrical power grids, global transportation network, brains in the living body, various economic and political social networks [1, 2]. As we all know that human beings and the environment they living are closely connected not only in natural science but also in some artificial societies. Dynamical systems are always used to describe this kind of connections and evolutions. For discussing a system, it should either explore the evolution rules of the individuals, or the influences affected by the other individuals. To smoothly investigate this kind of relationships, algebraic graph theory has been presented in [3], and therefore, the relationships among different individuals could be described as a complex dynamical network. In the graph, every node stands for an individual, and each path denotes the connections between different nodes. From the viewpoint of mathematics, complex networks can be written as an array of coupled ordinary differential equations. During the past three decades, the study on the complex networks has attracted lots of scientists' attention from different research fields.

In fact, synchronization [3–6] could be understood as the regulation on an array of coupled dynamical systems in order to force them to behave uniformly by regulating the coupling weights, coupling strengths or other system parameters of the complex networks. The phenomena of synchronization are common in daily life. Until now, different synchronization patterns of complex networks, such as (complete) synchronization, lag synchronization, phase synchronization, anti-synchronization, cluster synchronization have been deeply and widely discussed by the scholars from all over world, and as a result, many outstanding achievements have been presented [7, 8]. For different purposes in practical application, different synchronization patterns should be flexibly introduced. For instance, in some special situations, sometimes, the systems in the complex dynamical networks are required to be separated into some subgroups, namely, clusters, according to the topological structures of the whole complex networks. In this situation, only those systems in the same clusters are requested to behave uniformly while it doesn't matter for the systems in different clusters [9–13]. To study this issue, the cluster synchronization was discussed in [9] for a kind of complex networks with time-varying delayed couplings and stochastic disturbance by designing the randomly occurring controllers. It should be noted that the networked dynamic systems naturally have time delays through various channels due to several reasons in hardware structure and signal processing [14]. Moreover, Liu and Chen discussed the finite-time and fixed-time problems for cluster synchronization in [10] for the complex networks by introducing the distributed control protocol in order to guarantee the final synchronization.

As it is known to us that the impulse signal input to dynamical systems represent a kind of outer discontinuous forces at the impulsive instants, which makes the system to be perturbed or promoted suddenly and discontinuously. From the working mechanism of the impulses, it could be found that impulsive control is a kind of representative discrete-time control schemes, which provides the instantaneous motion only at the impulsive instants. From the viewpoint of control time, comparing with general continuous control methods, impulsive control could save the control costs [16, 22]. However, when an impulse signal is introduced to a dynamical system, it could be either the control motion or some kinds of stochastic disturbances [15], [17–20]. To study the impulsive control problems, the key factor is the impulsive effect, which determines the strength of the generated impulse signal. In this regard, to utilize the impulsive control method, not only the advantageous impulse but also the disadvantageous impulse should be considered [28]. Although authors in [21] extended our previous work [15] from complete synchronization to projective synchronization, the impulsive effects did not well discussed, and the results were not applicable to general coupled neural networks.

However, to the best of our knowledge, the cluster synchronization problem of nonidentically coupled Lur'e networks with derivative coupling and multiple time-varying delays, until now, has not drawn much attention. Both the significance in theoretical research and the importance in practical applications stimulate us to proceed with this problem in this chapter. This chapter is mainly concerned with the global and exponential synchronization on a kind of derivative coupled stochastic complex networks consisting of nonidentical Lur'e systems by jointly inducting the impulsive control and pinning feedback control strategies.

The main findings of this chapter can be outlined as follows:

(1) In order to describe more general complex network models, this chapter considers a nonidentically coupled Lur'e networks with three different kinds of coupling methods, namely, general state coupling, delay state coupling, and derivative state coupling;

(2) Considering the cluster-tree topological structure of the coupled Lur'e networks, a novel pinning feedback controller is designed, and it will be imposed only on these Lur'e systems in current cluster which have direct connections with the Lur'e systems in the other clusters;

(3) By considering the derivative coupling in networks, a novel Lyapunov functional is constructed, which is closely related to the derivative coupling strength and derivative coupling matrix;

(4) Based on the general comparison principle and parameter variation formulas for impulsive differential equations, sufficient criteria for global and exponential cluster synchronization are derived by discussing different functions of the impulsive effects.

The rest parts of this chapter are arranged as follows. In the second section, the nonidentically coupled Lur'e networks model with derivative coupling and stochastic disturbances will be introduced and meanwhile, some definitions, lemmas, and assumptions will be presented as well. Then, the global exponential synchronization of the derivative coupled Lur'e networks will be studied in the third section. In Sect. 4, we give some numerical simulations to verify the and theoretical analysis and the control strategy proposed in this chapter. We will draw the conclusion of this chapter in the final section.

Notations. H^T denotes the transpose of the matrix H. $\mathbb{R}^n$ denotes the n-dimensional Euclidean space. $\mathbb{R}^{n\times n}$ is the set of $n \times n$ real matrices. $diag\{\cdots\}$ stands for a diagonal matrix. The symbol $\|\cdot\|$ stands for the Euclid norm of the matrix or the vector. $\lambda_{\max}(H)$ stands for the largest eigenvalue of matrix H, and $\max_{1\leq i\leq N}\{\cdot\}$ denotes to take the maximum value. $A=[a_{ij}]\in\mathbb{R}^{n\times n}$ denotes the matrix with elements a_{ij}. If a matrix H is positive definite (semi-definite), then we denote it as $H>0(H\geq 0)$. I_n stands for the identity matrix with n dimension. $H=(h_{ij})$ denotes a $\mathbb{R}^{N\times N}$ matrix H with elements h_{ij} for $i,j=1,2,\ldots,N$. $Trace(H)$ denotes the trace of matrix H. The set $\mathcal{N}^+$ denotes the positive integers. $D^+u(t)=\bar{\lim}_{t->0^+}\frac{u(t+h)-u(t)}{h}$. $\mathcal{N}^+=\{1,2,\ldots\}$ Let $\mathcal{C}([-h,0],\mathbb{R}^n)$ be the continuous functions class, where the function maps from the interval $[-h,0]$ to the filed $\mathbb{R}^n$. The dimension of these vectors and matrices will be cleared in the context.

2 Model Description and Preliminaries

2.1 Network Structure Statement

Firstly, to proceed with this section, some statements on the nonidentically derivative coupled Lur'e networks with stochastic disturbance and time-varying delay are presented in this subsection.

Consider that the nonidentically derivative coupled Lur'e networks consist of N nonidentical Lur'e systems, and meanwhile, it could be split into l clusters, which satisfies $2\leq l<N$. If the Lur'e system i included int the cluster k, then define a function $\varphi_i=k$. Define the set that all Lur'e systems in cluster k is $\mathcal{W}_k$ and the set that all Lur'e systems in cluster j which have directed connections with the Lur'e systems in the other clusters is $\bar{\mathcal{W}}_j$. In summary, there satisfies the conclusion that (a) $\mathcal{W}_k\bigcap\mathcal{W}_q=\emptyset$, for $k\neq q$ and $k,q=1,2,\ldots,l$; (b) $\bigcup_{k=1}^{l}\mathcal{W}_k=\{1,2,\ldots,N\}$ based on the discussions above.

2.2 Problem Formulation

Let discuss the following nonidentically coupled Lur'e networks with derivative couplings and stochastic disturbance

$$dz_i(t)=[\Xi^A_{\varphi_i}z_i(t)+\Xi^B_{\varphi_i}f_{\varphi_i}(Cz_i(t))+\epsilon_1\sum_{j=1}^{N}g_{ij}\Gamma z_j(t)+\epsilon_2\sum_{j=1}^{N}\upsilon_{ij}\Gamma z_j(t-h(t))$$
$$+\epsilon_3\sum_{j=1}^{N}\rho_{ij}\Gamma\dot{z}_j(t)+u_i(t)]dt+\tilde{p}(t,z_i(t),z_i(t-h(t)))d\omega(t), \quad (1)$$

where the state vector of the i-th Lur'e system is $z_i(t) = [z_i^1(t), z_i^2(t), \ldots, z_i^n(t)]^T \in \mathbb{R}^n$ for $i = 1, 2, \ldots, N$, which is an n dimension vector; Three matrices $\Xi_{\varphi_i}^A \in \mathbb{R}^{n\times n}$, $\Xi_{\varphi_i}^B \in \mathbb{R}^{n\times m}$ and $C \in \mathbb{R}^{m\times n}$ are constant parameter matrices; The nonlinear Lur'e function $f_{\varphi_i} : \mathbb{R}^m \to \mathbb{R}^m$ is a memoryless vector-valued function, which is a continuous differentiable function on all real number fields, and it satisfies $f_{\varphi_i}(0) = 0$; The three positive constants $\epsilon_1, \epsilon_2, \epsilon_3$ are the coupling strengths for general coupling, time-varying delay coupling and derivative coupling terms, respectively; Positive definite diagonal matrix $\Gamma = diag\{r_1, \ldots, r_n\} \in \mathbb{R}^{n\times n}$ denotes the inner linking connections within the Lur'e systems; In this chapter, without loss of generality, simply let $\Gamma = I_n$; Constant function $h(t)$ is the communication transmission time-varying delay satisfying $0 \le h(t) \le h$; Two square matrices $G = [g_{ij}] \in \mathbb{R}^{N\times N}$ and $\Upsilon = [\upsilon_{ij}] \in \mathbb{R}^{N\times N}$ stand for two different outer communication methods among different Lur'e systems, and they are called outer-coupling matrices, which are totally determined by the topological structures of the coupled Lur'e networks. Furthermore, two outer-connected coupling matrices are assumed to satisfy with the following condition $g_{ii} = -\sum_{j\neq ij=1}^{N} g_{ij}$ and $\upsilon_{ii} = -\sum_{j\neq ij=1}^{N} \upsilon_{ij}$, where $g_{ij} > 0$ and $\upsilon_{ij} > 0$ mean that if there is a directed connection from the Lur'e system j to the Lur'e system i $(i \neq j)$, otherwise for $g_{ij} = 0$ and $\upsilon_{ij} = 0$; In addition, the matrix $\varrho = [\rho_{ij}] \in \mathbb{R}^{N\times N}$ describes the derivative coupling method among different Lur'e systems, and for simplicity in this chapter, we consider that ϱ is a symmetrical matrix satisfying $\rho_{ij} = \rho_{ji} > 0$ if there a directed connection between the Lur'e system j and the Lur'e system i $(i \neq j)$, otherwise for $\rho_{ij} = \rho_{ji} = 0$; The noise intensity function $\tilde{p} : \mathbb{R}^+ \times \mathbb{R}^n \times \mathbb{R}^n \to \mathbb{R}^{n\times m}$ is a matrix-valued function satisfying the initial values $\tilde{p}(t, 0, 0) = \mathbf{0}_{n\times m}$; Constant m-dimensional vector-valued function $\omega(t) \in \mathbb{R}^m$ describes the Brownian motion; And $u_i(t)$ means a kind of control inputs, it will be discussed in detail later. For the nonidentically derivative coupled Lur'e networks (1), the initial conditions could be given as follows $z_i(t) = \zeta_i(t) \in \mathcal{C}([-h, 0], \mathbb{R}^n)$.

Remark 1 In view of the concept on the derivative outer-coupling matrix ϱ, the maximum eigenvalue of matrix ϱ is zero, while other eigenvalues of matrix are less than zero. Therefore, all of the eigenvalues of matrix ϱ can be arrayed in the following sequence $0 = \lambda_1 \ge \lambda_2 \ge \cdots \ge \lambda_N$. Then, the minimum eigenvalues of the positive definite matrix $I_N - \epsilon_3\varrho$ is $\lambda_{\min}(I_N - \epsilon_3\varrho) = 1$, while the maximum eigenvalue $\lambda_{\max}(I_N - \epsilon_3\varrho) \ge 1$. The feature of this structured matrix $I_N - \epsilon_3\varrho$ will play an essential role in construction of the Lyapunov functional.

Consider the isolated Lur'e systems with time-varying delay and stochastic disturbance in different clusters

$$ds_{\varphi_i}(t) = [\Xi_{\varphi_i}^A s_{\varphi_i}(t) + \Xi_{\varphi_i}^B f_{\varphi_i}(C s_{\varphi_i}(t))]dt + \tilde{p}(t, s_{\varphi_i}(t), s_{\varphi_i}(t - h(t)))d\omega(t), \tag{2}$$

where $s_{\varphi_i}(t) = [s_{\varphi_i}^1(t), s_{\varphi_i}^2(t), \ldots, s_{\varphi_i}^n(t)]^T \in \mathbb{R}^n$ is the state vector of the target Lur'e system in the φ_i-th cluster for $i = 1, 2, \ldots, N$.

In this chapter, our objective is to study the cluster synchronization between the nonidentically derivative coupled Lur'e networks (1) and the isolated Lur'e systems

(2). In fact, the problem studied in this chapter could be viewed as special leaders and followers problems in each cluster. In particular, the trajectory $s_{\varphi_i}(t)$ of the above Lur'e system (2) could be viewed as the leader for the φ_i-th cluster while the Lur'e systems in the φ_i-th cluster could be regarded as its followers.

Let $\delta z_i(t) = z_i(t) - s_{\varphi_i}(t)$ be the error vector, where $\delta z_i(t) = [\delta z_i^1(t), \delta z_i^2(t), \ldots, \delta z_i^n(t)]^T \in \mathbb{R}^n$ for $i = 1, 2, \ldots, N$.

To realize the cluster synchronization of the nonidentically derivative coupled Lur'e networks, we propose the following impulsive pinning controllers

$$u_i(t) = \bar{u}_i(t) + \tilde{u}_i(t), \tag{3}$$

where

$$\bar{u}_i(t) = \sum_{s=1}^{+\infty}(\eta\delta z_i(t) + \mu\delta z_i(t - \tau(t)))\partial(t - t_s),$$

is the impulsive control input with another time-varying delay and furthermore, if the Lur'e system i satisfies that $i \in \bar{\mathcal{W}}_{\phi_i}$, let

$$\tilde{u}_i(t) = -\vartheta_i\delta z_i(t) - \epsilon_1\sum_{j=1}^{N} g_{ij}s_{\varphi_j}(t) - \epsilon_3\sum_{j=1}^{N}\rho_{ij}\dot{s}_{\varphi_j}(t) - \epsilon_2\sum_{j=1}^{N}\upsilon_{ij}s_{\varphi_j}(t - \tau(t)),$$

otherwise, just let $\tilde{u}_i(t) = 0$ for the Lur'e system satisfies $i \in \mathcal{W}_{\phi_i} - \bar{\mathcal{W}}_{\phi_i}$. Constant function $\tau(t)$ is the time-varying delay in the controlled state satisfying $0 \le \tau(t) \le \bar{\tau}$. Furthermore, for the two different time-varying delays, let $\tau = \max\{h, \bar{\tau}\}$. Two different scalars η and μ are called impulsive effects for the general error state $\delta z_i(t)$ and the time-varying delayed error state $\delta z_i(t - \tau(t))$ of the i-th Lur'e system; Function $\partial(\cdot)$ is the Dirac impulsive function; Let's define a impulse sequence $\zeta = \{t_1, t_2, \ldots\}$ with the impulsive instants t_k, which is a strictly increasing sequence of time series, that is, it satisfies the condition $t_{k-1} < t_k$ and $\lim_{k\to+\infty} t_k = +\infty$ for $k \in \mathcal{N}^+$. The constant $\vartheta_i \ge 0$ is called the negative feedback control gain for $i = 1, 2, \ldots, N$. Denote $D_\vartheta = diag\{\vartheta_1, \vartheta_2, \ldots, \vartheta_n\}$ for later usage.

Based on the design of the controller and the definition of the error vector, the nonidentically derivative coupled error Lur'e networks with stochastic disturbance and multiple time-varying delays could be represented as

$$\begin{cases} \delta\dot{z}_i(t) = [\Xi^A_{\varphi_i}\delta z_i(t) + \Xi^B_{\varphi_i}\Phi_{\varphi_i}(C\delta z_i(t)) + \epsilon_1\sum_{j=1}^{N} g_{ij}\delta z_j(t) \\ \qquad +\epsilon_2\sum_{j=1}^{N}\upsilon_{ij}\delta z_j(t - h(t)) + \epsilon_3\sum_{j=1}^{N}\rho_{ij}\delta\dot{z}_j(t) - \vartheta_i\delta z_i(t)]dt \\ \qquad +p(t, \delta z_i(t), \delta z_i(t - h(t)))d\omega(t) \\ \Delta\delta z_i(t_k) = \eta\delta z_i(t_k^-) + \mu\delta z_i(t_k^- - \tau(t_k^-)), \\ \delta z_i(t) = \varpi_i(t), \qquad -\tau \le t \le 0, \end{cases} \tag{4}$$

where the nonlinear structured function $\Phi_{\varphi_i}(C\delta z_i(t)) = f_{\varphi_i}(Cz_i(t)) - f_{\varphi_i}(Cs_{\varphi_i}(t))$, and the matrix-valued function $p(t, \delta z_i(t), \delta z_i(t-h(t))) = \tilde{p}(t, z_i(t), z_i(t-h(t))) - \tilde{p}(t, s_{\varphi_i}(t), s_{\varphi_i}(t-h(t))), i = 1, 2, \ldots, N$. Additionally, function $\varpi_i(t)$ is the initial condition of the above error networks for $-\tau \leq t \leq 0$.

2.3 Preliminaries

To derive the main results in this chapter for the cluster synchronization of the nonidentically derivative coupled Lur'e networks, we are now presenting some preliminaries including definitions, lemmas and assumptions.

Definition 1 (*Global Exponential Synchronization*) Consider the nonidentically derivative coupled error Lur'e networks (4). If there are two positive scalars $\tilde{F}$, δ, and a time instant $\bar{T} > 0$, such that for any time $t > \bar{T}$ and any initial values $\chi_i(\cdot)$ satisfying the following inequality

$$\mathbb{E}[\|\delta z_i(t)\|^2] \leq \tilde{F} \sup_{k\in[t-\tau,t]} \mathbb{E}[\|\chi_i(s)\|^2]e^{-\delta t}, \tag{5}$$

then it is said that global exponential synchronization the between the nonidentically derivative coupled Lur'e networks (1) and the target isolated Lur'e systems (2) is eventually realized in a mean-square. Furthermore, the positive scalar δ is named as the convergence velocity of the exponential synchronization.

Definition 2 (*Cluster Synchronization* [27]) Consider the nonidentically derivative coupled Lur'e networks (1) and the isolated target Lur'e systems (2). For the Lur'e systems in the same cluster, i.e., $\varphi_i = \varphi_j$ there holds the condition $\lim_{t\to+\infty} \|\delta z_i(t)\| = 0$, while for Lur'e systems in different clusters, namely, $\varphi_i \neq \varphi_j$, there holds that $\lim_{t\to+\infty} \|\delta z_i(t)\| \neq 0$ with $i, j = 1, 2, \ldots, N$, then it is said that the cluster synchronization between the nonidentically derivative coupled Lur'e networks (1) and the isolated target Lur'e systems (2) is eventually realized.

Definition 3 (*Average Impulsive Interval* [28]) Define the impulsive sequence $\zeta = \{t_1, t_2, \ldots\}$ during the time interval on (t_1, t_2). If one could find two constants $T_0 > 0$ and $T_a > 0$ satisfying

$$\frac{t_2 - t_1}{T_a} - T_0 \leq T_\zeta(t_2, t_1) \leq \frac{t_2 - t_1}{T_a} + T_0, \ \forall\, T \geq t \geq 0, \tag{6}$$

where $T_\zeta(t_2, t_1)$ means how many times that the impulses occur on impulsive sequence ζ, then it could be estimated that the average impulsive interval is not larger than T_a.

Lemma 1 ([29]) *Suppose that if there exists a nonlinear function $W(t) > 0$, such that the impulsive differential inequalities satisfied*

$$\begin{cases} D^+W(t) \le -\psi W(t) + \varsigma \sup_{k\in[t-\tau,t]}\{W(k)\}, \\ W(t) = \varphi(t), \qquad t \in [t_0-\tau, t_0], \end{cases} \tag{7}$$

then one could drive the conclusion that

$$W(t) \le \tilde{W}(t)e^{-\delta(t-t_0)}, \quad t \ge t_0, \tag{8}$$

where ψ *and* ς *are two scalars satisfying* $\psi > \varsigma \ge 0$*, the function* $\varphi(t)$ *is piecewisely continuous, the notation* $\tilde{W}(t) = \sup_{k\in[t-\tau,t]}\{W(k)\}$ *and the constant* $\delta > 0$ *is the unique solution to the system parameter equation* $\delta - \psi + \varsigma e^{\delta\tau} = 0$.

Lemma 2 ([30]) *Suppose that there are two scalars* ψ *and* ς *satisfying* $\psi > \varsigma \ge 0$*. Similar to Lemma 1, for the impulsive differential inequalities (7), if there exists a nonlinear function* $W(t) > 0$ *such that for any* $t > t_0 \ge 0$*, there holds*

$$\mathbb{E}[W(t)] \le \mathbb{E}[\tilde{W}(t_0)]e^{-\lambda(t-t_0)}, \tag{9}$$

where the constant $\delta > 0$ *is the unique solution to the system parameter equation* $\delta - \psi + \varsigma e^{\delta\tau} = 0$ *and* $\tilde{W}(t_0) = \sup_{k\in[t_0-\tau,t_0]}\{W(k)\}$.

Lemma 3 ([31]) *Let's discuss a nonlinear impulsive system with stochastic disturbance*

$$\begin{cases} dY(t) = F(t, C(t))dt + \Theta(t, Y(t))db(t), \ t \ne t_k, \\ Y(t_k^+) - Y(t_k^-) = u_k(Y(t_k^-)), \ k \in \mathcal{N}^+, \end{cases} \tag{10}$$

where $Y(t)$ *is the state vector,* $F(\cdot,\cdot), \Theta(\cdot,\cdot), C(\cdot), u_k(\cdot)$ *are constant functions. If there exist a positive function* $V(t, Y)$ *and two different nonlinear functions* ϖ, χ_k *with* $\varpi(t, 0) = \chi_k(0) = 0$ *satisfying the conditions*

(a) There exist two positive scalars l_1, l_2 *satisfying* $l_1 \cdot \|Y(t)\| \le V(t, Y) \le l_2 \cdot \| Y(t) \|$ *with* $t \ge t_0$*;*

(b) Define a continuous concave function as $\varpi : \mathbb{R}^+ \times \mathbb{R}^+ \to \mathbb{R}$*. Denote* $V_Y(t, Y) = [\frac{\partial V(t,Y)}{\partial Y_1}, \frac{\partial V(t,Y)}{\partial Y_2}, \ldots, \frac{\partial V(t,Y)}{\partial Y_n}]$ *and* $V_Y(t, Y) = (\frac{\partial^2 V(t,Y)}{\partial Y_i \partial Y_j})_{n\times n}$*. For the above conditions, there satisfies* $\mathcal{L}V(t, Y) = V_t(t, Y) + V_Y(t, Y)\varpi(t, Y) + \frac{1}{2}trace[\Theta^T(t, Y) V_{YY}\Theta(t, Y)]$*;*

(c) If there exist continuous concave functions $\chi_k : \mathbb{R}^+ \to \mathbb{R}^+$ *which satisfies the inequality* $V(t_k^+, Y(t_k^+)) \le \chi_k(V(t_k^-, Y(t_k^-)))$*,*
then the solution of the following comparison system

$$\begin{cases} \dot{w}(t) = \phi(t, w(t)), \ t \ne t_k, \\ w(t_k^+) = \chi_k(w(t_k^-)), \ k \in \mathcal{N}^+, \\ w(t_0) = \mathbb{E}[V(t_0, x_0)] \end{cases} \tag{11}$$

is exponentially stable, which directly implies that the exponential stability of the trivial solution of the nonlinear impulsive system with stochastic disturbance (10) is obtained, where $\mathbb{E}(\cdot)$ means the mathematical expectation.

Assumption 1 Consider the nonlinear Lur'e system function $f_{\varphi_i}(\cdot)$. Suppose that the function $f_{\varphi_i}(\cdot)$ satisfies with the Lipschitz condition with the Lipschitz constant $h_{\varphi_i} > 0 (i = 1, 2, \ldots, N)$. That is, there exists a positive constant h_{φ_i} for any two n-dimensional vectors $x, y \in \mathbb{R}^n$ satisfying

$$\|f_{\varphi_i}(x) - f_{\varphi_i}(y)\| \leq h_{\varphi_i}\|x - y\|, \quad i = 1, 2, \ldots, N.$$

□

Assumption 2 Consider the matrix-valued function $p(t, \cdot, \cdot) \in \mathbb{R}^{n\times m}$. Suppose that for two constant matrices Q, $Q_h \in \mathbb{R}^{n\times n}$ and any vectors $x(t), x(t - h(t)) \in \mathbb{R}^n$, the function $p(t, \cdot, \cdot)$ could be calculated as

$$trace((p(t, x(t), x(t - h(t))^T p(t, x(t), x(t - h(t))) \leq \|Qx(t)\|^2 + \|Q_h x(t - h(t))\|^2,$$

where $trace$ denotes the trace on the matrices. That is, the function $p(t, \cdot, \cdot)$ is a uniformly Lipschitz function in the form of matrix norm. □

3 Main Results

In this section, we will discuss the cluster synchronization of a kind of nonidentically derivative coupled Lur'e networks (1) with multiple time-varying delays and stochastic disturbance by applying an impulsive pinning controller. According to the extended comparison principle of the impulsive differential equations, the definition of the average impulsive interval and the mathematical taxonomy on impulsive parameters, the achievement of the cluster synchronization will be eventually guaranteed. In addition, the exponential convergence velocity of the derivative coupled Lur'e networks will be precisely estimated.

Theorem 1 *Consider the nonidentically derivative coupled error Lur'e networks (4) under the control strategy (3) with Assumptions 1 and 2. Suppose that for the impulse time series $\zeta = \{t_1, t_2, \ldots\}$, the average impulsive interval is less than T_a, which has been defined in Definition 3. If there exist a negative feedback control gain matrix $D_\vartheta = diag\{\vartheta_1, \vartheta_2, \ldots, \vartheta_N\}$ with at least one $\vartheta_i > 0 (i = 1, 2, \ldots, N)$ and three positive scalars α, β, h_{φ_i}, such that for the two different impulsive effects η and μ defined in controller (3) satisfying*

(i) The matrix-valued inequality

$$\Pi = \begin{pmatrix} \epsilon_1 G^s + \alpha I_N - D_\vartheta & \frac{1}{2}\epsilon_2 \Upsilon \\ \frac{1}{2}\epsilon_2 \Upsilon^T & -\beta I_N \end{pmatrix} \leq 0; \tag{12}$$

(ii) *The following inequality if there exists a positive constant* ι

$$\max\{\bar{a} + \bar{b}e^{\tau\lambda}, e^{\tau\lambda}\} \leq \iota, \tag{13}$$

where $\tau = \max\{h, \bar{\tau}\}$, $\bar{a} = \lambda_{\max}(I_N - \epsilon_3\varrho)(1+\eta)(1+\eta+\mu)$, $\bar{b} = \lambda_{\max}(I_N - \epsilon_3\varrho)(1+\eta+\mu)\mu$, *and the parameter* $\lambda > 0$ *is the unique solution to the following constructed parameter equation*

$$\lambda - \psi + \varsigma e^{\lambda\tau} = 0$$

satisfying the requirement $\psi > \varsigma \geq 0$, *and parameters are defined as follows*

$$\psi = -\frac{\min_{1\leq i\leq N}\{\lambda_{\min}(2\Xi^A_{\varphi_1} + \Xi^B_{\varphi_i}\Xi^{B\,T}_{\varphi_i} + h_{\varphi_i}CC^T + QQ^T - 2\alpha I_n)\}}{\lambda_{\max}(I_N - \epsilon_3\varrho)},$$
$$\varsigma = \max_{1\leq i\leq N}\{\lambda_{\max}(Q_h^T Q_h + 2\beta I_n)\};$$

(iii) *The exponentially convergence velocity in mean-square*

$$\frac{\ln\iota}{T_a} - \lambda < 0, \tag{14}$$

then the trajectory of the nonidentically derivative coupled error Lur'e networks (4) is exponentially stable. That is, the global and exponential synchronization between the nonidentically derivative coupled Lur'e networks (1) and the target isolated Lur'e systems (2) is finally realized by imposing the impulsive pinning controller (3).

Proof Based on the discussion in Remark 1, we construct the Lyapunov functional by thinking the coupling strength ϵ_3 and the derivative coupling matrix ϱ

$$V(t, \delta z(t)) = \frac{1}{2}\delta z^T(t)((I_N - \epsilon_3\varrho)\otimes I_n)\delta z(t), \tag{15}$$

and define the vector as $\delta z(t) = [\delta z_1^T(t), \delta z_2^T(t), \ldots, \delta z_N^T(t)]^T \in \mathbb{R}^{nN}$.

To study the impulsively controlled Lur'e networks, the $V(t, \delta z(t))$ of the impulse instants t_k and non-impulsive time interval $[t_{k-1}, t_k)$ should be discussed, respectively. At first, let's consider the situation that $V(t, \delta z(t))$ on impulsive time instant $t = t_k, k \in \mathcal{N}^+$.

Considering the impulsive relationship $\Delta\delta z_i(t_k) = \eta\delta z_i(t_k^-) + \mu\delta z_i(t_k^- - \tau(t_k^-))$ in the impulsively controlled error Lur'e networks (4) gives

$$\delta z_i(t_k^+) = (1+\eta)\delta z_i(t_k^-) + \mu\delta z_i(t_k^- - \tau(t_k^-)).$$

Recalling the truth in Remark 1 $\lambda_{\max}(I_N - \epsilon_3\varrho) \geq \lambda_{\min}(I_N - \epsilon_3\varrho) = 1$, it yields the following conclusion

$$
\begin{aligned}
V(t_k^+,\delta z(t_k^+)) &= \frac{1}{2}\delta z^T(t_k^+)((I_N-\epsilon_3\varrho)\otimes I_n)\delta z(t_k^+)\\
&\le \frac{1}{2}\lambda_{\max}(I_N-\epsilon_3\varrho)\delta z^T(t_k^+)\delta z(t_k^+)\\
&= \frac{1}{2}\lambda_{\max}(I_N-\epsilon_3\varrho)((1+\eta)\delta z^T(t_k^-)\\
&\quad + \mu\delta z^T(t_k^- - \tau(t_k^-)))((1+\eta)\delta z(t_k^-) + \mu\delta z(t_k^- - \tau(t_k^-)))\\
&= \frac{1}{2}\lambda_{\max}(I_N-\epsilon_3\varrho)((1+\eta)^2\delta z^T(t_k^-)\delta z(t_k^-)\\
&\quad + 2(1+\eta)\mu\delta z^T(t_k^-)\delta z(t_k^- - \tau(t_k^-))\\
&\quad + \mu^2\delta z^T(t_k^- - \tau(t_k^-))\delta z(t_k^- - \tau(t_k^-)))\\
&\le \frac{1}{2}\lambda_{\max}(I_N-\epsilon_3\varrho)((1+\eta)^2\delta z^T(t_k^-)\delta z(t_k^-)\\
&\quad + (1+\eta)\mu\delta z^T(t_k^-)\delta z(t_k^-)\\
&\quad + (1+\eta)\mu\delta z^T(t_k^- - \tau(t_k^-))\delta z(t_k^- - \tau(t_k^-))\\
&\quad + \mu^2\delta z^T(t_k^- - \tau(t_k^-))\delta z(t_k^- - \tau(t_k^-)))\\
&\le \frac{\lambda_{\max}(I_N-\epsilon_3\varrho)}{2\lambda_{\min}(I_N-\epsilon_3\varrho)}((1+\eta)(1+\eta+\mu)\delta z^T(t_k^-)\\
&\quad \times ((I_N-\epsilon_3\varrho)\otimes I_n)\delta z(t_k^-) + (1+\eta+\mu)\mu\delta z^T(t_k^- - \tau(t_k^-))\\
&\quad \times ((I_N-\epsilon_3\varrho)\otimes I_n)\delta z(t_k^- - \tau(t_k^-)))\\
&= \lambda_{\max}(I_N-\epsilon_3\varrho)((1+\eta)(1+\eta+\mu)V(t_k^-)\\
&\quad + (1+\eta+\mu)\mu V(t_k^- - \tau(t_k^-)))\\
&= \bar{a}V(t_k^-) + \bar{b}V(t_k^- - \tau(t_k^-)),
\end{aligned}
\tag{16}
$$

where we define the parameters $\bar{a} = \lambda_{\max}(I_N - \epsilon_3\varrho)(1+\eta)(1+\eta+\mu)$, $\bar{b} = \lambda_{\max}(I_N - \epsilon_3\varrho)\mu(1+\eta+\mu)$.

For the index $k \in \mathcal{N}^+$, calculating the mathematical expectation of the inequality (16) gives

$$
\mathbb{E}[V(t_k^+,\delta z(t_k^+))] \le \bar{a}\mathbb{E}[V(t_k^-,\delta z(t_k^-))] + \bar{b}\mathbb{E}[V(t_k^- - \tau(t_k^-),\delta z(t_k^-))]. \tag{17}
$$

After finishing the analysis of the impulsive instants t_k, let's consider the derivative on $V(t)$ during the non-impulsive time interval $t \in [t_{k-1}, t_k)$ for $k \in \mathcal{N}^+$.

Based on the definition of the Itô differential equation on the nonlinear impulsive system with stochastic disturbance, calculating the derivative values by applying $\mathcal{L}V(t,\delta z(t))$ in Lemma 3 with considering the impulsively pinning controlled error Lur'e networks (4) gives the following results

$$
\begin{aligned}
\mathcal{L}V(t,\delta z(t)) &= \delta z^T(t)((I_N-\epsilon_3\varrho)\otimes I_n)\delta\dot{z}(t)\\
&\quad+\frac{1}{2}\sum_{i=1}^{N} trace(p(t,\delta z_i(t),\delta z_i(t-h(t)))^T p(t,\delta z_i(t),\delta z_i(t-h(t))))\\
&\le \delta z^T(t)\delta\dot{z}(t)-\epsilon_3\delta z^T(t)(\varrho\otimes I_n)\delta\dot{z}(t)+\frac{1}{2}\sum_{i=1}^{N}\delta z_i^T(t)Q^TQ\delta z_i(t)\\
&\quad+\frac{1}{2}\sum_{i=1}^{N}\delta z_i^T(t-h(t))Q_h^TQ_h\delta z_i(t-h(t))\\
&=\sum_{i=1}^{N}\delta z_i^T(t)(\Xi_{\varphi_1}^A\delta z_i(t)+\Xi_{\varphi_i}^B\Phi_{\varphi_i}(C\delta z_i(t))+\epsilon_1\sum_{j=1}^{N}g_{ij}\delta z_j(t)\\
&\quad+\epsilon_2\sum_{j=1}^{N}\upsilon_{ij}\delta z_j(t-h(t))+\epsilon_3\sum_{j=1}^{N}\rho_{ij}\dot{\delta} z_j(t)-\vartheta_i\delta z_i(t))\\
&\quad-\epsilon_3\delta z^T(t)(\varrho\otimes I_n)\delta\dot{z}(t)+\frac{1}{2}\sum_{i=1}^{N}\delta z_i^T(t)Q^TQ\delta z_i(t)\\
&\quad+\frac{1}{2}\sum_{i=1}^{N}\delta z_i^T(t-h(t))Q_h^TQ_h\delta z_i(t-h(t))\\
&\le\frac{1}{2}\sum_{i=1}^{N}\delta z_i^T(t)(2\Xi_{\varphi_1}^A+\Xi_{\varphi_i}^B{\Xi_{\varphi_i}^B}^T+h_{\varphi_i}CC^T+QQ^T-2\alpha I_n)\delta z_i(t)\\
&\quad+\frac{1}{2}\sum_{i=1}^{N}\delta z_i^T(t-h(t))(Q_h^TQ_h+2\beta I_n)\delta z_i(t-h(t)))\\
&\quad+\sum_{k=1}^{n}\tilde{\delta} z_k^T(t)(\epsilon_1G-D_\vartheta+\alpha I_N)\tilde{\delta} z_k(t)+\epsilon_2\sum_{k=1}^{n}\tilde{\delta} z_k^T(t)\Upsilon\tilde{\delta} z_k(t-h(t))\\
&\quad-\beta\sum_{k=1}^{n}\tilde{\delta} z_k^T(t-h(t))\tilde{\delta} z_k(t-h(t))\\
&\le\min_{1\le i\le N}\{\lambda_{min}(2\Xi_{\varphi_1}^A+\Xi_{\varphi_i}^B{\Xi_{\varphi_i}^B}^T+h_{\varphi_i}CC^T+QQ^T-2\alpha I_n\}\\
&\quad\times\frac{1}{2}\sum_{i=1}^{N}\delta z_i^T(t)\delta z_i(t)\\
&\quad+\max_{1\le i\le N}\{\lambda_{\max}(Q_h^TQ_h+2\beta I_n)\}\\
&\quad\times\frac{1}{2}\delta z^T(t-h(t))((I_N-\epsilon_3\varrho)\otimes I_n)\delta z(t-h(t))\\
&\quad+\sum_{k=1}^{n}\mathcal{Y}^T(t)\Pi\mathcal{Y}(t)\\
&\le-\psi V(t)+\varsigma V(t-h(t)),
\end{aligned}
\tag{18}
$$

where define the vector as $\mathcal{Y}(t) = (\tilde{\delta z}_k^T(t), \tilde{\delta z}_k^T(t-h(t)))^T$, $\tilde{\delta z}_k(t) = (\delta z_1^k(t), \delta z_2^k(t), \ldots, \delta z_N^k(t))^T$ and denote $\psi = -\frac{\sigma}{\lambda_{\max}(I_N - \epsilon_3 \varrho)}$ with $\sigma = \min_{1\leq i\leq N}\{\lambda_{\min}(2\Xi_{\varphi_1}^A + \Xi_{\varphi_i}^B \Xi_{\varphi_i}^{B\,T} + h_{\varphi_i} CC^T + QQ^T - 2\alpha I_n)\}$, and $\varsigma = \max_{1\leq i\leq N}\{\lambda_{\max}(Q_h^T Q_h + 2\beta I_n)\}$.

Based on the above discussion, for the non-impulsive time interval $t \in [t_{k-1}, t_k)$, one could get the following conclusion by applying Lemmas 1 and 2

$$\mathbb{E}[V(t, \delta z(t))] \leq \mathbb{E}[\tilde{V}(t_{k-1}, \delta z(t_k^-))]e^{-\lambda(t-t_{k-1})}, \tag{19}$$

where denote the initial conditions as

$$\tilde{V}(t_{k-1}, \delta z(t_{k-1})) = \sup_{s\in[t_{k-1}-\tau, t_{k-1}]} V(s, \delta z(s)).$$

In order to derive the criteria on global exponential synchronization of the non-identically derivative coupled Lur'e networks, in the next procedure, based on the mathematical induction method, the following conclusion will be verified for a given positive constant ι and for any time satisfying $t > t_0 > 0$

$$\mathbb{E}[V(t, \delta z(t))] \leq \iota^{k-1}\mathbb{E}[\tilde{V}(t_0, \delta z(t_0))]e^{-\lambda(t-t_0)}, \tag{20}$$

where the initial conditions for $V(t, \delta z(t))$ on time instant t_0 is defined as $\tilde{V}(t_0, \delta z(t_0)) = \sup_{s\in[t_0-\tau, t_0]} V(s, \delta z(s))$.

In the first step, let's discuss the time during the interval $[t_{k-1}, t_k)$ with $k = 1$. According to the conclusion in inequality (19), there exists a scalar $\iota > 0$ such that

$$\begin{aligned}
\mathbb{E}[V(t, \delta z(t))] \leq& \mathbb{E}[\tilde{V}(t_0, \delta z(t_0))]e^{-\lambda(t-t_0)} \\
=& \iota^0\mathbb{E}[\tilde{V}(t_0, \delta z(t_0))]e^{-\lambda(t-t_0)} \\
=& \iota^{k-1}\mathbb{E}[\tilde{V}(t_0, \delta z(t_0))]e^{-\lambda(t-t_0)}.
\end{aligned}$$

In the next step, suppose that the conclusion (20) is satisfied for $k = s(s > 1)$. Then, in the following, by applying the mathematical induction method, we will prove that the conclusion (20) is still satisfied for $k = s + 1$. That is, for the situation $k = s(s > 1)$ and the time $t \in [t_{s-1}, t_s)$, one obtains the following results by considering the condition (13) in Theorem 1:

$$\begin{aligned}
\mathbb{E}[V(t_s, \delta z(t_s))] \leq& a\mathbb{E}[V(t_s^-, \delta z(t_k^-))] + b\mathbb{E}[V(t_s^- - \tau(t_s), \delta z(t_s^-))] \\
\leq& a \cdot \iota^{s-1}\mathbb{E}[\tilde{V}(t_0, \delta z(t_0))] \cdot e^{-\lambda(t_s-t_0)} \\
&+ b \cdot \iota^{s-1}\mathbb{E}[\tilde{V}(t_0, \delta z(t_0))]e^{-\lambda(t_s-\tau(t_s)-t_0)} \\
\leq& (\bar{a} + \bar{b}e^{\lambda\tau})\iota^{s-1}\mathbb{E}[\tilde{V}(t_0, \delta z(t_0))]e^{-\lambda(t_s-t_0)} \\
\leq& \iota^s\mathbb{E}[\tilde{V}(t_0, \delta z(t_0))]e^{-\lambda(t_s-t_0)}.
\end{aligned}$$

Therefore, we will derive the conclusion for $k = s + 1$ on the basis of $k = s$. Namely, for the time $t \in [t_s, t_{s+1})$, one has

$$\begin{aligned}
\mathbb{E}[V(t,\delta z(t))] &\leq \mathbb{E}[\tilde{V}(t_s, \delta z(t_s))]e^{-\lambda(t-t_s)} \\
&= \sup_{\sigma\in[t_s-\tau,t_s]} \mathbb{E}[V(\sigma, \delta z(\sigma))]e^{-\lambda(t-t_s)} \\
&= \max\{\sup_{\sigma\in[t_s-\tau,t_s)} \mathbb{E}[V(\sigma, \delta z(\sigma))], \mathbb{E}[V(t_s, \delta z(t_s))]\}e^{-\lambda(t-t_s)} \\
&\leq \max\big\{\iota^{s-1}\mathbb{E}[\tilde{V}(t_0, \delta z(t_0))]e^{-\lambda(t_s-\tau-t_0)}, \\
&\qquad \iota^{s}\mathbb{E}[\tilde{V}(t_0, \delta z(t_0))]e^{-\lambda(t_s-t_0)}\big\}e^{-\lambda(t-t_s)} \\
&= \max\{e^{\lambda\tau}, \iota\}\iota^{s-1}\mathbb{E}(\tilde{V}(t_0))e^{-\lambda(t-t_0)} \\
&\leq \iota^{s}\mathbb{E}[\tilde{V}(t_0, \delta z(t_0))]e^{-\lambda(t-t_0)}.
\end{aligned}$$

Above all, based on the Mathematical Induction Method, we have proved that there satisfies

$$\mathbb{E}[V(t, \delta z(t))] \leq \iota^{k-1}\mathbb{E}[\tilde{V}(t_0, \delta z(t_0))]e^{-\lambda(t-t_0)}$$

for any time $t \in [t_{k-1}, t_k)$ with $k \in \mathcal{N}^+$.

It should be noted that a positive scalar ι is introduced in the conclusion (20) to show the effectiveness of the inequality. However, an important problem is how to select the value ι such that (20) is satisfied for different impulsive effects. Therefore, by introducing the concept of an average impulsive interval, the global and exponential cluster synchronization issue on the nonidentically derivative coupled Lur'e networks will be studied by clarifying different values of the scalar $\iota > 0$.

(a) Considering the scalar satisfies $\iota \in (0, 1)$, then it gives the following results for the time $t \in [t_{k-1}, t_k)$, $k \in \mathcal{N}^+$

$$\begin{aligned}
\mathbb{E}[V(t, \delta z(t))] &\leq \iota^{k-1}\mathbb{E}[\tilde{V}(t_0, \delta z(t_0))]e^{-\lambda(t-t_0)} \\
&\leq \iota^{\frac{t-t_0}{T_a}-T_0}\mathbb{E}[\tilde{V}(t_0, \delta z(t_0))]e^{-\lambda(t-t_0)} \\
&= \iota^{-T_0}\mathbb{E}[\tilde{V}(t_0, \delta z(t_0))]\iota^{\frac{t-t_0}{T_a}}e^{-\lambda(t-t_0)} \\
&= \iota^{-T_0}\mathbb{E}[\tilde{V}(t_0, \delta z(t_0))]e^{(\frac{\ln\iota}{T_a}-\lambda)(t-t_0)}.
\end{aligned} \tag{21}$$

(b) For the positive scalar satisfying $\iota \in (1, +\infty)$, one could obtain the following results for the time $t \in [t_{k-1}, t_k)$, $k \in \mathcal{N}^+$

$$\begin{aligned}
\mathbb{E}[V(t, \delta z(t))] &\leq \iota^{k-1}\mathbb{E}[\tilde{V}(t_0, \delta z(t_0))]e^{-\lambda(t-t_0)} \\
&\leq \iota^{\frac{t-t_0}{T_a}+T_0}\mathbb{E}[\tilde{V}(t_0, \delta z(t_0))]e^{-\lambda(t-t_0)} \\
&= \iota^{T_0}\mathbb{E}[\tilde{V}(t_0, \delta z(t_0))]\iota^{\frac{t-t_0}{T_a}}e^{-\lambda(t-t_0)} \\
&= \iota^{T_0}\mathbb{E}[\tilde{V}(t_0, \delta z(t_0))]e^{(\frac{\ln\iota}{T_a}-\lambda)(t-t_0)}.
\end{aligned} \tag{22}$$

(c) For the special situation that the scalar $\iota = 1$, we could get the following results for the time $t \in [t_{k-1}, t_k)$, $k \in \mathcal{N}^+$

$$\begin{aligned}\mathbb{E}[V(t, \delta z(t))] \leq& \iota^{k-1}\mathbb{E}[\tilde{V}(t_0, \delta z(t_0))]e^{-\lambda(t-t_0)}\\ =&\mathbb{E}[\tilde{V}(t_0, \delta z(t_0))]e^{(\frac{\ln \iota}{T_a}-\lambda)(t-t_0)}.\end{aligned} \tag{23}$$

Based on the discussions above, it could be found that, for different values of the scalar ι, similar conclusions (21), (22) and (23) have been derived. According to Definition 1, the global exponential synchronization could be verified, that is, one could find two constants $\delta > 0$ and $\tilde{F} > 0$ such that the inequality (5) satisfied for any initial condition $\chi_i(t) \in \mathcal{C}([-\tau, 0], \mathbb{R}^n)$.

Furthermore, considering the construction of the Lyapunov function (15), it gives the following results

$$\|\delta z_i(t)\|^2 \leq \frac{2V(t, \delta z(t))}{\lambda_{\min}(I_N - \epsilon_3 \varrho)} = 2V(t, \delta z(t)). \tag{24}$$

By taking the mathematical expectation on inequality (24) gives

$$\mathbb{E}[\|\delta z_i(t)\|^2] \leq \tilde{F}\, \mathbb{E}[\tilde{V}(t_0, \delta z(t_0))]e^{(\frac{\ln \iota}{T_a}-\lambda)(t-t_0)}, \tag{25}$$

where $\tilde{V}(t_0, \delta z(t_0)) = \sup_{s\in[t_0-\tau,t_0]} V(s, \delta z(s))$. It further implies that the trajectories of the nonidentically coupled error Lur'e networks (4) are globally and exponentially stable with the convergence velocity $\frac{1}{2}(\frac{\ln \iota}{T_a} - \lambda)$. That is, the global exponential cluster synchronization between the nonidentically derivative coupled Lur'e networks (1) and the target isolated Lur'e systems (2) is finally realized in mean-square by imposing the impulsive pinning controller (3). By now, we completely finish the proof. □

Remark 2 Actually, the impulsive interval is one of the most important factors in impulsive control issues. In some related references like [22, 23], the impulsive interval T_a was roughly selected as $T_a = \min_k\{t_k - t_{k-1}\}$ or $T_a = \max_k\{t_k - t_{k-1}\}$. With similar steps in these references, if we deliberately let $T_a = \min_k\{t_k - t_{k-1}\}$ briefly in this chapter, then it gives a smaller exponential convergence velocity from the discussion in Theorem 1.

Correspondingly, a bigger impulsive effect μ and pinning control gains d_i are required for the achievement of the quasi-synchronization, which tremendously increase the requirements on the controller designing and the control costs.

Similarly, if take $T_a = \max_k\{t_k - t_{k-1}\}$ in this chapter, similar problems would occur. From this point of view, both $T_a = \min_k\{t_k - t_{k-1}\}$ and $T_a = \max_k\{t_k - t_{k-1}\}$ result in imprecise synchronization errors and convergence velocities. To avoid this situation, in this chapter, the concept of average impulsive interval has been introduced [22, 28].

From the definition in Definition 2, it gives that how many times that the impulses occur during the time interval (t, T) could be evaluated by jointly using the time interval (t, T), the freedom indicator T_0 and the positive constant T_a, which generally satisfies $\min_k\{t_k - t_{k-1}\} \leq T_a \leq \max_k\{t_k - t_{k-1}\}$. Based on the discussion above, different from previous works as we mentioned before, the introduction of average impulsive interval in this chapter would efficaciously reduce the conservatism in synchronization conditions and the requirements on impulsive effect and control gains.

Remark 3 For simplicity of the calculations, the constant impulsive effects η and μ have been considered in this chapter. In fact, due to the complicated working environment and uncertainties of the control input, in many practical applications, the impulsive effects on both general state and delayed state are time-varying as the impulses signal generates. As a result, in order to make the main theorem is applicable to this situation, the impulsive pinning feedback controller designed in this chapter should be substituted by the time-varying impulsive effects η_k and μ_k rather than constant ones. Correspondingly, the condition **(ii)** in Theorem 1 should be replaced as:

The following inequality is satisfied if there exists a positive constant ι

$$\max\{\bar{a}_k + \bar{b}_k e^{\lambda\tau}, e^{\lambda\tau}\} \leq \iota, \quad k = 1, 2, \ldots \tag{26}$$

where $\tau = \max\{h, \bar{\tau}\}$ $\bar{a}_k = \lambda_{\max}(I_N - \varepsilon_3\rho)(1 + \eta_k)(1 + \eta_k + \mu_k)$, $\bar{b}_k = \lambda_{\max}(I_N - \varepsilon_3\rho)(1 + \eta_k + \mu_k)\mu_k$, $k \in \mathcal{N}^+$, and the parameter $\lambda > 0$ is the unique solution to the following constructed parameter equation

$$\lambda - \psi + \varsigma e^{\lambda\tau} = 0$$

satisfying the requirement $\psi > \varsigma \geq 0$, and parameters are defined as follows:

$$\psi = -\frac{\min_{1\leq i\leq N}\{\lambda_{\min}(2\Xi^A_{\varphi_1} + \Xi^B_{\varphi_i}\Xi^{B\,T}_{\varphi_i} + h_{\varphi_i}CC^T + QQ^T - 2\alpha I_n)\}}{\lambda_{\max}(I_N - \epsilon_3\varrho)},$$

$$\varsigma = \max_{1\leq i\leq N}\{\lambda_{\max}(Q_h^T Q_h + 2\beta I_n)\}.$$

4 Numerical Simulation

In this section, a numerical simulation will be presented to verify the effectiveness of the theoretical analysis and the control strategy.

In the simulation, assume that there are totally 6 Lur'e systems in the coupled complex dynamical networks, which could be separated into two different clusters, i.e., $N = 6$ and let $\varphi_1 = \varphi_2 = \varphi_3 = 1$, $\varphi_4 = \varphi_5 = \varphi_6 = 2$.

For different clusters in the Lur'e complex, let's discuss the following isolated Chua's circuits with time-varying delay and stochastic disturbance

$$dz_{\varphi_i}(t) = [\Xi^A_{\varphi_i} z_{\varphi_i}(t) + \Xi^B_{\varphi_i} f_{\varphi_i}(C_{\varphi_i} z_{\varphi_i}(t))]dt \\ + \tilde{p}(t, z_{\varphi_i}(t), z_{\varphi_i}(t-h))d\omega(t), \quad (27)$$

where $\varphi_i = 1$ for $i = 1, 2, 3$ while $\varphi_i = 2$ for $i = 4, 5, 6$, and $z_{\varphi_i}(t) = (z_{\varphi_i}(t)^1, z_{\varphi_i}(t)^2, z_{\varphi_i}(t)^3)^T \in \mathbb{R}^3$ denotes the state vector of the i-th Lur'e system in the φ_i cluster, the time-varying delay $h(t) = 1 + 0.1\sin(0.1t)$, the system matrices are given as follows

$$\Xi^A_1 = \begin{bmatrix} -\frac{30}{7} & 15 & 0 \\ 1 & -1 & 0 \\ 0 & -25.5 & 0 \end{bmatrix}, \quad \Xi^B_1 = \begin{bmatrix} \frac{45}{7} \\ 0 \\ 0 \end{bmatrix}, \quad C_1 = \begin{bmatrix} 1 & 0 & 0 \end{bmatrix},$$

$$\Xi^A_2 = \begin{bmatrix} -3.2 & 10 & 0 \\ 1 & -1 & 0 \\ 0 & -15 & -0.0385 \end{bmatrix}, \quad \Xi^B_2 = \begin{bmatrix} 5.9 \\ 0 \\ 0 \end{bmatrix}, \quad C_2 = \begin{bmatrix} 1 & 0 & 0 \end{bmatrix},$$

and the nonlinear vector-valued function

$$f_1(C_1 z(t)) = f_2(C_2 z(t)) = \frac{1}{2}(\|z^1(t) + 1\|) - (\|z^1(t) - 1\|),$$

the stochastic matrix

$$\tilde{p}(t, z(t), z(t - h(t))) = 0.5 \cdot \|z(t)\| I_3,$$

and let $\omega(t)$ be the 3-dimensional Brownian motion. The phase graphs of the Lur'e systems in two different clusters are given in Fig. 1.

As stated in this chapter previously, two different clusters will be considered in the coupled Lur'e networks with three kinds of couplings, namely, general state coupling, derivative coupling, time-varying delayed state coupling. The topological structure of the time-varying delay coupled Lur'e networks is presented in Fig. 1b, where the 3-th Lur'e system in the $\varphi_i = 1$-th cluster and the 4-th Lur'e system in the $\varphi_i = 2$-th cluster are directed linked with each other. That is, the pinning feedback control term $\tilde{u}_i(t)$ in the impulsive pinning controller (3) will be directly imposed on these two Lur'e systems with $\tilde{u}_3(t)$ and $\tilde{u}_4(t)$.

To describe three kinds of couplings for the coupled Lur'e networks (1), let the coupling matrices as follows

$$G = \begin{pmatrix} -2 & 0 & 2 & 0 & 0 & 0 \\ 0 & -3 & 3 & 0 & 0 & 0 \\ 2 & 3 & -8 & 3 & 0 & 0 \\ 0 & 0 & 3 & -5 & 2 & 0 \\ 0 & 0 & 0 & 2 & -5 & 3 \\ 0 & 0 & 0 & 0 & 4 & -4 \end{pmatrix}, \quad \Upsilon = \begin{pmatrix} -5 & 1 & 0 & 2 & 0 & 2 \\ 1 & -5 & 3 & 0 & 1 & 0 \\ 0 & 3 & -5 & 1 & 0 & 2 \\ 2 & 0 & 0 & -5 & 1 & 2 \\ 2 & 1 & 1 & 0 & -5 & 1 \\ 2 & 2 & 1 & 0 & 2 & -5 \end{pmatrix},$$

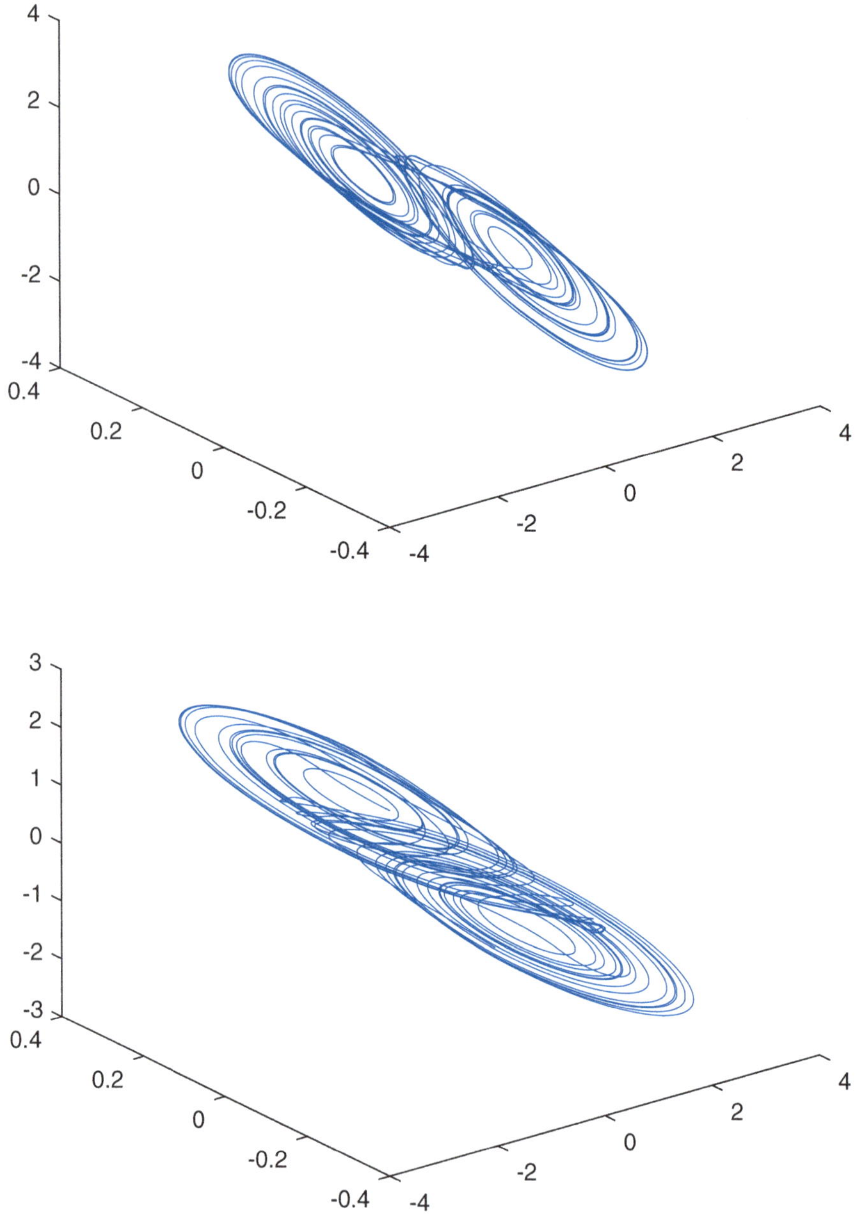

Fig. 1 The phase graph of two different Lur'e systems in (27)

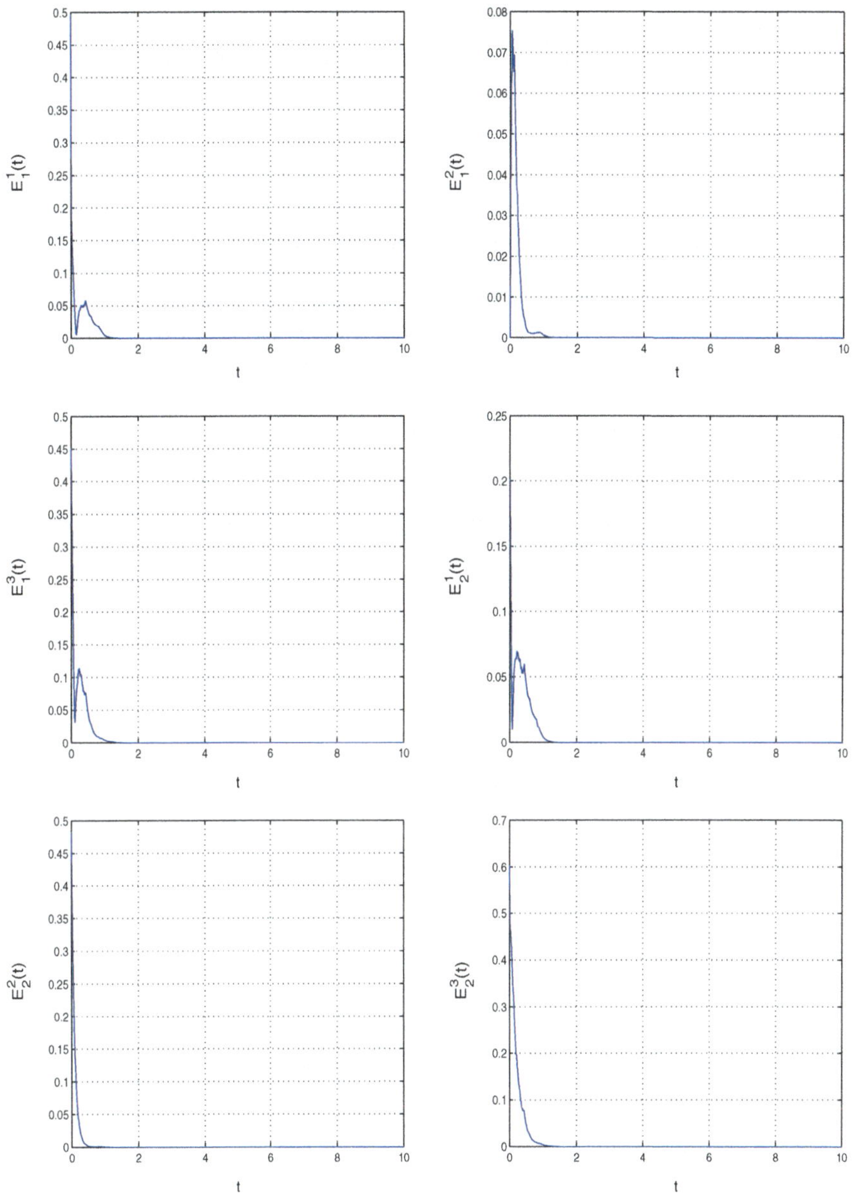

Fig. 2 The synchronization error curves of three states in two different clusters

$$\rho = \begin{pmatrix} -2 & 1 & 0 & 1 & 0 & 0 \\ 1 & -5 & 2 & 1 & 1 & 0 \\ 0 & 2 & -4 & 2 & 0 & 0 \\ 1 & 1 & 2 & -8 & 2 & 2 \\ 0 & 1 & 0 & 2 & -4 & 1 \\ 0 & 0 & 0 & 2 & 1 & -3 \end{pmatrix},$$

let coupling strengths are set with some positive constants $\varepsilon_1 = 0.5, \varepsilon_2 = 0.2, \varepsilon_3 = 0.2$, feedback control strength $d_1 = 0, d_2 = 0, d_3 = 0.8, d_4 = 0.5, d_5 = 0, d_6 = 0$, and the time-varying coupling delay as $\tau(t) = 0.2\sin(0.2t)$. In view of the construction of the Lyapunov function, it could be obtained that $\lambda_{\max}(I_6 - \varepsilon_3\rho) = 2.9792$.

In order to design the impulsive control input terms $\bar{u}_i(t)$ in the impulsive pinning feedback controller (3), take the nonuniform distribution of impulses into account [28]. Let the impulsive sequence as $\zeta = \{c, 2c, \ldots, (T_0 - 1)c, T_0T_a, T_0T_a + c, T_0T_a + 2c, \ldots, T_0T_a + (T_0 - 1)c, 2T_0T_a, \ldots\}$. It further denotes that $t_k - t_{k-1} = c$ if $\mathrm{mod}(k, T_0) \neq 0$ and $T_0(T_a - z) + c$, if $\mathrm{mod}(k, T_0) = 0$, where $0 \leq c \leq T_a$. Set the average impulsive interval is no larger than $T_a = 0.02$, and the free adjust constants $T_0 = 2$, $c = 0.01$. By the parameters, it could be verified that $\psi > \varsigma > 0$. It also could be calculated that the solution to the nonlinear system parameter equation $\lambda - \psi + \varsigma e^{\lambda\tau} = 0$ is $\lambda = 3.8912$.

To reflect the synchronization procedure of the nonidentically derivative coupled Lur'e networks (1), the synchronization error for each state in different clusters should be defined. Let $E_1^j(t) = \sqrt{\frac{1}{3}(\delta z_1^j(t)^2 + \delta z_2^j(t)^2 + \delta z_3^j(t)^2)}$, $j = 1, 2, 3$ be the synchronization error of each state in the first cluster and $E_2^j(t) = \sqrt{\frac{1}{3}(\delta z_4^j(t)^2 + \delta z_5^j(t)^2 + \delta z_6^j(t)^2)}$, $j = 1, 2, 3$.

In Fig. 2, the synchronization error curves of three states in the first cluster $E_1^j(t)$ and in the second cluster $E_2^j(t)$ are presented, where the error approach to zero within about two seconds. From the figures, it can be found that the cluster synchronization between the nonidentically derivative coupled Lur'e networks and the target isolated Lur'e system could be achieved within about two seconds.

5 Conclusion

In this chapter, the global exponential synchronization issue of a class of nonidentically derivative coupled Lur'e dynamical networks with time-varying delays and stochastic disturbance has been investigated. To be more applicable to general complex network models, general state coupling, delayed state coupling and derivative state coupling have been considered. By thinking that the cluster topological structure of the Lur'e networks, a novel impulsive pinning controller has been proposed, which has been imposed on the Lur'e systems in the clusters which have directed connections with the Lur'e systems in the other clusters. In view of the extended comparison principle of the impulsive differential equations, the concept of average impulsive interval and the mathematical induction method, sufficient criteria for the cluster synchronization of the coupled Lur'e networks have been obtained. Meanwhile, the exponential convergence velocity of the cluster synchronization has been precisely estimated. Furthermore, a numerical simulation has been given to demonstrate the availability of the theoretical analysis.

Acknowledgements This work was supported in part by the National Natural Science Foundation of China with Grant No. 61803180, in part by the Natural Science Foundation of Jiangsu Province with Grant No. BK20180599, in part by the National Key Research and Development Program of China with Grant No. 2018YFB1701903, and in part by the 111 Project with Grant No. B12018. The work of J. H. Park was supported by the National Research Foundation of Korea (NRF) grant funded by the Korea government (Ministry of Science and ICT) (No. 2019R1A5A8080290 11).

References

1. Chen, G., Wang, X., Li, X., Lu, J.: Some recent advances in complex networks synchronization. Stud. Comput. Intell. **254**, 3–16 (2009)
2. Wang, X.F., Chen, G.: Complex networks: small-world, scale-free and beyond. IEEE Circuits Syst. Mag. **3**(1), 6–20 (2003)
3. Li, C., Chen, G.: Synchronization in general complex dynamical networks with coupling delays. Phys. A: Stat. Mech. Appl. **343**(1–4), 263–278 (2004)
4. Liu, X., Chen, T.: Synchronization of complex networks via aperiodically intermittent pinning control. IEEE Trans. Autom. Control **60**(12), 3316–3321 (2015)
5. Park, J.H.: Synchronization of Genesio chaotic system via backstepping approach. Chaos, Solitons Fractals **27**(5), 1369–1375 (2006)
6. Bao, H., Park, J.H., Cao, J.: Adaptive synchronization of fractional-order memristor-based neural networks with time delay. Nonlinear Dyn. **82**(3), 1343–1354 (2015)
7. Lopes, L.M., Fernandes, S., Gacio, C.: Complete synchronization and delayed synchronization in couplings. Nonlinear Dyn. **79**(2), 1615–1624 (2014)
8. Tang, Z., Feng, J.W.: Adaptive cluster synchronization for nondelayed and delayed coupling complex networks with nonidentical nodes. Abst. Appl. Anal. **2013**, 1–13 (2013)
9. Tang, Z., Park, J.H., Lee, T.H., Feng, J.W.: Random adaptive control for cluster synchronization of complex networks with distinct communities. Int. J. Adapt. Control Signal Process. **30**(3), 534–549 (2016)
10. Liu, X.W., Chen, T.P.: Finite-time and fixed-time cluster synchronization with or without pinning control. IEEE Trans. Cybern. **48**(1), 240–252 (2018)
11. Liu, X.W., Chen, T.P.: Cluster synchronization in directed networks via intermittent pinning control. IEEE Trans. Neural Netw. **22**(7), 1009–1020 (2011)
12. Sorrentino, F., Pecora, L.M., Hagerstrom, A.M., Murphy, T.E., Roy, R.: Complete characterization of the stability of cluster synchronization in complex dynamical networks. Sci. Adv. **2**(4), e1501737:1-8 (2016)
13. Tang, Z., Park, J.H., Shen, H.: Finite-time cluster synchronization of Lur'e networks: a nonsmooth approach. IEEE Trans. Syst., Man, Cybern.: Syst. **48**(8), 1213–1224 (2018)
14. Park, J.H., Lee, T.H., Liu, Y., Chen, J.: Dynamic Systems with Time Delays: Stability and Control. Springer-Nature, Singapore (2019). https://doi.org/10.1007/978-981-13-9254-2
15. Tang, Z., Park, J.H., Feng, J.W.: Impulsive effects on quasi-synchronization of neural networks with parameter mismatches and time-varying delay. IEEE Trans. Neural Netw. Learn. Syst. **294**, 908–919 (2018)

16. He, W.L., Qian, F., Cao, J.D.: Pinning-controlled synchronization of delayed neural networks with distributed-delay coupling via impulsive control. Neural Netw. **85**, 1–9 (2017)
17. Tang, Z., Park, J.H., Zheng, W.X.: Distributed impulsive synchronization of Lur'e dynamical networks via parameter variation methods. Int. J. Robust Nonliear Control **28**(3), 1001–1015 (2018)
18. Zhang, W.B., Tang, Y., Wu, X.T., Fang, J.A.: Synchronization of nonlinear dynamical networks with heterogeneous impulses. IEEE Trans. Circuits Syst. I: Reg. Pap. **61**(4), 1220–1228 (2014)
19. Tang, Y., Gao, H.J., Zhang, W.B., Kurths, J.: Leader-following consensus of a class of stochastic delayed multi-agent systems with partial mixed impulses. Automatica **53**, 346–354 (2015)
20. Lu, J.Q., Ho, D.W.C., Cao, J.D., Kurths, J.: Exponential synchronization of linearly coupled neural networks with impulsive disturbances. IEEE Trans. Neural Netw. **22**(2), 329–336 (2011)
21. Kumar, R., Sarkar, S., Das, S., Cao, J.D.: Projective synchronization of delayed neural networks with mismatched parameters and impulsive effects. IEEE Trans. Neural Netw. Learn. Syst. To be published (2019). https://doi.org/10.1109/TNNLS.2019.2919560
22. Lu, J.Q., Wang, Z.D., Cao, J.D., Ho, D.W.C., Kurths, J.: Pinning impulsive stabilization of nonlinear dynamical networks with time-varying delay. Int. J. Bifurc. Chaos **22**(7), 1250176 (12 pages) (2012)
23. Zhang, W.B., Tang, Y., Miao, Q.Y., Fang, J.A.: Synchronization of stochastic dynamical networks under impulsive control with time delays. IEEE Trans. Neural Netw. Learn. Syst. **25**(10), 1758–1768 (2014)
24. Xu, Y.H., Zhou, W.N., Fang, J.A., Xie, C.R., Tong, D.B.: Finite-time synchronization of the complex dynamical network with non-derivative and derivative coupling. Neurocomputing **173**(3), 1356–1361 (2016)
25. Zheng, S.: Pinning and impulsive synchronization control of complex dynamical networks with non-derivative and derivative coupling. J. Frank. Inst. **354**(14), 6341–6363 (2017)
26. Deng, L.P., Wu, Z.Y., Wu, Q.C.: Pinning synchronization of complex network with non-derivative and derivative coupling. Nonlinear Dyn. **73**(1–2), 775–782 (2013)
27. Wu, W., Zhou, W.J., Chen, T.P.: Clsuter synchronization of linearly copuled complex networks under pinning control. IEEE Trans. Circuits Syst. I: Reg. Pap. **56**(4), 829–839 (2009)
28. Lu, J.Q., Ho, D.W.C., Cao, J.D.: A unified synchronization criterion for impulsive dynamical networks. Automatica **46**(7), 1215–1221 (2010)
29. Halanay, A.: Differential Equations: Stability, Oscillations, Time Lags. Academic, New York (1996)
30. Tang, Z., Park, J.H., Lee, T.H., Feng, J.W.: Mean square exponential synchronization for impulsive coupled neural networks with time-varying delays and stochastic disturbances. Complexity **21**(5), 190–202 (2016)
31. Liu, B.: Stability of solutions for stochastic impulsive systems via comparison approach. IEEE Trans. Autom. Control **53**(9), 2128–2133 (2008)

Finite-Time Synchronization Control for Markovian Jump Memristive Neural Networks with Reaction-Diffusion Terms

Xiaona Song, Jingtao Man, and Shuai Song

Abstract This chapter investigates the issue of finite-time synchronization for a class of memristive neural networks, where the reaction-diffusion items and Markovian jump parameters are both considered. Specifically, the drive and response systems proposed in this chapter follow the inconsistent Markov chains, which might be in accordance with realistic applications better. Meanwhile, a novel discontinuous controller is designed, so that the difficulty in deriving the synchronization criterion of time-space related second-order differential systems (SODSs) can be solved. In the process of dealing with the derivative of Lyapunov–Krasovskii functional (LKF), a canonical Bessel–Legendre inequality (BLI), which converts the limited interval required in traditional BLI to a general one, is employed. The main work is reflected as follows: first, the original SODSs is converted into a first-order differential system by employing a suitable variable substitution and then the error system in a compact first-order differential form is established; second, a less conservative criterion of finite-time synchronization is obtained by constructing a new LKF and utilizing the canonical BLI and free-weighting matrix methods. Ultimately, a numerical example and an application study are exploited to illustrate the feasibility and practicability of this chapter, thus the acquired theoretical results can be well supported.

Keywords Finite-time synchronization · Reaction-diffusion memristive neural networks · Inconsistent Markov chains · Discontinuous controller · Canonical Bessel–Legendre inequality

---

X. Song (✉) · J. Man
School of Information Engineering, Henan University of Science and Technology, Luoyang 471023, People's Republic of China
e-mail: xiaona_97@163.com

J. Man
e-mail: mjt546@163.com

S. Song
School of Automation, Nanjing University of Science and Technology, Nanjing 210094, People's Republic of China
e-mail: songshuai_1010@163.com

J. H. Park (ed.), *Recent Advances in Control Problems of Dynamical Systems and Networks*, Studies in Systems, Decision and Control 301,
https://doi.org/10.1007/978-3-030-49123-9_21

1 Introduction

The past few decades have witnessed the rapid development of neural networks (NNs), which have been applied to various areas such as image processing, pattern recognition, combinatorial optimization, associate memory design, etc [1–3]. With growing interest, an extraordinary volume of results on NNs have emerged (see e.g. [4–6] and references therein). However, in the hardware implementation of NNs, there is a common challenge that if we use traditional silicon-based semiconductor devices to simulate the synaptic electrical characteristics in biological neurons, only one synapse in a neuron will need dozens of transistors, resistors, capacitors, etc., which means that the implementation of complex NNs is impractical [7]. Fortunately, the invention of memristor solves this problem well. Because memristors are variable resistors and exhibit memory characteristics and their information storage and processing are very similar to human brain synapses. These characteristics show that memristive devices have many application prospects, such as device modeling and signal processing. Among them, the function of imitating synaptic behavior is not available in three types of basic circuit elements: resistance, capacitance and inductance [8].

Based on what has been discussed above, an increasing number of scholars have been focusing on memristive NNs (MNNs) and many relative results have been published. Based on the excitatory and inhibitory of memristive synaptic weights and by employing a new convex combination technique, [9] obtained a stabilization criterion for a class of T–S fuzzy MNNs via the switched fuzzy sampled-data control scheme. To reduce the complexity of available methods in dealing with the state-dependent memristive connection weights and conservatism of existing event-trigger mechanisms, [10] designed an exponential-attenuation-based switching event-trigger scheme to realize the global stabilization of delayed MNNs. Moreover, a class of MNNs with discontinuous activation functions and mixed time-varying delays were proposed in [11], where an adaptive state-feedback controller and a state-feedback controller were proposed, respectively, to synchronize the drive and response systems within a finite time.

As we know, reaction-diffusion NNs have become a research hotspot. The one reason is that diffusion phenomena will be obvious when electrons of NNs move in an electromagnetic field that is nonuniform. Another reason is they have been widely applied to biology, neuroscience, chemistry and some other fields. Therefore, considering reaction-diffusion terms in the model of MNNs is reasonable [12–14]. On the other hand, Markovian jumping phenomena often occur in real systems due to some factors such as sudden environmental changes and component failures [15, 16]. As a result, it is necessary to set the parameters as Markovian jumping ones during modeling and some results on Markovian reaction-diffusion MNNs have been appeared [17–19].

Nevertheless, all of the aforementioned papers about the issue of synchronization for Markovian systems suppose that the drive and response systems share the same Markov chain. While strictly speaking, in real systems, the drive and response systems are more likely to follow different Markovian jumping laws. Thus, in this chapter, we investigate the Markovian systems subject to inconsistent Markov chains. Then, the challenge of dealing with the inconsistent parameters in the error system must be faced. To solve this problem, a novel discontinuous controller is designed in this chapter, so that the influence of the inconsistent parameters can be avoided.

This chapter's main target is to propose an effective finite-time synchronization criterion with the designed control scheme to guarantee the response system can be synchronized with the established drive system.

To sum up, this chapter mainly contributes to the following three aspects:

(1) This chapter proposes a new class of drive and response MNNs, where the Markovian jumping parameters and reaction-diffusion terms are both contained. Note that the biggest innovation of this system model is that the drive and response systems subject to two different Markov chains, which is more practical than existing results on the synchronization of Markovian systems that suppose the drive and response systems share the same Markovian jumping law such as [20–22]. Moreover, to cope with the challenge caused by inconsistent parameters in the error system, inspired by [23], a novel discontinuous controller is proposed, such that the aforementioned problem can be solved.

(2) A novel LKF is constructed, where the triple and quadruple integral items are involved to capture more information of the time-varying delay, so that the conservatism of obtained results can be reduced. Additionally, by combining canonical BLI and free weighted matrix (FWM) methods, the conservatism of synchronization criterion proposed in this chapter can be further reduced.

(3) A numerical example and an application example are provided to illustrate the effectiveness and practicability of the main results. What should be emphasized is that the application is realized with a cryptosystem established by the Markovian reaction-diffusion MNNs, which is the first attempt in secret communication.

Notations: Throughout this chapter, $\mathbb{E}$ denotes the mathematical expectation operator; $\mathbb{R}^n$ represents the n-dimensional Euclidean space; $\mathtt{col}\{\cdots\}$ means expressing the elements in $\{\cdots\}$ as column vectors; $\dagger_x = [0_{n,(x-1)n}, I_n, 0_{n,(13-x)n}]$, $(x = 1, 2, \ldots, 13)$ is a block entry matrix; For an arbitrary matrix A, $Sym\{A\} = A + A^T$; I represents the identity matrix with appropriate dimensions; Additionally, for convenience, this chapter adopts some abbreviations as follows:

$$\S_\delta = \S_\delta(\upsilon, t)\ (\S = m, s, p, u),$$
$$\aleph_{lh} = \aleph_l(\upsilon, t - h(t))\ (\aleph = m, s, p),$$

where $\upsilon = (\upsilon_1, \upsilon_2, \ldots, \upsilon_q)^T \in \Omega \subset \mathbb{R}^q$ with $\Omega = \{\upsilon\, ||\upsilon_k| \leq \ell_k\}$ and $\ell_k > 0$, $(k = 1, 2, \ldots, q)$ are constants, and the meaning of m, s, p and u will be explained hereinafter.

2 Preliminaries

Suppose $\{r_t\}$ is a right-continuous Markov chain on the probability space that takes values in a finite-state space $S = \{1, 2, \ldots, \mathcal{N}\}$, and

$$\Pr\{r_{t+\Delta t} = \mu \,|\, r_t = \sigma\} = \begin{cases} \pi_{\sigma\mu}\Delta t + o(\Delta t), & \sigma \neq \mu \\ 1 + \pi_{\sigma\sigma}\Delta t + o(\Delta t), & \sigma = \mu, \end{cases}$$

where Pr is the transition probability from mode σ at time t to mode μ at time $t + \Delta t$; $\Delta t > 0$, $\lim_{\Delta t \to 0} [o(\Delta t)/\Delta t] = 0$, and $\pi_{\sigma\mu} > 0$ for $\sigma \neq \mu$, and $\pi_{\sigma\sigma} = -\sum_{\mu \in S, \sigma \neq \mu} \pi_{\sigma\mu}$.

Consider the following Markovian reaction-diffusion MNNs as the drive system

$$\begin{aligned} \frac{\partial m_\delta}{\partial t} =& \sum_{k=1}^{q} \frac{\partial}{\partial \upsilon_k}\left(\mathfrak{r}_{\delta k}\frac{\partial m_\delta}{\partial \upsilon_k}\right) - a_\delta m_\delta + \sum_{l=1}^{n} c_{\delta l\sigma}(m_\delta) g_l(m_l) \\ &+ \sum_{l=1}^{n} d_{\delta l\sigma}(m_\delta) g_l(m_{lh}), \end{aligned} \tag{1}$$

where m_δ is the δth neuron's state, $\delta = 1, 2, \ldots, n$; $\mathfrak{r}_{\delta k} > 0$ represents the transmission diffusion operator; $a_\delta > 0$ is a given constant; and $c_{\delta l\sigma}(m_\delta)$ and $d_{\delta l\sigma}(m_\delta)$ correspond to the feedback connection strength and delayed feedback connection strength, respectively, and

$$c_{\delta l\sigma}(m_\delta) = \begin{cases} c^{\flat}_{\delta l\sigma}, & |m_\delta| \leq \mathcal{Y}_\delta, \\ c^{\sharp}_{\delta l\sigma}, & |m_\delta| > \mathcal{Y}_\delta, \end{cases} \quad d_{\delta l\sigma}(m_\delta) = \begin{cases} d^{\flat}_{\delta l\sigma}, & |m_\delta| \leq \mathcal{Y}_\delta, \\ d^{\sharp}_{\delta l\sigma}, & |m_\delta| > \mathcal{Y}_\delta. \end{cases}$$

Additionally, $g_l(\cdot)$ is the activation function and $h(t)$ is time-varying delay, which satisfy the following assumptions:

$$\begin{aligned} &|g_l(\tau)| \leq \bar{g}_l, \;\; \varepsilon_l^- \leq \frac{g_l(\tau_1) - g_l(\tau_2)}{\tau_1 - \tau_2} \leq \varepsilon_l^+, \\ &0 < h_1 \leq h(t) \leq h_2, \;\; \frac{dh(t)}{dt} \leq h < \infty, \end{aligned}$$

where τ, τ_1, τ_2, ε_l^-, ε_l^+, h_1, h_2 and h are constants, $\tau_1 \neq \tau_2$, and $g_l(m_{lh}) = g_l(m_l(\upsilon, t - h(t)))$.

For a bounded and continuous function $\phi_\delta(\upsilon, o)$, the boundary and initial conditions of (1) are defined as

$$\begin{aligned} &m_\delta(\upsilon, t) = 0, (\upsilon, t) \in \partial\Omega \times [-h_2, +\infty), \\ &m_\delta(\upsilon, o) = \phi_\delta(\upsilon, o), (\upsilon, o) \in \Omega \times [-h_2, 0]. \end{aligned}$$

Now, we introduce another Markov chain $\{\kappa_t\}$ that takes values in $T = \{1, 2, \ldots, \mathcal{M}\}$ and depends on r_t with the following conditional probability:

$$Pr\{\kappa_t = \varpi | r_t = \sigma\} = \varrho_{\sigma\varpi},$$

where $\varrho_{\sigma\varpi} \geq 0$ for all $\sigma \in S$ and $\varpi \in T$, and $\sum_{\varpi \in T} \varrho_{\sigma\varpi} = 1$.

With this Markovian jumping law, the response system can be modeled as:

$$\begin{aligned}\frac{\partial s_\delta}{\partial t} =& \sum_{k=1}^{q} \frac{\partial}{\partial \upsilon_k}\left(\mathfrak{r}_{\delta k}\frac{\partial s_\delta}{\partial \upsilon_k}\right) - a_\delta s_\delta + \sum_{l=1}^{n} c_{\delta l \varpi}(s_\delta) g_l(s_l) \\ &+ \sum_{l=1}^{n} d_{\delta l \varpi}(s_\delta) g_l(s_{lh}) + u_\delta, \end{aligned} \quad (2)$$

where u_δ is the control input, and

$$c_{\delta l \varpi}(s_\delta) = \begin{cases} c^{\flat}_{\delta l \varpi}, & |s_\delta| \leq \mathcal{Y}_\delta, \\ c^{\sharp}_{\delta l \varpi}, & |s_\delta| > \mathcal{Y}_\delta, \end{cases} \quad d_{\delta l \varpi}(s_\delta) = \begin{cases} d^{\flat}_{\delta l \varpi}, & |s_\delta| \leq \mathcal{Y}_\delta, \\ d^{\sharp}_{\delta l \varpi}, & |s_\delta| > \mathcal{Y}_\delta. \end{cases}$$

Moreover,

$$\begin{aligned} &s_\delta(\upsilon, t) = 0, (\upsilon, t) \in \partial\Omega \times [-h_2, +\infty), \\ &s_\delta(\upsilon, o) = \psi_\delta(\upsilon, o), (\upsilon, o) \in \Omega \times [-h_2, 0], \end{aligned}$$

where $\psi_\delta(\upsilon, o)$ has the same characteristic as $\phi_\delta(\upsilon, o)$.

Define $p_\delta = s_\delta - m_\delta$, then, the error system can be described as

$$\begin{aligned}\frac{\partial p_\delta}{\partial t} =& \sum_{k=1}^{q} \frac{\partial}{\partial \upsilon_k}\left(\mathfrak{r}_{\delta k}\frac{\partial p_\delta}{\partial \upsilon_k}\right) - a_\delta p_\delta + \sum_{l=1}^{n} c_{\delta l \varpi}(s_\delta) f_l(p_l) \\ &+ \sum_{l=1}^{n} d_{\delta l \varpi}(s_\delta) f_l(p_{lh}) + \sum_{l=1}^{n} \left(c_{\delta l \varpi}(s_\delta) - c_{\delta l \sigma}(m_\delta)\right) g_l(m_l) \\ &+ \sum_{l=1}^{n} \left(d_{\delta l \varpi}(s_\delta) - d_{\delta l \sigma}(m_\delta)\right) g_l(m_{lh}) + u_\delta, \end{aligned} \quad (3)$$

where $f_l(p_l) = g_l(s_l) - g_l(m_l)$ and $f_l(p_{lh}) = g_l(s_{lh}) - g_l(m_{lh})$.

Obviously, $\varepsilon_l^- \leq \frac{f_l(\tau')}{\tau'} \leq \varepsilon_l^+$, $\tau' \in \mathbb{R}$, and

$$\begin{aligned} &p_\delta(\upsilon, t) = 0, (\upsilon, t) \in \partial\Omega \times [-h_2, +\infty), \\ &p_\delta(\upsilon, o) = \psi_\delta(\upsilon, o) - \phi_\delta(\upsilon, o), (\upsilon, o) \in \Omega \times [-h_2, 0]. \end{aligned}$$

For simplicity, we give the following compact form of error system (3):

$$\frac{\partial p}{\partial t} = \sum_{k=1}^{q} \frac{\partial}{\partial \upsilon_k}\left(\Lambda_k \frac{\partial p}{\partial \upsilon_k}\right) - \mathcal{A}p + \mathcal{C}_{\varpi}(s) f(p) + \mathcal{D}_{\varpi}(s) f(p_h) + \Im + u, \tag{4}$$

where $p = (p_1, p_2, \ldots, p_n)^T, u = (u_1, u_2, \ldots, u_n)^T, \Lambda_k = diag\{\mathfrak{r}_{1k}, \mathfrak{r}_{2k}, \ldots, \mathfrak{r}_{nk}\}$, $\mathcal{A} = diag\{a_1, a_2, \ldots, a_n\}, \mathcal{C}_{\varpi}(s) = (c_{\delta l \varpi}(s))_{n\times n}, \mathcal{D}_{\varpi}(s) = (d_{\delta l \varpi}(s))_{n\times n}, p_h = p(\upsilon, t - h(t)), f(\cdot) = (f_1(\cdot), f_2(\cdot), \ldots, f_n(\cdot))^T$; $\Im$ is short for $\Im(\upsilon, t)$ and $\Im = (\mathcal{C}_{\varpi}(s) - \mathcal{C}_{\sigma}(m))$ $g(m) + (\mathcal{D}_{\varpi}(s) - \mathcal{D}_{\sigma}(m))\, g(m_h)$, $\mathcal{C}_{\sigma}(m) = (c_{\delta l \sigma}(m))_{n\times n}$, $\mathcal{D}_{\sigma}(m) = (d_{\delta l \sigma}(m))_{n\times n}$, and $g(\cdot) = (g_1(\cdot), g_2(\cdot), \ldots, g_n(\cdot))^T$.

From aforementioned switching rules, the number of combination of $\mathcal{C}_{\varpi}(s)$ and $\mathcal{D}_{\varpi}(s)$ is 2^n at mode ϖ. Then, the 2^n cases at mode ϖ can be ordered as $(\mathcal{C}_{\varpi 1}, \mathcal{D}_{\varpi 1}), (\mathcal{C}_{\varpi 2}, \mathcal{D}_{\varpi 2}), \ldots, (\mathcal{C}_{\varpi 2^n}, \mathcal{D}_{\varpi 2^n})$. At any υ and t, the combination of $\mathcal{C}_{\varpi}(s)$ and $\mathcal{D}_{\varpi}(s)$ must be one of the above 2^n cases. Namely, there exist some φ such that $\mathcal{C}_{\varpi}(s) = \mathcal{C}_{\varpi\varphi}$, and $\mathcal{D}_{\varpi}(s) = \mathcal{D}_{\varpi\varphi}$.

We define

$$\xi_{\varpi\varphi}(\upsilon, t) = \begin{cases} 1, & \mathcal{C}_{\varpi}(s) = \mathcal{C}_{\varpi\varphi}, \mathcal{D}_{\varpi}(s) = \mathcal{D}_{\varpi\varphi}, \\ 0, & otherwise. \end{cases}$$

Obviously, $\sum_{\varphi=1}^{2^n} \xi_{\varpi\varphi}(\upsilon, t) \equiv 1$.

As a result, the error system (4) can be rewritten as:

$$\frac{\partial p}{\partial t} = \sum_{k=1}^{q} \frac{\partial}{\partial \upsilon_k}\left(\Lambda_k \frac{\partial p}{\partial \upsilon_k}\right) - \mathcal{A}p + \tilde{\mathcal{C}}_{\varpi} f(p) + \tilde{\mathcal{D}}_{\varpi} f(p_h) + \Im + u, \tag{5}$$

where $\tilde{\mathcal{C}}_{\varpi} = \tilde{\mathcal{C}}_{\varpi}(\upsilon, t) = \sum_{\varphi=1}^{2^n} \xi_{\varpi\varphi}(\upsilon, t)\mathcal{C}_{\varpi\varphi}$ and $\tilde{\mathcal{D}}_{\varpi} = \tilde{\mathcal{D}}_{\varpi}(\upsilon, t) = \sum_{\varphi=1}^{2^n} \xi_{\varpi\varphi}(\upsilon, t)\mathcal{D}_{\varpi\varphi}$.

To achieve the target of synchronization, we propose a discontinuous controller u as:

$$u = -K_{\varpi} p - H(\upsilon, t) sign(\tilde{p}) \tag{6}$$

where K_{ϖ} is the controller gains to be designed. $sign(\cdot)$ is a standard symbolic function; $sign(\tilde{p}) = \left[sign(\tilde{p}_1), sign(\tilde{p}_2), \ldots, sign(\tilde{p}_n)\right]^T$, $\tilde{p}_\delta = p_\delta + \dot{p}_\delta$; $\dot{p}_\delta = \partial p_\delta / \partial t$, $H(\upsilon, t) = diag\{H_1(\upsilon, t), H_2(\upsilon, t), \ldots, H_n(\upsilon, t)\}$, and

$$H_\delta(\upsilon, t) = \begin{cases} \mathfrak{h}_{1\delta}, & |m_\delta| \le \mathcal{Y}_\delta \ and \ |s_\delta| \le \mathcal{Y}_\delta, \\ \mathfrak{h}_{2\delta}, & |m_\delta| \le \mathcal{Y}_\delta \ and \ |s_\delta| > \mathcal{Y}_\delta, \\ \mathfrak{h}_{3\delta}, & |m_\delta| > \mathcal{Y}_\delta \ and \ |s_\delta| \le \mathcal{Y}_\delta, \\ \mathfrak{h}_{4\delta}, & |m_\delta| > \mathcal{Y}_\delta \ and \ |s_\delta| > \mathcal{Y}_\delta, \end{cases}$$

where $\mathfrak{h}_{i\delta} = \sum_{l=1}^{n} \left(\tilde{c}_{i\delta l} + \tilde{d}_{i\delta l}\right)\bar{g}_l$, $(i = 1, 2, 3, 4)$, $\tilde{c}_{1\delta l} = \left|c^{\flat}_{\delta l\varpi} - c^{\flat}_{\delta l\sigma}\right|$, $\tilde{c}_{2\delta l} = \left|c^{\flat}_{\delta l\sigma} - c^{\sharp}_{\delta l\varpi}\right|$, $\tilde{c}_{3\delta l} = \left|c^{\sharp}_{\delta l\sigma} - c^{\flat}_{\delta l\varpi}\right|$, $\tilde{c}_{4\delta l} = \left|c^{\sharp}_{\delta l\sigma} - c^{\sharp}_{\delta l\varpi}\right|$, $\tilde{d}_{1\delta l} = \left|d^{\flat}_{\delta l\varpi} - d^{\flat}_{\delta l\sigma}\right|$, $\tilde{d}_{2\delta l} = \left|d^{\flat}_{\delta l\sigma} - d^{\sharp}_{\delta l\varpi}\right|$, $\tilde{d}_{3\delta l} = \left|d^{\sharp}_{\delta l\sigma} - d^{\flat}_{\delta l\varpi}\right|$, and $\tilde{d}_{4\delta l} = \left|d^{\sharp}_{\delta l\sigma} - d^{\sharp}_{\delta l\varpi}\right|$.

Remark 1 As we all know, the invention of memristor greatly reduced the cost in the implementation of NNs, making the application of complex NNs possible. Therefore, MNNs are a class of system that is closely integrated with reality and it is necessary to consider various possible problems in the actual systems during the modeling of MNNs. For one thing, when studying the multilayer cellular NNs or electrons shift in a non-uniform electromagnetic field, the diffusion phenomena will be unavoidable. For another, due to the random failures, repairs of the network components, and/or sudden environment switching, the random mode switching would occur in NNs.

Above discussions imply that it is significant to consider reaction-diffusion terms and Markovian jump parameters on the basis of the existing MNNs models. What should be emphasized is that in existing results on the synchronization of Markovian jump systems, drive and response systems' jumping rules are usually set to be the same such as [24, 25]. However, it seems more reasonable for drive and response systems to jump by following different Markov chains.

From where we stand, the reason that existing papers don't consider the situation of inconsistent Markovian jumping laws is that it is a big challenge to deal with the inconsistent parameters in error systems. To solve this problem, an innovative part $-H(\upsilon, t)sign(\tilde{p})$ is tactfully employed in the controller designed in this chapter. Then, the aforementioned challenge is overcome. More details will be shown in subsequent sections.

Definition 1 ([26]) If the Markovian reaction-diffusion MNNs (5) satisfies

$$\mathbb{E}(\|p(\upsilon, t_1)\|^2) \leq \hbar_1 \Longrightarrow \mathbb{E}(\|p(\upsilon, t_2)\|^2) \leq \hbar_2, \quad t_1 \in [-h_2, 0], \quad t_2 \in (0, T^*],$$

where $0 \leq \hbar_1 < \hbar_2$, then, the system (5) is bounded within an FTI with respect to $(\hbar_1, \hbar_2, T^*)$.

Lemma 1 ([27]) *Let $\omega(\upsilon)$ be a real-valued function belonging to $C^1(\Omega)$ and satisfies $\omega(\upsilon)\mid_{\partial\Omega} = 0$. Then*

$$\int_{\Omega} \omega^2(\upsilon)d\upsilon \leq \ell_k^2 \int_{\Omega} \left|\frac{\partial\omega(\upsilon)}{\partial\upsilon_k}\right|^2 d\upsilon.$$

where Ω is a cube $|\upsilon_k| < \ell_k (k = 1, 2, \ldots, q)$.

Lemma 2 ([28]) *For given scalars c and d, $d > c$, if there exist an integer $N \geq 0$, an $n \times n$ positive constant matrix P, a matrix B with the appropriate dimensions, and a differentiable function $\omega : [c, d] \to \mathbb{R}^n$ that is vector-valued, the following inequality holds:*

$$-\int_c^d \dot{\omega}^T(x)P\dot{\omega}(x)dx \leq \varsigma_N^T\left[\tilde{X}_N^T\tilde{Y}_N^T B + B^T\tilde{Y}_N\tilde{X}_N + (d-c)B^T\mathcal{P}_N^{-1}B\right]\varsigma_N,$$

where

$$\varsigma_N = \texttt{col}\{\omega(d), \omega(c), \gamma_1, \ldots, \gamma_N\}, \ \mathcal{P} = diag\{P, 3P, \ldots, (2N+1)P\},$$

$$\tilde{X}_N = \begin{bmatrix} I & -I & 0 & 0 & \cdots & 0 \\ 0 & -I & I & 0 & \cdots & 0 \\ 0 & -I & 0 & 2I & \cdots & 0 \\ \vdots & \vdots & \vdots & \vdots & \ddots & \vdots \\ 0 & -I & 0 & 0 & 0 & NI \end{bmatrix}, \quad \tilde{Y}_N = \begin{bmatrix} I & 0 & \cdots & 0 \\ I & Y_{N1} & \cdots & 0 \\ \vdots & \vdots & \ddots & \vdots \\ I & Y_{N2} & \cdots & Y_{N3} \end{bmatrix},$$

with

$$\gamma_i = \int_c^d \frac{(d-x)^{i-1}}{(d-c)^i}\omega(x)dx, \quad (i = 1, 2, \ldots, N),$$

$$Y_{N1} = (-1)^1\binom{1}{1}\binom{1+1}{1}I,$$

$$Y_{N2} = (-1)^1\binom{N}{1}\binom{N+1}{1}I,$$

$$Y_{N3} = (-1)^N\binom{N}{N}\binom{N+N}{N}I.$$

Lemma 3 ([29]) *For any constant matrix $\mathcal{B} > 0$ and scalars $h_2 > h_1 > 0$, the following estimation holds:*

$$-\int_{-h_2}^{-h_1}\int_{t+\alpha}^{t}\omega^T(s)\mathcal{B}\omega(s)dsd\alpha$$
$$\leq -\frac{2}{h_2^2 - h_1^2}\int_{-h_2}^{-h_1}\int_{t+\alpha}^{t}\omega^T(s)dsd\alpha\mathcal{B}\int_{-h_2}^{-h_1}\int_{t+\alpha}^{t}\omega(s)dsd\alpha.$$

3 Main Results

This section proposes a finite-time synchronization criterion of the drive and response Markovian reaction-diffusion MNNs.

Before deriving our main results, we give some necessary donations below:

$$\chi^T = \left[p^T, f^T(p)\right], \ \chi_h^T = \left[p_h^T, f^T(p_h)\right], \ \chi_{h_1}^T = \left[p_{h_1}^T, f^T(p_{h_1})\right],$$
$$\chi_{h_2}^T = \left[p_{h_2}^T, f^T(p_{h_2})\right], \ h_{21} = h_2 - h_1, \ p_{h_1} = p(\upsilon, t - h_1),$$
$$p_{h_2} = p(\upsilon, t - h_2), \ \tilde{\Lambda} = diag\left\{\sum_{k=1}^{q} \frac{\mathfrak{r}_{1k}}{\ell_k^2}, \sum_{k=1}^{q} \frac{\mathfrak{r}_{2k}}{\ell_k^2}, \ldots, \sum_{k=1}^{q} \frac{\mathfrak{r}_{nk}}{\ell_k^2}\right\},$$
$$\vartheta_i = \int_{t-h_1}^{t} \frac{(t-\alpha)^{i-1}}{h_1^i} p(\upsilon, \alpha) d\alpha,$$
$$\tilde{\vartheta}_i = \int_{t-h_2}^{t-h_1} \frac{(t-h_1-\alpha)^{i-1}}{h_{21}^i} p(\upsilon, \alpha) d\alpha, (i = 1, 2, \ldots, N),$$
$$\wp = \mathtt{col}\{p, \epsilon_1, \epsilon_2, \ldots, \epsilon_N\}, \ \epsilon_i = \mathtt{col}\{\vartheta_i, \tilde{\vartheta}_i\},$$
$$\nabla = col\Big\{p, \ \dot{p}, \ p_{h_1}, \ p_h, \ p_{h_2}, \ f(p), \ f(p_{h_1}), \ f(p_h), \ f(p_{h_2}),$$
$$\int_{t-h_1}^{t} p(\upsilon, \alpha) d\alpha, \ \int_{t-h_2}^{t-h_1} p(\upsilon, \alpha) d\alpha, \ \int_{t-h_2}^{t-h_1} \dot{p}(\upsilon, \alpha) d\alpha, \ \epsilon_1, \ \epsilon_2, \ldots, \epsilon_N\Big\}.$$

Theorem 1 *For given scalars $\hbar_1 > 0$, $\iota > 0$, and $T^* > 0$, and an integer $N \geq 0$, the synchronization error system (5) is bounded over an FTI if there exist a positive scalar $\hbar_2$; symmetric matrices $P_\sigma^N = diag\{P_{1\sigma}^N, P_{2\sigma}^N, \ldots, P_{(N+1)\sigma}^N\} > 0$, $\Psi > 0$, $T' = \begin{bmatrix} T'_{11} & T'_{12} \\ * & T'_{22} \end{bmatrix} > 0$, $T'' = \begin{bmatrix} T''_{11} & T''_{12} \\ * & T''_{22} \end{bmatrix} > 0$, $T''' = \begin{bmatrix} T'''_{11} & T'''_{12} \\ * & T'''_{22} \end{bmatrix} > 0$, $Q > 0$, $R > 0$, $F_1 > 0$, $F_2 > 0$; arbitrary matrices Υ_j $(j = 1, 2, 3, 4, 5)$, M, and $\tilde{M}$; and diagonal matrices $\Theta_m (m = 1, 2, 3)$, such that the following LMIs hold:*

$$\begin{bmatrix} \Sigma & h_1 \lambda_1^T M & h_{21} \lambda_2^T \tilde{M} \\ * & -\mathcal{Q}_N & 0 \\ * & * & -\mathcal{R}_N \end{bmatrix} < 0, \tag{7}$$

$$\hbar_1 \varsigma e^{\iota T^*} < \min_{\sigma \in S} \left(\lambda_{\min}\left(P_{1\sigma}^N\right)\right) \hbar_2, \tag{8}$$

where

$$\Sigma = \Sigma_1 + \Sigma_2 + \Sigma_3 + \Sigma_4 + \Sigma_5, \ \lambda_1 = \mathtt{col}\{\dagger_1, \dagger_3, \bar{E}_1 \dagger_{13}, \ldots, \bar{E}_1 \dagger_{N+12}\},$$
$$\lambda_2 = \mathtt{col}\{\dagger_3, \dagger_5, \bar{E}_2 \dagger_{13}, \ldots, \bar{E}_2 \dagger_{N+12}\}, \ \mathcal{Q}_N = diag\{Q, 3Q, \ldots, (2N+1)Q\},$$
$$\mathcal{R}_N = diag\{R, 3R, \ldots, (2N+1)R\},$$
$$\varsigma = \lambda_{\max}\left((2N+1)I\right) \max_{\sigma \in S}\left(\lambda_{\max}\left(P_\sigma^N\right)\right) + \lambda_{max}\left(\sum_{k=1}^{q} \Lambda_k \Psi\right) + \frac{h_1^3}{2} \lambda_{\max}(Q)$$
$$+ h_2 \lambda_{max}\left(I + (E^+)^T E^+\right) \lambda_{\max}\left(T'\right) + \frac{h_{21}^3}{2} \lambda_{\max}(R) + \frac{h_1^3}{6} \lambda_{\max}(F_1)$$

$$+ h_1\lambda_{max}\left(I + (E^+)^T E^+\right)\lambda_{\max}\left(T''\right) + \frac{h_{21}^3}{6}\lambda_{\max}\left(F_2\right)$$
$$+ h_{21}\lambda_{max}\left(I + (E^+)^T E^+\right)\lambda_{\max}\left(T'''\right),$$

with

$$\begin{aligned}
\Sigma_1 =& Sym\{\mathfrak{X}^T P_\sigma^N \mathfrak{Y}\} + \mathfrak{X}^T \sum_{\mu\in S}\pi_{\sigma\mu}P_\mu^N \mathfrak{X} - 2\dagger_2^T\Psi\dagger_2 \\
& - \dagger_1^T\left(2\tilde{\Lambda}\Psi + 2\Psi\mathcal{A} + 2\Psi K_\varpi\right)\dagger_1 + Sym\{\dagger_1^T\Psi\tilde{\mathcal{C}}_\varpi\dagger_6\} \\
& - Sym\{\dagger_1^T(\Psi + \mathcal{A}^T\Psi^T + K_\varpi^T\Psi^T)\dagger_2\} + Sym\{\dagger_1^T\Psi\tilde{\mathcal{D}}_\varpi\dagger_8\} \\
& + Sym\{\dagger_2^T\Psi\tilde{\mathcal{C}}_\varpi\dagger_6\} + Sym\{\dagger_2^T\Psi\tilde{\mathcal{D}}_\varpi\dagger_8\}, \\
\Sigma_2 =& \dagger_1^T(T'_{11} + T''_{11})\dagger_1 + Sym\{\dagger_1^T(T'_{12} + T''_{12})\dagger_6\} \\
& - (1-h)\left[\dagger_4{}^T T'_{11}\dagger_4 + \dagger_8^T T'_{22}\dagger_8 + Sym\{\dagger_4{}^T T'_{12}\dagger_8\}\right] \\
& + \dagger_6^T(T'_{22} + T''_{22})\dagger_6 - \dagger_5^T T'''_{11}\dagger_5 - \dagger_3^T(T''_{11} - T'''_{11})\dagger_3 \\
& - Sym\{\dagger_3^T(T''_{12} - T'''_{12})\dagger_7\} - \dagger_7^T(T''_{22} - T'''_{22})\dagger_7 \\
& - \dagger_5^T T'''_{12}\dagger_9 - \dagger_9^T T'''_{22}\dagger_9, \\
\Sigma_3 =& h_1\lambda_1^T\left[\tilde{X}_N^T\tilde{Y}_N^T M + M^T\tilde{Y}_N\tilde{X}_N\right]\lambda_1 + \dagger_2^T(h_1^2 Q + h_{21}^2 R)\dagger_2 \\
& + h_{21}\lambda_2^T\left[\tilde{X}_N^T\tilde{Y}_N^T\tilde{M} + \tilde{M}^T\tilde{Y}_N\tilde{X}_N\right]\lambda_2, \\
\Sigma_4 =& \dagger_2^T\left(\frac{h_1^2}{2}F_1 + \frac{h_2^2 - h_1^2}{2}F_2\right)\dagger_2 - 2\dagger_1^T F_1\dagger_1 - \frac{2}{h_1^2}\dagger_{10}^T F_1\dagger_{10} \\
& + \frac{2}{h_1}Sym\{\dagger_1^T F_1\dagger_{10}\} - \frac{2h_{21}^2}{h_2^2 - h_1^2}\dagger_1^T F_2\dagger_1 \\
& + \frac{2h_{21}}{h_2^2 - h_1^2}\dagger_1^T F_2\dagger_{11} - \frac{2}{h_2^2 - h_1^2}\dagger_{11}^T F_2\dagger_{11}, \\
\Sigma_5 =& Sym\{\left(\dagger_6^T - E^-\dagger_1^T\right)\Theta_1\left(E^+\dagger_1 - \dagger_6\right)\} \\
& + Sym\{\left(\dagger_8^T - E^-\dagger_4^T\right)\Theta_2\left(E^+\dagger_4 - \dagger_8\right)\} \\
& + Sym\{\left[\dagger_6^T - \dagger_8^T - E^-\left(\dagger_1^T - \dagger_4^T\right)\right]\Theta_3\left[E^+(\dagger_1 - \dagger_4) - \dagger_6 + \dagger_8\right]\} \\
& + Sym\{\left(\dagger_1^T\Upsilon_1 + \dagger_2^T\Upsilon_2 + \dagger_6^T\Upsilon_3 + \dagger_8^T\Upsilon_4 + \dagger_{12}^T\Upsilon_5\right)(\dagger_3 - \dagger_5 - \dagger_{12})\},
\end{aligned}$$

and $\mathfrak{X} = \mathtt{col}\{\dagger_1, \dagger_{13}, \ldots, \dagger_{N+12}\}$, $\mathfrak{Y} = \mathtt{col}\{\dagger_2, \dagger_N\}$, $\dagger_N = \mathtt{col}\{\ddagger_1, \ddagger_2, \ldots, \ddagger_N\}$, $\ddagger_i = \mathtt{col}\{-\frac{1}{h_1}\dagger_3, -\frac{1}{h_{21}}\dagger_5\}$, $\bar{E}_1 = [I, 0]$, $\bar{E}_2 = [0, I]$.

Proof Consider the following LKF candidate for error system (5):

$$V(p,t) = \sum_{i=1}^{4} V_i(p,t), \tag{9}$$

where

$$
\begin{aligned}
V_1(p,t) &= \int_{\Omega}\left\{\wp^T P_{\sigma}^N \wp + \sum_{k=1}^{q}\left(\frac{\partial p}{\partial \upsilon_k}\right)^T \Lambda_k \Psi \left(\frac{\partial p}{\partial \upsilon_k}\right)\right\} d\upsilon,\\
V_2(p,t) &= \int_{\Omega}\left\{\int_{t-h(t)}^{t} \chi^T(\upsilon,\theta) T' \chi(\upsilon,\theta) d\theta\right.\\
&\quad + \int_{t-h_1}^{t} \chi^T(\upsilon,\theta) T'' \chi(\upsilon,\theta) d\theta\\
&\quad \left. + \int_{t-h_2}^{t-h_1} \chi^T(\upsilon,\theta) T''' \chi(\upsilon,\theta) d\theta\right\} d\upsilon,\\
V_3(p,t) &= \int_{\Omega}\left\{h_1 \int_{-h_1}^{0}\int_{t+\theta}^{t} \dot{p}^T(\upsilon,\alpha) Q \dot{p}(\upsilon,\alpha) d\alpha d\theta\right.\\
&\quad \left. + h_{21}\int_{-h_2}^{-h_1}\int_{t+\theta}^{t} \dot{p}^T(\upsilon,\alpha) R \dot{p}(\upsilon,\alpha) d\alpha d\theta\right\} d\upsilon,\\
V_4(p,t) &= \int_{\Omega}\left\{\int_{-h_1}^{0}\int_{\alpha}^{0}\int_{t+\beta}^{t} \dot{p}^T(\upsilon,\theta) F_1 \dot{p}(\upsilon,\theta) d\theta d\beta d\alpha\right.\\
&\quad \left. + \int_{-h_2}^{-h_1}\int_{\alpha}^{0}\int_{t+\beta}^{t} \dot{p}^T(\upsilon,\theta) F_2 \dot{p}(\upsilon,\theta) d\theta d\beta d\alpha\right\} d\upsilon.
\end{aligned}
$$

Then, the time derivative of $V(p,t)$ can be obtained as

$$
\mathcal{L}V(p,t) = \mathcal{L}V_1(p,t) + \mathcal{L}V_2(p,t) + \mathcal{L}V_3(p,t) + \mathcal{L}V_4(p,t),
$$

where

$$
\begin{aligned}
\mathcal{L}V_1(p,t) = &\int_{\Omega}\left\{2\wp^T P_{\sigma}^N \dot{\wp} + \wp^T \sum_{\mu\in S} \pi_{\sigma\mu} P_{\mu}^N \wp\right.\\
&\left. + 2\sum_{k=1}^{q}\left(\frac{\partial p}{\partial \upsilon_k}\right)^T \Lambda_k \Psi \left(\frac{\partial \dot{p}}{\partial \upsilon_k}\right)\right\} d\upsilon.
\end{aligned}
$$

Based on the error system (5), the following equation is obviously true:

$$
\begin{aligned}
0 = &2\tilde{p}\Psi\left[-\frac{\partial p}{\partial t} + \sum_{k=1}^{q}\frac{\partial}{\partial \upsilon_k}\left(\Lambda_k \frac{\partial p}{\partial \upsilon_k}\right) - \mathcal{A}p + \tilde{\mathcal{C}}_{\varpi} f(p) + \tilde{\mathcal{D}}_{\varpi} f(p_h)\right.\\
&\left. + \Im - K_{\omega} p - H(\upsilon,t) sign(\tilde{p})\right].
\end{aligned}
$$

One can conclude from the expression of $H(\upsilon,t)$ that $2\tilde{p}\Psi\left[\Im - H(\upsilon,t) sign(\tilde{p})\right] \leq 0$ for all cases.

Moreover, by using Lemma 1 and boundary conditions, we have

$$\int_{\Omega} 2\tilde{p}\Psi \sum_{k=1}^{q} \frac{\partial}{\partial \upsilon_k}\left(\Lambda_k \frac{\partial p}{\partial \upsilon_k}\right) d\upsilon = \int_{\Omega} 2\left(p+\dot{p}\right)\Psi \sum_{k=1}^{q} \frac{\partial}{\partial \upsilon_k}\left(\Lambda_k \frac{\partial p}{\partial \upsilon_k}\right) d\upsilon$$
$$\leq -2\int_{\Omega}\left\{p^T\tilde{\Lambda}\Psi p + \sum_{k=1}^{q}\left(\frac{\partial p}{\partial \upsilon_k}\right)^T \Lambda_k \Psi\left(\frac{\partial \dot{p}}{\partial \upsilon_k}\right)\right\} d\upsilon.$$

As a result,

$$\mathcal{L}V_1(p,t) \leq \int_{\Omega}\left\{2\wp^T P_{\sigma}^N \dot{\wp} + \wp^T \sum_{\mu\in S}\pi_{\sigma\mu}P_{\mu}^N \wp - 2p^T\tilde{\Lambda}\Psi p + 2(p+\dot{p})\Psi \right.$$
$$\left.\times\left[-\frac{\partial p}{\partial t} - \mathcal{A}p + \tilde{\mathcal{C}}_{\varpi} f(p) + \tilde{\mathcal{D}}_{\varpi} f(p_h) - K_{\varpi} p\right]\right\} d\upsilon.$$

One can readily obtain the following inequality based on some simple estimation:

$$\mathcal{L}V_2(p,t) \leq \int_{\Omega}\left\{\chi^T\left(T' + T''\right)\chi - (1-h)\chi_h^T T'\chi_h \right.$$
$$\left. -\chi_{h1}^T\left(T'' - T'''\right)\chi_{h1} - \chi_{h2}^T T'''\chi_{h2}\right\} d\upsilon.$$

Additionally, we have

$$\mathcal{L}V_3(p,t) = \int_{\Omega}\left\{h_1^2\dot{p}^T Q\dot{p} + h_{21}^2\dot{p}^T R\dot{p} - h_1\int_{t-h_1}^{t}\dot{p}^T(\upsilon,\alpha)Q\dot{p}(\upsilon,\alpha)d\alpha \right.$$
$$\left. -h_{21}\int_{t-h_2}^{t-h_1}\dot{p}^T(\upsilon,\alpha)R\dot{p}(\upsilon,\alpha)d\alpha\right\} d\upsilon.$$

On the other hand, the B–L inequality provided in Lemma 2 can yield that

$$-h_1\int_{t-h_1}^{t}\dot{p}^T(\upsilon,\alpha)Q\dot{p}(\upsilon,\alpha)d\alpha$$
$$\leq h_1\eta_N^T\left[\tilde{X}_N^T\tilde{Y}_N^T M + M^T\tilde{Y}_N\tilde{X}_N + h_1 M^T\mathcal{Q}_N^{-1}M\right]\eta_N,$$

where $\eta_N = \mathrm{col}\{p,\ \ p_{h_1},\ \ \vartheta_1,\ldots,\vartheta_N\}$.

Similarly,

$$-h_{21}\int_{t-h_2}^{t-h_1}\dot{p}^T(\upsilon,\alpha)R\dot{p}(\upsilon,\alpha)d\alpha$$
$$\leq h_{21}\tilde{\eta}_N^T\left[\tilde{X}_N^T\tilde{Y}_N^T\tilde{M} + \tilde{M}^T\tilde{Y}_N\tilde{X}_N + h_{21}\tilde{M}^T\mathcal{R}_N^{-1}\tilde{M}\right]\tilde{\eta}_N,$$

where $\tilde{\eta}_N = \mathtt{col}\{p_{h_1},\ \ p_{h_2},\ \ \tilde{\vartheta}_1, \ldots, \tilde{\vartheta}_N\}$.

Furthermore, it is straightforward that

$$
\begin{aligned}
\mathcal{L}V_4(p,t) = \int_{\Omega} \Bigg\{ & \frac{h_1^2}{2}\dot{p}^T F_1 \dot{p} + \frac{h_2^2 - h_1^2}{2}\dot{p}^T F_2 \dot{p} \\
& - \int_{-h_1}^{0} \int_{t+\alpha}^{t} \dot{p}^T(\upsilon,\theta) F_1 \dot{p}(\upsilon,\theta) d\theta d\alpha \\
& - \int_{-h_2}^{-h_1} \int_{t+\alpha}^{t} \dot{p}^T(\upsilon,\theta) F_2 \dot{p}(\upsilon,\theta) d\theta d\alpha \Bigg\} d\upsilon.
\end{aligned}
$$

By recalling Lemma 3, the following inequalities hold:

$$
\begin{aligned}
& - \int_{-h_1}^{0} \int_{t+\alpha}^{t} \dot{p}^T(\upsilon,\theta) F_1 \dot{p}(\upsilon,\theta) d\theta d\alpha \\
& \leq -\frac{2}{h_1^2} \int_{-h_1}^{0} \int_{t+\alpha}^{t} \dot{p}^T(\upsilon,\theta) d\theta d\alpha F_1 \int_{-h_1}^{0} \int_{t+\alpha}^{t} \dot{p}^T(\upsilon,\theta) d\theta d\alpha \\
& = -\frac{2}{h_1^2} \begin{bmatrix} h_1 p \\ \int_{t-h_1}^{t} p(\upsilon,\alpha) d\alpha \end{bmatrix}^T \mathfrak{F}_1 \begin{bmatrix} h_1 p \\ \int_{t-h_1}^{t} p(\upsilon,\alpha) d\alpha \end{bmatrix},
\end{aligned}
$$

and

$$
\begin{aligned}
& - \int_{-h_2}^{-h_1} \int_{t+\alpha}^{t} \dot{p}^T(\upsilon,\theta) F_2 \dot{p}(\upsilon,\theta) d\theta d\alpha \\
& \leq -\frac{2}{h_2^2 - h_1^2} \begin{bmatrix} h_{21} p \\ \int_{t-h_2}^{t-h_1} p(\upsilon,\alpha) d\alpha \end{bmatrix}^T \mathfrak{F}_2 \begin{bmatrix} h_{21} p \\ \int_{t-h_2}^{t-h_1} p(\upsilon,\alpha) d\alpha \end{bmatrix},
\end{aligned}
$$

where

$$
\mathfrak{F}_1 = \begin{bmatrix} F_1 & -F_1 \\ * & F_1 \end{bmatrix}, \quad \mathfrak{F}_2 = \begin{bmatrix} F_2 & -F_2 \\ * & F_2 \end{bmatrix}.
$$

The FWM method is applied to the following equation to further reduce the conservatism:

$$
\begin{aligned}
2 \Bigg(& p^T \Upsilon_1 + \dot{p} \Upsilon_2 + f(p) \Upsilon_3 + f(p_h) \Upsilon_4 + \int_{t-h_2}^{t-h_1} \dot{p}(\upsilon,\theta) d\theta \Upsilon_5 \Bigg) \\
& \times \left(p_{h_1} - p_{h_2} - \int_{t-h_2}^{t-h_1} \dot{p}(\upsilon,\theta) d\theta \right) = 0.
\end{aligned}
$$

By considering the activation function of this chapter, we have

$$0 \le 2\big[f(p) - E^- p\big]^T \Theta_1 \left[E^+ p - f(p)\right],$$
$$0 \le 2\big[f(p_h) - E^- p_h\big]^T \Theta_2 \left[E^+ p_h - f(p_h)\right],$$
$$0 \le 2\big\{f(p) - f(p_h) - E^- [p - p_h]\big\}^T \Theta_3 \left\{E^+ [p - p_h] - f(p) + f(p_h)\right\}.$$

It is obvious that if (16) holds, $\mathcal{L}V(p,t) < 0$. That is, there exists a constant $\iota > 0$, such that

$$\mathcal{L}V(p,t) < -\iota V(p,t). \tag{10}$$

Pre- and post-multiplying inequity (10) by $e^{-\iota t}$, and using Dynkin's formula, it follows that

$$\begin{aligned}
&\mathbb{E}\{V(p,t)\} < e^{\iota t}\mathbb{E}\{V(p,0)\} \\
&\le e^{\iota t}\Bigg\{\lambda_{\max}((2N+1)I)\max_{\sigma\in S}\left(\lambda_{\max}\left(P_\sigma^N\right)\right) + \lambda_{max}\left(\sum_{k=1}^{q}\Lambda_k\Psi\right) \\
&\quad + h_2\lambda_{max}\left(I + (E^+)^T E^+\right)\lambda_{\max}\left(T'\right) \\
&\quad + h_1\lambda_{max}\left(I + (E^+)^T E^+\right)\lambda_{\max}\left(T''\right) \\
&\quad + \frac{h_1^3}{2}\lambda_{\max}(Q) + h_{21}\lambda_{max}\left(I + (E^+)^T E^+\right)\lambda_{\max}\left(T'''\right) \\
&\quad + \frac{h_{21}^3}{2}\lambda_{\max}(R) + \frac{h_1^3}{6}\lambda_{\max}(F_1) + \frac{h_{21}^3}{6}\lambda_{\max}(F_2)\Bigg\} \\
&\quad \times \sup_{-h_2\le t\le 0}\left\{\int_\Omega (p^T p)\,d\upsilon, \int_\Omega (\dot{p}^T\dot{p})\,d\upsilon, \sup_{1\le k\le q}\left\{\int_\Omega \left(\frac{\partial p}{\partial \upsilon_k}\right)^T\left(\frac{\partial p}{\partial \upsilon_k}\right)d\upsilon\right\}\right\} \\
&\le \int_\Omega e^{\iota T^*}\varsigma\hbar_1 d\upsilon,
\end{aligned}$$

where $\hbar_1 = \sup_{-h_2\le t\le 0}\left\{(p^T p), (\dot{p}^T\dot{p}), \sup_{1\le k\le q}\left\{\left(\frac{\partial p}{\partial \upsilon_k}\right)^T\left(\frac{\partial p}{\partial \upsilon_k}\right)\right\}\right\}$.
Since $\min_{\sigma\in S}\left(\lambda_{\min}\left(P_{1\sigma}^N\right)\right)\mathbb{E}\left\{\|p\|^2\right\} \le \mathbb{E}\left\{p^T P_{1\sigma}^N p\right\}$, then

$$\mathbb{E}\left\{\|p\|^2\right\} < \frac{\hbar_1\varsigma e^{\iota T^*}}{\min_{\sigma\in S}\left(\lambda_{\min}\left(P_{1\sigma}^N\right)\right)}. \tag{11}$$

Therefore, from Definition 1, one can deduce that $\mathbb{E}\left\{\|p\|^2\right\} < \hbar_2$ if (17) holds, namely, the drive system (1) and response system (2) are synchronized over an FTI. □

Remark 2 The main target of this chapter is not only to obtain the finite-time synchronization criterion of Markovian reaction-diffusion MNNs, but also to reduce the conservatism of the obtained results. For one thing, we construct a novel LKF candidate that involves the triple integral term $V_3(p,t)$ and the quadruple integral term $V_4(p,t)$, with which more information of $h(t)$ can be captured to help to reduce the conservatism. For another, during handing the time derivative of LKF, the utilization of canonical BLI and FWM methods can further reduce the conservatism of obtained results.

Remark 3 The canonical BLI has been proven to be an effective tool in reducing the conservatism of main results represented by linear matrix inequalities [28]. But canonical BLI also has some limitations. As mentioned in [30], if ϑ_i and $\tilde{\vartheta}_i$ are not involved in $V(p,t)$, they will not arise in $\mathcal{L}V(p,t)$. Thus the BLI employed in this chapter will not be effective in reducing the conservatism of the main results.

If the reaction-diffusion phenomenon is not considered during the modeling, the new system model of Markovian MNNs that follow two different Markov chains can be described below:

$$\begin{aligned}\dot{m}_\delta(t) =& -a_\delta m_\delta(t) + \sum_{l=1}^{n} c_{\delta l\sigma}(m_\delta)g_l(m_l(t))\\ &+\sum_{l=1}^{n} d_{\delta l\sigma}(m_\delta)g_l(m_l(t-h(t))),\end{aligned} \tag{12}$$

$$\begin{aligned}\dot{s}_\delta(t) =& -a_\delta s_\delta(t) + \sum_{l=1}^{n} c_{\delta l\varpi}(s_\delta)g_l(s_l(t))\\ &+\sum_{l=1}^{n} d_{\delta l\varpi}(s_\delta)g_l(s_l(t-h(t))) + u_\delta(t).\end{aligned} \tag{13}$$

And the error system of (12) and (13) in a compact form can be obtained as

$$\dot{p}_\delta(t) = -\mathcal{A}p(t) + \tilde{\mathcal{C}}_\varpi f(p(t)) + \tilde{\mathcal{D}}_\varpi f(p(t-h(t))) + \Im(t) + u(t), \tag{14}$$

where

$$\begin{aligned}\Im(t) =& (\mathcal{C}_\varpi(s(t)) - \mathcal{C}_\sigma(m(t)))\,g(m(t))\\ &+ (\mathcal{D}_\varpi(s(t)) - \mathcal{D}_\sigma(m(t)))\,g(m(t-h(t))),\end{aligned}$$

and the controller $u(t)$ is designed as

$$u = -K_\varpi p(t) - H(t)sign(\tilde{p}(t)) \tag{15}$$

with $\tilde{p}_\delta(t) = p_\delta(t) + \dot{p}_\delta(t)$; $H(t) = diag\{H_1(t), H_2(t), \ldots, H_n(t)\}$, and

$$H_\delta(t) = \begin{cases} \mathfrak{h}_{1\delta}, & |m_\delta(t)| \le \mathcal{Y}_\delta \ and \ |s_\delta(t)| \le \mathcal{Y}_\delta, \\ \mathfrak{h}_{2\delta}, & |m_\delta(t)| \le \mathcal{Y}_\delta \ and \ |s_\delta(t)| > \mathcal{Y}_\delta, \\ \mathfrak{h}_{3\delta}, & |m_\delta(t)| > \mathcal{Y}_\delta \ and \ |s_\delta(t)| \le \mathcal{Y}_\delta, \\ \mathfrak{h}_{4\delta}, & |m_\delta(t)| > \mathcal{Y}_\delta \ and \ |s_\delta(t)| > \mathcal{Y}_\delta. \end{cases}$$

For the above systems, the following corollary can be obtained:

Corollary 1 *For given scalars $\tilde{\hbar}_1 > 0$, $\tilde{\iota} > 0$, and $\tilde{T}^* > 0$, and an integer $\tilde{N} \ge 0$, the synchronization error system (14) is bounded over an FTI if there exist a positive scalar $\tilde{\hbar}_2$; symmetric matrices $\tilde{P}_\sigma^N = diag\{\tilde{P}_{1\sigma}^N, \tilde{P}_{2\sigma}^N, \ldots, \tilde{P}_{(N+1)\sigma}^N\} > 0$, $\tilde{\Psi} > 0$, $\tilde{T}' = \begin{bmatrix} \tilde{T}'_{11} & \tilde{T}'_{12} \\ * & \tilde{T}'_{22} \end{bmatrix} > 0$, $\tilde{T}'' = \begin{bmatrix} \tilde{T}''_{11} & \tilde{T}''_{12} \\ * & \tilde{T}''_{22} \end{bmatrix} > 0$, $\tilde{T}''' = \begin{bmatrix} \tilde{T}'''_{11} & \tilde{T}'''_{12} \\ * & \tilde{T}'''_{22} \end{bmatrix} > 0$, $\tilde{Q} > 0$, $\tilde{R} > 0$, $\tilde{F}_1 > 0$, $\tilde{F}_2 > 0$; arbitrary matrices $\tilde{\Upsilon}_j$ $(j = 1, 2, 3, 4, 5)$, $\hat{M}$, and $\tilde{\hat{M}}$; and diagonal matrices $\tilde{\Theta}_m (m = 1, 2, 3)$, such that the following LMIs hold:*

$$\begin{bmatrix} \tilde{\Sigma} & h_1 \mathcal{X}_1^T \hat{M} & h_{21} \mathcal{X}_2^T \tilde{\hat{M}} \\ * & -\tilde{\mathcal{Q}}_{\tilde{N}} & 0 \\ * & * & -\tilde{\mathcal{R}}_{\tilde{N}} \end{bmatrix} < 0, \tag{16}$$

$$\tilde{\hbar}_1 \tilde{\varsigma} e^{\tilde{\iota} \tilde{T}^*} < \min_{\sigma \in S} \left(\lambda_{\min} \left(\tilde{P}_{1\sigma}^N \right) \right) \tilde{\hbar}_2, \tag{17}$$

where

$$\begin{aligned} \tilde{\Sigma} &= \tilde{\Sigma}_1 + \tilde{\Sigma}_2 + \tilde{\Sigma}_3 + \tilde{\Sigma}_4 + \tilde{\Sigma}_5, \\ \tilde{\mathcal{Q}}_{\tilde{N}} &= diag\{\tilde{Q}, 3\tilde{Q}, \ldots, (2\tilde{N}+1)\tilde{Q}\}, \\ \tilde{\mathcal{R}}_{\tilde{N}} &= diag\{\tilde{R}, 3\tilde{R}, \ldots, (2\tilde{N}+1)\tilde{R}\}, \\ \tilde{\varsigma} &= \lambda_{\max}\left((2\tilde{N}+1)I\right) \max_{\sigma \in S} \left(\lambda_{\max}\left(\tilde{P}_\sigma^N\right)\right) + \frac{h_1^3}{2} \lambda_{\max}\left(\tilde{Q}\right) \\ &\quad + h_2 \lambda_{max}\left(I + (E^+)^T E^+\right) \lambda_{\max}\left(\tilde{T}'\right) + \frac{h_{21}^3}{2} \lambda_{\max}\left(\tilde{R}\right) \\ &\quad + h_1 \lambda_{max}\left(I + (E^+)^T E^+\right) \lambda_{\max}\left(\tilde{T}''\right) + \frac{h_{21}^3}{6} \lambda_{\max}\left(\tilde{F}_2\right) \\ &\quad + h_{21} \lambda_{max}\left(I + (E^+)^T E^+\right) \lambda_{\max}\left(\tilde{T}'''\right) + \frac{h_1^3}{6} \lambda_{\max}\left(\tilde{F}_1\right), \end{aligned}$$

with

$$
\begin{aligned}
\tilde{\Sigma}_1 =& Sym\{\mathfrak{X}^T \tilde{P}_\sigma^N \mathfrak{Y}\} + \mathfrak{X}^T \sum_{\mu \in S} \pi_{\sigma\mu} \tilde{P}_\mu^N \mathfrak{X} - 2\dagger_2^T \Psi \dagger_2 - \dagger_1^T \left(2\tilde{\Psi} \mathcal{A} + 2\tilde{\Psi} K_\varpi\right) \dagger_1 \\
&+ Sym\{\dagger_1^T \tilde{\Psi} \tilde{\mathcal{C}}_\varpi \dagger_6\} - Sym\{\dagger_1^T (\tilde{\Psi} + \mathcal{A}^T \tilde{\Psi}^T + K_\varpi^T \tilde{\Psi}^T) \dagger_2\} \\
&+ Sym\{\dagger_1^T \tilde{\Psi} \tilde{\mathcal{D}}_\varpi \dagger_8\} + Sym\{\dagger_2^T \tilde{\Psi} \tilde{\mathcal{C}}_\varpi \dagger_6\} + Sym\{\dagger_2^T \tilde{\Psi} \tilde{\mathcal{D}}_\varpi \dagger_8\}, \\
\tilde{\Sigma}_2 =& \dagger_1^T (\tilde{T}'_{11} + \tilde{T}''_{11}) \dagger_1 + Sym\{\dagger_1^T (\tilde{T}'_{12} + \tilde{T}''_{12}) \dagger_6\} \\
&- (1-h) \left[\dagger_4{}^T \tilde{T}'_{11} \dagger_4 + \dagger_8^T \tilde{T}'_{22} \dagger_8 + Sym\{\dagger_4{}^T \tilde{T}'_{12} \dagger_8\}\right] \\
&+ \dagger_6^T (\tilde{T}'_{22} + \tilde{T}''_{22}) \dagger_6 - \dagger_5^T \tilde{T}'''_{11} \dagger_5 - \dagger_3^T (\tilde{T}''_{11} - \tilde{T}'''_{11}) \dagger_3 \\
&- Sym\{\dagger_3^T (\tilde{T}''_{12} - \tilde{T}'''_{12}) \dagger_7\} - \dagger_7^T (\tilde{T}''_{22} - \tilde{T}'''_{22}) \dagger_7 \\
&- \dagger_5^T \tilde{T}'''_{12} \dagger_9 - \dagger_9^T \tilde{T}'''_{22} \dagger_9, \\
\tilde{\Sigma}_3 =& h_1 \lambda_1^T \left[\tilde{X}_{\tilde{N}}^T \tilde{Y}_{\tilde{N}}^T \hat{M} + \hat{M}^T \tilde{Y}_{\tilde{N}} \tilde{X}_{\tilde{N}}\right] \lambda_1 + \dagger_2^T (h_1^2 \tilde{Q} \\
&+ h_{21}^2 \tilde{R}) \dagger_2 + h_{21} \lambda_2^T \left[\tilde{X}_{\tilde{N}}^T \tilde{Y}_{\tilde{N}}^T \tilde{\hat{M}} + \tilde{\hat{M}}^T \tilde{Y}_{\tilde{N}} \tilde{X}_{\tilde{N}}\right] \lambda_2, \\
\Sigma_4 =& \dagger_2^T (\frac{h_1^2}{2} \tilde{F}_1 + \frac{h_2^2 - h_1^2}{2} \tilde{F}_2) \dagger_2 - 2\dagger_1^T \tilde{F}_1 \dagger_1 - \frac{2}{h_1^2} \dagger_{10}^T \tilde{F}_1 \dagger_{10} + \frac{2}{h_1} Sym\{\dagger_1^T \tilde{F}_1 \dagger_{10}\} \\
&- \frac{2h_{21}^2}{h_2^2 - h_1^2} \dagger_1^T \tilde{F}_2 \dagger_1 + \frac{2h_{21}}{h_2^2 - h_1^2} \dagger_1^T \tilde{F}_2 \dagger_{11} - \frac{2}{h_2^2 - h_1^2} \dagger_{11}^T \tilde{F}_2 \dagger_{11}, \\
\Sigma_5 =& Sym\{\left(\dagger_6^T - E^- \dagger_1^T\right) \tilde{\Theta}_1 \left(E^+ \dagger_1 - \dagger_6\right)\} \\
&+ Sym\{\left(\dagger_8^T - E^- \dagger_4^T\right) \tilde{\Theta}_2 \left(E^+ \dagger_4 - \dagger_8\right)\} \\
&+ Sym\{\left[\dagger_6^T - \dagger_8^T - E^- \left(\dagger_1^T - \dagger_4^T\right)\right] \tilde{\Theta}_3 \left[E^+ (\dagger_1 - \dagger_4) - \dagger_6 + \dagger_8\right]\} \\
&+ Sym\{\left(\dagger_1^T \tilde{\Upsilon}_1 + \dagger_2^T \tilde{\Upsilon}_2 + \dagger_6^T \tilde{\Upsilon}_3 + \dagger_8^T \tilde{\Upsilon}_4 + \dagger_{12}^T \tilde{\Upsilon}_5\right) (\dagger_3 - \dagger_5 - \dagger_{12})\}.
\end{aligned}
$$

4 Numerical Examples

In order to prove the validity and practicability of the main results obtained in this chapter, this section provides two simulation examples.

First, for simplicity, we simplified the original systems' models, that is, we set $n = 1, 2, \sigma = 1, 2, \varpi = 1, 2$ and $q = 1$, then the drive and response systems can be described as

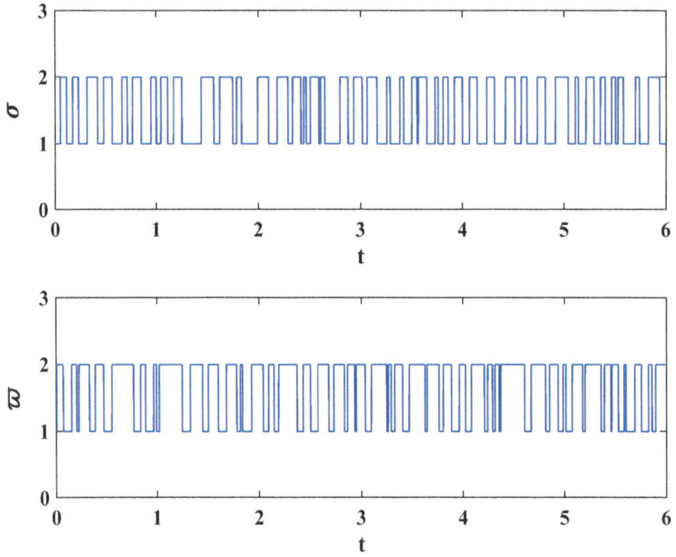

Fig. 1 The Markov stochastic processes of the drive and response systems

$$\frac{\partial m_\delta}{\partial t} = \sum_{k=1} \frac{\partial}{\partial \upsilon_k}\left(\mathfrak{r}_{\delta k}\frac{\partial m_\delta}{\partial \upsilon_k}\right) - a_\delta m_\delta + \sum_{l=1}^{2} c_{\delta l \sigma}(m_\delta) g_l(m_l) + \sum_{l=1}^{2} d_{\delta l \sigma}(m_\delta) g_l(m_{lh}), \tag{18}$$

and

$$\frac{\partial s_\delta}{\partial t} = \sum_{k=1} \frac{\partial}{\partial \upsilon_k}\left(\mathfrak{r}_{\delta k}\frac{\partial s_\delta}{\partial \upsilon_k}\right) - a_\delta s_\delta + \sum_{l=1}^{2} c_{\delta l \varpi}(s_\delta) g_l(s_l) + \sum_{l=1}^{2} d_{\delta l \varpi}(s_\delta) g_l(s_{lh}) + u_\delta. \tag{19}$$

For drive system (18) and response system (19), by choosing $\pi_{11} = -0.5$, $\pi_{12} = 0.5$, $\pi_{21} = 0.8$ and $\pi_{22} = -0.8$, $\varrho_{11} = 0.8$, $\varrho_{12} = 0.2$, $\varrho_{21} = 0.4$, $\varrho_{22} = 0.6$, the Markovian jumping laws of system (18) and system (19) can be shown as in Fig. 1.

Example 1 We set $\mathfrak{r}_{11} = \mathfrak{r}_{21} = 0.01$, $a_1 = 10.9$, $a_2 = 11.2$, $\mathcal{Y}_1 = 0.2$ and $\mathcal{Y}_2 = 0.15$, and the other parameters depend on the systems' modes:

Mode 1: $c^{\flat}_{111} = 0.2$, $d^{\flat}_{111} = -0.4$, $c^{\sharp}_{111} = 0.3$, $d^{\sharp}_{111} = -0.5$, $c^{\flat}_{121} = 0.15$, $d^{\flat}_{121} = -0.2$, $c^{\sharp}_{121} = 0.1$, $d^{\sharp}_{121} = -0.16$, $c^{\flat}_{211} = -0.4$, $d^{\flat}_{211} = 0.6$, $c^{\sharp}_{211} = -0.2$, $d^{\sharp}_{211} = 0.1$, $c^{\flat}_{221} = 0.5$, $d^{\flat}_{221} = -0.3$, $c^{\sharp}_{221} = 0.4$, $d^{\sharp}_{221} = -0.2$.

Mode 2: $c^{\flat}_{112} = -0.06$, $d^{\flat}_{112} = -0.5$, $c^{\sharp}_{112} = -0.05$, $d^{\sharp}_{112} = -0.6$, $c^{\flat}_{122} = 0.05$, $d^{\flat}_{122} = -0.1$, $c^{\sharp}_{122} = 0.1$, $d^{\sharp}_{122} = -0.15$, $c^{\flat}_{212} = -0.4$, $d^{\flat}_{212} = 0.5$, $c^{\sharp}_{212} = -0.5$, $d^{\sharp}_{212} = 0.6$, $c^{\flat}_{222} = 0.2$, $d^{\flat}_{222} = -0.5$, $c^{\sharp}_{222} = 0.15$, $d^{\sharp}_{222} = -0.4$.
Moreover,

$$
\begin{aligned}
m_1(\upsilon, o) &= -1.2sin(\upsilon), (\upsilon, o) \in \Omega \times [-h_2, 0],\\
m_2(\upsilon, o) &= 1.4sin(\upsilon), (\upsilon, o) \in \Omega \times [-h_2, 0],\\
s_1(\upsilon, o) &= -2.1sin(\upsilon), (\upsilon, o) \in \Omega \times [-h_2, 0],\\
s_2(\upsilon, o) &= 1.3sin(\upsilon), (\upsilon, o) \in \Omega \times [-h_2, 0].
\end{aligned}
$$

With considering $h(t) = \frac{2e^t}{15+5e^t}$ and $u = 0$, the trajectories of the error system can be described with Fig. 2. It is obvious that the error of drive system (18) and response system (19) is unbounded, that is, the drive and response systems cannot achieve synchronization without controller.

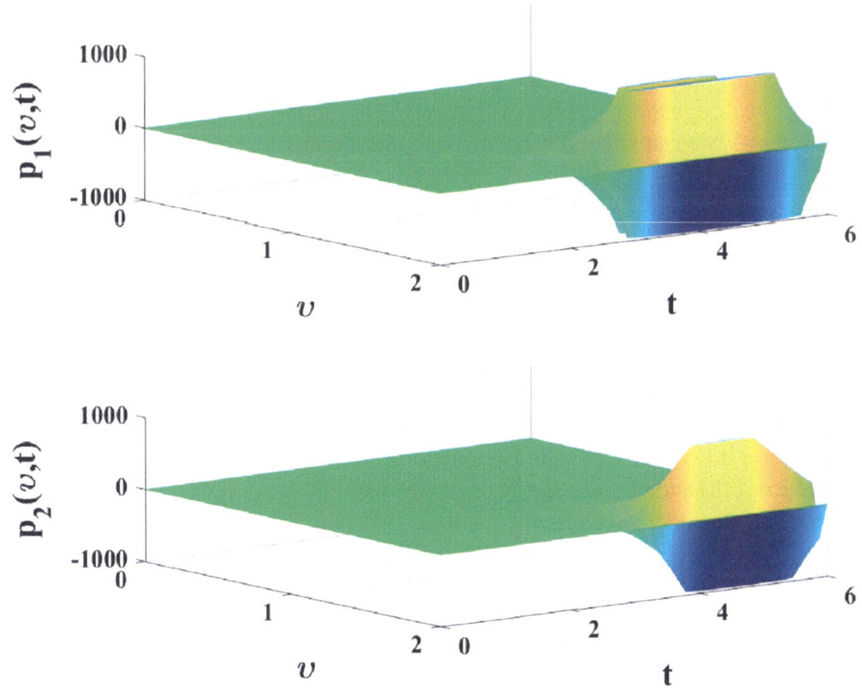

Fig. 2 The dynamic behaviors of the error system without controller

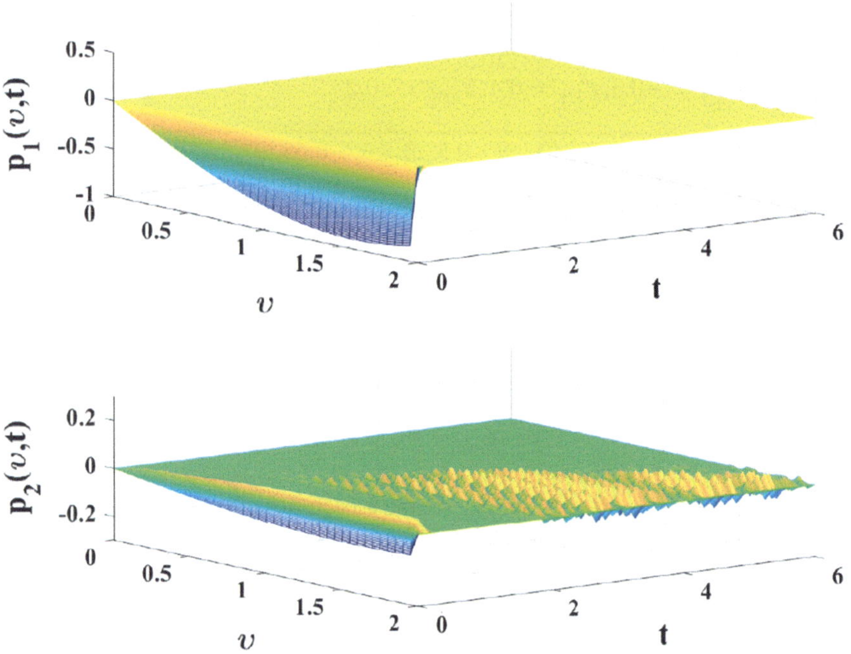

Fig. 3 The dynamic behaviors of the error system with the controller (6)

Then, we design the controller in (19) as (6), by solving the inequalities proposed in Theorem 1, we can obtain the following controller gains:

$$K_1 = \begin{bmatrix} 6.1195 & -7.5517 \\ 7.5164 & 6.4396 \end{bmatrix}, \quad K_2 = \begin{bmatrix} 6.0916 & -6.9876 \\ 7.6044 & 7.0124 \end{bmatrix}.$$

Furthermore, with these controller gains, the dynamic behaviors of the error system with the controller (6) are shown in Fig. 3, from which we can conclude that the error system of (18) and (19) is bounded, that is, the drive and response systems can achieve synchronization with the controller (6).

Additionally, we can obtain that $\varsigma = 16.71$ and $\hbar_1 = 0.82$ by some calculations, then, set $\iota = 0.1$, $T^* = 4$, from the inequation (17), there exists $\hbar_2 \geq 3.1485$, so that the systems (18) and (19) are synchronous over an FTI.

Example 2 Inspired by the applications of chaotic systems in Pseudo Random Number Generators (PRNG) proposed in [31–33], this example aims to apply our system to the field of PRNG. Note that Markovian reaction-diffusion MNNs are ideal tools for generating chaos because of their internal complexity. Therefore, applying Markovian reaction-diffusion MNNs to Pseudo Random Number Generators (PRNG) is reasonable.

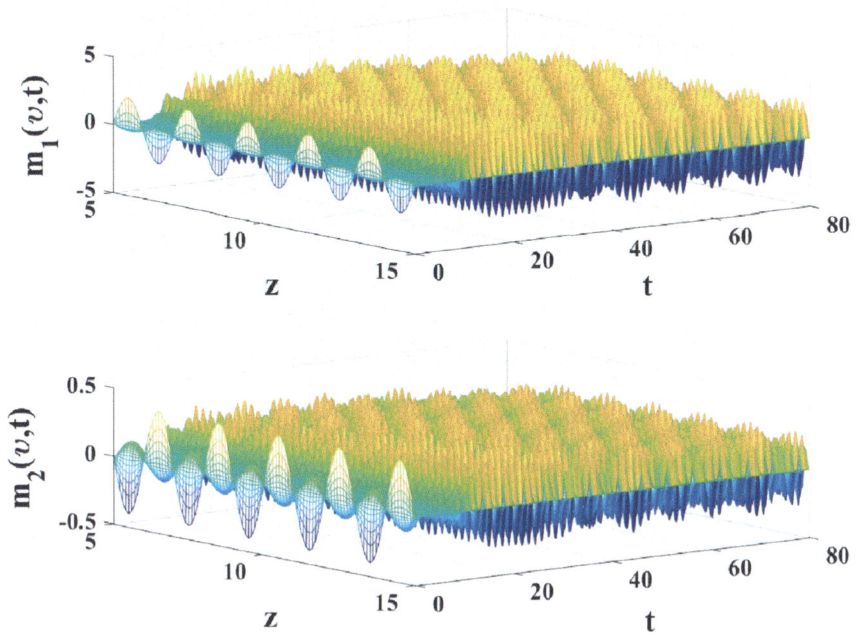

Fig. 4 The chaotic behaviors of the drive system (18)

This example realizes the encryption and decryption functions by complex dynamics of Markovian reaction-diffusion MNNs proposed in this chapter.

The following parameters are chosen for system (18): $a_1 = 6.6$, $a_2 = 6.5$, $\mathcal{Y}_1 = 1.6$, $\mathcal{Y}_2 = 0.5$.

When $MODE = 1$: $c^{\flat}_{111} = 2$, $d^{\flat}_{111} = -1.4$, $c^{\flat}_{121} = 20.1$, $d^{\flat}_{121} = 0.7$, $c^{\flat}_{211} = 0.1$, $d^{\flat}_{211} = 0.1$, $c^{\flat}_{221} = 2.05$, $d^{\flat}_{221} = -1.3$, $c^{\sharp}_{111} = 2.1$, $d^{\sharp}_{111} = -1.6$, $c^{\sharp}_{121} = 20.5$, $d^{\sharp}_{121} = 0.9$, $c^{\sharp}_{211} = 0.1$, $d^{\sharp}_{211} = 0.1$, $c^{\sharp}_{221} = 2.2$, $d^{\sharp}_{221} = -1.1$.

When $MODE = 2$: $c^{\flat}_{112} = 2.1$, $d^{\flat}_{112} = -1.5$, $c^{\flat}_{122} = 21.1$, $d^{\flat}_{122} = 0.6$, $c^{\flat}_{212} = 0.2$, $d^{\flat}_{212} = 0.14$, $c^{\flat}_{222} = 2.15$, $d^{\flat}_{222} = -1.2$, $c^{\sharp}_{112} = 2.0$, $d^{\sharp}_{112} = -1.5$, $c^{\sharp}_{122} = 20.5$, $d^{\sharp}_{122} = 1$, $c^{\sharp}_{212} = 0.1$, $d^{\sharp}_{212} = 0.11$, $c^{\sharp}_{222} = 2.21$, $d^{\sharp}_{222} = -1.3$.

And the other parameters are set the same as those in Example 1.

With these parameters, the state trajectories of the system (18) is shown in Fig. 4, where a typical spatiotemporal chaotic attractor can be found.

Now, we define a pseudo random number sequence in $t \in [t_{start}, t_{end}]$:

$$x(t) = \omega(\mathfrak{p}(t), \mathfrak{q}(t)) = \begin{cases} 1, & \mathfrak{p}(t) \leq \mathfrak{q}(t), \\ 0, & \mathfrak{p}(t) > \mathfrak{q}(t), \end{cases}$$

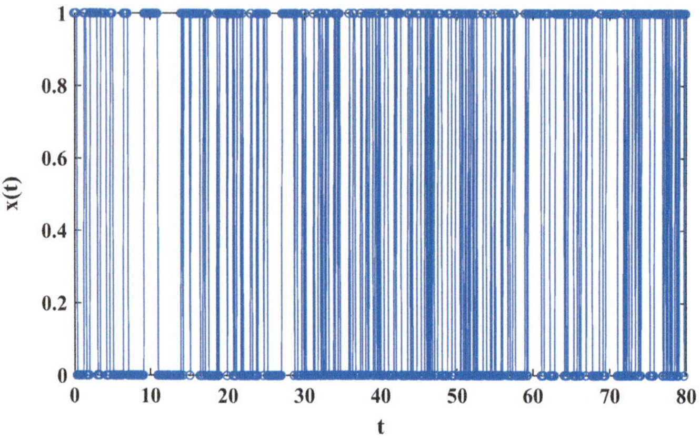

Fig. 5 PRNG produced by reaction-diffusion IMNNs (18)

with

$$\mathfrak{p}(t) = \frac{\hat{m}_1(t)}{\max\limits_{t \in [t_{start}, t_{end}]} \{\hat{m}_1(t)\}}, \quad \mathfrak{q}(t) = \frac{\hat{m}_2(t)}{\max\limits_{t \in [t_{start}, t_{end}]} \{\hat{m}_2(t)\}},$$

and $\hat{m}_1(t) = \int_\Omega m_1(v, t)dv$, $\hat{m}_2(t) = \int_\Omega m_2(v, t)dv$.
The PRNG can be produced by above function, which is described with Fig. 5. It can be seen from Fig. 5 that the sequence generated by Markovian reaction-diffusion MNNs (18) is random.

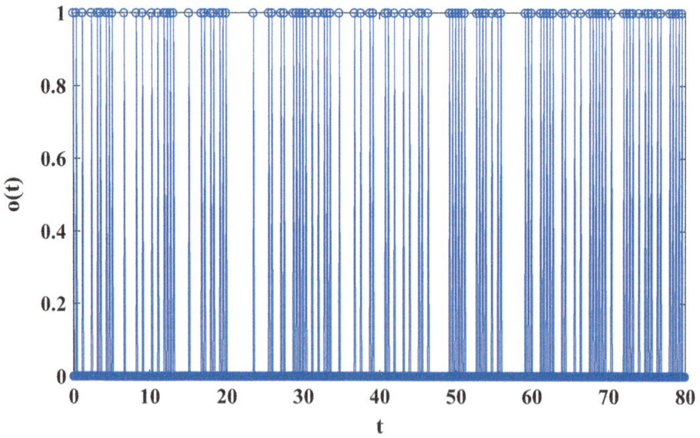

Fig. 6 Original signals $o(t)$

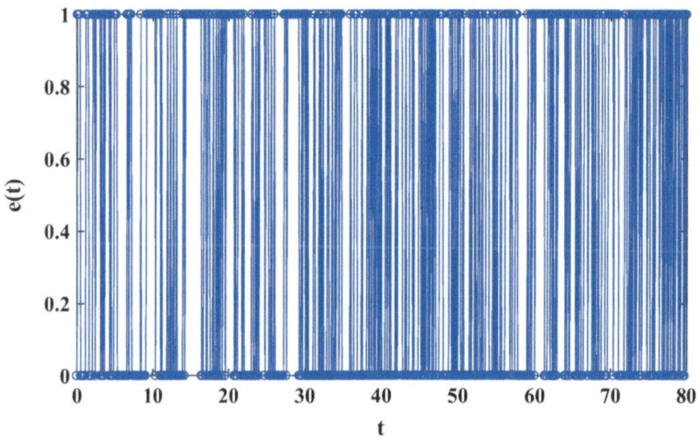

Fig. 7 Encrypted signal generated by $o(t)$ and $x(t)$

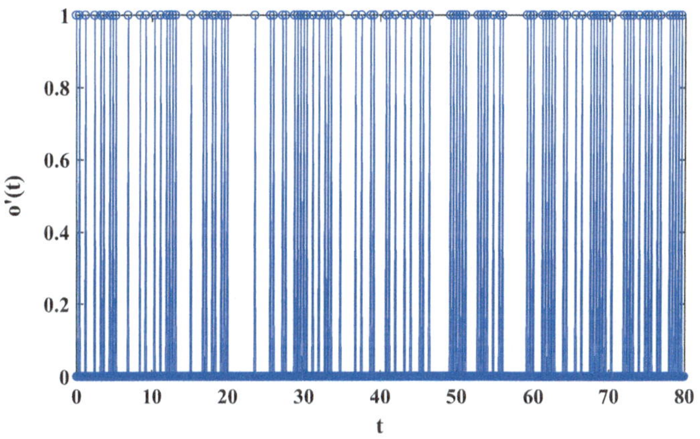

Fig. 8 Decrypted signal generated by $e(t)$ and $x(t)$

Choose $o(t)$ as the original signal that is represented by Fig. 6, and we define $e(t) = o(t) \oplus x(t)$ as the encrypted signal. Then, the trajectory of $e(t)$ can be described with Fig. 7. It is obvious that the encrypted signal and original signal are quite diverse by comparing Figs. 6 and 7, namely, the encryption algorithm we designed is very efficient.

The decryption algorithm is the inverse of encryption algorithm, thus, $o'(t) = e(t) \oplus x(t)$. The combination of PRNG in Fig. 5 and the encrypted signal in Fig. 7 can generates the decrypted signal, which can be represented by Fig. 8. By the comparison of Figs. 6 and 8, it can be seen that the decrypted signal and original signals are identical, which means that the signal is successfully decrypted.

5 Conclusion

This chapter aims to solve the problem of synchronization for Markovian reaction-diffusion MNNs over an FTI. In the process of modeling, the drive and response systems are set to obey inconsistent Markovian jumping laws, which is more practical. However, the challenge of handling the inconsistent parameters in the error system comes with it. To deal with this problem, a discontinuous controller is designed, so that the influence of the inconsistent parameters can be avoided. Moreover, by combining the novel LKF with the canonical BLI and FWM methods, a less conservative result on synchronization for reaction-diffusion MNNs over an FTI is obtained. Finally, a numerical example and an application study are provided to support the main results.

Acknowledgements This work supported by the National Natural Science Foundation of China (Nos. 61976081, U1604146), and Foundation for the University Technological Innovative Talents of Henan Province (No. 18HASTIT019).

References

1. Park, J.H., Kwon, O.M., Lee, S.M.: LMI optimization approach on stability for delayed neural networks of neutral-type. Appl. Math. Comput. **196**(1), 236–244 (2008)
2. Park, J.H.: Robust stability of bidirectional associative memory neural networks with time delays. Phys. Lett. A **349**(6), 494–499 (2006)
3. Bao, H., Park, J.H., Cao, J.: Synchronization of fractional-order complex-valued neural networks with time delay. Neural Netw. **81**, 16–28 (2016)
4. Li, D., Liu, Y., Tong, S., et al.: Neural networks-based adaptive control for nonlinear state constrained systems with input delay. IEEE Trans. Cybern. **49**(4), 1249–1258 (2018)
5. Ding, S., Wang, Z., Zhang, H.: Event-triggered stabilization of neural networks with time-varying switching gains and input saturation. IEEE Trans. Neural Netw. Learn. Syst. **29**(10), 5045–5056 (2018)
6. Xiao, Q., Huang, T., Zeng, Z.: Stabilization of nonautonomous recurrent neural networks with bounded and unbounded delays on time scales. IEEE Trans. Cybern. (2019). https://doi.org/10.1109/TCYB.2019.2922207
7. Bao, H., Park, J.H., Cao, J.: Non-fragile state estimation for fractional-order delayed memristive BAM neural networks. Neural Netw. **119**, 190–199 (2019)
8. Mathiyalagan, K., Anbuvithya, R., Sakthivel, R., et al.: Non-fragile H_∞ synchronization of memristor-based neural networks using passivity theory. Neural Netw. **74**, 85–100 (2016)
9. Wang, X., Park, J.H., Zhong, S., et al.: A switched operation approach to sampled-data control ttabilization of fuzzy memristive neural networks with time-varying delay. IEEE Trans. Neural Netw. Learn. Syst. (2019). https://doi.org/10.1109/TNNLS.2019.2910574
10. Fan, Y., Huang, X., Shen, H., et al.: Switching event-triggered control for global stabilization of delayed memristive neural networks: an exponential attenuation scheme. Neural Netw. **117**, 216–224 (2019)
11. Li, X., Zhang, W., Fang, J., et al.: Finite-time synchronization of memristive neural networks with discontinuous activation functions and mixed time-varying delays. Neurocomputing **340**, 99–109 (2019)
12. Wang, S., Guo, Z., Wen, S., et al.: Finite/fixed-time synchronization of delayed memristive reaction-diffusion neural networks. Neurocomputing **375**, 1–8 (2020)

13. Wang, S., Guo, Z., Wen, S., et al.: Global synchronization of coupled delayed memristive reaction-diffusion neural networks. Neural Netw. **123**, 362–371 (2019)
14. Wang, Z., Cao, J., Lu, G., et al.: Fixed-time passification analysis of interconnected memristive reaction-diffusion neural networks. IEEE Trans. Netw. Sci. Eng. (2019). https://doi.org/10.1109/TNSE.2019.2954463
15. Shen, Y., Wu, Z., Shi, P., et al.: Model reduction of Markovian jump systems with uncertain probabilities. IEEE Trans. Autom. Control **65**(1), 382–388 (2019)
16. Jun, C., Park, J.H., Zhao, X., et al.: Quantized nonstationary filtering of network-based Markov switching RSNSs: a multiple Hierarchical structure strategy. IEEE Trans. Autom. Control (2019). https://doi.org/10.1109/TAC.2019.2958824
17. Song, X., Man, J., Song, S., et al.: Integral sliding mode synchronization control for Markovian jump inertial memristive neural networks with reaction-diffusion terms. Neurocomputing **378**, 324–334 (2020)
18. Song, X., Man, J., Song, S., et al.: Finite-time nonfragile time-varying proportional retarded synchronization for Markovian inertial memristive NNs with reaction-diffusion items. Neural Netw. **123**, 317–330 (2020)
19. Song, X., Man, J., Ahn, C.K., et al.: Finite-time dissipative synchronization for Markovian jump generalized inertial neural networks with reaction-diffusion terms. IEEE Trans. Syst., Man, Cybern.: Syst. (2020). https://doi.org/10.1109/TSMC.2019.2958419
20. Cheng, J., Park, J.H., Karimi, H.R., et al.: A flexible terminal approach to sampled-data exponentially synchronization of Markovian neural networks with time-varying delayed signals. IEEE Trans. Cybern. **48**(8), 2232–2244 (2017)
21. Wei, Y., Park, J.H., Karimi, H.R., et al.: Improved stability and stabilization results for stochastic synchronization of continuous-time semi-Markovian jump neural networks with time-varying delay. IEEE Trans. Neural Netw. Learn. Syst. **29**(6), 2488–2501 (2017)
22. Shen, H., Jiao, S., Cao, J., et al.: An improved result on sampled-data synchronization of Markov jump delayed neural networks. IEEE Trans. Syst., Man, Cybern.: Syst. (2019)
23. Shen, H., Wang, T., Cao, J., et al.: Nonfragile dissipative synchronization for Markovian memristive neural networks: a gain-scheduled control scheme. IEEE Trans. Neural Netw. Learn. Syst. **30**(6), 1841–1853 (2018)
24. Liu, X., Tay, W.P., Liu, Z., et al.: Quasi-Synchronization of Heterogeneous networks with a generalized Markovian topology and event-triggered communication. IEEE Trans. Cybern. (2019). https://doi.org/10.1109/TCYB.2019.2891536
25. Guo, Y., Chen, B., Wu, Y.: Finite-time synchronization of stochastic multi-links dynamical networks with Markovian switching topologies. J. Franklin Inst. **357**(1), 359–384 (2019)
26. Fan, X., Zhang, X., Wu, L., et al.: Finite-time stability analysis of reaction-diffusion genetic regulatory networks with time-varying delays. IEEE/ACM Trans. Comput. Biol. Bioinf. **14**(4), 868–879 (2016)
27. Lu, J.: Global exponential stability and periodicity of reaction-diffusion delayed recurrent neural networks with Dirichlet boundary conditions. Chaos, Solitons Fractals **35**(1), 116–125 (2008)
28. Zhang, X., Han, Q., Zeng, Z.: Hierarchical type stability criteria for delayed neural networks via canonical Bessel-Legendre inequalities. IEEE Trans. Cybern. **48**(5), 1660–1671 (2017)
29. Sun, J., Liu, G., Chen, J., et al.: Improved delay-range-dependent stability criteria for linear systems with time-varying delays. Automatica **46**(2), 466–470 (2010)
30. Shen, H., Jiao, S., Xia, J., et al.: Generalised state estimation of Markov jump neural networks based on the Bessel-Legendre inequality. IET Control. Theory Appl. **13**(9), 1284–1290 (2018)
31. Cao, Y., Cao, Y., Wen, S., et al.: Passivity analysis of delayed reaction-diffusion memristor-based neural networks. Neural Netw. **109**, 159–167 (2019)
32. Xiao, Q., Huang, T., Zeng, Z.: Passivity and passification of fuzzy memristive inertial neural networks on time scales. IEEE Trans. Fuzzy Syst. **26**(6), 3342–3355 (2018)
33. Wen, S., Zeng, Z., Huang, T., et al.: Exponential adaptive lag synchronization of memristive neural networks via fuzzy method and applications in pseudorandom number generators. IEEE Trans. Fuzzy Syst. **22**(6), 1704–1713 (2013)

Index

J. H. Park (ed.), *Recent Advances in Control Problems of Dynamical Systems and Networks*, Studies in Systems, Decision and Control 301,
https://doi.org/10.1007/978-3-030-49123-9

The manufacturer's authorised representative in the EU is Springer Nature Customer Service Centre GmbH, Europaplatz 3, 69115 Heidelberg, Germany. If you have any concerns regarding our products, please contact ProductSafety@springernature.com

Printed and bound by CPI Group (UK) Ltd, Croydon, CR0 4YY
30/09/2026
02225651-0017